Communications in Computer and Information Science 901

More information about this series at http://www.springer.com/series/7899

Qinglei Zhou · Yong Gan
Weipeng Jing · Xianhua Song
Yan Wang · Zeguang Lu (Eds.)

Data Science

4th International Conference of Pioneering Computer Scientists, Engineers and Educators, ICPCSEE 2018 Zhengzhou, China, September 21–23, 2018 Proceedings, Part I

Editors
Qinglei Zhou
Zhengzhou University
Zhengzhou, Henan
China

Yong Gan
Zhengzhou University of Light Industry
Zhengzhou, Henan
China

Weipeng Jing
Northeast Forestry University
Harbin, China

Xianhua Song
Harbin University of Science
and Technology
Harbin, China

Yan Wang
Zhengzhou Institute of Technology
Zhengzhou, China

Zeguang Lu
National Academy of Guo Ding
Institute of Data Science
Beijing, China

ISSN 1865-0929 ISSN 1865-0937 (electronic)
Communications in Computer and Information Science
ISBN 978-981-13-2202-0 ISBN 978-981-13-2203-7 (eBook)
https://doi.org/10.1007/978-981-13-2203-7

Library of Congress Control Number: 2018951433

This Springer imprint is published by the registered company Springer Nature Singapore Pte Ltd.
The registered company address is: 152 Beach Road, #21-01/04 Gateway East, Singapore 189721, Singapore

Preface

As the general and program co-chairs of the 4th International Conference of Pioneer Computer Scientists, Engineers and Educators 2018 (ICPCSEE 2018, originally ICYCSEE), it is our great pleasure to welcome you to the proceedings of the conference, which was held in Zhengzhou, China, September 21–23, 2018, hosted by Henan Computer Federation and Zhengzhou Computer Federation and Zhengzhou University and Henan Polytechnic University and National Academy of Guo Ding Institute of Data Science. The goal of this conference is to provide a forum for computer scientists, engineers, and educators.

The call for papers of this year's conference attracted 470 paper submissions. After the hard work of the Program Committee, 125 papers were accepted to appear in the conference proceedings, with an acceptance rate of 26.5%. The major topic of this conference was data science. The accepted papers cover a wide range of areas related to Basic Theory and Techniques for Data Science including Mathematical Issues in Data Science, Computational Theory for Data Science, Big Data Management and Applications, Data Quality and Data Preparation, Evaluation and Measurement in Data Science, Data Visualization, Big Data Mining and Knowledge Management, Infrastructure for Data Science, Machine Learning for Data Science, Data Security and Privacy, Applications of Data Science, Case Study of Data Science, Multimedia Data Management and Analysis, Data-Driven Scientific Research, Data-Driven Bioinformatics, Data-Driven Healthcare, Data-Driven Management, Data-driven eGovernment, Data-Driven Smart City/Planet, Data Marketing and Economics, Social Media and Recommendation Systems, Data-Driven Security, Data-Driven Business Model Innovation, Social and/or Organizational Impacts of Data Science.

We would like to thank all the Program Committee members, 319 coming from 121 institutes, for their hard work in completing the review tasks. Their collective efforts made it possible to attain quality reviews for all the submissions within a few weeks. Their diverse expertise in each individual research area has helped us to create an exciting program for the conference. Their comments and advice helped the authors to improve the quality of their papers and gain deeper insights.

Great thanks should also go to the authors and participants for their tremendous support in making the conference a success. We thank Dr. Lanlan Chang and Jane Li from Springer, whose professional assistance was invaluable in the production of the proceedings.

Besides the technical program, this year ICPCSEE offered different experiences to the participants. We hope you enjoy the conference proceedings.

June 2018

Qinglei Zhou
Yong Gan
Qiguang Miao

Organization

The 4th International Conference of Pioneering Computer Scientists, Engineers and Educators (ICPCSEE, originally ICYCSEE) 2018 (http://2018.icpcsee.org) was held in Zhengzhou, China, during September 21–23 2018, hosted by Henan Computer Federation and Zhengzhou Computer Federation and Zhengzhou University and Henan Polytechnic University and National Academy of Guo Ding Institute of Data Science.

ICPCSEE 2018 General Chair

Qinglei Zhou — Zhengzhou University, China

Program Chairs

Yong Gan — Zhengzhou Institute of Technology, China
Qiguang Miao — Xidian University, China

Program Co-chairs

Qingxian Wang — Information Engineering University, China
Fengbin Zheng — Henan University, China
JiuCheng Xu — Henan Normal University, China
Jiexin Pu — Henan University of Science and Technology, China
ZongPu Jia — Henan Polytechnic University, China
Zhanbo Li — Zhengzhou University, China

Organization Chairs

Yangdong Ye — Zhengzhou University, China
WANG Yan — Zhengzhou Institute of Technology, China
Dong Liu — Henan Normal University, China
Junding Sun — Henan Polytechnic University, China
Zeguang Lu — National Academy of Guo Ding Institute of Data Science, China

Organization Co-chairs

Jianmin Wang — Zhengzhou University, China
Haitao Li — Zhengzhou University, China
Song Yu — Zhengzhou University, China
Song Wei — Zhengzhou University, China
Sun Yi — Zhengzhou University, China

Yan Gao	Henan Polytechnic University, China
Zhiheng Wang	Henan Polytechnic University, China
Fan Zhang	Zhengzhou Institute of Technology, China

Publication Chairs

Hongzhi Wang	Harbin Institute of Technology, China
Weipeng Jing	Northeast Forestry University, China

Publication Co-chairs

Xianhua Song	Harbin University of Science and Technology, China
Wei Xie	Harbin University of Science and Technology, China
Liuyuan Chen	Henan Normal University, China
Hui Li	Henan Polytechnic University, China
Xiaopeng Chang	Henan Finance University, China

Education Chairs

Shenyi Qian	Zhengzhou University of Light Industry, China
Miaolei Deng	Henan University of Technology, China

Industrial Chairs

Zheng Shan	Information Engineering University, China
Zhiyongng Zhang	Henan University of Science and Technology, China

Demo Chairs

Tianyang Zhou	Information Engineering University, China
Shuhong Li	Henan University of Economics and Law, China

Panel Chairs

Bing Xia	Zhongyuan University of Technology, China
Huaiguang Wu	Zhengzhou University of Light Industry, China

Poster Chairs

Guanglu Sun	Harbin University of Science and Technology, China
Liu Xia	Sanya Aviation and Tourism College, China

Expo Chairs

Shuaiyi Zhou	Henan Smart City Planning and Construction Specialized Committee, China
Junhao Jia	Henan King Source Information Technology Co., Ltd., China

Expo Co-chairs

Liang Bing	Henan Smart City Planning and Construction Specialized Committee, China
Dandan Jia	Henan Skylark Marketing Data Services Ltd., China

Registration/Financial Chair

Chunyan Hu	National Academy of Guo Ding Institute of Data Science, China

ICPCSEE Steering Committee

Jiajun Bu	Zhejiang University, China
Wanxiang Che	Harbin Institute of Technology, China
Jian Chen	Paratera, China
Xuebin Chen	North China University of Science and Technology, China
Wenguang Chen	Tsinghua University, China
Xiaoju Dong	Shanghai Jiao Tong University, China
TIAN Feng	Institute of Software Chinese Academy of Sciences, China
Qilong Han	Harbin Engineering University, China
Yiliang Han	Engineering University of CAPF, China
Yinhe Han	Institute of Computing Technology, Chinese Academy of Sciences, China
Hai Jin	Huazhong University of Science and Technology, China
Weipeng Jing	Northeast Forestry University, China
Wei Li	Central Queensland University, Australia
Min Li	Central South University, China
Junyu Lin	Institute of Information Engineering, Chinese Academy of Sciences, China
Yunhao Liu	Michigan State University, America
Zeguang Lu	National Academy of Guo Ding Institute of Data Science, China
Rui Mao	Shenzhen University, China
Qiguang Miao	Xidian University, China
Haiwei Pan	Harbin Engineering University, China
Pinle Qin	North University of China, China
Zhaowen Qiu	Northeast Forestry University, China
Zheng Shan	The PLA Information Engineering University, China
Guanglu Sun	Harbin University of Science and Technology, China

Jie Tang	Tsinghua University, China
Hongzhi Wang	Harbin Institute of Technology, China
Tao Wang	Peking University, China
Xiaohui Wei	Jilin University, China
Lifang Wen	Beijing Huazhang Graphics & Information Co., Ltd., China
Yu Yao	Northeastern University, China
Xiaoru Yuan	Peking University, China
Yingtao Zhang	Harbin Institute of Technology, China
Yunquan Zhang	Institute of Computing Technology, Chinese Academy of Sciences, China
Liehuang Zhu	Beijing Institute of Technology, China
Min Zhu	Sichuan University, China

ICPCSEE 2018 Program Committee Members

Chunyu Ai	University of South Carolina Upstate, America
Jiyao An	Hunan University, China
Xiaojing Bai	TsingHua University, China
Ran Bi	Dalian University of Technology, China
Yi Cai	South China University of Technology, China
Zhipeng Cai	Georgia State University, America
Cao Cao	State Key Laboratory of Mathematical Engineering and Advanced Computing, China
Zhao Cao	Beijing Institute of Technology, China
Baobao Chang	Peking University, China
Richard Chbeir	LIUPPA Laboratory, France
Che Nan	Harbin University of Science and Technology, China
Wanxiang Che	Harbin Institute of Technology, China
Bolin Chen	Northwestern Polytechnical University, China
Chunyi Chen	Changchun University of Science and Technology, China
Hao Chen	Hunan University, China
Quan Chen	Guangdong University of Technology, China
Shu Chen	Xiangtan University, China
Wei Chen	Beijing Jiaotong University, China
Wenliang Chen	Soochow University, China
Wenyu Chen	University of Electronic Science and Technology of China, China
Xuebin Chen	North China University of Science and Technology, China
Zhumin Chen	Shandong University, China
Ming Cheng	Zhengzhou University of Light Industry, China
Siyao Cheng	Harbin Institute of Technology, China
Byron Choi	Hong Kong Baptist University, China
Xinyu Dai	Nanjing University, China
Lei Deng	Central South University, China
Vincenzo Deufemia	University of Salerno, Italy
Jianrui Ding	Harbin Institute of Technology, China

Qun Ding	Heilongjiang University, China
Xiaofeng Ding	Huazhong University, China
Hongbin Dong	Harbin Engineering University, China
Xiaoju Dong	Shanghai Jiao Tong University, China
Zhicheng Dou	Renmin University of China, China
Jianyong Duan	North China University of Technology, China
Lei Duan	Sichuan University,China
Xiping Duan	Harbin Normal University, China
Junbin Fang	Jinan University, China
Xiaolin Fang	Southeast University, China
Guangsheng Feng	Harbin Engineering University, China
Jianlin Feng	Sun Yat-Sen University, China
Weisen Feng	Sichuan University, China
Guohong Fu	Heilongjiang University, China
Jianhou Gan	Yunnan Normal University, China
Jing Gao	Dalian University of Technology, China
Daohui Ge	Xidian University, China
Lin Ge	Zhengzhou University of Aeronautics, China
Dianxuan Gong	North China University of Science and Technology, China
Lila Gu	Xinjiang University
Yu Gu	Northeastern University, China
Hongjiao Guan	Harbin Institute of Technology, China
Tao Guan	Zhengzhou University of Aeronautics, China
Chunyi Guo	Zhengzhou University, China
Jiafeng Guo	Institute of Computing Technology, Chinese Academy of Sciences, China
Longjiang Guo	Heilongjiang University, China
Yibo Guo	Zhengzhou University, China
Yuhang Guo	Beijing Institute of Technology, China
Meng Han	Georgia State University, America
Meng Han	Kennesaw State University, America
Qi Han	Harbin Institute of Technology, China
Xianpei Han	Chinese Academy of Sciences, China
Yingjie Han	Zhengzhou University, China
Zhongyuan Han	Harbin Institute of Technology, China
Tianyong Hao	Guangdong University of Foreign Studies, China
Jia He	Chengdu University of Information Technology, China
Qinglai He	Arizona State University, America
Shizhu He	Chinese Academy of Sciences, China
Liang Hong	Wuhan University, China
Leong Hou	University of Macau, China
Yifan Hou	State Key Laboratory of Mathematical Engineering and Advanced Computing, China
Chengquan Hu	Jilin University, China
Wei Hu	Nanjing University, China
Zhang Hu	Shanxi University, China

Hao Huang	Wuhan University, China
Kuan Huang	Utah State University, America
Lan Huang	Jilin University, China
Shujian Huang	Nanjing University, China
Jian Ji	Xidian University, China
Ruoyu Jia	Sichuan University, China
Yuxiang Jia	Zhengzhou University, China
Bin Jiang	Hunan University, China
Feng Jiang	Harbin Institute of Technology, China
Hailin Jiang	Harbin Institute of Technology, China
Jiming Jiang	King Abdullah University of Science & Technology, Saudi Arabia
Wenjun Jiang	Hunan University, China
Xiaoheng Jiang	Zhengzhou University, China
Peng Jin	Leshan Normal University, China
Weipeng Jing	Northeast Forestry University, China
Shenggen Ju	Sichuan University, China
Fang Kong	Soochow University, China
Hanjiang Lai	Sun Yat-Sen University, China
Wei Lan	Central South University, China
Yanyan Lan	Institute of Computing Technology, Chinese Academy of Sciences, China
Chenliang Li	Wuhan University, China
Dawei Li	Nanjing Institute of Technology, China
Dun Li	Zhengzhou University, China
Faming Li	University of Electronic Science and Technology of China, China
Guoqiang Li	Norwegian University of Science and Technology, Norway
Hua Li	Changchun University, China
Hui Li	Xidian University, China
Jianjun Li	Huazhong University of Science and Technology, China
Jie Li	Harbin Institute of Technology, China
Kai Li	Harbin Institute of Technology, China
Min Li	Central South University, China
Mingzhao Li	RMIT University, Australia
Mohan Li	Jinan University, China
Moses Li	Jiangxi Normal University, China
Peng Li	Shaanxi Normal University, China
Qingliang Li	Changchun University of Science and Technology, China
Qiong Li	Harbin Institute of Technology, China
Rong-Hua Li	Shenzhen University, China
Ru Li	Shanxi University, China
Sujian Li	Peking University, China
Wei Li	Georgia State University, America
Xiaofeng Li	Sichuan University, China
Xiaoyong Li	Beijing University of Posts and Telecommunications, China

Xuwei Li	Sichuan University, China
Yunan Li	Xidian University, China
Zheng Li	Sichuan University, China
Zhenghua Li	Soochow University, China
Zhijun Li	Harbin Institute of Technology, China
Zhixu Li	Soochow University, China
Zhixun Li	Nanchang University, China
Hongfei Lin	Dalian University of Technology, China
Bingqiang Liu	Shandong University, China
Fudong Liu	State Key Laboratory of Mathematical Engineering and Advanced Computing, China
Guanfeng Liu	Soochow University, China
Guojun Liu	Harbin Institute of Technology,China
Hailong Liu	Northwestern Polytechnical University, China
Ming Liu	Harbin Institute of Technology, China
Pengyuan Liu	Beijing Language and Culture University, China
Shengquan Liu	XinJiang University, China
Tiange Liu	Yanshan University, China
Yan Liu	Harbin Institute of Technology, China
Yang Liu	Peking University, China
Yang Liu	TsingHua University, China
Yanli Liu	Sichuan University, China
Yong Liu	Heilongjiang University, China
Binbin Lu	Sichuan University, China
Junling Lu	Shaanxi Normal University, China
Wei Lu	Renmin University of China, China
Zeguang Lu	Sciences of Country Tripod Institute of Data Science, China
Jianlu Luo	Officers College of PAP, China
Jiawei Luo	Hunan University, China
Jizhou Luo	Harbin Institute of Technology, China
Zhunchen Luo	China Defense Science and Technology Information Center, China
Huifang Ma	NorthWest Normal University, China
Jiquan Ma	Heilongjiang University, China
Yide Ma	Lanzhou University, China
Hua Mao	Sichuan University, China
Xian-Ling Mao	Beijing Institute of Technology, China
Jun Meng	Dalian University of Technology, China
Hongwei Mo	Harbin Engineering University, China
Lingling Mu	Zhengzhou University, China
Jiaofen Nan	Zhengzhou University of Light Industry, China
Tiezheng Nie	Northeastern University, China
Haiwei Pan	Harbin Engineering University, China
Fei Peng	Hunan University, China
Jialiang Peng	Norwegian University of Science and Technology, China
Wei Peng	Kunming University of Science and Technology, China

Suge Wang	Shanxi University, China
Xiao Wang	Zhengzhou University of Light Industry, China
Xin Wang	Tianjin University, China
Yingjie Wang	Yantai University, China
Yongheng Wang	Hunan University, China
Yunfeng Wang	Sichuan University, China
Zhenyu Wang	State Key Laboratory of Mathematical Engineering and Advanced Computing, China
Zhifang Wang	Heilongjiang University, China
Zhewei Wei	School of Information, Renming University, China
Zhongyu Wei	Fudan University, China
Bin Wen	Yunnan Normal University, China
Huaiguang Wu	Zhengzhou University of Light Industry, China
Huayu Wu	Institute for Infocomm Research, China
Rui Wu	Harbin Institute of Technology, China
Xiangqian Wu	Harbin Institute of Technology, China
Yan Wu	Changchun University, China
Yufang Wu	Peking University, China
Zhihong Wu	Sichuan University, China
Guangyong Xi	Zhengzhou University of Light Industry, China
Rui Xia	Nanjing University of Science and Technology, China
Min Xian	Utah State University, America
Degui Xiao	Hunan University, China
Sheng Xiao	Hunan University, China
Tong Xiao	Northeastern University, China
Yi Xiao	Hunan University, China
Minzhu Xie	Hunan Normal University, China
Deyi Xing	Soochow University, China
Dan Xu	University of Trento, Italy
Jianqiu Xu	Nanjing University of Aeronautics and Astronautics, China
Jing Xu	Changchun University of Science and Technology, China
Pengfei Xu	Xidian University, China
Ruifeng Xu	Harbin Institute of Technology, China
Ying Xu	Hunan University, China
Yaohong Xue	Changchun University of Science and Technology, China
Mingyuan Yan	University of North Georgia, America
Shaohong Yan	North China University of Science and Technology, China
Xuexiong Yan	State Key Laboratory of Mathematical Engineering and Advanced Computing, China
Bian Yang	Norwegian University of Science and Technology, Norway
Chunfang Yang	State Key Laboratory of Mathematical Engineering and Advanced Computing, China
Donghua Yang	Harbin Institute of Technology, China
Gaobo Yang	Hunan University, China
Lei Yang	Heilongjiang University, China
Ning Yang	Sichuan University, China

Yajun Yang	Tianjin University, China
Bin Yao	Shanghai Jiao Tong University, China
Yuxin Ye	Jilin University, China
Dan Yin	Harbin Engineering University, China
Meijuan Yin	State Key Laboratory of Mathematical Engineering and Advanced Computing, China
Minghao Yin	Northeast Normal University, China
Zhongxu Yin	State Key Laboratory of Mathematical Engineering and Advanced Computing, China
Zhou Yong	China University of Mining and Technology, China
Jinguo You	Kunming University of Science and Technology, China
Bo Yu	National University of Defense Technology, China
Dong Yu	Beijing Language and Culture University, China
Fei Yu	Harbin Institute of Technology, China
Haitao Yu	Harbin Institute of Technology, China
Lei Yu	Georgia Institute of Technology, America
Yonghao Yu	Harbin Institute of Technology, China
Zhengtao Yu	Kunming University of Science and Technology, China
Lingyun Yuan	Yunnan Normal University, China
Ye Yuan	Harbin Institute of Technology, China
Ye Yuan	Northeastern University, China
Kun Yue	Yunnan University, China
Yue Yue	SUTD, Singapore
Hongying Zan	Zhengzhou University, China
Boyu Zhang	Utah State University, America
Dongxiang Zhang	University of Electronic Science and Technology of China, China
Fan Zhang	Wuhan University of Light Industry, China
Haixian Zhang	Sichuan University, China
Huijie Zhang	Northeast Normal University, China
Jiajun Zhang	Institute of Automation, Chinese Academy of Sciences, China
Kejia Zhang	Harbin Engineering University, China
Keliang Zhang	PLAUFL, China
Kunli Zhang	Zhengzhou University, China
Liancheng Zhang	State Key Laboratory of Mathematical Engineering and Advanced Computing, China
Lichen Zhang	Shaanxi Normal University, China
Liguo Zhang	Harbin Engineering University, China
Meishan Zhang	Heilongjiang University, China
Meishan Zhang	Singapore University of Technology and Design, Singapore
Peipei Zhang	Xidian University, China
Ping Zhang	State Key Laboratory of Mathematical Engineering and Advanced Computing, China
Tiejun Zhang	Harbin University of Science and Technology, China
Wenjie Zhang	The University of New South Wales, Australia
Xiao Zhang	Renmin University of China, China

Xiaowang Zhang	Tianjin University, China
Yangsen Zhang	Beijing Information Science and Technology University, China
Yi Zhang	Sichuan University, China
Yingtao Zhang	Harbin Institute of Technology, China
Yonggang Zhang	Jilin University, China
Yongqing Zhang	Chengdu University of Information Technology, China
Yu Zhang	Harbin Institute of Technology, China
Yuhong Zhang	Henan University of Technology, China
Bihai Zhao	Changsha University, China
Hai Zhao	Shanghai Jiao Tong University, China
Jian Zhao	Changchun University, China
Qijun Zhao	Sichuan University, China
Xin Zhao	Renmin University of China, China
Xudong Zhao	Northeast Forestry University, China
Wenping Zheng	Shanxi University, China
Zezhi Zheng	Xiamen University, China
Jiancheng Zhong	Hunan Normal University, China
Changjian Zhou	Northeast Agricultural University, China
Fucai Zhou	Northeastern University, China
Juxiang Zhou	Yunnan Normal University, China
Tianyang Zhou	State Key Laboratory of Mathematical Engineering and Advanced Computing, China
Haodong Zhu	Zhengzhou University of Light Industry, China
Jinghua Zhu	Heilongjiang University, China
Min Zhu	Sichuan University, China
Ruijie Zhu	Zhengzhou University, China
Shaolin Zhu	Xinjiang Institute of Sciences and Chemistry of the Chinese Academy of Sciences, China
Yuanyuan Zhu	Wuhan University, China
Zede Zhu	Hefei Institutes of Physical Science, Chinese Academy of Sciences, China
Huibin Zhuang	Henan University, China
Quan Zou	Tianjin University, China
Wangmeng Zuo	Harbin Institute of Technology, China
Xingquan Zuo	Beijing University of Posts and Telecommunications, China

Contents – Part I

Contents – Part II

Development of Scientific Research Management in Big Data Era

Bin Wang(✉) and Zhaowen Liu

Tianjin Normal University, Tianjin 300071, China
sdwb2004@126.com

Abstract. The advent of the big data era has brought tremendous development opportunities to scientific research management in universities, and led the development of scientific research management. However, the era of big data also raised new requirements for scientific research management. This paper discusses the scientific research management in universities, at the same time, put forward the innovative coping strategies.

Keywords: Big data era · Scientific research · Scientific research management

1 Introduction

At present, humans are entering an era of mass production, consumption and application of big data. Large-scale scientific research and the rapid development of the Internet have brought human into the "Big Data Era" in recent years. With the development of science and technology, each country invests more in scientific research than ever before. Scientific research has produced a large amount of resources for scientific technological achievements and scientific research information. With the development of information technology, how to efficiently manage these scientific information has become top priority [1].

As a part of education in the future, education and scientific research management face the age proposition of innovation and development. How to keep pace with the era of big data, use big data to promote the reform of education and scientific research management, improve management efficiency and promote the development of education are the tasks. Promoting education development is the main task of education scientific research management and practice [2]. Efficient management has a great significance about scientific research projects and scientific research achievements.

In the background of modern big data era, it has the leading role in the development of scientific research management, including the easier access to the original data; the active push of researchers can better improve the database, and can do scientific research management of data automatic acquisition, these are undoubtedly big data for scientific research in universities to promote the development, however, big data for the management of universities also issued new requirements [3]. On this basis, this article discusses the strategy of innovation in scientific research management in the era of big data, which can effectively help the scientific research management in universities play

Q. Zhou et al. (Eds.): ICPCSEE 2018, CCIS 901, pp. 1–7, 2018.
https://doi.org/10.1007/978-981-13-2203-7_1

its own role. With the help of the development of big data era, it will provide certain reference to the innovation of scientific research management in universities.

2 Overview of Big Data Era

2.1 The Definition of Big Data Era

Big data is a new term that people demand for data increases under the requirements of the development of the times. It was used in the IT industry. As the Internet has become more and more common in our daily life, the data storage and computing has a new way, the data is a valuable intangible assets, in business which it is the point of business management and development, and its high real-time data satisfaction, be able to put these data for people to use from time to time, people's lives and services have brought great benefits, and now the term big data era has more than just exist in the IT industry in physics, biology and other disciplines, as well as in the current hot industry in military affairs and finance has caused great concern in people. At the same time, the big data era has brought a certain level of innovation to people's lives.

2.2 The Characteristics of Big Data Era

Big data has four significant features, first of all, the large amount of data, the amount of the area has strict requirements, the starting unit of measurement of big data is a thousand T, only this large-capacity data can win the use of businesses, individuals is called the big data era in turn. In addition to the second largest amount of data in many aspects, and in terms of the type of data is very large, including network logs, audio and video, etc., of course, which on multiple types of data processing speed and processing effect. The third aspect is that the value density is low. Now that the Internet has a wide coverage and a high penetration rate, although the quantity and type of information can be said to be massive, its value density is relatively low. The refinement method can not make the data higher-level purification, which is also one of the big problems in the era of big data.

Finally, its processing speed has higher requirements in terms of timeliness, which is the biggest difference between big data and traditional data, which is a significant improvement in the collection of traditional data, it brings people convenience and benefits,which also bring new requirements to people about data mining and extraction. It can also be said that the potential of big data has not been fully exploited.

2.3 The Value of Big Data

Big data was originally proposed in the foreign IT industry. Now the well-known brands and websites of some developed countries in the West have made us aware of the potential value of the data, some of the more popular data is an important part of practical experience in the production of life carrier, we can purify some valuable information through data recording and storage for later use in production and life. This is the biggest value of big data.

This empirical introduction directly allows the big data age has some predictability, through the data collection, according to the laws and trends in the existing data to predict what will happen.

3 Big Data Lead the Development of Scientific Research Management

Big data plays a leading role in the scientific research management. This mainly depends on the acquisition of big data in the scientific research management and facilitates the progress of scientific research. Therefore, the management of scientific research is also affected by the big data and is constantly evolving. Specifically including the original data to facilitate the initiative to send scientific research personnel, completing the database, scientific research management of data collection automatically can help scientific research summary.

3.1 The Original Data Acquisition, Facilitate the Progress of Scientific Research

Since the establishment of college information, most of the schools have their own internal research management system, the relevant information and scientific research put into the system, to continuously collect the original data for scientific research, including the establishment of scientific research topics, research contract funding, scientific research and other information. Under the impetus of big data, through the simple statistics and sorting of scientific research managers, it is convenient for people to inquire about the scientific research achievements of their own since the past years as well as the scientific research achievement of others, and can provide certain guiding significance for people's scientific research in the next step. And under the influence of the big data era, it has effectively helped the researchers to establish their own scientific research information, set a reliable basis for the reform of the management policies and can effectively help the development of scientific research management.

3.2 The Active Push of Scientific Research Personnel, Improve the Database

In recent years, under the influence of big data, the data of the original contents of research users continue to rise and the rate of increase is very high, which is an important source of current research data. First of all, the rapid development of the new social networking platform represented by Weibo has made the desire of users to generate data. This is also in the background of big data, which enables researchers to push forward actively and scientific research. The establishment of the database has promoted the further development of scientific research management. In addition, since the increase in media users, smart phones can appear anytime and anywhere, expressing their opinions on the Internet, publishing their own scientific research has become more convenient, others view is also very convenient, online communication can span time geographical, which makes scientific research in academic exchanges to achieve real-time sharing, it also contributed to the establishment of scientific research team to facilitate the development of scientific research management.

3.3 Scientific Research Management Data Collection, Can Help Research Summary

Under the background of big data, automatic collection of scientific research management data can be realized, which facilitates the collection of scientific research information. With the development of big data technology, man has been able to start to manufacture some sensors with processing functions and is used widely. The collection and monitoring of scientific research data and scientific research forms of these devices are automatically generated to meet the needs of the present era. It can also be said that under the background of big data, it undoubtedly facilitates the automatic collection of scientific research management systems and realizes. The relationship between university scientific research management system and other university digital platforms. In order to achieve the teaching management system, personnel management systems, research management systems, financial systems, such as data sharing between the systems to facilitate automatic matching to collect scientific research personnel a variety of information.

4 The Requirements on Scientific Research Management in Big Data Era

4.1 Emphasis on Data, Establish a Big Data Awareness

Compared with the traditional research and management work, we should pay full attention to this awareness of application data. Nowadays, many electricity suppliers are gradually realizing that data are the new oil in the world today. It is a very valuable resource and an important source of generating all values. Therefore, if we want to do a good job in scientific research management and full innovation in the era of big data, we must attach importance to the data and implement the idea of big data in our daily work. We should constantly expand the channels for data collection and enrich the database. Take an effective way to effectively accumulate and sort the data for people to use and access.

4.2 Mining the Intrinsic Value of the Data

Data has a great intrinsic value, it also has great potential, the reason why big data to use a large character to describe, does not mean that it is a large number of a wide range of more describes the integration of large amounts of data. And the analysis of the process, the size of its large, and through effective exchange and integration, but also be able to create new data, discover new areas of science and technology, to create greater value, that is, big data can bring knowledge and technology. Profit and growth, which is also the core value of big data, and these core values are often hidden behind the data and require people to dig, largely relying on the technology, the application of some algorithms to the massive data, can predict the possibility of things effectively.

4.3 Use of the Thinking of Big Data Era

In the process of scientific research management, the way of thinking in the era of big data should be used to better expand the space for people's thinking through better collection and utilization of data, and to think in many aspects, which is to require scientific research. Managers in the data, data management process, to have an open mind, good at exploring new methods, new ideas. Different from traditional scientific research management, big data needs to sort out the relationship of things effectively, not only the data of causality, but also the hidden links between more interdisciplinary data, which is also a way of thinking that is required in the big data age helps to analyze the internal relations and the hidden relations between things.

5 Innovation Scientific Research Management Strategy in the Big Data Era

5.1 Innovative Scientific Research Management Concepts

At the innovation of scientific research management in universities, the core content should be the management, that is to say, it's essence lies in serving scientific researchers and serving other researchers. In the concrete implementation process, we should abandon the past scientific research management workers passively record and organize data, so that they can accomplish some work mechanically without realizing the quality of scientific research management innovation in colleges and universities. Therefore, it is necessary to start with intensifying education and training efforts and strengthening the service consciousness of scientific research managers at universities. By letting university scientific research managers understand the trends of big data and the significance they propose, so as to effectively enable them to exert their own values, promote scientific research management innovation steady, passive and proactive, and secondly to upgrade their sense of service, so that they can forward the service, take the initiative to analyze the mining. The data can be obtained, and pay for action to innovative scientific research management concepts.

5.2 Improve the Quality of Scientific Research Managers

The quality of scientific research managers directly affects the quality and process of scientific research management innovation in universities, because only the professional qualifications of scientific research managers reach a certain level, can accurately grasp and implement various management skills, through their own technology to help data efficient acquisition and analysis can effectively screen out some of the data and quickly convert reliable and valid data into standardized, shared information. Therefore, universities should also start preparations for improving the quality of scientific research managers, improve their barriers to entry, set some exemplary figures for research management team, and improve the quality of scientific research managers as a whole to improve the effectiveness of research management.

5.3 Strengthen Data Collection, Promote Data Sharing

In the background of big data, it is inseparable from the data collection and analysis, then in order to improve the quality of data collection, we must strengthen data collection efforts, while promoting data sharing. In this specific work, universities must strengthen the basic information construction, to maintain the coverage of the information collected is broad, rich in content, and can be accurate. On this basis, to realize its application value and management value, coupled with the construction of technical information, the existing data can once again be deep mining and analysis. In the university research work, there will be multidisciplinary or interdisciplinary problems arise, with the increase in the number of scientific research, this multidisciplinary issue is properly classified and collected for truly multidisciplinary sharing. Universities should also study in depth the delivery and sharing of big data technology management work to meet the actual needs of each university research and development management.

5.4 Improve the Relevant Mechanism About Scientific Research Management

The related mechanism of scientific research management in universities is a rigid management law of scientific research management content and work quality, which can set a standard for the specific work of scientific research managers and also provide a basis for their management innovation. In the era of big data, universities need to have some skills to deeply study the actual situation of their own scientific research management. At the same time, learn advanced management experience of other universities or other institutions, including exchanges with other countries, and be able to selectively absorb and apply them to promote their own research and development management mechanism. Only in this way can we provide a guarantee for the smooth development of the next scientific research management and to standardize the contents and working standards of the scientific research managers at a certain extent so as to lay a foundation for further innovation in scientific research management.

5.5 Scientific Research in Universities for Intelligent Management

The university scientific research management system involves many aspects, including research projects, research funds, scientific research results, technical support, intellectual property rights, research teams, scientific research incentives, academic exchanges and other subsystems. This is a huge workload for managers. Therefore, an intelligent management method will greatly reduce the work pressure of staff and increase work efficiency.

Genetic algorithms are widely used, so it can be considered to apply genetic algorithms to project management issues of scientific research management. Genetic algorithms are often used to generate effective solutions to find the optimal solution to solve the problem. The development of a project needs to consider many factors, such as the type of project declared, financial support, technical support, and expected results. Through these factors, it is judged whether the project is reliable and whether it can be carried out. The project category and declaration project index are input, and then the intelligent detection system performs a comparison and judgment based on the optimal solution generated by the genetic algorithm. Whether or not to pass the project declaration, the declaration result is finally output; if it is not successful, feedback is provided. For scientific researchers, the scientific research personnel will make amendments and re-apply. This method can replace the traditional manual judgment, saving a lot of manpower and time.

6 Summary

In short, under the background of the big data era, all walks of life have been affected to varying degrees, so in the management of scientific research in universities, it should also comply with the requirements of the big data era of thinking and theoretical requirements in the management process. We must pay great attention to the significance of big data, with the current situation in-depth study of innovation in scientific research management, the real work of scientific research management can be applied to the development of universities and the development of scientific research.

Acknowledgment. This paper is supported by one projects:

1. Tianjin Science and technology Development Fund projects in colleges and universities (No. 20110818).

References

1. Zhang, F., Zhang, D.: Analysis of research management data center construction in Chinese academy of sciences in age of big data. Stud. Sci. Sci. **34**(2), 166–170 (2016)
2. Li, Z.: Research on the countermeasures of university library information resources construction in big data era. J. Acad. Library Inf. Sci. **35**(1), 36–40 (2018)
3. Ren, C.X., Han, C.L., Wu, L.: The role of national scientific research platform in promoting two innovative talents. Univ. Educ. **1**, 137–139 (2017)
4. Cong, P.M.: The impact of big data on research management and decision-making-a policy study view. E-sci. Technol. Appl. **4**(6), 29–35 (2013)
5. Du, Y.H.: Analysis of research management data center construction in chinese academy of sciences in age of big data. Comput. Syst. Appl. **24**(1), 79–85 (2015)
6. Yang, W.R.: Study on scientific research management innovation of colleges and universities in big data era. Sci. Technol. Manage. Res. **14**, 1–4 (2015)
7. Zhang, Y.N.: A study of big data applications in the field of education based on the practice in america. East China Normal University (2016)
8. Cui, P.: The analysis of universities' sciences research management information construction. J. Central Univ. Financ. Econ., 81–83 (2014)
9. Qian, X.H., Xie, S.P.: Research on the construction and improving path of university scientific research management information system. Res. High. Educ. Eng. **1**, 107–112 (2015)

The Competence of Volunteer Computing for MapReduce Big Data Applications

Wei Li(✉) and William Guo

School of Engineering and Technology, Central Queensland University, Rockhampton, QLD 4702, Australia
{w.li,w.guo}@cqu.edu.au

Abstract. It is little to find off-the-shelf research results in the current literature about how competent Volunteer Computing (VC) performs big data applications. This paper explores whether VC scales for a large number of volunteers when they commit churn and how large VC needs to scale in order to achieve the same performance as that by High Performance Computing (HPC) or computing grid for a given big data problem. To achieve the goal, this paper proposes a unification model to support the construction of virtual big data problems, virtual HPC clusters, computing grids or VC overlays on the same platform. The model is able to compare the competence of those computing facilities in terms of speedup vs number of computing nodes or volunteers for solving a big data problem. The evaluation results have demonstrated that all the computing facilities scale for the big data problem, with a computing grid or a VC overlay being in need of more or much more computing nodes or volunteers to achieve the same speedup as that of a HPC cluster. This paper has confirmed that VC is competent for big data problems as long as a large number of volunteers is available from the Internet.

1 Introduction

Big Data is coined as it cannot be handled by a single computer in a reasonable amount of time. Big data is a result of accumulated business transactions or social media events. Big data is an enterprise's legacy and asset in support of smart decision, predicting business trends, deepening customer engagement or optimizing operations. However, the computational analysis of big data makes incompetent of the traditional data processing by a single computer and demands parallel or distributed approaches and tools [21]. Traditionally parallel and distributed computing is to solve compute-intensive problems, with no or little data exchange between the parallel or distributed tasks. Big data problem challenges the traditional parallel or distributed computing paradigms by a large amount of inter-task data exchange in the course of computing. Thus the needs, such as distributing the data, parallelizing the computation and synthesizing results, demand a newer programming paradigm to handle. In this area, MapReduce [7] has been a successful model, originated from Google to process the large data set of crawled documents, inverted indices and web request logs. Thereafter, MapReduce has been applied to more extensive big data problems by the big enterprises like Yahoo, Facebook

Q. Zhou et al. (Eds.): ICPCSEE 2018, CCIS 901, pp. 8–23, 2018.
https://doi.org/10.1007/978-981-13-2203-7_2

and Microsoft. Nowadays medium or even small enterprises also face big data processing problems.

The MapReduce computing paradigm is abstracted as in Fig. 1. Among the 3 steps, the second step *shuffle* involves a large amount of data exchange between computing nodes. Thus naturally MapReduce software tools such as Hadoop [14] has been designed to utilize High Performance Computing (HPC) clusters, where the high speed network supports fast data exchange between cluster nodes via Distributed File Systems (DFS) like the Hadoop HDFS. The problem is that the high cost of HPC facility makes it inaccessible to every enterprise, especially small or medium enterprises. The extensive needs of big data analysis enables the exploring of cheaper distributed facilities. Among them, computing grid, making use existing corporate computers connected by high speed intranet or broadband wan, or Volunteer Computing (VC) [24], making use of the donated computing cycles and temporary storage from millions of the Internet volunteers, are extremely attractive.

Map Step

1. The original big dataset is divided into a number of small datasets.
2. The datasets with processing code are distributed as *map tasks* onto a HPC cluster.
3. The datasets will be computed in parallel by the same *map function* by the entire cluster.
4. Each computer will emit a number of *<key, value>* pairs with some keys being the same.

Shuffle Step

1. All the *<key, value>* pairs with the same key will be merged together in the form of *<key, a list of values>*.
2. These pairs will be further sorted by the keys into a number of *reduce tasks* with no any 2 keys being the same.
3. The reduce tasks are re-distributed onto the HPC cluster.

Reduce Step

1. The reduce tasks are computed in parallel by the same *reduce function* by the entire cluster.
2. Each computer will emit a number of *<key, value>* pairs as the final results.

Fig. 1. The MapReduce algorithm

Computing grids and particularly VC overlays exhibit more dynamics and opportunism than HPC clusters in terms of heterogeneous compute-capacity, storage and communication bandwidth and churn. Whether they are competent to solve big data problems has not been confirmed by the current literature. By competence, we are meant whether MapReduce on computing grids or VC overlays is scalable to a large number of computing nodes or volunteers and for a given big data problem, how large a computing grid or a VC overlay needs to scale in order to achieve the same performance as that by a High Performance Computing (HPC) cluster. Although computing grid or VC has been successfully applied to large-scale scientific projects such as SETI@Home [16], FiND@Home [12], ATLAS@home [2], Climateprediction.net [3] and DrugDiscovery@Home [10], those projects are normally compute-intensive. Consequently, it is necessary to study their competence to solve big data problems, with intensive data exchange between computing nodes or volunteers during the course of computing, before we really take the effort or invest to build such a hardware and software facility in dynamic and opportunistic environments. This paper addresses the above issues via the following approaches.

1. Proposing *Standard Step* to universally model the compute-capacity of a HPC cluster, a computing grid or a VC overlay, the cost to search for tasks or exchange data, and the compute-load of a big data problem.
2. Modelling churn as opportunistic behaviour of volunteers, who can join, leave or crash at any time and in a large number during the course of computing.
3. Proposing a unification model to support the construction of virtual HPC cluster, computing grid or volunteer overlay on the same platform.
4. Evaluating the competence of HPC cluster, computing grid or VC overlay for solving the given big data problem by inspection of the speedup and scalability vs a large number of computing nodes or volunteers.

The proposed models have been fully implemented as a simulator, where the dynamics and stabilization of VC overlay has been built on the Distributed Hash Table (DHT) protocol Chord [26]. The initial evaluation results are promising from 2 aspects. First, HPC cluster, computing grid and VC all scale to the big data problem, particularly, VC scales to 40K volunteers in the evaluation. Second, for the given big data problem, a computing grid or a VC overlay can achieve the same speedup as that of a HPC cluster by simply adding more nodes or volunteers, particularly, the VC overlay can achieve the same speedup in the face of *30%* or *50%* volunteers committing churn in the course of computing.

2 Related Work

In the current literature, some research have promoted VC to solve MapReduce big data problems. Tang et al. [27] developed a model to apply MapReduce to a desktop grid environment, where computing nodes were heterogeneous in compute-capacity and could fail. The architecture was not a pure P2P structure as stable nodes are introduced as the master nodes and the shuffle step of MapReduce was centralized. The experimental environment was of *356* computing nodes in a Gigabit Ethernet desktop grid. A word count application of *2.5* TB data was processed by *50,000* map and reduce tasks. The systems scaled for throughput (*MB/s*) vs the number of computing nodes. However, whether VC scaled for larger computing nodes or stronger churn was not confirmed.

Monsalve et al. [18] developed a complete simulator (named SimGrid) of BOINC [1] to analyze the performance of VC to solve data-intensive problems. The number of volunteer nodes, CPU power, node availability and unavailability vary for the simulations. The bottleneck problem of the data servers had been confirmed by the simulation. For the given number of servers (*1* to *64*), the size of input files (*4* KB, *256* KB, *1* MB, *16* MB and *64* MB) and the number of volunteer nodes (*3,000* and *30,000*), the servers were saturated at a certain number of volunteer nodes, causing a severe deceleration of throughput and validated results. The new model of [19] had improved the previous model in 2 aspects, using volunteer nodes as data server and data access locality. The new model outperformed the previous one for scaling throughput vs up to *10,000* volunteer nodes with *30%* as data servers.

The GiGi-MR model of Costa et al. [4, 5] introduced DHT to the original BOINC structure to allow data exchange between volunteer nodes in the shuffle step of a MapReduce application. The evaluation was conducted on *50* homogeneous volunteer nodes connected by an intranet of *700* KB/s download speed. The result showed that GiGi-MR was much better (half of the turnaround time) than the original BOINC although there was a large number of intermediate file exchanges. A volunteer node in [4, 5] communicated with another volunteer node directly, rather than by an overlay, to request a data set. The process could stop if the target volunteer committed churn. Costa et al. [6] was their following work, but improvement seemed little on the above issues.

Ryden et al. [23] was similar to Monsalve et al. [18] in that volunteer nodes were used as a part of system storage (data servers). As a result, the data exchange of the shuffle step of a MapReduce application could happen between volunteer nodes. The partitioned key-value pairs from the map step were pulled by the volunteers that were assigned reduce tasks. The disadvantage was that the pulling needed to contact other *n* volunteers. Thus once a reduce task was aborted, redoing the same task on another volunteer needed to contact the *n* volunteers again. The evaluation results showed a better performance of [23] than the original BOINC because of the removal of data bottleneck and the introduced locality-awareness of data access. The results also showed that the system still worked on *70%* node failure. The results were still unable to demonstrate the scalability of the system because of the use of small amount of data (*2* GB) or small number (*52*) of volunteer nodes.

MOON [17] was an extension to the MapReduce middleware Hadoop [14] by bringing Hadoop into the opportunistic VC environment. To cope with the unavailability of volunteer nodes, MOON made use of a small number of dedicated nodes and exploited data replication and task replication. For the real world applications *sort* and *word count*, MOON scaled against the unavailability of volunteer nodes, working effectively at different unavailable rate and working better at a lower unavailable rate. The drawback of the experimental results of MOON was that the problem size or volunteer number was still small when the data size was *24* GB and the overlay comprised of *60* volatile plus *6* dedicated nodes, which were homogeneous in compute-capacity and connected by 1 Gbps Ethernet.

The above studies cannot confirm how competent VC performs a big data problem, in comparison with the traditional big data computing facilities like HPC cluster or computing grid, when the problem size is fairly large, volunteers are heterogeneous in compute-capacity and bandwidth and committing churn at a large rate and particularly when there are a large amount intermediate data exchange between volunteers. This paper aims at filling up the gap by confirming the competence of VC for solving big data problems to clarify the above issues.

3 The Conceptual Unification of 3 Computing Facilities

This section conceptually unifies 3 computing facility HPC cluster, computing grid and VC overlay, along with the related communication cost and computing load of big data problems, by proposing the concept of virtual *Base Capacity* and *Standard Step*. Such a conceptual unification enables the modelling and sound comparison of different computing facilities on the same platform.

First, we need to measure the compute-capacities of HPC cluster, computing grid or VC overlay. The start point of such a measurement is a general personal computer (PC). The test results from TechPowerUp [28] have shown that the popular Intel Core i3, i5 and i7 processors vary from *10s* to *170s GFlops* with the average of *60s GFlops*. An investigation of some HPC clusters in Australia has been summarized in Table 1. On this basis, the simple equivalence by flops between a HPC cluster and the numbers of PC is also listed in Table 1.

Table 1. The compute-capacities (flops) of some Australia HPC clusters

HPC cluster	Flops	Equivalence (PCs)
Raijin [22], Australia National Computational Infrastructure	1.37 PFlops	23,000
Fujin [13], Australia National Computational Infrastructure	22.7 TFlops	378
Isaac Newton Cluster [15], Central Queensland University	12.35 TFlops	206
Tinaroo [30], University of Queensland	234.2 TFlops	3,900
The SGI Altix XE Cluster [25], Queensland University of Technology	127 TFlops (Double Precision)/271 TFlops (Single Precision)	2,117/4,517
The Fawkes Cluster [11], University of South Queensland	27.1 TFlops	452

The compute-capacities in terms of flops of the world top 500 supercomputing supremacy are officially listed on Top500 [29]. The general conclusion from the above investigation is that the compute-capacity is relative between any 2 machines. If the slowest computer is chosen as the *Base Capacity* (shorten as *BC*), all others' compute-capacities can be simplified as a times of the base capacity, e.g. *1.5BCs*, *2BCs* etc.

Second, we also need to measure the communication cost of HPC cluster, computing grid or VC overlay. When the broadband standards depend on countries, Australia National Broadband [20] provides 3 speed plans for download and upload (Mbps): *25/5*, *50/20* and *100/40*. Some speed test results by using the popular internet speed test tools including the Australian government recommended *speedtest.net* showed that the internet speed of Central Queensland University (the authors' affiliation) corporate network compiles with the *100/40* and one of the authors' home internet connection complies with the 25/5 standards. Thus for a given dataset, e.g. *50* MB, the data exchange speed of *16/80* s for an internet volunteer and *4/10* s for a corporate network is a safe assumption. In this paper the communication cost is unified by *Standard Step* like *4/10*, representing the download and upload speed of *4* standard steps and *10* standard steps for a *50* MB dataset. The internal network between computing nodes of a HPC cluster is significantly fast. For a *50* MB data exchange, it costs less than 1 s (1 standard step in this paper) in the current available test reports [8, 9].

Third, we need to measure the compute-load of a map or reduce task. The compute-load of a task depends on the processing algorithm, the size of a big data problem, the split numbers of the original data set and the compute-capacity of a computer. However, once the above are determined, the compute-load of a task is determined. As described previously in this section, the compute-capacity is relative for any 2 computing facilities. Thus the s*tandard step* is also used to represent the workload of a task for a single *BC* computer to complete it. For example, if the standard steps of a task is *48K*, it means that a *BC* computer needs *48K* stand steps to complete it, but a *2BCs* computer needs only *24K* standard steps to complete the same task.

Based on the above analysis, the conceptual unification of 3 types of computing facility can be summarized as follows. A HPC cluster consists of reliable and homogeneous computing nodes, having the same compute-capacity, e.g. *6BCs*. The data exchange speed between the computing nodes of a HPC cluster is significantly fast, for example a standard step for a *50* MB dataset. A computing grid consists of reliable but heterogeneous computing nodes, having different compute-capacity *mBCs*, where *m* is a positive number, e.g. *3* or *6* etc. However, the difference in compute-capacity is not too much because the grid is based on the corporate computers. The computing nodes of a computing grid solely work for the big data problem during the course of computing. The communication network between the computing nodes of a grid are reliable and the data exchange speed between the nodes is fast. A volunteer overlay consists of heterogeneous volunteers, having different compute-capacity *mBCs*, where *m* is a positive number and can be significantly different, e.g. *1*, *1.2*, *3* or *6* etc. The volunteers are opportunistic because they can join a big data computing at any time, leave or crash even before completing the assigned tasks. Both the reliability of communication network and the data exchange speed between volunteers are assumed sound.

4 The DHT Based MapReduce Model

This section proposes a DHT-based model to bind the most complex case of volunteer overlay modelling and MapReduce modelling together. On such a model, the computing grid can be treated as a special case of VC with computing nodes behaving reliably (no churn). Furthermore, the HPC cluster can also be treated as a special case of VC as well with all computing nodes behaving reliably and being homogeneous in compute-capacity. The benefit of such a modelling unifies the 3 computing facility structures to make performance comparison between them sound on the same platform.

The DHT-based modelling incorporates the following considerations. For a big data problem, it is allowed the problem to be split into any *number of map or reduce tasks*. Each task's *compute-load* is represented by the number of standard steps of a single *BC* computer. For data exchange speed, it is allowed in standard steps of the setting of *search time*, *download time* and *upload time* for a task to be located and downloaded and to return the result of the task. For a computing facility, it is allowed the setting of the *number of computing nodes* and the *compute-capacity* of each computing node. For volunteers, it is allowed the setting of the *number of volunteers*, the *join time* of every volunteer and the *leave* or *crash time* of each churn volunteer. Based on the considerations, the computing progress of a task can be trigged by a logical clock. If we assume that each clock cycle is n standard steps, a task will progress $n \times m$ standard steps in a clock cycle, where the task is assigned to an *mBC* computer and m is a positive number. A task will be certainly completed by the computing node to which it is assigned in a HPC cluster or a computing grid, but a task can be left unfinished by a volunteer which leaves or crashes. The DHT-based modelling allows the progress of an unfinished task of a leave volunteer to be check-pointed so that the unfinished task can be picked up by another volunteer to continue the progress. However, the modelling does not check-point the progress of an unfinished task of a crash volunteer, thus, the task must be completely redone by another volunteer.

The DHT-based modelling firstly considers the most complex condition of MapReduce on a volunteer overlay. To model MapReduce on DHT, any object like a task or a result is represented by a key-value pair. DHT structure naturally matches the data structure of MapReduce as depicted in Fig. 1 of Sect. 1 of this paper, and the modelling of this paper extends the key-value pairs to represent a MapReduce task as a finite state machine. In the Map step, each map task changes between the following states as illustrated in Fig. 2.

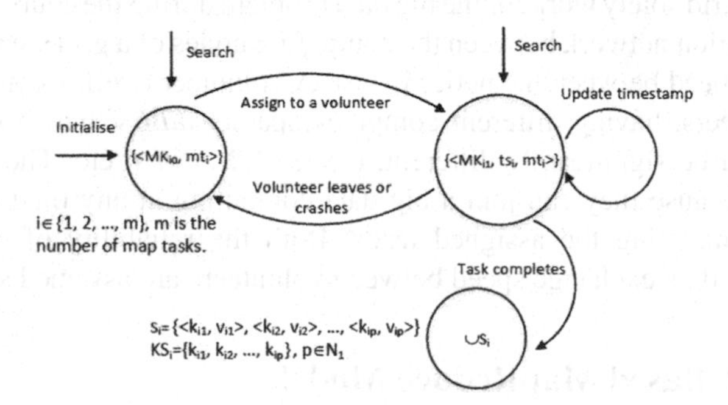

Fig. 2. The state change of map tasks

The original big data problem is split into $m \in N_1 = \{1, 2, \ldots\}$ map tasks $<MK_{i0}, mt_i>$, which are uploaded into the VC overlay, where MK_{i0} is the key and mt_i is the task and $i \in \{1, 2, \ldots, m\}$. When a volunteer gets a task, it will put the task into the *execution state* by changing MK_{i0} to MK_{i1}. If the volunteer leaves before completing the task, it will change MK_{i1} back to MK_{i0}. An active volunteer needs to update the timestamp ts_i of the task within an agreed time period in order to inform that it is still active on the overlay. Otherwise, the task will be treated as dead (from a crash volunteer). Under such assumption, a volunteer has 2 search points: search of an available task $<MK_{i0}, mt_i>$ or search of a dead task $<MK_{i1}, ts_i, mt_i>$ which timestamp ts_i has not been updated for longer than the agreed time period. A map task mt_i is a self-satisfied object, which includes the executable code and data that are encapsulated in the data structures that are appropriate to a particular MapReduce application. A method call on mt_i such as $mt_i.execute()$ will perform the map task and emit a set of key-value pairs $S_i = \{<k_{i1}, v_{i1}>, <k_{i2}, v_{i2}>, \ldots, <k_{ip}, v_{ip}>\}$. Furthermore we assume $KS_i = \{k_{i1}, k_{i2}, \ldots, k_{ip}\}$ and $p \in N_1$ from the output of this step.

In the Shuffle step as depicted in Fig. 3, the output of the last step is re-distributed into a number of reduce tasks $<RK_{j0}, rt_j>$ by a hash function hf, where RK_{jo} is the key of an available reduce task rt_j, $j = \{1, 2, \ldots, r\}$ and $r \in N_I$ is the total number of reduce tasks. The shuffle procedure is to feed data to each reduce task by the following rule:

1. At the beginning, $rt_j = \phi$, where $j = \{1, 2, \ldots, r\}$ and r is the total number of reduce tasks
2. For any $k \in KS_i$ (see the output of last step), $hf(k) = RK_{j0}$, where $i \in \{1, 2, \ldots, m\}$ and $m \in N_1$ is the total number of map tasks
3. For any $<k_1, v_1>, <k_2, v_2> \in \cup S_i \cup rt_j$ (see the output of last step) and $hf(k_1) = hf(k_2) = RK_{j0}$, if $k_1 \neq k_2, rt_j = rt_j \cup \{<k_1, v_1>, <k_2, v_2>\}$. Otherwise if $k_1 = k_2 = k, rt_j = rt_j - \{<k, list_k>\} \cup \{<k, list_k \cup \{v_1, v_2\}>\}$, where $list_k$ is the value list of k.

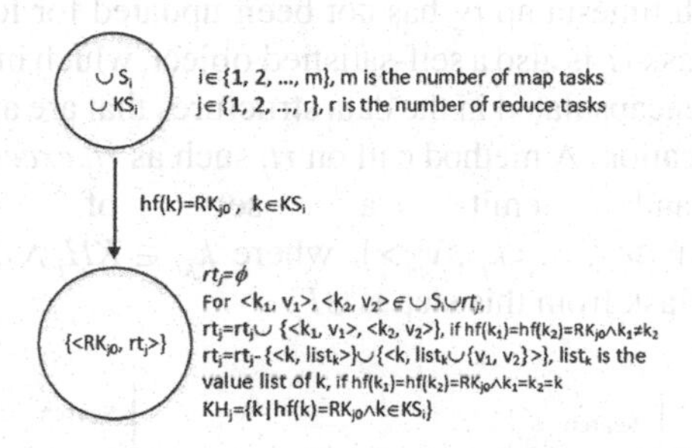

Fig. 3. The redistribution of the output data from map step into reduce tasks

That is, the shuffle procedure will classify the output key-value pairs from the map step into r sets, of which the same key will represent a single set only and multiple values for the same key comprise the value list of the key. Furthermore we assume $KH_j = \{k | hf(k) = RK_{j0} \wedge k \in KS_i\}$.

The shuffle step incurs a large amount of data exchanges between volunteer nodes. For the result set $S_i = \{<k_{i1}, v_{i1}>, <k_{i2}, v_{i2}>, \ldots, <k_{ip}, v_{ip}>\}$ from the map task mt_i, where $i \in \{1, 2, \ldots, m\}$ and $m \in N_I$ is the total number of map tasks and $p \in N_I$, we define a *re-distribution factor* rf_i, the total number of the other volunteers that the volunteer of mt_i needs to redistribute S_i. That is, if S_i is divided by hf into $S_i = SS_1 \cup SS_2 \cup \ldots \cup SS_t$ subsets where $t = rf_i$, there will be rf_i other volunteer nodes to be contacted for the redistribution of S_i. For each re-distribution, the data size is different. However the total data size of the re-distribution still remains DS_i, which is the data size of S_i, because for any SS_a and SS_b, $SS_a \cap SS_b = \phi$, where $a, b \in \{1, 2, \ldots, rf_i\}$. Furthermore we define a *re-distribution cover times* as:

$$\frac{\text{the total number of map result redistribution}}{\text{the total number of reduce tasks}}$$

For example, if *we* assume $m = 10{,}000$, $r = 5{,}000$, $rf_i = 200$ and $DS_i = 50$ MB, each S_i needs *200* result re-distributions, of which each re-distribution will upload a *250* KB (*50* MB/*200*) data set in average. For total *10,000* result sets, there will be *2,000,000* (*10,000* × *200*) data re-distributions. Consequently the redistribution cover times will be *400* (*2,000,000*/*5,000*). Furthermore, the lookup performance of the volunteer node of storing a reduce task is Chord provable performance $O({}_{log}n)$, where $n \in N_1$ is the total number of volunteers. Thus the overall lookup performance for the re-distribution of each result set S_i will be $rf_i \times O({}_{log}n)$.

The reduce step is similar to the map step as illustrated in Fig. 4. In the step, there are $r \in N_1$ available reduce tasks $<RK_{j0}, rt_j>$ from the shuffle step, where $j = \{1, 2, \ldots, r\}$. When a volunteer gets a task, it will put the task into execution state by changing RK_{j0} to RK_{j1}. If the volunteer leaves before completing the task, it will change RK_{j1} back to RK_{j0}. An active volunteer needs to update the timestamp ts_j of the task within an agreed time period in order to inform that it is still active on the overlay. Otherwise, the task will be treated as dead (from a crash volunteer). Under such assumption, a volunteer has 2 search points: search of an available reduce task $<RK_{j0}, rt_j>$ or search of a dead task $<RK_{j1}, ts_j, mt_j>$, which timestamp ts_j has not been updated for longer than the agreed time period. A reduce task rt_j is also a self-satisfied object, which includes the executable code and data that are encapsulated in the data structures that are appropriate to a particular MapReduce application. A method call on rt_j such as *rt*$_j$*.execute()* will perform the reduce task and emit a set of key-value pairs $T_j = \{<k_{j1}, v_{j1}>, <k_{j2}, v_{j2}>, \ldots, <k_{jd}, v_{jd}>\}$, where $k_{jd} \in KH_j \wedge d \in N_1$. The result of the entire MapReduce task from this step is $\cup T_j$.

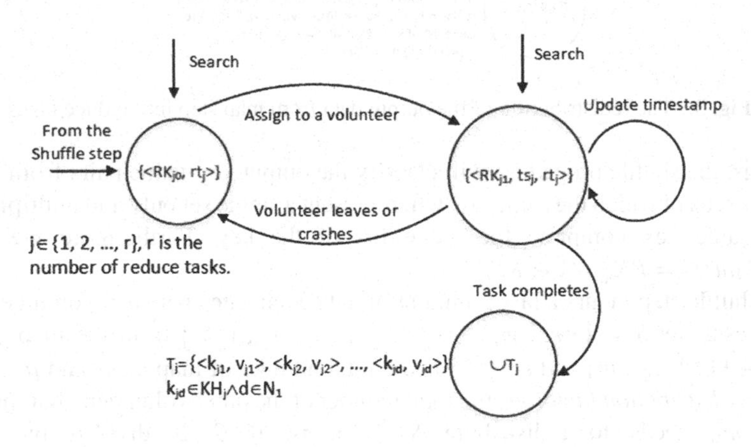

Fig. 4. The state change of reduce tasks

5 The Evaluation Environment

Written in Java, the simulator of the proposed DHT-based MapReduce model has been implemented on the Chord DHT protocol [26]. Once a volunteer overlay is constructed by the simulator, it will have the following certain reliability and performance from the Chord.

1. Chord overlay transparently replicates the uploaded key-value pairs to tolerate volunteer crashes to provide built-in reliability.
2. Chord stabilisation protocol periodically updates the routing information to reflect churn; Chord consistent hashing ensures the storage load balance among volunteer nodes.
3. Given a key, searching the value identified by the key is done in logarithmic performance $O(_{log}n)$, where $n \in N_1$ is the number of peers on the Chord overlay.

Based on the above condition, a MapReduce problem can be completed by a Chord DHT overlay as long as there are a number of volunteers continuously active on the overlay.

Instead of using a real big data problem, a virtual problem is proposed to generalize any big data problems as long as they comply with the MapReduce programming paradigm as defined by [7]. This virtual problem is able to demonstrate the generality of the simulation model in the volunteer environments, thus the runtime behaviors of this virtual problem could be applied to a wide range of MapReduce problems. In the virtual problem, a safe assumption of the size of map or reduce tasks is *50* MB. Such assumption has real world application support, e.g. each SETI@Home [16] work unit is only *0.34* MB. The simulator has no restriction on the number of map or reduce tasks, but in this evaluation, both the number of map and reduce tasks is randomly chosen as equally 1.4 million. Thus the big data problem size is $1.4 \times 2 \times 50$ MB = *140* TB, which is big enough for the effectiveness evaluation of the model.

For the evaluation, the compute-load of each map or reduce task is randomly chosen as *8,000* standard steps by a single *BC* computer. This compute-load assumption is big enough for a single *BC* computer to process the entire big data problem in $1.4 \times 2 \times 8{,}000 = 22.4 \times 10^9$ standard steps.

The download time and upload time of a *50* MB map or reduce task is determined by the general assumption of a volunteer's internet speed of *25/5* Mbps. Thus they are *16/80* s (standard steps in the evaluation) respectively. For a computing grid, the internet speed is assumed as *100/40* Mbps, thus, the download time and upload time are *4/10* s (standard steps in the evaluation) respectively. For a HPC cluster using a distributed file system, the speed of download and upload is significantly fast. In the evaluation, we set them as *1/1* because a standard step is the minimum granularity of communication cost.

For a volunteer to search for a task, the search time is calculated on the provable performance of Chord lookup services of $O(_{log}n)$ [26], where $n \in N_1$ is the number of volunteers, because the simulator is built on the Chord protocol. If we assume that each lookup message needs *4* standard steps to turnaround, the search time for a maximum *40,000* peers overlay will be $4 \times log(40{,}000) \approx 60$ *standard steps*. The real situation could be better when these messages are somehow processed concurrently and when the volunteers are actually less than *40,000* due to churn. For a computing grid or a HPC cluster, we assume the search time of 0. The reason is that because the computing nodes are reliable in a HPC cluster or a computing grid, the map or reduce tasks can be pre-assigned to certain nodes.

For the volunteer overlay, we represent the heterogeneity of volunteers by assuming that they have different compute-capacities from 6 randomly chosen levels between *BC* and *6BCs*. This assumption of difference is sound for modern PCs, of which a high-end PC can be several times faster than a low-end one. The assumption of a computing grid is that it has computing nodes with small heterogeneity in compute-capacity. Thus in the evaluation, we set 2 levels of compute-capacity as having *3BC* and *6BCs* computing nodes for the computing grid. The HPC cluster has homogeneous and high-end computing nodes, thus, setting all the computing nodes as having *6BCs* compute-capacity is a safe assumption.

Table 2. The scenario setting of evaluation for 3 computing facilities

Scenario variable	Value		
	Volunteer overlay	Computing grid	HPC cluster
The number of map tasks	1,400,000	1,400,000	1,400,000
The number of reduce tasks	1,400,000	1,400,000	1,400,000
The compute-load in standard steps of a map or reduce task for a single *BC* computer	8,000	8,000	8,000
The turnaround to lookup a particular task on the overlay	4	0	0
The result re-distribution factor and re-distribution cover times	200 and 400	0	0
The search time for a map or reduce task in standard steps	$4 \times O(_{log}cn)$, where $cn \in N_1$ *is the current number of volunteers on the overlay*	0	0
The download time for a map or reduce task in standard steps	16	4	1
The upload time for a map or reduce result in standard steps	80	10	1
The number of volunteers/computing nodes	1,000 to 40,000	1,000 to 12,000	1,000 to 8,000
The compute-capacity	Evenly in *BC, 2BCs, 3BCs, 4BCs, 5BCs, 6BCs*	Evenly in *3BCs* and *6BCs*	*6BCs*
The churn rate	30%, 50% of total volunteers	0	0
The peer join interval	20 standard steps	All joins at time 0	All joins at time 0

Another assumption is about how churn occurs in the evaluation. Assume that there are n volunteers, of which c commit churn and $n, c \in N_1$. For volunteer *1* to n, they are divided into c segments, of which each segment has n/c volunteers and a churn volunteer is randomly chosen for each segment. The churn time for each volunteer is also randomly chosen so that the volunteer could commit churn when it is searching for, downloading or computing a map or reduce task or uploading a result. For all churn volunteers, a half leaves and the other half crashes.

In the evaluation scenario, volunteers join the VC overlay in an interval of *20* (randomly chosen) standard steps and the churn would happen to any volunteers if it is randomly chosen for. All computing nodes of a HPC cluster or a computing grid join at the same time when the computing starts and they never commit churn. The evaluation scenario is set as in Table 2.

6 The Evaluation of Competence

In this section, the evaluation results and analysis are reported. The performance of a computing facility is measured by speedup, that is,

$$\frac{\text{the total standard steps to complete the entire problem by a single BC computer}}{\text{the total standard steps to complete the entire problem by a computing facility}}$$

The evaluation of the 3 computing facilities has been conducted by testing the speedup against the number of computing nodes or volunteers starting from 1,000 for an increasing step of 1,000 each time. The results have been reported in 2 ways. First (Way 1), the speedup has been reported by the *average compute-capacity* (in the measurement of *BCs*) of a computing facility, which is calculated as follows and where the compute-capacity types are evenly distributed among computing nodes or volunteers, *cn* is the current number of computing nodes or volunteers and the churn rate is 0 for the HPC cluster or the computing grid.

$$\frac{\text{The sum of compute} - \text{capacity of each type}}{\text{The total number of compute} - \text{capacity types}} \times (1 - \text{the churn rate}) \times \text{cn}$$

Based on the above formula, the average compute-capacity of the computing facilities in Table 2 is $3.5\ (i.e.,\ 1+2+3+4+5+6)/6) \times (70\%\ or\ 50\%) \times cn$ for the volunteer overlay, $4.5\ (i.e.,\ (3+6)/2) \times cn$ for the computing grid and $6 \times cn$ for the HPC cluster respectively. The speedup results have been illustrated in Figs. 5, 6 and 7 respectively for each computing facility in Table 2. From the results, any of the 3 computing facility's speedup scales for the big data problem when the average compute-capacity increases, particularly for the HPC cluster or computing grid, the speedup is close to linear growth. If we define *speedup/compute-capacity* as the measurement of each *BC* contribution to the speedup acceleration, the *average speedup acceleration* of the HPC cluster, the computing grid and volunteer overlay is listed in Table 3. The differences between the maximum or minimum speedup acceleration and the average speedup acceleration are also listed in Table 3.

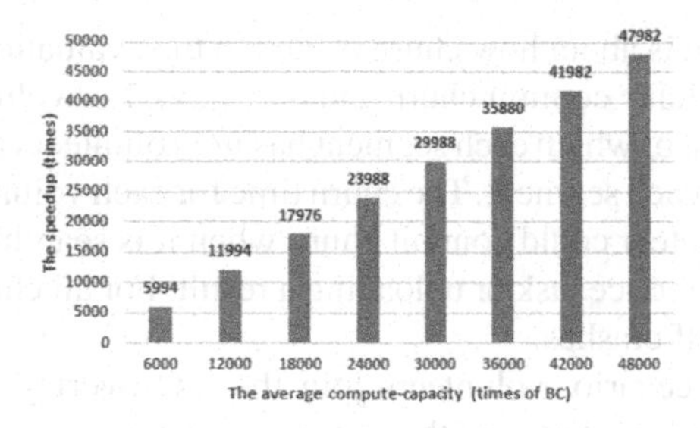

Fig. 5. The speedup of the HPC cluster vs the average compute-capacity

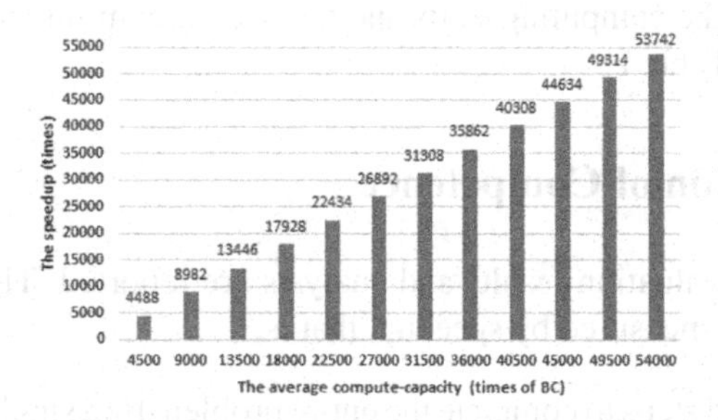

Fig. 6. The speedup of the computing grid vs the average compute-capacity

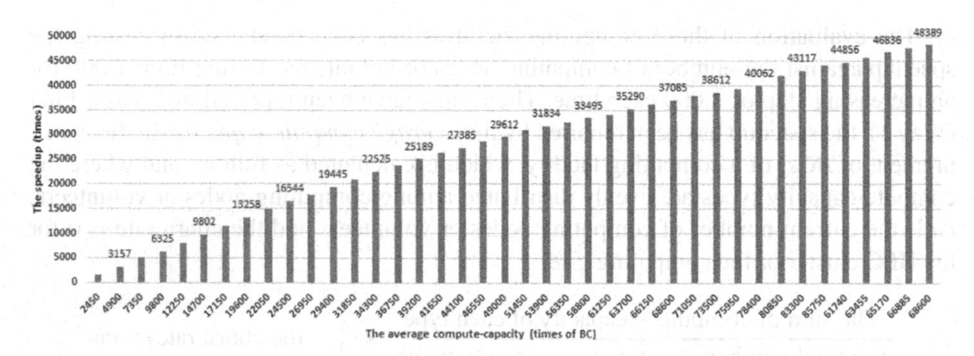

Fig. 7. The speedup of the VC overlay (with *30%* churn rate) vs the average compute-capacity

Table 3. The average speedup acceleration, the differences between the maximum or minimum speedup acceleration and the average speedup acceleration

Computing facility	Average speedup acceleration	Maximum acceleration difference	Minimum acceleration difference
HPC	0.99902	0.00235	0.00002
Computing grid	0.99576	0.00389	0.00025
Volunteer overlay	0.619708	0.106823	0.00127

Comparing the results in Table 3, the computing grid has almost reached the speedup acceleration of the HPC cluster for each more *BC* contribution. The volunteer overlay has a lower speedup acceleration because of the waste of computing from peer crashes, the search for reduce tasks to re-distribute map results and higher cost of communication. Either the maximum or the minimum acceleration difference is in the ascending order of the HPC cluster, the computing grid and volunteer overlay. That property complies with stability of the 3 computing facilities.

Second (Way 2), the speedup of 3 computing facilities has been contrasted for the same number of computing nodes or volunteers in a single chart in Fig. 8, where the VC overlay has *30%* and *50%* churn rate respectively. The results have demonstrated that any of the 3 facilities can reach the same speedup for different number of computing nodes or volunteers, complying with the conclusion in Way 1. It can be inferred that the computing grid needs more computing nodes to reach the same speedup as that of the HPC cluster because of the heterogeneous compute-capacity of the computing nodes and higher communication cost between computing nodes in the computing grid. Furthermore the volunteer overlay needs much more volunteers to reach the same speedup as that of the HPC cluster because it has more heterogeneous compute-capacities of volunteers, much higher communication cost than that of the HPC cluster, the need of search for reduce tasks to re-distribute map results and particularly the peer churn.

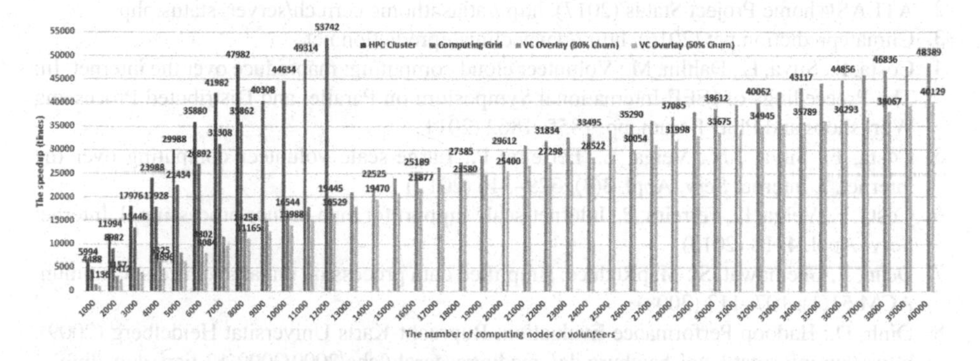

Fig. 8. The speedup contrast of the 3 types of computing facility

7 Conclusion

The 3 most important types of parallel or distributed computing facility, HPC cluster, computing grid and VC, have been successfully modelled onto a single simulation platform with computing grid and HPC cluster being treated as the special cases of VC. By configuration, a computing grid is just a VC overlay without churn or the need of search for reduce tasks to re-distribute map results and with only small heterogeneity of compute-capacity between computing nodes. Furthermore on computing grid configuration, if all computing node are homogeneous in compute-capacity, it is configured as a HPC cluster. As all 3 facilities have been unified onto a single platform, a sound comparison of their performance to solve MapReduce big data problems has become possible.

The competence of computing facilities has been quantitatively evaluated in terms of overall speedup by using a virtual big data problem. The evaluation results have confirmed that any computing facility's speedup scales for the 140 TB big data problem in the evaluation and any computing facility can reach the given speedup as long as the computing nodes or volunteers have been increased to a certain number, where the average compute-capacity of the facility is big enough for the required speedup. The main difference among the 3 computing facilities is the actual compute-capacity that they can provide in the same number of computing nodes or volunteers in consideration of the heterogeneity and possible churn of computing nodes or volunteers.

When an optimized simulation algorithm for a higher time efficiency in simulation is developed, the future work includes more intensive evaluations by using even bigger data problems and larger number of volunteers such as millions of volunteers to further confirm the competence of VC.

References

1. Anderson, D.P.: BOINC: a system for public-resource computing and storage. In: The Proceedings of 5th IEEE/ACM International Conference on Grid Computing, pp. 4–10 (2004)
2. ATLAS@home Project Status (2017). http://atlasathome.cern.ch/server_status.php
3. Climateprediction.net (2016). http://www.climateprediction.net
4. Costa, F., Silva, L., Dahlin, M.: Volunteer cloud computing: mapreduce over the internet. In: The Proceedings of IEEE International Symposium on Parallel and Distributed Processing Workshops and Ph.D Forum, pp. 1855–1862 (2011)
5. Costa, F., Silva, J.N., Veiga, L., Ferreira, P.: Large-scale volunteer computing over the internet. J. Internet Serv. Appl. **3**(3), 329–346 (2012)
6. Costa, F., Veiga, L., Ferreira, P.: Internet-scale support for map-reduce processing. J. Internet Serv. Appl. **4**, 18 (2013)
7. Dean, J., Ghemawat, S.: MapReduce: simplified data processing on large clusters. Commun. ACM **51**(1), 107–113 (2008)
8. Dinh, D.: Hadoop Performance Evaluation, Ruprecht-Karls Universitat Heidelberg (2009). https://wr.informatik.uni-hamburg.de/_media/research/labs/2009/2009-12-tien_duc_dinh-evaluierung_von_hadoop-report.pdf
9. Donvito, G., Marzulli, G., Diacono, D.: Testing of several distributed file-systems (HDFS, Ceph and GlusterFS) for supporting the HEP experiments analysis. J. Phys. Conf. Ser. **513**(4), 1–7 (2014)
10. DrugDiscovery@Home (2017). http://www.drugdiscoveryathome.com/
11. Fawkes Cluster (2017). https://www.usq.edu.au/research/support-development/development/eresearch/hpc/hardware
12. FiND@Home (2017). http://findah.ucd.ie
13. Fujin (2017). http://nci.org.au/systems-services/national-facility/peak-system/fujin/
14. Hadoop (2014). https://wiki.apache.org/hadoop/ProjectDescription
15. Isaac Newton Cluster (2017). https://my.cqu.edu.au/web/eresearch/hpc-systems
16. Korpela, E.J.: SETI@home, BOINC, and volunteer distributed computing. Ann. Rev. Earth Planet. Sci. **40**, 69–87 (2012)
17. Lin, H., Ma, X., Archuleta, J., Feng, W.C., Gardner, M., Zhang, Z.: Moon: MapReduce on opportunistic environments. In: The Proceedings of the 19th ACM International Symposium on High Performance Distributed Computing, pp. 95–106 (2010)

18. Monsalve, S.A., Carballeira, F.G., Mateos, A.C.: Analyzing the performance of volunteer computing for data intensive applications. In: The Proceedings of International Conference on High Performance Computing and Simulation, pp. 597–604 (2016)
19. Alonso-Monsalve, S., García-Carballeira, F., Calderón, A.: A new volunteer computing model for data-intensive applications. Concur. Comput. Pract. Exp. **24**, e4198 (2017)
20. Australia National Broadband (2017). http://www.nbnco.com.au/learn-about-the-nbn/speed.html
21. Oracle: An Enterprise Architect's Guide to Big Data - Reference Architecture Overview, Oracle Enterprise Architecture White Paper (2016)
22. Raijin (2017). http://nci.org.au/systems-services/national-facility/peak-system/raijin/
23. Ryden, M., Oh, K., Chandra, A., Weissman, J.: Nebula: distributed edge cloud for data intensive computing. In: The Proceedings of IEEE International Conference on Cloud Engineering, pp. 57–66 (2014)
24. Sarmenta, L.: Volunteer Computing, Ph.D thesis, Massachusetts Institute of Technology (2001)
25. SGI Altix XE Cluster (2017). http://www.itservices.qut.edu.au/researchteaching/hpc/hpc_infrastructure.jsp
26. Stoica, I., et al.: Chord: a scalable peer-to-peer lookup protocol for internet applications. IEEE/ACM Trans. Netw. **11**(1), 17–32 (2003)
27. Tang, B., Moca, M., Chevalier, S., He, H., Fedak, G.: Towards mapreduce for desktop grid computing. In: The Proceedings of International Conference on P2P, Parallel, Grid, Cloud and Internet Computing, pp. 193–200 (2010)
28. TechPowerUp (2017). https://www.techpowerup.com/forums/threads/processor-gflops-compilation.94721/
29. Top500 (2017). https://www.top500.org/lists/2017/11/
30. Tinaroo (2017). https://rcc.uq.edu.au/tinaroo

Research on the Security Protection Scheme for Container-Based Cloud Platform Node Based on BlockChain Technology

Xiaolan Xie[1,2], Tao Huang[1(✉)], and Zhihong Guo[1]

[1] College of Information Science and Engineering, Guilin University of Technology, Guilin, Guangxi Zhuang Autonomous Region, China
228152339@qq.com

[2] Guangxi Universities Key Laboratory of Embedded Technology and Intelligent System, Guilin University of Technology, Guilin, China

Abstract. With the development of science and technology, container-based cloud platforms have been widely used. For the security issues for container-based cloud platforms, this paper lists the current threats to container-based cloud platforms and demonstrates the principle of BlockChain technology. The aim of this paper is to demonstrate a security protection scheme for container-based cloud platform node based on BlockChain technology. With the characteristic of tamper-resistant and traceability of BlockChain, the Merkle tree is utilized in the scheme to protect vital documents in child nodes. The effectiveness of this technology for file security protection has been verified.

Keywords: BlockChain · Merkle tree · Container-based cloud platform Hash encryption · Node protection

1 Introduction

Container technology has the characteristics of continuous deployment and testing, high resource utilization and isolation, ease of use and easy understanding, and it has been rapidly developed and popularized. However at the same time, the security of containers and nodes in container-based cloud platforms has also become increasingly prominent. Attacks such as node attacks, kernel exploits, malicious mirroring, and key acquisition have emerged in an endless stream, seriously threatening the security of container-based cloud platforms, and constraining the development of container technology. The existing security protection methods have certain limitations and cannot protect the security of the nodes in the container-based cloud platform well. With the advent of BlockChain technology, a new idea for the protection of container-based cloud platform node key file information is proposed. Characteristics of BlockChain such as tamper-resistance, forge-resistance, distribution and so on can be applied to the container-based cloud platform node security protection, thereby greatly improving the node's security performance.

Q. Zhou et al. (Eds.): ICPCSEE 2018, CCIS 901, pp. 24–32, 2018.
https://doi.org/10.1007/978-981-13-2203-7_3

BlockChains are mainly divided into Public BlockChains, private BlockChains, and Consortium BlockChains. The Public BlockChain is a BlockChain where everyone can access and read transaction information and users are not interfered with by developers. In the Public BlockChain, all data is public by default. It is mainly applied to Bitcoin, Ethereum, Hyperledger and other fields. The characteristic of the private BlockChain is that its write permission is only in the hands of an organization or individual, and the read permission and openness are limited. Compared to the Public BlockChain, the private BlockChain has better protection in terms of file privacy, and the transaction speed is also faster than Public BlockChains. The Consortium BlockChain can be considered as "partial decentralization", and the public can consult and trade, but can not verify the transaction, either publish smart contracts, unless the permission from Consortium is obtained [1].

In this paper, the container-based cloud platform is considered as a private BlockChain. The master node is the only node that has the key file write permission for all child nodes. When creating a child node, hash encrypt the information of all tamper-resisitance key files in the node and generate two Merkle trees. Save the hash digests of all root nodes to the master node. By continuously verifying the consistency of the hash value of the root node (Fig. 1), the key file of the container-based cloud platform node cannot be tampered with, and the overall security of the container-based cloud platform is improved.

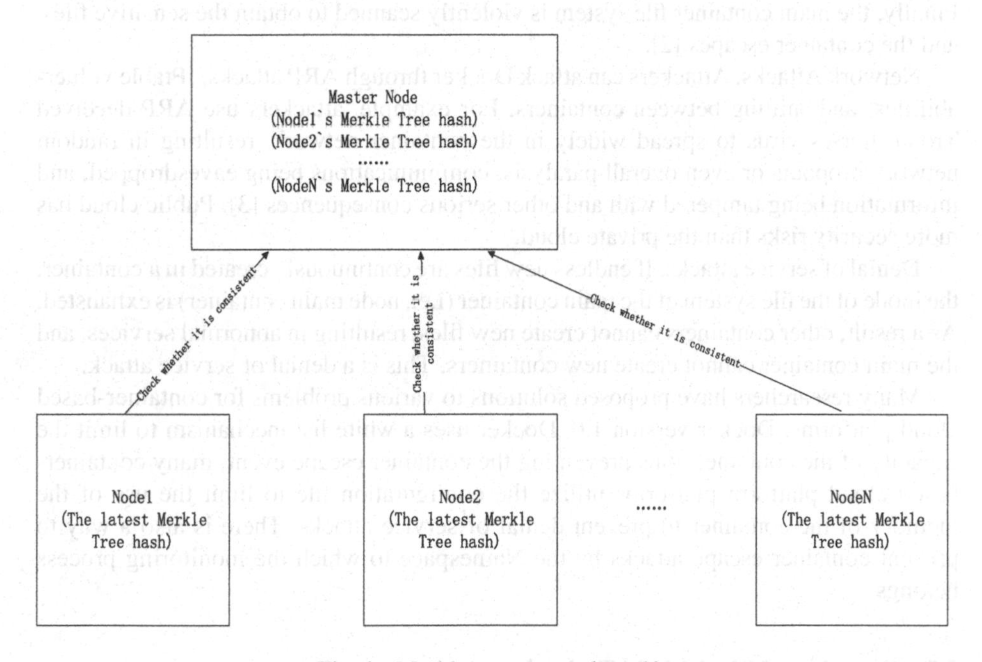

Fig. 1. Merkle tree root node verify

2 Overview of Container-Based Cloud Platform Security and BlockChain Technology

2.1 Overview of Container-Based Cloud Platform Security

Container technology is a lightweight virtualization technology with the advantages of rapid integrated deployment, high suitability, version controllability, high utilization, and cross-platform capability. Its vigorous development has also brought many security issues:

Remote attack problem. Due to the user's negligence in configuration, the remote access port was opened, resulting in a malicious attacker gaining access to the remote container through the remote access interface. When the container-based cloud platform is started, the hacker tampers with the root file, writes its own ssh private key, and finally gains remote control of the main container.

Container escape problem. The Docker operating system is virtualized to share the kernel, memory, CPU, and disk. However, when the container escapes, it poses great threats to all the containers connected to the node as well as the node itself. For example, if the operator calls the open_by_handle_at function, then the operator will obtain the inode index number of the target file by constructing a file_handle, and then uses the index to construct the file_handle again, and calls the open_by_handle_at function. Finally, the main container file system is violently scanned to obtain the sensitive files, and the container escapes [2].

Network Attacks. Attackers can attack Docker through ARP attacks, IPtable vulnerabilities, and sniffing between containers. For example, attackers use ARP-deceived Trojan horses virus to spread widely in the container network, resulting in random network dropouts or even overall paralysis, communications being eavesdropped, and information being tampered with and other serious consequences [3]. Public cloud has more security risks than the private cloud.

Denial of service attacks. If endless new files are continuously created in a container, the inode of the file system of the main container (i.e., node main container) is exhausted. As a result, other containers cannot create new files, resulting in abnormal services, and the main container cannot create new containers. This is a denial of service attack.

Many researchers have proposed solutions to various problems for container-based cloud platforms. Docker version 1.0, Docker uses a white list mechanism to limit the capacity of the container, thus preventing the container escape event; many container-based cloud platform platforms utilize the configuration file to limit the use of the memory of the container to prevent denial of service attacks. There is also a way to prevent container escape attacks by the Namespace to which the monitoring process belongs.

2.2 Overview of BlockChain Technology

BlockChain is a distributed shared ledger that can hardly be tampered. It has the following features:

(1) Decentralization. The BlockChain utilizes the p2p network. Each node has the same authority and all obey the same transaction rule. This rule is based on cryptographic algorithms.
(2) Safe and credible. Asymmetric encryption is used to encrypt transaction content, and a one-way hashing algorithm is used to add time irreversibility, enabling traceability to any intrusion tampering.
(3) Smart contracts. Smart contracts can not only exert the advantages of smart contracts in terms of cost, but also avoid the interference of malicious behaviors on the normal execution of contracts. The smart contract is written into the BlockChain in a digital form. The features of the BlockChain technology ensure that the whole process of storage, reading, and execution is transparent and can not be traced. At the same time, a state machine system is constructed from the consensus algorithm of the BlockChain, which makes smart contracts run efficiently [4].

BlockChains generally fall into three categories: private BlockChains, Public BlockChains, and Consortium BlockChains.

(1) Public BlockChain: Public BlockChain refers to a BlockChain where anyone can access the BlockChain system at any time to read data, send confirmable transactions, and compete for billing. The Public BlockChain is completely decentralized because no individual or institution can control or tamper with the data.
(2) private BlockChain: private BlockChain refers to the BlockChain whose write permission is controlled by an organization or institution. It is characterized by fast transactions, privacy protection and low transaction costs.
(3) Consortium BlockChain: A consortium BlockChain refers to a BlockChain that is jointly managed by several organizations. Each organization runs one or more nodes. The data in the BlockChain is only allowed to read, write, and send transactions by different organizations in the system, and record transaction data together.

3 Tentative Scheme for the Security Protection for Container-Based Cloud Platform Node Based on BlockChain Technology

3.1 The Basic Idea

In this scheme, the secure and tamper-resistant characteristics of the BlockChain is integrated into the security mechanism of the container-based cloud platform. A Merkle tree is established at each node, which combines the tamper-resistance and traceability features of the key configuration files in the container-based cloud platform nodes, thereby improving the security performance of the nodes.

3.2 Build Merkle Tree

Hash Algorithm
The Hash algorithm can transform data of any length into a fixed-length data, and it is irreversible, and it is impossible to inversely derive the original data according to a series

of hash data. The Hash algorithm is also sensitive to the content of the input. Even if the content of the file changes only a little, the resulting calculations will be very different. Because of the sensitivity and irreversibility of its input content, Hash algorithm has been widely used in cryptography, and it is also the foundation of establishing Merkle tree.

Merkle Tree Construction Process

Hash algorithm is used to convert the tamper-resistant configuration file of the child node into a Hash digest, and then merge the two adjacent hash summaries into a string. This string is converted into a new digest using the Hash algorithm, and the two new digests are merged and hashed. This is a reverse-generation process of the tree. At the end, we will get a root node. This root node is also a fixed-length Hash digest. as shown in Fig. 2.

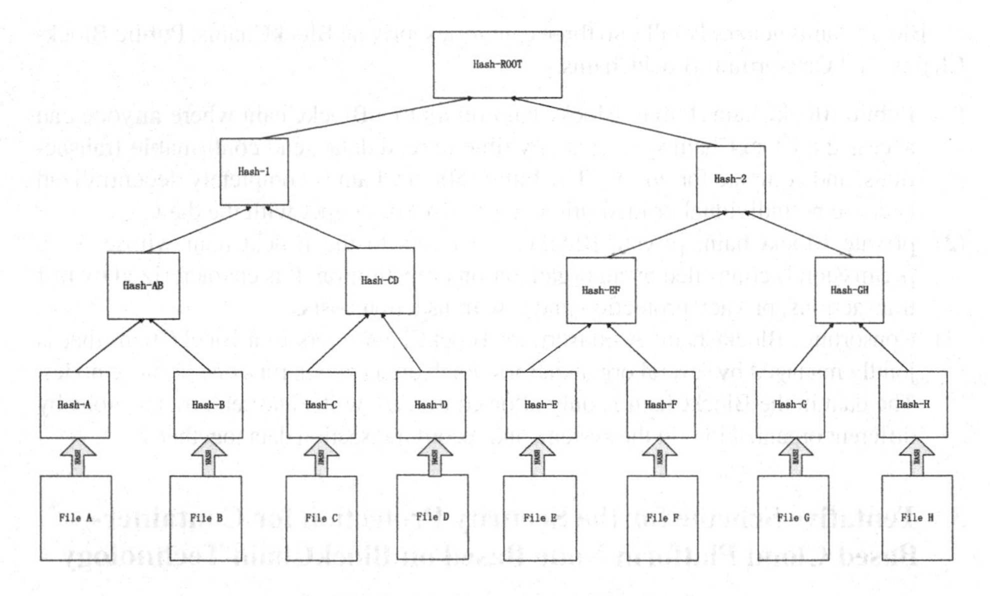

Fig. 2. Merkle tree on child node

If the leaf node's file content is tampered with, the Hash digest content of the Merkle tree root node will change too. If it is found that the content of the Merkle tree root node is incorrect, each leaf node will be compared from the top to the bottom to trace the specific tampered configuration file.

3.3 Program Implementation Process

(1) At the beginning of the creation of the child node, the configuration file to be protected is sent to the master node to be saved as a copy.
(2) Hash algorithm is utilized to generate Hash digest on child nodes. Combining the Hash digests in pairs and recalculating them to build a whole Merkle tree. This Merkle tree and leaf nodes are also stored on the master node with the configuration

file mapping table. The information of Merkle tree on original child node is shown in Fig. 3.

```
Node: node1,
Timestamp:1525508241151,
Files List: test1/test2/test3/test4,
Merkle Tree:
ee42b4e06c5e2ece49a9eb8e75343cc7b88ff021c544a6b8e46b5eacc6898f88
7e33af6e0a922f8198bae6f8549dda031adbce3a8784284cbf6a0ed40ceaa75b / 1a225eed96853fe74ee7
5bbd2c374d9625cb7027443c8c13761320798b0cc80b
3b14a1cf5773020f3758a10af721e8bc5c4d659ee62bdff3ffe105cad1a5923d / 9cf067e9a30e24d89d7f
179e8ab08102c9ed004e0f2fd83c40674a0588c1b3d1 / 783316f171e60dd78bd7d565f4c7890ef48b6591
0c682b5961658191630d8639 / 579de136cde9f489b50230caebabf7ff370daba291cb72724b84c1a95645
3422
~
~
~
~
~
~
~
~
```

Fig. 3. The information of Merkle tree on original child node

(3) The Merkle tree of the child node is recalculated every period of time and the calculation result is sent to the master node for review. The contents of the audit file has the Merkle tree root node and timestamp.
(4) Compared master nodes, the Hash digest of the root node of the Merkle tree is found to be inconsistent with the original one, and the master node requests the child node for the complete Merkle tree.
(5) When the request reaches the Merkle tree, the child node's container-based cloud platform service is stopped. By comparing two Merkle trees, the specific tampered leaf nodes are locked. According to the leaf node and the configuration file mapping table, it is easy to find the tampered configuration file. As shown in Fig. 4, analyzing the Merkle tree sent from the child node in conjunction with Fig. 3, it can be known that the test2 file has been modified. In fact, only 1 byte of data has been tampered with, but the difference in Hash values is obvious.
(6) Send the original copy of the configuration file saved at the master node to the child node, and restart the container-based cloud platform services.

```
Node: node1,
Timestamp: 1525508652083,
Merkle Tree:
d68a57658dee5a5fe245c7d29930a142fb00c9ed7015330d16ca5713eff4694c
21b4aa47cd27067dd0f0555c2def897cc4d3e8345f82b465311bf8d3b07ed8e7 / 1a225eed96853fe74ee7
5bbd2c374d9625cb7027443c8c13761320798b0cc80b
3b14a1cf5773020f3758a10af721e8bc5c4d659ee62bdff3ffe105cad1a5923d / 7c796e8f914acdf83fc0
9e7d634f216eb55ea2bb2a7d12e4bd0086adcb92b0dd / 783316f171e60dd78bd7d565f4c7890ef48b6591
0c682b5961658191630d8639 / 579de136cde9f489b50230caebabf7ff370daba291cb72724b84c1a95645
3422
~
~
~
~
~
~
```


Fig. 4. Merkle tree information for tracking tampered files

3.4 Partial Audit

Partial Audit Principle

The process of constructing and calculating a Merkle tree requires a large number of Hash operations, which requires a large amount of calculation and a long calculation time, and also increases the I/O consumption of the child nodes. Storj is an open source decentralized storage platform that uses a partial audit approach. This approach reduces the computational overhead by auditing a subset of the data and reduces the node's I/O consumption.

This method has two important parameters: a set of byte index x within the file and a set of byte whose length is b. The child node holds a set of triples (s, x, b), where s represents a specific configuration file. When the leaf node i is generated, a piece of data of length b_i in the file is found through the index x_i, and the leaf node is generated using this data. The entire process does not affect the generation of Merkle trees.

Partial Audit Mistake Probability

If only audit a part of the data, it may cause erroneous judgments, but in fact this method has a very low probability of making mistakes. Storj utilizes the confidence of each partial audit to explain the possibility of false positives: $P_{rfalsepositive}(N, K, n)$ is the probability of false positives for partial auditing of n bytes of an N byte file. The probability of tampering with K byte in the file is a hypergeometric distribution of $k = 0$ [5]. The formula is as follows:

$$P_{rfalsepositive}(N, K, n) = \frac{\binom{N-K}{n}}{\binom{N}{n}} \tag{1}$$

In theory, the probability of false positives is very small, even if only a small part is tampered with.

In order to verify the reliability of some audits, in this article the following experiments has been done: randomly select 512 bytes from a text file of size 8192 bytes for Hash encryption, then randomly modify K bytes, re-encrypt, compare before and after twice whether the encryption result is the same. If the result is the same, the test fails. Carry out 100,000 experiments for each set of data. The results are shown in Table 1.

Table 1. Experimental results

N	K	n	$P_{rfalsepositive}(N, K, n)$
8192	50	512	0.03995
8192	100	512	0.00169
8192	150	512	0.00004

The experimental results show that even if only a small part of the document is tampered with, partial audit can have a very low error rate.

3.5 Partial Audit Methods in the Scheme

The partial audit approach adopted in this scheme is to generate two triples for the N-byte file. The two triples divide the content of the file into two parts. The collection of these two parts of data is the complete file data. Therefore, each node will generate 2 Merkle trees, and the master node only needs to save a set of hash values in root node. Each partial audit of a child node is equivalent to a full audit. The Merkle tree number is added to the audit result sent to the master node for distinguishing different Merkle Trees. This saves computing costs, reduces I/O consumption for child nodes, and eliminates the possibility of false positives. As long as the interval between two consecutive audits is short enough, the time cost of growth can be ignored.

4 Conclusion

This article focuses on the combination of BlockChain technology and container-based cloud platform node security, and utilizes the Hash encryption algorithm and Merkle tree in the BlockChain to design a preliminary scheme for container-based cloud platform node profile protection. Through the hash algorithm, the key configuration files of the child nodes are encrypted into hash abstracts, and the unidirectionality and input sensitivity of the hash algorithm are used to ensure that the master node can respond in time and accurately track when the contents of the files are tampered with. By improving partial audit method, I/O consumption of child nodes can be reduced, and computational cost can be saved. This paper verifies the reliability of partial audits. However how to design and implement monitoring and protection systems applied to container-based cloud platforms, and well reduce the computational cost needs to be further researched.

References

1. Zhejiang University SEL Laboratory: Docker container and container cloud. People Post Press`(2016)
2. Xin, S.M., Feng, W., Wang, J.Q.: Research on secure digital document protection scheme based on blockchain technology. J. Inf. Secur. Res. **3**(10), 884–892 (2017)
3. Shawn, W.: Storj a peer-to-peer cloud storage network
4. Xu, Q., Jin, C., Rasid, M.F.B.M., et al.: Blockchain-based decentralized content trust for docker images. Multimed. Tools Appl. (2017)
5. Brinckman, A., et al.: A comparative evaluation of blockchain systems for application sharing using containers. In: 2017 IEEE 13th International Conference on e-Science (2017)
6. Niranjanamurthy, M., Nithya, B.N., Jagannatha, S.: Analysis of blockchain technology: pros, cons and SWOT. Cluster Comput. **20**, 1–15 (2018)
7. Cha, B., Kim, J., Moon, H., et al.: Global experimental verification of docker-based secured mVoIP to protect against eavesdropping and DoS attacks. J Wirel. Commun. Netw. **2017**, 63 (2017)
8. Singh, I., Lee, S.-W.: Comparative requirements analysis for the feasibility of blockchain for secure cloud. In: Kamalrudin, M., Ahmad, S., Ikram, N. (eds.) APRES 2017. CCIS, vol. 809, pp. 57–72. Springer, Singapore (2018). https://doi.org/10.1007/978-981-10-7796-8_5
9. Nakamoto: Bitcoin: a peer-to-peer electronic cash system (2008)
10. Raval, S.: Decentralized Applications: Harnessing Bitcoin's Blockchain Technology. O'Reilly Media, Inc., Newton (2016)
11. Swan, M.: Blockchain: Blueprint for a New Economy. O'Reilly Media Inc, Newton (2015)
12. Olleros, F.S., Zhegu, M.: Blockchain technology: principles and applications. In: Research Handbook on Digital Transformations. Edward Elgar Publishing (2016)

SeCEE: Edge Environment Data Sharing and Processing Framework with Service Composition

Yasu Zhang, Haiquan Wang(✉), Jiejie Zhao, and Bo An

State Key Laboratory of Software Development Environment, Beihang University, Beijing 100191, China
{susan940910,whq,zjj,anbo_software}@buaa.edu.cn

Abstract. A centralized computing paradigm, such as cloud computing, cannot satisfy the explosive growth in the amount of data and computing needs. Therefore, a computing paradigm for the edge environment is proposed to enable real-time data analysis with large volumes of data by decentralizing heavy computational loads and reducing the consumption of network bandwidth. Data ownership can provide substantial commercial interests for data owners. However, traditional data processing exposes data on the Internet and incurs the risks of data value reduction and privacy issues. By applying the computing paradigm in the edge environment to data sharing and processing, people can build data processing applications without providing the whole original dataset. However, existing work on lightweight methods of building applications and decomposing computation tasks is still lacking. In this paper, we present SeCEE, a framework for data sharing and processing in the edge environment. This framework utilizes geographically distributed datasets to analyze data without programming and comprises (i) a hierarchical task network-based approach that describes datasets and corresponding services from different stakeholders, on the basis of which the features and relationships among datasets and services are recorded; (ii) a service composition method that instantiates an abstract process model for multiple data flows in a dynamic environment; and (iii) an execution engine that coordinates the computing process by dispatching computing tasks to edge servers and collects results for combination and further processing. A case study of a data processing application for electronic toll collection demonstrates the effectiveness of the proposed framework.

Keywords: Edge environment · Data sharing · Data processing
Service composition

1 Introduction

In the last decade, cloud computing has been the most common solution for big data processing and storage that is adopted by different domains, such as business, transportation, education, and health care. Cloud-based data analytics requires data collection, and all computations are conducted using cloud resources. However, cloud

Q. Zhou et al. (Eds.): ICPCSEE 2018, CCIS 901, pp. 33–47, 2018.
https://doi.org/10.1007/978-981-13-2203-7_4

computing and other centralized computing paradigms cannot satisfy the explosive increase in data volumes and computing needs caused by the rapid growth in the number of data producer and service consumer nodes (e.g., sensors and user apps). Decentralized computing paradigms, such as fog computing [1] and edge computing [2], have been presented to address this new challenge in the edge environment. By transferring data storage and computing tasks as closely as possible to the edge near data producers, the consumption of network bandwidth and centralized computing resources by data transfer and computation are considerably reduced, thereby enabling real-time data processing. In addition, a decentralized computing paradigm can avoid the duplication of data without permission. Such a paradigm prevents data from exposure outside the data owner's systems, thus ensuring data privacy and commercial value. Hence, a decentralized computing paradigm in the edge environment is an effective substitute to cloud computing and other centralized computing paradigms.

Computing in the edge environment has been discussed by several existing works but rarely involving data sharing. Complex data processing logic, which is severely restricted by data analytics to only a single data source, can be implemented with multiple datasets shared by different stakeholders. Nevertheless, data sharing and processing in the edge environment encounters the following challenges:

(1) Decomposing a data processing task into several subtasks that can be executed separately;
(2) Coordinating the execution of subtasks on distributed computing resources; and
(3) Accessing data that are not directly provided by data owners due to commercial competition and privacy concerns.

Zhang et al. [3] provided a feasible data sharing solution in the edge environment. A set of privacy preserving functions was provided by data owners, and these functions were bound to original datasets. The functions generated several intermediate results without sensitive information. With a combination of several functions, various complex data processing applications with multiple datasets shared by different data owners can be formed by reusing software. On the one hand, the aforementioned solution simplifies task decomposition by reducing the coupling of data processing application using predefined functions. On the other hand, this solution ensures that data will not be exposed on the Internet. Consequently, an urgent requirement of service composition is raised for data analytics in the edge environment. Selecting services and composing them together by manual programming is effective but prone to error and time consuming; the quality of data analytics may be substandard.

Service composition has had achievements in the academia and the industry [4]. However, traditional service composition approaches can hardly handle multiple dataflows due to the special geographical distribution of data and services in the edge environment. For cloud computing, data are stored in a centralized data center, where the query on a dataset can be accomplished by a single service. Thus, existing approaches need to consider only the composition of workflow; i.e., only one service is bound to each activity in the process. However, in the edge environment, a whole dataset comprises several small ones owned by different data owners and stored separately for the geographical distribution of business systems. A query on that dataset requires several services cooperating with one another. Most existing service composition approaches fail to achieve this cooperation.

Figure 1 illustrates the incapability of traditional approaches to facilitate service composition in the edge environment. Every chain bookstore of the company in the figure is assumed to record every book it sells. The data are not collected and reported to the headquarters. The sales of a specific book in all chain stores are added to produce the total sales. In this process, two or more services are utilized to produce the final result. Hence, in addition to vertical service composition along the direction of workflow, horizontal composition on a geographically distributed service is essential. However, the latter is neglected by existing service composition approaches.

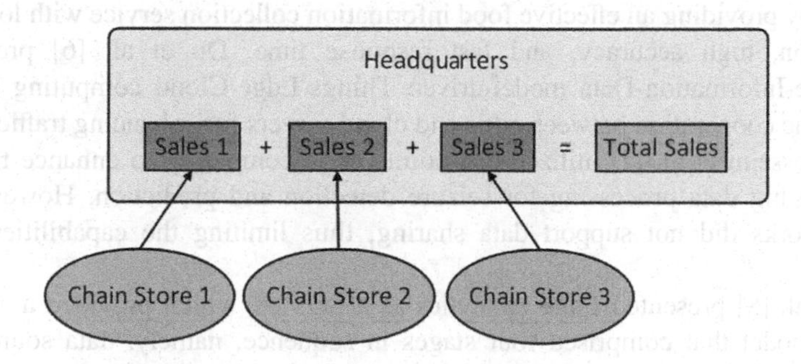

Fig. 1. Service composition example of chain stores.

Therefore, SeCEE, a data sharing and processing framework for the edge environment, is presented to address the aforementioned problems. End users can build data processing applications by service composition without programming, thereby simplifying the construction of an application and improving the quality of data processing. The main contributions of this study are as follows:

(1) A Hierarchical Task Network (HTN)-based service/data register method presented by an ontology, which describes the services and data and the relations among them (especially the possibility of service composition);
(2) A multi-dataflow service composition approach for transforming abstract process models into executable composite services despite the special data distribution in the edge environment; and
(3) An execution engine that coordinates the execution of composite services and handles the exception whenever it occurs.

The remainder of this paper is organized as follows. In Sect. 2, existing works on data sharing and processing and service composition are reviewed. Section 3 discusses the users and architecture of SeCEE. In Sect. 4, a detailed description of the implementation of the three main components in the framework is given. Section 5 demonstrates the effectiveness of the proposed framework through a real and simple case study. In Sect. 6, the work is concluded and future directions are identified.

2 Related Works

In this section, an overview of related works on data sharing and processing in the edge environment is presented. Furthermore, existing methods of service composition are discussed as an effective data sharing method in the proposed framework.

2.1 Data Sharing and Processing in Edge Environment

Data processing in the edge environment is still in its early stage. Hence, few studies have focused on this topic. Liu et al. [5] applied a deep learning-based algorithm and edge computing paradigm to design and implement a real-time food recognition system, thereby providing an effective food information collection service with low energy consumption, high accuracy, and fast response time. Du et al. [6] presented a Knowledge-Information-Data model-driven Things-Edge-Cloud computing paradigm to enable the cooperation between edge and cloud servers for enhancing traffic data as a service. Hosseini et al. [7] utilized autonomic edge computing to enhance traditional health care big data processing for seizure detection and prediction. However, most existing works did not support data sharing, thus limiting the capabilities of data analytics.

Xu et al. [8] presented Edge Analytics as a Service, which provided a rule-based analytics model that comprised four stages in sequence, namely, data source, transformation, rule, and action. In the first two stages, data sources were parsed and transformed into a user-designable format. In the third stage, data tuples were emitted to action stage or not according to a specific rule and rule-hit mode. In the fourth stage, services were activated on an Internet of Things (IoT) cloud platform or on a local gateway. The model application that ran on Edge Analytics Agent in the IoT gateway enables data analytics in the edge environment. However, the model was used to reduce computing workload in the action stage but did not providing complete data for analytics and was incapable of data analytics with multiple data sources. The main data processing logic was implemented in the action stage by utilizing other services. The construction of a data analytics application required programming.

Zhang et al. [3] presented Firework, a computing paradigm for distributed big data processing and sharing in the edge environment. Firework nodes provided various types of data that could be accessed only using privacy-preserving functions that were bound to datasets. A Firework manager kept a view defining datasets and their corresponding functions and coordinated the interaction of nodes in data processing jobs that consisted of operations on different data nodes. When a query was started, a user job was transformed into a directed acyclic graph, which was subsequently split into several tasks on different data nodes. The tasks were then dispatched to various nodes and executed after scheduling. Finally, the results were returned to the manager and fused. Firework could provide data processing on multiple datasets with the consideration of data privacy and processed data with complex logic. However, the service execution plan is generated manually and cannot define a data processing logic without programming.

2.2 Web Service Composition

Web service composition is an effective technique of processing data with multiple data sources. Existing service composition approaches exploited workflow [9, 10], graph [11, 12], HTN [13], Petri net [14], machine learning [15, 16], and other techniques to compose web services from functional and non-functional perspectives. However, existing works focused on task-centric composition scenarios [17], which were not applicable to data-centric composition. Meanwhile, most studies involved only combining single tasks into a complete data analytics process (i.e., vertical composition) and ignored the composition of multiple services in the same task (i.e., horizontal composition), which was common in the edge environment. Although some approaches [18, 19] allowed binding multiple services for a task, they aimed to yield only improved service utility on QoS and did not address the functional requirement of covering all related datasets.

Therefore, a data sharing and processing framework in the edge environment is proposed to address the above problems. This framework enables data processing on multiple datasets by composing services of geographically distributed datasets. End users can easily define abstract data process logic without programming. A multi-dataflow service composition approach automatically instantiates a composite service for different datasets. A service set is integrated into the framework to provide continuously updating algorithms and models designed and implemented by domain specialists.

3 Framework Architecture

SeCEE requires end users to define a data processing model in as detailed a manner as possible for effective and accurate service composition and data processing with complex logic. The problem in this study can be summarized as service presentation, service composition, and composite service execution in the edge environment. A detailed definition is as follows.

On the basis of the registered service information provided by business systems, we aim to bind services for every activity in the abstract process model defined by end users to form an executable composite service, which is subsequently executed by coordinating servers on the edge (i.e., business systems) to yield an outcome. Several services with the same processing logic may be bound to a single activity according to the data coverage constraints specified by the end user given that data is geographically distributed on different systems in the edge environment.

3.1 Framework Users

Figure 2 shows the architecture of the proposed framework, which considers three types of users, **algorithm providers**, **data providers,** and **end users**. Algorithm providers are specialists in a specific domain who construct an algorithm set that consists of algorithms and models and update them continuously to satisfy end users' data analytics demand. Data providers are administrators of mutually independent business systems who choose the data that can be accessed by end users and then register them to the cloud. Data providers select available parts from the new algorithms and models in the algorithm set and bind these parts to one or more datasets. Then, the functions are deployed on servers as services that will be registered to the cloud. End users can then check data, service registration, and abstract actions by invoking application APIs provided by the cloud and describe data processing logic in a Business Process Execution Language (BPEL) or through the graphical interface. The description will be sent to the cloud as a data processing request and web service composition requirement. The composite service, which can be modified through a graphical API, will be returned to the end user and executed after verification.

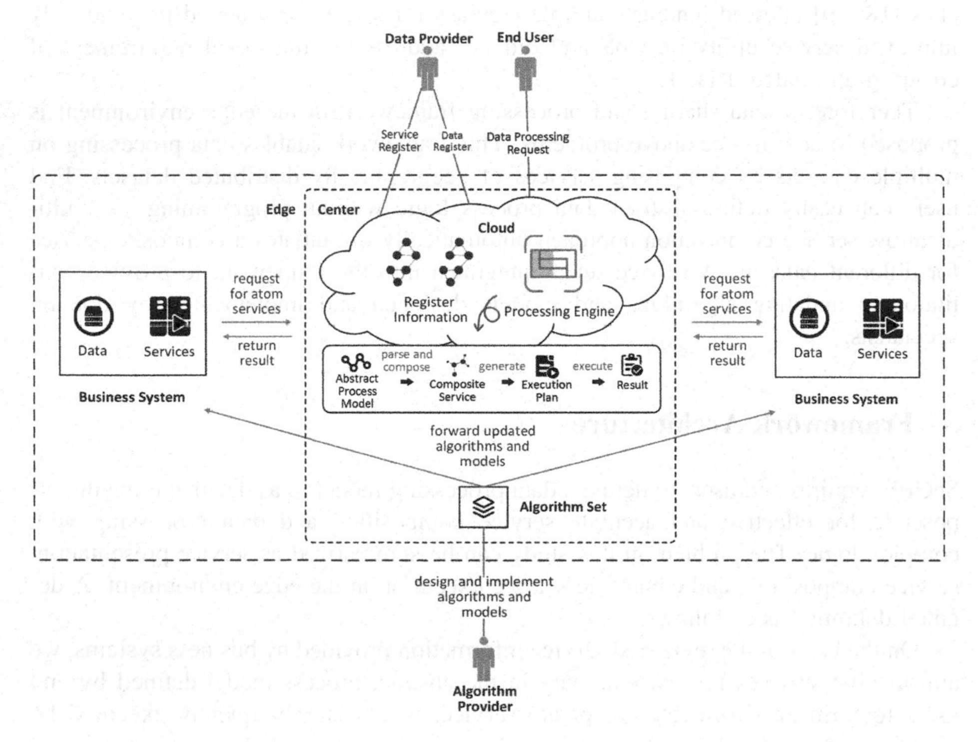

Fig. 2. Framework architecture.

3.2 Components

The main components of the proposed framework are the **cloud**, **business systems,** and an **algorithm set**. The cloud is the center of the framework and manages, schedules, and coordinates computations. Upon receiving a data processing request from the end user, the cloud splits the process into several computing tasks and assigns these tasks to the business systems for execution. Therefore, data are stored and computations are implemented on the servers of the business systems instead of the cloud. This computing paradigm is similar to distributed systems. However, servers of business systems that execute computing tasks are more autonomous in comparison with nodes in distributed systems because the former are not directly managed and scheduled by the cloud.

The cloud is responsible for managing data and service registration and responding to data processing requests. The key component of the cloud is the **processing engine**, which parses a data processing logic description in BPEL (i.e., **abstract process model**) and generates an executable composite service from the description by the service composition approach. The composite service will be sent to an execution engine and executed in a distributed manner. Moreover, the cloud provides several graphical APIs for registering data and services, checking the registered data and services, describing data processing logic, and monitoring the execution status of data processing tasks and the status of servers.

Business systems are mutually independent and geographically distributed systems. Several types of data, such as real-time stream data and historical data that is generated by sensors, and static basic information data, are collected and accumulated inside the business systems, which are responsible for facilitating data analytics regularly for decision making. All business systems enable data access for data fusion and computing with other datasets to maximize the value of data. However, original datasets must not be exposed outside the business system on account of commercial competition or privacy preservation. Therefore, business systems adopt specific services to provide mediate results that are based on datasets instead of sending them to the cloud for centralized computing. These services, i.e., computation tasks, run on business systems on the edge. Therefore, business systems provide computing resources in addition to data and services.

An algorithm set includes implementations of generalized or domain-specified algorithms and models that can be applied in data analytics. The algorithm set may be extended and updated due to changes in requirements or software defects. These new contents will be forwarded to the business systems, and functions related to the data inside the business systems will then be deployed on servers or substitute the previous version. Unlike in [3], the data access services in the present framework is designed by domain experts instead of data provider themselves. Therefore, these services are more suitable for professionals and can satisfy end user requirements more easily than those in [3].

4 Framework Implementation

4.1 Service/Data Register and Representation

In this paper, services and their relations are modeled by using the HTN to present the possibility of composition between services. Inspired by [13], we use an ontology to describe the register information for the following reasons.

(1) The **concept layer** of the ontology provides a vocabulary from which users select formalized classes as activity name and combine these classes to form an abstract process model, thereby allowing the service composition approach to resolve the model correctly.

(2) The **instance layer** of the ontology is a graph that can describe the register information and relations between services that represent how tasks are decomposed (or bottom–up service composition).
(3) The correspondence between services and activities can be built by instantiating a class in a concept layer to instances in an instance layer.

The service ontology comprises two layers, i.e., concept and instance layers. The concept layer contains a series of classes that denotes the activities that registered services can accomplish. For each activity from a correctly defined abstract process model, a corresponding class can be found in the ontology classes. Meanwhile, the instance layer of the service ontology, an HTN described by a graph, comprises nodes and edges that denote the services and relations between them. The two types of services, i.e., **entity** and **abstract services**, are equal to the **compound** and **primitive tasks** in an HTN, respectively. The definitions are as follows.

Definition 1. Entity Service. Entity service is an executable and indivisible service provided by business systems.

Definition 2. Abstract Service. Abstract service is a non-executable service that can be instantiated to a single entity service or a composite service composed of several entity services.

Figure 3 shows part of the services and data registered in one of the applications of the proposed framework in transportation data processing. The blue and white circles depict the abstract and entity services, respectively.

The relations between services indicate how an abstract service is decomposed into several specific services. In HTN, the corresponding term is **method**. Only two types of service relations, i.e., **instantiation** and **composition relations**, are considered because we focus only on which services can be combined and the new features of the composite service. Instantiation relation is a relation between an abstract service and one or more entity services and describes how to obtain a result for the abstract service by combining and executing entity services. Meanwhile, the composition relation is a relation between one abstract service and others, which shows that the result of the former can be obtained by the combination of the latter. The difference between the two types of relation is that the object of instantiation relation is entity service, whereas that of the composition relation is abstract service. During service composition, the object abstract services of the composition relation still need further instantiation until it generates an executable composite service. Different instantiation and composition relations of an abstract service represent various solutions of a compound task that the service describes. In Fig. 3, the two relations are depicted by a set of solid lines with an arc crossing through. The abstract service that calculates Electronic Toll Collection (ETC) turnover, for instance, can be decomposed into two abstract services that calculate ETC turnover inside province and cross-provinces according to the composition relation between them. An instantiation relation exists between the service that counts ETC users and the combination of services that count ETC users in Beijing, Tianjin, and Hebei.

In addition, every instantiation or composition relation has a weight that represents the priority of the decomposition solution. The higher the priority, the better the

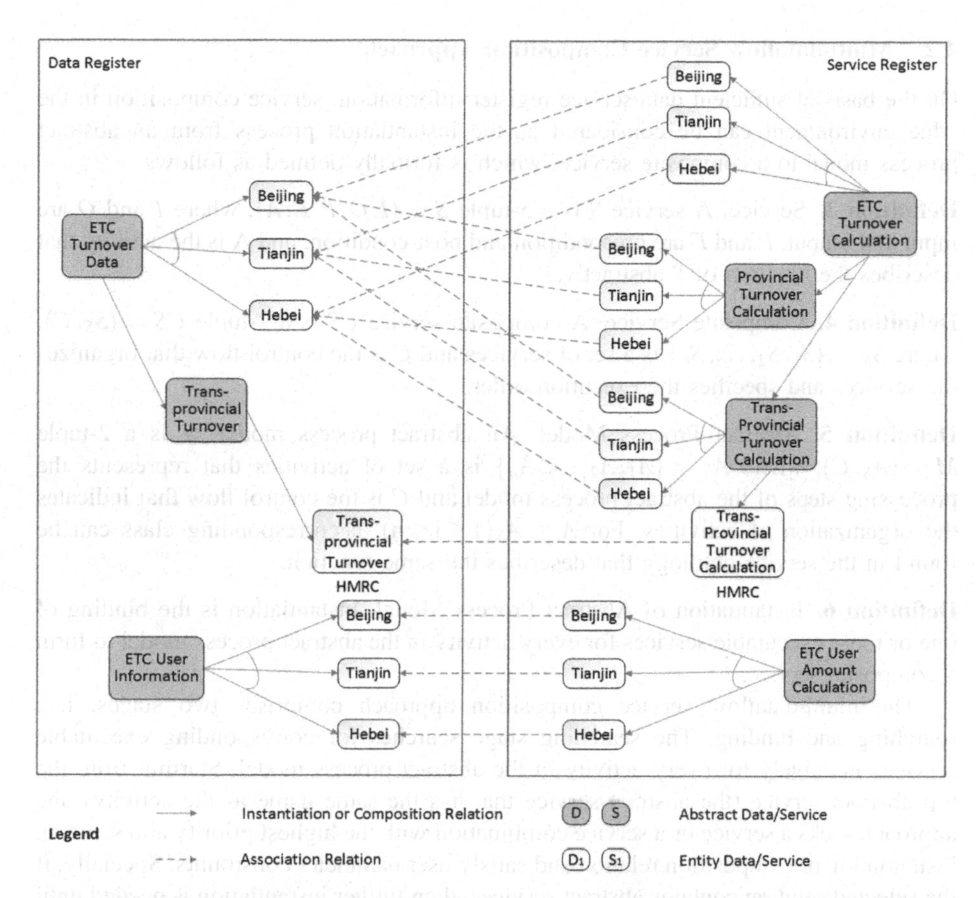

Fig. 3. Example of SeCEE application in transportation data processing. Only part of the service/data register information is shown. The result combination functions, such as average() and sum(), are omitted here.

solution. Therefore, when several candidate decomposition solutions exist, the one with highest priority will be chosen; if the selected solution contains unavailable atom services, then the solution with the lower priority will be employed until an available solution is found.

Although data from business systems technically cannot be modeled as HTN, data ontology can be easily created in nearly the same way as service ontology because of the similarity between data and service in the modeling of concepts (i.e., representing activities of services and types of data), entities (i.e., representing services and data), and relations. The construction of data ontology is omitted here.

The **association relation** is used to involve the corresponding relationship between a service and a dataset, i.e., describing which dataset is utilized by the service an. In Fig. 3, the dotted line represents the association relation. The service that counts ETC users in Beijing are associated with the user information dataset of Beijing, which means that the service retrieves a result that is based on that dataset.

4.2 Multi-dataflow Service Composition Approach

On the basis of sufficient data/service register information, service composition in the edge environment can be considered as the instantiation process from an abstract process model to a composite service, which is formally defined as follows.

Definition 3. Service. A service S is a 5-tuple $S = (I, O, P, E, A)$, where I and O are input and output, P and E are precondition and post-condition, and A is the activity that describes the function of S abstractly.

Definition 4. Composite Service. A composite service CS is a 2-tuple $CS = (S_S, C)$, where $S_S = \{S_1, S_2, \ldots, S_n\}$ is a set of services and C is the control flow that organizes the services and specifies the execution order.

Definition 5. Abstract Process Model. An abstract process model M is a 2-tuple $M = (A_S, C)$, where $A_S = \{A_1, A_2, \ldots, A_n\}$ is a set of activities that represents the processing steps of the abstract process model and C is the control flow that indicates the organization of activities. For $A_i \in A_S(1 \leq \mathrm{i} \leq \mathrm{n})$, a corresponding class can be found in the service ontology that describes the same function.

Definition 6. Instantiation of Abstract Process Model. Instantiation is the binding of one or more executable services for every activity in the abstract process model to form a composite service.

The multi-dataflow service composition approach comprises two stages, i.e., searching and binding. The searching stage searches for corresponding executable services iteratively for every activity in the abstract process model. Starting from the top abstract service (the abstract service that has the same name as the activity), the approach seeks a service or a service combination with the highest priority and share an instantiation or composition relation and satisfy user parameter constraints. Specially, if the selected solution contains abstract services, then further instantiation is needed until all services involved are entity services.

In the binding stage, services are bound to activities in the abstract model. The activities are substituted by the outcome of the searching stage. If more than one entity service is bound to a single activity, then a result combination service should be involved according to the service register information. Algorithms 1 and 2 are the detailed algorithms. Algorithm 1 briefly specifies the two stages, and Algorithm 2 further describes how an abstract service is instantiated by iteratively searching for entity services.

Algorithm 1 Service-Composition(*m*)

Input: *m*: the abstract process model
Output: *m*: the abstract process model after instantiation
1: **for each** abstract activity *A* in *m* **do**
2: **if** *A.name* in classes from service ontology **then**
3: find the instance *R* with the same name
4: *ins* = *Find-Possible-Composition(R, A.params)*
5: **end if**
6: **if** *ins.services.count*>1 **then**
7: add result combination to *ins* according to instances relations
8: **end if**
9: substitute *A* with *ins*
10: **end for**
11: **return** *m*

Algorithm 2 Find-Possible-Composition (*node, params*)

Input: *node*: the abstract process model
params: the parameters of the activity
Output: *sol*: the instantiation of the activity
1: *solutionSet* = ∅
2: *solutionSet*= *solutions* ∪ *node.hasInstantiation* ∪ *node.hasComponents*
3: find *sol, where sol*∈*solutionSet, and for* ∀*sol'*∈*solutionSet, sol.priority*≥*sol'.priority*
4: **if** *sol* contains abstract service **then**
5: **for each** abstract service *C* in *sol.nodes*
6: substitute *Find-Possible-Composition(c, params)* for *C*
7: **end for**
8: **else**
9: delete unrelated services in *sol* using filter with parameter set *A*, where $A \subset \boldsymbol{params}$
10: bind parameter remaining *B* to services in *sol*, where $B = \boldsymbol{params} - A$
11: **end if**
12: **return** *sol*

The parameters given by users can be classified into service properties and function parameters. Therefore, in Algorithm 2, some of the parameters are used to screen the entity services (Line 9). Meanwhile, the rest of the parameters are assigned to entity services as function parameters (Line 10).

Although HTN is utilized to model the composition possibility of services, the service composition approach is not exactly identical to HTN planning methods. In HTN planning, the decomposition solution of a method is fixed. Regardless of the constraints specified by a user, the subtasks decomposed from a compound task in a method do not change. However, in the proposed service composition approach, only part of the services from a service relation may be involved according to the constraints given by the user at runtime. According to the example in Sect. 4.1, if an activity of an

ETC user quantity calculation is constrained by district "Beijing", then only the service that counts ETC user in Beijing will be bound to it and services from other districts will be eliminated.

4.3 Composite Service Execution

Composite service execution engine executes atom services in a specific order and continuously combines the results due to the dependencies between the atom services. Meanwhile, the execution engine is responsible for exception handling during the execution of the composite service. The common exceptions and the handling mechanism are specified below. Section 6 shows the detailed handling process.

Service Failure. Any invalid service or service exception may cause an inaccurate final result or even a fatal error that hinders the execution of the composite service. Once a service fails, the execution engine will request a new composite service without the failing service. If no other composition solution can be found, then the execution will be aborted and the execution log is returned to the end user.

Lack of Services. A lack of services may cause defects in the service composition that contradict real-world business logic. These services cannot be found in the service ontology. However, they may exist in certain business systems that are waiting to be involved. As shown in Fig. 3, a user request of the ETC user calculation nationwide is hard to satisfy due to the lack of counting services in other districts, except Beijing, Tianjin, and Hebei. The execution engine can detect the lack of atom services in the composite service and inquire whether to resume the execution or not.

By composing services implemented by different techniques, such as Spark, Storm, MapReduce, or even such traditional data analytics techniques as SQL, the barriers between heterogeneous business systems are eliminated. All the heterogeneous services at the edge can be involved in the same computing process without data collection, thereby enabling data analytics with multiple datasets in the edge environment and substantially improving the efficiency of data processing.

5 Case Study

In this chapter, the usability of SeCEE is demonstrated using a simple case of transportation data processing. An end user must collect data about average ETC consumption per user in different months of a year to estimate the development trend of ETC. Two statistics, i.e., monthly ETC turnover and total number of ETC users, are required to obtain the average consumption.

In China, ETC systems of different provinces are relatively independent and owned by different agencies. Therefore, the transaction data and user information are distributed to different systems (specifically, all of the data of a user are stored in a system owned by the agency from which the user applied for the ETC account) and not shared with other systems. Meanwhile, the Highway Monitoring & Response Center (HMRC), an intermediary organization responsible for settling trans-provincial transactions, collect all such records. A trans-provincial transaction takes place in a province different from where the ETC account is from. For example, a user has an ETC account applied for in Beijing but pays the highway tolls for the same account in Henan. Concentrating all distributed data on a platform for centralized data processing is impractical due to the large data volume and privacy concern. Therefore, a computing paradigm in the edge environment is adopted to enable decentralized data analytics by sharing geographically distributed datasets. In this scenario, only business systems from Beijing, Tianjin, and Hebei provide services for accessing transaction and user data, as shown in Fig. 3. With the variety of services, multiple composite services can be generated on the basis of the same requirements.

By visiting the graphical interface provided by the cloud, end users design an abstract process model (Fig. 4) that will be translated to a BPEL file. They can also provide a BPEL-represented abstract process model directly. For every activity, the end user specifies a set of parameters that will be used to screen services or assigned to functions. In the "total ETC turnover" activity example, a time period is specified (e.g., January). Meanwhile, without a specific geographical range, the activity aims to calculate a nationwide ETC turnover.

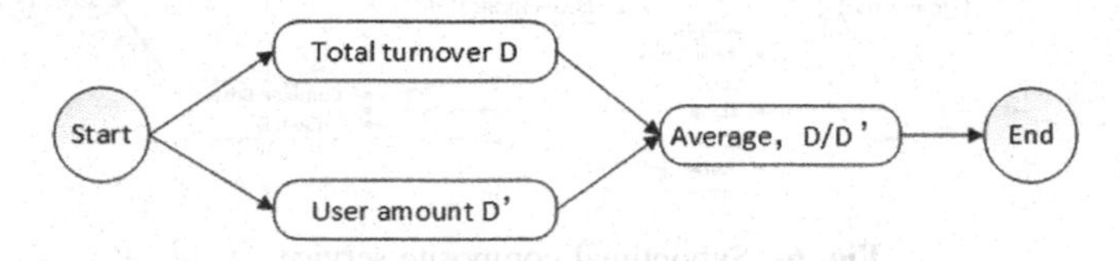

Fig. 4. Abstract process model of average ETC turnover calculation.

The processing engine then checks and resolves the abstract process model. On the basis of the registered services, three candidate composition solutions can be found. Figure 5 shows the optimal composite service among them. The end user can modify the atom services or control the flow through a graphical interface. After user verification, the execution engine will execute the composite service. In this scenario, the registered services (services from only the business systems in Beijing, Tianjin, and Hebei are available) cannot provide the total ETC turnover nationwide. Therefore, the processing engine will also report the absence of the services and ask the end user to confirm whether to resume service execution.

An additional service composition is required if any atom service is invalid. For example, if the service that calculates total ETC turnover of Beijing is unavailable, then a suboptimal solution should be adopted to substitute the former one. Figure 6 illustrates the new solution. All transactions can be classified into provincial and trans-provincial transactions. Therefore, the turnover can be obtained by calculating the provincial and trans-provincial turnovers using the services provided by ETC agencies and HMRC. Furthermore, if the trans-provincial turnover calculation service in HMRC is invalid, then it can be replaced by similar services from ETC agencies. The details are omitted here.

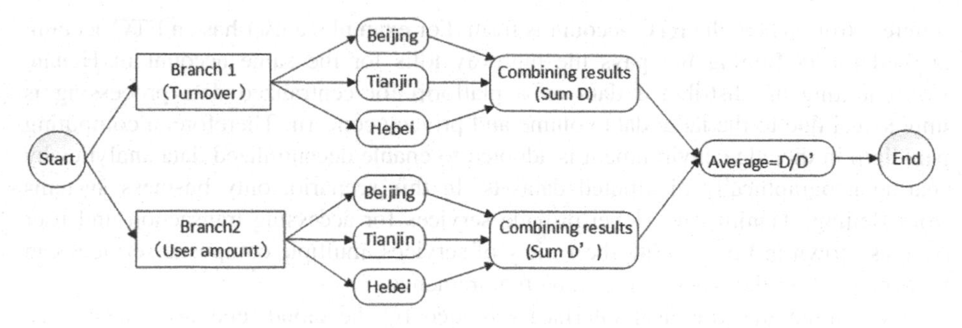

Fig. 5. Optimal composite service.

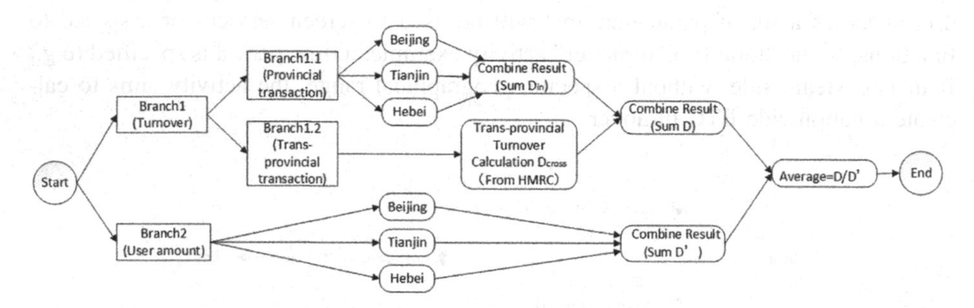

Fig. 6. Suboptimal composite service.

6 Conclusion

In this paper, we present SeCEE, a big data sharing and processing framework in the edge environment. According to end user data processing requirements, the framework can compose a series of related services provided by different business systems, execute the composite service, and handle exceptions whenever they occur. The service composition is based on the register information of data and services, which is modeled as an HTN and presented by an ontology. The case study of a real transportation data processing application demonstrates the effectiveness of SeCEE.

In future work, we will improve the automation level of our framework such that the end user is no longer required to provide a detailed processing model with considerable domain knowledge. Meanwhile, we plan to adjust the priority of the service composition plan in the service ontology using non-functional attributes, such as composite service execution time, to improve the quality of service composition.

Acknowledgments. This research was supported by the National Key R&D program under Grant No. 2016YFC0801700, Beijing Municipal Science and Technology Project No. Z171100000917016, the National Natural Science Foundation Project under Grant No. U1636208.

References

1. Bonomi, F., Milito, R., Zhu, J., Addepalli, S.: Fog computing and its role in the internet of things. In: Proceedings of the First Edition of the MCC Workshop on Mobile Cloud Computing, pp. 13–16. ACM (2012)
2. Shi, W., Cao, J., Zhang, Q., Li, Y., Xu, L.: Edge computing: vision and challenges. IEEE Internet Things J. **3**(5), 637–646 (2016)
3. Zhang, Q., Zhang, X., Shi, W.: Firework: big data processing in collaborative edge environment. In: IEEE/ACM Symposium on Edge Computing (SEC), pp. 81–82. IEEE (2016)
4. Ahmed, T., Mrissa, M., Srivastava, A.: MagEl: a magneto-electric effect-inspired approach for web service composition. In: 2014 IEEE International Conference on Web Services (ICWS), pp. 455–462. IEEE (2014)
5. Liu, C., et al.: A new deep learning-based food recognition system for dietary assessment on an edge computing service infrastructure. IEEE Trans. Serv. Comput. **PP**(99), 1 (2017)
6. Du, B., Huang, R., Xie, Z., Ma, J., Lv, W.: KID model-driven things-edge-cloud computing paradigm for traffic data as a service. IEEE Netw. **32**(1), 34–41 (2018)
7. Hosseini, M.P., Tran, T.X., Pompili, D., Elisevich, K., Soltanian-Zadeh, H.: Deep learning with edge computing for localization of epileptogenicity using multimodal rs-fMRI and EEG big data. In: 2017 IEEE International Conference on Autonomic Computing (ICAC), pp. 83–92. IEEE (2017)
8. Xu, X., Huang, S., Feagan, L., Chen, Y., Qiu, Y., Wang, Y.: EAaaS: edge analytics as a service. In: 2017 IEEE International Conference on Web Services (ICWS), pp. 349–356. IEEE (2017)
9. Lécué, F., Gorronogoitia, Y., Gonzalez, R., Radzimski, M., Villa, M.: SOA4All: an innovative integrated approach to services composition. In: 2010 IEEE International Conference on Web Services (ICWS), pp. 58–67. IEEE (2010)
10. Lee, C.H., Hwang, S.Y., Yen, I.L.: A service pattern model for flexible service composition. In: 2012 IEEE 19th International Conference on Web Services (ICWS), pp. 626–627. IEEE (2012)
11. Tong, H., Cao, J., Zhang, S., Li, M.: A distributed algorithm for web service composition based on service agent model. IEEE Trans. Parallel Distrib. Syst. **22**(12), 2008–2021 (2011)
12. Rodriguez-Mier, P., Mucientes, M., Lama, M.: Automatic web service composition with a heuristic-based search algorithm. In: 2011 IEEE International Conference on Web Services (ICWS), pp. 81–88. IEEE (2011)
13. Ko, R.K., Lee, E.W., Lee, S.G.: Business-OWL (BOWL)—a hierarchical task network ontology for dynamic business process decomposition and formulation. IEEE Trans. Serv. Comput. **5**(2), 246–259 (2012)
14. Tang, X., Jiang, C., Zhou, M.: Automatic web service composition based on horn clauses and petri nets. Expert Syst. Appl. **38**(10), 13024–13031 (2011)
15. Peng, S., Wang, H., Yu, Q.: Estimation of distribution with restricted boltzmann machine for adaptive service composition. In: 2017 IEEE International Conference on Web Services (ICWS), pp. 114–121. IEEE (2017)
16. Wang, H., Wu, Q., Chen, X., Yu, Q., Zheng, Z., Bouguettaya, A.: Adaptive and dynamic service composition via multi-agent reinforcement learning. In: 2014 IEEE International Conference on Web Services (ICWS), pp. 447–454. IEEE (2014)
17. Chen, X., Wu, T., Xie, Q., He, J.: Data flow-oriented multi-path semantic web service composition using extended SPARQL. In: 2017 IEEE International Conference on Web Services (ICWS), pp. 882–885. IEEE (2017)
18. Llinás, G.A.G., Nagi, R.: Network and QoS-based selection of complementary services. IEEE Trans. Serv. Comput. **8**(1), 79–91 (2015)
19. Liang, X., Qin, A.K., Tang, K., Tan, K.C.: QoS-aware web service composition with internal complementarity. IEEE Trans. Serv. Comput. **PP**(99), 1 (2016)

Research on Pricing Model of Offline Crowdsourcing Based on Dynamic Quota

Lu Yuan[1], Yan Zhou[2], Jia-run Fu[1], Ling-yu Yan[1(✉)], and Chun-zhi Wang[1]

[1] School of Computer Science, Hubei University of Technology, Wuhan 430068, Hubei, China
yuanluemail2@163.com

[2] Hubei Entry-Exit Inspection and Quarantine Bureau, Wuhan 430050, Hubei, China

Abstract. Crowdsourcing is a business model that relies on computer networks. It was first introduced in 2006 and more and more people began to participate in this business model. At present, how to improve the completion rate of crowdsourcing mode task, while reducing the cost of the publisher has became an urgent problem. Based on the nonstandard form, we implement experiments on the dataset given by CUMCM and propose a time sharing + dynamic pricing model to solve the problem of offline tasks in crowdsourcing. Experiment results show that the model task completion rate improved to 83.5%, 26.6% higher than the traditional mode, the average task publication costs 69 yuan, 4 yuan lower than the traditional. Finally, according to the experiment results, we make some reasonable suggestions to the future crowdsourcing platform design.

Keywords: Non standard assignment problem · Timesharing assignment
Dynamic quota · Crowdsourcing

1 Introduction

In June 2006, American journalist Geoff Howe [1] introduced the concept of crowdsourcing for the first time in the Wired. Crowdsourcing: "a company or an institution outsourced the tasks which previously performed by employees to a non specific mass network in a free and voluntary way". The biggest change between crowdsourcing and outsourcing is that crowdsourcing is assigning tasks to non specific groups, outsourcing is assigning tasks to specific groups which emphasizing highly specialized [2]. At present, crowdsourcing is an increasingly important approach to pursuing innovation but the research on crowdsourcing mainly stays in the definition and the model [3–5]. With the development of Internet technology [6], there are many commercial crowdsourcing platforms, like Amazon Mechanical Turk (Mturk), Crowd-Flower, Samasource CloudCrowd and so on. The well-known crowdsourcing platforms in home market, such as Brain Store, Zhu Bajie, San Da Ha. These commercial crowdsourcing platforms provide the corresponding services according to different needs of task publishers and recipients [7, 8], and charge some management fees to the publisher.

Q. Zhou et al. (Eds.): ICPCSEE 2018, CCIS 901, pp. 48–59, 2018.
https://doi.org/10.1007/978-981-13-2203-7_5

After users registered members, they can publish or receive tasks through platforms. As for platforms, only to ensure the task completion rate can improve the influence, and attract more members to profit; As for task publishers, wish the published tasks were completed on time and with high quality, while reducing the price as low as possible. As for the task recipients, can quickly get tasks through the platforms according to their locations and free time to get paid.

Crowdsourcing, as a new type of economic model, carries out and accepts tasks through the Internet [9–11], shapes online communication and cooperation patterns, subverts the traditional mode of solving problems, provides new impetus for the development of enterprises and networks, creates great value, realizes a win-win situation, and realizes a kind of social resources [12]. There are a wide variety of crowdsourcing tasks [13], including market research, trademark design, business planning, online translation and even website design. The tasks are mainly online and offline. Online tasks mainly refer to: the task members can finish tasks without going outdoors, such as designing a logo, creating a professional website, etc.; offline tasks mainly refer to: the task members are required to finish tasks to the specify location, such as to distribute leaflets to streets, classify the goods in the malls. Crowdsourcing bid problem has been plaguing the crowdsourcing platforms and task publishers [14], reduced people's feeling sense and easy-using to crowdsourcing patterns, so we propose our model to solve it.

2 Related Work

At present the study of crowdsourcing bid is mainly about the nature of the strategy, and function crowdsourcing bids are rarely found. So it is a key problem to design a reasonable price which can improve the completion rate and reduce the cost of pricing. In recent years, Enrique Estellees Arolas, Bayus, Vuculescu have studied crowd packet mode [15], public package task and public packet network platform from management or technical point of view. Mason and others have found that the high or low price of a task will have a certain impact on the completion of the task [16]. If the price is too high can attract more workers to answer the task, but will not improve the quality to complete the task, instead of increasing the burden of task request money; in addition, the high price of the tasks are easier to attract fraudsters and lead a low quality [17]; and if the task price is too low, members will have no interest in it so it will be difficult to finish the tasks on time. The main idea to research crowdsourcing is to apply game theory and reverse auction theory to deduce the theory [18] (Dominic Di Palantino, Milan Vojnovic, 2011), and to study the expression of crowdsourcing task pricing is very few.

In order to solve this problem and combine the practical significance of crowdsourcing (economic benefits), we have read many excellent papers, and we find that most of them is just discussing the conception, pattern and the development prospect. However, the papers about pricing for the crowdsourcing tasks are seldom, finally we propose a time sharing + dynamic pricing model. This include distribution method and pricing method.

(1) The method of distribution. We put the coordinate of tasks and members on the map and get their distance from each other, then we calculate the number of tasks and members in every period. Then we use 0-1 distribution method to account finished and on finished tasks and the members receive tasks and not. We find that the number of members receiving the task is different in each time period, if all the tasks are published in one time, the task will be accepted by the members and not completed in time. Therefore, the public package platform can distribute the task according to the time period according to the acceptance task time of the member, and form a model of hunger marketing. Improve the timeliness of tasks, such as reducing the release of tasks at 8:00 to 9:30 at the low activity peak period, increases the number of missions during the rush hour 20:21:30. Taking the information in the packet as an example, we calculates the number of members in each time period, and the average tasks of each member. By comparing the impact of the one-time release task and the time sharing task on the membership evaluation task, the effectiveness of the allocation method is judged.
(2) The method of pricing. For the new dynamic pricing model the price should be given by crowdsourcing platforms by obtaining the user's task attributes (such as ease, geographical location, completion time) and access to the task of membership attributes (such as membership reputation, membership geographical location) to give a reference value *l_price*. In theory, reference price is the minimum value to ensure tasks members can be profitable while answering tasks. According to the reference price publishers will give the finally price *r_price*. If the *r_price* > *l_price* means the pricing is reasonable and the platform can push such pricing tasks to the appropriate members. However if *r_price* < *l_price*, it means that the pricing was unreasonable, the risk of failed to complete the task is higher, and the platforms would not push such tasks to the members, This kind of task can only be selected by members actively screening information. By obtaining the data of membership attribute and task attribute, we first analyze the main factors that affect the pricing, take the original price in the data packet as Y, then calculate the weight of each parameter by multiple regression and fuzzy mathematics, and bring the result into metadata inspection, get its error, and finally compare the new task completion rate and total task pricing with traditional pricing methods(price is determined by the publisher, the crowdsourcing platforms just the spread tasks news and manage them).

In this paper, we take the most common offline task "take photos to make money" as the research object. CUMCM provides two data packages for reference. Data package one is the data of finished tasks, including the latitude and longitude of each task, pricing and completion ("1" means completion, "0" means incompletion), the number of tasks in packet one is 835; packet two is member information, and there are 1848 members in total, including latitude and longitude of members, credit value, and the start booking time and booking limit, in principle, the higher the member credit, the more priority to start the selection of tasks, the greater the predetermined limit.

3 Analysis of the Unfinished Task

According to the latitude and longitude we spread the members and tasks to the global open source image basemap, and make member distribution thermodynamic diagram, and superimposed task price information on it. Figure 1 shows the distribution of membership hotspot and task price overlay, light circles indicate low prices, dark blue circles indicate high price tasks, the blue background shows the distribution of members, and the deeper the background color indicates that members are densely distributed. Figure 2 shows the hotspot chart of members' distribution and the completion of the task. The red circles indicate the successful tasks. The blue circles indicate the unfinished tasks. Here are the two figures.

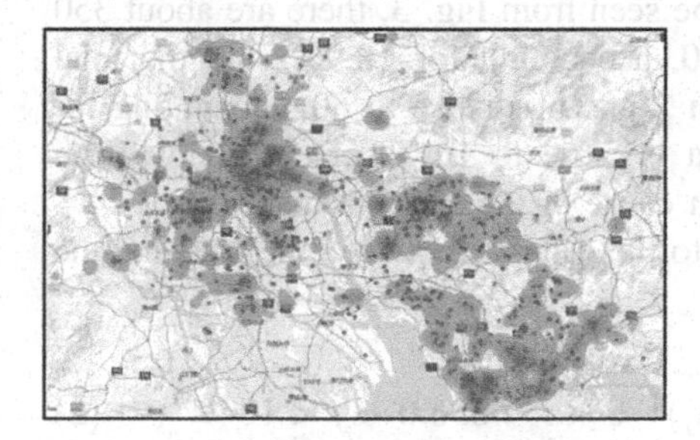

Fig. 1. Membership distribution and price (Color figure online)

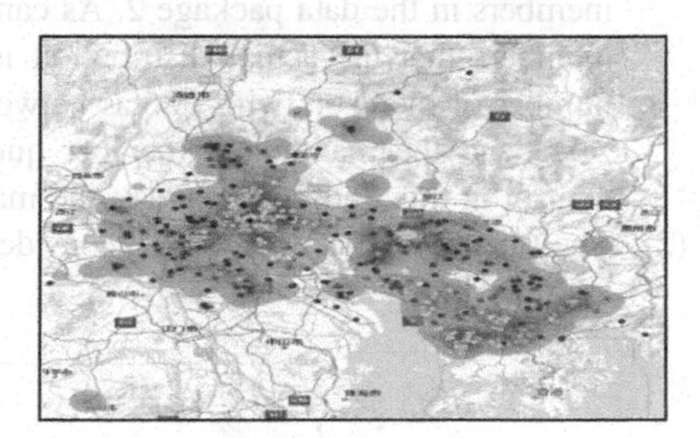

Fig. 2. Membership and task completion (Color figure online)

3.1 Task Pricing Rules and Task Completion

(1) **A significant negative correlation between task pricing and member reservation limit.**

Figure 1, in the high quota area mainly distributed the low price tasks, and the area surrounding the city, members of the actual amount of allotment is relatively low but the price of them are high.

(2) The higher the distance to the membership, the higher the price is.

Figure 1, the distribution of scattered tasks on the map is often with high price, around the task the number of members is very small and the credit level of the members is low away from the membership cluster near the task, its price is often low.

(3) The completion of different regional tasks is different.

Figure 2, the completion rate of tasks near Dong Guan is the highest, the completion rate of tasks around Guang Zhou and Fo Shan is higher, and the completion rate of tasks in Shenzhen is the lowest, mainly because each city has different economic development and economically developed. The urban residents tend to have high expectation on the price of the task.

3.2 Reasons for Pricing Failure

(1) Calculate the actual distribution of each member in the package

$$allot_j = floor(\frac{reservlimit_j}{\sum reservlimit_j} \times T + 0.5) \quad (j = 1, 2, 3\ldots1848) \quad (1)$$

floor indicates rounding down. $allot_j \geq 0.5$, $allot_j = 1$, $allot_j \leq 0.5$, $allot_j = 0$. Making MATLAB programs and draw Fig. 3 the actual distribution of all members in the data package 2. As can be seen from Fig. 3, there are about 350 members that the actual distribution is 0, accounting for 18.9%, and the total number of members with quotas between 2 and 6 is close to 50, accounting for 2.7% and the maximized member quota is close to 16. There is no platform control in this quota way and is the main cause of the task unfinished.

(2) The pricing does not make full consider to the distance between the task and the membership.

$$dist_{ij} = \sqrt{(x_i - x_j)^2 + (y_i - y_j)^2} \quad 0 < i < 836.0 < j < 1849 \quad (2)$$

i: task number, j: member number, we calculate a 1848 * 835 matrix, import the data into the MATLAB, we draw the picture 4 the distance between members i and j.

From Fig. 4, for the same tasks, the distance costs for different members to take the task are different (assuming all the members have access to the task). Therefore, when the exact distance between the task and the membership increases, the costs of time and the transportation will increase of course the price of the tasks should be increased too.

(3) Did not consider the impact of membership credit. The higher the credit of members are the number of the task that they can get is more.

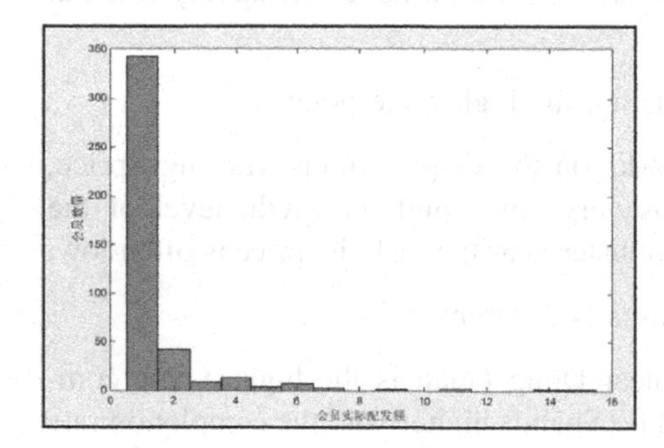

Fig. 3. Actual quota of members

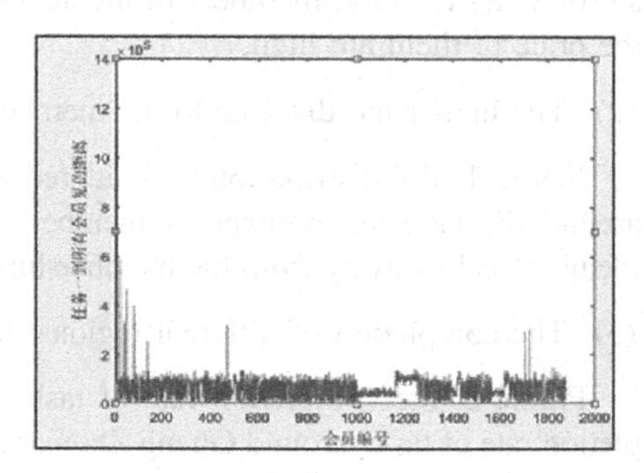

Fig. 4. Distance between tasks and members

(4) Not considering the difficulty of the task, the completion time of the task and the weather and other external non-control factors λ. The task more difficult, more urgent, price should be higher.

4 Time-Sharing Assignment Model

The model of "take photos to make money" is essentially a non-standard assignment problem. In this paper, j members $M_1, M_2...M_j$ are assigned to complete i tasks $T_1, T_2...T_j$, requiring each task to be completed by at most k members ($k > 0$), and the number of each members complete tasks is different, members will choose the most profitable tasks priorly to maximize their benefits. The benefit that member M_j to complete task T_j is $benefit_{ij}$. It is considered that the assignment task is the most efficient in the case of the maximum income of the member, according this we establish the following model.

$$x_{ij} = \begin{cases} 0 \ member\ M_j\ don't\ answer\ task\ T_i \\ 1 \ member\ M_j\ answer\ task\ T_{ii} \\ i = 1, 2, \cdots, n, j = 1, 2, \cdots, m \end{cases}$$

$$\max \ Benefit = \sum_{i=1}^{n}\sum_{j=1}^{m} benefit_{ij}x_{ij}$$

$$s.t. \begin{cases} \sum_{i=1}^{n} x_{ij} \le allot_j \\ \sum_{j=1}^{m} x_{ij} \le 1 \\ x_{ij} = 0/1 \end{cases}$$

Time-sharing assignment model is mainly composed of the following four steps:

4.1 Data Preparation

Calculate the distance matrix for each member from each member $dist_{ij}$, the information *task_price* that the data package one.

$$benefit_{ij} = price_i - \lambda * dist_{ij} \tag{3}$$

$$dist_{ij} = \sqrt{(x_i - x_j)^2 + (y_i - y_j)^2}$$

$$price_i = f(i, j) = R_i \odot j \tag{4}$$

benefit$_{ij}$: the benefit of member j finish task i; *price*$_i$: the price of task i; λ: taking into account the degree of difficulty of the tasks, the time to complete the task and other factors cost per km distance cost. R: factor matrix, consider the Relationship strength $\odot$ between i and j, we should calculate the weight in the matrix. Statistic the time members start to answer tasks *start_time*, We divide the time into four stages 6:30–7:00, 7:00–7:30, 7:30–8:00 and after 8:00 we receive The total number of members receiving the task at each period, specific results are shown in Table 1.

Table 1. Distribution of members

Time	Member	Actual quota > 0	Idea number of task
6:30–7:00	377	190	401
7:00–7:30	274	177	129
7:30–8:00	207	72	72
After 8:00	1059	47	47

It can be seen from Table 1 that the total number of theoretical distribution tasks in the four time periods is 649, which is smaller than the total number of tasks in packet one 877. There are a large number of non-qualified members in each period, with the largest number of members after 8:00, the quota is the least, so it is necessary to carry out time-sharing dynamic distribution of tasks.

4.2 Randomize the Membership to Get the Task

Each time, send tasks to qualified members. In real life, for each member to meet the conditions, they get the equal opportunity of the task, we use the Excel function random to randomize member number *idx_member_qualified.*

4.3 Dispatch Tasks by Time, Until the End

According to the new membership number *idx_member_qualified* to assign tasks. For each task i, the provisions of the task number is k, when the number of members to complete the task $count_{ij} = k$ means the task has been completed, its state value $state_{ij} = -1$, and delete this task in assignment table, and calculated the $benefit_{ij}$ when finishing this task. Once a member fails to receive the task in the period, he will be canceled the qualification to receive the task in the subsequent time period. The members will receive the task according to the size of the income.

4.4 Make Computer Programs to Simulate, the Model and Get a Reasonable λ

There is an unknown parameter λ, in order to get a value that is consistent with the actual comparison, we make experiments that make λ values between 1–15. The specific results are shown in Fig. 5 and Table 2.

Combining Table 2 and Fig. 4, in Table 2 When the value is 12, the number of tasks assigned by simulation is 519, which is the closest to the real data 522 in the packet 1, comparison of task distribution with original results in Dong Guan, Guang Zhou, Fo Shan and Shen Zhen by SPSS, there are four significant results, which are 0.102, 0.042, 0.033 and 0.051 respectively, all of which are significant in four regions.

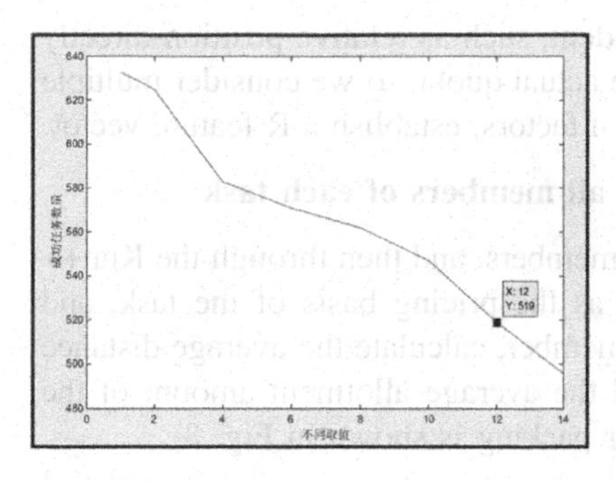

Fig. 5. Relationship between λ and tasks

Table 2. Tasks number with λ

Completed tasks	λ	Completed tasks	λ
633	1	562	8
625	2	547	10
583	4	519	12
571	6	496	14

The highest price in packet 1 is 84 yuan, it is assumed that the maximum allowable membership activities range of 7 km.

4.5 Model Verification

In order to verify the correctness of λ, we take the original $price_i$ into expression (3), when $benefit_{ij} \leq 0$, it means task i is successfully completed, on the contrary when

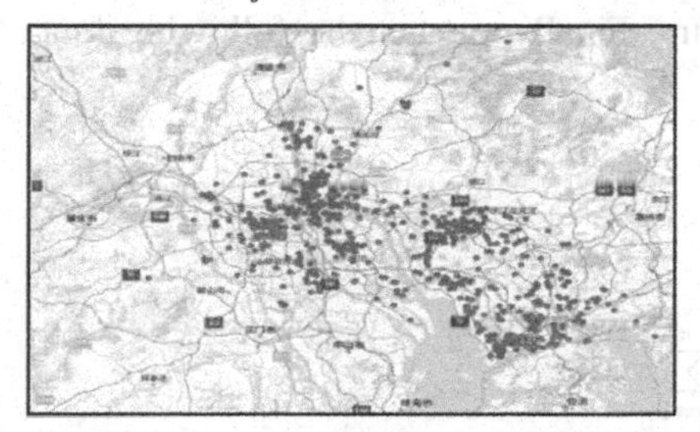

Fig. 6. Simulated distribution results

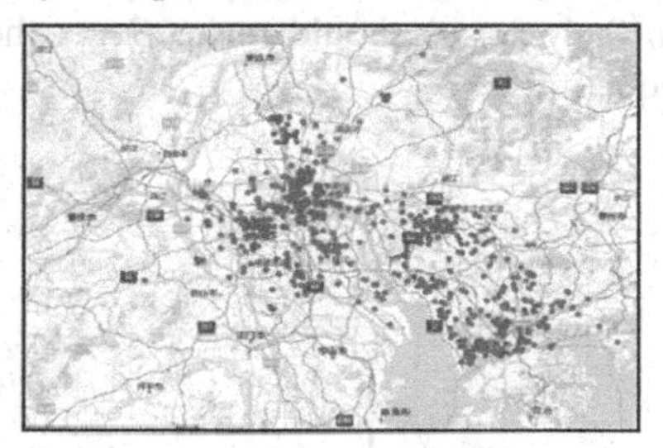

Fig. 7. Actual distribution in packet 1

$benefit_{ij} > 0$, it means task i is successfully completed, on the contrary, if $benefit_{ij} < 0$, it means task allocation failed. The results of the simulation and the actual results in the data packet 1 are respectively shown in Figs. 6 and 7.

As can be seen from the comparison of the figure, the distribution of tasks completed and unfinished is basically the same on the map, which indirectly proves that the established task allocation model is correct, and $\lambda = 12$ is reasonable.

5 Dynamic Pricing Model

(1) The establishment of feature vector R

The study found that the main factors affecting the price is the relative position of task and members (x, y), and actual quota allot of the members, the membership degree of credibility credit. But each factor is not independent, such as relative position directly affects the distance, credibility directly affect the actual quota, so we consider multiple linear relation between factors the combination of factors, establish a R feature vector.

(2) **Using the Knn++ algorithm to calculate all members of each task**

For each task to calculate the distance of all the members, and then through the Knn++ algorithm, get the nearest member, Package it as the pricing basis of the task, and calculate the geographic location (*x, y*) of each member, calculate the average distance between the members to the task *avg_dist*, and the average allotment amount of the member *avg_allot*. The classification effect after packing is shown in Fig. 8.

(3) **Eigenvector solution**

Simplified the expression Price (4):

$$y = a0 + a1x1 + a2x2 + a3x3 + \ldots a_n x_n$$

Then the fitted values y for the follow:

$$\hat{y} = \hat{a}0 + \hat{a}1x_1 + \hat{a}2x_2 + \hat{a}3x_3 + \ldots \hat{a}nx_n$$

According to the principle of least square method $a_i(0, 1, 2\ldots n)$ Then $\hat{a}_i(0, 1, 2\ldots n)$ should make Q get the minimum. Finally we get the following four equations.

$$Q = \sum_{a=1}^{n} (y_a - \hat{y}_a)^2 = \sum_{a=1}^{n} [y_a - (b_0 + b_1x_{1a} + b_2x_{2a} + \ldots + b_kx_{ka})]^2$$

$$\begin{cases} nb_0 + (\sum_{a=1}^{n} x_{1a})b_1 + \ldots + (\sum_{a=1}^{n} x_{ka})b_k = \sum_{a=1}^{n} y_a \\ (\sum_{a=1}^{n} x_{1a})b_0 + (\sum_{a=1}^{n} x_{1a}^2)b_1 + \ldots + (\sum_{a=1}^{n} x_{1a}x_{ka})b_k \\ (\sum_{a=1}^{n} x_{2a})b_0 + (\sum_{a=1}^{n} x_{1a}x_{2a})b_1 + \ldots + (\sum_{a=1}^{n} x_{2a}x_{ka})b_k \\ (\sum_{a=1}^{n} x_{ka})b_0 + (\sum_{a=1}^{n} x_{1a}x_{ka})b_1 + \ldots + (\sum_{a=1}^{n} x_{ka}^2)b_k \end{cases}$$

Using Lingo obtain the parameter values. The specific results are shown in Table 3.

(4) **Model verification**

Bring the data in packet 1 into the model and we obtain the price of each task, then use the income matrix formula to calculate the number of tasks that delivered successfully in the dynamic pricing + assigned scenario. Calculate the completion rate of the task and the total cost of the published. Compared with packet 1 whether the total cost is

Table 3. Pricing model coefficient

a0	6.496	a6	−0.126
a1	0.01	a7	−2.998
a2	0.019	a8	−0.007
a3	2.404	a9	−0.14
a4	2.875	a10	0.002
a5	−0.108	a11	0.64

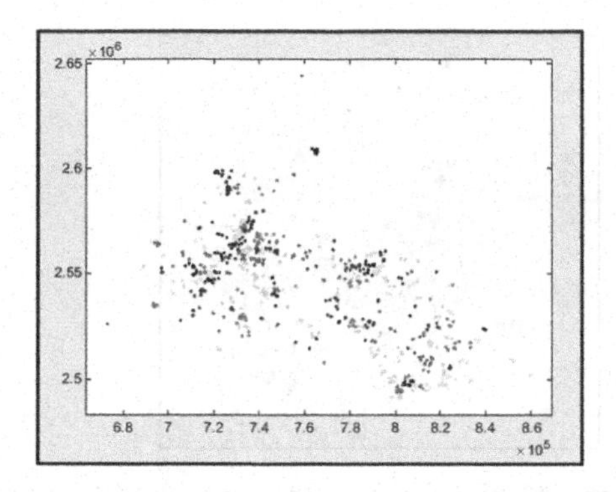

Fig. 8. Packaging classification effect

Table 4. Results of Simulation

Total cost of the task (yuan)	48298.00
Number of successful assignments	700.00
The average cost of completing the task (yuan)	69.00
The average distance (km)	2.48
Number of members to completed the task	330.00
The average number members receive tasks	2.12
The average benefit members completed tasks	35.74

Table 5. The real data in data packet 1

Total cost of the task (yuan)	36446.00
Number of successful assignments	499
The average cost of completing the task (yuan)	65
The average distance (km)	1.95
Number of members to completed the task	278
The average number members receive tasks	1.71
The average benefit members completed tasks	76.72

effectively reduced. Finally, count the benefit of every member and the number of members to complete the task. Take the calculated data into our program and get Table 4 and the Fig. 9 (Table 5).

It can be seen from the results (the red represents the task has been completed, the blue represents the unfinished), under the dynamic quota pricing + time sharing assignment scheme, though the average benefit members completed the tasks has declined about 35 yuan for each member, it means the benefit is distributed to more people, in other words it there are more people taking into crowdsourcing business model than before. The total number of successfully assigned tasks is 700 compared with the traditional fixed quota program increased by 195, task completion rate in the new model is 83.5% higher than the tradition completion rate 59.7%, average cost of completing the task is 69 yuan of each task, compared with the previous average cost of 73 yuan of each task, with an average cost reduction of 4 yuan. The membership distance to complete the task is from 1.95 km expand to 2.48 km, effectively expanding the membership of the task to complete the range (Fig. 10).

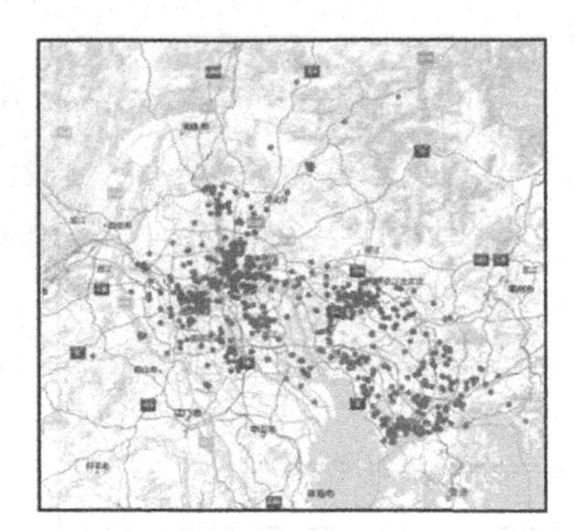

Fig. 9. Simulation results of model (Color figure online)

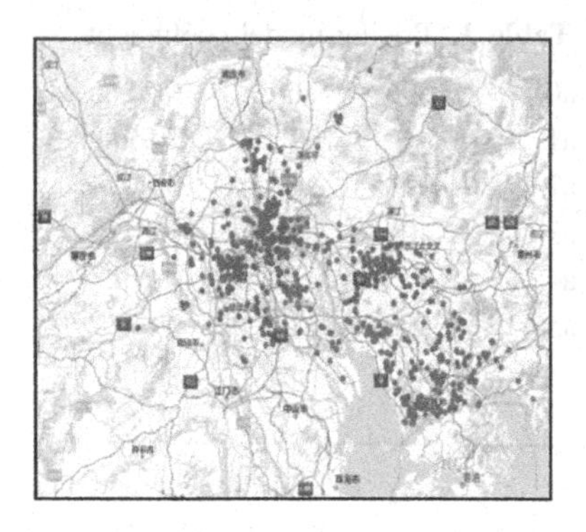

Fig. 10. Actual distribution in packet

6 Conclusion and Suggestion

Through mathematical modeling, we get a new dynamic pricing formula (4). Compared with the existing traditional pricing methods, we find that dynamic pricing is beneficial to reduce the cost of the publisher and improve the task completion rate, which is more conducive to the popularization of crowdsourcing economy and development.

The future crowdsourcing platforms should equipped with intelligent push function, the system provides member information and task information, and according the information pushed different profitable tasks to different members. The task's access is two-way selection between members and systems, that is, members can receive messages directly from the system, and can also search information by themselves to complete tasks. In order to promote the development of crowdsourcing economy and let more people participate in crowdsourcing business, we give the following suggestions.

1. When registering for membership, you need to submit an idle time period, as well as your own acceptable range of activities, as a reference for pushing tasks.
2. Members must submit the task content, geographical location, the required completion time, the number of tasks expected to complete, the initial quotation of the task (in principle, the more urgent the time, the higher the task difficulty coefficient, the price must be higher) to facilitate the platform calculate the ideal pricing of these tasks for different tasks.
3. All members have the same chance of selecting tasks, and members with high credit standing can get additional platform rewards when the task is completed, and the higher the credit, the more rewarded they get.
4. For members who do not complete or fail to receive tasks for a long time, they must deduct their member's reputation value. When the member's credit value is reduced to 0, the platform will no longer push the task to the member.

Acknowledgement. Project supported by the National Natural Science Foundation (61502155, 61772180); Education cooperation and cooperative education project (201701003076); Research start-up fund of Hubei university of technology (BSQD029); University student innovation and entrepreneurship project of Hubei university of technology (201710500047).

References

1. Howe, J.: The rise of crowdsourcing. Wired **14**(6), 176–183 (2006)
2. Saxton, G.D., Kishore, R.: Rules of crowdsourcing: models, issues, and systems of control. Inf. Syst. Manag. **30**(1), 2–20 (2013)
3. Yan, T., Kumar, V., Ganesan, D.: CrowdSearch: Exploiting crowds for accurate real time image search on mobile phones. In: Proceedings of the International Conference on Mobile Systems, Applications, and Services, San Francisco, USA, pp. 77–90 (2010)
4. Alonso, O., Rose, D.E., Stewart, B.: Crowdsourcing for relevance evaluation. J. SIGIR Forum (SIGIR) **42**(2), 9–15 (2008)
5. Alonso, O., Mizzaro, S.: Can we get rid of TREC assessors? Using Mechanical Turk for relevance assessment. In: Proceedings of the SIGIR Workshop on the Future of IR Evaluation, Boston, Massachusetts, USA, pp. 15–16 (2009)
6. Lease, M., Carvalho, V.R., Yilmaz, E.: Crowdsourcing for search and data mining. J. SIGIR Forum (SIGIR) **45**(1), 18–24 (2011)
7. Thrift, N.: Re-inventing invention: new tendencies in capitalist commodification. Econ. Soc. **35**(2), 279–306 (2006)
8. Penin, J., Burger-Helmchen, T.: Crowdsourcing of Inventive activities: definition and limits. Int. J. Innov. Sustain. Dev. **5**(2/3), 246–263 (2011)
9. Yuen, M.C., King, I., Leung, K.S.: A survey of crowdsourcing systems. In: Proceedings of the IEEE Third International Conference on Social Computing, SocialCom, pp. 766–773. IEEE, Boston, October 2012
10. Kittur, A., Smus, B., Khamkar, S., Kraut, R.E.: CrowdForge: crowdsourcing complex work. In: Proceedings of the 24th Annual ACM Symposium on User Interface Software and Technology, UIST 2011, USA, pp. 43–52, October 2011
11. Wang, J., Kraska, T., Franklin, M.J., et al.: CrowdER: crowdsourcing entity resolution. Proc. VLD Endow. **5**(11), 1483–1494
12. Sakamoto, Y., Tanaka, Y., Yu, L., Nickerson, J.V.: The crowdsourcing design space. In: Schmorrow, D.D., Fidopiastis, C.M. (eds.) FAC 2011. LNCS, vol. 6780, pp. 346–355. Springer, Heidelberg (2011). https://doi.org/10.1007/978-3-642-21852-1_41
13. Feng, J., Li, G., Wang, H., Feng, J.: Incremental quality inference in crowdsourcing. In: Bhowmick, S.S., Dyreson, C.E., Jensen, C.S., Lee, M.L., Muliantara, A., Thalheim, B. (eds.) DASFAA 2014. LNCS, vol. 8422, pp. 453–467. Springer, Cham (2014). https://doi.org/10.1007/978-3-319-05813-9_30
14. Pu, L., Chen, X., Xu, J., Fu, X.: Crowdlet: optimal worker recruitment for self-organized mobile crowdsourcing. In: Proceedings of the 35th Annual IEEE International Conference on Computer Communications, IEEE INFOCOM 2016, USA, April 2016
15. He, Z., Cao, J., Liu, X.: High quality participant recruitment in vehicle-based crowdsourcing using predictable mobility. In: Proceedings of the 34th IEEE Annual Conference on Computer Communications and Networks, IEEE INFOCOM 2015, pp. 2542–2550, May 2015
16. Mason, W.A., Watts, D.J.: Financial incentives and the "performance of crowds". In: Proceedings of the ACM SIGKDD Workshop on Human Computation, Paris, France, pp. 77–85 (2009)
17. Han, Y., Wu, H.: Minimum-cost crowdsourcing with coverage guarantee in mobile opportunistic D2D networks. IEEE Trans. Mob. Comput. **16**(10), 2806–2818 (2017)
18. Yang, K., Zhang, K., Ren, J., Shen, X.: Security and privacy in mobile crowdsourcing networks: challenges and opportunities. IEEE Commun. Mag. **53**(8), 75–81 (2015)

Research on Hybrid Data Verification Method for Educational Data

Lin Dong[1(✉)], Xinhong Hei[1], Xiaojiao Liu[2], Ping He[1], and Bin Wang[1]

[1] Xi'an University of Technology, Xi'an 710048, China
1132665427@qq.com
[2] Xi'an Educational Television, Xi'an 710001, China

Abstract. With the development of educational informatization, the problem of data quality has become the main problem that restricts the development of educational informatization. It is particularly important to manage and improve the data quality at the life cycle of the data by intelligent means. This paper presents a data verification framework orienting education data based on rule base, and gives the hybrid data verification method. The method is applied to the education statistics foundation database platform in the data verification process, and improves the precision of data.

Keywords: Education data · Data verification · Rule base

1 Introduction

Education statistics directly affects the scientificity, rationality and effectiveness of government policy making and regulation management. Therefore, it is very important to standardize the whole life cycle of the collection, storage, processing, use and sharing of educational data so as to ensure the authenticity and accuracy of the data.

In the current educational data management system, due to the lack of effective data prevention and detection methods for the data sources and the relationship between data, so that it can not detect and correct the possible data quality problems, thus increasing the difficulty and complexity of data reporting work and data analysis. In the field of data verification, Han realized data verification based on knowledge library [1], but the drawback is that the construction of knowledge base is highly dependent on computer capabilities. Xiao checks the input data by reading and analyzing the rules dynamically [2]. At present, a complete verification mechanism has not been formed according to its own characteristics in the field of education data verification. Liu and others introduced some constraints of data quality problems on the basis of a depth understanding of the existing verification tools, they combine Excel and rules to solve the related technical questions [3]. Huang puts forward a method of data quality detection based on multi-dimensional check rules, aiming at each item of the target data [4]. In addition, there are still some literatures about data verification methods [5–10].

Therefore, this paper proposes a hybrid data verification method combining client and server data verification, and giving the specific verification model and verification

Q. Zhou et al. (Eds.): ICPCSEE 2018, CCIS 901, pp. 60–73, 2018.
https://doi.org/10.1007/978-981-13-2203-7_6

method. In this paper, this method is applied to the verification process of educational statistics, and improves the accuracy of the data. The innovation of this paper is for the first time to study the verification method of educational data, using SQL rules and other algorithms to implement a large number of rules to verify the data, separate the business part from the code, and reduce the coupling degree.

2 Principle of Data Verification

Data verification exists in all stages of data quality control, according to different standards, it is divided into different types. From the check time, it is divided into verification before data acquisition and verification after collecting data. From the check place, it includes client verification, server verification and database verification [11]. From the check content, it's composed of business relation verification, logical relation verification and empirical relation verification and so on. The main process of data verification is shown in Fig. 1.

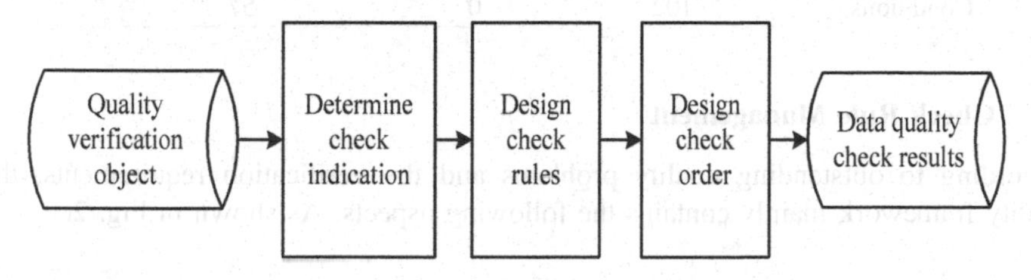

Fig. 1. Data quality check main flow

In order to realize educational data verification system, firstly, we need to research and build a data verification model. Check the data by designing different check rules for different indicators to obtain quality check results.

2.1 Educational Statistics

The educational statistics mainly includes organization basic information, business information, educational user data, school expansion information, school area and conditions etc. The organization basic information is consist of organization name, code, nature, detail address, e-mail, postal and historical evolution and so on. Business information is consist of the party leader information, the executive director information and persons information in charge of statistics. Education user data includes teacher and student information. Expand information includes washing facilities information, water supply situation and toilet facilities information. School area and conditions includes green area, sports area, the number of books, computers and classrooms. In addition, educational statistics is also related to school construction information and other related information.

2.2 Verification Indicator

The educational statistics involves many fields and various kinds of information, and relevant departments hope to analyze and explore some rules deeply from these data. However, many records in the education information database have some obvious quality problems, which are not conducive to data analysis. This paper makes a statistical analysis of some outstanding quality problems on the historical educational data of a city in 2016. The results are shown in Table 1.

Table 1. Outstanding quality problems (unit: records)

Description	Problem data	Duplicate records	Unreasonable data
Student	155	7628	10052
Teacher	213	288	567
Schoolhouse	87	65	234
School	472	146	0
Expansion information	447	0	131
Conditions	102	0	57

2.3 Check Rule Management

According to outstanding quality problems and the verification requirements, this quality framework mainly contains the following aspects. As shown in Fig. 2.

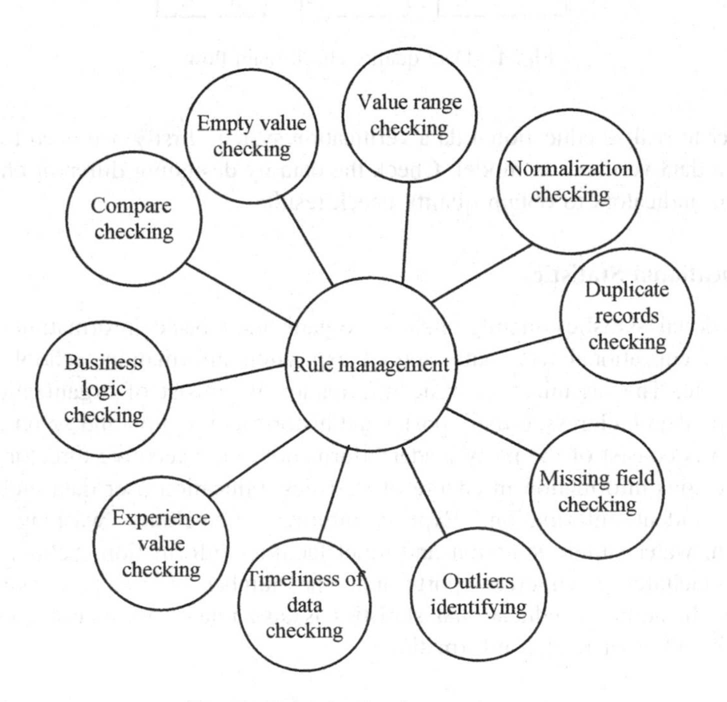

Fig. 2. Data check rule management

According to the analysis of education data, data verification rules of this paper mainly are divided into empty value checking, value range checking, normalization checking, duplicate records checking and missing field checking, outliers identifying, experience value checking and timeliness of data checking, business logic checking and standard information comparison checking.

3 A Hybrid Data Verification Framework

The framework is divided into client verification and server verification.

Client verification checks input data for business logic checking, duplicate checking, outliers identifying and missing field checking according to all kinds of rules. The unqualified data don't enter the process of server verification. Server verification includes logic checking and experience value checking and so on. The results are divided into error and warn information, both results will be displayed, and the user checks the warning information again. Data verification framework is as shown in Fig. 3.

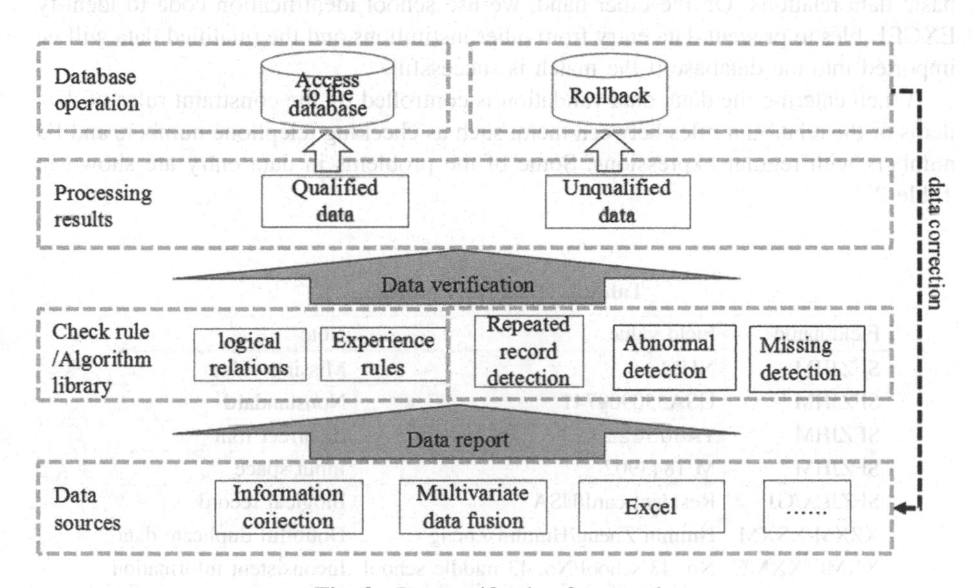

Fig. 3. Data verification framework

The verification framework manages the whole life cycle of the data, including the data collection, storage and use process, Aiming at various problems, this framework try to find out a set of extensive education data checking rule base.

In particular, the data verification includes the following steps.

① Design and analyze verification indicator: Determined verification Indicators through the error rate of the statistical indicator and the expert knowledge.
② Define data verification rules and work flow: According to the verification indicators obtained from the Step 1, we take a comprehensive consideration of the number

of data sources and dependencies between data source to define validation regulations and arrange the verification place, verification time and verification order.
③ Execute data verification rule: Execute the process on the source data determined in the previous step by rule execution engine, then display the verification result to the user.
④ Verify and solve problem data: According to the result of data verification, we need to iterate and revise. During the actual data checking process, there are rule analysis, design and verification process for many times before we get satisfactory data verification rules and result.
⑤ Repeat the step 3 to step 4 until the check result is passed.

3.1 Client Verification Method

The client verification includes three methods, which are EXCEL file checking, and the data verification when data is inputting and after data input.

When we import data by using batch import, on the one hand, we need to satisfy the basic data relations. On the other hand, we use school identification code to identify EXCEL files to prevent data entry from other institutions and the qualified data will be imported into the database if the match is successful.

When entering the data, data validation is controlled by the constraint rules of data items or the relational rules between them, such as checking telephone numbers and ID numbers with regular expressions. Some of the problems in data entry are shown in Table 2.

Table 2. Data entry problems

Field name	Field value	Note
SFZJHM	NULL	Missing value
SFZJHM	USA530909741	Nonstandard
SFZJHM	YA8054213	Incorrect font
SFZJHM	M 1843902	Input space
SFZJLX/GJ	Resident card/USA	Illogical record
XSXM/XSXM	Huimin Zheng/Huimin Zheng	Doubtful duplicate data
XXMC/XXMC	No. 43 school/No. 43 middle school	Inconsistent information

Duplicate data detection is also a very important process in data verification. Multi-source heterogeneous data fusion or manual entry may lead to similar duplicate records. We can use the existing matured similar duplicate record detection algorithms, such as SNM algorithm, to detect duplicate or conflict records [12, 13].

3.2 Server Verification Method

The quality check rule set is composed of different verification rules, which are derived from a careful analysis of the report file. The server verification is mainly to check the basic table data of the school by single or multiple tables, and you can also view all the logical and experience rules of the reports. As shown in Fig. 4:

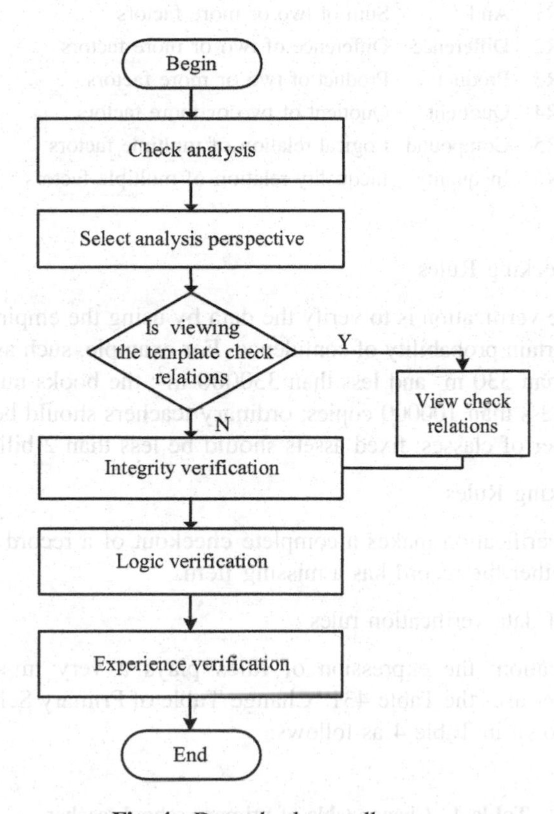

Fig. 4. Data check overall process

(1) Design rule base

The rule base is composed of many data checking rules. These check rules are derived from the comprehensive analysis of the reports in the data reporting system, and combined with the fact database and expert knowledge experience. A complete and extensive rule base will detect bad data to the greatest extent and reflect the data detection ability and efficiency.

The educational data verification rules can be divided into logical checking rule, experience checking rule and integrity checking rule. The general expression of the rule is defined as shown in Table 3:

- Logic Checking Rules

The logic check formula indicates that the relationship between data tables by written in a conditional form, and the logical check is to determine whether the relationship of different tables is consistent.

Table 3. Checking rule expressions

No.	Rule name	Rule description
R1	And	Sum of two or more factors
R2	Difference	Difference of two or more factors
R3	Product	Product of two or more factors
R4	Quotient	Quotient of two or more factors
R5	Compound	Logical relation of multiple factors
R6	Inequality	Inequality relation of multiple factors

- Experience Checking Rules

The experience verification is to verify the data by using the empirical value of the indicator with a certain probability of confidence. For example, such as the school area should be larger than 330 m^2 and less than 330000 m^2; the books numbers of school library should be less than 100000 copies; ordinary teachers should be greater than or equal to the number of classes; fixed assets should be less than 2 billion yuan.

- Integrity Checking Rules

The integrity verification makes a complete checkout of a record in the database, which checks whether the record has a missing item.

(2) Expression of data verification rules

In data verification, the expression of rules plays a very important role. For example, This paper uses the Table 431 "Change Table of Primary School Teacher" as an example, as shown in Table 4 as follows:

Table 4. Change table of primary school teacher

B	C	D	E	F	G	H	I..
	No.	Full-time teacher numbers of last year	Increasing numbers of teacher				
			Total	Number of graduate teachers		Transferred	Changes in campus
				Total	Among them: normal students		
		1	2	3	4	5	6
Primary school	01						
Female	02						

The verification rules are expressed as follows:

Error Rule 1:

[Table 431]. [primary school full-time teachers] should be greater than or equal to [Table 912]. [primary school full-time teachers]

Expression:

[Table 431]. (line 01, column 13) ≥ [Table 912]. (line 07, column 09)

Error Rule 2:

[Table 431]. [Full-time teacher Numbers of the school year in primary school] should be greater than or equal to [Table 4412]. [the number of primary school of the Communist Youth League]

Expression:

[Table 431]. (line 01, column 13) ≥ [Table 4412]. (line 03, column 2)

Error Rule 3:

[Table 431]. [Number of Graduate Teachers] should be greater than or equal to [among them: normal students]

Expression:

[Table 431]. column 3 ≥ [Table 431]. column 4

Warn Rule 4:

The Cell "others" has a value in Increasing Numbers of Teacher. Please explain the situation.

Expression:

The Cell "others" has a value in Increasing Numbers of Teacher. Please explain the situation.

Error Rule 5:

[Table 431]. [Numbers of full-time teachers in primary school at the beginning of the school year] should be greater than or equal to [Table 4412]. [sum of Communist Party, the Communist Youth League and the democratic parties in Primary school]

Expression:

[Table 431]. (line 01, column 13) ≥ [Table 4412].{(line 03, column 1) + (line 03, column 2) + (line 03, column 3)}

Error Rule 6:

[Table 431]. [Numbers of female full-time teachers in special school at the beginning of the school year should be greater than or equal to [Table 4412]. [the female teacher sum of Communist Party and the Communist Youth League and the democratic parties in special school]

Table 5. Formula expression for rules

Formula	Description	Condition
D1:I1 ≥ #2	line 01 ≥ line 02	None
F1:F8 ≥ #G	column 3 ≥ column 4	None
F1 ≥ F2 − G2 + G1	(line 01, column 3)–(line 01, column 4) ≥ (line 02, column 3)–(line 02, column 4)	None
F5 ≥ F6 − G6 + G5	(line 01, column 3)–(line 02, column 3) ≥ (line 01, column 4)–(line 02, column 4)	None

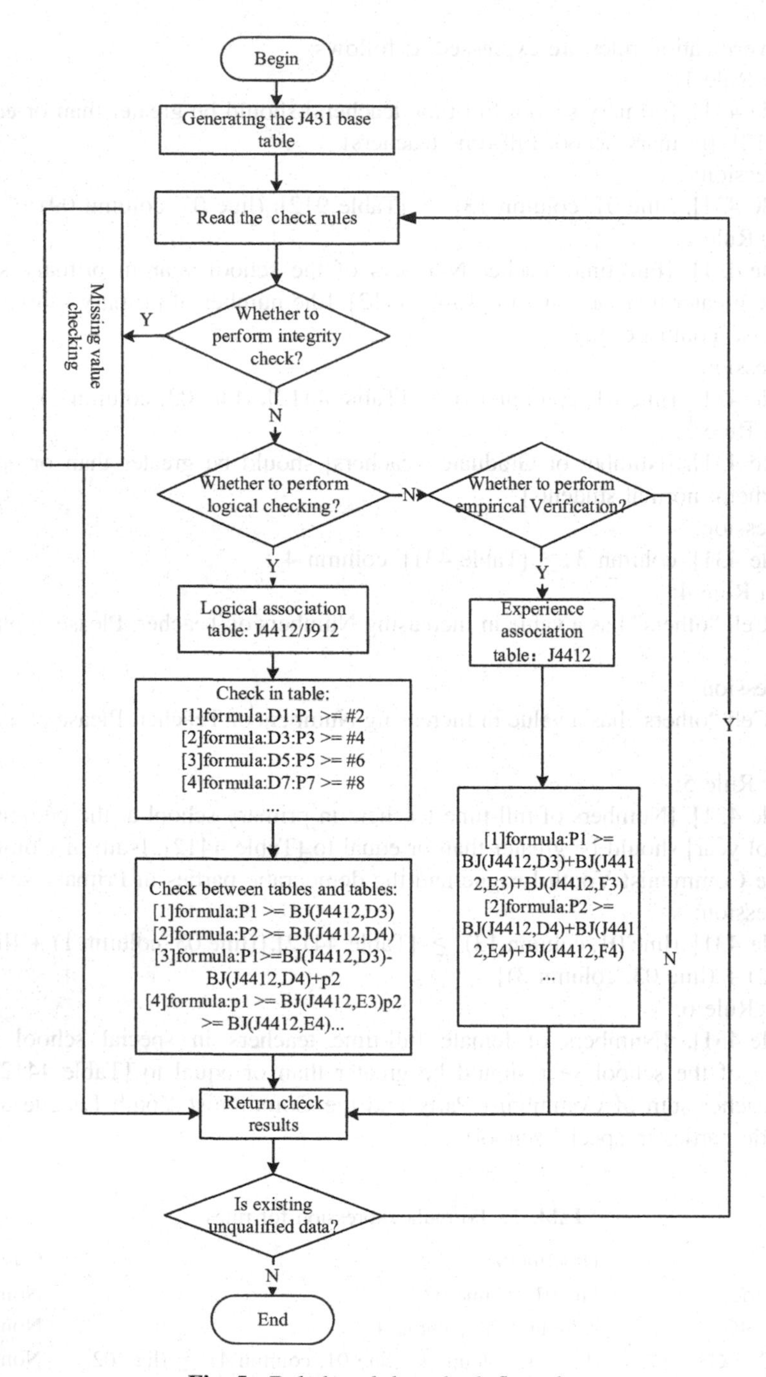

Fig. 5. Rule-based data check flow chart

Expression:

[Table 431]. (line 08, column 13) $\geq$ [Table 4412].{(line 10, column 1) + (line 10, column 2) + (line 10, column 3)}

Then, we take out a few examples, and the following are the forms that are expressed in a formula. As shown in Table 5.

(3) The execution process of data verification

The flow chart of the server's data check process is shown in Fig. 5. Firstly, we need to generate base table data. Secondly, query association tables and data. Thirdly, determine the categories and check sequences to check.

4 Realization and Experimental Results Analysis

4.1 Experimental Data Source

The educational data that is used in data verification is obtained through the educational statistics basic database platform of a city. The distribution of data sets as shown in Table 6.

Table 6. Data set samples

Number of data sheets	Rule number
19	758

4.2 Experimental Code Implementation

The pseudo code implemented by rule checking algorithms is as follows (Table 7):

4.3 Experimental Results Analysis

After several iterations of the verification process, we get the result of Fig. 6 below. The number of problem and unreasonable data tends to be zero. But the suspect duplicate data is almost unchanged at a stable value. It is normal to have some suspected data, these data needs to verify artificially.

After the accuracy meets the requirements of the target, the check time of this paper method has not greatly increased under the number of different rules. The result of check time is as shown in Fig. 7.

Table 7. Algorithm description of the rule checking

Input: recordYear ← string, checkModel ← Model, edudataDBName ← string
Output: verification result;

```
1. Problem data process implementation
2. for example:
3.     select *,case when (sfzjlx is null or sfzjlx=' ') then 'The type of
4.     identity card is empty' else'' end as Q1, case when (sfzjhm is null
5.     or sfzjhm =' ')
6.     then 'The ID card number is empty' else'' end as Q2 into #tempA from
7.     #tempJS_TSJY where (sfzjlx is null or sfzjlx =' ') or (sfzjhm is
8.     null or sfzjhm =' ')
9. connect database
10. delete old check results;
11. Generation data process implementation
12. for example:
13.     declare @DID2 int = 1
14.     declare @A2 int =6
15.     declare @E2 int ,@F2 int ,@G2 int,@H2 int,@I2 int,@J2 int,@K2 int
16.     select @E2 = count(ID) from #tempTeacherYEY where GWLB=@GWLB1
17.     and SFJRBXJZG = @SFJRBXJZG1
18.     select @F2 =... ; select @G2 = ...
19.     insert into #J411(DID,A,E ,F, G ,H , I, J, K)
20.     values(@DID2,@A2,@E2,@F2,@G2,@H2,@I2,@J2,@K2)
21. Data verification process implementation
22. checkinfo.checkType = "warn";
23. query listTemplates;
24. foreach(string current in listTemplates)
25. Update progress value
```

```
26. If listTemplates > 0 then
27.    execution rule formulas;
28.    invoking the SQL stored procedure;
29.    for example:
30.      --▲公式:E1 < 5
31.      insert into #checkTmp(instanceID,DID,expressID )
32.          select instanceID,-1,2 from #J411_R_TMP
33.                   where DID=1
34.                    and E >= 5
35.      insert check result to the checkout table;
36. End if
37. checkinfo.checkType = "error";
38. execution step 9 to 22 for error rules;
39. If rule set = 0 then
40.    output check result;
41. End if
```

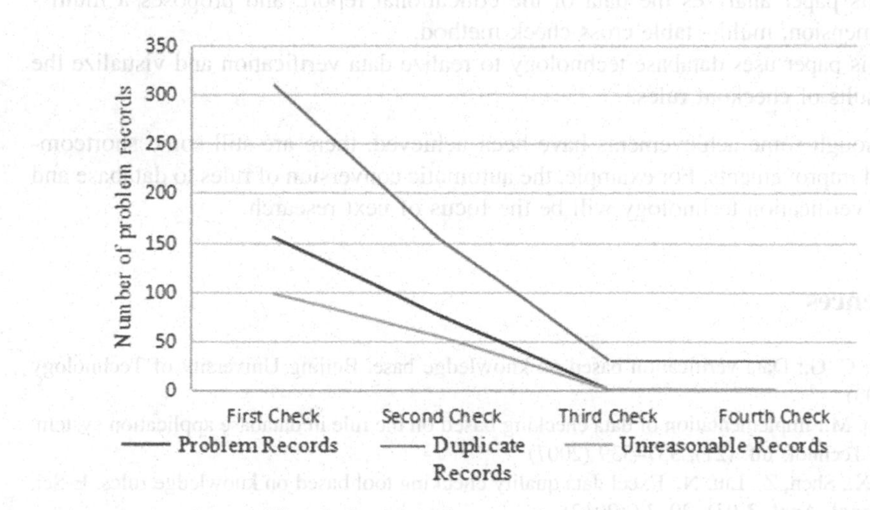

Fig. 6. Iterative diagram of verifying results

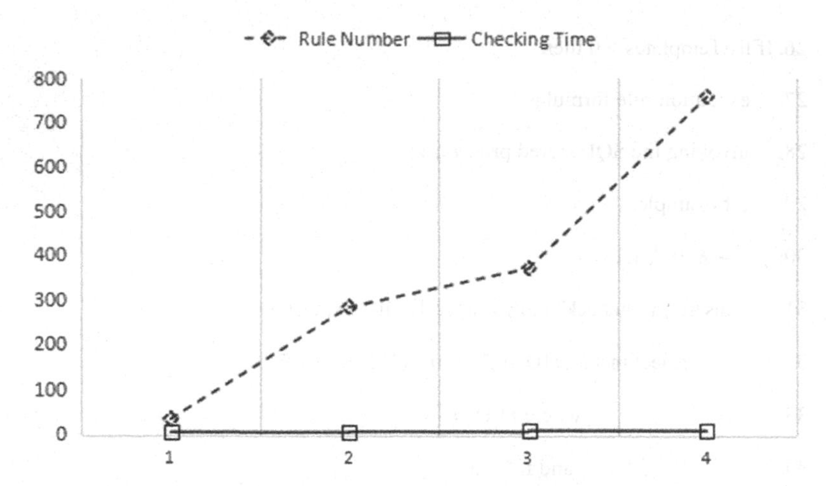

Fig. 7. Data check time contrast under different rules

5 Conclusions

This paper proposes a data verification framework model based on rule base, giving the hybrid data verification method for education data and forming a closed-loop process control of data quality. To a great extent, the framework maximizes the effectiveness of data quality.

The innovative points of this article are as follows:

(1) Put forward a perfect data checking framework. By using the respective characteristics of each stage of the life cycle of the educational data.
(2) This paper analyzes the data of the educational report, and proposes a multi - dimension, multi - table cross check method.
(3) This paper uses database technology to realize data verification and visualize the results of checkout rules.

Although some achievements have been achieved, there are still some shortcomings and improvements. For example, the automatic conversion of rules to database and parallel verification technology will be the focus of next research.

References

1. Han, C.-G.: Data verification based on knowledge base. Beijing University of Technology (2003)
2. Xiao, M.: Implementation of data checking based on the rule in database application system. Sci. Technol. Inf. (21), 337–339 (2007)
3. Su, X., Shen, Z., Liu, N.: Excel data quality checking tool based on knowledge rules. E-Sci. Technol. Appl. **3**(03), 29–37 (2012)
4. Lin, X., Shen, D., Shi, Y., Qiao, D.: Configurable data checking method. Comput. Syst. Appl. **24**(05), 161–166 (2015)

5. Huang, H.: A data quality detection method based on multi-dimensional check rules. Shandong CN106528828A 22, March 2017
6. Xu, H., Wang, J., Wu, L., Sun, R., Li, L.: Research and application of multi-dimension check for running data of new energy unit. N. China Electr. Power (10), 7–12+18 (2017)
7. Wang, T.: Discussion on data checking mechanism of rule-based power dispatching plan. Electr. Prod. (12), 263 (2015)
8. Liu, Y., Shen, X., Xu, L., et al.: A MapReduce based parallel algorithm for CIM data verification. In: International Conference on Fuzzy Systems and Knowledge Discovery, pp. 704–709. IEEE (2014)
9. Liu, B., Gen, Y.: The mining method of data quality detecting rules. Pattern Recogn. Artif. Intell. **25**(05), 835–844 (2012)
10. Ebaid, A., Elmagarmid, A., Ilyas, I.F., et al.: NADEEF: a generalized data cleaning system. Proc. VLDB Endow. **6**(12), 1218–1221 (2013)
11. Liu, C., Zhang, K., Chen, J.: Research on data validation scheme of mixed mode. Comput. Eng. Des. **34**(01), 366–371 (2013)
12. Zhou, Y.: Research and application of data cleaning algorithm. Qingdao University (2005)
13. Sheng, X., Hun, S.: Similar duplicate records elimination based on improved SNM algorithm. J. Chongqing Univ. Technol. (Nat. Sci.) **30**(04), 91–96 (2016)

Efficient User Preferences-Based Top-k Skyline Using MapReduce

Linlin Ding, Xiao Zhang, Mingxin Sun, Aili Liu, and Baoyan Song(✉)

School of Information, Shenyang, China
bysong@lnu.edu.cn

Abstract. As an important variant of skyline query, top-k skyline can find the best k points as the final results. In order to generate the results meeting the needs of users for massive data, we propose an efficient user preferences-based top-k skyline combining partially and totally ordered domains in MapReduce, named P/T_SKY_MR. The whole course contains two main phases, partially ordered domains processing and totally ordered domains processing. In partially ordered domains processing, we propose the binary encoding and a pruning strategy to present the precedence relationship about the partially ordered domains and different user preferences. Meanwhile, in totally ordered domains processing, for finding the final results, a defined ranking criterion is also proposed in order to reduce the calculation cost and minimize the response time. A large number of experiments show that our method is effective, flexible and scalable.

Keywords: Top-k skyline · User preferences
Partially ordered domains · Totally ordered domains · MapReduce

1 Introduction

Skyline computation has the extensive application fields, such as multi-objective decision making, mine microseism warning and data visualization. However, with the increasing of data volume, the size of the skyline result may also very large, leading to the users to making decisions difficultly. To identify the best points among the skyline results, top-k skyline has been proposed.

At present, top-k skyline algorithms usually require a user to define the ranking function over the data objects collection, and return the k top ranked data objects. However, some data dimensions (for example, in the forms of hierarchies, intervals, and preferences) are partially ordered in many applications. Besides, it sometimes requires us to adopt a user-specific metric to narrow down skylines into the user-specified retrieval size k. It means that different attributes will take on different degrees of importance for different users.

In order to generate the results meeting the needs of users for massive data, this paper presents the study on user preferences combining the partially and totally ordered domains for efficient top-k skyline computation using

Q. Zhou et al. (Eds.): ICPCSEE 2018, CCIS 901, pp. 74–87, 2018.
https://doi.org/10.1007/978-981-13-2203-7_7

MapReduce, named P/T_SKY_MR. There are mainly two phases MapReduce computations of the whole course, respectively the partially ordered domains processing, P_SKY_MR, and totally ordered domains processing, T_SKY_MR. In the P_SKY_MR phase, we deal with the partially ordered domains by employing the binary encoding to present the order of different candidate choices. Then a pruning strategy is applied to filter out a part of results which can't be the final results in advance avoiding the redundancy of computation and storage. When it comes to the second phase T_SKY_MR, we handle the numeric attributes with a defined ranking criterion to order the candidates from the P_SKY_MR phase for further filtering. Then, the final k results can be gained by the ranking of combing the partially and totally ordered domains.

The remainder of this paper is organized as follows. Section 2 reviews the related work. Section 3 shows the problem statement and the whole distributed processing course. The partially ordered domains processing is proposed in Sect. 4. Section 5 presents the totally ordered domains processing. Section 6 reports on the experimental studies. Section 7 concludes the paper.

2 Related Work

2.1 User Preferences Skyline

The user preferences skyline can return the data objects that meeting the requirements of users best. Lee et al. [1] proposed the personalized skyline queries as identifying "truly interesting" objects based on the user-specific preference and retrieval size k. Cheong Fung et al. [2] defined a novel concept called k-dominate p-core skyline to decide whether a point is an interesting one or not and proposed an effective tree structure LBM tree to process the preference skyline. Paper [3] introduced a new operator, namely the most desirable skyline object query, to identify the manageable size of truly interesting skyline objects. Paper [4] introduced a novel concept called collective skyline to deal with the problem of multiple users preferences, then developed a suitable algorithm based on optimization techniques for efficiently computing the collective skyline. Paper [5] proposed an incremental method for calculating the skyline points related to several dimensions associated with dynamic preferences, which improved notably the execution time and storage size of queries. Paper [6] presented the first study on processing top-k representative skyline queries in uncertain databases, based on user-defined references, regarding the priority of individual dimensions, then developed two novel algorithms and several pruning conditions. Li et al. [7] proposed a system model in a mobile and distributed environment to perform distributed skyline queries in any subspace according to user preferences. An efficient parallel algorithm for processing the Subspace Skyline Query (SSQ) using MapReduce is applied to the system model.

2.2 Top-k Skyline

Top-k skyline is a variant of skyline, which can control the number of skyline query results. Paper [8] defined a novel type of skyline query to find

k combinations of data points according to a monotonic preference function, and proposed two efficient query algorithms to incrementally and quickly generate the combinations of possible query results. Paper [9] proposed an efficient method based on the quad-tree index and used four pruning strategies to solve the top-k skyline query. Paper [10] proposed k-objects selection function that selects various k objects that are preferable for all users who may have the different scoring functions by applying the idea of skyline queries using MapReduce. Paper [11] proposed a distributed top-k skyline method in MapReduce of big data. Paper [12] considered a method for computing the top-k dominating query and k-skyband query in a parallel distributed framework called MapReduce, which is a popular framework to handle big data. Siddique et al. [13] found the conventional algorithms for computing k-dominant skyline queries are not well suited for the parallel and distributed environments, so they considered an efficient parallel algorithm to process k-dominant skyline query in MapReduce framework.

3 Problem Statement and Distributed Processing Course

3.1 Problem Statement

Definition 1 *(Top-k Skyline). For any point p in data set P, $p \in P$, the scoring function is F(), then we call p_1, p_2,..., p_k is the top-k skyline set of P, only when $F(p_1) + F(p_2) + \ldots + F(p_k)$ is the max.*

Definition 2 *(User Preferences-based top-k skyline). For any point p in data set P, $p \in P$, the scoring function is F(), then we call p_1, p_2,..., p_k is the top-k skyline set of P, only when $F(p_1) + F(p_2) + \ldots + F(p_k)$ is the max. F() is defined by the users.*

Different dimensions have different importance. In one dimension, different values also have different importance.

3.2 Distributed Processing Course

The main course of our user preferences-based top-k skyline includes the following two phases: partially ordered domains processing, named P_SKY_MR, and totally ordered domain processing, T_SKY_MR. The two phases are all processed in MapReduce framework, involving mappers and reducers. The overall flow of top-k skyline is illustrated in Fig. 1.

- P_SKY_MR: We focus on the processing of partially ordered domains in one MapReduce computation. In the Map phase, we identify the binary bitstring for every data object by the binary bitstring coding method according to the user preferences, and combine the objects with the same bitstring, and then compute the virtual min and max points. In Reduce phase, a pruning strategy is applied to filter the data objects not meeting the conditions, avoiding the redundancy of computation and storage. The P_SKY_MR will be presented in Sect. 4 in detail.

- *T_SKY_MR*: We focus on the processing of totally ordered domains and calculate the user preferences-based top-k skyline results in one MapReduce computation. After receiving the output of *P_SKY_MR*, in Map phase, we use our criterion to order the candidate data objects. In Reduce phase, we obtain the final results by our ranking criterion combining the partially and totally ordered domains. The *T_SKY_MR* will be presented in Sect. 5 in detail.

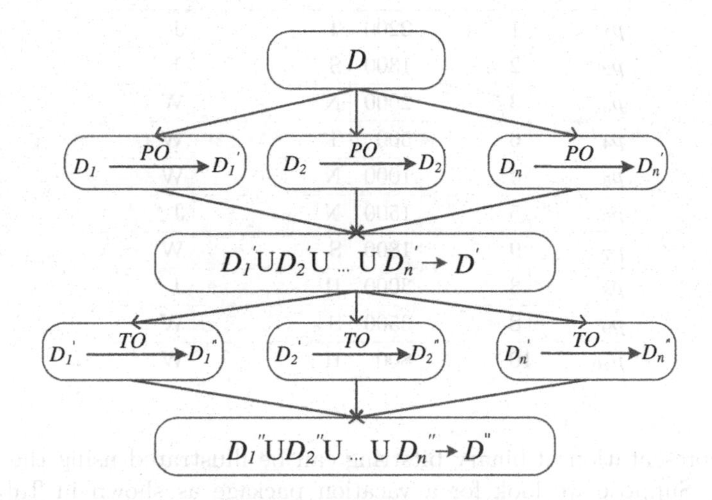

Fig. 1. The overflow of user preference top-k skyline in MapReduce

4 Partially Ordered Domains Processing

4.1 Binary Bitstring and Pruning Strategy

In order to present the user preferences and filter the data objects not meeting the user preferences, we propose the following binary bitstring strategy and the pruning strategy.

Given a query with a user-specific preference over partially ordered domains, we need to use a binary bitstring strategy below to present the precedence relationship of the different candidate objects.

Definition 3 *(Binary Bitstring). A bitstring is assigned to each data object in the candidate sets based on its value in partially ordered domains (PO). In each partially ordered domain (PO), there are v attributes given by the user preferences priority. In the w-dimensional partially ordered domains, each of which consists of v attributes and the user takes more interest in the PO placed at the front in default. The binary bitstring of the object in the candidate set is represented by* $id = a_{w \times v} \ldots a_2 a_1$*, where each v bits will be* $V_i = a_v \ldots a_i \ldots a_2 a_1$*. We encode* V_i *as a v-bit binary bitstring where each* a_i *corresponds to ith most significant bit related to the priority in PO. Similarly, we can simply obtain the* *binary bitstring of* $V_i (1 \le i \le w)$*. Finally we merge the* w *binary bitstrings to form the whole binary bitstring of* $w \times v$ *bits for gaining the final bitstring id. Note that we can get a great variety of binary bitstrings which are equal to* v^w*.*

Table 1. An example of user preference-based top-k skyline

Package	Distance	Price	Hotel service	Airline
p_1	1	2200	T	J
p_2	2	1300	S	J
p_3	3.5	2000	N	W
p_4	6	500	T	W
p_5	4	1000	N	W
p_6	5	1500	N	J
p_7	9	1800	S	W
p_8	8	3000	H	J
p_9	8	3500	S	W
p_{10}	10	800	H	W

The representation of binary bitstring can be illustrated using the example in Table 1. Suppose we look for a vacation package as shown in Table 1. We know that the attribute Distance and the attribute Price are numeric attributes where a package with shorter distance from beach or a cheaper one is more preferable. However, the attribute Hotel service and the attribute Airline are partially ordered domains. Consider a user-specific preference: compared with the attribute Airline, the attribute Hotel service is more important for users. For each PO, a user provides us with some partially order relations expressing his/her preferences: PO1: $N < H, T < S$; PO2: $J < W$. We can transform all the partially ordered domains of each object in Table 1. According to the definition of binary bitstring and the user-specific preference, the partially ordered domains of p_1 on the attributes Hotel service and Airline of package are T, J respectively, so the *id* of PO1 is 01 and the *id* of PO2 is 01. The whole *id* of p_1 is 0101. Following the same principle, the *ids* of p_2 and p_3 are 1001, 0110.

Top-k skyline operation is expensive when the large scale data sets are encountered. Due to the characteristic that the top-k skyline result set is much smaller than the original data set, the pruning strategy can eliminate the impossible objects in advance.

Given several clusters, each cluster consists of plenty of data points. We extract from each cluster to form the two artificial d-dimensional points as the virtual max point and virtual minimum point. Virtual max point is denoted by $vxp(k) = max_{p \in cluster} p(k)$ with $1 \le k \le d$. Virtual minimum point is denoted by $vmp(k) = min_{p \in cluster} p(k)$ with $1 \le k \le d$.

Theorem 1 *(Pruning Strategy). If the $vxp(k)$ in cluster 1 dominates the $vmp(k)$ in cluster 2, we can safely remove all the data points in cluster 2 because all the data points in cluster 1 dominate all the data points in cluster 2.*

Proof. Since $vxp(k)$ in cluster 1 dominates $vmp(k)$ in cluster 2, $vxp(k) \leq vmp(k)$ for each k with $1 \leq k \leq d$ and there exists k such that $vxp(k) < vmp(k)$. By the definition of the virtual max point $vxp(k)$ and virtual minimum point $vmp(k)$, all points p in cluster 1, $p(k) \leq vxp(k)$, holds for each k and all points q in cluster 2, $vmp(k) \leq q(k)$, holds for each k. Thus, we also have $p(k) \leq q(k)$ for each k and there exists at least one k such that $p(k) < q(k)$. In other words, all points p in cluster 1 dominate all points q in cluster 2.

4.2 Computing of Candidates Results

We next present the parallel algorithm *P_SKY_MR* that calculates the candidate objects based on the user preferences about the partially ordered domains in large data sets using MapReduce.

Map Phase: Each map function is called with an object p in subset D_i. We take the key-value pair (p, <(*numeric attributes*), (*partially ordered attributes*)>) as input. The map function next computes the binary bitstring of each object according to user preferences-based with the partially ordered attributes. Then we identify the virtual minimum point $vmp(k)$ and virtual maximum point $vxp(k)$ of each map to report to the master. The key-value pair of Map phase is (*id*, <*p*, (*numeric attributes*)>).

Reduce Phase: In accordance with the Theorem 1, if the $vxp(k)$ of one map dominates the $vmp(k)$ of another map, the latter map can be pruned in advance. Subsequently, the reduce function is called and combines the objects with same binary bitstring as the output in terms of the bitstring in ascending order. Algorithm 1 formally describes the query algorithm of *P_SKY_MR*.

Algorithm 1. *P_SKY_MR*

```
 1: class MAPPER
 2:     p[i]=ExtractPartiallyOrderedAttributes(p);
 3:     p[j]=ExtractNumericAttributes(p);
 4:     id=GetBinaryBitstring(p[i]);
 5:        if p.pruned==false then
 6:           emit(p, < p[j], p.bitstring >);
 7: class REDUCER
 8:     for each emitted p do
 9:        combine p with the same bitstring;
10:     output (id, list of (p, p[j])) in ascending order of p.bitstring;
```

Example: Consider the candidate package set in Table 1, Figs. 2 and 3 show the data flow of *P_SKY_MR*. Each map function is invoked with an object p in D_i as illustrated in Fig. 2. For instance, (0101, <p1, 1, 2200>) is emitted

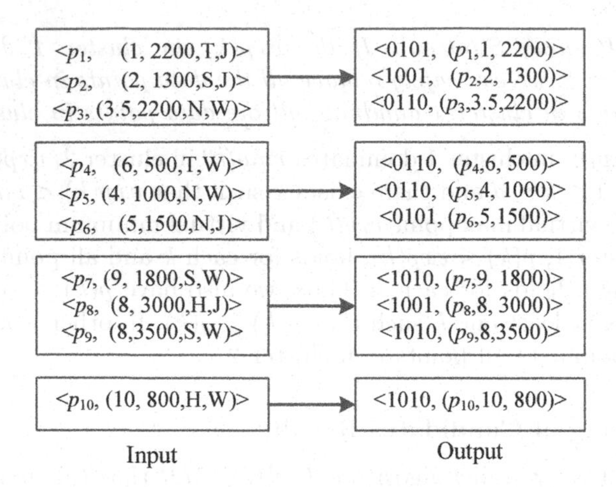

Fig. 2. Map phase of P_SKY_MR

since the binary bitstring of p_1 can be represented by 0101. In addition, each mapper calculates its own virtual max point $vxp(k)$ and virtual minimum point $vmp(k)$ and sends them to the master. The $vxp(k)$ and $vmp(k)$ of mapper1, mapper2, mapper3 and mapper4 can be $(<3.5, 2200, 1001>, <1, 1300, 0101>)$, $(<6, 1500, 0110>, <4, 500, 0101>)$, $(<9, 3500, 1010>, <8, 1800, 1001>)$, $(<9, 800, 1010>, <9, 800, 1010>)$ respectively. Because the $vxp(k)$ of mapper2 (6, 1500, 0110) dominates the $vmp(k)$ (8, 1800, 1001) of mapper3, so the whole mapper3 is not emitted to the reducer and can be discarded safely. Next, each reduce function combines the key-value pairs grouped by each distinct binary bitstring in ascending order provided in Fig. 3.

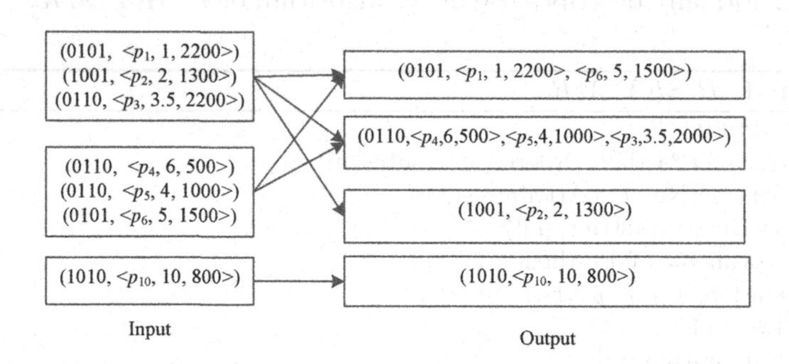

Fig. 3. Reduce phase of P_SKY_MR

5 Totally Ordered Domains Processing

5.1 Scores and Ranking Criterion

Definition 4 *(Dominating Score). Suppose D be the data set, in the totally ordered domains of D, for an object $p \in D$, the dominating score of p, denoted by v(p), could be defined as: $v(p) = |q \in D | p \prec q|$.*

In other words, the score $v(p)$ is the number of objects dominated by p. The higher the $v(p)$ is, the better p is. Consider the Table 1, for instance, p_2 is more preferable than p_1 as $v(p_2) = 5$ and $v(p_1) = 2$. Hence, a natural ordering of objects can be derived based on the dominating score. However, when two objects share the same dominating score, the tie has to be broken.

Definition 5 *(Dominated Score). Let D be the data set, in the totally ordered domains of D, for an object $p \in D$, the dominated score of p, denoted by v'(p), could be defined as: $v'(p) = \frac{1}{v(p) = |q \in D | p \prec q|}$.*

Intuitively, the dominated score of an object takes the number of objects dominating it into consideration. In fact, an object may have the larger power if it is dominated by as few objects as possible. As stated in Table 1, the power of p_3 is greater than that of p_6, since $v'(p3) = 1$ and $v'(p6) = 1/2$. Hence, we define the B-dominating score of an object: an object is more preferable if it dominates as many objects with higher dominated scores as possible.

Definition 6 *(B-Dominating Score). Let D be the data set, in the totally ordered domains of D, for an object $p \in D$, the B-dominating score of p, denoted by $\mu(p)$, could be defined as: $\mu(p) = \sum_{q=q \in D | p \prec q} v'(q)$.*

In Table 1, for instance, p_4 is more preferable than p_5 as $\mu(p_4) = 131/84$ and $\mu(p_5) = 89/84$ although $v(p_4) = v(p_5) = 4$. In fact, the B-dominating score of an object considers the accumulated power of all objects dominated by it. Based on the Definitions 4, 5 and 6, the ranking criterion is formalized as follows:

Definition 7 *(Ranking Criterion). Let D be the data set, in the totally ordered domains of D, for two objects p and q, p is more preferable than q, denoted by p q, if and only if:*

$$p \lhd q \Leftrightarrow \begin{cases} v(p) > v(q) \\ v(p) = v(q) \wedge \mu(p) > \mu(q) \end{cases}$$

5.2 Computing the Final Results

The procedure *T_SKY_MR* computes the final k objects in every non-empty list with the same bitstring independently using MapReduce.

Map Phase: Each map processes the current totally ordered domains of data set. The input key-value pair is (*id*, *<p, numeric attributes>*). In the map function called with each distinct key, it calculates the dominating score and B-dominating score and then ranks the objects on the basis of the ranking criterion as the output. The output key-value pair of Map is (id, $<p, v(p), \mu(p)>$).

Reduce Phase: Each reduce function next selects the object ranking the first in every mapper. Based on a scoring function, every candidate object will be given a score. Eventually it returns the first k objects to users as the final results. If the candidate results is smaller than k, the Reducer can compute by the same method again for the remainder candidate results. Algorithm 2 formally describes the query algorithm of T_SKY_MR.

Algorithm 2. T_SKY_MR

```
1: class MAPPER
2:    for p in each distinct bitstring do
3:    compute the dominating score v(p) and the B-dominating score(p) about p[j];
4:    emit (p, < v(p), μ(p) >);
5: class REDUCER
6:    for each p coming from the same bitstring do
7:       rank p according to the ranking criterion;
8:       select the top object;
9:    rank all top objects according to the score function;
10:   output the top k objects;
```

Example: The behavior of T_SKY_MR is illustrated in Figs. 4 and 5. Assume the value of k is 3 in our example. For instance, the map function with p_1 outputs (0101, $<p1, 2, 13/42>$) since the dominating score and B-dominating score of p_1 are 2 and 13/42 respectively. Note that in mapper2, the objects p_4, p_5 have the same dominating score, but the B-dominating score of p_4 is much larger than p_5. Based on the ranking criterion, p_4 is more preferable and sort p_4 ahead of p_5. The reduce function is called with all mappers by selecting the first object of every mapper. We think over both the aspects of totally ordered domains and partially ordered domains. Thus, we convert the binary bitstring into decimal number of every candidate object. Then we put the reciprocal of decimal number as horizontal ordinate and dominating score as vertical coordinate. We know the rectangle composed of the horizontal and vertical ordinate is the dominating region. So the area formula of the rectangle represents the candidate's overall controlling ability. As a consequence, horizontal ordinate is multiplied by vertical ordinate to get the final scoring function. In this way, we can get the top 3 objects which are p_4, p_6, p_2.

We now analyze the space and time complexity involved in our query. In the w-dimensional partially ordered domains and t-dimensional numeric attributes, each of the partially ordered domains consists of v attributes. Suppose P be the data set, the space and time complexity are $(w \times v + t) \times vw$, $log(P) + vw$ respectively since we utilize R-tree to store all the candidate objects.

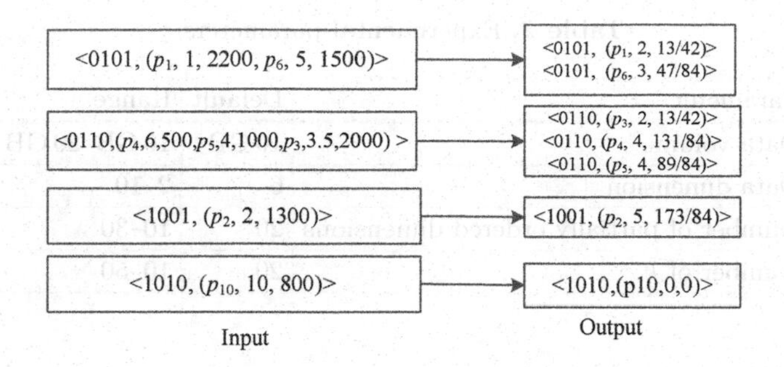

Fig. 4. Map phase of T_SKY_MR

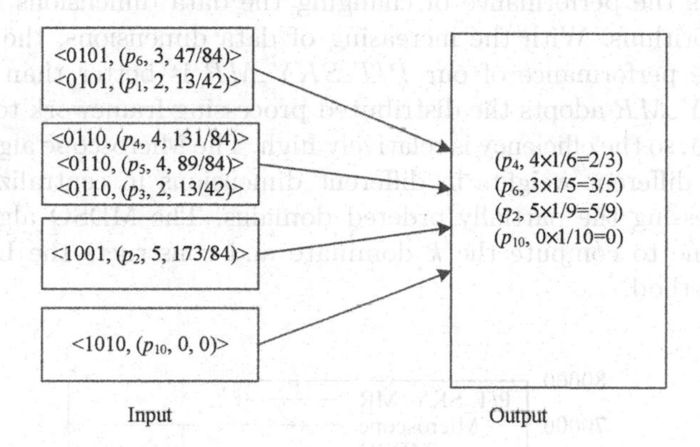

Fig. 5. Reduce phase of T_SKY_MR

6 Experimental Evaluation

6.1 Experimental Setup

This section reports our experimental results to validate the effectiveness and efficiency over our proposed method. We evaluate the performance of P/T_SKY_MR by comparing it with the approaches Microscope algorithm [1,15] and MDSO algorithm [2].

Extensive experiments have been conducted on a homogeneous cluster consisting of 8 nodes of Intel(R) Core(TM) i5-4210U CPU 2.4 GHz machines with 4 GB of main memory running Linux. The machines are connected by a 100 Mb/s LAN. We use the operating system Ubuntu 10.10 to run the machines, and Hadoop 1.0.3 to build the MapReduce environment on the cluster.

For the data set, we use the synthetic data sets of anti-correlated distributions. In the experiments, we adopt the data sets generated by [14], which contains both numeric attributes and partially ordered attributes. The partially ordered attributes are generated according to the Zipfian distribution. The default values of the experimental parameters are shown in Table 2.

Table 2. Experimental parameters

Parameter	Default	Range
Data volume	10 GB	10 GB–20 GB
Data dimension	6	2–10
Number of partially ordered dimensions	20	10–30
Number of k	20	10–50

6.2 Experimental Results

Figure 6 shows the performance of changing the data dimensions to illustrate the three algorithms. With the increasing of data dimensions, the query time increases. The performance of our P/T_SKY_MR is better than the others. The P/T_SKY_MR adopts the distributed processing framework to handle the numerous data, so the efficiency is relatively high. The Microscope algorithm only considers the different weights in different dimensions in centralized method, without processing the partially ordered domains. The MDSO algorithm also uses much time to compute the k dominate and constructs the LMB tree in centralized method.

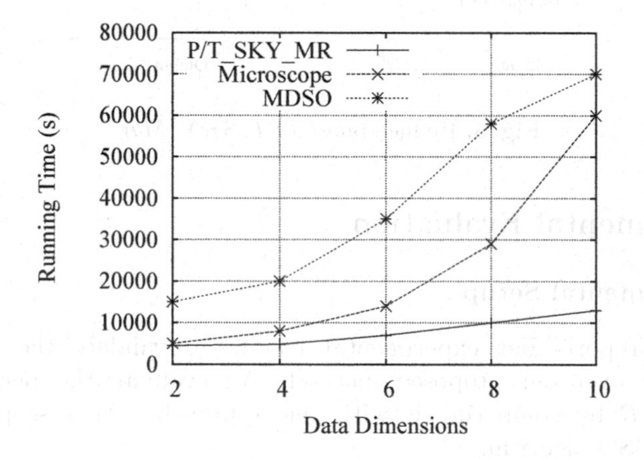

Fig. 6. Performance of changing dimensions

Figure 7 shows the performance of space storage overhead by changing the data size. With the increasing of data size, the space storage overhead also increases. The performance of P/T_SKY_MR is better than the others. When the data size increases, the data dimensions also increase, so the skyline results would be large and the dominate computing would be increase. The space storage overhead is increasing. The Microscope algorithm retrieves the skycube and compute the subspace skyline results. When the number of data size increases, the number of subspace and the results of subspace skyline increase too, so the space storage overhead is large. The MDSO algorithm uses LMB tree to process the query, without any compressing of LMB tree. So, when the data size increases, the space storage overhead increases.

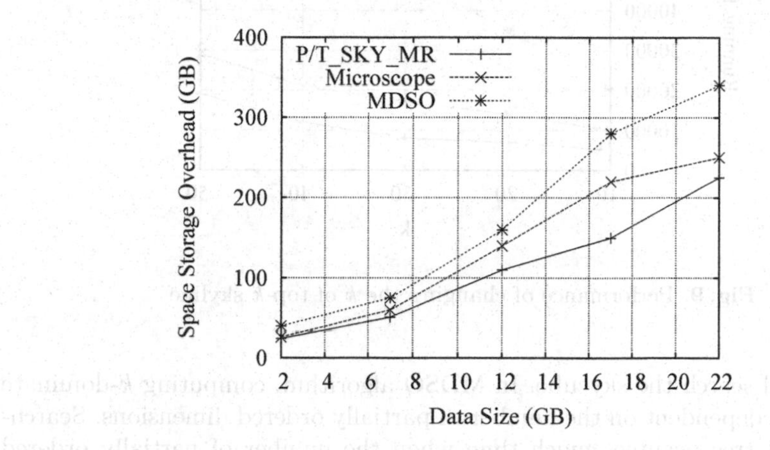

Fig. 7. Performance of changing data size

Figure 8 shows the performance of changing the number of partially ordered dimensions. With the increasing of the number of partially ordered dimensions, the running time also grows. The P/T_SKY_MR has obvious advantages. The main time on processing the partially ordered dimensions is spent on gaining the coding *ids* to stand for the user preferences. The Microscope algorithm uses the skycube to compute the top-k skyline. When the number of partially ordered dimensions increases, the Microscope algorithm would spend numerous time to

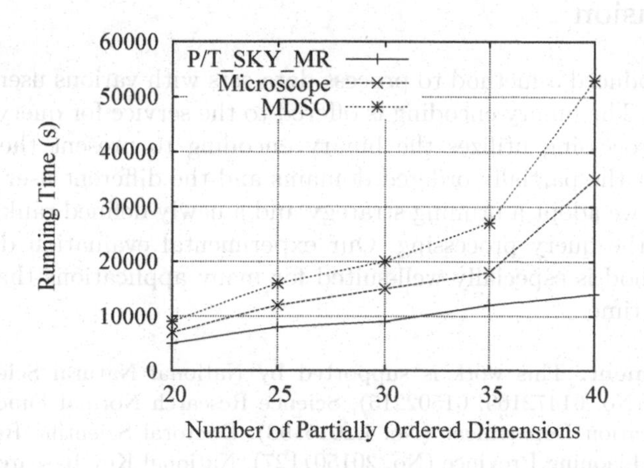

Fig. 8. Performance of changing the number of partially ordered dimensions

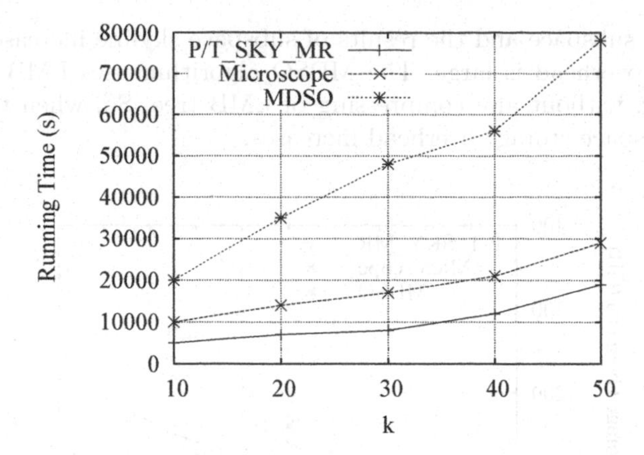

Fig. 9. Performance of changing the k of top-k skyline

construct and search the skycube. In MDSO algorithm, computing k-dominate core points is dependent on the number of partially ordered dimensions. Searching the LMB tree occupies much time when the number of partially ordered dimensions increases.

Figure 9 shows the performance of changing k of the three algorithms. With the increasing of k, the running time of the three algorithms all increase. The P/T_SKY_MR algorithm has the better performance than the other two algorithms. The P/T_SKY_MR algorithm uses MapReduce framework to process the user preferences top-k skyline combining the partially and totally ordered domains. The Microscope algorithm needs to scan the skycube for the final results. The MDSO constructs the LMB tree without any compressing, so the running time is large.

7 Conclusion

We have introduced a method to process data sets with various user preferences on attributes. The binary encoding is offered to the service for query processing. The query processing utilizes the binary encoding to present the precedence relation about the partially ordered domains and the different user preferences. Furthermore, we adopt a pruning strategy and a newly defined ranking criterion to speed up the query processing. Our experimental evaluation demonstrates that our method is especially well-suited for many applications that requires a fast response time.

Acknowledgment. This work is supported by National Natural Science Foundation of China (No. 61472169, 61502215), Science Research Normal Fund of Liaoning Province Education Department (No. L2015193), Doctoral Scientific Research Start Foundation of Liaoning Province (No. 201501127), National Key Research and Development Program of China (No. 2016YFC0801406).

References

1. Lee, J., You, G., Hwang, S.: Personalized top-k skyline queries in high-dimensional space. Inf. Syst. **34**(1), 45–61 (2009)
2. Cheong Fung, G.P., Lu, W., Yang, J., Du, X., Zhou, X.: Extract interesting skyline points in high dimension. In: Kitagawa, H., Ishikawa, Y., Li, Q., Watanabe, C. (eds.) DASFAA 2010. LNCS, vol. 5982, pp. 94–108. Springer, Heidelberg (2010). https://doi.org/10.1007/978-3-642-12098-5_7
3. Gao, Y., Hu, J., Chen, G., Chen, C.: Finding the most desirable skyline objects. In: Kitagawa, H., Ishikawa, Y., Li, Q., Watanabe, C. (eds.) DASFAA 2010. LNCS, vol. 5982, pp. 116–122. Springer, Heidelberg (2010). https://doi.org/10.1007/978-3-642-12098-5_9
4. Benouaret, K., Benslimane, D., HadjAli, A.: Selecting skyline web services for multiple users preferences. In: ICWS, Hawaii, pp. 635–636 (2012)
5. Bouadi, T., Cordier, M.-O., Quiniou, R.: Computing skyline incrementally in response to online preference modification. In: Hameurlain, A., Küng, J., Wagner, R., Liddle, S.W., Schewe, K.-D., Zhou, X. (eds.) Transactions on Large-Scale Data- and Knowledge-Centered Systems X. LNCS, vol. 8220, pp. 34–59. Springer, Heidelberg (2013). https://doi.org/10.1007/978-3-642-41221-9_2
6. Nguyen, H.T.H., Cao, J.: Preference-based top-k representative skyline queries on uncertain databases. In: Cao, T., Lim, E.-P., Zhou, Z.-H., Ho, T.-B., Cheung, D., Motoda, H. (eds.) PAKDD 2015. LNCS, vol. 9078, pp. 280–292. Springer, Cham (2015). https://doi.org/10.1007/978-3-319-18032-8_22
7. Li, Y., Qu, W., Li, Z., et al.: Skyline query based on user preference with MapReduce. In: DASC, Dalian, pp. 153–158 (2014)
8. Jiang, T., Zhang, B., Lin, D.: Incremental evaluation of top-k combinatorial metric skyline query. Knowl.-Based Syst. **74**, 89–105 (2015)
9. Jheng, R.S., Wang, E.T., Chen, A.L.P.: Determining top-k candidates by reverse constrained skyline queries. In: DATA, pp. 101–110 (2015)
10. Siddique, M.A., Morimoto, Y.: Efficient selection of various k-objects for a keyword query based on MapReduce skyline algorithm. In: Madaan, A., Kikuchi, S., Bhalla, S. (eds.) DNIS 2014. LNCS, vol. 8381, pp. 40–52. Springer, Cham (2014). https://doi.org/10.1007/978-3-319-05693-7_3
11. Song, B., Liu, A., Ding, L.: Efficient top-k skyline computation in MapReduce. In: IEEE WISA, pp. 67–70 (2015)
12. Siddique, M.A., Tian, H., Morimoto, Y.: Selecting representative objects from large database by using K-skyband and top-k dominating queries in MapReduce environment. In: Luo, X., Yu, J.X., Li, Z. (eds.) ADMA 2014. LNCS, vol. 8933, pp. 560–572. Springer, Cham (2014). https://doi.org/10.1007/978-3-319-14717-8_44
13. Siddique, Md.A., Tian, H., Morimoto, Y.: k-dominant skyline query computation in MapReduce environment. IEICE Trans. Inf. Syst. **98–D**(5), 1027–1034 (2015)
14. Borzsony, S., Kossmann, D., Stocker, K.: The skyline operator. In: 17th International Conference on Data Engineering, pp. 421–430. IEEE (2001)
15. Lee, J., You, G.W., Sohn, I.C., et al.: Supporting personalized top-k skyline queries using partial compressed skycube. In: Proceedings of the 9th Annual ACM International Workshop on Web Information and Data Management, pp. 65–72. ACM (2007)

An Importance-and-Semantics-Aware Approach for Entity Resolution Using MLP

Yaoli Xu[1,2](✉), Zhanhuai Li[1,2], and Wanhua Qi[3]

[1] School of Computer Science and Engineering, Northwestern Polytechnical University, Xi'an 710072, Shaanxi, People's Republic of China
yaolixu@mail.nwpu.edu.cn, lizhh@nwpu.edu.cn
[2] Key Laboratory of Big Data Storage and Management, Northwestern Polytechnical University, Ministry of Industry and Information Technology, Xi'an, People's Republic of China
[3] Anyang Institute of Technology, Anyang 455000, Henan, People's Republic of China
qiwanhua@outlook.com

Abstract. Entity resolution (ER), as the process of identifying records which depict the same real-world entity, plays a fundamental role in data integration and data cleaning tasks. Although deep learning techniques of data science have transformed various applications, there are few efforts to leverage these techniques to deal with entity resolution. We also observe the importance of overlapped tokens and the semantic similarity from pre-trained word vectors can benefit ER. To this end, we propose a deep learning based framework for ER, which can leverage the state-of-the-art techniques in deep neural network communities. We also propose an importance-and-semantics-aware approach for ER using a multilayer perceptron (MLP), to combine the importance of overlapped tokens, semantic similarity and textual similarity of corresponding attribute values of pairs. Comparative experiments demonstrate that our method outperforms the traditional method.

Keywords: Data integration · Entity resolution
Semantic similarity · Textual similarity · Importance

1 Introduction

In data cleaning and data integration fields, Entity Resolution (ER), also known as entity matching, record deduplication or record linkage, is a long-standing challenge. With the advent of the era of big data, there are enormous published data sources which may store comprehensive descriptions of overlapped entities. Since misspellings, inconsistent digital conventions, and ongoing changes exist, ER has been paid more and more attention to, and plenty of methods have been

Q. Zhou et al. (Eds.): ICPCSEE 2018, CCIS 901, pp. 88–100, 2018.
https://doi.org/10.1007/978-981-13-2203-7_8

proposed in order to achieve data with higher quality and outstanding value. For data science, ER is a fundamental problem since data in a consistent form can bring more knowledge or insights.

Deep learning architectures (e.g., the feedforward artificial neural networks, recurrent neural networks, and convolutional neural networks) have been broadly applied in analyzing visual imagery, speech recognition, natural language process and so forth. Most of these models leverage the strategy of distributed representations. There are many amazing semantic vector space models proposed (e.g., GloVe [1] and word2vec [2]) which can generate word representations (i.e., a kind of distributed representation). Such representation makes it possible to feed semantic information derived from local contexts of words or the statistics of global corpus (e.g., the global co-occurrence counts), into deep neural networks. There are many available pre-trained word vectors, which model the fine-grained semantic and syntactic regularities from large datasets (e.g., Wikipedia or Twitter). However, there are few studies to leverage such semantic vectors to deal with ER. In our work, we leverage such pre-trained word vectors to construct the semantic similarity features of record pairs.

Table 1. Records in D_1 and D_2.

rID	Title	Authors
$r_{1,1}$	Broadcast disks: data management for asymmetric communications environments	S Acharya, R Alonso, M Franklin, S Zdonik
$r_{2,1}$	Broadca. rt l) ikr: data management for uymmetrir communications environment	S Acharya, R Alonso, MJ Franklin, S Zdonik
$r_{1,2}$	PSoup: a system for streaming queries over streaming data	S Chandrasekaran, M Franklin
$r_{2,2}$	Streaming queries over streaming data	S Chandrasekaran, MJ Franklin

Table 2. Record pairs.

pID	rID	rID	GT
$p_{1,1}$	$r_{1,1}$	$r_{2,1}$	P
$p_{1,2}$	$r_{1,1}$	$r_{2,2}$	N
$p_{2,1}$	$r_{1,2}$	$r_{2,1}$	N
$p_{2,2}$	$r_{1,2}$	$r_{2,2}$	N

Most existing ER approaches [3,4] depend on the similarity of attribute values. Various similarity metrics (e.g., the Jaccard coefficient and Levenshtein distance) only quantify the overall word-level or letter-level concurrences among corresponding attribute values for each pair [5]. However, for some challenging pairs, the similarity may fail. For instance, in Table 1 each record describes the title (Title), authors (Authors) and identity (rID) of an article, and the first subscript 1 (or 2) of rID implies that this record is from D_1 (or D_2). In Table 2, each pair consists of two records (i.e., one in D_1, and the other in D_2), where P indicates the match, N indicates the non-match and the values at the column of GT indicate the ground-truth status. For a text, there are two kinds of items: word and letter. A token is an item or a contiguous sequence of n items (a.k.a. n-gram). For pairs (e.g., $p_{1,1}$ or $p_{2,2}$), there are indicative tokens (e.g., "system" in $r_{1,2}$), and unrelated tokens (e.g.,"rt" in $r_{2,1}$), as shown in Tables 1 and 2, but these hints can not be explicitly presented by similarity measures. In addition, for conventional one hot representation, one record is encoded by a vector, where each dimension corresponds to a token and the column value indicates whether the token appears in the record. It turns out the vector for each record (or each record pair) has a great number of dimensions, as for any entity matching task, the number of tokens may be far more than ten thousands. The deep neural network using such vector, may suffer from the curse of dimensionality. Therefore, it is challenging to construct pair vectors which can incorporate the importance of tokens and prevent from such curse. We observe that for a matched pair (e.g., $p_{1,1}$), the importance of overlapped tokens for related records is similar. For instance, an overlapped token (e.g., "management") in $r_{1,1}$ with higher importance, is also important to the other $r_{2,1}$. We provide an importance-aware feature generator, and leverage the TFIDF weighting scheme to quantify the token importance.

The traditional training set for ER is constructed by a threshold-Random strategy [6]. However, in practice, it is challenging to find a generally appropriate threshold for different datasets. In addition, the imbalance between matched and non-matched pairs is severe, especially for a Clean-Clean ER [7]. Although there are many blocking techniques [7–10] in order to reduce massive pair comparisons, the imbalance is still serious. For instance, after we employed the state-of-art blocking technique (e.g., BLAST [9]), the ratio of non-matched to matched pairs was still up to 171, in DblpScholar datasets which consist of a wealth of literature records. Therefore, we propose a novel training selection method, which leverages machine learning models to automatically produce the training set and to alleviate the imbalance, by filtering non-matched pairs.

In summary, the main contributions of our work are the following.

1. We propose a deep learning based framework for imbalanced ER in order to leverage the state-of-the-art strategies within deep neural network communities. We also implement a greedy forward search algorithm for optimal features.
2. We present an importance-and-semantics-aware approach for ER using a multilayer perceptron (MLP) which combines semantic similarity, textual similarity and the importance of overlapped tokens. To the best of our knowledge, we are the first to combine three kinds of information via a multilayer perceptron. We also propose a novel training selection method, which leverages machine learning models to automatically produce the training set, by filtering non-matched pairs.

3. We conducted extensive experiments, which demonstrate our method outperforms the traditional approach.

The rest of this paper is organized as follows. We review extensive related work in Sect. 2. We introduce our deep learning based framework for imbalance ER, and detail our importance-and-semantics-aware approach in Sect. 3. We present our evaluation results in Sect. 4. Finally, we conclude this paper and present our remarks concerning future work in Sect. 5.

2 Related Work

Entity resolution is also known as entity matching or record linkage. Since it is a fundamental component in data integration and data cleaning systems, various approaches [3,11–17] and tools (e.g., Febrl [18]) have been proposed. These approaches can be broadly classified into three kinds: learning-based ER, non-learning based ER and efficiency.

Learning-Based ER. There are a few learning-based methods [6,12,18]. Both the SuppVecMachine classifier in Febrl [18], and MARLIN [12] model ER as a binary classification problem which is solved by a supervised support vector machine (SVM). However, they only leverage textual similarity or use TFIDF as the weight of the textual similarity. In our work, the pair features also include the semantic similarity. We implemented the method proposed in [6], and consider it as our baseline method.

A few researchers [11,17] utilize various similarity metrics (e.g., Jaccard coefficient and Levenshtein distance) to generate entity matching rules. Our pair features based on semantic similarity or token importance can also be integrated in the process of mining these rules.

Non-learning Based ER. Some researchers [3,13] focus on how to combine the textual similarity of record pairs to a match measure. The footrule distance [13] is the sum of the absolute differences of the positions of tuples of a relational table in the two rankings. ODetec [3] utilizes Principal Component Analysis to compute a centrifugal distance as a match measure. However, it is non-trivial to determine appropriate thresholds for these match measures, since different dataset has different distribution of the textual similarity. Our learning-based method can avoid how to determine appropriate thresholds for different datasets.

Efficiency. Since candidate pairs are enormous, there are a great number of blocking techniques [7–10] to reduce pair comparisons. They focus on how to construct an indexing key or a composition key, so that only records in the same blocking are needed to be compared. Wang et al. [16] proposed an approximate algorithm, which formalize the group-wise entity resolution as the partition of the vertexes in a weighted graph into cohesive subgraphs. For efficiently performing entity resolution on a large dataset, they also have provided a heuristic algorithm. Several works [14,15] focus on how to resolve entity resolution task in pay-as-you-go fashion, so as to resolve big datasets efficiently. Different from these approaches, our method focuses on the effectiveness, not the efficiency.

3 An Importance-and-Semantics-Aware Approach

In this section, we first overview a deep learning based framework for imbalanced entity resolution, and then present its implementation, an importance-and-semantics-aware approach in detail.

Given datasets D_1 and D_2, labeled pairs P_{train}, and non-labeled pairs P_{test}, the problem is to predict whether a pair $p \in P_{test}$ is matched or non-matched. To resolve this problem, we propose a deep learning based framework for imbalanced entity resolution, as shown in Fig. 1. Its main idea is to employ state-of-the-art deep neural networks to combine different informative hints (e.g., semantic similarity, token importance and traditional textual similarity). Its procedure is the following. Given datasets, D_1 and D_2, it first employs some blocking model to generate candidate record pairs P_c which consists of P_{train} and P_{test}. Since P_c has a severe imbalance between matched and non-matched pairs, a training selection model is utilized to generate a more valid training (or test) set P^*_{train} (or P^*_{test}) in order to rectify such imbalance. And then, it quantifies a variety of hints, and regards quantified values as features. Finally, it incrementally selects optimal features using a greedy feature selection algorithm by selecting and evaluating, and outputs a deep neural network model. Since there are various kinds of tokens (e.g., word-level and letter-level tokens), enumerating all these tokens is computationally expensive and does not always bring the best result. It is worthwhile to select optimal features that benefit ER. Once the optimal features and trained model are achieved, pairs in P^*_{test} are resolved. In the following, we present the main components (i.e., the training selection model, semantic similarity, token importance, textual similarity, optimal feature selection, and a deep neural network (MLP)) of our importance-and-semantics-aware approach in detail.

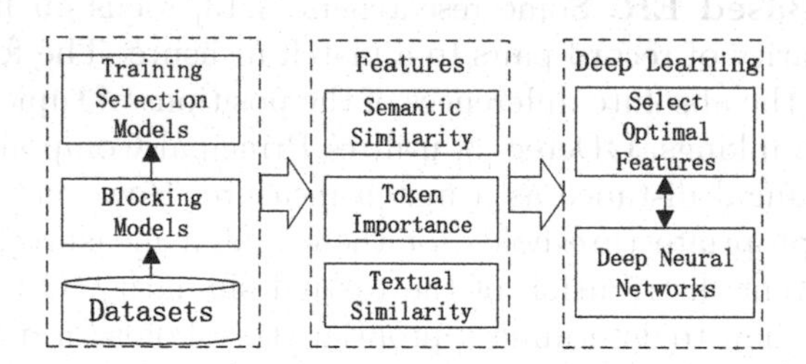

Fig. 1. Framework overview

3.1 A Novel Training Selection Model

The traditional training set for ER is constructed by a Threshold-Random strategy [6]. However, in practice, it is challenging to find generally appropriate thresholds for various datasets. To this end, we propose a novel training selection method to automatically output more valid training set, by filtering non-matched pairs.

There are unsupervised models (e.g., Isolation Forest Algorithm, or One-Class SVM) which assume matched pairs belong to the same distribution [19]. The main idea of One-Class SVM is to estimate a rough, close frontier from observations (e.g., matched pairs in the training set), and then to compute the distance between the frontier to a pair to be determined. Such frontier determines the contour of these observations' distribution. If the pair lies in the frontier's subspace, it is matched, otherwise non-matched. The farther distance implies that the prediction is with higher confidence. The main idea of Isolation Forest algorithm is to recursively partition observations by randomly selecting a feature and a splitting value to construct a few trees. The averaged path length among these random threes, is a measure to predict whether a pair is matched or not. Pairs with shorter averaged path length, are more likely to be non-matched. To avoid mistakenly filtering a truly matched pair, our training selection model first leverages One-Class SVM and Isolation Forest algorithms to predict whether a pair is matched or not, and then derives more valid training (or test) set through filtering consistently non-matched pairs and retaining pairs within some constraint (e.g., the similarity is larger than 0.5) from the original counterpart.

3.2 Semantic Similarity

Semantic vector space models have been studied thoroughly to generate distributed representations. Several amazing models (e.g., word2vec and GloVe) for distributed representations of words have been proposed. Their main ideas are to capture the semantic and systematic patterns among a great number of unstructured texts, and quantify such information as low-dimensional vectors after matrix factorization or local context analyses. For ER, each attribute value is a natural language snippet with typos and various errors, not a complete sentence. Therefore, it is non-trivial to directly extract these patterns from datasets. Fortunately, both word2vec and GloVe provide pre-trained word vectors derived from some gigantic corpora such as Wikipedia or Twitter. These massive external sources can also provide better representations for words appeared in ER tasks. Based on this intuition, we leverage pre-trained semantic vectors generated by GloVe, to compute the semantic similarity features for each pair.

Pre-trained vectors only contain a great number of distributed representations (i.e., low-dimensional vectors with numeric values) for words, not for record pairs. Therefore, we first define the semantic representation for each attribute value of a record as the sum of distributed representations of words appearing in the attribute value, given by Eq. 1.

$$SR(r_{i,j}, an_k) = \sum_{w \in r_{i,j}[an_k]} v(\mathbb{V}, w) \tag{1}$$

where $r_{i,j}$ is the j-th record of the i-th database, an_k is the k-th attribute name, $r_{i,j}[an_k]$ is the attribute value of an_k of $r_{i,j}$, w indicates a word, and $v(\mathbb{V}, w)$ is the distributed representation of w in pre-trained vectors $\mathbb{V}$. Notice that if w does not appear in $\mathbb{V}$, $v(\mathbb{V}, w)$ is a zero vector with the same dimension.

In addition, some informative words may not appear in corresponding attribute values. We regard all attribute values of a record as a document, and compute the semantic representation of the document in a similar way, given by Eq. 2.

$$SR(r_{i,j}, doc(r_{i,j})) = \sum_{w \in doc(r_{i,j})} v(\mathbb{V}, w) \quad (2)$$

where $doc(r_{i,j})$ is the concatenation of all attribute values of $r_{i,j}$.

Pair features derived from the semantic similarity, include the cosine similarity of corresponding attribute representations, given by Eq. 3, and the cosine similarity of the semantic representation for record documents, given by Eq. 4. Assuming there are k attributes in D_1 and D_2, for $p_{i,j}$, there are k+1 features in terms of the semantic similarity.

$$SS(p_{i,j}, an_k) = cos(SR(r_{1,i}, an_k), SR(r_{2,j}, an_k)) \quad (3)$$

$$SS(p_{i,j}, doc) = cos(SR(r_{1,i}, doc(r_{1,i})), SR(r_{2,j}, doc(r_{2,j}))) \quad (4)$$

where $cos(\cdot)$ is a similarity measure between two vectors.

Example 1. Consider Tables 1 and 2. Given $\mathbb{V}$, a distributed representation of "data" $v(\mathbb{V}, \text{"data"})$ is $[0.5748, 0.0356, 0.4859, \ldots, -0.3627]$. According to Eq. 1, $SR(r_{1,1}, \text{"Title"})$ is the sum of $v(\mathbb{V}, \text{"data"})$, ..., $v(\mathbb{V}, \text{"environments"})$. Given $SS(p_{1,1}, \text{"Title"}) = 0.8896$, $SS(p_{1,1}, \text{"Authors"}) = 1.0$, $SS(p_{1,1}, doc) = 0.9214$, semantic features of $p_{1,1}$ include $SS(p_{1,1}, \text{"Title"})$, $SS(p_{1,1}, \text{"Authors"})$ and $SS(p_{1,1}, doc)$. □

3.3 Token Importance

We observe that for a matched pair $p_{i,j}$, the importance of overlapped tokens is similar, which means when a token t is with higher importance in $r_{1,i}$, t is also with higher importance in $r_{2,j}$, if t exists in $r_{2,j}$. Based on this, we propose our importance-aware features.

TFIDF, a numeric statistic metric, is used to reflect how important a word is to certain document. It is the most popular and has been widely used in information retrieval and document classification. In general, it can make the weight of frequent and trivial tokens be lower. TFIDF is a product of TF and IDF, given by Eq. 5. TF, short for term frequency, is the number of occurrences of a token t in a document d, and IDF, short for inverse document frequency, quantifies how much information the token provides, given by Eq. 6.

$$\mathsf{TFIDF}(t, d, D) = \mathsf{TF}(t, d) * \mathsf{IDF}(t, D) \quad (5)$$

$$\mathsf{IDF}(t, D) = log\frac{|D|}{|\{d \in D\} : t \in d| + 1} \quad (6)$$

where t is a token, d is a document, and D is a corpus which consists of a host of documents. We utilize TFIDF to quantify the importance of overlapped tokens of a pair. But directly utilizing this hint is challenging. Firstly, most of deep neural network models require that the input has the same dimension. However, different record has different number of tokens. Secondly, the traditional one hot representation may bring more than ten thousands features, because it regards each token as a feature, and in general the number of tokens on ER datasets is enormous. To this end, we propose an importance-aware feature generator (IAFG), to construct importance features of pairs using TFIDF of overlapped tokens. Its core idea is first to compute the summed TFIDF for each overlapped token in each pair, sort these TFIDF in descending order, and regard these sorted TFIDF as the importance-aware features. The procedure of how to construct such features, as shown in Algorithm 1, takes as input D_1, D_2, $P_{train} \cup P_{test}$, token types TT, and outputs the importance-aware features of pairs IF. There are many token types (e.g., biletter or triletter). In this work, we utilized six kinds of tokens (i.e., biletter, triletter, word, biword, triword, and four-word).

Algorithm 1. IAFG

```
   input : D1, D2, Ptrain ∪ Ptest, and TT
   output: IF
 1 Convert each record r ∈ D1 ∪ D2 into a document by concatenating its attribute
   values, and regard all these documents as a corpus;
 2 Tokenize each document into tokens according to TT and compute TFIDF for
   each token in r;
 3 maxLen ← 0;
 4 IF ← ∅ ;
 5 for p ∈ Ptrain ∪ Ptest do
 6     r1,i,r2,j ← p;
 7     Select overlapped tokens T appeared in r1,i, and r2,j;
 8     if |T| > maxLen then
 9         maxLen = |T|
10     end
11     ifp ← ∅ ;
12     for ot ∈ T do
13         temp = TFIDF(r1,i, ot) + TFIDF(r2,j, ot);
14         ifp ← ifp ∪ temp;
15     end
16     Sort ifp in descending order;
17     IF ← IF ∪ (p, ifp);
18 end
19 for (p, ifp) ∈ IF do
20     if len(ifp) < maxLen then
21         Pad ifp with zeros, update (p, ifp) ∈ IF;
22     end
23 end
```

3.4 Textual Similarity

There are various similarity metrics (e.g., Levenshtein distance or Jaccard coefficient) to measure the similarity of two strings. Traditional approaches leverage these metrics to compute the textual similarity of corresponding attribute values of a pair. For instance, the textual features of $p_{1,1}$ in Table 2, are $S(p_{1,1},$ "Title"$)$ and $S(p_{1,1},$ "Authors"$)$, where $S(\cdot)$ stands for some similarity metric. In our work, we employ Jaccard coefficient.

3.5 A Greedy Feature Selection Algorithm

For traditional attribute-based approaches for ER, the feature number is quite finite, and linearly scales to the number of attributes in datasets. However, once importance-aware features are leveraged, pair features are quite a lot, since there are many kinds of tokens (e.g., word, biword, triword, and so on). Leveraging all these tokens is computationally expensive and does not always bring the best result. To this end, we propose a greedy feature selection algorithm, which first ranks these features according to their scores (i.e., the mean of a feature), and then adds the feature with the highest score at each iteration, finally evaluates the performance of the trained deep neural network. If the effectiveness of the trained model is increased after adding a feature, we reserve the feature, otherwise we drop it. When all features are evaluated, we can obtain optimal features, and a deep learning model with the best effectiveness.

3.6 A Multilayer Perceptron

In the big data era, various deep neural networks and related powerful techniques (e.g., normalization, regularization, or optimization techniques [20]) have been proposed. All of these state-of-the-art techniques can be leveraged in our framework. In this work, we formalize ER as a binary classification task, which takes as input labeled pairs in P_{train}, and outputs labels of pairs in P_{test}. We leverage a multilayer perceptron (MLP) as our training model with a modified error function. MLP, an optimization algorithm, learns an objective function with the minimum error through estimating weights of perceptron units. Such error function is the cross-entropy error function. Although the effectiveness of ER focuses on truly matched pairs, rather than accuracy, the conventional error function does not take class weights into account. We incorporate class weights into the error function, as Eq. 7 shows.

$$E(p) = -cw(t_p) * (t_p ln \hat{y_p} + (1 - t_p) ln(1 - \hat{y_p})) \tag{7}$$

where t_p is 1 if p is truly matched, otherwise 0; $cw(t_p)$ is the class weight of pairs with t_p in the training set; and $\hat{y_p}$ is the estimated probability that trained MLP model predicates it to be matched. Plenty of novel regularization techniques (e.g., dropout or early stopping) have been proposed to mitigate overfitting. We employed dropout.

4 Experimental Results

In this section, we conducted extensive experiments on real-world datasets to demonstrate the effectiveness of our method in terms of Recall, Precision and F_1. We evaluated the importance-and-semantics-aware approach for entity resolution using MLP (MLP-ER), comparing with SVM-ER [6], and the impact of different information (i.e., textual similarity, token importance and semantic similarity) on MLP-ER.

4.1 Experimental Settings

Environment. We used ubuntu 16.04 64 bit as the operating system and python 3.5 as the runtime environment. All experiments were conducted on a machine with a 2.50 GHz Intel(R) Core(TM) i7-4710MQ processor and 16 GB of RAM.

Evaluation Measures. We use Recall, Precision, and F_1 to measure the effectiveness of our method. Positive pairs are predicted to be matched, denoted as P_{pos}. Ground truth is composed of truly matched pairs, denoted as P_{GT}. Truly positive pairs, denoted as P_{TPos}, are predicted to be matched and belong to ground truth. Recall is the ratio of truly positive pairs to ground truth, i.e., $\text{Recall} = \#P_{TPos}/\#P_{GT}$. Precision is the ratio of truly positive pairs to positive pairs, i.e., $\text{Precision} = \#P_{TPos}/\#P_{pos}$. F_1 is the harmonic mean of Precision and Recall, i.e., $F_1 = 2 \cdot \text{Recall} \cdot \text{Precision}/(\text{Recall} + \text{Precision})$.

Datasets. We utilized two real-world datasets in bibliographic domain exploited in existing literature [3,4,9,12]. Cora is a collection of 1,295 bibliographic records which involve title, publisher, author, address, etc. Cora contains 17,184 duplicate pairs. DS, short for DblpScholar, is composed of two relational tables (i.e., DBLP and Scholar), where DBLP contains 2,616 citations and Scholar contains 64,263 citations. Both of them consist of title, authors, venue, and year. The number of duplicate pairs is 5,347.

4.2 Comparison with Baseline Method

In the first set of experiments, we compared MLP-ER with SVM-ER. The experimental results were showed in Table 1. The top values are in bold. We make the following observations. (i) MLP-ER outperforms SVM-ER in all of evaluation measures (i.e., Recall, Precision and F_1). Recall, Precision, and F_1 of MLP-ER on DS are improved by 5.35%, 2.53% and 3.97% respectively, compared with those of SVM-ER. Recall, Precision, and F_1 of MLP-ER on Cora are increased by 7.56%, 4.26% and 5.96% respectively, compared to those of SVM-ER. (ii) Recall of MLP-ER achieves the highest improvement among all of measures. It means semantic and importance features make more matched pairs be identified.

4.3 Impact of Different Information

Three kinds of information (i.e., semantic similarity, token importance, and textual similarity) can be used as the hints of whether a pair matches or not. In

Table 3. Comparative result

Evaluation measure	SVM-ER(DS)	MLP-ER(DS)	SVM-ER(Cora)	MLP-ER(Cora)
Recall (%)	88.15	**93.50**	85.81	**93.37**
Precision (%)	91.37	**93.90**	90.00	**94.26**
F_1 (%)	89.73	**93.70**	87.86	**93.82**

this section, we conducted experiments to evaluate the impact of different information on MLP-ER. The experimental results were showed in Fig. 2, where TS, SS, IT, Hybrid are separately short for textual similarity, semantic similarity, the importance of tokens, and all of previous information. We can achieve the following. (i) Different information on different datasets has its ad-hoc advantage. On DS, compared with TS and SS, IT achieves the highest Recall, while Precision of TS is the highest compared to those of IT and SS. On Cora, Recall of SS is the highest, compared to those of TS and IT. (ii) Their combination can achieve better quality than any of them. For both Cora and DS, Recall, Precision, and F_1 of Hybrid are higher, comparing to their counterparts in IT, SS or TS (Fig. 2).

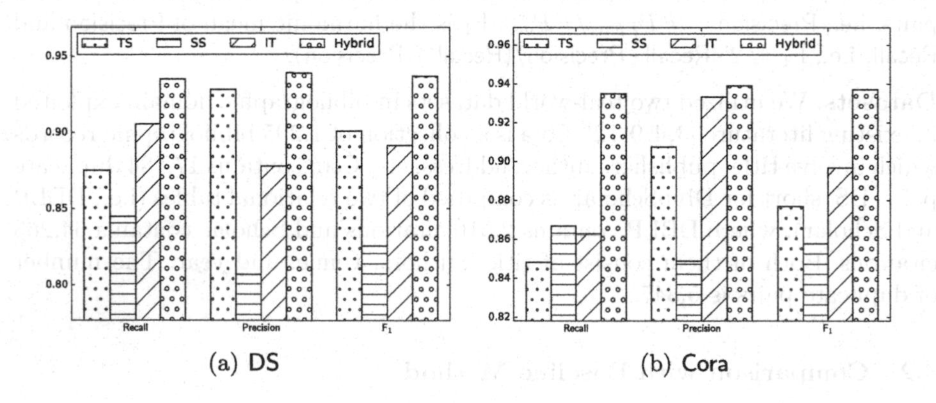

(a) DS (b) Cora

Fig. 2. Different information

5 Conclusion and Future Work

In this paper, we made the first step to combine the semantic similarity, textual similarity and importance of tokens using a deep feedforward network (i.e., MLP) to improve the effectiveness of ER. We also provided a greedy feature selection algorithm to search optimal features. We also proposed a novel automatic training selection approach to alleviate the imbalance and to reduce computation cost.

In the future, we would like further study other fundamental problems in data science, such as how to utilize novel deep neural networks to address data repairing problem [21]. We also would like to investigate how to learn more informative vectors for records from dirty datasets.

Acknowledgments. This work was supported by the Ministry of Science and Technology of China, National Key Research and Development Program (Project Number: 2016YFB1000703), and the National Natural Science Foundation of China under Grant No. 61732014, No. 61332006, No. 61472321, No. 61502390 and No. 61672432.

References

1. Pennington, J., Socher, R., Manning, C.D.: Glove: global vectors for word representation. In: Proceedings of the 2014 Conference on Empirical Methods in Natural Language Processing, pp. 1532–1543. ACL, Stroudsburg (2014)
2. Mikolov, T., Chen, K., Corrado, G., Dean, J.: Efficient estimation of word representations in vector space. CoRR abs/1301.3781 (2013). http://arxiv.org/abs/1301.3781
3. Fan, F., Li, Z., Chen, Q., Liu, H.: An outlier-detection based approach for automatic entity matching. Chin. J. Comput. **40**(10), 2197–2211 (2017). https://doi.org/10.11897/SP.J.1016.2017.02197
4. Köpcke, H., Thor, A., Rahm, E.: Evaluation of entity resolution approaches on real-world match problems. PVLDB **3**(1), 484–493 (2010). https://doi.org/10.14778/1920841.1920904
5. Cohen, W.W., Ravikumar, P., Fienberg, S.E.: A comparison of string distance metrics for name-matching tasks. In: Proceedings of IJCAI-03 Workshop on Information Integration on the Web, pp. 73–78. AAAI Press, Palo Alto (2003)
6. Köpcke, H., Rahm, E.: Training selection for tuning entity matching. In: Proceedings of the International Workshop on Quality in Databases and Management of Uncertain Data, pp. 3–12 (2008)
7. Papadakis, G., Koutrika, G., Palpanas, T., Nejdl, W.: Meta-blocking: taking entity resolutionto the next level. IEEE Trans. Knowl. Data Eng. **26**(8), 1946–1960 (2014). https://doi.org/10.1109/TKDE.2013.54
8. Wang, Q., Cui, M., Liang, H.: Semantic-aware blocking for entity resolution. IEEE Trans. Knowl. Data Eng. **28**(1), 166–180 (2016). https://doi.org/10.1109/TKDE.2015.2468711
9. Simonini, G., Bergamaschi, S., Jagadish, H.V.: BLAST: a loosely schema-aware meta-blocking approach for entity resolution. PVLDB **9**(12), 1173–1184 (2016). https://doi.org/10.14778/2994509.2994533
10. Efthymiou, V., Papadakis, G., Papastefanatos, G., Stefanidis, K., Palpanas, T.: Parallel meta-blocking for scaling entity resolution over big heterogeneous data. Inf. Syst. **65**, 137–157 (2017). https://doi.org/10.1016/j.is.2016.12.001
11. Li, L., Li, J., Gao, H.: Rule-based method for entity resolution. IEEE Trans. Knowl. Data Eng. **27**(1), 250–263 (2015). https://doi.org/10.1109/TKDE.2014.2320713
12. Bilenko, M., Mooney, R.J.: Adaptive duplicate detection using learnable string similarity measures. In: Proceedings of the 9th ACM SIGKDD International Conference on Knowledge Discovery and Data Mining, pp. 39–48. ACM, New York (2003). https://doi.org/10.1145/956750.956759
13. Guha, S., Koudas, N., Marathe, A., Srivastava, D.: Merging the results of approximate match operations. In: Proceedings of the 30th International Conference on Very Large Data Bases, pp. 636–647. VLDB Endowment, USA (2004)

14. Whang, S.E., Marmaros, D., Garcia-Molina, H.: Pay-as-you-go entity resolution. IEEE Trans. Knowl. Data Eng. **25**(5), 1111–1124 (2013). https://doi.org/10.1109/TKDE.2012.43
15. Efthymiou, V., Stefanidis, K., Christophides, V.: Minoan ER: progressive entity resolution in the web of data. In: Proceedings of the 19th International Conference on Extending Database Technology, pp. 670–671. OpenProceedings.org, Konstanz (2016). https://doi.org/10.5441/002/edbt.2016.79
16. Wang, H., Li, J., Gao, H.: Efficient entity resolution based on subgraph cohesion. Knowl. Inf. Syst. **46**(2), 285–314 (2016). https://doi.org/10.1007/s10115-015-0818-7
17. Wang, J., Li, G., Yu, J.X., Feng, J.: Entity matching: how similar is similar. PVLDB **4**(10), 622–633 (2011). https://doi.org/10.14778/2021017.2021020
18. Christen, P.: Febrl: a freely available record linkage system with a graphical user interface. In: Proceedings of the 2nd Australasian Workshop on Health Data and Knowledge Management, pp. 17–25. Australian Computer Society Inc, Darlinghurst (2008)
19. Pedregosa, F., et al.: Scikit-learn: machine learning in Python. JMLR **12**, 2825–2830 (2011)
20. Goodfellow, I., Bengio, Y., Courville, A.: Deep Learning. MIT press, Cambridge (2016)
21. Li, Z., Wang, H., Shao, W., Li, J., Gao, H.: Repairing data through regular expressions. PVLDB **9**(5), 432–443 (2016). https://doi.org/10.14778/2876473.2876478

Integration of Big Data: A Survey

Jingya Hui, Lingli Li(✉), and Zhaogong Zhang

Department of Computer Science and Technology,
Heilongjiang University, Harbin, China
ahhjy0807@163.com, lilingli_grace@163.com,
zhaogong.zhang@qq.com

Abstract. Data integration provides users a uniform interface for multiple heterogonous data sources. This problem has attracted a large amount of attention from both research and industry areas. In this paper, we overview the state-of-art approaches in data integration which are roughly divided into five parts: schema matching, entity resolution, data fusion, integration system, and new problems arisen.

Keywords: Data integration · Schema matching · Entity resolution
Data fusion

1 Introduction

Data integration systems offer users a uniform interface to a set of data sources. For instance, first the user submits a query based on a mediated schema; secondly, the system reformulates the query over the relevant sources based on schema matching; thirdly, all the answers of the query are combined from the relevant sources. Due to the importance of data integration, it has attracted significant research attention. On one hand, challenges of the basic operations, such as schema matching, entity resolution, data fusion, continue to appear in the new contexts; on the other hand, some new problems have attracted interest, such as, integrating extremely large data sets, extremely large numbers of data sets, and extremely heterogeneous data sets, such as data lake.

The general data integration system can be roughly divided into three steps: (1) schema matching: generating correspondences between elements of two given schemas; (2) entity resolution: identifying duplicated records which represent the same real-world entity; and (3) data fusion: resolving conflicts and finding the truth from different data sources.

1.1 Schema Matching

Schema matching is the problem of finding correspondences between elements of two schemas. It is usually the first step in data integration. The process of schema matching often consists of the following steps: (1) Mediate Schema which provides a unified view for different data sources; (2) Attribute Matching which matches the attributes in each source schema to the corresponding attributes in mediate schema; and

Q. Zhou et al. (Eds.): ICPCSEE 2018, CCIS 901, pp. 101–121, 2018.
https://doi.org/10.1007/978-981-13-2203-7_9

(3) Schema Mapping in which the semantic relationship between the content of the data source and the content of the mediate data are illustrated.

Schema matching is the first and most important step in data integration. Even though traditional schema matching can improve data quality, there are still several shortcomings, such as lacking support for extremely large table corpus, heterogeneous data, online query and so on.

1.2 Entity Resolution

The goal of entity resolution (ER) is to identify the records referring to the same real-world entity from multiple data sources. ER is a central component in data integration and has attracted much attention in recent years. The process of ER can be divided into the following steps.

(1) Blocking: a block function is established on one or more attribute values, and the entity is divided into several blocks by using this function. The goal of blocking is to filter most of the unlikely matching pairs so that the number of comparison between records can be reduced.
(2) Record Matching is to compare tuples based on a given similarity function (or rules) and determine how similar they are.
(3) Clustering is to partition records where the records that are determined to represent the same entity are put the same group.

Entity resolution is a hotspot in data integration and has attracted much research attention for decades. The focus of entity resolution has shifted to some new applications, such as ER on large-scale data, crowdsourcing ER and temporal ER.

1.3 Data Fusion

The third step in data integration is data fusion. The task of data fusion is to fuse multiple records representing the same real-world entity into a single, consistent, and clean representation. The most critical component in data fusion is truth discovery, which is to identify the true values of data items from multiple sources.

Recent work on data fusion can be roughly divided into three categories. One is data fusion for special data, such as probability data, time data and graph data. The second is to optimize the performance of traditional fusion technologies by using data features (e.g. the long tail effect), characteristics of data sources (the relevance and reliability of data sources) and specific domain knowledge, etc. The third is the emerging problems in data fusion, such as knowledge fusion, data fusion based on query, etc. We will introduce these three technologies respectively.

The rest of the paper is structured as follows. Section 2 describes and classifies the new techniques for schema matching. Section 3 presents technologies for entity resolution, Sect. 4 gives an overview of integrated information systems and Sect. 5 present new problems emerged in data integration.

2 Schema Matching

Schema matching is the first and most important step in data integration. Even though traditional schema matching can improve data quality, there are still several shortcomings, such as lacking the Synthesizing Mapping Relationships in table corpus, the online query data system do not have high efficiency, lacking viable technologies on heterogeneous data. In this section, we will propose solutions to the above problems.

2.1 Schema Mapping Based on Table Corpus

Wang et al. [1] studies the problem of automatically synthesizing mapping relationships from existing mapping tables (also referred to as "bridge tables"). Mapping tables are versatile data assets with many applications in data integration and data cleaning. They formalize this as an optimization problem that maximizes positive compatibility between tables while respecting constrains of negative compatibility imposed by functional dependencies. It shows that this optimization problem is NP-hard. A greedy heuristic is developed which can scale to large table corpus. Figure 1 shows the main steps of the solution.

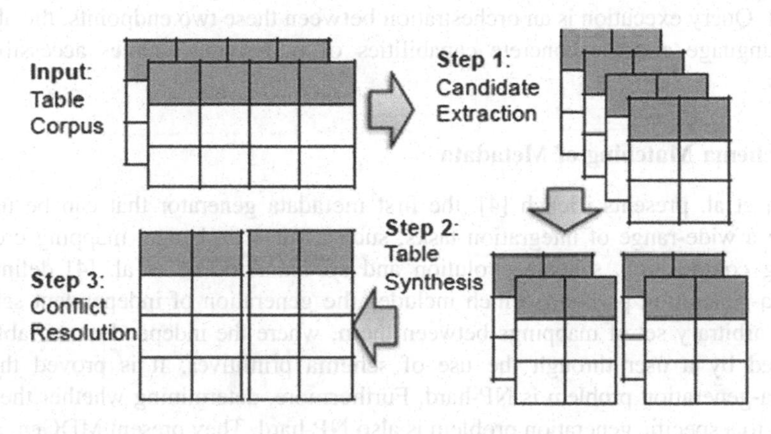

Fig. 1. solution overview [1]

Step 1. Candidate Extraction. In this step, candidate tables for synthesis are generated from all matching tables in the corpus. Low-quality tables are pruned based on co-occurrence statistics. Specifically, Point-wise Mutual Information (PMI) [2] is used to quantify the coherence of a column C, and C is filtered out if its coherence is lower than a threshold.

Step 2. Table Synthesis. In this step, they synthesize two-column tables that describe the same relationship and are compatible with each other by partitioning. However, this process has a high computational complexity $O(N^2)$ where N is the total number of candidate tables. To address this problem, they use inverted-index-like regrouping in a Map-Reduce round, so that only similar tables can be partitioned into the same group.

Step 3. Conflict Resolution. After partitioning tables, this step is to find out the largest subset P_T of P such that no two tables in P_T conflict with each other, where P is the partition results of Step 2. This step assumes that most of tables in the partition should agree with the ground-truth. The objective is to include as many value pairs as possible, where no pairs of tables in $P\text{-}_T$ have conflict. This problem is shown to be NP-hard, and they use a greedy filtering strategy to find the solution.

2.2 Heterogeneous Data Source Querying

To answer a specific research question, heterogeneous in-situ data from different sources need to be integrated. However, classic Extract-Transform-Load approach is too complex and time-consuming and unable to handle the heterogeneous in-situ data sets. To address these problems, Chamanara et al. presents QUIS [3], a system for an open source heterogeneous in-situ data querying. QUIS consists of three main components: a query language, a set of adapters, and a query execution engine. Users write their queries and applications using the query language and submit them to the execution engine. The engine chooses the best adapter to transform and execute each query. The queries may be optimized before sending them to the adapters. Furthermore, to complement any lacking features of the adapters, query rewriting might also be required. Query execution is an orchestration between these two endpoints; the abstract query language and the concrete capabilities of underlying sources accessible via adapters.

2.3 Schema Matching of Metadata

Arocena et al. presents iBench [4], the first metadata generator that can be used to evaluate a wide-range of integration tasks, such as data exchange, mapping creation, mapping composition, schema evolution and so on. Arocena et al. [4] defines the metadata-generation problem, which includes the generation of independent schemas with an arbitrary set of mappings between them, where the independent variables are controlled by a user through the use of schema primitives. It is proved that the metadata-generation problem is NP-hard. Furthermore, determining whether there is a solution to a specific generation problem is also NP-hard. They present MDGen, a best-effort, randomized algorithm, that solves the metadata-generation problem. The main innovation behind MDGen is in permitting flexible control over independent variables describing the metadata in a fast, scalable way.

Data Lakes (DLs) provides a centralized data store accommodating all forms of data, such as structured data and semi-structured data. However, current DL systems lack details about the required metadata management features. Constance [5] is a DL system which manages structural and semantic metadata, provides means to enrich the metadata with schema matching and schema summarization techniques. Figure 2 show the architecture of Constance.

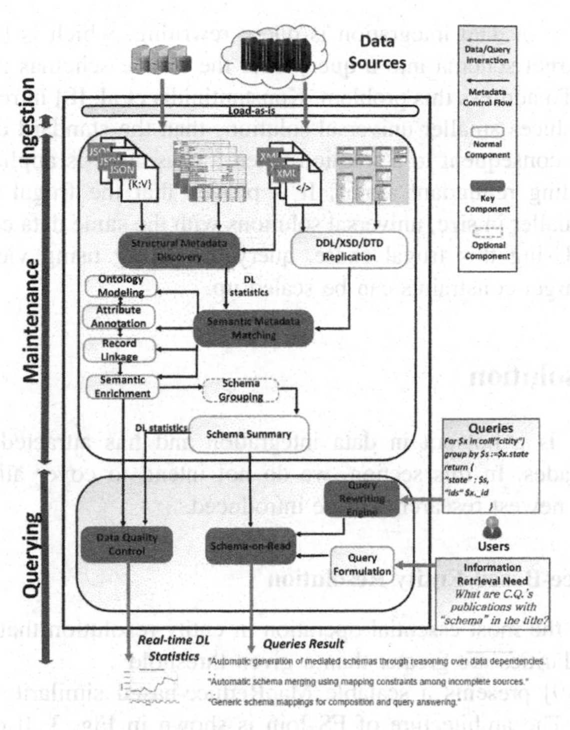

Fig. 2. Constance system [5]

The system can be divided into three layers: ingestion, maintenance, and querying, where the second layer Maintenance is in charge of the metadata management. More specifically, this layer consists of the following components: (1) the Structural Metadata Discovery component extracts metadata from the sources; (2) the Metadata Matching component consists of four steps as shown in Fig. 2. The output is a graph representation of the extracted metadata and their relationships; (3) the Schema Grouping component clusters the schemas and picks up the core of each cluster as its presentation; and (4) the Schema Summary component extracts a compact structure of the currently schemata, namely skeleton [6].

2.4 Other Problems

Kolaitis et al. [7] studies the fundamental reasoning tasks and structural properties of nested tgds in schema mapping. They show the following problems are decidable: the implication problem of nested tgds, and the problem of deciding whether a given nested GLAV mapping is equivalent to some GLAV mapping. They also show that whether a given plain SO tgd is equivalent to some GLAV mapping is undecidable as soon as a single key dependency is allowed in the source schema.

A key problem in data integration is query rewriting, which is how to rewrite a query over the target schema into a query over the source schemas that provides the certain answers. To address this problem, Konstantinidis et al. [8] introduces the frugal chase, which produces smaller universal solutions than the standard chase. Instead of adding the entire consequent to a solution when a chase rule is applicable, the frugal chase avoids adding redundant atoms. It is proven that the frugal chase results in equivalent, but smaller in size, universal solutions with the same data complexity as the standard chase. Using this frugal chase, query answering using views under LAV weakly acyclic target constraints can be scaled up.

3 Entity Resolution

Entity resolution is a hotspot in data integration and has attracted much research attention for decades. In this section, we do not intend to cover all aspects in ER, instead, only the newest research will be introduced.

3.1 MapReduce-Based Entity Resolution

Similarity join is the most essential operation in entity resolution that finds all record pairs whose similarities are greater than a given threshold

Rong et al. [9] presents a scalable MapReduce-based similarity join algorithm, namely FS-Join. The architecture of FS-Join is shown in Fig. 3. It consists of three major steps: ordering, filtering and verification.

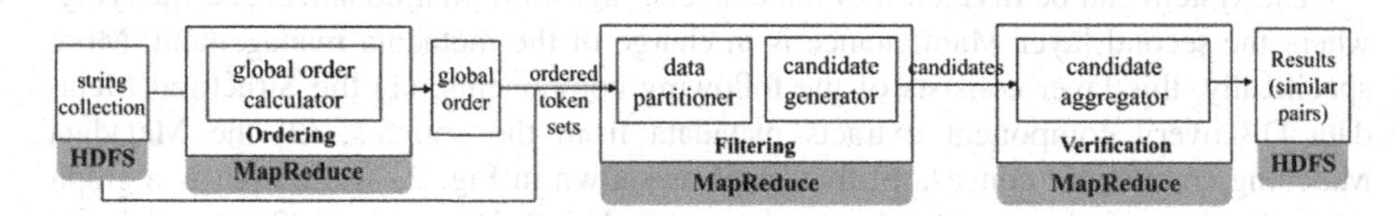

Fig. 3. Architecture of FS-join [9]

Step1. Ordering. A global order O is obtained by applying the ordering algorithm proposed in [10]. In particular, the tokens are sorted in an ascending order based on their frequencies. One MapReduce job is required to compute the global order for the string collection in HDFS.

Step2. Filtering. In step 2, one MapReduce job is required to generate candidate string pairs by pruning dissimilar pairs without computing their accurate similarity scores. This is the core operation of FS-Join and has two phases: partitioning and join. More precisely, the *partitioner* first selects a number of pivots from O returned from Step 1. Next, it sorts all the tokens in each received strings according to the global order. Then, the data *partitioner* partitions each string into several segments based on a set of chosen pivots. Finally, FS-Join treats the partition id as the key and the segment in the corresponding partition as the value in the Map phase.

The segments belonging to the same partition are shuffled to the same reduce node. In each reducer, the candidate generator of FS-Join generates candidates by using a prefix-based inverted index and several filtering methods.

Step3. Verification. One MapReduce job is used to verify the candidates generated in Step 2. If the number of common tokens of a string pair is larger than a given threshold, then it is determined to be an answer. Otherwise, it is not.

3.2 Hybrid Human-Machine Framework

Li [11] presents a hybrid human-machine entity resolution framework. It first uses rules (i.e., similarity-based rules and knowledge-based rules) to identify candidate matching pairs and then utilizes the crowd to verify these candidate results. To improve the efficiency of the first step, they build a distributed in-memory system and use a signature-based method such that if two records refer to the same entity, they must share a common signature. To save the cost of the second step, a selection-inference-refine framework is proposed. First, some "beneficial" tasks are selected to ask the crowd, and then the answers of unasked tasks are inferred based on the crowd results; finally, the inferred answers with high uncertainty are refined.

3.3 Temporal Entity Resolution

To construct a high-quality integrated knowledge repository from different data sources, a big challenge is to understand how facts across different sources are related to one another over time. This is called temporal record linkage problem. Li et al. [12] presents a framework, called MAROON, for temporal record linkage problem. temporal record linkage problem is more challenging than traditional entity resolution problem, because how facts across different sources are related to one another over time should be understood. To address this problem, two algorithms are proposed. First, a transition model learns the probabilities of the transitions between different values over time. This transition model enables to answer the following question: if an attribute of an entity currently has a value v, what is the probability that this value will change to a value v? after Δt time? After that, a matching algorithm is proposed to link temporal records to the entities in the right time period by jointly considering the value transition probability and the freshness of the data sources.

4 Data Fusion

The third step in data integration is data fusion. When different data sources provide values for the same property of the same entity, because of classification error, incorrect calculation, information expiration, semantic interpretation inconsistency or false information can lead to conflict.

Recent work on data fusion can be roughly divided into three categories. One is data fusion for special data, such as probability data, time data and graph data. The second is to optimize the performance of traditional fusion technologies by using data features (e.g. the long tail effect), characteristics of data sources (the relevance and reliability of data sources) and specific domain knowledge, etc. The third is the emerging problems in data fusion, such as knowledge fusion, data fusion based on query, etc. We will introduce these three technologies respectively.

4.1 Coping with Data Diversity

Probability-Based Data Fusion. Probabilistic data can be modeled in various formalisms, such as pc-tables, Bayesian networks, and stochastic automata. Olteanu et al. [13] presents an integration system, namely *Πgora,* which provides a uniform relational interface to the heterogeneous probabilistic sources. The components of the system are shown in Fig. 4.

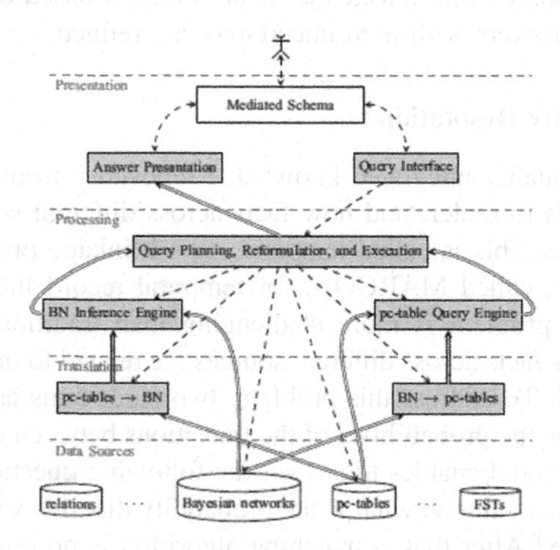

Fig. 4. The architecture of Πgora [13]

The system works as follows. Each local source is registered to the system with a relational schema. The user's query is expressed as a SQL query extended with an exact and approximate probability computation aggregate and with a *given* clause, which allows to formulate conditionals and ask for the probability of an event given another event.

Data Sources. To handle heterogeneous probabilistic sources, MayBMS is used to manage pc-tables [14], FSTs are managed by Staccato [15] and Bayesian networks are represented in the XML-based BayesNets Interchange Format.

Translation Layer. The query evaluation strategy requires to translate all sources into one formalism and executes the query using one dedicated engine. However, this translation has a considerable overhead for networks that are not tree-shaped. To address this problem, they allow *let definitions* in the pc-table formalism, whereby events can be named and re-used several times.

Inference and Query Engines. Πgora is implemented in Java on top of the probabilistic management system MayBMS [14] with the SPROUT query engine [16] for queries on pc-tables, SMILE [17] for Bayesian inference, and Staccato for selection queries on FSTs [15].

Query Planning, Reformulation, and Execution. The task of this component is to evaluate query. The default strategy is to identify subqueries that are naturally supported by the formalisms of the sources. They then use the engines associated with the formalism of the data sources to answer the subqueries. A more advanced strategy is to convert all data sources by the query into either the pc-tables or Bayesian networks formalism, followed by evaluation using either a query or an inference engine respectively, after reformulating the query over the corresponding formalism.

Time-Based Data Fusion. Abedjan et al. [18] studies the rule discovery problem for temporal web data. Such a discovery process is very challenging because the temporal web data are (1) sparse over time, (2) reported with delays, and (3) often reported with errors. Machine learning techniques, such as association measures and outlier detection, are used to identify approximate temporal functional dependencies and their durations. This duration is then used as the time window for the rule to identify temporal outliers. Alexe et al. [19] presents an entity integration framework, called PRAWN, that integrates temporal data and resolves conflicts based on user-defined preference rules.

Graph-Based Data Fusion. Petermann et al. [20] presents a graph-based data integration system, namely BIIIG (Business Intelligence with Integrated Instance Graphs). A largely automatic pipeline (shown in Fig. 5) is used for integrating metadata and instance data. More specifically, metadata from heterogeneous sources are integrated in a so-called Unified Metadata Graph (UMG) while instance data is combined in a single integrated instance graph (IIG).

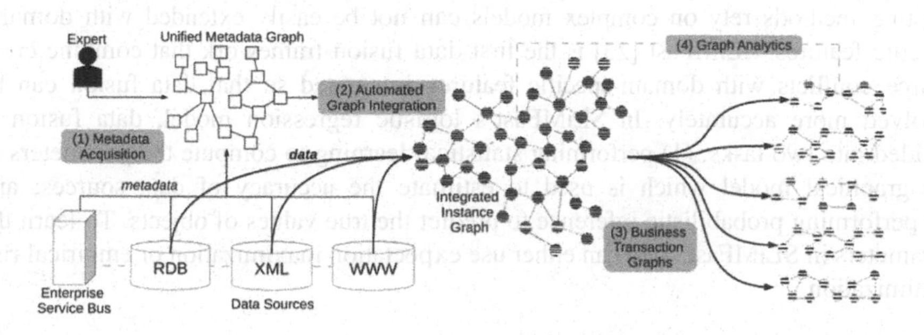

Fig. 5. Conceptual overview of BIIIG evaluate [20]

4.2 Coping with Data Truth

In data fusion, how to improve the accuracy has always been the focus of research. Actually, there are some important factors should be considered: the correlation between data sources, the reliability of data sources, the long tail of data, domain-specific information and human involvement. With these consideration, the performance of data fusion can be greatly improved.

Source Reliability. Identifying the reliability of sources can greatly improve the accuracy of truth discovery. Therefore, it is essential to automatically identify trustworthy information and sources from multiple conflicting data sources. As heterogeneous data is ubiquitous, a joint estimation on various data types can lead to better estimation of truths and sources reliability. However, existing conflict resolution work either regards all the sources equally reliable, or model different data types individually. Li et al. [21] proposes a general optimization framework, namely CRH, for truth discovery. In this framework, conflicts from heterogeneous data sources are resolved by incorporating source reliability estimation. The conflict resolution problem is formalized as an optimization problem that targets at minimizing overall weighted difference between truths and input data, where the weights in the objective function correspond to source reliability degrees. Various loss functions can be plugged into the framework to characterize different types of data.

Long-Tail Data. Traditional truth discovery approaches do not consider the effect of the long-tail phenomenon. A confidence-aware truth discovery (CATD) method that can automatically detect truth from conflicting data with long-tail phenomenon is proposed by [22] This method not only estimates source reliability, but also considers the confidence interval of the estimation. The overall goal is to minimize the weighted sum of the variances to obtain a reasonable estimate of the source reliability degrees. By optimizing the source weights, they can assign high weights to reliable sources and low weights to unreliable sources when the sources have sufficient claims. When a source only provides very few claims, the weight is mostly dominated by the chi-squared probability value so that the source reliability degree is automatically smoothed and small sources will not affect the trustworthiness estimation heavily.

Domain-Specific Information. Current data fusion methods only use the conflicting observations across sources and do not consider domain knowledge. Moreover, existing methods rely on complex models can not be easily extended with domain-specific features. SLiMFast [23] is the first data fusion framework that combine cross source conflicts with domain-specific features integrated so that data fusion can be resolved more accurately. In SLiMFast's logistic regression model, data fusion is divided into two tasks: (1) performing statistical learning to compute the parameters of the graphical model which is used to estimate the accuracy of data sources; and (2) performing probabilistic inference to predict the true values of objects. To learn the parameters in SLiMFast, they can either use expectation maximization or empirical risk minimization.

Human Involvement. Most of the existing fusion techniques automatically identify correct claims for data items. However, for crucial data, automated data fusion is not enough and human involvement should be integrated to achieve a higher. Sometimes, validation from experts is also expected.

In order to improve the confidence of data, a novel crowdsourcing-based machine-crowd hybrid system, namely *CrowdFusion*, is designed by [24] to select a set of tasks to ask crowds. However, since data are correlated and crowds may provide incorrect answers, how to select a proper set of tasks to ask the crowd is a very challenging problem. Chen et al. [24] has proved that, given n possible tasks, selecting k of them to reach the highest value of utility function is an NP-hard problem. An greedy algorithm of approximation ratio $1 - 1/e$ is proposed to address this problem. In each iteration, the most uncertain variable given the ones selected. To further improve the efficiency, a pruning strategy and a preprocessing method are also presented.

User feedback is leveraged by [25] to validate data conflicts and improve the fusion result. It is the first work to leverage user feedback in Bayesian-based data fusion models. They formalize the problem of ordering user feedback for data fusion and propose a decision-theoretic ranking strategy to evaluate data items. In this model, the utility function obtained by validating a data item is based on the concept of VPI [26]. In each iteration, a validating which maximizes the gain in utility function is selected. However, this algorithm has a prohibitively expensive computational cost. To scale up to large-scale datasets, an approximation algorithm is proposed by analytically estimating the impact of a validation on other unvalidated data items, and selecting a claim that has the maximum utility gain over the estimates.

Correlation Between Data Sources. Traditional approaches discover the copying relationship between sources based on the intuition that common mistakes are strong evidence of copying [27, 28]. However, correlation between data sources is much broader than copying. Previous approaches are effective in detecting positive correlation on false data, but are not effective in detecting positive correlation on true data or negative correlation. Moreover, their models often rely on the single-truth assumption.

Pochampally et al. [29] presents a novel technique that models correlations between sources. They present two quality metrics of sources based on conditional probability, joint precision and joint recall, to measure the correlation between a subset of sources. From the precision and recall of the sources, the probability of a knowledge triple being true can be derived using Bayesian analysis.

4.3 Data Fusion: New Problem

Query-Based Data Fusion. Yu et al. [30] presents a framework, namely FuseM, to address the query-centric data fusion problem on Web markup data. It includes two major steps: entity retrieval and data fusion. The first step is to obtain a pool of candidate facts, denoted *F*. The second step is to obtain the complete and correct facts *F'* from *F*, which consists of a classification step and a diversification step. In the classification step, each fact is classified into one of the two classes {'correct', 'incorrect'} based on their 11 quality features, i.e. 3 relevance features, 4 clustering features and 4 quality features.

State-of-the-art classification algorithms, such as SVM, kNN, can be applied. After classification, in order to improve the diversity of F', the duplicated facts are removed in the diversification step.

5 Advanced Data Integration Systems

In this section, we will introduce systems about data integration with some feature. For instance, someone builds a web application to implement a knowledge intensive application that identifies those people that may have been affected due to natural disasters or man-made disasters at any geographical location and notify them with safety instructions. someone builds a system for extracting relational databases from dark data. someone demonstrates a new system, called BIIIG, it is a new system for graph-based data integration and analysis. And other people also demonstrates Data Civilizer to ease the pain faced in analyzing data in the numerous data sources, etc.

5.1 Safe Check System

Pandey et al. [31] build a web application called Safety Check which identifies those people that may been affected by disasters. This involves extraction of data from various sources for emergency alerts, weather alerts, and contacts data. However, disaster data are extremely heterogeneous, both structurally and semantically, which brings a big challenge for data integration and ingestion. To solve this problem, [31] adopt semantic web technologies, which are best suited to handle data with high - volume, velocity, and variety [32]. Specifically, Apache Jena (or Jena in short), an open source Java framework for building Semantic web and Linked Data application [33] is been used. The framework of Safety Check that uses semantic technology is shown in Fig. 6.

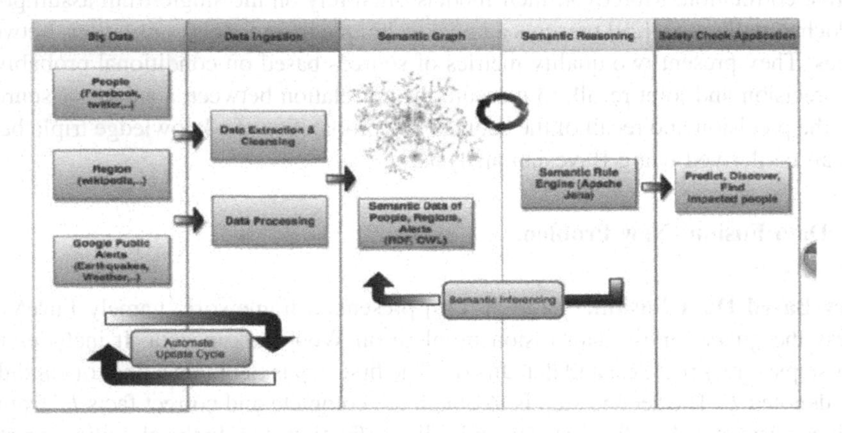

Fig. 6. Semantic technology for information data analysis for emergency management system [31]

In first step, Data Extraction and Cleansing - gathering data from multiple sources. In the 2nd step, Data Aggregation processes and translates data into Semantic Graph with RDF/XML format. Note that, to support data update, there is an automate update cycle between Phase I and II. In 3rd step, Semantic Reasoning - discovering new information using "computing engines" that operates on semantic graph. In the 4th step, transformed data can be accessed and shared using the SPARQL Semantic Web query language.

5.2 Dark Data

Dark data is the mass of the text, tables, and images which are collected and stored but cannot be queried using traditional relational database tools. Because the information embedded in dark data can be massively valuable, institutions have hired human beings to manually read documents and populate a relational database by hand [35].

However, this practice is expensive, incredibly slow, and surprisingly error-prone. DeepDive [34] can extract structured databases from dark data accurately and vastly more quickly than human beings. And its ability to produce databases of extremely high accuracy with reasonable amounts of human engineering effort.

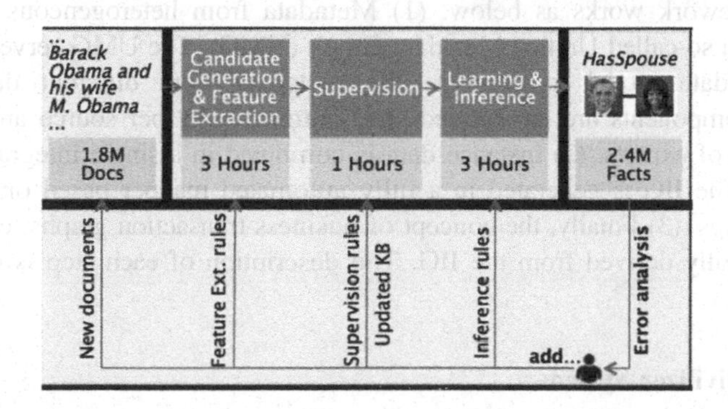

Fig. 7. DEEPDIVE takes as input dark data—such as unstructured documents—and outputs a structured database [34]

DeepDive's high quality is possible because of the unique design based around probabilistic inference and a specialized engineering development cycle; they are underpinned by several technical innovations for efficient statistical training and sampling. DeepDive consists of the following phrases (shown in Fig. 7): extraction, integration, and cleaning.

1. In candidate generation, DeepDive applies a user-defined function to each document to generate candidate extractions. The candidate are intended to be high-recall, low-precision.
2. In supervision, DeepDive applies user-defined distant supervision rules to provide labels for some of the candidates. Supervision rules are intended to be low-recall, high-precision.

3. In learning and inference, they construct a graphical model that represents all of the labeled candidate extractions, train weights, then infer a correct probability for each candidate.

5.3 BIIIG System

Petermann et al. [20] propose a new system, BIIIG (Business Intelligence with Integrated Instance Graphs), for graph-based data integration and analysis. It aims at improving business analytics compared to traditional OLAP approaches by comprehensively tracking relationships between entities and making them available for analysis.

In contrast to data warehouses, BIIIG does not require defining a global schema for data integration such as a star or snowflake schema. While such an approach serves many OLAP queries, it is often too inflexible as it can only evaluate facts according to the predefined dimensions and relationships. BIIIG thus aims at supporting the analysis of relationships between business entities in addition to standard analysis tasks. For this purpose, they present a bottom-up data integration approach that combines metadata and instance data from relevant data sources in flexible and generic graph models that preserve existing relationships for later analysis.

The framework works as below. (1) Metadata from heterogeneous sources are integrated in a so-called Unified Metadata Graph (UMG). The UMG serves BIIIG as a generic metadata model to combine the metadata from different data sources. The UMG components are determined semi-automatically per source and integrated with the help of experts. (2) Instance data is combined in a single integrated instance graph (IIG) The IIG is generated in a fully automated manner based on the source-UMG mappings. (3) Finally, the concept of business transaction graphs, called BTGs, are automatically derived from the IIG. The description of each step is described as following.

5.4 Data Civilizer System

One difficult problem of in data integration is to find relevant data from numerous data sources for specific tasks. In practice, data scientists are routinely reporting that the majority (more than 80%) of their effort is spent finding, cleaning, integrating, and accessing data of interest to a task at hand.

Motivated by this, Fernandez et al. [36] builds an end-to-end big data management system, Data Civilizer, with components for: (1) Discovering datasets relevant to the task at hand; (2) obtaining access to these datasets; (3) integrating datasets and deduplicating the result; (4) stitching together datasets through the best join paths; (5) cleaning datasets under a limited budget; and (6) querying datasets that live across different systems. Figure 8 give an overview of the DATA CIVILIZER system.

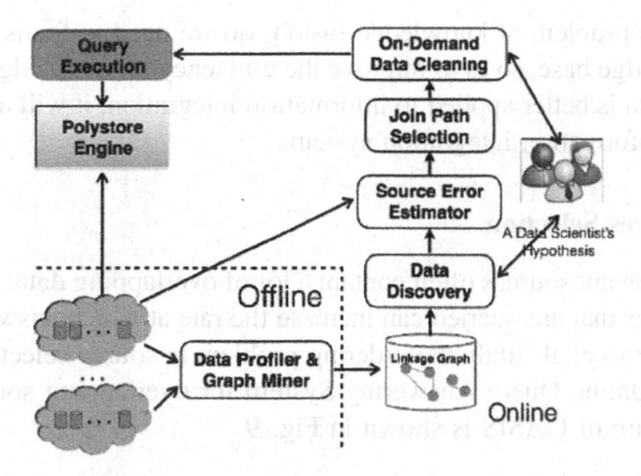

Fig. 8. Data civilizer system [37]

Each module of Data Civilizer is described as below:

Data Profiler and Graph Miner. This offline module creates the metadata required by the subsequent online modules. It mainly summarizes data contents into an index and relationships among data elements into a linkage graph.

Data Discovery. This module allows users to capture their intuition about the desired datasets into a Source Retrieval Query Language (SRQL). SRQL queries are based on a discovery algebra for expressing different discovery needs.

Join Path Selection. Given multiple datasets obtained from the previous module, users are usually interested in finding ways to stitch them together and ask meaningful queries. We propose a join path selection algorithm that permits users to choose the most appropriate join path. The algorithm uses the source error estimator to determine the cleanliness of the datasets.

On-Demand Cleaning. Cleaning all source tables is infeasible with large volumes of data. Data Civilizer incorporates a cleaning module that, given a budget, suggests the best cells to clean for the desired view. This module uses the source error estimator in selecting the cells to clean.

Data Query. Analysts want to get the necessary data sets, and they may want to query multiple data sets from different systems.

6 Data Integration: New Problem

In this section, we present some new problems and technologies. These problems and technologies are critical to the data integration, determine whether data integration is successful. For instance, when we discover the cost of data integration, we need to consider the potential benefits of integrating new data sources. Then, based on the potential benefits, determine which data sources are worth integrating. When providing basic services to users, help them find the data source of their domain knowledge.

After defining the problem of knowledge fusion, do we need to focus on building and managing knowledge base, so as to improve the efficiency of knowledge fusion. Last, if heterogeneous data is better applied to information integration, it will also contribute to the progress of information integration system.

6.1 Data Sources Selection

Since different relevant sources often contain a lot of overlapping data, choosing a good ordering of sources that are queried can increase the rate at which answers are returned. To this end, Salloum et al. study the ordering problem in source selection and presents OASIS [37], an Online Query Answering System for overlapping sources.

The architecture of OASIS is shown in Fig. 9.

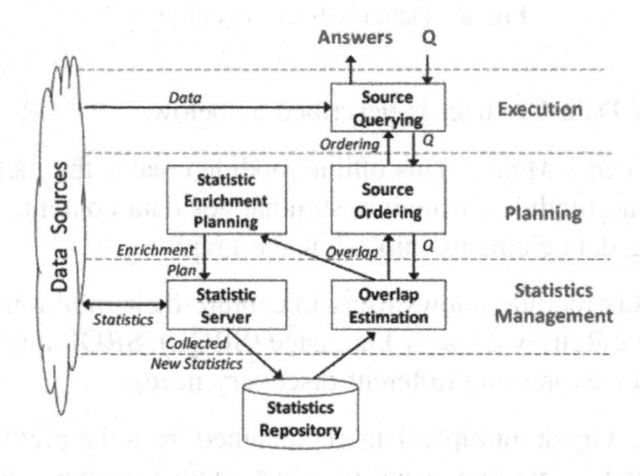

Fig. 9. OASIS system [37]

1. The Statistics management is at the bottom and consists of two components. The Statistics Server component is to collect statistics, such as source coverage and overlap information. Since the collected statistics may be incomplete, the Overlap Estimation component estimates upon receiving a query the overlap statistics that are unavailable under the Maximum Entropy principle.
2. The Planning layer contains two components. The Source Ordering component decides the ordering of sources based on the statistics provided by the Statistics management layer and the statistics collected during query answering. The Statistics Enrichment component selects a set of overlaps to enrich and sends the request to the Statistics Server.
3. The Execution layer receives a user query, queries the sources according to the ordering decided by Source Ordering, and returns the retrieved answers.

Rekatsinas et al. [38] studies the problem of source selection considering dynamic data sources whose content change over time. They propose the time perception source selection problem, which is to reason the profit of the acquisition and integration of dynamic source, so as to select the best subset of the source to be integrated. A set of time-dependent metrics are defined, including coverage, freshness and accuracy,

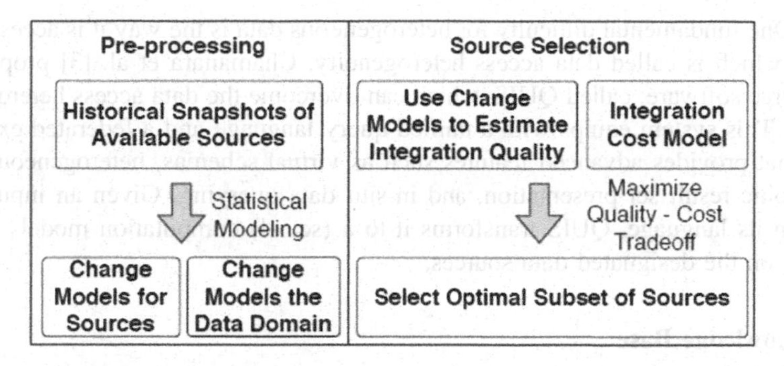

Fig. 10. The Framework for solving time-aware source selection [39]

to characterize the quality of integrated data. The main components of their proposed framework for solving time-aware source selection are shown in Fig. 10. In this framework, it is assumed that knowledge of data changes over a past time window *T*. In the pre-processing phase, historical snapshots of sources are used to build change models for sources and data domains. In the source selection step, they use the change models to estimate the integration quality and cost for an arbitrary set of sources.

6.2 Heterogeneous Data Integration

There are different forms of data in different data sources. We call them heterogeneous data. In data integration, the first step is to access data, and the most important step in accessing data is to parse the data. To parse the data, we must understand the structure of the data.

In this section, we will introduce how to resolve conflicts among multiple sources of heterogeneous data types.

CRH Framework. Li et al. [21] develops a conflict resolution framework, namely CRH, which discovers the truths from conflicting heterogeneous sources. The conflict resolution problem on heterogeneous data sources is modeled as a general optimization problem. In this optimization model, truth is defined as the value that incurs the smallest weighted deviation from multi-source input in which weights represent source reliability degrees. A two-step iterative procedure is derived, including the computation of truths and source weights as a solution to the optimization problem. In order to characterize different data types and weight distributions effectively, various loss and regularization functions are plugged in the framework.

TATOOINE. Bonaque et al. [39] demonstrates a lightweight data integration prototype for data journalism, called TATOOINE, which allows journalists to exploit heterogeneous data sources of different data models based on key-word search. Specifically, for each source, a digest is built that comprises: (i) its schema and (ii) the domain of each attribute. Digests can be viewed as directed graphs (e.g., for a relational database, there is one node per attribute, one edge per key-foreign key constraint, etc.). Given the search keywords, the engine looks up the keywords in the digest; and the corresponding structured query is generated from the keywords.

QUIS. One fundamental difficulty for heterogeneous data is the way it is accessed and queried, which is called data access heterogeneity. Chamanara et al. [3] proposes an open source software, called QUIS, which can overcome the data access heterogeneity problem. This system equips with a unified query language and a federated execution engine that provides advanced features such as virtual schemas, heterogeneous joins, polymorphic result set presentation, and in-situ data querying. Given an input query written in its language, QUIS transforms it to a (set of) computation models that are executed on the designated data sources.

6.3 Knowledge Base

A knowledge base (KB) typically contains a set of concepts, instances, and relations. Examples of KBs include DBLP, Google Scholar, and Freebase. In recent years, many well-known KBs have been built, and this topic has received growing attention in both industry and research community. KBs can be distinguished as two classes: ontology-like KBs, which attempt to capture all relevant concepts, instances, and relationships in a domain, and source-specific KBs, which integrate a set of given data sources.

Deshpande et al. [40] describes an end-to-end process of how to build, maintain, curate, and use a global KB at Kosmix and later at WalmartLabs. The first step, building a source-specific KB, is in essence a data integration problem, where data from a given set of data sources must be integrated. Given a KB *B*, the process of integrating a source *S* into *B* is as follows. First, the data instances and the taxonomy *T* over the instances are extracted from *S*. Second, the categories of the taxonomy *T* are matched to those in the KB *B*, using a state-of-the-art matcher, then clean and add such matches to a concordance table. Finally, the instances of S are matched and moved over to the KB *B*.

Integrating knowledge bases (KBs) has attracted interests in the last several years. Knowledge bases may differ greatly in their schema, using different named and different granularity relations and properties. This brings challenges for KB fusion. Rodríguez [41] introduces a knowledge base system, called SigmaKB, that can incorporate both strong, high-precision KBs and weaker noisy KBs into a single, cohesive master KB. Rather than integrate all data sources into a single, monolithic KB, SigmaKB chooses to remain modular, querying over each KB individually and fusing the results on-the-fly. In order to combine different ontologies of KBs into a single mediated ontology, all KBs are unioned and relations that are semantically equivalent are canonicalized. Rather than simply take the union results from all individual KBs, a reasoning component is used to combine duplicate and conflicting entries into a cohesive, singular entry. Specifically, aggregation across individual KBs is handled using a state-of-the-art Consensus Maximization Fusion framework in which complementary and conflicting data can be leveraged to present users with a probabilistic interpretation of their results.

7 Conclusion

Data become more and more important in modern life. In practice, data are often distributed in multiple heterogonous data sources, which bring the challenges of data integration. Facing big data era, many new techniques have been proposed to cope data integration problems in new environment. We believe that it is time for data integration to eliminate the end-to-end obstacles. Then the original data should be accessible at all levels of the system.

In this paper, we overview state-of-art techniques. Even though researchers make great progress in this area, following problems are still not solved, which is left for researchers to do.

1. How to make use of our mobile devices to combing structured and unstructured data?
2. How to focus on the semantics of the data to make data fusion more intelligent?
3. How to involve data quality issues into data integration?
4. How to scale data integration on massive data sources?

Acknowledgment. This work was supported by NSFC61602159, 61370222 and Program for Group of Science Harbin technological innovation 2015RAXXJ004.

References

1. Wang, Y., He, Y.: Synthesizing mapping relationships using table corpus, pp. 1117–1132. ACM (2017)
2. Church, K.W., Hanks, P.: Word association norms, mutual information, and lexicography. Comput. Linguist. **16**(1), 22–29 (1990)
3. Chamanara, J., König-Ries, B., et al.: QUIS: InSitu heterogeneous data source querying. VLDB **10**, 1877–1880 (2017)
4. Arocena, P.C., Glavic, B., Ciucanu, R., et al.: The iBench intergration metadata generator. VLDB **9**, 108–119 (2015)
5. Hai, R., Geisler, S., Quix,C.: Constance: an intelligent data lake system, pp. 2097–2100. ACM (2016)
6. Wang, L., et al.: Schema management for document stores. PVLDB **8**(9), 922–933 (2015)
7. Kolaitis, P.G., Pichler, R., Sallinger, E., et al.: Nested dependencies: structure and reasoning, pp. 176–187. ACM (2014)
8. Konstantinidis, G., Ambite, J.L.: Optimizing the chase: scalable data integration under constraints. VLDB **7**, 1869–1880 (2014)
9. Rong, C., Lin, C., Silva, Y.N., et al.: Fast and scalable distributed set similarity joins for big data analytics, pp. 1059–1070. IEEE (2017)
10. Vernica, R., Carey, M., Li, C.: Efficient parallel set-similarity joins using MapReduce. In: SIGMOD, pp. 495–506. ACM (2010)
11. Li, G.: Human-in-the-loop data integration. VLDB **10**, 2006–2017 (2017)
12. Li, F., Lee, M.L., Hsu, W., et al.: Linking temporal records for profiling entities, pp. 593–605. ACM (2015)

13. Olteanu, D., Papageorgiou, L., van Schaik, S.J.: Πgora: an integration system for probabilistic data, pp. 1324–1327. IEEE (2013)
14. Huang, J., Antova, L., Koch, C., Olteanu, D.: MayBMS: a probabilistic database management system. In: SIGMOD (2009)
15. Kumar, A., Ré, C.: Probabilistic management of OCR data using an RDBMS. PVLDB **5**(4), 322–333 (2011)
16. Olteanu, D., Huang, J., Koch, C.: Approximate confidence computa- tion in probabilistic databases. In: ICDE (2010)
17. Druzdzel, M.: SMILE: structural modeling, inference, and learning engine and GeNIe: a development environment for graphical decision - theoretic models. In: AIII (1999)
18. Abedjan, Z., Akcora, C.G., Ouzzani, M., et al.: Temporal rules discovery for web data cleaning. VLDB **9**, 336–347 (2015)
19. Alexe, B., Roth, M., Tan, W.-C.: Preference-aware integration of temporal data. VLDB **8**, 365–376 (2014)
20. Petermann, A., Junghanns, M., Müller, R., et al.: Graph-based data integration and business intelligence with BIIIG. VLDB **7**, 1577–1580 (2014)
21. Li, Q., Li, Y., Gao, J., et al.: Resolving conflicts in heterogeneous data by truth discovery and source reliability estimation, pp. 1187–1198. ACM (2014)
22. Li, Q., Li, Y., Gao, J., et al.: A confidence-aware approach for truth discovery on long-tail data. VLDB **4**, 425–436 (2014)
23. Joglekar, M., Rekatsinas, T., Garcia-Molina, H., et al.: SLiMFast: guaranteed results for data fusion and source reliability, pp. 1399–1414. ACM (2017)
24. Chen, Y. Chen, L., Zhang, C.J.: CrowdFusion: a crowdsource approach on data fusion refinement, pp. 127–130. IEEE (2017)
25. Pradhan, R., Bykau, S., Prabhakar, S.: Staging user feedback toward rapid conflict resolution in data fusion, pp. 603–618. ACM (2017)
26. Russell, S.J., Norvig, P.: Articial Intelligence: A Modern Approach, 2nd edn. Prentice Hall, Upper Saddle River (2003)
27. Dong, X.L., Berti-Equille, L., Hu, Y., Srivastava, D.: Global detection of complex copying relationships between sources. PVLDB **3**(1–2), 1358–1369 (2010)
28. Dong, X.L., Berti-Equille, L., Srivastava, D.: Integrating conflicting data: the role of source dependence. PVLDB **2**(1), 550–561 (2009)
29. Pochampally, R., Das Sarma, A., Dong, X.L., et al.: Fusing data with correlations, pp. 433-444. ACM (2014)
30. Yu, R., Gadiraju, U., Fetahu, B., et al.: FuseM: query-centric data fusion on structured web markup, pp. 179–182. IEEE (2017)
31. Pandey,Y., et al.: Safety check – a semantic web application for emergency management. ACM (2017)
32. Hristidis, V., et al.: Survey of data management and analysis in disaster situations. J. Syst. Softw. **83**(10), 1701–1714 (2010)
33. McBride, B.: Jena: a semantic web toolkit. IEEE Internet Comput. **6**, 55–59 (2002)
34. Zhang, C., Shin, J., et al.: Extracting databases from dark data with DeepDive, pp. 847–859. ACM (2016)
35. Peters, S.E., et al.: A machine reading system for assembling synthetic paleontological databases. PloS One **9**, e113523 (2014)
36. Fernandez, R.C., Deng, D., Mansour, E., et al.: A demo of the data civilizer system, pp. 1639–1642. ACM (2017)

37. Salloum, M., Dong, X.L., Srivastava, D., et al.: Online ordering of overlapping data source. VLDB **7**, 133–144 (2014)
38. Rekatsinas, T., Dong, X.L., Srivastava, D.: Characterizing and selecting fresh data sources, pp. 919–930. ACM (2014)
39. Bonaque, R., Cao, T.D., Mendoza, O., et al.: Mixedinstance querying: a lightweight integration architecture for data journalism. VLDB **9**, 1513–1516 (2016)
40. Deshpande, O., Lamba, D.S., Tourn, M., et al.: Building, maintaining, and using knowledge bases: a report from the trenches, pp. 1209–1220. ACM (2013)
41. Rodríguez, M., Goldberg, S., Wang, D.Z.: SigmaKB: multiple probabilistic knowledge base fusion. VLDB **9**, 1577–1580 (2016)

Scene-Based Big Data Quality Management Framework

Xinhua Dong[1(✉)], Heng He[2], Chao Li[1], Yongchuan Liu[1], and Houbo Xiong[1]

[1] Hubei University of Technology, Wuhan 430068, Hubei, China
xhdong@hust.edu.cn, {lich.mail,5070708}@qq.com,
hboxiong99@163.com

[2] Wuhan University of Science and Technology, Wuhan 430065, Hubei, China
heheng@wust.edu.cn

Abstract. After the rise of big data to national strategy, the application of big data in every industry is increasing. The quality of data will directly affect the value of data and influence the analysis and decision of managers. Aiming at the characteristics of big data, such as volume, velocity, variety and value, a quality management framework of big data based on application scenario is proposed, which includes data quality assessment and quality management of structured data, unstructured data and data integration stage. In view of the current structured data leading to the core business of the enterprise, we use the research method to extend the peripheral data layer by layer on the main data. Big data processing technology, such as Hadoop and Storm, is used to construct a big data cleaning system based on semantics. Combined with JStorm platform, a real-time control system for big data quality is given. Finally, a big data quality evaluation system is built to detect the effect of data integration. The framework can guarantee the output of high quality big data on the basis of traditional data quality system. It helps enterprises to understand data rules and increase the value of core data, which has practical application value.

Keywords: Data quality · Big data · Quality management · Scene
Data cleaning

1 Introduction

Data quality reflects the degree of satisfaction of data to specific applications [1]. Obtaining high-quality data helps managers to make optimal decisions efficiently, while low quality data will greatly affect decision makers' judgment. With the advent of the era of big data, big data technology is gradually applied in all walks of life, and more enterprises take big data as a means of asset appreciation. Big data specific "4V" features: volume, velocity, variety and value. Because of the 4V features of big data, the quality of big data is also faced with the problem of extracting high quality data from massive, fast changing, and complex data.

At present, there are mainly four challenges. First, the diversity of data sources brings rich data types and complex data structures, which increases the difficulty of data

Q. Zhou et al. (Eds.): ICPCSEE 2018, CCIS 901, pp. 122–139, 2018.
https://doi.org/10.1007/978-981-13-2203-7_10

integration. Second, the amount of data is huge, which puts forward higher requirements for the technology of mass data processing. Third, the speed of data change is fast and the data "timeliness" is very short. It is difficult to judge the quality of the data in an operable time. Fourth, there are no unified big data quality standards at home and abroad, which is little research on the quality of big data at present.

Existing technologies have resolved partly the issues of big data quality from a different perspective, but they did not consider the data quality of the basic units under the big data characteristics. However, big data quality management is a systematic closed loop. In this paper, in view of data diversity, we adopt the scene-based or topic approach to reduce the scale of data, and propose a scene-based quality management framework for big data. For the huge volume of data, we consider the Hadoop distributed software framework for parallel computing and cleaning. In view of the fast changing speed of big data and the short timeliness of data, a more efficient Spark or Storm system is selected. Aiming at the lack of unified data quality standard, a bottom-up data quality research method is put forward, and the quality standard system of big data is formed in the promotion and application of many industries.

The rest of this paper is organized as follows. Section 2 describes the related work. Section 3 proposes a systematic big data quality management framework based on scene. Section 4 introduces the quality management of structured data. Section 5 gives the unstructured data quality management scheme. Section 6 describes quality management of the data integration stage, and gives the framework of big data quality evaluation system. The conclusion is given in Sect. 7.

2 Related Work

For data quality management, a number of research institutions and scholars at home and abroad have made fruitful explorations and researches, mainly focusing on analysis of data quality problems, structured data quality, unstructured data quality, data quality evaluation and data quality monitoring.

2.1 Analysis of Data Quality Problems

The research of data quality in the computer field began in the 90s of the last century. Considering the number of data sources, it can be divided into single data source data quality problems and multiple data source quality problems. Single data sources are mainly lack of integrity constraints and data recording errors, such as data missing, duplicate data and inaccuracy of data expression. In addition to all the quality problems with single data sources, multiple data sources also have structural conflicts and naming conflicts. Data quality can also be divided according to different stages of data processing. First, in data collection phase, manual input may have low accuracy and low efficiency. But automatic acquisition also has problems such as integrity and data reliability caused by hardware and software failures. Second, in the stage of data integration, due to the inconsistent data format, structure and semantics, there are some problems such as erroneous relationship mapping and incomplete data record. Third, in the data application stage, if only the data is read, it may not produce data quality problems.

In order to get more accurate and comprehensive data, enterprises first build data centers, and then use data cleaning technology [2] to ensure accuracy, integration, consistency, timeliness and entity identity of the data quality requirements [3]. Data quality management [4] runs through every stage of data usage cycle. It is a series of management activities that implements the identification, measurement, monitoring and early warning of various data quality problems that may arise from data acquisition, storage, analysis and application. MIT led the Total Data Quality Management project (TDQM) [5], which put forward the cycle of data quality continuous improvement, including four stages of definition, evaluation, analysis and improvement. The Six Sigma quality management [6] process is divided into five stages: definition, measurement, analysis, improvement and control.

2.2 Improving the Quality of Structured Data

For improving the quality of traditional structured data, data cleaning technology [7] occupies an important position, and related research mainly includes **(1) abnormal data detection.** The traditional method of abnormal data detection is data audit [8], also known as data quality mining [9], which first generalizes the data to obtain the overall distribution of the data, and then excavates and processes the data quality problems. Christen et al. [10] based on q-gram set and probabilistic language model technology, using word features to train a one-class support vector machine classifier, can automatically discover abnormal textual values. D'Urso [11] proposed a new method to analyze the quality of data sets, the proposed method is based on proper revisions of different approaches for outlier detection that are combined to boost overall performance and accuracy. **(2) Missing data detection.** Missing data detection is a means to detect the lack of data in the key data fields and can be processed automatically. It includes the single filling method, the multiple filling methods, the maximum expected filling method and the maximum likelihood estimation method [12]. **(3) Detection and processing of inconsistent data** [13]. In the integration of multiple data sources, the method of sorting and fusion [14] is often used to deal with the inconsistent data. **(4) Duplicated data detection.** Duplicated data detection is the detection and processing of duplicate records and semi-structured duplicate records in the relational database. The traditional sorting and merging algorithm is commonly used in duplicate data detection. After sorting, each data is compared with other data records, and then duplicate records are found, which has the characteristics of high accuracy. The basic neighbor algorithm [15] sets the sliding window on the basis of sorting the data set according to the extraction field, and asks for every new entry window's record to compare with other records of the window to find out duplicate records. The advantage is that the search speed is fast, but the search precision is not high. An improved K- nearest neighbor algorithm [16] further improves the precision of the search.

2.3 Improvement of Unstructured Data Quality

As with structured data, the quality problems of unstructured data are more complex. Wang [17] sums up the problems in the quality management of big data, and gives three ways of quality management, which are low-quality data query processing, data error discovery and data error repair. The query processing of low-quality data mainly includes inconsistent data and empty value data. Using Hadoop MapReduce architecture, Yu et al. [18] proposed an automatic discovery algorithm of inconsistent Web data based on hierarchical probability. Clark et al. [19] carried out an experimental comparison of two kinds of incomplete data of the empty value and the attribute concept value. For the discovery of data errors, the research is focused on three aspects: entity recognition, error discovery based on rules and main data. Entity recognition is the key technology to guarantee data quality. Researchers put forward various entity recognition algorithm [20], including how to improve the efficiency of identifying entities [21], but at present, the computational complexity of entity recognition is relatively high. Rule-based error discovery refers to using given rules to catch errors in the data, the main task is to find out the tuples of violate rules, including conditional function constraints, accuracy constraints, conditions include constraints, function constraint of describing consistency and timing constraints of describing timeliness [22]. Master data refers to the data shared between computer systems, which are a high-quality data set, which can provide a synchronous view of the core business entities for many applications. Otto et al. [23] described the design process of the main data quality management function reference model. Data error repair refers to modifying or supplemented wrong data, including rule-based repair, true value discovery, and machine learning based repair. Bohannon et al. [24] proposed a function dependency based Greedy_repair repair algorithm, which uses a heuristic method to repair string data. Li et al. [25] combined the rules and statistical methods to repair stale data. In order to detect the accuracy of big data, Berti-Equille et al. [26] proposed a cross-modal and cross-lingual truth discovery scheme. The repair methods based on machine learning technology mainly include decision tree, neural network and Bayesian network. In recent years, new technologies have been gradually applied to data quality. Storm is a real-time streaming data processing platform for Twitter [27], which can be used to deal with flow data, real-time micro-blog and real-time GPS data. In view of the inadequacy of Storm task allocation, Alibaba Company launched a Java-enhanced JStorm based on Storm, which has been adopted by a large number of enterprises [28]. From the statistics and management perspective, Yu [29] studies and constructs a framework for the quality assurance of non-traditional data in Chinese government statistics from three aspects of data source conditions, metadata and data. On the basis of analysis of big data processing flow, Mo [30] constructs big data quality impact model, and puts forward suggestions and measures for big data quality assurance.

2.4 Data Quality Evaluation and Monitoring

The effect of data quality can be evaluated and monitored [31], that is, the quality dimension results of data can be evaluated through data quality analysis, so as to provide basis for other applications. Data quality assessment refers to quantitative or qualitative analysis [34] of each data quality dimension [32], that is, accuracy [33], integrity, consistency and timeliness. Karvounarakis et al. [35] proposed a provenance query language ProQL based on tuple and semi-ring provenance, which can solve the storage and query problems of data provenance. Geisler et al. [36] proposed a data quality framework based on ontology. Chapman et al. [37] described open research issues in quick and dirty information quality assessment. From the perspective of accuracy, Huang et al. [38] discussed the data quality assessment framework of Internet data sources, and gave the practice and key points of single source quality assessment, multi-source integration evaluation and event information assistant evaluation. Zhao et al. [39] proposed a quality assessment method of Web data source. It establishes a unified data model and data quality standard model for multi-source Internet platform, and then gives quality standard measurement and representation methods for full sample data analysis of big data, and finally achieves unified quality metrics of Web data source by comprehensive evaluation of multidimensional data quality.

As far as we know, few researches systematically focus on the quality management of big data. Considering the data quality problem of the basic unit under the big data characteristics, we discussed quality management of structured data, unstructured data and data integration phase, and data quality assessment.

3 Big Data Quality Management Framework Based on Scene

Because of the potential data increment brought by big data, and the quality questions of all kinds of data will bring serious consequences, we need to further study the quality system of big data on the basis of traditional data quality. In order to facilitate data provenance, we propose a big data extension method based on scene, and a big data quality management framework based on scene.

3.1 Big Data Expansion Method Based on Scene

The advent of the era of big data brings great imagination to enterprises and users, but the specific use of big data should have a real foothold. For small and medium-sized enterprises and institutions, the structured main data (points) of their own are used as core data to expand (lines) according to the application scenario. For large enterprises and institutions, we can get more insights and valuable data information by further analyzing and applying the extended results from multiple application scenarios (surfaces). Data expansion based on application scenario is the basic unit of big data analysis and sharing. Following the "step by step" strategy, it expands from inside to outside, as shown in Fig. 1.

Figure 1 shows extended form of core data, peripheral related data, conventional channel data and socialized media data from inside to outside. The core data is the main data, which is the most important business database (such as CRM) of the enterprise or institution. The second layer is peripheral related data, which are structured data of other business departments inside the enterprise, and also include structured data that is related to core data after the special collection. The conventional channel data on the third layer are structured data outside the enterprise, and the enterprise can combine

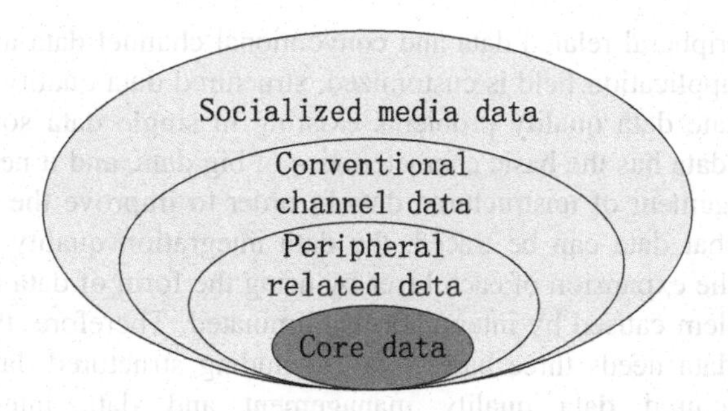

Fig. 1. General form of data extension based on application scene

with its own main data for marketing. Fourth layers socialized media data are unstructured data that enterprises can collect and can be used for promotion and innovation. When switching to another application scenario, the data sources involved are different, so the core data need to be coupled with different outer data to customize the new application scenario.

For example, an application scenario is the credit investigation of a college teacher. The core data is the basic information base of the college teachers. The peripheral related data may be teacher's salary records of school finance department. Conventional channel data may be derived from the record of the bank's registered income and loan (repayment). Socialized media data may be a number of online shopping records or the evaluation of a circle of friends, or even the rating data of sesame credit.

3.2 Big Data Quality Management Framework Based on Scene

According to the general form of data extension based on the application scenario, the corresponding data quality management framework is shown in Fig. 2.

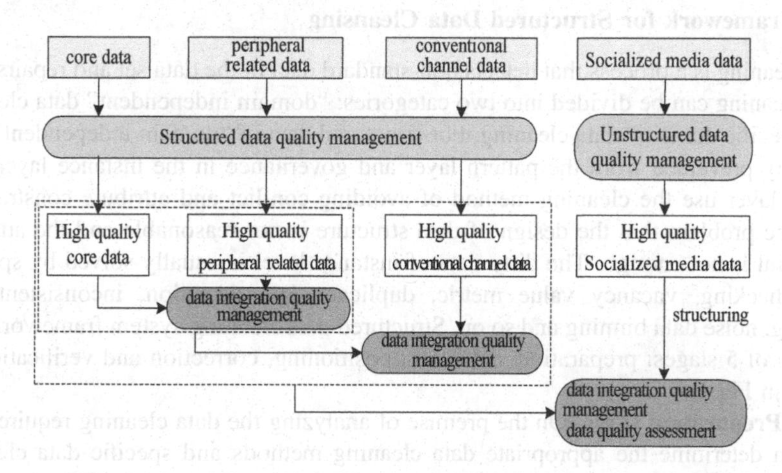

Fig. 2. Big data quality management framework based on scene

Core data, peripheral related data and conventional channel data are all structured data. Before the application field is customized, structured data quality management is needed to eliminate data quality problems existing in single data source itself. The socialized media data has the basic characteristics of big data, and it needs to carry out the quality management of unstructured data in order to improve the data quality. In order to ensure that data can be traced, the data integration quality management is carried out after the expansion of each layer by using the form of data integration. The data quality problem caused by integration is eliminated. Therefore, the quality management of big data needs three basic tasks, including structured data quality management, unstructured data quality management and data integration quality management. Finally, the data after data integration is standardized, data profiling and quality evaluation.

4 Structured Data Quality Management

For the quality management of structured data, it is only necessary to use the traditional data cleaning scheme under the scientific data collection mechanism.

4.1 Scientific Data Collection Mechanism

There are two ways to obtain the basic data, one is manual entry, and the other is the automatic collection of system. For manual entry, the accuracy and integrity of the data should be taken into consideration. In particular, the lack of data fields causes the data to be disconnected. For the data collected automatically, it is checked by a special tool and assisted by manual testing. Incomplete data are forbidden to appear in the standard database, preventing all kinds of low-level errors from the entry, so that errors in subsequent processing are constantly magnified.

4.2 Framework for Structured Data Cleansing

Data cleaning is a process that detects non-standard data in the data set and repairs data. Data cleaning can be divided into two categories: "domain independent" data cleaning and "specific domain" data cleaning. For structured data of "domain independent", it is generally prevented from the pattern layer and governance in the instance layer. The pattern layer use the cleaning method of avoiding conflict and attribute constraint to solve the problem that the design of data structure is not reasonable and the attribute constraint is not enough. The dirty data of instance level is usually solved by spelling error checking, vacancy value metric, duplicate data detection, inconsistent data cleaning, noise data binning and so on. Structured data cleaning system framework [40] consists of 5 stages: preparation, detection, positioning, correction and verification, as shown in Fig. 3.

(1) Preparation stage. On the premise of analyzing the data cleaning requirement, we then determine the appropriate data cleaning methods and specific data cleaning tasks. After setting up the data interface and other basic configurations, a complete set of data cleaning schemes have been formed. **(2) Detection stage.** After data

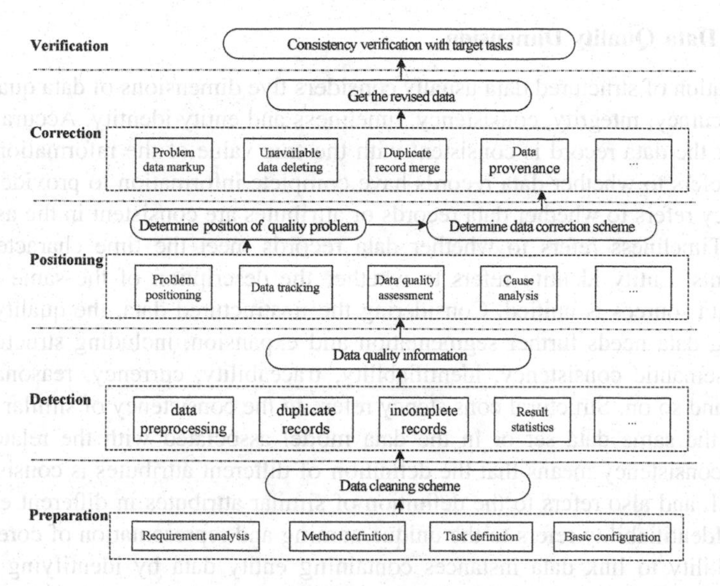

Fig. 3. Framework of structured data cleaning

preprocessing, we need to complete the detection of all kinds of problem data, such as duplicate records, incomplete records, logical errors and exception data, and statistically detect the results, in order to get comprehensive data quality information. **(3) Positioning stage.** The quality problem data are located and tracked, and data quality is evaluated based on detect results, and then we find out the causes of data quality problems, so as to determine the location of quality problems, and finally give the modification plan. **(4) Correction stage.** Based on the modification plan, we label problem data, delete the unavailable data, merge duplicate records, estimate and fill vacancy data, and conduct data provenance management. **(5) Verification stage.** Whether the revised data is consistent with the target task, if it is not consistent, it needs to return to the positioning stage for further positioning analysis and correction. The whole five processes can start from different stages, and can be used for customization.

5 Unstructured Data Quality Management

After obtaining relevant unstructured data based on scene, entity recognition and data link of unstructured data must first be carried out, so as to further reduce the scope of processing data. Considering that there are more quality dimensions in big data, the big data cleaning system based on semantics is used to clean and repair data in specific application scenarios, and the data after cleaning are structured. Finally, a real time control scheme for large data quality is given.

5.1 Big Data Quality Dimension

The evaluation of structured data usually considers five dimensions of data quality, that is, data accuracy, integrity, consistency, timeliness and entity identity. Accuracy refers to whether the data record is consistent with the true value of the information source. Integrity refers to whether data records have complete information to provide queries. Consistency refers to whether data records or attributes are consistent in the associated data set. Timeliness refers to whether data records meet the time characteristic of requirements. Entity identity refers to whether the description of the same entity in various data sources is unified. Considering the unstructured data, the quality dimension of big data needs further segmentation and expansion, including structural consistency, semantic consistency, identifiability, traceability, currency, reasonableness, authority and so on. Structural consistency refers to the consistency of similar attribute values in the same data set or in the data model associated with the related table. Semantic consistency means that the definition of different attributes is consistent in a data model, and also refers to the definition of similar attributes in different enterprise data sets. Identifiability refers to the unique naming and representation of core objects, and the ability to link data instances containing entity data by identifying attribute values. Traceability is used to describe the origin of data. Currency is used to measure the freshness of information. It can be measured by the expected refresh frequency function of data element, that is, whether the information is correct in the case of changing time and date value. Reasonableness refers to a comprehensive review of data consistency or rationality expectation in a time series. Authority refers to the official data sources issued by the government or the trustworthy enterprise. The suitable data quality dimension can be selected according to your own business needs.

5.2 Parallel Entity Recognition and Data Link Based on Hadoop

In order to implement data quality management for massive unstructured uncertain data, it is necessary to solve the problem of large-scale computing. The solution is to use parallel technology to process massive amounts of data. The Hadoop and MapReduce parallel computing frameworks are the first choice for current unstructured data processing. The entity recognition algorithm is used to parse and standardize the data values, and then compares them to determine whether the two records are directed to the same entity. For parallel entity recognition, we first calculate the similarity between records by attribute values, and then identify entity based on clustering method, and output the final result.

Data link is basically a matching operation. First, we select multiple attributes that can be used to uniquely identify entities or attributes, and then search all records with the same attributes. These records can form a candidate set that provides data for the new information. When a new record is merged into an existing set of data, a higher value can be obtained by combining multiple data sets. The focus of data consolidation is to view duplicate records and check the tightness of the other attributes of the record. Link are established only when the selected attributes are properly matched.

5.3 Big Data Cleaning System Based on Semantics

The vast majority of socialized media data are from Web pages, which belong to unstructured data. The big data cleaning system based on semantics mainly aims at these data. Its basic management process includes data grabbing, preprocessing, cleaning, quality assessment and monitoring feedback. First, we analyze the content of web pages, and then we dig out the semantic information expressed by web pages through massive data comparison and machine learning [41], so as to determine the quality and sensitivity of data. The system framework is shown in Fig. 4.

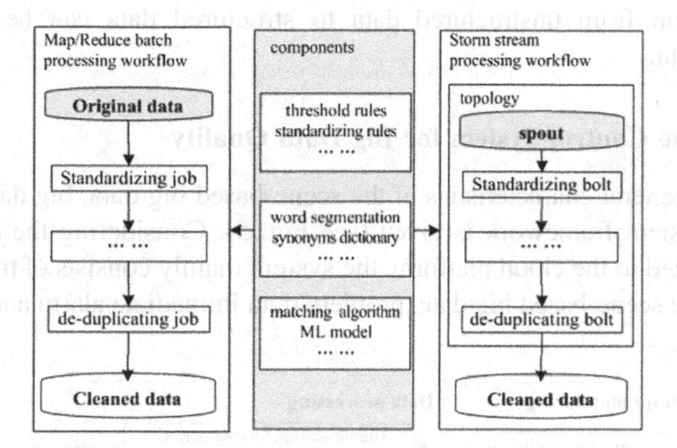

Fig. 4. Framework of big data cleaning system based on semantics

The semantic-based big data cleaning system can support real-time and non-real-time big data cleaning. The non-real-time big data cleaning mainly runs on the MapReduce batch processing workflow, while the real-time big data cleaning runs on the Storm flow data framework, which is mainly used to clean log files, Internet data and video streams. The non-real-time big data of various sources is stored in HDFS. The MapReduce data cleaning workflow can access the HDFS to obtain the original massive data. For real-time applications, the semantic-based big data cleaning system builds a topology to connect to these applications and uses spout to obtain persistent original stream data.

In semantic-based big data cleaning system, the rules of discovering and cleaning abnormal data are managed by Drools engine [42], and the relevant rules have threshold rules and standardization rules. The pattern matching and rule activation mechanism in Drools is implemented by the Rete algorithm [43]. The system can detect abnormal data such as error data, inconsistent data, and duplicate data. In order to improve the results of data duplication detection, the system uses a semantic-based keyword matching algorithm, which can complete the cleaning of repeated values by keyword extraction, keyword semantic similarity calculation and keyword matching. The parallel cleaning of big data is performed by a series of MapReduce or Bolt functions. During the whole data cleaning process, the log of the data changes is tracked by the audit. Finally, as a result of the output, the massive data after the cleaning is stored in HDFS or Storm framework.

5.4 Data Structure Based on Tetrahedron Model

For the description of unstructured data, tetrahedron data model [44] is proposed by Beihang University in 2010. The model consists of a vertex, four surfaces and edge between surfaces. The vertex represents the unique identifier of the unstructured data, the bottom represents the original data, and the three sides represent the basic attributes, semantic features and underlying features. The edges of each surface represent the relationship between the elements on each surface. The model provides unified, integrated and related descriptions for unstructured data of different types, and can support intelligent data services such as retrieval and data mining. Based on this data model, the direct conversion from unstructured data to structured data can be accomplished through metadata.

5.5 Real-Time Control System for Big Data Quality

In view of the general characteristics of the scene-based big data, big data quality real-time control system framework is given (see Fig. 5). Considering the collection data has been uploaded to the cloud platform, the system mainly consists of three parts: real-time audit of the scene-based big data, problem data immediate alarm and problem data

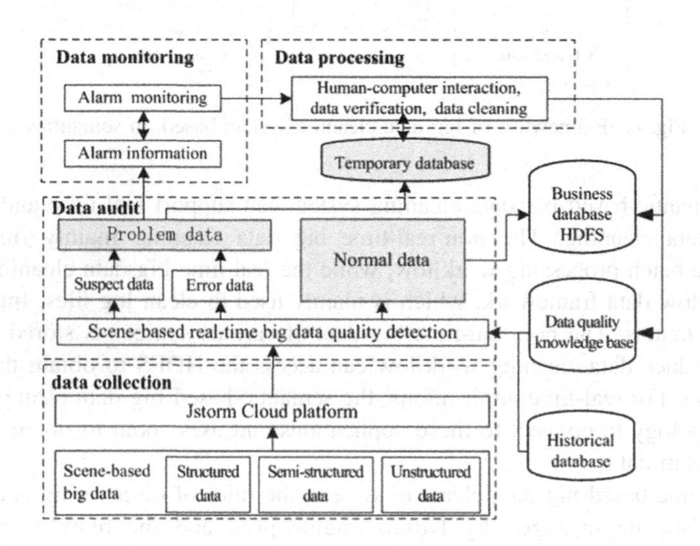

Fig. 5. Framework of real-time control system for scene-based big data quality

processing.

In Fig. 5, the big data real-time audit section can be used to detect the data that imported into the JStorm cloud computing platform. The normal data after real-time auditing is directly stored in the business database or HDFS, and the suspicious data and erroneous data identified will retain their records and store them in the temporary database, so as to perform data cleaning and other operations. The data quality dimension and audit rules required in the process of detection come from the data quality knowledge base. Real-time audits are used to examine the accuracy, integrity and consistency of the data. If the big data value of some scenes is difficult to obtain, the statistical results in the historical database can also be used as the actual value.

The immediate alarm of the problem data means that if the system monitors the problem data, it will automatically trigger the alarm function. The problem data will be saved to the temporary database waiting for processing, and the data quality control personnel are also informed by e-mail or SMS.

Problem data processing is a key part of improving the data quality. When the data quality control personnel receive the alarm information, the problem is extracted from the temporary database for verification and processing. If the data is repeated, it is deleted directly; if it is wrong data, it is corrected or deleted; if it is the other problem, it is further analyzed and processed. In general, data cleaning software is used to automatically processing, and manual method is used when necessary. Data that is only processed and consistent with the quality standards will be transferred from the temporary database to the business database or HDFS. In order to retain the historical experience, the solutions and experiences formed in the whole process will also be written to the data quality knowledge base.

6 Data Integration Quality Management, Data Profiling and Quality Evaluation

The quality management of data integration phase and later stage includes the construction of data quality model, standardization of integrated data, data profiling and data quality evaluation.

6.1 Data Quality Evaluation Model in Integration Stage

Data integration is a logical or physical centralization of data from different sources, formats, and properties to provide comprehensive data sharing for the enterprise. At the stage of data integration, data quality assessment should ensure that the data satisfy the specified data quality rules. Only when data are passed through these rules one by one, can we say that the integrated data is of high quality. The input of data quality assessment model is data quality dimension of data object, data quality rule and data object, and output is data quality dimension evaluation result. As shown in Fig. 6.

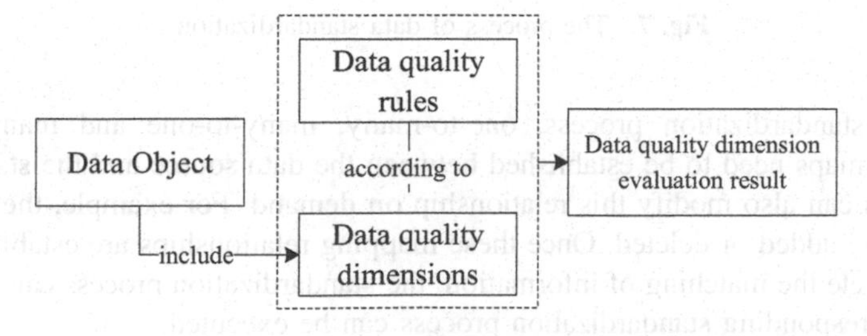

Fig. 6. The data quality assessment model

Data object refers to the data to be processed after integration, and its quality dimension includes data integrity, accuracy, consistency, timeliness and entity identity. These dimensions are the constraints of data objects on the data quality. Only when these constraints are satisfied can the quality of the data object be guaranteed. Data quality rules are the corresponding rules formulated to satisfy the data quality dimension. For example, for integrity constraints, data quality rules give nonempty rules and domain integrity rules. If the data object can satisfy these two rules, it indicates that the data quality dimension of the data object satisfies the integrity. The results of the data quality dimension assessment indicate the degree of satisfaction of the data in the quality dimension. Therefore, the data quality evaluation model can deal with the quality of the data object through data quality rules, and help to improve data quality.

6.2 Standardized Management of Integrated Data

In order to implement standardize data management, it is necessary to convert data of data sources into a unified standardized form, including standardized rules and standardized processes. The standardization rule contains all kinds of standardized rule information sets, and the standardized process is the combination of data source and standardization rule. The data standardization process is shown in Fig. 7.

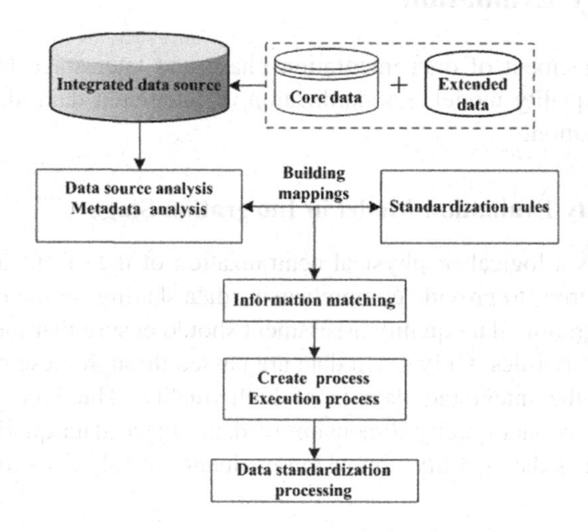

Fig. 7. The process of data standardization

In the standardization process, one-to-many, many-to-one and many-to-many association maps need to be established between the data source and the standardized rules. Users can also modify this relationship on demand. For example, the rules of a data field are added or deleted. Once these mapping relationships are established, that is, to complete the matching of information, the standardization process can be created and the corresponding standardization process can be executed.

6.3 Data Profiling

Data profiling can empirically examine potential data problems. Data profiling is mainly used in analysis of exception data and discovery of data quality rule. In order to realize the availability of data sets, we must first establish measurement datum of data set quality through data profiling. Exception analysis is to examine all data elements of the data set, check the frequency distribution and the relationship between data columns to reveal defective data value, data elements and records. The dependencies that data integration must adhere to are the data quality rules. Data profiling can be used to check data sets to identify and extract embedded business rules. However, these rules can be combined with predefined data quality expectation indicators to be used as a target for data quality audit and monitoring.

Through the results of data profiling, column exceptions, cross column exceptions, and cross table exceptions can be identified. The purpose of data profiling is to find rules consistent with data business expectations, including column-oriented rules, cross-column rule discovery and cross-table rule discovery.

6.4 Big Data Quality Evaluation System

In order to verify the quality of data after standardization, standardized data sets need to pass the corresponding quality dimensions, evaluation models and rules, so as to analyze the overall situation of the corresponding data sets in the data quality dimension. To evaluate the data quality, a scene-based standardized big data quality assessment system (see Fig. 8) was established.

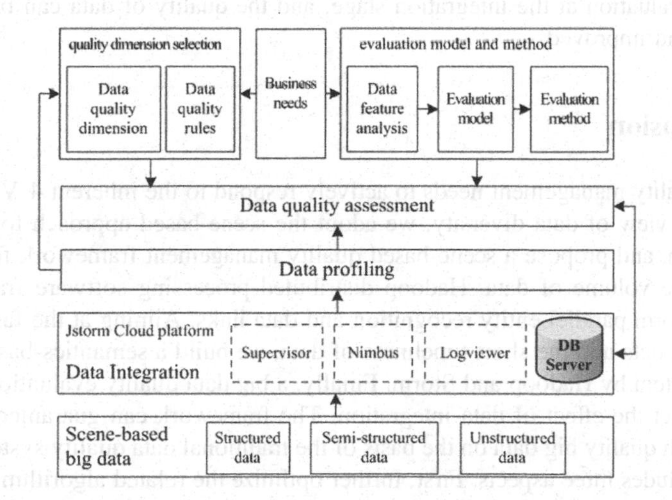

Fig. 8. Framework of scene-based big data quality assessment system

The data integration module collects big data from different scenarios and stores it in the database and file system of the cloud computing platform. Therefore, a JStorm cluster is built to provide data integration and processing capabilities, and the related data sources are stored in the MySQL server of JStorm cluster.

Data profiling module is responsible for statistics and data quality related information, including cardinal analysis, range analysis, type detection, fluctuation inspection and data distribution and so on. According to the results of data profiling, the appropriate data quality assessment standards and quality benchmarks can be selected for different data sources.

The selection of the quality dimensions is based on the specific requirements of a particular scene and the results of data profiling. The quality dimension selection module is responsible for determining the quality dimensions of the various data related to the scene in the evaluation.

The evaluation model and method module is responsible for analyzing the characteristics of all kinds of data in a scene. Based on the analysis results and the selected quality dimensions and their evaluation indicators, the evaluation model is established.

The data quality assessment module performs evaluation operations on the JStorm cloud platform based on the selected quality dimensions, evaluation models and evaluation methods. The data quality assessment results will be compared with the quality benchmark. If the results meet the quality benchmark, data quality is acceptable and subsequent processing is performed. Otherwise, data cleaning operation will be performed. After data cleaning is completed, quality evaluation will be performed again. If the benchmark is met, data will be retained for further data persistence. Otherwise, new data will be selected from the data source and continue to be evaluated.

In a word, the original data from various data sources will perform quality management of structured and unstructured data, and then perform data quality management and evaluation at the integration stage, and the quality of data can be gradually improved and improved.

7 Conclusion

Big data quality management needs to actively respond to the inherent 4 V features of big data. In view of data diversity, we adopt the scene-based approach to reduce the scale of data, and propose a scene-based quality management framework for big data. For the huge volume of data, Hadoop distributed processing software framework is used to perform parallel entity recognition and data links. Aiming at the fast changing speed of big data and the short timeliness of data, we build a semantics-based big data cleaning system by Hadoop and Storm. Finally, a big data quality evaluation system is built to detect the effect of data integration. The framework can guarantee the acquisition of high quality big data on the basis of the traditional data quality system. Further research includes three aspects. First, further optimize the related algorithm of big data cleaning system. Second, we verify the framework and compare with other systems through a large number of experiments in different datasets. Third, we should apply the quality management framework to more industries, and use the bottom-up data quality research method to explore the general quality standard system for big data gradually.

Acknowledgment. This work is supported by the National Natural Science Foundation of China 61602351, 61572221, U1401258 and 61433006, the Science and Technology Support Program of Hubei Province under grant 2015AAA013, the Natural Science Foundation of Hubei Province 2017CFB326, the Scientific Research Fund of Hubei Provincial Department of Education Q20141410, and the Innovation Fund of Hubei University of Technology BSQD2016019. We sincerely thank the anonymous reviewers for their very comprehensive and constructive comments.

References

1. Yang, W.L., Strong, D.M.: Knowing-why about data processes and data quality. J. Manag. Inf. Syst. **20**(3), 13–39 (2003)
2. Brüggemann, S., Appelrath, H.J.: Context-aware replacement operations for data cleaning. In: ACM Symposium on Applied Computing, pp. 1700–1704. ACM, New York (2011)
3. Han, J., Xu, L., Dong, Y.: An overview of data quality research. Comput. Sci. **35**(2), 1–5 (2008)
4. Kwon, O., Lee, N., Shin, B.: Data quality management, data usage experience and acquisition intention of big data analytics. Int. J. Inf. Manag. **34**(3), 387–394 (2014)
5. Moges, H.-T., Dejaeger, K., Lemahieu, W., et al.: A total data quality management for credit risk: new insights and challenges. Int. J. Inf. Qual. **3**(1), 1–27 (2012)
6. Zhang, G., Xi, Y., Liu, N.: Research on the evaluation index system of quality management in service industry based on lean six sigma. J. Hunan Univ. (Soc. Sci.) **30**(6), 79–84 (2016)
7. Zhimao, G., Zhou, A.: Research on data quality and data cleaning: a survey. J. Softw. **13** (11), 2076–2082 (2002)
8. Yu, J., Xu, B., Shi, Y.: The domain knowledge driven intelligent data auditing model. In: 2010 IEEE/WIC/ACM International Conference on Web Intelligence and Intelligent Agent Technology (WI-IAT), pp. 199–202. IEEE, Piscataway (2010)
9. Januzaj, E., Januzaj, V.: An application of data mining to identify data quality problems. In: Third International Conference on Advanced Engineering Computing and Applications in Sciences, ADVCOMP 2009, pp. 17–22. IEEE, Piscataway (2009)
10. Christen, P., Gayler, R.W., Tran, K.N., et al.: Automatic discovery of abnormal values in large textual databases. J. Data Inf. Qual. **7**(1–2), 1–31 (2016)
11. D'Urso, C.: Experience: glitches in databases, how to ensure data quality by outlier detection techniques. J. Data Inf. Qual. **7**(3), 1–22 (2016)
12. Beirami, M.H.N., Ghavifekr, M.H.N., Khajei, R.P.: Predicting missing attribute values using cooperative particle swarm optimization. J. Basic Appl. Sci. Res. **3**(1), 885–890 (2013)
13. Wu, A.: A framework of annotation-based query process over inconsistent data. J. Shanghai Maritime Univ. **34**(1), 84–89 (2013)
14. Meng, X., Du, Z.: Research on the big data fusion: issues and challenges. J. Comput. Res. Dev. **53**(2), 231–246 (2016)
15. Karakasidis, A., Verykios, V.S.: A sorted neighborhood approach to multidimensional privacy preserving blocking. In: 2012 IEEE 12th International Conference on Data Mining Workshops, pp. 937–944. IEEE, Piscataway (2012)
16. Agrawal, R.: A modified K-nearest neighbor algorithm using feature optimization. Int. J. Eng. Technol. **8**(1), 28–37 (2016)
17. Wang, H.: Big data quality management: problems and progress. Sci. Technol. Rev. **32**(34), 78–84 (2014)

18. Yu, W., Li, S., Yang, S., et al.: Automatically discovering of inconsistency among cross-source data based on web big data. J. Comput. Res. Dev. **52**(2), 295–308 (2015)
19. Clark, P.G., Grzymalabusse, J.W.: Mining incomplete data with lost values and attribute-concept values. In: IEEE International Conference on Granular Computing, pp. 49–54. IEEE, Piscataway (2014)
20. Tang, Z., Jiang, L., Yang, L., et al.: CRFs based parallel biomedical named entity recognition algorithm employing MapReduce framework. Cluster Comput. **18**(2), 493–505 (2015)
21. Christen, P.: A survey of indexing techniques for scalable record linkage and deduplication. IEEE Trans. Knowl. Data Eng. **24**(9), 1537–1555 (2012)
22. Fan, W., Geerts, F., Wijsen, J.: Determining the currency of data. In: Thirtieth ACM Sigmod-Sigact-Sigart Symposium on Principles of Database Systems, pp. 71–82. ACM, New York (2011)
23. Otto, B., Österle, H.: Toward a functional reference model for master data quality management. Inf. Syst. e-Business Manag. **10**(3), 395–425 (2012)
24. Bohannon, P., Fan, W., Flaster, M., et al.: A cost-based model and effective heuristic for repairing constraints by value modification. In: Proceedings of the 2005 ACM SIGMOD International Conference on Management of Data, pp. 143–154. ACM, New York (2005)
25. Li, M., Li, J.: Algorithms for improving data currency. J. Comput. Res. Dev. **52**(9), 1992–2001 (2015)
26. Berti-Equille, L., Ba, M.L.: Veracity of big data: challenges of cross-modal truth discovery. J. Data Inf. Qual. **6**(6), 144 (2016)
27. Li, X., Wang, F.: Research on data provenance's security model. J. Shandong Univ. Technol. (Nat. Sci. Edn.) **24**(4), 56–60 (2010)
28. Peiyu, J., Jun, C., Xin, X., et al.: Data traceability of large-scale sensor networks. J. Suzhou Univ. Sci. Technol. (Nat. Sci.) **30**(4), 55–59 (2013)
29. Yu, F.: The international experience of the quality assessment for the non-traditional data and their reference to China. Stat. Res. **34**(12), 15–23 (2017)
30. Mo, Z.: Analysis on influence of data quality in the procedure of big data processing. Modern Inf. **37**(3), 69–72 (2017)
31. Fürber, C., Hepp, M.: Towards a vocabulary for data quality management in semantic web architectures. In: Proceedings of the 1st International Workshop on Linked Web Data Management, pp. 1–8. ACM, New York (2011)
32. Sidi, F., Panahy, P.H.S., Affendey, L.S., et al.: Data quality: a survey of data quality dimensions. In: International Conference on Information Retrieval and Knowledge Management, pp. 300–304. IEEE, Piscataway (2012)
33. Zhou, N., Sheng, W., Liu, K., et al.: WR approach: determining accurate attribute values in big data integration. J. Comput. Res. Dev. **53**(2), 449–458 (2016)
34. Xiaoou, D., Hongzhi, W., Xiaoying, Z., et al.: Association relationships study of multi-dimensional data quality. J. Softw. **27**(7), 1626–1644 (2016)
35. Karvounarakis, G., Ives, Z.G., Tannen, V.: Querying data provenance. In: ACM SIGMOD International Conference on Management of Data, pp. 951–962. ACM, New York (2010)
36. Geisler, S., Quix, C., Weber, S., et al.: Ontology-based data quality management for data streams. J. Data Inf. Qual. **7**(4), 18 (2016)
37. Chapman, A.P., Rosenthal, A., Seligman, L.: The challenge of "quick and dirty" information quality. J. Data Inf. Qual. **7**(1–2), 1–4 (2016)
38. Huang, H., Tao, R., Fu, D.: Business register database revision: internet data sources and data quality assessment. Stat. Res. **34**(1), 12–22 (2017)
39. Zhao, X., Li, S., Yu, W., et al.: Research on web data source quality assessment method in big data. Comput. Eng. **43**(2), 48–56 (2017)

40. Jianjun, C., Xinngchun, D., Shuang, C., et al.: Data cleaning and its general system framework. Comput. Sci. **39**(s3), 207–211 (2012)
41. Da, H., Hui, C., Haitao, W., et al.: Discussion about quality management issues of web big data. Inf. Commun. Technol. **11**(4), 60–64 (2017)
42. Ordóñez, A., Eraso, L., Ordóñez, H., et al.: Comparing drools and ontology reasoning approaches for automated monitoring in telecommunication processes. Procedia Comput. Sci. **95**(2016), 353–360 (2016)
43. Forgy, C.L.: Rete: a fast algorithm for the many pattern/many object pattern match problem. Artif. Intell. **19**(1), 17–37 (1982)
44. Li, W., Lang, B.: A tetrahedral data model for unstructured data management. Sci. China Inf. Sci. **53**(8), 1497–1510 (2010)

The Construction Approach of Statutes Database

Linxia Yao[1,2], Haojie Huang[1,2], Jidong Ge[1,2](✉), Simeng Zhao[1,2], Peitang Ling[1,2], Ting Lei[1,2], Mengting He[1,2], and Bin Luo[1,2]

[1] State Key Laboratory for Novel Software Technology, Nanjing University, Nanjing 210093, China
gjdnju@163.com

[2] Software Institute, Nanjing University, Nanjing 210093, China

Abstract. The storage of statutes is a preliminary and essential work for Chinese text processing, which plays an important role in legal area. Nowadays, there are not specific databases for statutes. Most existing database of statutes focus on using the traditional methods and not put the focus-point on the judgment documents itself, which language features and writing formats are unique. In this paper, we proposed a strategy to construct database in the basis of statutes. We defined a series of rules under the study of structure on statutes. We chose the statutes with good formats and do some preprocessing before constructing database and proposed a method to solve the problem of granularity and wrong classification based on statutes. We defined several parameters according to the research on the judgement documents. Moreover, we combined the dependencies between statutes and found inner relationship between single statute. The experiments on the three types of law show that the result of the database we constructed obtains a good consequence and can store enough legal statutes.

Keywords: Chinese statutes · Database · Compiler theory

1 Introduction

Nowadays, the number of statutes are increasing at an exponential level. Along with the development of science and technology on data processing, which provides opportunities to the study on statutes. Researchers on the domain can have rich data of statutes and can easily access. In 2013, China Judgment Online System officially opened. Up to now, it has recorded more than 45 million electronic judgment documents and became the largest judgment document sharing website around the world.

The storage of statutes is a basic work and it is an essential preprocessing step for upper application of natural language processing on statutes in legal area. Nowadays, there is no specific technology of data storage on statutes which has its own language feature and writing format. So the work on how to store legal documents and what should be stored based on statutes have to be solved seriously.

In this paper, we proposed an approach integrating the language feature of statutes to construct database in the basis of statutes.

Q. Zhou et al. (Eds.): ICPCSEE 2018, CCIS 901, pp. 140–150, 2018.
https://doi.org/10.1007/978-981-13-2203-7_11

The contribution of this paper can be summarize as follows:

(1) Propose an method to do text proposing based on statutes, including making the laws standardization and converting laws into XML file.
(2) Collect a series of statutes for different cases.
(3) Combine the dependencies between statutes and inner relationship between single statutes.
(4) Propose a method to solve the problem of granularity and wrong classification based on statutes.
(5) Define several parameters according to the research on the statutes. Realize the methods we proposed and can run successfully with right result and obtain powerful storage capacity.

2 Related Work

The application and design of database plays an important role in natural language processing and there are many works on it. Batini et al. presented an introduction to database design, data modeling concepts, methodologies for conceptual design, view design, functional analysis for database design and so on [1]. Let us know more about database design. Navathe et al. addressed the vertical partitioning of a set of logical records or a relation into fragments and the rationale behind vertical partitioning is to produce fragments, groups of attribute columns [2]. Zilio et al. claimed their work is the very first industrial-strength tool that covers the design of as many as four different features, a significant advance to existing tools, which support no more than just indexes and materialized views [3]. Goelman et al. created a tool provides a visual introduction to important concepts in database design using Entity Relationship Diagrams as the primary visual design model, relating these same concepts to other models, such as UML diagrams and the crow's feet notation used in MySQL Workbench [4]. Rodriguez et al. explored modern technologies to implement geographical objects (polygons, points, polylines) that were to contain real-time information about geographical objects around our user (buildings, floors, classrooms). And we implemented a Node.js server to retrieve data from our MongoDB according to our user's current location and then handle that GIS information using the Google Maps API [5]. Brossier et al. used the existing bedside information system and network architecture of their PICU, implemented an ongoing high-fidelity prospectively collected electronic database, preventing the continuous loss of scientific information, which offers the opportunity to develop research on clinical decision support systems and computational models of cardiorespiratory physiology for example [6]. Alotaibi et al. proposed a novel normalization forms for relational database design that match the related data attribute, who proposed approach called Matching Related Data Attribute Normal Form (MRDANF) and that A civil registration database system is used as a case study to validate the proposed approach, which results show that using their proposed approach has the positive impact on database quality and performance as the data redundancy will be reduced [7]. Alzarka et al. claimed the HRFDCC Clinical Database they constructed is designed to capture the clinical variability within and among this set of disorders and ultimately to serve as a resource of well-phenotyped patients for future interventional studies [8]. Fleri et al. invented tools which include validated and benchmarked methods to predict MHC class I and class II binding and which predictions from these tools can be combined with tools predicting antigen processing, TCR recognition, and B cell epitope prediction [9]. Bouzeghoub et al. presented the implementation of SECSI, an expert system for database design written in Prolog and starting from an application description given with either a subset of the natural language, or a formal language, or a graphical interface, the system generates a specific semantic network portraying the application [10].

3 Approach

3.1 Overview

In this section, we describe the approach we proposed to construct database for statutes that we take on the statutes in detail. Subsection 3.1 presents an overview of the workflow of our approach. Subsection 3.2 describes the preprocessing work of statutes. Subsection 3.3 introduces the database we constructed and what it is. Subsection 3.4 describes our method on how to construct the database.

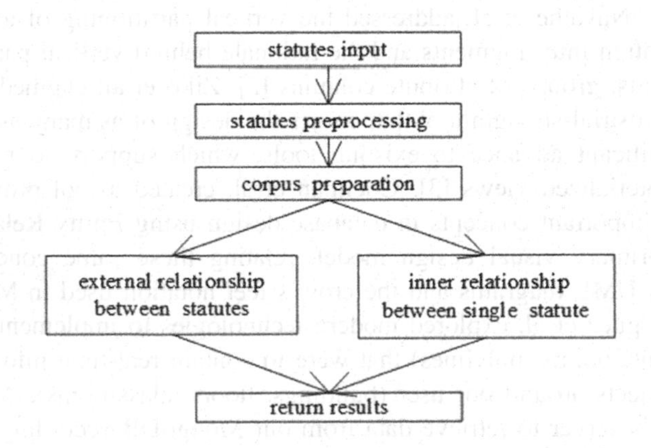

Fig. 1. Overview of the workflow of our approach on statutes

Figure 1 presents an overview of workflow for our work on statutes. The first work that we have to do is to collect statutes, which should have structured information with wide coverage and include different categories of cases. The original statutes are pure text format, in order to have a good result during the work followed, we have to do some preprocessing work on the statutes. In Subsect. 3.2 we will introduce the preprocessing work. In Subsect. 3.3 we will describe the content we constructed on statutes. In Subsect. 3.4 we present the methods we used in detail.

3.2 Preprocessing

In this section, we mainly focus on the methods we used to preprocess statutes for further research.

3.2.1 Prepare Corpus on Statutes

It is important to choose corpus that have good formats and with wide area. Statutes can be classified into three types: civil, criminal and administrative based on their case nature. Moreover, statutes can be classified into certain types by trail procedure. For example, first instance, second instance and so on. For this reason, we set up six kinds of XML templates for all types of statutes and prepare a series of statutes for different cases.

3.2.2 Make Statutes Standardization

In law, statutes are a decision of a court regarding the rights and liabilities of parties in a legal action or proceeding. Statutes documents also generally provide the court's explanation of why it has chosen to make a particular court order. According to the different formats of judgment documents, we convert the format of statutes into the formats of UTF-8 in unite. Moreover, we do some deal on the formats of the judgment documents, because the formats of documents may be different. Moreover, the formats of the judgement documents should comply with the following principles: first, in each chapter, the interpretations of legal clause have to be appeared on the top of the documents. And before the character "第", it should be with a space on each legal clause (see Fig. 2). Second, for the chaos format, we change the wrong format into right format on artificial way (see Fig. 3).

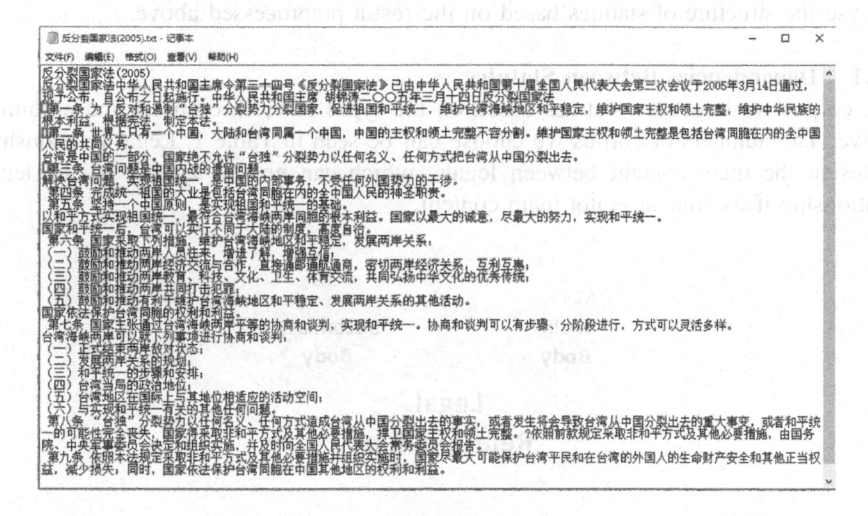

反分裂国家法(2005).txt - 记事本

文件(F) 编辑(E) 格式(O) 查看(V) 帮助(H)

反分裂国家法(2005)
反分裂国家法中华人民共和国主席令第三十四号《反分裂国家法》已由中华人民共和国第十届全国人民代表大会第三次会议于2005年3月14日通过，现予公布，自公布之日起施行。中华人民共和国主席 胡锦涛二〇〇五年三月十四日反分裂国家法
□第一条 为了反对和遏制"台独"分裂势力分裂国家，促进祖国和平统一，维护台湾海峡地区和平稳定，维护国家主权和领土完整，维护中华民族的根本利益，根据宪法，制定本法。
□第二条 世界上只有一个中国，大陆和台湾同属一个中国，中国的主权和领土完整不容分割。维护国家主权和领土完整是包括台湾同胞在内的全中国人民的共同义务。
台湾是中国的一部分。国家绝不允许"台独"分裂势力以任何名义、任何方式把台湾从中国分裂出去。
□第三条 台湾问题是中国内战的遗留问题。
解决台湾问题，实现祖国统一，是中国的内部事务，不受任何外国势力的干涉。
 第四条 完成统一祖国的大业是包括台湾同胞在内的全中国人民的神圣职责。
 第五条 坚持一个中国原则，是实现祖国和平统一的基础。
以和平方式实现祖国统一，最符合台湾海峡两岸同胞的根本利益。国家以最大的诚意，尽最大的努力，实现和平统一。
国家和平统一后，台湾可以实行不同于大陆的制度，高度自治。
 第六条 国家采取下列措施，维护台湾海峡地区和平稳定，发展两岸关系：
（一）鼓励和推动两岸人员往来，增进了解，增强互信；
（二）鼓励和推动两岸经济交流与合作，直接通邮通航通商，密切两岸经济关系，互利互惠；
（三）鼓励和推动两岸教育、科技、文化、卫生、体育交流，共同弘扬中华文化的优秀传统；
（四）鼓励和推动两岸共同打击犯罪；
（五）鼓励和推动有利于维护台湾海峡地区和平稳定、发展两岸关系的其他活动。
国家依法保护台湾同胞的权利和利益。
 第七条 国家主张通过台湾海峡两岸平等的协商和谈判，实现和平统一。协商和谈判可以有步骤、分阶段进行，方式可以灵活多样。
台湾海峡两岸可以就下列事项进行协商和谈判：
（一）正式结束两岸敌对状态；
（二）发展两岸关系的规划；
（三）和平统一的步骤和安排；
（四）台湾当局的政治地位；
（五）台湾地区在国际上与其地位相适应的活动空间；
（六）与实现和平统一有关的其他任何问题。
 第八条 "台独"分裂势力以任何名义、任何方式造成台湾从中国分裂出去的事实，或者发生将会导致台湾从中国分裂出去的重大事变，或者和平统一的可能性完全丧失，国家得采取非和平方式及其他必要措施，捍卫国家主权和领土完整。?依照前款规定采取非和平方式及其他必要措施，由国务院、中央军事委员会决定和组织实施，并及时向全国人民代表大会常务委员会报告。
 第九条 依照本法规定采取非和平方式及其他必要措施并组织实施时，国家尽最大可能保护台湾平民和在台湾的外国人的生命财产安全和其他正当权益，减少损失；同时，国家依法保护台湾同胞在中国其他地区的权利和利益。

Fig. 2. Right format of statutes

Moreover, there are other wrong formats on statutes, such as the input method is not uniform, some are full-width and some are half-width. Under this circumstances, we make it standardization with the formats of half-width.

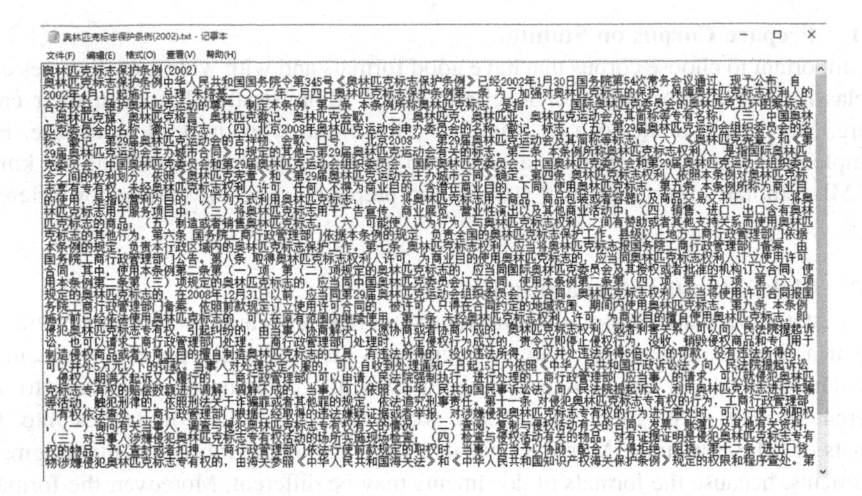

Fig. 3. Wrong format of statutes

3.3 Construct What

Nowadays, it has great achievements on natural language processing, in this section we present the content we decided to construct on legal area and the method we chose to analyse the structure of statutes based on the result preprocessed above.

3.3.1 Dependencies Between Statutes

The corpus we select are statutes based on the types of civil, criminal and administrative. The numbers of corpus we choose can be seen in Table 1. Legal relationship represent the main content between legals, which can not be a party to the legal relationship if its content is not main content.

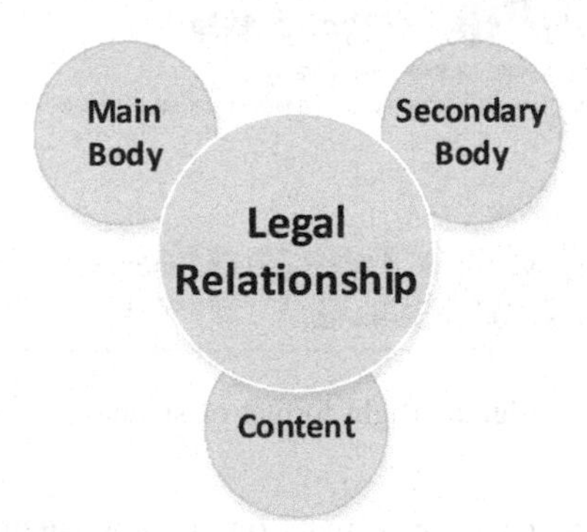

Fig. 4. The relationship between statutes

Things can not become the main body of law, because the main body of law must have the will to enjoy the benefits provided by the law and to fulfill its obligations under the law. Therefore, the relationship between people and things in property rights relationship is not a legal relationship, but the relationship between rights subject and rights object. Property relationship is the relationship between man and the right holder of the right to enjoy the property. Based on the principle, we defined the relationship between judgement documents as following construction principles which can be seen in Fig. 4.

From Fig. 4, we can see that legal relationship consists of main body, secondary body and content. Personally, we defined main body is consist of citizens, legal persons and other organizations. The content of the legal relationship includes rights and obligations. The object of legal relationship means the object to which the rights and obligations of subjects of legal relationship are directed. It can be divided into: things, behavior, personality interests, intellectual achievements (see Table 1).

Table 1. Position vector for every character

Main body	Content	Secondary body
Citizens Legal persons Other organizations countries	Rights Obligations	Things Behaviors Personality interests intellectual achievements

3.3.2 Inner Relationship Between Single Statutes

Statutes are the forms of legal rules, which are the content of statutes. The relationship between legal rules and statutes corresponds generally. It may have the following principles:

Table 2. Key elements on statutes

Number	Main element	Secondary element
1	Legal category	Legal category name
		Documents release time
		Documents note
2	Law article	Legal category
3	Legal chapter	Chapter order
		Chapter name
		Legal code id
4	Legal section	Section order
		Section name
		Legal chapter id
		Legal section id
5	Statutes	Clause order
		Clause name
		Chapter
		Section
6	Legal funds	Payment order
		Payment name
		Clause id

(1) A complete legal rule is expressed by several statutes.
(2) The contents of legal rules are respectively expressed by the statutes of different normative legal documents.
(3) A clause expresses different legal rules or their elements.
(4) The law provides only one element or several elements of the rule of law.

Based on the principles above and the study we do on the statutes, we combined a series of key elements on statutes. The results can be seen in Table 2.

3.4 How to Construct the Database We Want

In this section, we introduced the method we used in detail based on statutes. In the basis of the description about the content we choose to introduce above, we propose several methods to construct it.

3.4.1 Introduce the Tools We Used to Construct Database

We choose MySQL database for further work, which is an open source relational database management system (RDBMS) for database management. Which use the most commonly used database management language, structured query language (SQL). MySQL is an open source, so anyone can download it under the general public license and make changes to it if needed.

MySQL is gaining traction because of its speed, reliability and adaptability. Most people think MySQL is the best way to manage content without the need for transactional processing [11]. We choose the method of mvn to do some configuration and use MyEclipse to import the project based on mvn. It makes the build process easier, provide a unified build system, high-quality project information, best practice guidelines for development and can seamlessly add new features.

And for the unique of the number of each statue, we use UUID to make a coding. Which is a standard for software construction and is also part of the open software foundation organization in the field of distributed computing environments [12]. Its purpose is to allow all elements in the distributed system to have unique identification information without the need to specify the identification information through the central control terminal.

3.4.2 The Solution on Granularity and Wrong Classification on Laws

The Chinese statutes are consist of parts, chapters, articles, paragraphs, subparagraphs and items. Nowadays, most methods to construct database on Chinese statutes are under the analyse and locking of articles, which have omissions when dealing with the upper natural language processing. The method we proposed is based on the items which is the smallest unit of statues, when scanning statues we do analysis on the items and proposed the method combining the idea of principles based on LL1 analysis of compiler theory and text markup language. The method can be summarized as follows.

(1) Read two lines when scanning statues and according to the beginning of the next line to determine the current state of the line.
(2) Distinguish chapters or articles: the end of a section must be a colon or period. The chapter's ending is Chinese characters.

(3) Distinguish articles or paragraphs: exclude subparagraphs and items, in the same articles are paragraphs.
(4) Distinguish subparagraphs or items: through the shortest match under the beginning of the subparagraphs and items.

We also defined special marks, such as @Para@, @Item@ and so on, which can be used to represent some words like the components of the statues, but actually they are just the ordinary vocabularies, the circumstances like these may produce ambiguities. So we make a mark for these words and then put them into the buffer pool.

3.4.3 Introduce the Method We Used in Detail

Based on the rules we proposed above, we decided to define some parameters to represent the relationships between documents (see Table 3). Based on the parameters we defined, we create database table.

Table 3. Parameters of key elements on statutes

Number	Main element	Main element definition	Secondary element	Secondary element definition
1	Legal category	pub_fllb	Legal category name	mc
			Documents release time	fbsj
			Documents note	bz
2	Law article	pub_flb	Legal category	lbid
3	Legal chapter	pub_flz	Chapter order	zsx
			Chapter name	zmc
			Legal code id	bid
4	Legal section	pub_flj	Section order	jsx
			Section name	jmc
			Legal chapter id	zid
			Legal section id	flz
5	Statutes	pub_fltw	Clause order	twsx
			Clause name	twmc
			Chapter	zid
			Section	jid
6	Legal funds	pub_flkx	Payment order	kxsx
			Payment name	kxmc
			Clause id	twid

4 Experiments

The corpus we select are statutes based on the types of civil, criminal and administrative. The numbers of corpus we choose can be seen in Table 4.

Table 4. The corpus we choose

Corpus type	Corpus counts
Civil	231067
Criminal	220890
Administrative	203487

Based on the documents processed before, we convert the statutes into database with a database table as shown in Table 5. And the judgement documents saved in the database is shown in Fig. 5.

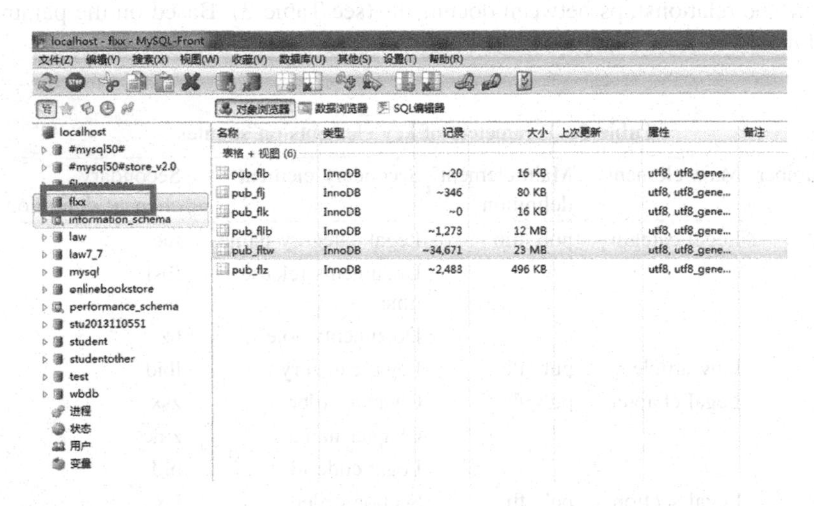

Fig. 5. The result presented

Table 5. Key elements on statutes

1	2	3	4	5	6
pub_fllb	pub_flb	pub_flz	pub_flj	pub_fltw	pub_flkx

The pseudocode we use the coding language of Java, the pseudocode can be seen in Table 6.

Table 6. The pseudocode of LawConvert

```
Code: LawConvert
input: Both path and name of the document which is ready to convert.
output: A complete segmentation of law and storage of DAO.
begin
1:   lawlist ← GETLIST(lawtext)
2:     for law : lawList do TRIM(law)
3:         i ← law:indexOf(firstSpace)
4:         maxLength law ← length
5:         title ← law.subString(0, i)
6:         content ← law.subString(i + 1, maxLength)
7:         if title ← contains("Part" OR "Chapter" OR "Section") then
8:             id ← UUIDGEN(())
9:             obj ← new DaoObj(id, title, content,rootId)
10:            Dao.save(obj)
11:        else if title.contains("Article") then
12:            id ← UUIDGEN(())
13:            obj ← new ArticleDaoObj(id, title, content)
14:            Dao.save(obj)
15:            if EXIST("Paragraph", content) then
16:                paraList[] ← PARSE("Paragraph", content)
17:            else
18:                paraList[0] ← law
19:            end if
20:            for para : paraList do
21:            if EXIST("SubParagraph", para) then
22:                subparaList[] PARSE("SubParagraph", para)
23:                for subpara : subparaList do
24:                    if EXIST("Item", subpara) then
25:                        itemList[] ← PARSE("Item", subpara)
26:                        for item : itemList do
27:                            obj ← new ItemObj(UUIDGEN(),
                               item.title, item.content, item.root)
28:                        end for
29:                    end if
30:                    obj ← new SubparaObj(UUIDGEN(),
                       subpara.title, subpara.content, subpara.root)
31:                end for
32:            end if
33:            obj ← new ParaObj(UUIDGEN(),
                 para.title, para.content, para.root)
34:           end for
35:        end if
36:     end for
end
```

5 Conclusion

In this paper, we proposed an approach to construct database in the basis of statutes. Natural language processing in statutes plays an important role in the legal area. On the basis of the language rules of statutes, We defined a series of rules of statutes. We do some preprocessing work before constructing database for a good corpus and defined several parameters according to the research on the statutes. In addition, we combined the dependencies between statutes and inner relationship between single statutes and proposed the solution on granularity and wrong classification on laws.

Moreover, experiments show that the result of the database we constructed obtains a good consequence and can store enough legal statutes.

Acknowledgements. This work was supported by the National Key R&D Program of China (2016YFC0800803).

References

1. Batini, C., Ceri, S., Navathe, S.B.: Conceptual Database Design: An Entity-Relationship Approach. Addison-Wesley, Boston (1991)
2. Navathe, S., Ceri, S., Wiederhold, G., et al.: Vertical partitioning algorithms for database design. ACM Trans. Database Syst. **9**(4), 680–710 (1984)
3. Zilio, D.C., Rao, J., Lightstone, S., et al.: DB2 design advisor: integrated automatic physical database design. In: Proceedings of the Thirtieth International Conference on Very Large Data Bases, VLDB Endowment, vol. 30, pp. 1087–1097. Morgan Kaufmann (2004)
4. Goelman, D., Dietrich, S.W.: A Visual introduction to conceptual database design for all. In: Proceedings of the 49th ACM Technical Symposium on Computer Science Education, pp. 320–325. ACM (2018)
5. Rodriguez, J., Malgapo, A., Quick, J., et al.: Distributed architecture of mobile gis application using NoSQL database. J. Comput. Sci. Coll. **33**(3), 68 (2018)
6. Brossier, D., El Taani, R., Sauthier, M., et al.: Creating a high-frequency electronic database in the PICU: the perpetual patient. Pediatr. Crit. Care Med. **19**, 189–198 (2018)
7. Alotaibi, Y., Ramadan, B.: A novel normalization forms for relational database design throughout matching related data attribute (2017)
8. Alzarka, B., Morizono, H., Bollman, J.W., et al.: Design and implementation of the hepatorenal fibrocystic disease core center clinical database: a centralized resource for characterizing autosomal recessive polycystic kidney disease and other hepatorenal fibrocystic diseases. Front. Pediatr. **5**, 80 (2017)
9. Fleri, W., Paul, S., Dhanda, S.K., et al.: The immune epitope database and analysis resource in epitope discovery and synthetic vaccine design. Front. Immunol. **8**, 278 (2017)
10. Bouzeghoub, M., Gardarin, G., Métais, E.: Database design tools: an expert system approach. In: Proceedings of the 11th International Conference on Very Large Data Bases, VLDB Endowment, vol. 11, pp. 82–95 (1985)
11. Chen, P.: The entity-relationship approach to logical database design (1991)
12. Leach, P.J., Mealling, M., Salz, R.: A universally unique identifier (UUID) URN namespace (2005)

Weighted Clustering Coefficients Based Feature Extraction and Selection for Collaboration Relation Prediction

Jiehua Wu[1,2](✉)

[1] School of Computer and Information Engineering, Guangdong Polytechnic of Industry and Commerce, Guangzhou 510510, China
kodakwu@126.com

[2] School of Computer Science and Engineering, South China University of Technology, Guangzhou 510641, China

Abstract. Existing methods of scientific collaboration prediction often use one-step weighted attribute of the common neighbor to construct associated feature. Such methods make no distinction between the different contributions of different neighbors and they cannot effectively deal with the redundant information between the features. Consequently, a general feature-based framework for weighted scientific network relation prediction is proposed. The framework is based on the heuristics of the Naive Bayes link prediction model. Firstly, we introduce weighted clustering coefficients to define several weighted Naive Bayes features, then use the mRMR feature selection method to deal with the information between the SE relevant features. Extensive experiments on real-world scientific network datasets demonstrate that the effectiveness of our proposed framework. Further experiments are conducted to understand the impacts of leveraging of extraction and selection of the proposed weighted features.

Keywords: Weighted network · Feature extraction
Feature selection · Relation prediction · Link prediction
Clustering coefficients

1 Introduction

Scientific Collaboration Network (SCN) [1], which may also be referred to academic network or bibliographic network, is a kind of complex network [2]. It analyzes various types of research academic cooperation from the perspective of network interconnection such as Google Scholar, CNKI Scholar Network and Arxiv preprint network etc. SCN is mainly divided into two types: research collaboration network and paper reference network. In research collaboration network, nodes is researcher and the links between them mean they published a paper with collaboration. In papers reference network, paper is node and their reference relation formulate a link. Thus, as a basic element, the relationship

Q. Zhou et al. (Eds.): ICPCSEE 2018, CCIS 901, pp. 151–164, 2018.
https://doi.org/10.1007/978-981-13-2203-7_12

between nodes plays an important role in the SCN analysis. One of the hottest research directions is the research relationship prediction (link prediction), which is of great value both in the applied and theoretical aspects [3]. Judging from the application level in the paper reference network, there are many commonly quoted papers between two papers that have not yet been quoted. The predictive algorithm can be used to judge whether there is any intentional non-quotation or false plagiarism between these two papers. In the research collaboration network, it is possible to find out the possibility of future cooperation between collaborators who do not yet have a cooperative relationship by the inferring and predicting technique. Such results can effectively narrow the distance between academic circles and communities and promote cooperation [4]. At theoretical level, since the evolution of research networks is driven by the establishment of relationships among entities, predicting relationships helps to grasp the changing patterns of SCN and discovers which areas of research are becoming more and more popular and which academic circles collaborate more and more closely [5].

In order to realize the above requirements, it is necessary to develop a efficient and effective prediction algorithm to determine whether there exists a node that has not currently generated a relationship (link) [6]. The relation prediction methods can be categorized into similarity based methods and learning based methods [7]. Our work is focused on learning based methods. The learning based methods try to incorporate machine leaning theory into link prediction, on of the most common one is the link classification method based on SCN features. The core idea of this algorithm [8] is that whether the potential links of the network are formed or not is treated as a binary classification problem.

However, many existing methods [8,18,24] are all based on the unweighted network. In reality, the links between network nodes are likely to have an interaction strength. We call this weighted network. The network of research collaborators is a classic weighted network. If we regard it as an unweighted network, the links only reflect whether the two scientists have the relationship of collaborating on published papers. If we regard it as weighted network, the links will reflect how many papers have been published in cooperation. Therefore, there is a growing trend recently to adopt the methods in weighted scene [9,10]. However, there still exist some potential issues that need further concerns. On one hand, researchers with a close research circle is very willing to create opportunities for cooperation intuitively. These features do not reflect the density information of the network's local structure. On the other hand, lacking feature selection to process the feature redundancy with similar structure information and choose more discriminatory features. To be specific, nearly all the features are based on common neighbors, and cannot be directly applied to the classification model. In contrast, feature selection methods which are able to achieve the above functions are extensively required in real-world applications.

In view of the above problems, this paper proposes a relational classification framework which jointly preserves the process of feature extraction and feature selection. The illustration of our framework is shown in Fig. 1. And our main contributions can be summarized as follows:

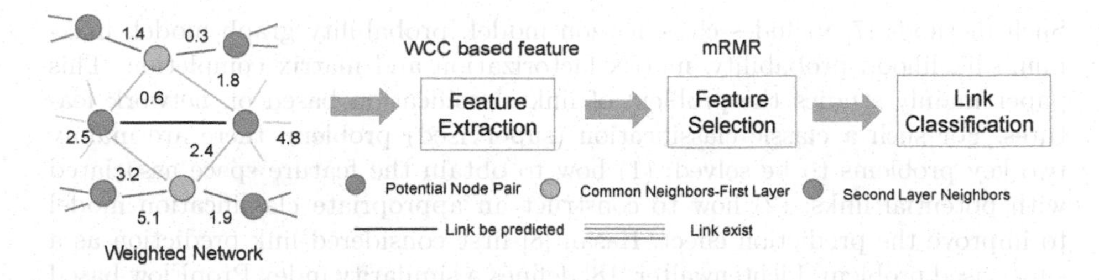

Fig. 1. Proposed link classification model. (Color figure online)

1. Novelly analyzing the SCN from the perspective of weight information, and we propose a general classification model for relation prediction.
2. At the stage of feature processing, the definition of weighted clustering coefficient is given first, and then based on this definition, a weighted naive bayes model based weighting feature is constructed to measure the weight density of local structure.
3. In the feature selection phase, mRMR (Max-Relevance Min-Redundancy) algorithm is introduced to process the redundant information between features to filter out the most discriminative features.
4. Experiments in several real SCN show that the proposed framework is effective and robust.

2 Related Works

Link prediction techniques are mainly divided into two kinds: similarity methods based on the network structure information and methods based on statistical learning. The idea of first kind of methods is to calculate the similarity between two potential nodes by analyzing the structure information of the network, and using the similarity as the probability of whether it generates a link. The greater the similarity is, the more likely it is to generate the link. It is mainly divided into several categories: (1) Common Neighbors based methods (Common Neighbors (CN) [6], Adamic Adar (AA) [6], Tree Augmented Naive Bayes (TAN) [11]). (2) Path based methods (Meta Path (MP) [29], Local Path (LP) [12]). (3) Random Walk based methods (Asymmetric Local Random Walk (ALRW) [14], Superposed Random Walk (SRW) [15]). (4) Global structural information based methods (Structural Perturbation Method (SPM) [16], Low Rank matrix completion (LR) [17]) etc. Their advantages are fast speed and high efficiency, so some progress has been made in recent years. However, as the research on similarity methods continues to deepen and the evolution and scale of such network structures continue to expand, how to deeply tap the influence of SCN structure on link formation is still a worthwhile study.

Another kind of methods are based on the idea of statistical learning which calculates the probability of link generation between two potential nodes just by using the model in the field of machine learning and data mining.

Such methods [7] includes classification model, probability graph model, maximum likelihood probability, matrix factorization and matrix completion. This paper mainly studies the problem of link classification based on network features. For such a classic classification (supervised) problem, there are mainly two key problems to be solved: (1) how to obtain the feature space associated with potential links; (2) how to construct an appropriate classification model to improve the prediction effect. Hasan [8] first considered link prediction as a supervised problem. Lichtenwalter [18] defines a similarity index PropFlow based on the flow information, and then builds HPLP (High Performance Link Prediction) classification model based on Bagging and Random Forest classifier to effectively solve the problem of network sparsity and sample imbalance. Scellato [19] found that 30% of the links generated in the location information network are related to their geographical location, and then they acquired various types of features such as social, location and global, and predicted the relationship between users and users and between users and locations. Similar works [20,21] were also presented.

However, the above algorithms aim to predict link in the unweighted network, but there is many weighted networks in the real world. De Sá [10] put forward a classification algorithm based on weighted attribute structure of topology with weighting. This algorithm builds the feature by integrating the weight information into the index of similarity of local co-adjacent nodes, For better results, Lu [22] points out that Weak-Ties theory plays an important role in the weighting of network players and proposes a method for defining the similarity index based on weighted information. On the basis of Weak-Ties theory, Lin [23] introduced a method to sort the authority of nodes, and BenefitRanks defined the similarity between nodes to achieve better results. In addition, Zhu [24] constructs a new link prediction index WMI based on Weighted Mutual Information, which can be effectively extended to many weighted similarity algorithms and achieves some improvement effects.

3 Proposed Model

3.1 Problem Definition

A weighted SCN can be defined as a weighted graph: $G = (V, E, W)$. It consisting of a node set or vertex set V, an edge set E, and a weighted value that with each edge associates two nodes. All values formulates a weighted matrix W. When nodes u and v are the endpoints of an edge, e_{uv} defines an edge and w_{uv} is its weighted.

In the link classification problem, a G is partitioned into a training network G_T and a predicting network G_P. There is a link mapping function $\psi : E \rightarrow Y$. It means each exist or non-exist links in G_T belongs to a positive (+1) or negative label (−1) respectively. Further, each link e_{uv} is associated with a feature vector $\theta(u, v)$ and all the feature in G_T is defined as Θ. Then the goal is to utilize a classification model to learn a function and predict the label in G_P:

$$f_{\Theta}(G_P|G_T) \rightarrow Y \tag{1}$$

3.2 Feature Extraction Based on Weighted Clustering Coefficient

A common neighbor node is a set of nodes that are connected with two nodes. Because of its direct connection with potential node pairs, many weighted network link classification algorithms are based on such similarity structure definition. The representation features in literature are (Weighted CN-WCN), (Weighted AA-WAA), (Weighted RA-WRA) etc. There are two problems in the above features: (1) The algorithm is only a superposition of the topological properties, and does not deeply reflect the generation and evolution trend of the local structure on the network. (2) The above algorithms are the first layer (one-step) common-node nodes of the candidate node pairs, and do not further expand the local influence range, it is possible to ignore the weight information of the neighbor nodes. The left part of Fig. 1 depicts the relevant methods of local structure diagram. Two potential nodes are blue and their neighboring nodes are yellow. Orange represents adjacent nodes of the node adjacency node (the second node). The above features are based on blue and yellow nodes' topological properties of the link weight (gray light line and green line), ignoring the nodes attributes, and orange and yellow orange node link weights (red line). It is essential to introduce a broader structure framework of computing the dense local similarity. Therefore, the concept of weighted clustering coefficient [25] is proposed to solve this problem.

The clustering coefficient c_u denotes the link cohesion of node u and its neighboring nodes, which is used to quantify the density of local structures in complex networks [26]. The clustering coefficient generally decreases with the increase of the vertex degree and is considered as a symbol of the network hierarchy. It is defined as the number of triangles formed by any node u and the maximum number of triangles that u may participate in forming ratio:

$$c_u = \frac{2t_u}{k_u \cdot (k_u - 1)} \tag{2}$$

Where k_u is the node degree of node u. If $c_u = 0$ indicates that the node has no neighbor, $c_u = 1$ indicates that all the neighboring nodes are closely connected. Many studies have pointed out that the trend of forming links between adjacent nodes is higher than the distance between nodes, and the triangular structure formed by local nodes is very useful in estimating the similarity between nodes. However, Eq. (2) is based solely on the non-ownership of the network, irrespective of the fact that weights may reflect that some of the neighbors are more important than others. In order to solve this problem, we combine the topology information with the weight distribution of the network to define a weighted clustering measure [17], that is, weighted clustering coefficient $c_u^{\mathbf{w}}$:

$$c_u^{\mathbf{w}} = \frac{1}{p_u(k_u - 1)} \sum_{(v,w)} \frac{\mathbf{w}_{vu} + \mathbf{w}_{uw}}{2} a_{vw} a_{vu} a_{uw} \tag{3}$$

Where $a_{uw} = 1$ indicates that there is a link between nodes u and v. This coefficient is a measure of the local cohesion of the network, which fully considers the importance of the local clustering structure, that is, based on the actual interaction strength and information flow of nodes on the local triangle structure. In this way, we consider not only the number of closed triplets near the vertices, but also the total relative weights of their intensities relative to the vertices. The normalization factor $p_u(k_u - 1)$ takes into account the weight of each edge times the maximum possible number of triples it can participate in, and guarantees $0 \leq c_u^{\mathbf{w}} \leq 1$ [27].

Then we can draw the conclusion that the similarity index of the weighted clustering coefficient of the integrated neighbor nodes is defined as:

$$WCC(u,v) = \sum_{\omega \in CN(u,v)} c_\omega^{\mathbf{w}} \tag{4}$$

3.3 Feature Extension

The purpose of weighted clustering coefficients is to introduce two-step weighting information, but not to differentiate to define the contribution of a neighbor node. Therefore, this section proposes to introduce weighted clustering coefficients into the newly proposed Local Naive Bayes (LNB) [28], and use a weighted LNB similarity definition method [32] for constructing more discriminative features. The present work is quite different from [32] as our proposed methods is a classification-based prediction method and it also incorporate a feature selection process.

The LNB assumes that the link generation probability depends conditionally on the local public neighbor. In such a model, two adjacent potentially coordinated links of the same size may have very different link generation possibilities, which may be defined as:

$$s_{u,v}^{LNBCN} = \frac{P(e_{uv})}{P(\overline{e_{uv}})} \prod_{\omega \in N(u,v)} \frac{P(\overline{e_{uv}})}{P(e_{uv})} \prod_{\omega \in N(u,v)} \frac{c_\omega}{1 - c_\omega} \tag{5}$$

Since c_ω is a non-weighted clustering coefficient, the definition of this formula is obviously based on the construction of unweighted network and can not calculate the local link weight information. In order to further expand the similarity in LNB to calculate the weighted network, c_ω is replaced by $c_\omega^{\mathbf{w}}$, so the formula of Naive Bayesian CN-based weighted network is as follows:

$$s_{u,v}^{LNBCN} = \frac{P(e_{uv})}{P(\overline{e_{uv}})} \prod_{\omega \in N(u,v)} \frac{P(\overline{e_{uv}})}{P(e_{uv})} \prod_{\omega \in N(u,v)} \frac{c_\omega^{\mathbf{w}}}{1 - c_\omega^{\mathbf{w}}} \tag{6}$$

Similarly, the WLNBCN model can be extended to AA and RA algorithms available:

$$s_{u,v}^{WLNBAA} = \mathcal{C} \sum_{\omega \in CN(u,v)} \frac{1}{\log |N(\omega)|} + \sum_{\omega \in CN(u,v)} \frac{\log \frac{c_\omega^{\mathbf{w}}}{1 - c_\omega^{\mathbf{w}}}}{\log |N(\omega)|} \tag{7}$$

$$s_{u,v}^{WLNBRA} = C \sum_{\omega \in CN(u,v)} \frac{1}{|N(\omega)|} + \sum_{\omega \in CN(u,v)} \frac{\log \frac{c_\omega^{w}}{1-c_\omega^{w}}}{|N(\omega)|} \quad (8)$$

3.4 mRMR Feature Selection

From the previous analysis, we can see that all the features in **F** are obtained based on the local structure composed of the common neighbors. Because of the correlation and redundancy between features [29], it is ineffective to have a better classification performance. Therefore, we use mRMR (Minimum Redundancy Maximum Dependency) [30] to process noise information between features and pick out more discriminative features.

The mRMR feature selection method selects the most redundant feature that has the highest correlation with the target category and is also the smallest, i.e., selects the features that are the largest dissimilar to each other. The smaller index of the feature indicates that it has a better trade-off between maximum correlation with the target and minimum redundancy. In mRMR, correlation and redundancy information is quantified by Mutual Information (MI), which functions to estimate the degree to which one vector relates to another, which is defined as:

$$I(x,y) = \int\int p(x,y) \log \frac{p(x,y)}{p(x) \cdot p(y) dx dy} \quad (9)$$

In this formula, x and y are feature vectors, $p(x,y)$ is the joint distribution probability, $p(x)$ and $p(y)$ are the marginal probability density. **F** represents the set of features, $\mathbf{F}_s$ represents the set of m selected features, $\mathbf{F}_t$ represents the set of n features to be selected. The correlation between the feature f in $\mathbf{F}_t$ and the classification target δ can be defined as:

$$D = I(f, \delta) \quad (10)$$

And the redundancy R between all features in $\mathbf{F}_t$ and all features in $\mathbf{F}_s$ can be expressed as:

$$R = \frac{1}{m} \sum_{f_i \in \mathbf{F}_s} I(f, f_i) \quad (11)$$

To make the feature f_j in $\mathbf{F}_t$ have the maximum correlation and the minimum redundancy, the definition of mRMR function can be obtained by combining Eqs. (12) and (13):

$$max_{f_j \in \mathbf{F}_s}[I(f_j, \delta) - \frac{1}{m} \sum_{f_i \in \mathbf{F}_s} I(f_j, f_i)] \quad (12)$$

When there are $N(m+n)$ features for a given feature set, the feature's evaluation of the function is executed N times and returns a set S: $S = \{f_1^{'}, f_2^{'}, f_3^{'} \cdots f_h^{'} \cdots f_N^{'}\}$. In feature set S, the subscript h of each feature indicates that the feature will be selected on the *hth* time. The smaller the value of h, the more satisfying the Eq. (14) is the discriminant feature.

Table 1. Statistics of networks.

Network	#Node	#Links	$\langle CC \rangle$	$\langle AV \rangle$	$\langle AW \rangle$
NetScience	1461	2742	0.8770	6.6312	0.7489
countryLevel	226	28869	0.8715	286.6814	27443
cond-mat-2003	5638	20912	0.4584	1.8841	3.2632
hep-th	8255	15751	0.6378	35.878	1.8334
astro-ph	7171	56547	0.4408	7.6529	2.8005

4 Experiments

4.1 Experimental Setting

To prove the validity of the method, partition ratio r is set to 0.9. The experimental results are used AUC for evaluation. The experimental platform is based on the complex network analysis package which is developed by Matlab. At the same time, the weighted features (called Weighted features) are composed of similarity metrics such as WCN, WAA and WRA, WLNBCN, WLNBAA and WLNBRA are used to construct the LNBweighted features, and the mRMR feature selection algorithm is used to obtain the eigenvectors.

Then we use three types of classical classification models [31]:

- LR (Logistic Regression). Logistic regression is the most commonly used method of machine learning in the industry. It is used to estimate the predictive possibility and can also be used to solve the dichotomy problem.
- RT (Regression Tree) Regression Tree classification. RT is a classification method of local data modeling by constructing decision tree, which can effectively divide the classification features into multiple easily-modeled data, and then use linear regression to fit.
- DA (Discriminant Analysis) Discriminant Analysis. In DA, a classification rule consisting of numerical indices is established as a discriminant function, and then the rule is applied to the samples of the unknown classification for prediction.

4.2 Scientific Collaboration Network

We conducted experiments in a number of real SCN: NetScience[1], hep-th[2], cond-mat-2003 (see footnote 2), astro-ph (see footnote 2), countryLevel[3]. The structural attributes of these datasets are shown in Table 1. Where V is the number of nodes, E is the number of edges, CC, AV and AW are the clustering coefficient, average degree and average weight, respectively.

[1] http://www.linkprediction.org/index.php/link/resource/data.
[2] http://www-personal.umich.edu/~mejn/netdata/.
[3] http://opsahl.co.uk/tnet/datasets/.

Table 2. Experimental results

Network	Weighted			LNBweighted			LNBweighted + FS		
	LR	RT	DA	LR	RT	DA	LR	RT	DA
NetScience	0.9977	0.9988	0.9933	0.9979	0.9989	0.9876	0.9994	**0.9996**	0.9908
countryLevel	0.8676	0.9894	0.8409	0.9419	0.9931	0.9196	0.9483	**0.9933**	0.9221
cond-mat-2003	0.9905	0.9955	0.9923	0.9937	0.9956	0.9925	0.9944	**0.9984**	0.9947
hep-th	0.9793	0.9913	0.9822	0.9877	0.9916	0.9847	0.9818	**0.9952**	0.9842
astro-ph	0.9843	0.9936	0.9860	0.9873	0.9936	0.9866	0.9877	**0.9937**	0.9867

4.3 Experimental Results

First, we tested the results of all the experiments under all datasets using weighted feature (Weighted), LNB based weighted feature (LNBweighted) and full model (LNBweighted + mRMR) as shown in Table 2. No matter using LR, RT or DA classification models, the LNBweighted + mRMR results were superior to those of LNBweighted and Weighted with 0.028%, 0.148%, 0.153% and 2.421%, 0.233% and 1.747% respectively. In addition, LNBweighted improve Weighted with 1.812%, 0.012% and 1.591% respectively. The performance shows that it is necessary to extract the features based on weighted clustering coefficients and select discriminative features in the task of classifying the collaboration in SCN. It can also be seen from the results that the effect of classification on the countryLevel dataset is enhanced by up to 5.804% on average. The reason may lie in the fact that the higher average degree and clustering coefficient of the dataset and the denser local structure of the corresponding co-located nodes. As such structure leads to more advantageous for calculating discriminative weighted naive Bayesian features. Finally, it is observed that the performance of RT classifiers are all above 0.99 (at least 3% above LR and DA). Such performance partly due to the fact that RT is an ensemble method that is more effective than other classifiers.

This part of the experiment aims to further explore the effectiveness of the proposed feature extraction process. We set training ratio r from 0.5 to 0.9 with a 0.02 step and report the classification performance corresponding to each step in Fig. 2. Specifically, countryLevel-LR means the results of countryLevel dataset with LR classification algorithm and weighted and LNBweighted is the proposed features. In addition, we have a similar effect in the experiment with all datasets, because the layout of the relationship, we only give two cases of data sets and have the following observations: (1) From the trend point of view, as the training set increases, the classification effects of the two features are gradually improved. For each classification model, the curves of the LNBweighted features represent the curves above, when compared with the Weighted features. This shows that the introduction of weighted coefficient can better measure the impact of more dense local structure, making the corresponding impact to improve prediction performance. (2) It can also be seen that the LNBweighted feature is more

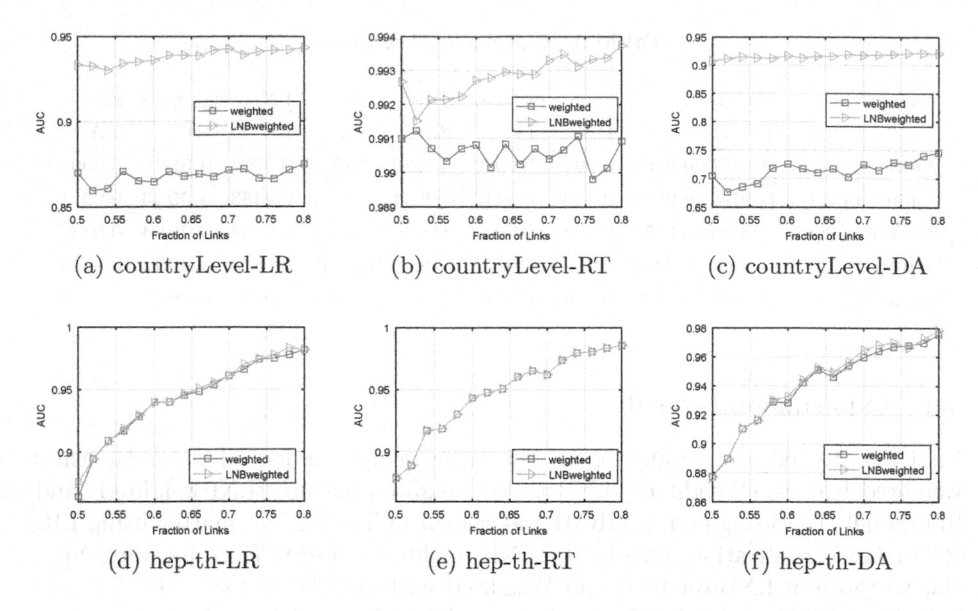

(a) countryLevel-LR (b) countryLevel-RT (c) countryLevel-DA

(d) hep-th-LR (e) hep-th-RT (f) hep-th-DA

Fig. 2. Performance of AUC by varying the percentage of training network

discriminative than the Weighted feature when the value of r is small. It means that LNBweighted features can play an important role when the network is very sparse and show that our proposed framework is stable. (3) We can find that the difference between the curves of is more obvious than that of hep-ph. Similar results are observed as in the Table 2. This is because the performance improvement in countryLevel is higher than hep-ph. But if we look carefully, we will also find the curves of the LNBweighted features are consistently lying on the top.

In addition, in order to further verify the validity of mRMR, this experimental section intends to introduce another statistical feature (distance measure) to evaluate the feature selection model - Relief for comparison. Relief algorithm is a feature selection algorithm based on feature weight weighting algorithms that help mark the importance of different features. The experiments only use LNBweighted feature to compare for convenient. The results are shown in Fig. 3. In particular, due to the limitation of the x-axis length, the network datasets in Table 1 are denoted by shortened as net, cL, cond, hep and astro. It can be clearly seen that the classification accuracy of mRMR model is higher than that of Relief model, no matter which classification method is adopted. The major reason is that mRMR can measure the nonlinear relationship between features. It demonstrates that mRMR is a better strategy for feature selection. And it also proves that mining information correlation between features is more effective than using distance measure and the computational efficiency is high.

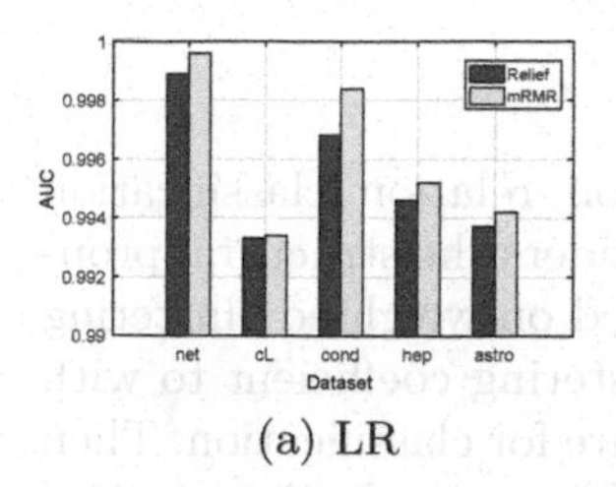

(a) LR

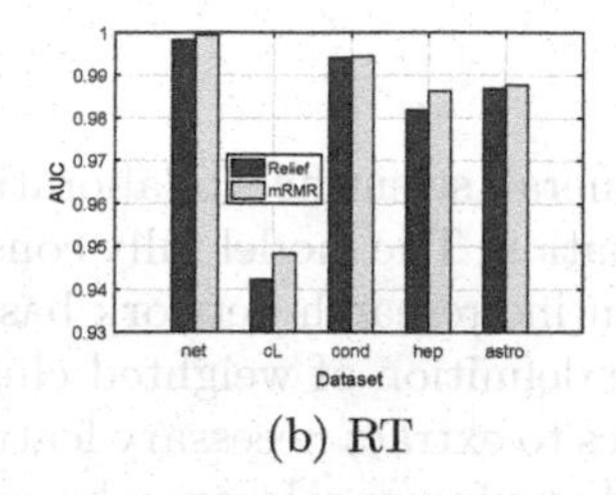

(b) RT

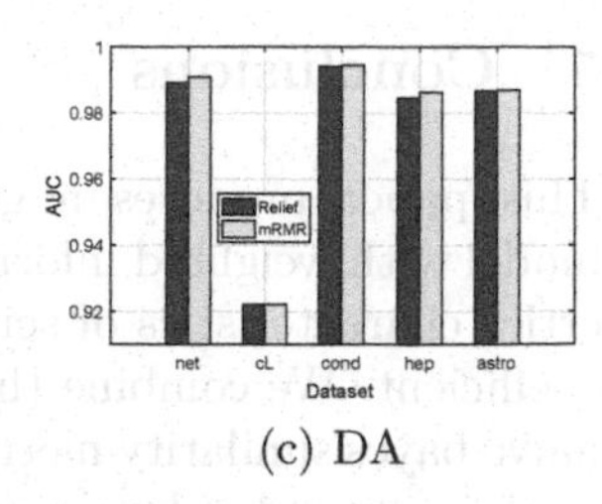

(c) DA

Fig. 3. Performance of AUC FS

4.4 Experimental Results in Unsupervised Methods

The similarity metric that constitutes LNBweighted eigenvector can also be used to predict the relationship of unsupervised method. In this section, WLNBCN, WLNBAA, WLNBRA and WCC are compared with some classic indicators to further validate the effectiveness of the weighted clustering coefficient definition. The comparison of indicators are WPA, WJaccard, WSalton and WSorenson and the corresponding results are shown in Table 3. The best classification results are highlighted in bold. It can be clearly seen that the method based on locally weighted clustering coefficients is better than the method based on weighted information of common neighbor nodes. The average of WLNBCN, WLNBAA and WLNBRA is 17.597% higher than that of WPA, WJaccard, WSalton and WSorenson with a margin by 12.393%, 9.411%, 9.287%, 19.349%, 14.067%, 11.041%, 10.915% and 16.947%, 11.772%, 8.807% and 8.683% respectively. This is due to the fact that scientific research networks usually have two characteristics: the aggregation of local structures and the different roles of the common neighbors. Other weighted similarity metrics based on neighborhood only utilize the weight information of the neighbor nodes and can not mine the implicit information represented by the above features, thus resulting in poor performance. In addition, we also find that the performance of simple metric WCC is better than WPA, WJaccard, WSalton and WSorenson with a margin by 14.964%, 10.863%, 8.651%, 9.993%. This results again show that our approach can also effectively integrate local weighted cluster coefficient to learn more effective common neighbors influence. We also observe that the comparison methods (WJaccard, WSalton and WSorenson) obtain very poor performance in countryLevel network. The major reason is that countryLevel is not a sparse network and these metrics can not capture the dense local structure.

Table 3. Experimental results with similarity index

Network	WCC	WLNBCN	WLNBAA	WLNBRA	WPA	WSalton	WJaccard	WSorenson
NetScience	0.9912	0.9994	**0.9996**	0.9908	0.4543	0.9767	0.9782	0.9791
countryLevel	0.9061	0.9438	**0.9933**	0.9221	0.9276	0.5851	0.6873	0.6931
cond-mat-2003	0.9476	0.9944	**0.9984**	0.9941	0.8977	0.9195	0.9328	0.9233
hep-th	0.9574	0.9818	**0.9952**	0.9862	0.9121	0.9116	0.9122	0.9137
astro-ph	0.9815	0.9877	**0.9937**	0.9868	0.9811	0.9731	0.9745	0.9809

5 Conclusions

This paper proposes a general scientific collaboration relation classification model with weighted information. The model fully considers the structural properties characteristics of scientific research network based on weighted clustering coefficient. We combine the definition of weighted clustering coefficient to with naive bayes similarity metrics to extract necessary feature for classification. Then a minimum-redundancy and maximum-relevancy based feature selection method is incorporated, which effectively overcome the problems of feature weighted identification and solves the redundancy and the similarity between features. In order to prove the validity and robustness of proposed model, we conducted experiments to evaluate in several real-world SCN. The results show that the model has obvious improvement when compared with the benchmark algorithms. In the next stage of work, we will focus on how to improve the capability of the model, so as to better solve the relationship classification of large-scale and heterogeneous scientific research networks.

Acknowledgements. Jiehua Wu is supported by Guangdong Provincial Higher outstanding young teachers Training Program (Nos. YQ2015177). This work is supported by Natural Science Foundation of Guangdong Province (No. 2017ZC0348). Guangdong Provincial major scientific research project (No. 2017GKTSCX009) and Major engineering technical and commercial services applied research project (No. GDGM2015-ZZ-C03).

References

1. Zhang, C., Bu, Y., Ding, Y., et al.: Understanding scientific collaboration: homophily, transitivity, and preferential attachment. J. Assoc. Inf. Sci. Technol. **69**(1), 72–86 (2018)
2. Martínez, V., Berzal, F., Cubero, J.C.: A survey of link prediction in complex networks. ACM Comput. Surv. (CSUR) **49**(4), 69 (2017)
3. Zhou, J., Zeng, A., Fan, Y., et al.: Identifying important scholars via directed scientific collaboration networks. Scientometrics **114**(3), 1327–1343 (2018)
4. Wang, W., Yu, S., Bekele, T.M., et al.: Scientific collaboration patterns vary with scholars' academic ages. Scientometrics **112**(1), 329–343 (2017)
5. Gupta, M., Gao, J., Han, J.: Community distribution outlier detection in heterogeneous information networks. In: Proceedings of 2013 European Conference on Machine Learning and Principles and Practice of Knowledge Discovery in Databases (ECMLPKDD 2013), Prague, Czech, September, pp. 557–573 (2013)
6. Liben Nowell, D., Kleinberg, J.: The link prediction problem for social networks. J. Assoc. Inf. Sci. Technol. **58**(7), 1019–1031 (2007)
7. Wang, P., Xu, B.W., Wu, Y.R.: Link prediction in social networks: the state-of-the-art. Sci. China Inf. Sci. **58**(1), 1–38 (2015)
8. Al Hasan, M., Zaki, M.J.: A survey of link prediction in social networks. In: Aggarwal, C. (ed.) Social Network Data Analytics, pp. 243–275. Springer, Boston (2011). https://doi.org/10.1007/978-1-4419-8462-3_9

9. Murata, T., Moriyasu, S.: Link prediction of social networks based on weighted proximity measures. In: Proceedings of the IEEE/WIC/ACM International Conference on Web Intelligence, pp. 85–88. IEEE Computer Society (2007)
10. De Sá, H.R., Prudêncio, R.B.C: Supervised link prediction in weighted networks. In: The 2011 International Joint Conference on Neural Networks (IJCNN), pp. 2281–2288. IEEE (2011)
11. Wu, J.H., Zhang, G.J., Ren, Y.Z.: Exploiting neighbors' latent correlation for link prediction in complex network. In: 2013 International Conference on Machine Learning and Cybernetics (ICMLC), vol. 3, pp. 1077–1082. IEEE (2013)
12. Lü, L., Jin, C.H., Zhou, T.: Similarity index based on local paths for link prediction of complex networks. Phys. Rev. E **80**(4), 046122 (2009)
13. Li, J., Ge, B., Yang, K.: Meta-path based heterogeneous combat network link prediction. Phys. A: Stat. Mech. Appl. **482**, 507–523 (2017)
14. Fan, C., Li, D., Teng, Y., Fan, D., Ding, G.: Exploiting non-visible relationship in link prediction based on asymmetric local random walk. In: Liu, D., Xie, S., Li, Y., Zhao, D., El-Alfy, E.-S.M. (eds.) ICONIP 2017. LNCS, vol. 10638, pp. 731–740. Springer, Cham (2017). https://doi.org/10.1007/978-3-319-70139-4_74
15. Liu, W., Lü, L.: Link prediction based on local random walk. EPL (Europhys. Lett.) **89**(5), 58007 (2010)
16. Lü, L., Pan, L., Zhou, T.: Toward link predictability of complex networks. Proc. Natl. Acad. Sci. **112**(8), 2325–2330 (2015)
17. Pech, R., Hao, D., Pan, L.: Link prediction via matrix completion. EPL (Europhy. Lett.) **117**(3), 38002 (2017)
18. Lichtenwalter, R.N., Lussier, J.T., Chawla, N.V: New perspectives and methods in link prediction. In: Proceedings of the 16th ACM SIGKDD International Conference on Knowledge Discovery and Data Mining, pp. 243–252. ACM (2010)
19. Scellato, S., Noulas, A., Mascolo, C.: Exploiting place features in link prediction on location-based social networks. In: Proceedings of the 17th ACM SIGKDD International Conference on Knowledge Discovery and Data Mining, pp. 1046–1054. ACM (2011)
20. Zhao, Y., Wu, Y.J., Levina, E.: Link prediction for partially observed networks. J. Comput. Graph. Stat. **26**(3), 725–733 (2017)
21. Li, Y., Luo, P., Fan, Z.: A utility-based link prediction method in social networks. Eur. J. Oper. Res. **260**(2), 693–705 (2017)
22. Lü, L., Zhou, T.: Link prediction in weighted networks the role of weak ties. EPL (Europhys. Lett.) **89**(1), 18001 (2010)
23. Lin, Z., Yun, X., Zhu, Y.: Link prediction using benefit ranks in weighted networks. In: 2012 IEEE/WIC/ACM International Conferences on Web Intelligence and Intelligent Agent Technology (WI-IAT), vol. 1, pp. 423–430. IEEE (2012)
24. Zhu, B., Xia, Y.: Link prediction in weighted networks: a weighted mutual information model. PloS one **11**(2), e0148265 (2016)
25. Barrat, A., Barthelemy, M., Pastor-Satorras, R.: The architecture of complex weighted networks. Proc. Natl. Acad. Sci. USA **101**(11), 3747–3752 (2004)
26. Kalna, G., Higham, D.J.: A clustering coefficient for weighted networks, with application to gene expression data. AI Commun. **20**(4), 263–271 (2007)
27. Nascimento, M.C.V.: A graph clustering algorithm based on a clustering coefficient for weighted graphs. J. Braz. Comput. Soci. **17**(1), 19–29 (2011)
28. Liu, Z., et al.: Link prediction in complex networks: a local naive Bayes model. EPL (Europhy. Lett.) **96**(4), 48007 (2011)
29. Li, J., Cheng, K., Wang, S., et al.: Feature selection: a data perspective. ACM Comput. Surv. (CSUR) **50**(6), 94 (2017)

30. Peng, H., Long, F., Ding, C.: Feature selection based on mutual information: criteria of max-dependency, max-relevance, and min-redundancy. IEEE Trans. Pattern Anal. Mach. Intell. **27**(8), 1226–1238 (2005)
31. Bishop, C.M.: Pattern recognition. Mach. Learn. **128**, 1–58 (2006)
32. Wu, J.H., Zhang, G.J., Ren, Y.Z., et al.: Weighted local Naive Bayes link prediction. J. Inf. Process. Syst. **13**(4) (2017)

A Representation-Based Pseudo Nearest Neighbor Classifier

Yanwei Qi(✉)

School of Computer Science and Telecommunication Engineering,
Jiangsu University, Zhenjiang 212013, China
qywujs@163.com

Abstract. K-nearest neighbor rule (KNN) is a simple and powerful classification algorithm. In this article, we develop a representation-based pseudo nearest neighbor rule (RPNN), which is based on the idea of pseudo nearest neighbor rule (PNN). In the proposed RPNN, a query point is represented as a linear combination of all the training samples in each class, and we use k largest representation coefficients to determine k nearest neighbors per class instead of Euclidean distance. Then, we design the pseudo nearest neighbor of the query point per class and compute the distance between the query point and the pseudo nearest neighbor. The query point belongs to the class which has the closest representation-based pseudo nearest neighbor among all classes. Experimental results on twelve data sets have demonstrated that the proposed RPNN classifier can improve the classification accuracy in the small sample size cases, compared to the traditional KNN-based methods.

Keywords: K-nearest neighbor rule · Pseudo nearest neighbor rule
Pattern classification

1 Introduction

Pattern classification has been a research hot spot in the field of artificial intelligence nowadays. Machine learning methods are widely employed in classification, for example, traffic classification [1, 2] and so on. As one of the top 10 algorithms in data mining [3], K-Nearest neighbor rule (KNN) [4] firstly proposed by Fix and Hodges, has been widely used in many practical applications for its superiorities. The basic rationale of KNN rule is that the class label of every query point is often determined by a simple majority vote among its *k* nearest neighbors, which are searched from the training set by means of Euclidean distance. The main characteristic of the KNN is its asymptotical classification performance in the Bayes sense [4, 5]. Besides, KNN has several prominent advantages, such as intuitiveness, simplicity and efficiency. However, the conventional KNN still exists some major disadvantages as follows. First of all, the choice of the neighborhood size *k* is sensitive, especially in the small sample size cases. When *k* is too small, the query point is very sensitive to the existing outliers and noise points [6]. Conversely, when *k* is too large, the neighborhood contains too many points from other classes which the query point does not belong to [7, 8]. Secondly, the identical weight

Q. Zhou et al. (Eds.): ICPCSEE 2018, CCIS 901, pp. 165–181, 2018.
https://doi.org/10.1007/978-981-13-2203-7_13

instead of different weights is assigned to k nearest neighbors in KNN. In general, each neighbor should have different weight according to the classification contribution. However, the uncorrelated or farther points are given the same weight and the representative ones don't play a significant role in classification. Clearly, this way is unreasonable in pattern classification. Thus, how to assign the appropriate weights to nearest neighbors is not an ignorable issue in KNN-based classification. Thirdly, the way of determining nearest neighbors is always to use Euclidean distance measure. It could not find proper nearest points, which can contribute more to classifying the query point. In fact, the last two issues can further aggravate the sensitivity of k.

To overcome these issues above, many variants of KNN have been introduced [9–16]. Aiming at reducing the influence of the sensitivity of k and finding the suitable values of k, a proposal for local k values for k-nearest neighbor rule has been put forward in [9]. This method can fix the adaptive value of k according to different distributions of points. In [10, 11], two schemes of dynamically choosing the numbers of k nearest neighbors are also designed for correctly classifying different query points. To address the problem that k neighbors of each query point are not given by the rational weights, a classical weighted voting method for KNN, called the distance-weighted k-nearest-neighbor rule (WKNN), was proposed in [12]. As a good extension of WKNN, the dual distance-weighted k-nearest neighbor classier (DWKNN) was proposed to reduce the sensitivity of k and improve the classification performance in [13]. In order to overcome the sensitivity of k caused by Euclidean distance as the similarity measure between every query point and neighbors, nearest centroid neighborhood (NCN) as an alternative neighborhood was introduced in [14, 15]. It not only considers the similarities of the neighbors, but also takes the geometrical distributions of the neighbors into account. Based on NCN, some related k-nearest centroid neighbor classifiers were developed for good classification in [14–18]. In addition, combining k-nearest neighbor and centroid neighbor classifier was proposed in [19] in order to get the fast and robust classification. To well improve the recognition rate of the nearest neighbor classification method, combining nearest neighbor classifiers using multiple feature subsets was proposed in [20]. Besides, the selection of k-nearest neighbor can be used as a nearest neighbor measurement, such as in multiview feature embedding for pattern classification [21] and multiple nearest neighbor feature matching [22].

The sensitivity of the neighborhood size k could be heavily aggravated in the small sample size cases with the existing outliers. To solve this issue, a local mean-based k-nearest neighbor rule (LMKNN) was proposed, which adopts the local mean vector of k nearest neighbors from each class [23]. In LMKNN, the query point is classified into the class with the closest local mean vector among all classes. On basis of the idea of LMKNN using local mean vector, some related methods were introduced to improve the KNN-based classification performance and overcome the negative influence of the existing outliers, especially in the case of the small training sample size cases [24–29]. As an extension of LMKNN, a new method has been proposed, which used the cosine distance to compute the distance between the test samples and the local mean-based vectors [24]. Among these methods, two promising ones are the pseudo nearest neighbor rule (PNN) [25] and the local mean-based pseudo nearest neighbor rule (LMPNN) [31]. In PNN, k nearest neighbors from each class are firstly found and the corresponding distances to the query point are calculated. Then the pseudo nearest neighbor per class is determined by the weighted sum of distances from neighbors to the query point. The query point is finally classified into the class with the closest pseudo nearest neighbor. Unlike PNN, k local mean vectors per class are used for finding pseudo nearest neighbor in LMPNN, and the categorical k local mean vectors are computed by the corresponding k nearest neighbors.

In this article we propose a representation-based pseudo nearest neighbor rule (RPNN) on the basis of the idea of PNN. Our purpose is to further improve the KNN-based classification performance and overcome the problem which the value of k is very sensitive in the case of the small training sample size. In RPNN, a query point is first represented as a linear combination of all the training samples from each class, and the representation coefficient associated with each training sample to represent the query point are solved. Then, the representation coefficients are employed as the similarity measure to seek k nearest neighbors. After that, k training samples corresponding to k largest coefficients are chosen as k nearest neighbors in each class. We design the pseudo nearest neighbor from each class using k nearest neighbors per class, the same as in PNN and LMPNN. Finally, RPNN classifies the query point into the class that has the closest pseudo nearest neighbor among all classes. Distinguished from traditional KNN-based classifiers, the proposed RPNN adaptively searches k nearest neighbors by k largest representation coefficients instead of Euclidean distance. So the representative neighbors of each query point can be found in RPNN. Moreover, the distance between the query point and the pseudo nearest neighbor per class is used as the classification decision rather than the simple majority vote among k nearest neighbors in RPNN. Thus, the proposed RPNN can overcome the sensitivity of k mainly caused by ways of determining neighbors and making the classification decision. To evaluate the classification performance of our method, we compare RPNN to the state-of-art methods including KNN, WKNN, LMKNN, PNN and LMPNN on eighteen numerical real data sets with the small numbers of the sample size. Experimental results have shown that the proposed RPNN performs very well, especially in the small sample size cases.

This article is organized as follows. In Sect. 2, we briefly make a summary about the related work. In Sect. 3, we elaborate the proposed RPNN method in detail. In Sect. 4, we present our experimental results and analysis. Finally, Sect. 5 draws the conclusions.

2 The Related Work

In this section, we briefly review two KNN-based methods that are related to our work. Both are the pseudo nearest neighbor rule (PNN) [25] and the local mean-based pseudo nearest neighbor rule (LMPNN) [31]. Some common symbols used in what follows are first denoted. $M = \{\mathrm{x_i} \in \mathrm{R^t}\}_{\mathrm{i=1}}^{\mathrm{n}}$ is a training set including n training samples in t-dimensional feature space, $C = \{1, 2, \ldots, Q\}$ is the set of the class labels that all training samples belong to, and the class label l_i of each sample x_i is $l_i \in C.M^j = \{x_i^j \in R^t\}_{i=1}^{n_j}$ is the training set from class j with n_j samples and y is a given query point.

2.1 PNN

In PNN, the pseudo nearest neighbor in each class is decided by the weighted sum of distances of k nearest neighbors per class to the query point, and the query point is classified into the class with the closest pseudo nearest neighbor among all classes.

Firstly, k nearest neighbors of the query point y from each class j in M^j are found by using Euclidean distances in Eq. (1).

$$d(y, x_i^j) = \sqrt{(y - x_i^j)^T (y - x_i^j)}. \quad (1)$$

Let k nearest neighbors corresponding to the first k smallest distances from class j and their distances be denoted as $\left\{ x_{1j}^{NN}, x_{2j}^{NN}, \ldots, x_{kj}^{NN} \right\}$ and $\left\{ d(y, x_{1j}^{NN}), d(y, x_{1j}^{NN}), \ldots, d(y, x_{1j}^{NN}) \right\}$, respectively.

Secondly, the weight W_{ij} of the i-th neighbor x_{ij}^{NN} from the class j is defined as

$$W_{ij} = \frac{1}{i}, i = 1, \ldots, k. \quad (2)$$

Thirdly, Let the pseudo nearest neighbor of the query point y in class j be x_j^{PNN} and its distance $d(y, x_j^{PNN})$ between y and x_j^{PNN} is designed as

$$d(y, x_j^{PNN}) = (W_{1j} \times d(y, x_{1j}^{NN}) + W_{2j} \times d(y, x_{2j}^{NN}) + \ldots + W_{kj} \times d(y, x_{kj}^{NN})). \quad (3)$$

Finally, the query point y is classified into the class j with the smallest $d(y, x_i^{PNN})$.

$$c = \arg\min_j d(y, x_j^{PNN}). \quad (4)$$

2.2 LMPNN

In LMPNN, k local mean vectors computed by using k nearest neighbors are employed to design the pseudo nearest neighbor in each class, and the query point is classified into the class with the closest pseudo nearest neighbor among all classes, the same as PNN.

Firstly, k nearest neighbors of the query point y from each class j are determined by Euclidean distance.

Secondly, k local mean vectors of the query point y are calculated by using nearest neighbors in each class. The i-th local mean vector $\bar{x}_{ij}^{NN}$ of the first i nearest neighbors of y in class j is calculated as

$$\bar{x}_{ij}^{NN} = \frac{1}{i} \sum_{i=1}^{i} x_{ij}^{NN}, i = 1, \ldots, k. \quad (5)$$

Let $\left\{ d(y, \bar{x}_{1j}^{NN}), d(y, \bar{x}_{2j}^{NN}), \ldots, d(y, \bar{x}_{kj}^{NN}) \right\}$ represent the Euclidean distances between y and k local mean vectors from class j.

Thirdly, the weight W_{ij} of the i-th local mean vector $\bar{x}_{ij}^{NN}$ from the class j is defined, the same as Eq. (2).

Fourthly, the local mean-based pseudo nearest neighbor $\bar{x}_j^{PNN}$ of the query point y from class j is determined. Its distance to y is computed as

$$d(y, \bar{x}_j^{PNN}) = (W_{1j} \times d(y, \bar{x}_{1j}^{NN}) + W_{2j} \times d(y, \bar{x}_{2j}^{NN}) + \ldots + W_{kj} \times d(y, \bar{x}_{kj}^{NN})). \quad (6)$$

Lastly, the query point y is assigned into the classwith the closest local mean-based pseudo nearest neighbor among all classes.

$$c = \arg\min_{j} d(y, \bar{x}_j^{PNN}). \quad (7)$$

Note that both PNN and LMPNN are the same as KNN when $k = 1$.

3 The Proposed RPNN

In this section, we detailedly describe our proposed RPNN method which is based on the idea of PNN.

3.1 Basic Rationale

As we know, the sensitivity of the neighborhood size k could heavily degrade the KNN-based classification performance [3, 9, 11, 13], because the noise points or outliers always exist in the k-neighborhood regions, especially in the small sample size cases. Thus, how to determine the representative neighbors of each query point plays an important role in the KNN-based classification. Moreover, the ways of determining nearest neighbors and making the classification decisions in many variants of KNN can aggravate the sensitiveness to the choice of the neighborhood size k. Generally, k nearest neighbors of each query point are often determined by Euclidean distance. Since Euclidean distance measure only considers the similarities of samples, the appropriate samples as the neighbors cannot be obtained. A simple majority vote among k nearest neighbors is always adopted as the classification decision to classify the query points. In fact, this way gives identical weight to each neighbor. However, such classification decision cannot provide more classification contribution when neighbors are more representative. As the good extensions of KNN, PNN can overcome the sensitivity of k, especially in the small sample size cases with existing outliers. However, the way of determining k nearest neighbors is to use Euclidean distance in PNN, in order that the representative neighbors could be not well determined. Therefore, to further reduce the sensitivity of k and improve the KNN-based classification performance, we propose a new KNN-based method, called a representation-based pseudo nearest neighbor rule (RPNN), an extension of PNN.

In the proposed RPNN, a new scheme of choosing k nearest neighbors instead of Euclidean distance is mainly introduced. For given query point, it is first represented by a linear combination of all training samples from each class. Then, the representation coefficients associated with the training samples are used as similarity measure for determining class-specific k nearest neighbors. After finding k nearest neighbors from each class, the pseudo nearest neighbor in each class is designed for the query sample by weighted sum of distances between the query sample and the neighbors, the same as what is in PNN. Finally, the query point is classified into the class with the closest pseudo nearest neighbor among all the classes. In the proposed RPNN, three properties are held: (1) the class-specific representation coefficients instead of Euclidean distances are used for seeking representative neighbors, (2) each class-specific neighbor is weighted in designing pseudo nearest neighbor in each class for classification, (3) the distance of pseudo nearest neighbor from each class to the query point is used for the classification decision.

3.2 The RPNN Method

The proposed RPNN method on basis of the rationale of PNN is presented in this section. In RPNN, each query point is represented as a linear combination of all the class-specific training samples to construct the similarities between the query point and the training samples, in order to determine k nearest neighbors from each class. The classification process of RPNN is carried out as follows.

(1) The given query point y can be represented as a linear combination of all the training samples from each class.

$$y = x_1^j a_1^j + x_2^j a_2^j + \ldots + x_{nj}^j a_{nj}^j = M^j a^j. \tag{8}$$

where $a^j = [a_1^j, a_2^j, \ldots, a_{nj}^j]^T$ is the representation coefficient vector associated with the training samples from class j. In fact, a^j can be solved in the following situations: If M^j is a nonsingular square matrix, a^j is directly solved as

$$a^j = (M^j)^{-1} y. \tag{9}$$

If M^j is a non-square matrix and $(M^j)^T M^j$ is a nonsingular matrix, a^j is easily solved as

$$a^j = ((M^j)^T M^j)^{-1} (M^j)^T y. \tag{10}$$

In general, $(M^j)^T M^j$ could be a singular matrix. To obtain the robust solution of a^j and address the singularity, a^j is always regularized with l_2-norm as

$$\min \|y - M^j a^j\|_2^2 + \gamma \|a^j\|_2^2. \tag{11}$$

According to the lagrange multiplier, a^j in Eq. (11) is achieved as

$$a^j = ((M^j)^T M^j + \gamma I)^{-1} (M^j)^T y. \quad (12)$$

Where I is an identity matrix. Note that the representation coefficient vector in Eq. (8) is resolved by Eq. (12) in RPNN.

(2) K nearest neighbors of the query point y from each class are determined the representation coefficients. The coefficients to represent the query point in Eq. (8) can well reflect the similarities between the training samples and the query point, and are considered as the similarity measure. The larger a_i^j denotes that sample x_i^j is more similar to y. Thus, the representation coefficients $\left\{ a_1^j, a_2^j, \ldots, a_{nj}^j \right\}$ are sorted in a decreasing order, and the training samples corresponding to the k largest coefficients in the coefficient vector a^j are regarded as k nearest neighbors from class j, denoted as $\left\{ x_{1j}^{NN}, x_{2j}^{NN}, \ldots, x_{kj}^{NN} \right\}$. Then, the Euclidean distances between y and k nearest neighbors in class j are calculated and denoted as $d(y, x_{1j}^{NN}), d(y, x_{2j}^{NN}), \ldots, d(y, x_{kj}^{NN})$.

(3) Each nearest neighbor is assigned with the weight. The weight W_{ij} of the neighbor x_{ij}^{NN} from the class j is also defined as in Eq. (2)

(4) The pseudo nearest neighbor in each class is designed. Let $\bar{\bar{x}}_j^{PNN}$ be the pseudo nearest neighbor of the query point y in class j. The distance between y and $\bar{\bar{x}}_j^{PNN}$ is defined as

$$d(y, \bar{\bar{x}}_j^{PNN}) = (W_{1j} \times d(y, x_{1j}^{NN}) + W_{2j} \times d(y, x_{2j}^{NN}) + \ldots + W_{kj} \times d(y, x_{kj}^{NN})). \quad (13)$$

(5) The query point y is classified in the class with closest pseudo nearest neighbor among all classes as follows

$$c = \arg\min_j d(y, \bar{\bar{x}}_j^{PNN}). \quad (14)$$

3.3 The RPNN Algorithm

As discussed in Subsect. 3.2, the proposed RPNN method is summarized in Algorithm 1 by means of pseudo codes.

Algorithm 1 A Representation-Based Pseudo Nearest Neighbor Classifier.

Require:

Error! Reference source not found.: a query point, ***Error! Reference source not found.***: the neighborhood size.

Error! Reference source not found.: a training subset from class ***Error! Reference source not found.***.

Step 1: Use the training samples from each class j to represent the query point y.

Error! Reference source not found.

Step 2: Calculate the representation coefficient vector**Error! Reference source not found.**associated with the training samples from class ***Error! Reference source not found.***.

Error! Reference source not found.

Step 3: Search ***Error! Reference source not found.*** nearest neighbors of the query point ***Error! Reference source not found.*** from class ***Error! Reference source not found.*** using the representation coefficient vector**Error! Reference source not found.** a^j. The training samples corresponding to ***Error! Reference source not found.*** largest representation coefficients in**Error! Reference source not found.** a^j are chosen as ***Error! Reference source not found.*** nearest neighbors from class ***Error! Reference source not found.***, which are denoted as **Error! Reference source not found.**.

Step 4: Compute Euclidean distances from the categorical ***Error! Reference source not found.*** nearest neighbors to the query point **Error! Reference source not found.** in each class.

For i=1 to k **do**

$$d(y, x_{ij}^{NN}) = \sqrt{(y - x_{ij}^{NN})^T (y - x_{ij}^{NN})}.$$

end for

Step 5: Allocate the weight**Error! Reference source not found.**to the ***Error! Reference source not found.***-th nearest neighbor**Error! Reference source not found.**.

For i=1 to k **do**

$$W_{ij} = \frac{1}{i}, i = 1, \ldots, k.$$

end for

Step 6: Design the pseudo nearest neighbor **Error! Reference source not found.**in each class j by calculating the weighted sum of distances between the query point and ***Error! Reference source not found.*** nearest neighbors.

$$d(y, \bar{\bar{x}}_j^{PNN}) = (W_{1j} \times d(y, x_{1j}^{NN}) + W_{2j} \times d(y, x_{2j}^{NN}) + \ldots + W_{kj} \times d(y, x_{kj}^{NN})).$$

Step 7: Classify the query point to the class c with the closest pseudo nearest neighbor among all classes.

$$c = \arg\min_j d(y, \bar{\bar{x}}_j^{PNN}).$$

4 Experimental Results

To verify the classification performance of the proposed RPNN, we compare RPNN to the competing methods including KNN, WKNN, LMKNN, PNN and LMPNN in terms of the classification accuracy. Our experiments are conducted on eighteen real data sets from UCI [32] or KEEL [33] repositories. Furthermore, since RPNN, PNN and LMPNN design the pseudo nearest neighbor from each class to classify the query points, we analyze the distances from their categorical pseudo nearest neighbors to the given query points, in order to further evaluate the proposed RPNN.

4.1 Data Sets

In the experiments, we use eighteen data sets that are taken from UCI or KEEL repositories. The information about the attributes, classes and total samples of these data sets is shown in Table 1. Among these data sets, the 'Climate Model Simulation', 'Led7digit', 'Hayes-roth' 'Hillvalley', 'Parkinsons', 'Wisconsin' and 'Transfusion' data sets are abbreviated as 'Climate', 'Led', 'Hayes', 'Hill', 'Park', 'Wis' and 'Tran', respectively. The total samples on each data set are randomly divided into the training and the testing sets ten times. The number of the testing samples on each data set is also displayed in Table 1. The final classification results of each method on all data sets are the averages of the classification accuracy rates on ten runs with 95% confidence. It should be noted that the chosen data sets used in the experiments have the small training sample sizes, and are used to well study the classification performance of the proposed RPNN in the small sample cases.

Table 1. The real data sets used in the experiments

Data	Total samples	Attributes	Classes	Testing samples
Wine	178	13	3	59
Seeds	210	7	3	60
Sonar	208	60	2	67
Musk	476	166	2	162
Titanic	2201	3	2	703
Vehicle	846	18	4	282
Led	500	7	10	151
Hayes	160	4	3	58
Bupa	345	6	2	169
Iris	150	4	3	56
Climate	540	18	2	230
Plrx	182	12	2	40
Hill	1212	100	2	600
Park	195	1	2	65
Tae	151	5	3	60
Wis	683	9	2	223
Glass	146	9	2	36
Tran	748	4	2	90

4.2 Experiment 1

In order to study the classification performance of the proposed RPNN method, we compare RPNN to KNN, WKNN, LMKNN, PNN and LMPNN by varying the neighborhood size k in terms of the classification accuracy rates. The value of k is varied from 1 to 15 with step 1 on each data set. The classification results of each method with increasing k are displayed in Figs. 1 and 2. As shown in Figs. 1 and 2, we can observe that the classification accuracy rates of the proposed RPNN method quickly increase at the smaller values of k at first, and then nearly keep stable when the values of k become larger on most of the data sets. It is clear that RPNN always performs better than the other competing methods at large values of k. The significant classification performance of RPNN can be obviously seen on Wine, Seeds, Titanic, Vehicle, Hill and Tran data sets. This fact implies that RPNN can use more representative nearest neighbors to well classify the query points. Meanwhile, the comparative classification results in Figs. 1 and 2 show that RPNN is less sensitive to k than KNN, WKNN, LMKNN, PNN and LMPNN at the large values of k. Thus, we can conclude that the representation coefficients as a similarity measure instead of Euclidean distance are very useful for selecting the representative neighbors for query samples.

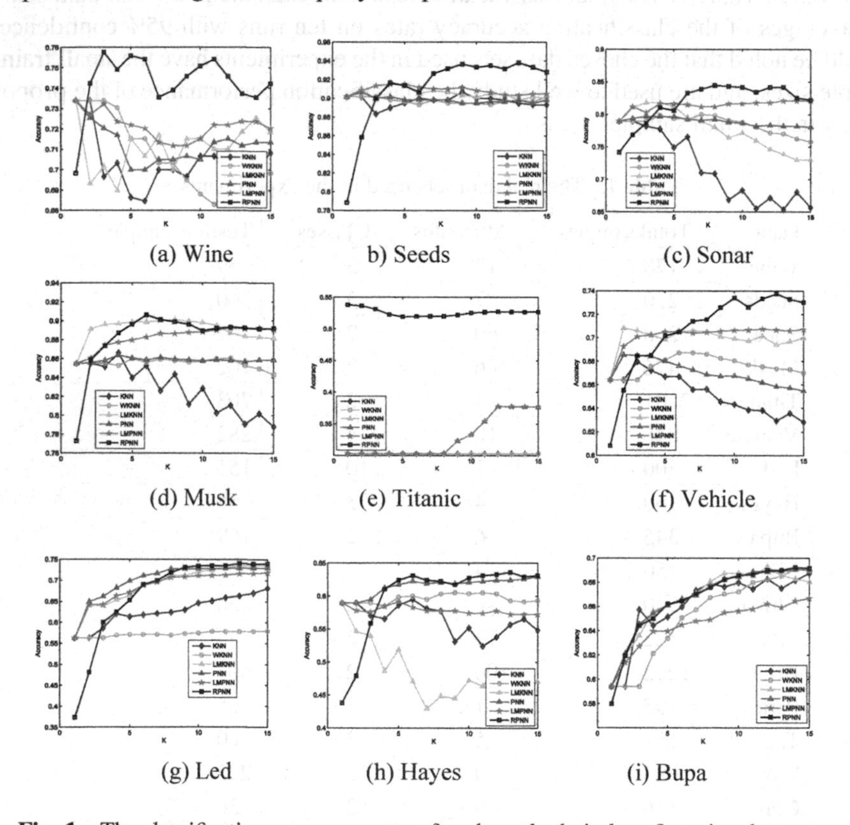

(a) Wine b) Seeds (c) Sonar

(d) Musk (e) Titanic (f) Vehicle

(g) Led (h) Hayes (i) Bupa

Fig. 1. The classification accuracy rates of each method via k on first nine data sets.

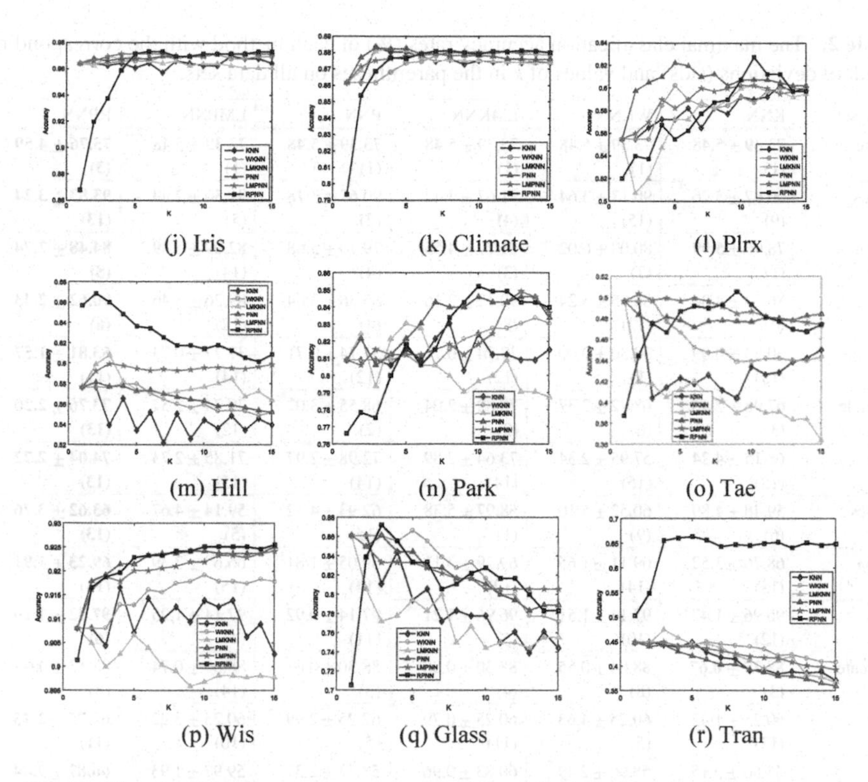

Fig. 2. The classification accuracy rates of each method via k on other nine data sets.

According to the classification results in Figs. 1 and 2, the maximal classification accuracy rates (%) of the competing methods with the corresponding standard deviations (stds) and values of k in the parentheses on each data set are shown in Table 2. Note that the maximal classification performance of each method is obtained among the rang of k from 1 to 15 with step 1, and the best performance on each data set among the competing methods is denoted in bold-face. It is obvious that the proposed RPNN nearly performs better than all competing methods, and the average of the maximal classification accuracy rates of RPNN on all data sets is larger than that of the other methods. Furthermore, our method can show distinguished performances on Wine, Seeds, Titanic, Vehicle, Hill and Tran data sets, as we can see from Figs. 1, 2 and Table 2. The reason may be that we use the way of representation coefficients to determine k nearest neighbors. The representation coefficients can well reflect the degree of similarity between training samples and testing samples, which is critical for classification. Therefore, the experimental results shown in Figs. 1, 2 and Table 2 illustrate that our proposed RPNN method is less sensitive to k, and it has more satisfactory classification performances than those of KNN, WKNN, LMKNN, PNN and LMPNN.

Table 2. The maximal classification accuracy rates (%) of each method with the corresponding standard deviations (stds) and values of k in the parentheses on all data sets.

Data set	KNN	WKNN	LMKNN	PNN	LMPNN	RPNN
Wine	73.39 ± 5.48 (1)	73.39 ± 5.48 (1)	73.39 ± 5.48 (1)	73.39 ± 5.48 (1)	73.39 ± 5.48 (1)	**75.76 ± 4.59 (3)**
Seeds	90.67 ± 3.26 (9)	90.17 ± 3.64 (15)	91.67 ± 3.42 (4)	90.67 ± 3.78 (2)	91.50 ± 2.99 (5)	**93.83 ± 3.24 (13)**
Sonar	78.96 ± 5.99 (1)	80.00 ± 6.02 (7)	81.79 ± 4.55 (3)	79.70 ± 5.68 (4)	82.09 ± 5.89 (14)	**84.48 ± 2.74 (8)**
Musk	86.60 ± 4.04 (4)	86.05 ± 3.24 (12)	90.12 ± 2.16 (8)	86.36 ± 3.64 (4)	89.26 ± 3.46 (12)	**90.62 ± 2.43 (6)**
Titanic	30.33 ± 0.09 (13)	30.30 ± 0.00 (1)	37.74 ± 0.71 (12)	37.74 ± 0.71 (12)	37.77 ± 0.73 (14)	**53.81 ± 1.57 (11)**
Vehicle	67.98 ± 2.46 (3)	68.72 ± 2.37 (6)	70.82 ± 2.04 (2)	68.55 ± 3.02 (2)	70.74 ± 2.54 (12)	**73.76 ± 2.20 (13)**
Led	68.15 ± 4.24 (15)	57.95 ± 2.34 (15)	73.64 ± 2.89 (14)	72.98 ± 2.97 (14)	71.85 ± 2.73 (15)	**74.04 ± 2.72 (13)**
Hayes	59.48 ± 4.89 (6)	60.52 ± 5.10 (9)	58.97 ± 5.38 (1)	62.93 ± 4.32 (15)	59.14 ± 4.67 (5)	**63.62 ± 3.76 (13)**
Bupa	68.76 ± 3.52 (15)	69.11 ± 1.65 (14)	69.29 ± 3.11 (12)	69.05 ± 1.81 (14)	66.69 ± 3.39 (15)	**69.23 ± 1.91 (14)**
Iris	96.96 ± 1.47 (12)	97.14 ± 1.51 (10)	96.96 ± 1.21 (3)	97.14 ± 0.92 (11)	97.14 ± 1.25 (4)	**97.32 ± 1.26 (6)**
Climate	**88.35 ± 0.67 (3)**	88.04 ± 0.55 (8)	88.30 ± 0.88 (4)	88.30 ± 0.63 (3)	88.09 ± 0.74 (14)	88.13 ± 0.65 (8)
Plrx	60.25 ± 1.42 (13)	60.25 ± 4.63 (5)	60.25 ± 0.79 (11)	62.25 ± 2.99 (5)	60.25 ± 3.22 (10)	**62.75 ± 2.75 (11)**
Hill	57.56 ± 3.15 (1)	58.00 ± 2.49 (5)	60.83 ± 0.96 (3)	58.13 ± 2.37 (2)	59.97 ± 1.93 (3)	**66.87 ± 7.44 (2)**
Park	83.08 ± 2.88 (8)	83.38 ± 2.28 (14)	83.08 ± 2.43 (5)	84.62 ± 1.54 (13)	84.92 ± 2.28 (13)	**85.23 ± 2.79 (10)**
Tae	49.67 ± 4.10 (1)	49.67 ± 4.10 (1)	49.67 ± 4.10 (1)	49.67 ± 4.10 (1)	50.33 ± 5.03 (2)	**50.58 ± 4.87 (9)**
Wis	91.66 ± 1.29 (4)	91.93 ± 1.01 (10)	90.81 ± 1.14 (1)	92.51 ± 1.16 (15)	92.42 ± 0.68 (15)	**92.56 ± 1.14 (15)**
Glsss	86.11 ± 2.78 (1)	86.67 ± 4.56 (5)	86.11 ± 2.78 (1)	86.11 ± 2.78 (1)	86.11 ± 2.78 (1)	**87.22 ± 3.17 (3)**
Tran	44.89 ± 2.97 (5)	45.00 ± 4.23 (6)	47.44 ± 3.99 (3)	44.89 ± 3.82 (1)	45.00 ± 4.23 (3)	**66.33 ± 3.48 (5)**
Average	71.27 ± 3.04	70.91 ± 3.07	72.83 ± 2.67	72.50 ± 2.87	72.59 ± 3.00	**76.45 ± 2.93**

It should be noted that, through extensive experiments, we found that our method can perform well under the circumstance of the small sample size. So we select many small datasets to verify the classification performance of the proposed RPNN method.

4.3 Experiment 2

To further explore the power of the pattern discrimination of the proposed RPNN, we compare RPNN to PNN, LMPNN on the Seeds, Wine and Iris data sets within three classes by considering their pseudo distances in the process of the classification decision. Since RPNN, PNN and LMPNN use pseudo distances between the given query point and categorical pseudo nearest neighbors to determine the class label of the query point, the differences among the pseudo distances play an important role in classification. If the differences among the pseudo distances are large, the corresponding methods can easily classify the query point. The pseudo distances between the given query points and the categorical pseudo nearest neighbors on these three data sets in RPNN, PNN and LMPNN are shown in Figs. 3, 4 and 5. Note that *ci* denotes the class *i* and the value of *k* is five. From the pseudo distances in RPNN, PNN and LMPNN, we can see that differences among the categorical pseudo distances on each data set in RPNN are larger than that in PNN and LMPNN. Moreover, RPNN correctly classifies the given query points, but PNN and LMPNN mistakenly classify the given query points. As described above, the pseudo distances in RPNN, PNN and LMPNN are designed by weighted sum of the distances from *k* nearest neighbors to the query points. The experimental results in Figs. 3, 4 and 5 implies that *k* nearest neighbors chosen by RPNN are more representative than ones by PNN and LMPNN.

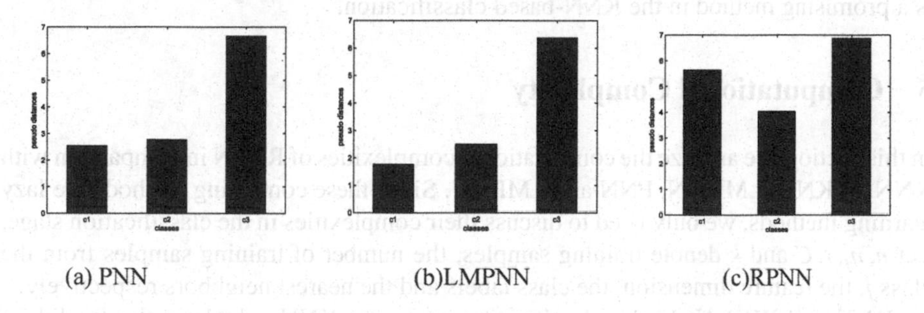

(a) PNN (b)LMPNN (c)RPNN

Fig. 3. The pseudo distances between the query point from class 2 and categorical pseudo nearest neighbors on seeds.

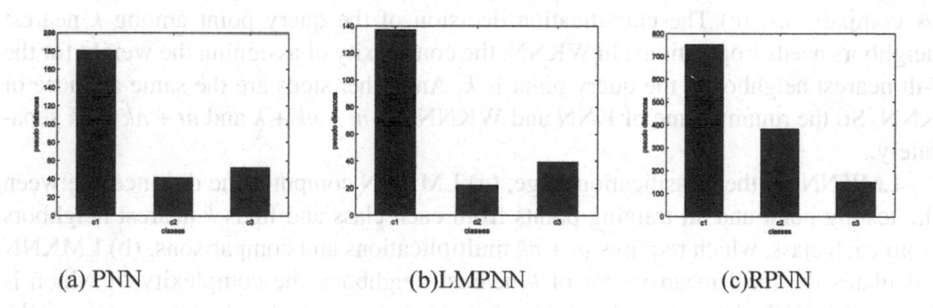

(a) PNN (b)LMPNN (c)RPNN

Fig. 4. The pseudo distances between the query point from class 3 and categorical pseudo nearest neighbors on Wine.

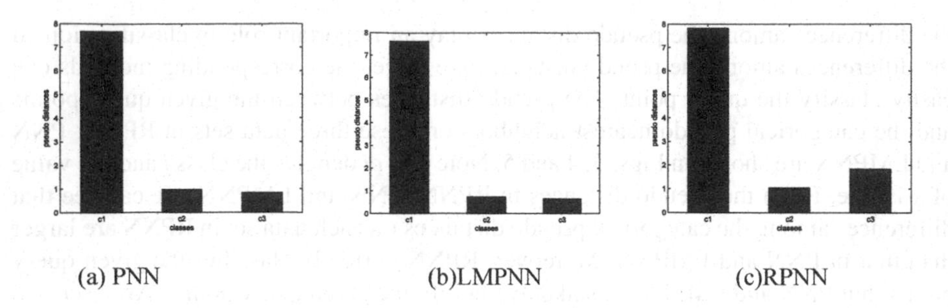

Fig. 5. The pseudo distances between the query point from class 2 and categorical pseudo nearest neighbors on Iris.

Through the experiments above, we can draw a conclusion that the proposed RPNN is a promising method in the KNN-based classification.

5 Computational Complexity

In this section, we analyze the computational complexities of RPNN in comparison with KNN, WKNN, LMKNN, PNN and LMPNN. Since these competing methods are lazy learning methods, we only need to discuss their complexities in the classification stage. Let n, n_j, t, C and k denote training samples, the number of training samples from the class j, the feature dimension, the class labels and the nearest neighbors respectively.

KNN and WKNN: In the classification stage, (a) KNN calculates the Euclidean distances between the query point and all training samples, in which the complexity nt is required. (b) KNN determines the k nearest neighbors according to the distances with nk comparisons. (c) The classification decision of the query point among k nearest neighbors needs k operations. In WKNN, the complexity of assigning the weight for the j-th nearest neighbor of the query point is k. And other steps are the same as those of KNN. So the running time of KNN and WKNN are $nt + nk + k$ and $nt + nk + 2k$ separately.

LMKNN: In the classification stage, (a) LMKNN computes the distances between the testing point and all training points from each class and finds k nearest neighbors from each class, which requires $nt + nk$ multiplications and comparisons. (b) LMKNN calculates the local mean vector of k nearest neighbors, the complexity of which is Ckt. (c) LMKNN determines the closest distance between the local mean vector and the query point and ultimately decides the classification result. In this step, $Ct + t$ operations are required. So the computational complexity of LMKNN is $nt + nk + Ckt + Ct + t$.

PNN: In the classification stage, (a) PNN determines k nearest neighbors from each class by computing the distances between the testing point and all training points from each class, which requires $nt + nk$ multiplications and comparisons. (b) PNN allocates the weight W_j to the j-th nearest neighbor and needs Ck divisions. (c) $Ck + C$ operations are required to design pseudo neighbor in each class and determine the class of query point. Totally, the computational complexity of PNN is $nt + nk + 2Ck + k$.

LMPNN: In the classification stage, (a) LMPNN needs the same step as PNN, which requires $nt + nk$ multiplications and comparisons. (b) LMPNN computing k local mean vectors of k nearest neighbors needs $2Ckt$ operations. (c) The weight W_j is allocated to the j-th local mean vector and the local mean-based pseudo neighbor is found, which needs $3Ck$. (d) Finally, deciding the class of query point requires C. So the running time of LMPNN is $nt + nk + 2Ck + 3Ck + C$.

RPNN: In the classification stage, (a) we first represent the query point with a linear combination of all the training samples from each class and determine the representation coefficients; (b) We find k nearest neighbors from each class; (c) assign the weight W_j to the j-th nearest neighbor; (d) Finally, we design pseudo neighbor in each class and determine the class which query point belongs to. In step (a), $nt^2 + t^3$ multiplications are required. In step (b), $n_1k + n_2k + \ldots + n_ck$ namely, nk comparisons are required. In step (c), Ck divisions are required. In step (d), $Ck + C$ operations are required. Thus, the running time of RPNN is $nt^2 + t^3 + nk + 2Ck + C$.

6 Conclusion

In this article, we propose a representation-based pseudo nearest neighbor (RPNN) rule. The aim is to further improve the KNN-based classification performance and mainly overcome the sensitivity of the neighborhood size k, especially in the small sample size cases. The RPNN is inspired by the idea of PNN. However, unlike PNN, the new way of choosing k nearest neighbors of the query sample is introduced. In RPNN, the query point is represented by the linear combinations of the class-specific training samples. The representation coefficients are used for determining categorical k nearest neighbors. Using the k nearest neighbors per class, we design the categorical pseudo nearest neighbor, whose distance to the query sample is the sum of weighted distances between the neighbors and the query sample. The query sample is finally classified into the class with the closest pseudo nearest neighbor among all classes. To demonstrate the effectiveness of the proposed RPNN, the experiments are carried out on eighteen real data sets with the small numbers of samples, in comparison with the competing methods. The experimental results show that the propose method perform very well. In our method, the values of k are set manually in the process of classification. In order to further improve the classification performance, we will extend the RPNN by adaptively determining the values of k. The basic idea of the proposed RPNN method can enhance the pattern discrimination in the small sample size cases, it may be useful in the classification tasks with big data. We will study this issue and use the idea of RPNN to design some extensions of the KNN-based classification to fit big data in the future.

Acknowledgements. This work was supported in part by National Natural Science Foundation of China (Grant No. 61502208), Natural Science Foundation of Jiangsu Province of China (Grant Nos. BK20150522, BK20140571, BE2017700) and Research Foundation for Talented Scholars of JiangSu University (Grant No. 14JDG037).

References

1. Sun, G., Chen, T., Su, Y., et al.: Internet traffic classification based on incremental support vector machines. Mob. Netw. Appl. **23**, 1–8 (2018)
2. Sun, G., Liang, L., Chen, T., et al.: Network traffic classification based on transfer learning. Comput. Electr. Eng. **69**, 1–8 (2018)
3. Wu, X., Kumar, V., Ross Quinlan, J.: Top 10 algorithms in data mining. Knowl. Inf. Syst. **14**, 1–37 (2008)
4. Cover, T.M., Hart, P.E.: Nearest neighbor pattern classification. IEEE Trans. Inform. Theory **13**, 21–27 (1967)
5. Covergence of the nearest neighbor rule. IEEE Trans. Inf. Theory, **17**, 566–571 (1971)
6. Fukunage, K.: Introduction to Statistical Pattern Recognition. Academic Press, Inc., San Diego (1990)
7. Yosipof, A., Senderowitz, H.: K-nearest neighbors optimization-based outlier removal. J. Comput. Chem. **36**, 493–506 (2015)
8. Wang, X.X., Ma, L.Y.: A compact k nearest neighbor classification for power plant fault diagnosis. J. Inf. Hiding Multimedia Sig. Process. **5**(3), 508–517 (2014)
9. Garciapedrajas, N., Del Castillo, J.A., Cerruelagarcia, G.: A proposal for local k values for k-nearest neighbor rule. IEEE Trans. Neural Netw. Learn. Syst. **28**(2), 470–475 (2015)
10. Zhang, S., Li, X., Ming, Z.: Efficient kNN classification with different numbers of nearest neighbors. IEEE Trans. Neural Netw. Learn. Syst. **99**, 1–12 (2017)
11. Bulut, F., Amasyali, M.F.: Locally adaptive k, parameter selection for nearest neighbor classifier: one nearest cluster. Pattern Anal. Appl. **20**, 1–11 (2015)
12. Dudani, S.A.: The distance-weighted k-nearest-neighbor rule. IEEE Trans. Syst. Man Cybern. **6**(4), 325–327 (1967)
13. Gou, J., Du, L., Zhang, Y.: A new distance-weighted k -nearest neighbor classier. J. Inf. Comput. Sci. **9**, 1429–1436 (2012)
14. Chaudhuri, B.B.: A new definition of neighborhood of a point in multi-dimensional space. Pattern Recogn. Lett. **17**(1), 11–17 (1996)
15. Sánchez, J.S., Marqués, A.I.: Enhanced neighborhood specification for pattern classification. In: Pattern Recognition and String Matching, vol. 13. Kluwer Academic Publishers (2002)
16. Sánchez, J.S., Pla, F., Ferri, F.J.: On the use of neighbourhood-based non-parametric classifiers 1. Pattern Recogn. Lett. **18**(11–13), 1179–1186 (1997)
17. Gou, J., Du, L.: Weighted k-nearest centroid neighbor classification. J. Comput. Inf. Syst. **8**, 851–860 (2012)
18. Li, P., Gou, J., Yang, H.: The distance-weighted k-nearest centroid neighbor classification. J. Intell. Inf. Hiding Multimedia Sig. Process. **8**(3), 611–622 (2017)
19. Chmielnicki, W.: Combining k-nearest neighbor and centroid neighbor classifier for fast and robust classification. In: Martínez-Álvarez, F., Troncoso, A., Quintián, H., Corchado, E. (eds.) HAIS 2016. LNCS (LNAI), vol. 9648, pp. 536–548. Springer, Cham (2016). https://doi.org/10.1007/978-3-319-32034-2_45
20. Bay, S.: Nearest neighbor classification from multiple feature subsets. Intell. Data Anal. **3**(3), 191–209 (1999)
21. Zhang, L.: Ensemble manifold regularized sparse low-rank approximation for multiview feature embedding. Pattern Recogn. **48**, 3102–3112 (2015)
22. Zamir, A.R., Shah, M.: Image Geo-Localization Based on MultipleNearest Neighbor Feature Matching UsingGeneralized Graphs. IEEE Computer Society (2014)
23. Mitani, Y., Hamamoto, Y.: A local mean-based nonparametric classifier. Pattern Recogn. Lett. **27**(10), 1151–1159 (2006)

24. Zhang, X.: A new local mean-based nonparametric classification method. In: International Conference on Electrical, Computer Engineering and Electronics (2015)
25. Zeng, Y.: Pseudo nearest neighbor rule for pattern classification. Expert Syst. Appl. **36**, 3587–3595 (2008)
26. Zeng, Y.: Nonparametric classification based on local mean and class statistics. Expert Syst. Appl. **36**, 8443–8448 (2009)
27. Gou, J., Yi, Z., Du, L.: A local mean-based k-nearest centroid neighbor classifier. Comput. J. **55**(9), 1058–1071 (2012)
28. Ma, H., Wang, X., Gou, J.: Pseudo nearest centroid neighbor classification. In: Hung, J., Yen, N., Li, K.C. (eds.) Frontier Computing. LNEE, vol. 375. Springer, Singapore (2016). https://doi.org/10.1007/978-981-10-0539-8_12
29. Pan, Z., Wang, Y.: A new k-harmonic nearest neighbor classifier based on multi-local means. Expert Syst. Appl. **67**, 115–125 (2017)
30. Yong, Z., Shu, H., Jiangping, H.U., et al.: Adaptive pseudo nearest neighbor classification based on bp neural network. J. Electron. Inf. Technol. (2016)
31. Gou, J., Zhan, Y., Hebiao, Y.: Improved pseudo nearest neighbor classification. Knowl.-Based Syst. **70**, 361–375 (2014)
32. UCI Machine Learning Repository. University of California, School of Information and Computer Science 2013, Irvine, CA. http://archive.ics.uci.edu/ml
33. Alcala-Fdez, J., et al.: Herrera KEEL data-mining software tool: data set repository, integration of algorithms and experimental analysis framework. J. Multiple-Valued Log Soft Comput. **17**(2–3), 255–287 (2011)

Research on Network Intrusion Data Based on KNN and Feature Extraction Algorithm

Shuai Dong(✉) and Xingang Wang(✉)

School of Information, Qilu University of Technology, Shandong, China
552862678@qq.com, wxg@qlu.edu.cn

Abstract. In order to solve the problem of high data dimension in network intrusion detection, the paper proposes KNN classifier and two kinds of effective feature selection algorithms that include Automatic encoder (Autoencoder) and Principal Component Analysis (PCA). The algorithms that the paper proposes will be applied in the field of network intrusion detection. First of all, we combine the KNN classifier with the effective feature selection algorithms to form a novel intrusion detection model. Secondly, we input the preprocessed data into the model. Finally, the experimental results show that the combination of KNN classifier and Autoencoder (KNN-Autoencoder) makes the accuracy of intrusion detection reach 93%, and the combination of KNN classifier and PCA (KNN-PCA) makes the accuracy of intrusion detection reach 91%. Apparently, the efficient feature selection algorithms can effectively reduce the impact of non-related attributes on the classification results. The performance of the automatic encoder is better than the principal component analysis. All experiments are done based on the KDD UP99 data set.

Keywords: Intrusion detection · KNN · Feature selection · KNN-autoencoder KNN- PCA · Autoencoder · PCA

1 Introduction

With the advent of the internet, the network has become an important part of people's life, an increasing number of people are beginning to notice the importance of network security. Therefore, for the network, the first important thing is to promise the network security [1], because the internet has been closely related to people's life, for example, most human tend to do online shopping or trading, so if an attacker steals personal information or property by invading their network, that will put people in a very insecure state. In order to solve these problems, it is necessary to carry out the detection and identification of intrusion of network [2–4].

Since 1970, machine learning and artificial intelligence have begun to rise, those began to become the most important tools to solve the classification problems. However, in recent years, with the rapid development in data dimension, it is difficult to solve this problem by using the traditional classification algorithms.

Q. Zhou et al. (Eds.): ICPCSEE 2018, CCIS 901, pp. 182–191, 2018.
https://doi.org/10.1007/978-981-13-2203-7_14

In the past, the researchers introduced different machine learning algorithms such as unsupervised learning self-organizing mapping algorithm (SOM) [5], artificial neural network algorithm (ANN) [6], support vector machine (SVM) [7], etc. Nevertheless, among those machine learning algorithms, the K Nearest Neighbor algorithm (KNN), which proposed by Cover and Hart in 1968, is the most popular one because it is simple and has a low error rate.

Even if the KNN classifier is used to classify the data easily, the time complexity will increase many times as the dimension increases. So the most important questions now are how to solve high-dimensional data, how to reduce the time complexity and how to improve the computational efficiency and accuracy.

With the continuous study of machine learning, the Deep Learning gradually begins to become popular. This paper presents two new feature selection algorithms, automatic encoders [8] and principal component analysis [9]. Both algorithms, which can be very effective in dimensionality reduction of high dimensional data, are attributed to unsupervised learning [10].

Recently, Lakhina proposes a method of combining PCA with ANN [11], which is applied to network intrusion detection. Kuang et al. use the combination of kernel genetic analysis (KPCA) and genetic algorithm for network intrusion detection, in which the feature selection of data is done by KPCA [12, 13]. Aung and Min proposes the combination of K-means and Random Tree to solve the problem of intrusion detection [14]. Yang, Li and Zhang et al. use the Random Forest to extract the characteristics of the original data and then import the extracted data into the SVM classifier to identify the normal and abnormal data [15]. However, those methods are not able to establish a mapping from high to low dimensions while automatic encoders and PCAs can overcome this shortcoming, these two methods are able to learn the optimal subset of feature subsets and accelerate the calculation. So far, there are few researchers use the combination of KNN and feature selection algorithms like Autoencoder and PCA in intrusion detection.

For the above problems, we use the combination of efficient KNN classifier and efficient feature extraction algorithms such as automatic encoder and PCA to do intrusion detection, and then test which combination can be more effective in the application of network intrusion detection. The second part of the paper introduces the KNN classifier, automatic encoder, and PCA. The third part shows that the proposed work. The fourth part reveals that the process and results of the experiment. The fifth part shows that the conclusion of those experiments we've done. The last part proposes the future directions.

2 Related Works

2.1 K-Nearest Neighbor Classifier

In the field of machine learning, the KNN algorithm is very popular in classification applications. In this paper, the training data and the corresponding training labels are placed in the KNN classifier to form a classification model and then drop the test data into the classifier, finally, we are able to get the accuracy of intrusion detection after comparing the predicted labels with the real labels. KNN classifier can be achieved by following steps:

1. Using vectors to store the training data. For instance, $Tr(A_{c1}, A_{c2}, A_{c3} \cdots, A_{cm} \ldots, A_{cn})$, m represents the weight of the attribute in the cth sample.
2. Calculating the distance between the testing sample and other samples, determining the K value and finding the K samples that are closest to the test sample.
3. Finding the main type of the samples from the K neighboring samples to determine the type of test sample.

2.2 Autoencoder

In order to solve the problem of high dimension in network intrusion detection, here is a very efficient data reduction method called automatic encoder and it is a branch of the Artificial Neural Network. The Autoencoder can convert high-dimensional data into low-dimensional data effectively. The automatic encoder is a feedforward neural network that contains one or more hidden layers and the automatic encoder that includes one or more hidden layers tries to reconstruct its input data according to its output. Therefore, the major difference between automatic encoder and traditional neural network is the size of output layer, the size of input layer always has the same size as the output layer. In addition, the size of output layer and input layer is always larger than the size of hidden layer. Figure 1 shows the workflow of the Automatic Encoder.

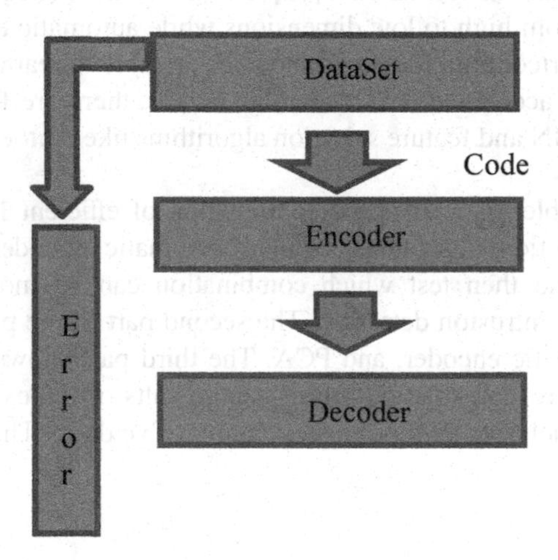

Fig. 1. The process of the automatic encoder

2.3 Principal Components Analysis

PCA is one of the most efficient methods, this method can extract the features among the 41 dimension-dataset, it is usually used for the linear transformation and unsupervised learning.

We assume that a given training dataset contains N samples, and each data sample is represented by an N-dimensional vector, $X \in R^n$ and m is the average.

$$X = [X_1, X_2, X_3 \ldots, X_n]^T \in R^n \tag{1}$$

The covariance matrix of the training set is defined as:

$$Z = \sum_{i=1}^{n} (X_i - m)(X_i - m)^T \tag{2}$$

Selecting j eigenvectors in Z and sort them according to the eigenvalues, the dimensionality-reduced feature subspace is expressed as: $P \in R^j$, d < < n.

$$P = Q^T X \tag{3}$$

$$P = \{p_{11}, p_{21}, p_{31}, \cdots p_{j1}\}^T \tag{4}$$

The most important characteristics of PCA are that the samples should be dispersed as much as possible in the low-dimensional space, the difference of the original samples should be kept in the original space, and the minimum mean square error should be controlled between the projection data and the original data.

3 Proposed Work

3.1 The Design of the System

After all the preparations are completed, the processed data is dimensionally reduced. The dimensionality reduction tools here use the Autoencoder and PCA, then we put the data, which dimensions have been reduced, into KNN classifier. The whole system will be shown in Fig. 2.

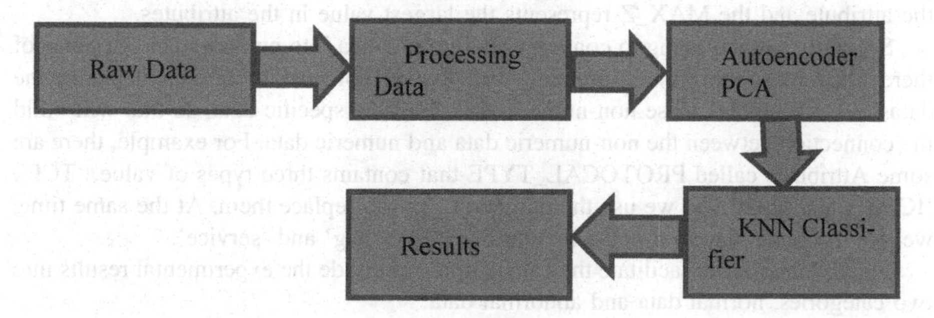

Fig. 2. Design of proposed system

3.2 Data Resources

In this paper, it's effective to use the KDD'99 [16] to do the experiment. The dataset was collected by the Lincoln laboratory at Massachusetts Institute of Technology, so that it is authoritative.

The KDD'99 data set has 494021 samples as training data and 311029 samples as test data, and there are four categories of attack ways: Denial of service (DOS), Remote-to-Local (R2L), User-to-Root (U2R) and Probing. Table 1 shows the number of each original type of data.

Table 1. The number of training and testing samples

Typle	Training_samples	Testing_samples
Normal	97278	60593
DoS	391458	229853
R2L	1126	16189
U2R	52	228
Probe	4107	4166

3.3 Data Preprocessing Phase

There are 41 categories of attributes involved in the Net Intrusion Dataset, and three attributes are non-numeric data and the rest are numeric, so the process of the treatment will be involved into two major steps.

First of all, we normalize the original data, which solves the problem of too much differences between different data value, and then improve the convergence rate. In this paper, we use the No.9 formula for normalization, which is able to reduce the value of the attribute to 0-1effectively.

$$Z' = (Z - MIN\_Z)/(MAX\_Z - MIN\_Z) \quad (5)$$

In this formula, Z is the value of attribute, MIN_Z shows that the minimum value of the attribute and the MAX_Z represents the largest value in the attributes.

Secondly, we are going to convert non-numeric data into numeric data. Because of there are three columns non-numeric data that is not able to be calculated in the datasheet. we replace these non-numeric data with the specific data, so that will build the connection between the non-numeric data and numeric data. For example, there are some Attributes called PROTOCAL_TYPE that contains three types of value, 'TCP', 'ICMP', and 'UDP', so we use the number 1, 2, 3 to replace them. At the same time, we use the same way to process the attributes like 'flag' and 'service'.

Finally, in order to facilitate the calculation, we divide the experimental results into two categories, normal data and abnormal data.

3.4 The Preparation of Training and Testing Samples

In the experimental data set, I would like to use the data from Table 1 and extract 6 groups of dataset randomly to make sure that the proportion between normal data and abnormal data is equal in each group of training samples. Apparently, training samples and testing samples in each group will be shown in Fig. 3.

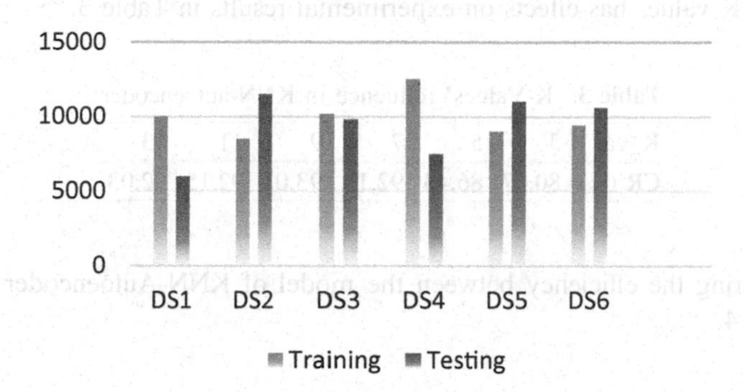

Fig. 3. The proportion of training and testing samples

3.5 The Standard of the Estimation

In order to test the efficiency of the models, we do the detection experiment by comparing the accuracy and wrong detection rate, so there are some definitions as follows:

$$CR = (CN + CP)/(CN + CP + WN + WP) \quad (6)$$

$$WR = WP/(CP + WP) \quad (7)$$

The formulas above introduce what the definitions are. CR indicates that the total accuracy, CN shows the number of abnormal samples which are detected rightly, CP means the quantities of normal sample that are detected rightly, WN introduces the number of the abnormal samples which are detected wrongly, WP indicates that the quantities of normal samples that have been detected wrongly.

4 Experiment and Analysis

At the beginning of the experiments, we use the KNN algorithm to do the intrusion detection, and then record the result. Furthermore, we use the KNN classifier with Autoencoder and PCA to do the intrusion detection, and record the result. There are four following groups of experiments.

Ex1: In the model of KNN-PCA classifier, we oversee if the K values has effects on the result of PCA and record the experimental result in Table 2.

Table 2. K-values' influence in KNN-PCA

K_val	3	5	7	9	11	13
CR (%)	80.23	83.21	88.43	91.23	90.43	90.01

Ex2: Analyzing the model of KNN-Autoencoder, and seeing if the number of selected features and K values has effects on experimental results in Table 3.

Table 3. K-Values' influence in KNN-autoencoder

K_val	3	5	7	9	11	13
CR (%)	80.67	86.43	92.12	93.01	92.15	92.03

Ex3: Comparing the efficiency between the model of KNN-Autoencoder and KNN-PCA in Fig. 4.

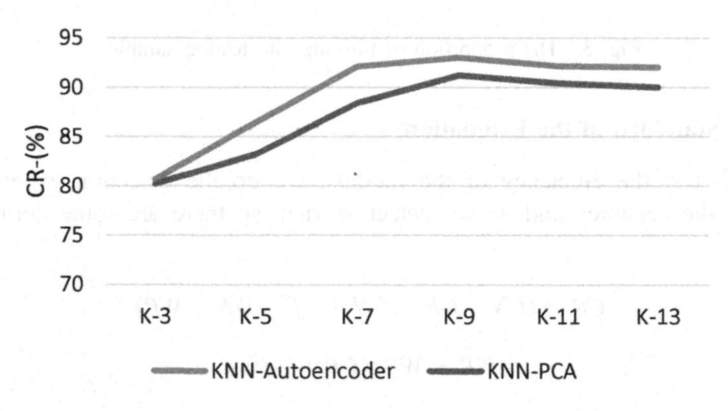

Fig. 4. Comparison between KNN-Autoencoder and KNN-PCA

Through Fig. 4, we find that when K-value is 9, both methods have got their best accuracy. However, when K-value is greater than 9, the accuracy intend to decline. We also find that the KNN-Autoender is better than KNN-PCA by comparing these two models.

Ex4: Analyzing the False Alarm Rate which calculated by KNN-Autoencoder and KNN-PCA in Table 4.

Table 4. False alarm rate recording

K_val	3	5	7	9	11	13
Auto-WR (%)	8.60	7.34	7.04	4.45	6.14	7.18
PCA-WR (%)	9.74	9.57	9.08	7.53	7.80	7.94

Ex5: Comparing the False Alarm Rate between the KNN-Autoender and KNN-PCA in Fig. 5.

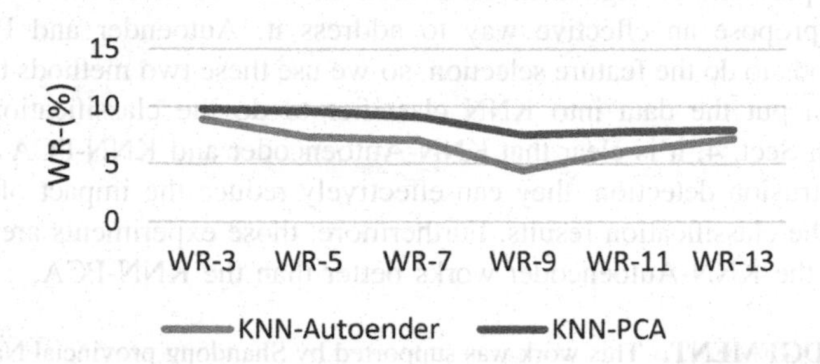

Fig. 5. Comparing the false alarm rate between KNN-autoencoder and KNN-PCA

In Fig. 5, the wrong rate of the detection declines from WR-3 to WR-9, but the wrong rate recovers from WR-9 to WR-13. Apparently, the WR of KNN-Autoencoder is lower than the KNN-PCA When the K-value reaches 9.

Ex6: Comparing the accuracy between the Auto-KNN and SVM in Fig. 6

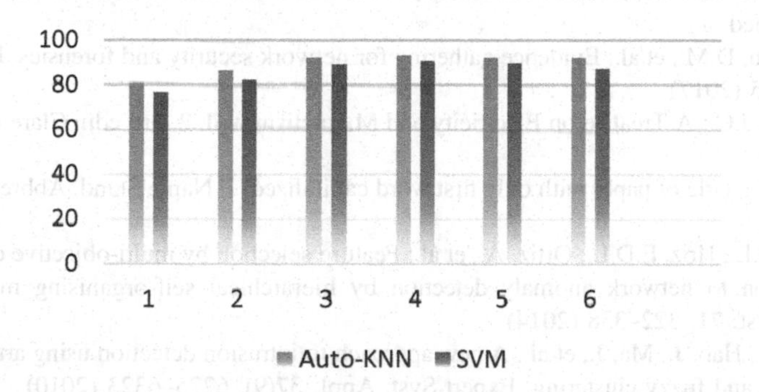

Fig. 6. The comparison between auto-KNN and SVM in accuracy

It is obvious that Auto-KNN classifier can perform better than the SVM classifier in accuracy of detecting intrusion. The experiment is divided into 6 groups, apparently, Auto-KNN has a good performance in each group.

5 Conclusions

To solve the problem of high data dimension in network intrusion detection, it is necessary to propose an effective way to address it. Autoender and PCA are the efficient methods to do the feature selection, so we use these two methods to extract the raw data, then put the data into KNN classifier to do the classification. After the experiments in Sect. 4, it is clear that KNN-Autoencoder and KNN-PCA can perform well in the intrusion detection, they can effectively reduce the impact of non-related attributes on the classification results. Furthermore, those experiments are detailed for us to see that the KNN-Autoencoder works better than the KNN-PCA.

ACKNOWLEDGEMENT. This work was supported by Shandong provincial Natural Science Foundation, China (************) and by Key Research and Development Plan of Shandong Province (*************).

References

1. Bloede, K., Mischou, G., Senan, A., Koontz, R.: The Internet of Things. Woodside Capital Partners (2015). http://www.woodsidecap.com/wp-content/uploads/2015/03/WCP-IOT-M_and_A-REPORT-2015-3.pdf. Accessed 27 Oct 2016. Elissa, K.: Title of paper if known, unpublished
2. Divakaran, D.M., et al.: Evidence gathering for network security and forensics. Digit. Invest. **20**, 56–65 (2017)
3. Maxwell, J.C.: A Treatise on Electricity and Magnetism, vol. 2, 3rd edn. Clarendon, Oxford (1892)
4. Nicole, R.: Title of paper with only first word capitalized. J. Name Stand. Abbrev., 68–73, in press
5. Hoz, E.D.L., Hoz, E.D.L., Ortiz, A., et al.: Feature selection by multi-objective optimisation: application to network anomaly detection by hierarchical self-organising maps. Knowl. Based Syst. **71**, 322–338 (2014)
6. Wang, G., Hao, J., Ma, J., et al.: A new approach to intrusion detection using artificial neural networks and fuzzy clustering. Expert Syst. Appl. **37**(9), 6225–6323 (2010)
7. Chitraker, R., Huang, C.: Selection of candidate support vectors in incremental SVM for network intrusion detection. Comput. Secur. **45**(3), 231–241 (2014)
8. Hinton, G.E., Salakhudinov, R.R.: Reducing the dimensionality of data with neural networks. Science **313**(5786), 504–507 (2006)
9. PCA indexing based feature learning and feature selection
10. Cheryadat, A.M.: Unsupervised feature learning for aerial scene classification. IEEE Trans. Geosci. Remote Sens. **52**(1), 439–451 (2014)
11. Lakhina, S., Joseph, S., Verma, B.: Feature reduction using principal component analysis for effective anomaly-based intrusion detection on NSL-KDD. Int. J. Eng. Sci. Technol. **2**(6), 1790–1799 (2010)
12. Kuang, F., Xu, W., Zhang, S.: A novel hybrid KPCA and SVM with GA model for intrusion detection. Appl. Soft Comput. **18**(4), 178–184 (2014)

13. Kuang, F., Xu, W., Wnag, Y.: A novel approach of KPCA and SVM for intrusion detection. J. Comput. Inf. Syst. **8**(8), 3237–3244 (2012)
14. Aung, Y., Min, M.: An analysis of random forest algorithm based network intrusion detection system. In: 2017 18th IEEE/ACIS International Conference on Software Engineering, Artificial Intelligence, Networking and Parallel/Distributed Computing (SNPD), Kanazawa, Japan, pp. 127–132 (2017)
15. Chang, Y., Li, W., Yang, Z.: Network intrusion detection based on random forest and support vector machine. In: 2017 IEEE International Conference on Computational Science and Engineering (CSE) and IEEE International Conference on Embedded and Ubiquitous Computing (EUC), Guangzhou, China, pp. 635–638 (2017)
16. Stolfo, S.J., Fan, W., Lee, W.K., et al.: Cost-based modeling for fraud and intrusion detection: results from the jam project [EB/OL]. http://kdd.ics.uci.edu/databases/kddcup99/kddcup99.html

PSHCAR: A Position-Irrelevant Scene-Aware Human Complex Activities Recognizing Algorithm on Mobile Phones

Boxuan Jia[1], Jinbao Li[1,2(✉)], and Hui Xu[1]

[1] School of Computer Science and Technology, Heilongjiang University, Harbin 150080, Heilongjiang, China
jiaboxuan1177@163.com, {jbli,xuhui}@hlju.edu.cn

[2] Key Laboratory of Database and Parallel Computing of Heilongjiang Province, Harbin 150080, Heilongjiang, China

Abstract. Recognizing human complex activities has become an essential topic in pervasive computing research area. With the growing popularity of mobile phones, more and more studies have been dedicated to identifying human complex activities using mobile phones in recent years. However, previous works often restrain the position and orientation of cell phones which limit the applicability of their methods. To overcome this limitation, we propose a novel position-irrelevant activities identification method named PSHCAR, which efficiently utilize information from multiple sensors on smartphones. Moreover, besides commonly-used features such as accelerometer and gyroscope, PSHCAR also employ the knowledge about scenes of activities, which is helpful but ignored by previous works, to identify complex activities of mobile phone users. Comparative experiments show that our method performs better than several strong baselines on the task of human complex activities recognition. In conclusion, our method achieves state-of-the-art performance without any limitation on position or orientation of mobile phones.

Keywords: Human complex activities · Activity recognition · Mobile phone Mobile computing · Pervasive computing · Data science

1 Introduction

Activity recognition is an essential task in pervasive computing research area. Identifying simple activities is well-studied, and now, more researchers are working at recognizing complex activities, which are used to detect human diseases, care old aged, predict human social behavior, etc. Many activities of daily living have more than one sub-activity. For example, people eating lunch may involve sitting on the chair, picking chopsticks, picking food, etc. We define these sub-activities as atomic activities which cannot be broken down further given application semantics. Complex activities in daily life consist of atomic activities which may interleave and occur concurrently or serially.

Moreover, in recent years, the percentage of the population owning a smartphone has increased significantly. Modern smartphones ship with lots of sensors such as

Q. Zhou et al. (Eds.): ICPCSEE 2018, CCIS 901, pp. 192–211, 2018.
https://doi.org/10.1007/978-981-13-2203-7_15

accelerometers and gyroscopes which enable detecting the activities of users. Thence smartphones have become a more and more popular platform for human activity recognition.

Human activities recognizing approaches usually include a series of steps, i.e., human motion data acquisition, features extraction and activities identification using statistical machine learning algorithms. Extracting features from raw data is a significant work in the recognition methods. In this paper, we extract features from multiple data sources, including audio information from microphone and motion information from the linear accelerometer, gravitational accelerometer, and gyroscope on mobile phones. In previous studies, some mobile phone recognition methods take the location and direction of the mobile phone into account [1, 2], but most of them are too complicated to run on a mobile phone. We design a lightweight mobile phone-based position-irrelevant feature extraction algorithm which can overcome the restriction of placement of phones. Also, the past activity recognition methods focused on how to identify activities more accurately but neglected the scenes where the activities happen. However, different activities often occur on distinct scenes and the probability of various activities occur in different scenes varies, especially for some complex activities. Moreover, the limitation of scenes is smaller for the occurrence of simple activities, but the occurrence of complex activities has certain particularity. For example, walking as a simple activity can be carried out in many places, while the complex activity of playing badminton will not be carried out in a canteen. Thus, the knowledge of scenes is more conducive to the identification complex activities. Taking this into consideration, we present a method named PSHCAR, which is short for 'Position-irrelevant Scene-aware Human Complex Activities Recognition' in this paper. It first employs the scenes detector to find where activities occur, and then use the position-irrelevant algorithm to collapse the variance of placement of phones, and finally use the scene-aware classifier to identify the activities of smartphone users.

The main contributions of this paper are as follows: First, the PSHCAR method overcomes the limitation of the placement of mobiles phones. Secondly, it could use the information of scenes to help identify a variety of activities, and effectively improves the accuracy of recognition.

The rest of this paper is organized as follows: The next section describes the related works of human activities recognition. The third section describes PSHCAR algorithm including the scenes detector, position-irrelevant algorithm, scene-aware activities classifier. The fourth section describes the data collection, data analysis process, and the experiments. The last section is the conclusion and future work.

2 Related Work

Recognizing human activities is an increasingly popular research area, attracting many researchers to enter. Human activities recognition has been applied to smart home [3, 4], health monitor [5, 6], sports training [7], etc. and other related fields. There are many ways to identify human activities so far, which are mainly divided into two categories, the environment sensor-based identification approaches and wearable sensor-based identification methods. For example, in [8], eight different sensors are deployed in the environment to identify complex activities such as cooking, work, bed toilet transition, et al. [9] uses wearable devices to identify 22 simple and complex activities, such as run, washing machine, drinking, et al. And their method effectively used background sounds.

However, besides the sensing functionality in the smart environment and body-worn sensors, smartphones are also convenient interaction platforms for persuasive applications to motivate healthier behavior, monitor disease [10], recognize activities [11, 12] and other domains. Smartphones that are integrated with a rich set of sensors, such as the accelerometer, gyroscope, GPS, microphone, camera, proximity and light sensors, Wi-Fi, and Bluetooth interfaces, provide a suitable platform for personal activity recognition systems. In summary, due to the diversity of cellphones' functions, the use of these sensors on mobile phones to study and explore human activities has become increasingly meaningful. For instance, in [13], Dernbach et al. used the accelerometer to monitor some complex activities, such as cleaning, cooking, meditating and so on. The activity recognition system was developed on the Android platform but lack accuracy. Alqassim et al. used microphone and accelerometer to distinguish breathing, movement, and no-sleep-apnea on Windows and Android phones [14]. Concone et al. presented a framework for human activity recognition using data captured using embedded triaxial accelerometer and gyroscope sensors on smartphones. Some statistics over the captured sensor data are computed to model each activity [15]. In [16], Lane et al. designed an automated well-being application named BeWell for the Android smartphones. BeWell system monitors sleep, physical activity, and social interactions and gave feedback to promote better health. GPS and accelerometer in Bewell were used to monitor a person's physical activities, such as driving, being in a stationary state, running, and walking, and microphone was used to recognize social interactions by identifying voicing and non-voicing states. Sleep durations were approximated by measuring phone usage patterns, such as phone recharging, movement, and ambient sound.

Many of the activity recognition studies, not necessarily in the field of mobile phone sensing, focused on the use of statistical machine learning techniques to infer information about the user activities from raw data. The learning phase can be supervised, unsupervised and semi-supervised approaches, but the last method is less used. There are many approaches to recognize human activities. HMM (Hidden Markov Model), SVM (Support Vector Machines) and RF (Random Forest) are typical supervised learning methods, which are often used in activity recognition. For example, Lee et al. [17] proposed a layered Hidden Markov Model to recognize both short-term activity and long-term activity in real time. Min and Cho [18] proposed a method to recognize activities by combining multiple classifiers such as support vector machine (SVM) and Bayesian network (BN) using accelerometer and wearable sensors. Yonggang Lu [19] employed an unsupervised method for recognizing physical activities using smartphone accelerometers. Features were extracted from the raw acceleration data collected by smartphones, then an unsupervised classification method which was a clustering algorithm based on density for activities recognition.

However, most of these above studies have a limitation of placement of smartphones or the position-irrelevant algorithm is complicated. They do not consider the scene of the activity, either. The occurrence of activities is related to some specific scenes, and the detection of scenes can better assist the recognition of activities. Take this into account, based on the previous research, we propose a method that gets rid of the limitation of mobile phone placement and devise an identification method that focuses on scene and activities to improve the recognition accuracy of complex activities.

3 PSHCAR Algorithm

3.1 Overview of the PSHCAR Algorithm

There are three components in the PSHCAR algorithm: the scenes detector which detects the location where activities occur; the position-irrelevant algorithm which collapses the variance of the placement and orientation of cell phones; the scene-aware activities classifier that combines knowledge of scenes with information of audio and motion to identify the activities of users.

Figure 1 shows the overview of the PSHCAR algorithm, which extracts features from multiple data sources, including audio information from microphone and motion information from the linear accelerometer, gravitational accelerometer, and gyroscope on mobile phones. Then it uses the position-irrelevant algorithm to process the accelerometer data. At the same time, it employs the scene-aware algorithms to detect the scenes and form a new feature based on Wi-Fi and time features. Finally, it ensembles these features to identify the activities of users.

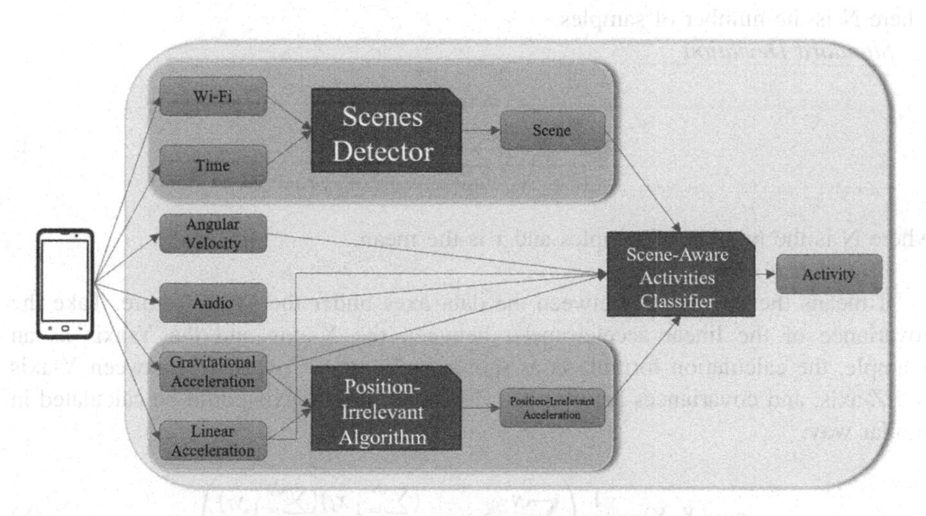

Fig. 1. The overview of the PSHCAR algorithm.

3.2 Features Selection

The PSHCAR algorithm uses multiple data sources on mobile phones, which comprise motion data from accelerometers, audio signals, Wi-Fi and time information.

Acceleration Data

Acceleration data from motion sensors includes three-dimensional linear acceleration, gravitational acceleration, and angular velocity. These are commonly used features in previous studies. However, raw readings from these motion sensors are relevant to the position and orientation of cell phones which is a restriction of previous studies on human complex activities recognition. We will use the position-irrelevant algorithm which will be introduced in Sect. 3.4 to collapse the variance of positions of mobile phones. We use the maximum, minimum, mean, standard deviation and covariance of the acceleration data as motion features.

Maximum

$$Max = Max(a_1, \cdots, a_N) \tag{1}$$

where N is the number of samples.

Minimum

$$Min = Min(a_1, \cdots, a_N) \tag{2}$$

where N is the number of samples.

Mean

$$Mean = \frac{1}{N}\sum_{i=1}^{N} a_i \tag{3}$$

where N is the number of samples.

Standard Deviation

$$S = \sqrt{\frac{1}{N}\sum_{i=1}^{N}(x_i - \bar{x})^2} \tag{4}$$

where N is the number of samples and $\bar{x}$ is the mean.

Covariance

It means the covariance between the data axes under the same feature. Take the covariance of the linear accelerometer between the X-axis and the Y-axis as an example, the calculation formula is as shown in Eq. (5). Covariance between Y-axis and Z-axis, and covariances between the Z-axis and the X-axis could be calculated in similar way.

$$\mathrm{cov}(X, Y) = \frac{1}{N}\left(\sum\nolimits_{i=1}^{N} x_i y_i - \frac{\left(\sum_{i=1}^{N} x_i\right)\left(\sum_{i=1}^{N} y_i\right)}{N}\right) \tag{5}$$

where N is the number of samples and x_i, y_i are the readings of the sensors under this dimension.

Audio Information

It uses the jAudio [20] package to extracts audio features from the microphone on smartphones. These extracted features are commonly used in audio information retrieval studies and are also useful to identify the activities of mobile phone users. The main audio features used in this paper are as follows:

Zero Crossing

Zero Crossing is calculated by counting the number of times that the time domain signal crosses zero within a given window. Mathematically, the zero crossing can be expressed as: $(X_n - 1 < 0$ and $X_n > 0)$ or $(X_n - 1 > 0$ and $X_n < 0)$ or $(X_n - 1 \neq 0$ and $X_n = 0)$.

RMS

RMS, Root Mean Square, is used to calculate the basis of a window, which can be expressed as Eq. (6)

$$\mathrm{RMS} = \sqrt{\frac{1}{N}\sum_{i=1}^{N} x_i^2} \tag{6}$$

where N is the number of samples in a window. RMS is used to calculate the amplitude of the window.

Fraction of Low Amplitude Frames

Low sub-frame score, this feature is used to extract the low-energy window ratio from each window, it can be a good measure of how much of the remaining signal is relatively quiet.

Spectral Flux

Spectral Flux is a good metric for the spectral change of a signal. It first calculates the difference between the amplitude of each frame in the current window and the corresponding amplitude in the previous window, and finally the square of all the differences summation.

Spectral Rolloff

Spectral Rolloff Point is a way to describe the spectral waveform and has been widely used to distinguish speech and music. It is often used to indicate the degree of tilt of the frequency in the current window.

Compactness

Tightness, the signal noise interference is a good indicator.

Method of Moments

Moment method, this function consists of the first five statistical moments of the spectrometer, including the area (zero order), mean (first order), Power Spectral Density (second order), Spectral Skew (third order), Spectral Kurtosis (fourth order). These features describe the shape of the spectrometer for a given window.

MFCC

MFCC is also known as Mel Frequency Cepstral Coefficients. It has been widely applied in the field of automatic speech recognition.

LPC

Linear prediction, LPC, is one of the most effective analysis methods in speech signal processing. It is mainly used in audio signal processing and automatic speech recognition to represent digital speech signal in compressed form according to the information of linear prediction model.

Beat Histogram

Beat Histogram, which builds a regular histogram of rhythms, is used as a base feature to determine the best rhythm match.

Wi-Fi Hotspots and Time Features

Wi-Fi signal intensity is various in different scenes. For example, there are often more Wi-Fi hotspots in offices than in gyms. Meanwhile, the probability that people in some locations is related to time. For instance, people are more probable in offices than in gyms during work hours. Therefore, Wi-Fi hotspots and time features are useful to distinguish the scenes where activities occur.

3.3 Scenes Detector

The features used in scenes detector are time features and the number of Wi-Fi hotspots features, which are conditionally independent and discretely distributed. Therefore, we select a Naive Bayesian model as the scenes detector.

Naive Bayes Classifier

Naive Bayes classifier is based on the Bayesian theorem and the hypothesis that features are conditionally independent. For a given training dataset, it first learns the joint probability distribution of inputs and outputs based on the hypothesis. For a given input x, the classifier uses the Bayesian theorem to find the output y with the highest posterior probability as the category of x [21].

Let the input space $\mathcal{X} \subseteq R^n$ be a collection of n-dimensional vectors and the output space $\mathcal{Y} = \{c_1, c_2, \ldots, c_k\}$ be a set of class labels. The training data set is $T = \{(x_1, y_1), (x_2, y_2), \ldots, (x_n, y_n)\}$. The input is feature vector $x \in \mathcal{X}$ and the output is class label $y \in \mathcal{Y}$. X is a random vector defined on input space $\mathcal{X}$, and Y is a random variable defined on output space $\mathcal{Y}$. $P(X, Y)$ is the joint probability distribution of X and Y.

Specifically, the conditional independence assumption can be expressed as Eq. (7).

$$\begin{aligned} P(X = x|Y = c_k) &= P(X^{(1)} = x^{(1)}, \ldots, X^{(n)} = x^{(n)}|Y = c_k) \\ &= \prod_{j=1}^{n} P(X^j = x^j|Y = c_k) \end{aligned} \tag{7}$$

The Naive Bayesian classifier's final decision is represented by Eq. (8).

$$y = \arg\max_{c_k} \frac{P(Y = c_k)\prod_j P(X^{(j)} = x^{(j)}|Y = c_k)}{\sum_k P(Y = c_k)\prod_j P(X^{(j)} = x^{(j)}|Y = c_k)} \tag{8}$$

In Eq. (8), the denominators are the same for all c_k, so we only need to consider Eq. (9).

$$y = \arg\max_{c_k} P(Y = c_k) \prod_j P(X^{(j)} = x^{(j)} | Y = c_k) \tag{9}$$

In this experiment, the Naive Bayesian model has two-dimensional features. $x^{(1)}$ is the number of Wi-Fi hotspots as the first-dimension feature and $x^{(2)}$ as the second-dimension feature. At the same time, c_k is one of the three kinds of scenes, namely the gymnasium, canteen, and office. Using the Naive Bayesian model, we can calculate the probability that each sample exists in each scene. Moreover, Bayesian classifier method is simple and costs little amount of computation, so it is suitable for mobile phones platforms.

Scenes Detector Algorithm

The scenes detector detects the location where activities occur which is shown in Algorithms 3.1 and 3.2:

Algorithm 3.1: Naive Bayes Algorithm.

Input: Dataset $T = \{(x_1, y_1), (x_2, y_2), \ldots, (x_n, y_n)\}$

Where $x_i = (x_i^{(1)}, x_i^{(2)}, \ldots, x_i^{(n)})^T$ **and** $x_i^{(j)}$ **means the** ***jth*** **feature of the** ***ith*** **sample,** $x_i^{(j)} \in \{a_{j1}, a_{j2}, \ldots, a_{jS_j}\}$ **and** a_{jl} **means the** ***jth*** **feature may get the** ***lth*** **value.**

$j = 1, 2, \ldots, n$, $l = 1, 2, \ldots, S_j$, $y_i \in \{c_1, c_2, \ldots, c_K\}$; **Sample** x .

Output: y **(category of sample** x **).**

1. Calculate the prior probability and the conditional probability.

$$P(Y = c_k) = \frac{\sum_{i=1}^{N} I(y_i = c_k)}{N}, k = 1, 2, \ldots, K$$

$$P(X^{(j)} = a_{jl} | Y = c_k) = \frac{\sum_{i=1}^{N} I(x^{(j)} = a_{jl}, y = c_k)}{\sum_{i=1}^{N} I(y = c_k)}$$

$$j = 1, 2, \ldots, n; l = 1, 2, \ldots, S_j; k = 1, 2, \ldots, K$$

2. For the given sample $x_i = (x_i^{(1)}, x_i^{(2)}, \ldots, x_i^{(n)})^T$ **, calculate**

$$P(Y = c_k) \prod_j P(X^{(j)} = x^{(j)} | Y = c_k), k = 1, 2, \ldots, K .$$

3. Determine the category of sample x **.**

$$y = \arg\max_{c_k} P(Y = c_k) \prod_j P(X^{(j)} = x^{(j)} | Y = c_k) .$$

Algorithm 3.2: Scenes Detector Algorithm.

Input: Dataset $T = \{x_i, y_i\}_{i=1}^{n}$, where $x_i = (w_i, t_i)^T$, w is the number of Wi-Fi hotspots and $w_i \in \{0,1,...,15\}$, τ is the time and $\tau \in \{6,7,...,23\}$, $y_i \in \{office, gymnasium, canteen\}$; test sample x.

Output: y (scene of x).

1. Call Algorithm 3.1 training model to get the $P(Y)$, $P(X \mid Y)$.
2. Compare $P(Y = c_k \mid X = x)$ for each c_k and choose the one with max probability as y, i.e., the scene of sample x.

3.4 Position-Irrelevant Algorithm

To make the activity recognition method applicable in most scenes, we relax the placement of the phone. Even for the same activity, different position and direction of the phone could cause different readings from the sensors, which makes the performance of recognition unstable. To overcome this challenge, it needs to find an invariant coordinate system during the recognition.

Projection of a Vector on Another Vector

Suppose there are two vectors $\vec{a}(a_x, a_y)$ and $\vec{b}(b_x, b_y)$ in 2D space, as Fig. 2 shown, α is angle between $\vec{a}$ and $\vec{b}$, as Eq. (10):

$$\vec{a} \cdot \vec{b} = |\vec{a}| \cdot \left|\vec{b}\right| \cdot \cos\alpha \tag{10}$$

While we have Eqs. (11) and (12):

$$\vec{a} \cdot \vec{b} = a_x \times b_x + a_y \times b_y \tag{11}$$

$$|\vec{a}| = \sqrt{a_x^2 + a_y^2} \tag{12}$$

To modify Eqs. (10) to (13):

$$\frac{\vec{a} \cdot \vec{b}}{|\vec{a}|} = \left|\vec{b}\right| \cdot \cos\alpha \tag{13}$$

As shown in Fig. 2 we have Eq. (14):

$$\left|\vec{b}\right| \cdot \cos\alpha = \left|\vec{b}\right| \cdot \frac{P_{ba}}{\left|\vec{b}\right|} = P_{ba} \tag{14}$$

If we combine these formulas, we have P_{ba}, which is the projection of a vector $\vec{b}$ on a vector $\vec{a}$:

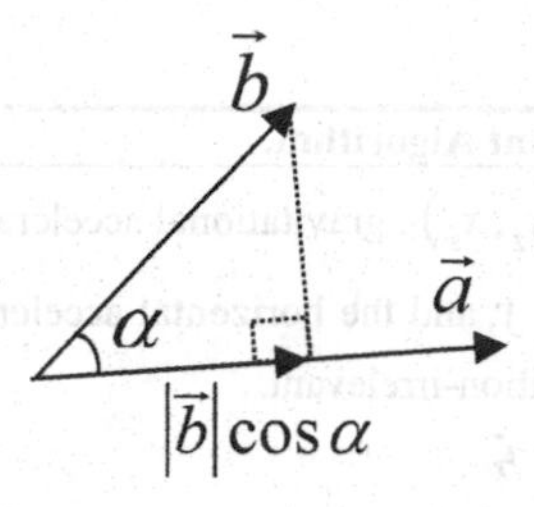

Fig. 2. Illustration of projection of a vector on another vector.

$$P_{ba} = \frac{\vec{a} \cdot \vec{b}}{|\vec{a}|} = \frac{a_x \times b_x + a_y \times b_y}{\sqrt{a_x^2 + a_y^2}} \tag{15}$$

Vectors in 3D space could be derived similarly. On Earth, the gravitational acceleration always keeps pointing vertically downward. Thus, we could decompose the linear acceleration along the direction of gravity and the horizontal direction. In a 3D space, the three base vectors can determine a spatial coordinate system. Since we have determined one base vector along the direction of gravity, we should choose the other two directions from the horizontal plane. One possible solution is along the cardinal directions, i.e. north, east, south, and west. However, human activities are not performed according to them, so it is unable to identify and classify the activities using these cardinal directions. Therefore, the linear acceleration is decomposed into the direction of gravity and the horizontal plane. Suppose we have an acceleration vector $\vec{a}(a_x, a_y, a_z)$ and the gravity vector $\vec{g}(g_x, g_y, g_z)$, and we could extend Eq. (15) to 3D space, i.e., the projection of a vector $\vec{a}$ on a vector $\vec{g}$ is Eq. (16):

$$P_{ag} = \frac{\vec{a} \cdot \vec{g}}{|\vec{a}|} = \frac{a_x \times g_x + a_y \times g_y + a_z \times g_z}{\sqrt{a_x^2 + a_y^2 + a_z^2}} \tag{16}$$

Note that we have Eq. (17):

$$a_{x'}^2 + a_{y'}^2 + a_{z'}^2 = |\vec{a}|^2 = a_x^2 + a_y^2 + a_z^2 \tag{17}$$

where $(a_{x'}, a_{y'}, a_{z'})$ is the vector that $\vec{a}$ after the normalization. Meanwhile $a_{z'}$ is P_{ag}, which is the projection of a vector $\vec{a}$ on a vector $\vec{g}$, then the remaining vector $\sqrt{a_{x'}^2 + a_{y'}^2}$ is the projection of $\vec{a}$ in the horizontal plane. Thus, the space vector $\vec{a}(a_x, a_y, a_z)$ will be broken down into two parts, one part is projected in the gravitational direction, the other is projected to the horizontal plane.

Position-Irrelevant Algorithm

The position-irrelevant algorithm collapses the variance of the placement and orientation of cell phones, which is described as below:

Algorithm 3.3: Position-Irrelevant Algorithm.

Input: Linear acceleration $\vec{a}(x_1, x_2, x_3)$, gravitational acceleration $\vec{b}(x_4, x_5, x_6)$.

Output: The vertical acceleration y_1 and the horizontal acceleration y_2 under the new coordinate system which is position-irrelevant.

1. Calculate $\vec{a}$ in the direction of $\vec{b}$.

$$y_1 = \frac{\vec{a} \cdot \vec{b}}{|\vec{b}|} = \frac{x_1 x_4 + x_2 x_5 + x_3 x_6}{\sqrt{x_4^2 + x_5^2 + x_6^2}}$$

2. Calculate the projection of $\vec{a}$ in the horizontal plane.

$$y_2 = \sqrt{x_1^2 + x_2^2 + x_3^2 - y_1^2}$$

We apply Algorithm 3.3 for each accelerometer reading. So, each acceleration has two more dimensions, i.e., horizontal projection and gravitational projection.

3.5 Scene-Aware Activities Classifier

Random Forest

Human activity recognition is essentially a classification process. The choice of the classifier will affect the classification results. Compared to the traditional model, Random Forest (RF) is an efficient statistical learning theory. It is a discriminant classification method. And many theoretical researches and practices have proved that RF has a high prediction accuracy. It has good tolerance to outliers and noise, and it is also not prone to over-fitting [22]. And in the previous work [23], the Random Forest classifier also shows good applicability. In addition, trees in RF is relatively independent, which make it easy to make a parallel implementation. It is suitable for the multi-core platform of the mobile phone. It uses the bootstrap sampling method to extract multiple samples from the original sample set to form a new sampling set. It models a decision tree for each new sampling set and then combines multiple decision trees to predict. The final forecast result is obtained through majority voting.

Random Forest classification model is composed of multiple decision trees $\{h(X, \theta_k), k = 1, \ldots\}$. And its parameters $\{\theta_k\}$ are random variables that independent identically distributed. For a given sample X, each decision tree makes its judgment. RF constructs a different training set to increase the variance between decision trees, then to obtain a combined classification model. The final classification results are generated by the majority vote method, as Eq. (18).

$$H(x) = \arg \max_{Y} \sum_{i=1}^{k} I(h_i(x) = Y) \tag{18}$$

where $H(x)$ represents the combined classification model, h_i is a single decision tree, Y represents the category, $I(\cdot)$ is an identity function.

Scene-Aware Activities Classifier

The scene-aware activities classifier combines knowledge of scenes with information of audio and motion to identify the activities of users.

Algorithm 3.4: Learning Algorithm of Random Forest Classifier.

Input: Data samples $\left\{\left(x_1^{t,n},\ldots,x_9^{t,n},y_1^{t,n},y_2^{t,n}\right)_{t=1}^{T_n}\right\}_{n=1}^{N}$, where N is the number of samples, T_n is the number of sampling from the nth example, $x_1,\ldots,x_9$ are three-dimensional linear acceleration, gravitational acceleration and angular velocity. y_1, y_2 are the linear acceleration components in the vertical and horizontal directions. $z_1,\ldots,z_z$ are the audio features.

Output: Classification Model F.

1. For n = 1 to N:
2. For j=1 to 9:
3. $Max_x^{j,n} = \max(x_j^{1,n}, x_j^{T_n,n})$
4. $Min_x^{j,n} = \min(x_j^{1,n}, x_j^{T_n,n})$
5. $Avg_x^{j,n} = avg(x_j^{1,n}, x_j^{T_n,n})$
6. $Sigma_x^{j,n} = sigma(x_j^{1,n}, x_j^{T_n,n})$
7. $\text{cov}_{1,2}^n = \text{cov}(\{x_1^{t,n}, x_2^{t,n}\}_{t=1}^{T_n})$
8. $\text{cov}_{2,3}^n = \text{cov}(\{x_2^{t,n}, x_3^{t,n}\}_{t=1}^{T_n})$
9. $\text{cov}_{1,3}^n = \text{cov}(\{x_1^{t,n}, x_3^{t,n}\}_{t=1}^{T_n})$
10. $\text{cov}_{4,5}^n = \text{cov}(\{x_4^{t,n}, x_5^{t,n}\}_{t=1}^{T_n})$
11. $\text{cov}_{5,6}^n = \text{cov}(\{x_5^{t,n}, x_6^{t,n}\}_{t=1}^{T_n})$
12. $\text{cov}_{4,6}^n = \text{cov}(\{x_4^{t,n}, x_6^{t,n}\}_{t=1}^{T_n})$
13. $\text{cov}_{7,8}^n = \text{cov}(\{x_7^{t,n}, x_8^{t,n}\}_{t=1}^{T_n})$
14. $\text{cov}_{7,9}^n = \text{cov}(\{x_7^{t,n}, x_9^{t,n}\}_{t=1}^{T_n})$
15. $\text{cov}_{8,9}^n = \text{cov}(\{x_8^{t,n}, x_9^{t,n}\}_{t=1}^{T_n})$
16. For j=1, 2 call Algorithm 3.3:
17. $Max_y^{j,n} = \max(y_j^{1,n}, y_j^{T_n,n})$
18. $Min_y^{j,n} = \min(y_j^{1,n}, y_j^{T_n,n})$
19. $Avg_y^{j,n} = avg(y_j^{1,n}, y_j^{T_n,n})$
20. $Sigma_y^{j,n} = sigma(y_j^{1,n}, y_j^{T_n,n})$
21. Form a training set of sample characteristics

$$S=\left\{\begin{array}{l} \max_x^{1,n},\ldots,\max_x^{9,n},\max_y^{1,n},\max_y^{2,n}, \\ \min_x^{1,n},\ldots,\min_x^{9,n},\min_y^{1,n},\min_y^{2,n}, \\ avg_x^{1,n},\ldots,avg_x^{9,n},avg_y^{1,n},avg_y^{2,n}, \\ sigma_x^{1,n},\ldots,sigma_x^{9,n},sigma_y^{1,n},sigma_y^{2,n}, \\ \mathrm{cov}_{1,2},\mathrm{cov}_{1,3},\mathrm{cov}_{2,3},\mathrm{cov}_{4,5},\mathrm{cov}_{4,6},\mathrm{cov}_{5,6}, \\ \mathrm{cov}_{7,8},\mathrm{cov}_{7,9},\mathrm{cov}_{8,9},z_1^n,\ldots,z_z^n \end{array}\right\}_{n=1}^{N}=\{s_n\}_{n=1}^{N}$$

22. For n=1 to N
23. Call Algorithm 3.2 call to calculate r_n (Scene of s_n).
24. Get the new training dataset $R=\{s_n,r_n\}_{n=1}^{N}$.
25. Train the classification model F from R.

After the scene-aware activities classifier is trained, we can use the model to identify the activities, as shown in Algorithm 3.5:

Algorithm 3.5: PSHCAR Activity Recognition Algorithm.

Input: Classification model F, dataset $\left\{\left(x_1^t,\ldots,x_9^t,y_1^t,y_2^t\right)_{t=1}^{T},z_1,\ldots z_z,w,\tau\right\}$, where N is the number of samples , T is the number of sampling from example , y_1,y_2 are the linear acceleration components in the vertical and horizontal direction. $z_1,\ldots,z_z$ are audio features. w is the number of Wi-Fi hotspots and $w\in\{0,1,\ldots,15\}$, τ is the time feature and $\tau\in\{6,7,\ldots,23\}$; Sample x.

Output: The activity of x.

1. Call Algorithm 3.2 and 3.4 to form the representation of sample x

$$x=\left\{\begin{array}{l} \max_x^1,\ldots,\max_x^9,\max_y^1,\max_y^2, \\ \min_x^1,\ldots,\min_x^9,\min_y^1,\min_y^2, \\ avg_x^1,\ldots,avg_x^9,avg_y^1,avg_y^2, \\ sigma_x^1,\ldots,sigma_x^9,sigma_y^1,sigma_y^2, \\ \mathrm{cov}_{1,2},\mathrm{cov}_{1,3},\mathrm{cov}_{2,3},\mathrm{cov}_{4,5},\mathrm{cov}_{4,6},\mathrm{cov}_{5,6}, \\ \mathrm{cov}_{7,8},\mathrm{cov}_{7,9},\mathrm{cov}_{8,9},z_1,\ldots,z_z,r \end{array}\right\}=\{s\ ,r\}$$

2. According to the classification model F to classify x.

4 Experiments and Analysis

4.1 Data Collection

In our experiment, there are three scenes: office, gymnasium, and canteen. Although there are only three kinds of experimental scenarios designed in this paper, the scenes detecting algorithm and the time and Wi-Fi feature can also be applied to other scenes. We designed eight different kinds of complex activities. Our experiments mainly focus on the complicated activities in the office, e.g., drinking, washing, sweeping, calling, talking. The specific description of each activity is as follows:

(a) Drinking
Experimental participants picked up the glass, stood up, walked to the side of the drinking fountain, bent over to received water, drank, and then returned to their seat to sat down.

(b) Washing
Experimental participants stood up, walked out of the office to the restroom, turned on the faucet, washed their hands, turned off the faucet after washing his hands, returned to the office and sat down.

(c) Sweeping
Experimental participants stood up, picked up the broom, swept the floor, dumped the trash into the trash, sent back the broom, returned to the seat and sat down.

(d) Calling
Experimental participants dialed up their phones, standing up, talked through the phone, hang up the phone, return to the seat, and sit down. They could walk around in a small area during the phone call.

(e) Talking
Experimental participants stood up and talked with their colleagues around, after the conversation returned to the seat, and sat down. They could walk around in a small area during the phone call.

In addition to these five complex activities at the office, we also collected two complex activities in the gymnasium and a complex activity in the canteen. In the

Fig. 3. Screenshot of the data collection App on Android phone.

gymnasium, we collected data of the ping pong and badminton activities of the experimental participants. In the canteen, we collected some complex activities related to eating. These activities include ordering food, eating, serving plates, etc. These activities collectively referred to like eating.

Figure 3 shows the mobile phone APP we used to collect the data. We used Android 4.4.2 based HTC mobile phone to collect data. We collected a total of 336 samples from all the complicated activities. There is no specific requirement on the location of the participants' mobile phones on their bodies, and there are no restrictions on the length of time. At the same time, no matter where the experimenter is located, we collected data from cell phone sensors and recorded the number of Wi-Fi hotspots at that moment and the time.

Figure 4 shows the number of Wi-Fi hotspots that can be received by the mobile phones carried by the experimental participants at different times and in different scenes. Figure 4 also shows the relationship between different scenes where participants are at different times and the number of samples.

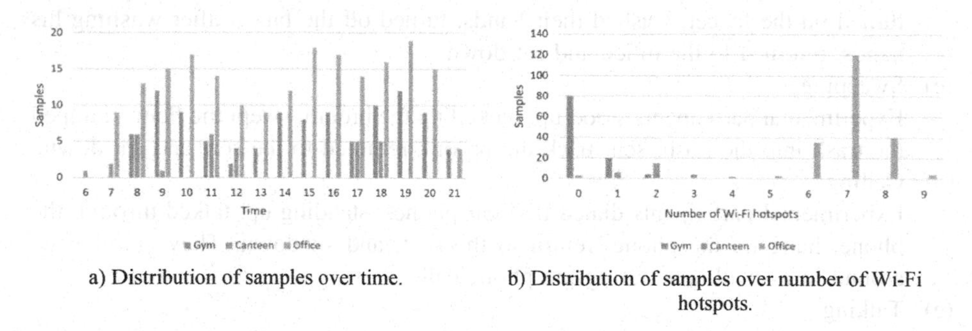

a) Distribution of samples over time.

b) Distribution of samples over number of Wi-Fi hotspots.

Fig. 4. Distribution of samples over time, number of Wi-Fi hotspots.

From there figures, we can see that in the office environment participants can receive the most Wi-Fi hotspots, such as the office's Wi-Fi signals, or the China Mobile Wi-Fi signal CMCC. As the gymnasium signal is not very strong, so the number of hotspots participants can collect is the least. Concerning time distribution, we can see that in the morning, noon and evening, many people are going to the canteen for meals in these three periods. At the same time, we can see that the number of people who appeared in the gymnasium at noon was relatively low. Most people appeared in the office from 8 o'clock to 11 o'clock and from 15 o'clock to 20 o'clock.

4.2 Results of Complex Activities Recognition

We used 10-fold cross-validation in our experiments. Table 1 shows the results of the scene detector using the Naive Bayes method. From the table, we can see that it performs best in the office scene, probably due to office Wi-Fi signals are more intensive which makes smartphones receive more Wi-Fi hotspots. Table 2 shows the classification result of the scene-aware activities classifier.

We constructed 70 trees in the random forest, and each tree randomly selected eight features for feature classification. Comparing Tables 1 and 2, we also find that the misclassified scenes are modified in the scene-aware activities classifier, which proves that the activity recognition method in this paper is not a hard classification method.

Table 1. Confusion matrix of the scenes detector.

	Office	Canteen	Gymnasium
Office	200.0	1.0	1.0
Canteen	0.0	21.0	9.0
Gymnasia	0.0	5.0	99.0

Table 2. Confusion matrix of the scene-aware activities classifier.

	Eating	Ping pong	Badminton	Calling	Drinking	Sweeping	Talking	Washing
Eating	27.0	1.0	1.0	0.0	0.0	0.0	1.0	0.0
Ping pong	0.0	51.0	1.0	0.0	0.0	0.0	0.0	0.0
Badminton	0.0	1.0	51.0	0.0	0.0	0.0	0.0	0.0
Calling	0.0	0.0	0.0	22.0	0.0	0.0	0.0	0.0
Drinking	0.0	0.0	0.0	0.0	51.0	0.0	0.0	1.0
Sweeping	0.0	0.0	0.0	0.0	0.0	52.0	0.0	0.0
Talking	0.0	0.0	0.0	1.0	0.0	0.0	23.0	0.0
Washing	0.0	0.0	0.0	0.0	1.0	0.0	0.0	51.0

Figure 5 shows the running results of PSHCAR. We observe that the Kappa statics is 0.97. The Kappa closer to 1, the classification performance is better. The average absolute error is 0.04, and the root mean square error is 0.096, indicating that the misclassification rate in this paper is rather low. The relative absolute error is 16.93%, and the root mean square error is 29.22%.

```
Correctly Classified Instances       328        97.619  %
Incorrectly Classified Instances       8         2.381  %
Kappa statistic                        0.9724
Mean absolute error                    0.0365
Root mean squared error                0.096
Relative absolute error               16.9269 %
Root relative squared error           29.22   %
Total Number of Instances            336
```

Fig. 5. The running results of PSHCAR.

4.3 Comparative Experiments About Scenes

This section describes the accuracy of the model with scene information and the model without scene information to identify the complex activities. We can conclude from Fig. 6 that the accuracy of the activity recognition using the PSHCAR approach is higher than that of the classifier direct recognizing activities, which shows the effectiveness of the integration of knowledge about scenes.

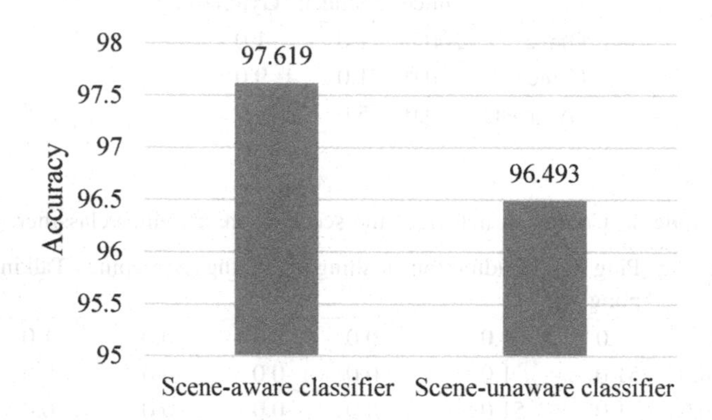

Fig. 6. Results of comparative experiments about scenes.

4.4 Ablation Experiments

Table 3 shows the ablation experimental results of different data feature sets. We remove each feature from the whole feature set to test the effectiveness of it. PI features mean the position-irrelevant accelerations. We can observe from the table that there is a performance regression after removing the position-irrelevant features, which verifies the usefulness of our position-irrelevant algorithm. The accuracy also decreases if we remove the audio features, which means they are a good complement to motion features for complex activities recognition. The fifth row, i.e., 'PSHCAR – PI features – Audio features', only uses the raw readings from motion sensors of smartphones. The last row only employs audio features. The last two rows obtain poor performance, showing that only using motion or audio features could not efficiently identify human complex activities.

Table 3. Results of ablation experiments.

Features	Accuracy (%)
PSHCAR (all features)	97.619
PSHCAR – PI features	96.030
PSHCAR – Audio features	88.614
PSHCAR – PI features – Audio features	86.634
Audio features only	84.555

4.5 Comparative Experiments

In this section, we explore the performance of different classifiers as the final activities identifiers. The experimental results are shown in Fig. 7, where PSHCAR means the proposed activity recognition algorithm, SVM (Support Vector Machines), J48 (a decision tree classifier), NB (Naive Bayes classifier) are baseline classifiers. We find that the PSHCAR algorithm using random forest outperforms all other baseline methods on human complex activities recognition task. At the same time, we can see that the proposed position-irrelevant algorithm for mobile phones can improve classification accuracy for various classifiers, showing the effectiveness of it.

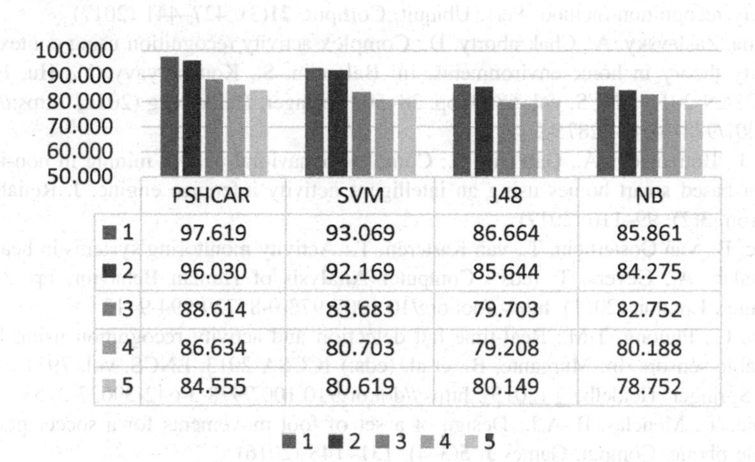

	PSHCAR	SVM	J48	NB
1	97.619	93.069	86.664	85.861
2	96.030	92.169	85.644	84.275
3	88.614	83.683	79.703	82.752
4	86.634	80.703	79.208	80.188
5	84.555	80.619	80.149	78.752

Fig. 7. Comparative experiment results: 1 represents PSHCAR (all features); 2 represents PSHCAR – PI features; 3 represents PSHCAR – Audio features; 4 represents PSHCAR – PI features – Audio features; 5 represents Audio features only.

5 Conclusion and Future Work

According to the characteristics of complex human activities, this paper presents a method PSHCAR to recognize complex activities. It combines features from multiple data sources of mobile phones and gets rid of the limitation of mobile phone placement using a lightweight algorithm. At the same time, the method effectively uses the relationship between scenes and activities by first detecting the scenes and then identify activities using the scene-aware classifier to improve the recognition accuracy. Furthermore, experiments show that the accuracy of the proposed method in identifying individual complex activities is higher than baseline methods. However, there is still room for improvement, e.g., due to the computational complexity is rather high, it might not be applied to low-end devices. We will continue to optimize the algorithm to reduce the complexity of the algorithm to apply to low-end mobile devices in the further work.

Acknowledgements. This work is supported by Basic Research Funds for Higher Education Institution in Heilongjiang Province (Fundamental Research Project, Grant No. KJCXYB201702).

References

1. Vathsangam, H., Zhang, M., Tarashansky, A., Sawchuk, A.A., Sukhatme, G.S.: Towards practical energy expenditure estimation with mobile phones. In: Asilomar Conference on Signals, Systems and Computers, pp. 74–79. IEEE, Pacific Grove (2013)
2. Shi, D., Wang, R., Wu, Y., Mo, X., Wei, J.: A novel orientation- and location-independent activity recognition method. Pers. Ubiquit. Comput. **21**(3), 427–441 (2017)
3. Saguna, Zaslavsky, A., Chakraborty, D.: Complex activity recognition using context driven activity theory in home environments. In: Balandin, S., Koucheryavy, Y., Hu, H. (eds.) NEW2AN 2011. LNCS, vol. 6869, pp. 38–50. Springer, Heidelberg (2011). https://doi.org/10.1007/978-3-642-22875-9_4
4. Hao, J., Bouzouane, A., Gaboury, S.: Complex behavioral pattern mining in non-intrusive sensor-based smart homes using an intelligent activity inference engine. J. Reliab. Intell. Environ. **3**(2), 99–116 (2017)
5. Kröse, B., van Oosterhout, T., van Kasteren, T.: Activity monitoring systems in health care. In: Salah, A., Gevers, T. (eds.) Computer Analysis of Human Behavior, pp. 325–346. Springer, London (2011). https://doi.org/10.1007/978-0-85729-994-9_12
6. Pham, C., Phuong, T.M.: Real-time fall detection and activity recognition using low-cost wearable sensors. In: Murgante, B., et al. (eds.) ICCSA 2013. LNCS, vol. 7971, pp. 673–682. Springer, Heidelberg (2013). https://doi.org/10.1007/978-3-642-39637-3_53
7. Lavoie, T., Menelas, B.-A.J.: Design of a set of foot movements for a soccer game on a mobile phone. Comput. Games J. **5**(3–4), 131–148 (2016)
8. Malazi, H.T., Davari, M.: Combining emerging patterns with random forest for complex activity recognition in smart homes. Appl. Intell. **48**(2), 315–330 (2018)
9. Zhan, Y., Kuroda, T.: Wearable sensor-based human activity recognition from environmental background sounds. J. Ambient Intell. Human. Comput. **5**(1), 77–89 (2014)
10. Capela, N.A., Lemaire, E.D., Baddour, N., Rudolf, M., Goljar, N., Burger, H.: Evaluation of a smartphone human activity recognition application with able-bodied and stroke participants. J. Neuro Eng. Rehab. **13**(5), 1–10 (2016)
11. Chatzaki, C., Pediaditis, M., Vavoulas, G., Tsiknakis, M.: Human daily activity and fall recognition using a smartphone's acceleration sensor. In: Röcker, C., O'Donoghue, J., Ziefle, M., Helfert, M., Molloy, W. (eds.) ICT4AWE 2016. CCIS, vol. 736, pp. 100–118. Springer, Cham (2017). https://doi.org/10.1007/978-3-319-62704-5_7
12. Bugdol, M.D., Mitas, A.W., Grzegorzek, M., Meyer, R., Wilhelm, C.: Human activity recognition using smartphone sensors. In: Piętka, E., Badura, P., Kawa, J., Wieclawek, W. (eds.) Information Technologies in Medicine. AISC, vol. 472, pp. 41–47. Springer, Cham (2016). https://doi.org/10.1007/978-3-319-39904-1_4
13. Dernbach, S., Das, B., Krishnan, N.C., Thomas, B.L., Cook, D.J.: Simple and complex activity recognition through smart phones. In: 2012 Eighth International Conference on Intelligent Environments, pp. 214–221. IEEE, Guanajuato (2012)
14. Alqassim, S., Ganesh, M., Khoja, S., Zaidi, M., Aloul, F., Sagahyroon, A.: Sleep Apnea monitoring using mobile phones. In: 2012 IEEE 14th International Conference on e-Health Networking, Applications and Services (Healthcom), pp. 443–446. IEEE, Beijing (2012)

15. Concone, F., Gaglio, S., Lo Re, G., Morana, M.: Smartphone data analysis for human activity recognition. In: Esposito, F., Basili, R., Ferilli, S., Lisi, F.A. (eds.) AI*IA 2017. LNCS, pp. 58–71. Springer, Cham (2017). https://doi.org/10.1007/978-3-319-70169-1_5
16. Lane, N., et al.: BeWell: a smartphone application to monitor, model and promote wellbeing. In: 5th International ICST Conference on Pervasive Computing Technologies for Healthcare, pp. 23–26. IEEE, Dublin (2011)
17. Lee, Y.-S., Cho, S.-B.: Layered hidden Markov models to recognize activity with built-in sensors on Android smartphone. Pattern Anal. Appl. **19**(4), 1181–1193 (2016)
18. Min, J.-K., Hong, J.-H., Cho, S.-B.: Combining localized fusion and dynamic selection for high-performance SVM. Expert Syst. Appl. **42**(1), 9–20 (2015)
19. Lu, Y., Wei, Y., Liu, L., Zhong, J., Sun, L., Liu, Y.: Towards unsupervised physical activity recognition using smartphone accelerometers. Multimed. Tools Appl. **76**(8), 10701–10719 (2017)
20. McKay, C.: jAudio: towards a standardized extensible audio music feature extraction system. Course Paper. McGill University, Canada (2009)
21. Russell, S.J., Norvig, P.: Artificial Intelligence: A Modern Approach, 3rd edn. Prentice Hall, Upper Saddle River (2009)
22. Ho, T.K.: Random decision forests. In: Proceedings of 3rd International Conference on Document Analysis and Recognition, pp. 278–282. IEEE, Montreal (1995)
23. Jia, B., Li, J.: Recognizing human activities in real-time using mobile phone sensors. In: Sun, L., Ma, H., Fang, D., Niu, J., Wang, W. (eds.) CWSN 2014. CCIS, vol. 501, pp. 638–650. Springer, Heidelberg (2015). https://doi.org/10.1007/978-3-662-46981-1_60

Visual-Based Character Embedding via Principal Component Analysis

Linchao He[1], Dejun Zhang[1(✉)], Long Tian[1], Fei Han[1], Mengting Luo[1], Yilin Chen[2], and Yiqi Wu[3]

[1] Sichuan Agricultural University, Yaan 625014, China
djz@sicau.edu.cn
[2] Wuhan University, Wuhan 430072, China
[3] China University of Geosciences, Wuhan 430074, China
https://github.com/djzgroup/Visual-basedCharacterEmbedding

Abstract. Most dense word embedding methods are based on statistics and semantic information currently. However, for hieroglyphs, these methods ignore the visual information underlaid in the characters, moreover this visual information in the expression of characters plays an extremely important role. Therefore, the visual information can be uncovered from the single character image through Convolutional Neural Network (CNN). Compared with the mainstream methods, the CNN method is inferior in efficiency and precision. In this study, we present a novel model called Img2Vec: using Principal Component Analysis (PCA) to generate word embedding vectors. Because the semantic and the visual information of the characters are complementary, we feed Word2Vec and Img2Vec embeddings into two different fusion models to implement text classification. Experiments show that our Img2Vec model has significant improvements in training time and precision. Finally, the visualizations of our Img2Vec character embedding prove that our model has a state-of-the-art representation of the visual information.

Keywords: Principal Component Analysis
Convolutional Neural Network · Word embedding · Text classification

1 Introduction

Images and texts are the essential components of Big Data era. Enormous works have done to analysis the information behind texts. Word embedding becomes an important part of representing texts which can directly impact the performances of tasks [12] like information retrieval [18], search query expansions [8] and representing semantics of words etc. In recent years, neural network methods [17] and statistical methods [1,7] are widely researched to generate word embedding effectively for Natural Language Processing tasks. In those works, bag-of-word model (BOW model) by Harris [4], Word2Vec [16,17] and GloVe [19] are widely recognized.

Q. Zhou et al. (Eds.): ICPCSEE 2018, CCIS 901, pp. 212–224, 2018.
https://doi.org/10.1007/978-981-13-2203-7_16

The radicals can affect the expression of characters while Word2Vec-like methods can't learn it from text. Meanwhile, the expressions of rare character embedding depend on the size of the corpus. In Fig. 1, radical structure information can determine the character semantics, and we can abstract the structure information from character image. We transform each character into an image, generating the corresponding embedding through the image without considering the rarity of the characters. However, compared with mainstream word embedding method (Word2Vec) [16,17], previous works which utilize visual information have significant weaknesses in the accuracy and train time.

(a) 口 + 乞 → 吃
mouth supplicate eat
(b) 火 + 火 → 炎
fire fire fire
(c) 疒 + 矢 → 疾
illness arrow illness or fast

Fig. 1. There are three kinds of paradigms of radical combinations: (a) We can see that the first radical which means mouth and the combined character which means eat; (b) Two same radicals which both mean fire can be combined into a character which also mean fire; (c) First radical means illness and second radical means arrow which leads to the character means disease or fast.

Inspired by the success of Principal Component Analysis (PCA) in the field of machine learning [20], we improve the Convolutional Neural Network (CNN) [11] feature extractor with PCA based on the departure point of effectiveness. Our Img2Vec model exploits PCA algorithm to concentrate on the global visual information, which can represent the whole character and the associations between radicals and characters. We propose the fusion models which can combine the statistical embedding from Word2Vec with the visual embedding from PCA to achieve better accuracy.

The rest of this paper is organized as follows. We present a brief review of the related work in Sect. 2 and our Img2Vec model and the highlights in Sect. 3. The two parts of Img2Vec model is introduced in Sects. 4 and 5, respectively. Next, two kinds of fusion models are described in Sect. 6. In Sect. 7, we discuss the evaluation results and compare it with representative models. Finally, Sect. 8 concludes the paper and demonstrates our prospect.

2 Related Work

The task of generating word embedding has been a popular topic in the research community for years. And a lot of neural networks are proposed to take advantages of varietal word embeddings. We will briefly outline connections and differences to four related work of research.

Word Embedding. Word Embedding methods that exploit neural networks to learn distributed representations of words or characters have been widely developed. BOW [4] is early referenced by Harris as an early approach to get word embedding. Word2Vec [16,17] is one of the mainstream word embedding methods which make use of the linguistic contexts of words by building two-layer neural networks. It does not only generate the word embedding, but also can produce a language model that can be exploited in other NLP applications. Furthermore, GloVe [19] has some similarities with Word2Vec, adding the global contexts information into its training process. Both models have great performances in tasks. Meanwhile, they are based on the statistic information of the corpus, which means they have a same feature that the bigger corpus is, the better representation they have. However they are based on the neural network, it takes too much time to train and produces large-size linguistic model lacking the interpretation of a single character embedding.

Character-Level-Embedding. Besides the word-level embedding, Zhang et al. [25] proposed a character-level embedding which decompose the word to character. Experiments demonstrate that their methods are state-of-the-art in represents of tweets vectors [2]. However, for hieroglyphics, we hold the opinion that the embedding of characters is based on the radical-level.

Long Short-Term Memory. Hochreiter and Schmidhuber et al. propose Long Short-Term Memory (LSTM) [5] which solves long-term dependency problem efficiently, meanwhile LSTM is widely used in machine translation [22], language modeling [9], and multilingual language processing [3]. Furthermore, LSTM also has a great performance in image captioning [24]. Thus, we choose LSTM as our topic classifier based on its excellent performance.

CNN Extract Image. Liu et al. [15] proposed a CNN Extract Image (CEI) model to generate word embedding, which exploits the visual characteristic of characters. CEI model concentrates on the visual data which mimics human action. CNN are employed to extract the visual information from character-level images in CEI model. The experiments show that the visual model can have almost identical performances with the look-up model. However, the CEI model has a huge disadvantage in training time. Based on the point of departure of reducing training time, we propose a character-level embedding model.

3 Overview

Our overall model has two main components: the PCA embedding layer and the RNN classifier, as illustrated in Fig. 2.

1. *The PCA embedding layer.* We apply PCA algorithm to transform the character images to fixed dimensional embedding vectors, which gives us an approach to extract the features from the 2-D image matrices. The embedding vectors generated from the PCA layer can be considered as the input vectors of the topic classifier.

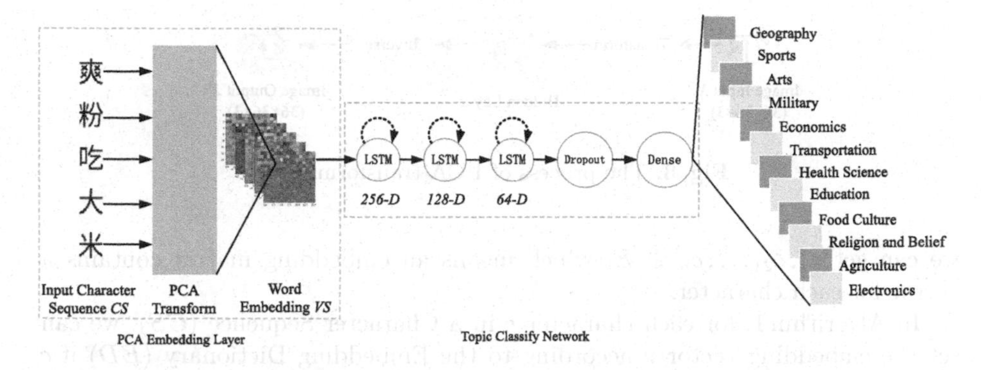

Fig. 2. The flowchart of Img2Vec model.

2. *The RNN classifier.* There are three layers of LSTM and one layer of fully-connected in the RNN network, and dropout [21] is adopted to make the Img2Vec model have a satisfactory generalization. For the RNN classifier, it regards minimize the loss function of classification as its target.

Our Img2Vec is a visual-based model with a better performance compared with previous works. The highlights of our model are summarized as follows:

1. *Better Performance.* Our Img2Vec takes several seconds to generate embedding vectors while Word2Vec-like method need a couple of hours. Simultaneously, our Img2Vec have a significant improvement in the accuracy. Comparison with the CEI model, our model has a same excellent performance to prior works by using less time to train the whole network.
2. *Exportable.* The embedding generated by the PCA embedding layer can be exported to a standalone embedding that can be utilized in other tasks. Moreover, it can initialize the embedding as a pretrained word embedding input like Word2Vec does.
3. *Visual Information.* We employ the visual information of characters to yield embedding, however, the mainstream methods take advantage of statistical analysis to generate embedding which ignore the visual information. Thus, we can coalesce our Img2Vec with Word2Vec to obtain better performance.

4 The PCA Embedding Layer

4.1 Layer Configuration

The PCA processing of our model is illustrated in Fig. 3. A character vocabulary C can be obtained by traversing the corpus. Then $c_1, c_2, \ldots, c_n \in c$ can be converted into $i_1, i_2, \ldots, i_n \in I$ which means image dataset. It should be noted that the input dimensionality and output dimensionality of the PCA layer can be alterable, which means the embedding dimensionality can be changed during training to adapt different applications. After PCA processing the image dataset,

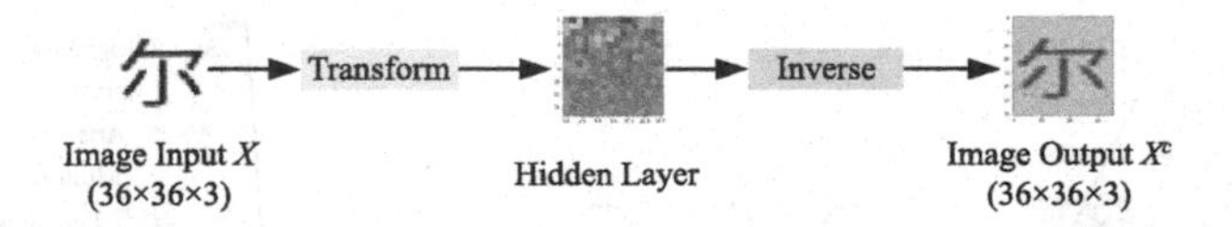

Fig. 3. The process of PCA transforming.

we can get $e_1, e_2, \ldots, e_n \in E$, which means an embedding matrix contains a vector for each character.

In Algorithm 1, for each character c in a Character Sequence (CS), we can set the embedding vector v according to the Embedding Dictionary (ED) if c belong to ED. Otherwise, we set a zero vector which has a same shape with embedding vector. Finally, the output Vector Sequence (VS) constructed by converted character vectors v is the input matrix of the RNN classifier.

Algorithm 1. Embedding layer algorithm.

Input: CS, ED
Output: VS
1 **for** *each c in CS* **do**
2   **if** $c \in ED$ **then**
3     $v \leftarrow c$
4   **else**
5     $v \leftarrow \mathbf{0}$
6   $VS \leftarrow VS \cup \{v\}$
7 **Return** VS

5 The RNN Classifier

An embedding vector dictionary VC which contains the corresponding vector of each characters can be gotten through the PCA embedding layer. We exploit VC

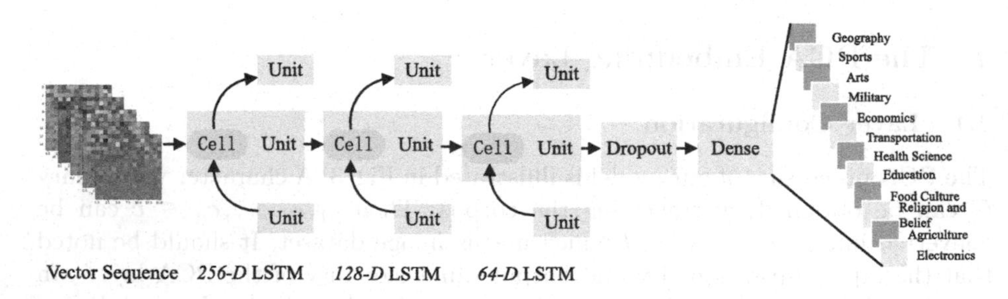

Fig. 4. We add a dropout layer and a fully-connected layer after three layers of LSTM.

to initialize the embedding layer of the neural network, transforming character to their corresponding embedding vectors. Then, we utilize a RNN network to implement a simplified Chinese text classification with three layers of LSTM as our experiment setup. The structure of topic classifier is illustrated in Fig. 4. We add a dropout layer and a fully-connected (FC or Dense) layer after three layers of LSTM.

6 Fusion Models

In this section, we describe two kinds of fusion models which are named early fusion model (Fig. 5 left) and late fusion model (Fig. 5 right) respectively. They are the supplementary models of Img2Vec model. Both of them can utilize the visual and statistic information to represent texts for better performance.

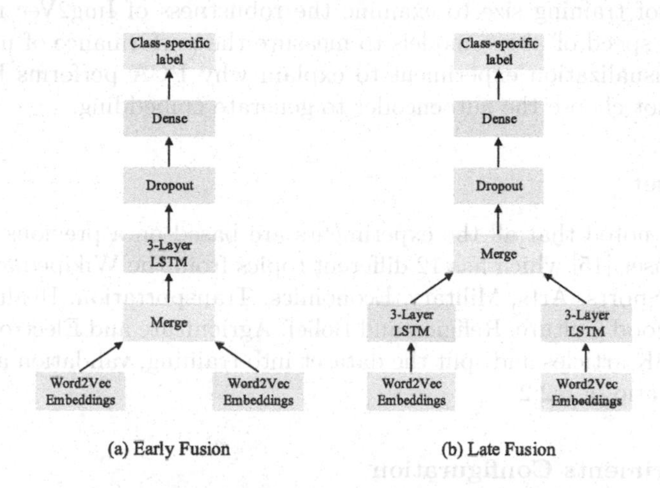

Fig. 5. The illustration of two fusion models.

6.1 Early Fusion Model

We describe each components of the early fusion model. First, two kinds of embeddings with fixed d-dimensions can be concatenated to a $2 \times d$-dimensions vector in the Merge layer. Then, 3-layer LSTM are applied to process the vector and reduce its dimensions to 64. Next, the dropout rate is set to $p = 0.5$ to prevent it from being overfitted. Finally, we construct a fully-connected layer with softmax function as our final classifier in the Dense layer.

6.2 Late Fusion Model

We now consider not to merge two kinds of embeddings at the beginning. Both embeddings are processed by 3-layer LSTM which final output vector dimensions are $d = 64$. Next, a merge layer concatenates the processed vectors whose output dimensions $2 \times d = 128$. Then we utilize the Dropout layer with the possibility of $p = 0.5$ to enhance generalization ability. Finally, a fully-connected layer is our classifier to output the labels. In this model, we don't concatenate the embeddings directly, instead, we concatenate the vectors after 3-layer LSTM processing to mix the visual and statistic information.

7 Experiments and Results

In this section, we introduce the dataset for our experiments. Then, we set an experiment configuration for a uniform standard, and discuss the trade-offs involved between different embedding dimensions. Next, a quantitative analysis is performed to evaluate the effects of models including our Img2Vec model, Word2Vec model, two kinds of fusion models and CEI model. We test different percentages of training size to examine the robustness of Img2Vec model and compare the speed of above models to measure the performance of models. We conduct a visualization experiment to explain why PCA performs better and why we do not choose the autoencoder to generate embedding.

7.1 Dataset

It should be noted that all the experiments are based on a previous Simplified Chinese dataset [15] which has 12 different topics from the Wikipedia web page: Geography, Sports, Arts, Military, Economics, Transportation, Health Science, Education, Food Culture, Religion and Belief, Agriculture and Electronics. They collected 593K articles and split the dataset into training, validation and testing sets with a ratio of 6:2:2.

7.2 Experiments Configuration

We sample images of characters with a configuration of a resolution of 36×36 pixels with 3 channels. Meanwhile, in the Img2Vec model, we need to use a conversion tool[1] to transform each character to corresponding image. Moreover, we use these images to train the PCA layer. After fitting the PCA layer, we use the processed low-dimensionality vector dictionary to initialize the visual embedding. And we deploy a pretrained Word2Vec model to initialize the Word2Vec embedding layer.

The batch size for entire models is set to $B = 4096$ and the dimensionality of the embedding d_c have 7 values containing the dimensionality of 32, 64, 128, 256, 512, 1024 and 2048. Moreover, Adam [10] with a learning rate $lr = 0.001$ is our stochastic optimization strategy. Meanwhile, to avoid overfitting, we use the dropout layer whose dropout rate is set to 0.5 to generalize models. All the sequences are padded to the identical length $l = 10$ which is same to previous works and the number of epochs are 50.

[1] https://github.com/djzgroup/Visual-basedCharacterEmbedding.

7.3 Experiment on Different Embedding Size

In these experiments, we test 7 kinds of visual embedding sizes to discuss the influence by dimension. The results are shown in Table 1. The table shows the different dimensionality does not have many associations with accuracy and F1-Scores. We consider 256 as our embeddings dimensions in the next experiments for achieving a balance between time and performance.

Table 1. The influences of different dimensionality.

Dimensions	Accuracy	F1-Scores
32	53.52%	**48.53%**
64	53.32%	47.97%
128	53.19%	47.74%
256	**56.31%**	48.04%
512	53.26%	47.89%
1024	53.40%	48.18%
2048	53.32%	48.17%

7.4 Comparison Between Models

We evaluate 5 kinds of model including Img2Vec model, Word2Vec model, the two kinds of fusion models and CEI model through performing a simplified Chinese text classification. We initialize our Word2Vec embedding layer with pretrained Word2Vec model[2] based on a WeChat[3] Simplified Chinese corpus. For a uniform standard, we set the batch size $B = 4096$, embedding size $d_c = 256$ and only use the standard three LSTM layers. What's more we adopt the same padding length with setups of Liu et al.

Table 2. The accuracy and F1-scores of Img2Vec, Word2Vec, early fusion, late fusion and CEI.

Accuracy/F1	Train	Test
Img2Vec	66.47%/56.59%	52.02%/47.04%
Word2Vec	56.74%/44.31%	**55.29%**/45.15%
Early fusion	68.38%/58.98%	54.75%/**49.65%**
Late fusion	**73.34%/64.04%**	53.94%/49.49%
CEI	–%/–%	55.07%/–

2 http://spaces.ac.cn/archives/4304/.
3 http://www.wechat.com/en/.

We exploit the above models to perform a simplified Chinese text classification to evaluate performances. The result is shown in Table 2. We can see our models are better than the Word2Vec model in training. But in testing, Word2Vec model has a little improve over others in accuracy. Meanwhile, all the models have almost identical performance in test. Thus, we attribute these problem to lack of generalization ability and the defects of datasets.

We can observe the two kinds of fusion models both have great performances in Table 2. And the late fusion model is better than the early fusion model in accuracy and F1-Scores. The late fusion model has 6.1M parameters while the early fusion model has 5.6M parameters. thus, more parameters make the late fusion model represent better.

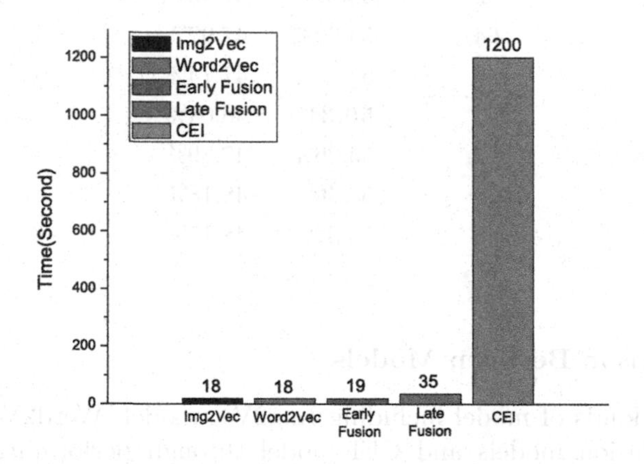

Fig. 6. Time comparison between Img2Vec, Word2Vec, early fusion, late fusion and CEI.

It should be noted that if we don't exploit the pretrain Word2Vec model, it occupies a long time to access the corpus and train the language model to obtain the Word2Vec model. Moreover, the train time of the CEI model is similar with the Word2Vec model. The CEI model needs to train the whole network which contains the CNN embedding layer and the RNN classifier, which leads them cost more time. Our Img2Vec model can be converge in 15 min, and it has a better performance of time compared to the Word2Vec model. The comparison of time costing among the above models is illustrated in Fig. 6. As we can see, the training process of CEI model has a poor efficiency. We blame CNN for learning global features from local features, whereas the PCA does not need this process.

7.5 The Result of Different Training Sizes

We test four training sizes, 25%, 50%, 75% and 100%, to indicate the influences. We find that there is about 5% of the gap in accuracy and 4% of the gap in F1-Scores between the full training size and the 25% training size.

The results of experiments which is illustrated in Fig. 7 meet our expectation and the results do not drop off too much which can suggest the visual information has a good representation of the character. Moreover, these experiments prove that our Img2Vec model has a high robustness.

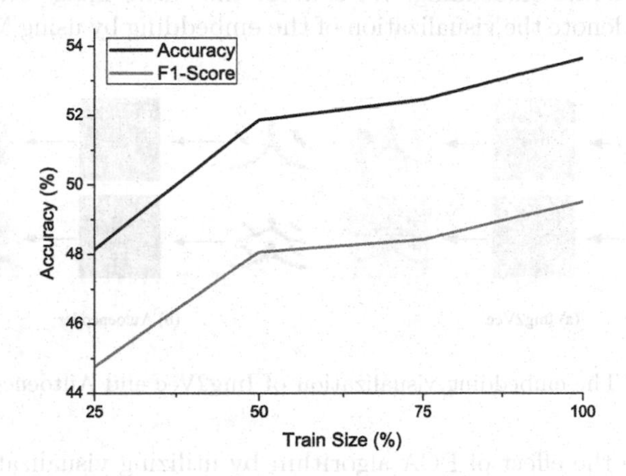

Fig. 7. The different size of training dataset comparison.

People exploit the pretrain embedding model like Word2Vec to initialize the weights of the embedding layer to improve convergence speed and avoid saddle point in many NLP applications. Img2Vec embedding and Word2Vec embedding are utilized to initialize the embedding layer of the RNN classifier and set to trainable separately. Table 3 shows the performances of different embeddings of initialization. The results show our Img2Vec embedding performs 7% better than Word2Vec embedding. Because Img2Vec learns embeddings from images, making similar texts have similar word vectors. We consider that the Img2Vec embedding has a specialty of keeping stabilization. No matter what kind of corpus is exploited, the embeddings remain unchanged, but the Word2Vec embedding lacks this stabilization. In theory, the bigger corpus is, the better Word2Vec embeddings performs, whereas it leads the training time increase if using big corpus.

Table 3. Pretrain embeddings comparison of Img2Vec and Word2Vec

Accuracy/F1	Img2Vec	Word2Vec
Train	**72.97%/64.86%**	65.63%/55.58%
Test	52.29%/**49.40%**	**54.95%**/48.30%

7.6 Embeddings Visualization

We perform embeddings visualization qualitatively to analysis the effects of representation of embeddings. Two kinds of visualization models are PCA-based and Autoencoder-based respectively. We analyze why our model can be an excellent approach to generate embedding. We convert the entire image sets to $256-D$ vectors and we denote the visualization of the embedding by using Matplotlib [6].

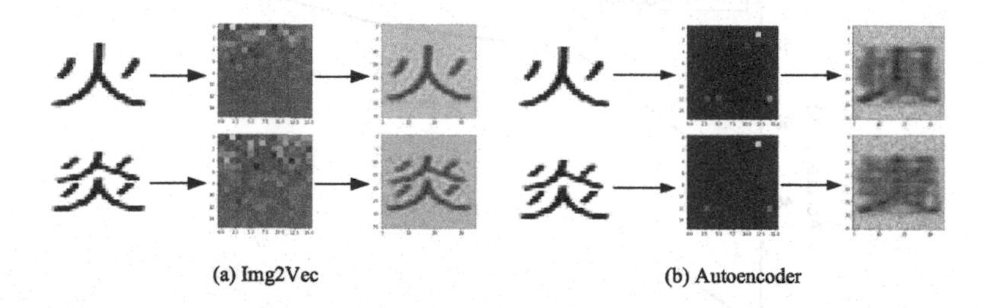

Fig. 8. The embedding visualization of Img2Vec and Autoencoder.

We evaluate the effect of PCA algorithm by utilizing visualization. As illustrated in Fig. 8 (left), the different characters have different embeddings. These characters with a same radical can denote an identical feature which is a serviceable attribute for character representation. Because characters which have one or more radicals often have semantic association with other characters [23]. This figure reflects that the PCA algorithm can convert the character into a vector with low loss of visual information. This method also works for the rare words in the corpus and the complex words. The PCA algorithm can be broken into some single radicals which appear in the other character. We can infer causal associations between the complex characters and the simple structure characters to represent the embedding for the complex characters. The smallest-grained items which are radicals can be represented in the visual information of the characters, thus word segmentation does not need to be applied in the Img2Vec model.

We construct an autoencoder [13,14] with two fully-connected layers which have a same purpose with the PCA model to reduce dimension. Moreover, we exploit the same datasets to train the autoencoder. We can find the relationships of radicals among the characters clearly, as illustrated in Fig. 8 (right). Compared with reconstruction images of embeddings from PCA algorithm, the images generated by the autoencoder model has disturbed by noise that it is hard to find out the characters. Thus the embedding vectors can not represent the visual information of the characters well. Consequently, the PCA algorithm can represent the global visual information of characters better than the autoencoder.

8 Conclusion and Future Work

We propose a novel method for extracting feature and improve pervious works via PCA. We peel off the train process from the whole process, thus the train processes can be standalone to be integrated in other applications. First, we propose a model called Img2Vec which contains an embedding layer based on PCA algorithm and a RNN classifier. Then, we construct two kinds of fusion models to utilize statistical information (Word2Vec [16,17]) and visual information (Img2Vec) simultaneously which both achieve better performance compared with models only using one kind of information. Next, we perform some experiments with different embedding size, model structure and training size. Finally, we construct two kinds of **visualization** to prove that Img2Vec model can generate visual information embedding effectively compared with autoencoder.

In the future, we will study about the influences of using different fonts of characters and the input image parameters. What's more, we will combine the local and global semantic information with the visual information to generate a new type of word embedding, which can perform well whether rareness the words are and can explain every radical on different grains.

Acknowledgments. This paper was supported by the National Science Foundation of China (Grant No. 61702350).

References

1. Blei, D.M., Ng, A.Y., Jordan, M.I.: Latent Dirichlet allocation. J. Mach. Learn. Res. **3**(10), 993–1022 (2003)
2. Dhingra, B., Zhou, Z., Fitzpatrick, D., Muehl, M., Cohen, W.W.: Tweet2vec: character-based distributed representations for social media, pp. 269–274, May 2016
3. Gillick, D., Brunk, C., Vinyals, O., Subramanya, A.: Multilingual language processing from bytes, November 2015
4. Harris, Z.S.: Distributional structure. In: Hiż, H. (ed.) Papers on Syntax. SLAP, vol. 14, pp. 3–22. Springer, Dordrecht (1981). https://doi.org/10.1007/978-94-009-8467-7_1
5. Hochreiter, S., Schmidhuber, J.: Long short-term memory. Neural Comput. **9**(8), 1735–1780 (1997)
6. Hunter, J.D.: Matplotlib: a 2D graphics environment. Comput. Sci. Eng. **9**(3), 90–95 (2007)
7. Iyyer, M., Manjunatha, V., Boyd-Graber, J., Daumé III, H.: Deep unordered composition rivals syntactic methods for text classification. In: Proceedings of the 53rd Annual Meeting of the Association for Computational Linguistics and the 7th International Joint Conference on Natural Language Processing, (Volume 1: Long Paper), vol. 1, pp. 1681–1691 (2015)
8. Jones, R., Rey, B., Madani, O., Greiner, W.: Generating query substitutions. In: Proceedings of the 15th International Conference on World Wide Web, pp. 387–396. ACM (2006)
9. Jozefowicz, R., Vinyals, O., Schuster, M., Shazeer, N., Wu, Y.: Exploring the limits of language modeling, February 2016
10. Kingma, D., Ba, J.: Adam: a method for stochastic optimization, December 2014
11. Krizhevsky, A., Sutskever, I., Hinton, G.E.: ImageNet classification with deep convolutional neural networks. In: Advances in Neural Information Processing Systems, pp. 1097–1105 (2012)

12. Le, Q., Mikolov, T.: Distributed representations of sentences and documents. In: Proceedings of the 31st International Conference on Machine Learning (ICML 2014), pp. 1188–1196 (2014)
13. Liou, C.Y., Cheng, W.C., Liou, J.W., Liou, D.R.: Autoencoder for words. Neurocomputing **139**, 84–96 (2014)
14. Liou, C.Y., Huang, J.C., Yang, W.C.: Modeling word perception using the elman network. Neurocomputing **71**(16), 3150–3157 (2008)
15. Liu, F., Lu, H., Lo, C., Neubig, G.: Learning character-level compositionality with visual features. In: Proceedings of the 55th Annual Meeting of the Association for Computational Linguistics (Volume 1: Long Papers), pp. 2059–2068. Association for Computational Linguistics, Vancouver, July 2017. http://aclweb.org/anthology/P17-1188
16. Mikolov, T., Chen, K., Corrado, G., Dean, J.: Efficient estimation of word representations in vector space, January 2013
17. Mikolov, T., Sutskever, I., Chen, K., Corrado, G.S., Dean, J.: Distributed representations of words and phrases and their compositionality. In: Advances in Neural Information Processing Systems, pp. 3111–3119 (2013)
18. Paşca, M., Lin, D., Bigham, J., Lifchits, A., Jain, A.: Names and similarities on the web: fact extraction in the fast lane. In: Proceedings of the 21st International Conference on Computational Linguistics and the 44th annual meeting of the Association for Computational Linguistics, pp. 809–816. Association for Computational Linguistics (2006)
19. Pennington, J., Socher, R., Manning, C.: Glove: global vectors for word representation. In: Proceedings of the 2014 Conference on Empirical Methods in Natural Language Processing (EMNLP), pp. 1532–1543 (2014)
20. Roweis, S.T., Saul, L.K.: Nonlinear dimensionality reduction by locally linear embedding. Science **290**(5500), 2323–2326 (2000)
21. Srivastava, N., Hinton, G.E., Krizhevsky, A., Sutskever, I., Salakhutdinov, R.: Dropout: a simple way to prevent neural networks from overfitting. J. Mach. Learn. Res. **15**(1), 1929–1958 (2014)
22. Sutskever, I., Vinyals, O., Le, Q.V.: Sequence to sequence learning with neural networks. In: Advances in Neural Information Processing Systems, pp. 3104–3112 (2014)
23. Taft, M., Zhu, X., Peng, D.: Positional specificity of radicals in Chinese character recognition. J. Memory Lang. **40**(4), 498–519 (1999)
24. Vinyals, O., Toshev, A., Bengio, S., Erhan, D.: Show and tell: a neural image caption generator. In: Proceedings of the IEEE Conference on Computer Vision and Pattern Recognition, pp. 3156–3164 (2015)
25. Zhang, X., Zhao, J., LeCun, Y.: Character-level convolutional networks for text classification. In: Advances in Neural Information Processing Systems, pp. 649–657 (2015)

A Novel Experience-Based Exploration Method for Q-Learning

Bohong Yang[1], Hong Lu[1(✉)], Baogen Li[2], Zheng Zhang[3], and Wenqiang Zhang[2]

[1] Shanghai Key Laboratory of Intelligent Information Processing, School of Computer Science, Fudan University, Shanghai, People's Republic of China
honglu@fudan.edu.cn
[2] Shanghai Engineering Research Center for Video Technology and System, School of Computer Science, Fudan University, Shanghai, People's Republic of China
[3] School of Computer Science, New York University Shanghai, Shanghai, People's Republic of China

Abstract. Reinforcement learning algorithms are used to deal with a lot of sequential problems, such as playing games, mechanical control, and so on. Q-Learning is a model-free reinforcement learning method. In traditional Q-learning algorithms, the agent stops immediately after it has reached the goal. We propose in this paper a new method—Experience-based Exploration method—in order to sample more efficient state-action pairs for Q-learning updating. In the Experience-based Exploration method, the agent does not stop and continues to search the states with high bellman-error inversely. In this setting, the agent will set the terminal state as a new start point, and generate pairs of action and state which could be useful. The efficacy of the method is proved analytically. And the experimental results verify the hypothesis on Grid-world.

Keywords: Inverse exploration · Reinforcement learning
Q-learning · Deep learning · Deep Q-learning Network

1 Introduction

The goal of reinforcement learning [1] is to learn good policies for agent to maximize their expected future rewards from interacting with an unknown environment. The idea on which reinforcement learning based is deeply rooted in psychological and neuroscientific perspectives on animal behaviour. This universal problem applies on learning control of mobile robot, optimizing operation sequence in factory, learning the board games, etc.

Supervised learning has some difficulties during dealing with reinforcement learning problems, since the reward signal in reinforcement learning (corresponds to the label in supervised learning) is normally sparse, noisy and delayed.

Q. Zhou et al. (Eds.): ICPCSEE 2018, CCIS 901, pp. 225–240, 2018.
https://doi.org/10.1007/978-981-13-2203-7_17

Another issue is that for supervised learning, the data is assumed that is subjects to a fixed underlying distribution. While for reinforcement learning, the data distribution normally changes when the algorithm is learning new behaviors. Whereas, Deep Q-learning Network (DQN) [2] demonstrates that neural network can overcome these challenges to perform better in controlling policy learning which reinforcement learning want to learn.

In some situations such as robot working space, traditional methods use the 4-tuple sampled in exploration to update the policy. But it wastes much information when resetting the environment. So the method converges slowly.

In this paper we propose a method to help the reinforcement learning algorithm to sample more useful state-action pairs to update its policy. We prove that our method could make the updating of policy more efficient. Experimental results demonstrate that the method can improve policy converge effectively and efficiently.

2 Background

Reinforcement learning [1] focuses on learning control policies for agents interacting with an unknown environment. The environment is usually modelled as a Markov decision process (MDP) which is completely described by a tuple(S, A, P, R), i.e. the set of states, the set of actions, the probability of transformation and the reward function. Many reinforcement learning algorithms learn from sequential experience in the form of transition tuples $(s_t, a_t, r_{t+1}, s_{t+1})$. At each time step t, the agent interacting with the MDP observes a state $s_t \in S$, and chooses an action $a_t \in A$ which determines next state $s_{t+1} \sim P(s_t, a_t)$ with the reward $r_{t+1} \sim R(s_t, a_t)$. The goal of agent is to improve its policy in order to maximize its discounted gain, i.e. $G_t = \sum_{i=t+1}^{T} \gamma^{i-(t+1)} r_i$, where γ is the discount coefficient of rewards. The discounted gain is a random variable of the agent's cumulative feature rewards starting from time t. A common objective of them is to learning the action-value function, $Q(s, a) = \mathbb{E}^{\pi}[G_t | s_t = s, a_t = a]$, which is defined as the expected gain of takeing action a in state s and following policy π thereafter. If an agent learns about the policy that it currently subjects to, we call it learning on-policy. Otherwise, if the agent learns from experience of another agent or another policy, we call it learning off-policy.

Q-learning [3] is a popular off-policy reinforcement learning method. It learns about the greedy policy which always chooses the action with the highest estimated value Q at each state. The bellman optimal equation will update the current Q-value estimation iteratively towards the observed reward r and the estimated utility of the resulting state s':

$$Q(s, a) = Q(s, a) + \alpha(r + \gamma \max_{a'} Q(s', a') - Q(s, a)) \tag{1}$$

A naive method to store the Q-value is using an $S \times A$ array to represent the value of each individual state-action pair. The array is called a look-up table. We could update the individual Q-values with Eq. (1) for state-action pairs incrementally. But in many challenging domains, such as Atari games, there are too many unique states. It would lead to "Curse of Dimensionality" if we try to maintain a separate estimate for each state-action pair. Deep Q-learning Network (DQN) [2] uses function approximator to approximate the Q-values. Specifically, it uses a neural network parameterized by weights and biases denoted as θ to represent the Q-value, and build an experience pool to store the agent's experiences at each time-step. Then it updates the network parameters θ in batch by minimizing a sequence of loss functions $L_i(\theta_i)$ that changes at each iteration i. The loss function is defined as below.

$$L_i(\theta_i) = \mathbb{E}_{s,a\sim\rho}[(y_i - Q(s,a;\theta_i))^2] \tag{2}$$

where $\rho(s,a)$ is a probability distribution over state s and action a that we refer to as the behaviour distribution.

In traditional methods, the agent gets s_t from the environment, takes a_t depended on its policy, gets r_{t+1} and s_{t+1} and updates its policy with the 4-tuple $(s_t, a_t, s_{t+1}, r_{t+1})$ repeatedly, until the agent reaches the terminal state.

3 Related Work

In the domain of game control, DQN [2] performs better than human players in Atari games. The network is composed of convolutional neural network (CNN) and fully-connected layers instead of linear function approximator. DQN has some problems such as overestimating and relying on experience pool. For overestimating, [4] proposes Double DQN inspired from [5] to reduce the observed overestimation. For experience pool, [6] uses priority experience pool to improve the efficiency of learning. In addition, [7] optimizes DQN controller with asynchronous gradient descending algorithm. There is much work on DQN. [8] defines a novel method, i.e. "Actor-Mimic" of multi-task and transfer learning. [9] presents a new dueling neural network architecture which has two separate estimators. [10] trains AlphaGO with the combined algorithm of reinforcement learning and Monte Carlo tree search program. In March 2016, AlphaGO won the match by 4 to 1 with S. Lee who is one of the top GO players. And in May 2017, AlphaGO won the match by 3 to 0 with J. Ke who is the best GO player in the world now.

For rigid body control problem, policy gradient algorithms are perhaps the most popular class of continuous action reinforcement learning algorithms. [11] presents the deterministic policy gradient algorithm instead of stochastic policy gradient algorithm. [12] solves more than 20 simulated physics tasks with deep deterministic policy gradient algorithm and deep neural network. In [13], reinforcement learning also directs the agent to shoot and performed pretty well, better than the world champion of 2012.

In other domains, [14] investigates the possibility of exploiting emotions in agent learning in order to facilitate the emergence of cooperation in social dilemmas. [15] proposes the first probably approximately correct algorithm for continuous deterministic systems without relying on any system dynamics. [16] proposes a method for the choice and adaptation of the smoothing parameter of the probabilistic neural network. [17] shows how self-organizing neural networks designed for online and incremental adaptation can integrate domain knowledge and reinforcement learning. [18] tries to address this challenge by introducing a recurrent deep neural network for real-time financial signal representation and trading.

4 Method

In robot working space, the cost of resetting the environment is expensive. The reason lies that resetting the environment means to control the robot go back from the terminal state to the start state. Traditional methods lose much information of resetting the environment. And the agents learn nothing during resetting process. Then we try to make the agents extract more useful information to learn in the process. In our method, we change the environment settings. Specifically, we remove the absorbing state, so the agent could still explore the whole space after it has reached the goal we set. And we need not move the agent from the terminal state to the start point. The episode is not be finished until the agent reaches the goal for the second time.

By observing the updating process of Q-learning, we could find that back propagation of reward signal from terminal state to start point through all states visited in the trajectory is the essence of Q-learning. A simple idea is to update the states' estimation following the inverse order of the trajectory, such as Monte-Carlo method, which updates the state's estimate according to the next state that is close to goal during one updating phase. However, it is too hard to store the whole trajectory if the trajectory is infinite. One solution is to explore inversely, which means the agent will continue exploring after it has reached the terminal state. The agent will take the original terminal state as a new start point. And the reward will be propagated forward along the trajectory. Then we propose a method which is shown in Algorithm 1. The method makes the agent inversely search and sample states after it reaches the terminal state.

In fact, our method includes two games and unify them together. Our method makes the agent play two games. One game is the original game, which has the long winding paths. It's hard for the agent to learn. The other is the inverse game, which has only one "straight" path and makes it easy for agent to learn. This inverse game makes the value function converges more efficiently and allows the agent to learn the original game. In order to unify these two games, we propose an algorithm as below.

We give the agent a coarse module P_R based on its own physical structure in the line 8 of Algorithm 1, such as "go straight" would make the agent move straight. This module could be different from environmental model. But the module could help the agent to explore the environment.

The module P_R is an alternative representation of the robot's system error. It can be obtained by pre-training. We train the module from the easy case which is only 3×3 Gridworld. The agent could use this module P_R to find the set of

Algorithm 1. Framework of Experience-Based Exploration Method.

1: Initialize, $inverse\_flag = 0$ and $last\_reward = 0$
2: Initialize the action-value function Q
3: **for** $t = 1$ to T **do**
4: //pick an action a based on state s
5: **if** $inverse\_flag == 0$ **then**
6: get an action a with $\epsilon - greedy$ strategy
7: **else**
8: bulid a set of neighbor states
$S' = \{s' | \exists a, s' * a \rightarrow s\}$
9: compute bellman-error
$e(s') = (y - Q_t(s', a))^2 | s' * a \rightarrow s$
where $y = r(s', a, s) + max_{a''} \gamma Q_t(s, a'')$
10: pick up an action with the action selection strategy (ASS)
$a' = ASS(e(s'))$
11: get an inverse action
$a = a'^{-1}$
12: **end if**
13: execute action a in emulator and observe reward r and next state s'
14: //update based on the tuple (s, a, s', r)
15: **if** $inverse\_flag == 0$ **then**
16: add sample tuple (s, a, s', r)
17: **else**
18: add sample tuple $(s', a^{-1}, s, last\_reward)$
19: **end if**
20: update action-value function Q with bellman optimal equation Eq. (1)
21: $last\_reward = r$
22: **if** $s \in terminalstates$ **then**
23: $inverse\_flag = 1 - inverse\_flag$
24: **end if**
25: $inverse\_flag = 0$ if $inverse\_flag == 1$ for a long time
26: **end for**

neighboring states it could reach in one step. Then we compute the bellman-error of all candidate states with bellman optimal equation Eq. (1). The larger the errors are, the fewer times the state has been visited. And the higher possibilities it should be visited in exploration.

As for action selection strategy, we select the action according to the possibilities for every action computed by $\epsilon - greedy$ strategy. We compare 3 kinds of action selection strategies in Sect. 5. The first one is greedy strategy. We choose the action with the largest bellman-error, i.e. $\pi = argmax_a Q(s, a)$. The second one is $\epsilon - greedy$ strategy. It is an exploration strategy which selects a random action with probability ϵ, otherwise selects the action with max Q-value. The last one is softmax strategy. We choose the action with the probabilities computed by softmax function, i.e. $\pi = P(a|s) = \frac{exp(Q(s,a))}{\sum_{a'} exp(Q(s,a'))}$. When the $inverse\_flag$ is false, we select action with $\epsilon - greedy$ strategy which is the same as the baseline—DQN's strategy, otherwise, we select action from those 3 strategies.

We could regard samples from trajectories searched inversely both as inverse samples and as normal samples. But in some special conditions, the samples from trajectory can not be reversed, such as the agent hitting the obstacles. In this case, the agent will take one action a and stay in the same state s. If we reverse this sample, the agent will take inverse action a^{-1} and stay in the same state. Then we would get incorrect states in tuples. Thus we remove all these inverse samples where $s = s'$ in 4-tuple (s, a, s', r).

As for the reward, we only care about the state after the agent has taken action. The reward function could be rewritten as $r_t(s_t, a, s_{t+1}) \sim r(s_{t+1})$, because the environment gives reward in accordance with the state instead of the action. The reward function in computing bellman-error could be calculated by

$$r_t(s', a, s) = \frac{\sum_{\tau=1}^{t} r_\tau \delta s', s_{\tau+1}}{\sum_{\tau=1}^{t} \delta s', s_{\tau+1} + \varepsilon} \tag{3}$$

where $\delta x, y$ is the Kronecker delta function, and the ε is a small number added with denominator to prevent divided by zero.

In inverse sampling, the last time-step reward is related to the state s'. Then this reward should be the reward of the reverse sample (s', a^{-1}, s). This time difference is caused by the environment with giving reward according to the state instead of the action And we set the value *last_reward*.

Inverse exploration starts when the agent reaches the terminal state and stops with finite steps according to the state space, or breaks when the agent reaches the terminal state again. *inverse_flag* indicates whether agent is inversely searching or not. Since terminal states are sparse in state space, the *inverse_flag* will not change frequently.

The whole process is like that a person resets the agent to where the agent always makes mistakes, but the resetting works automatically. If the part of experience based exploration is removed, the process degenerates into imitation learning. We propose to make the agent inversely explore and reverse the sample for updating policy. And the method enables the reward information be propagated from the terminal state. It is similar to update with Monte-Carlo method step by step.

Our idea comes from the bellman-equation showing in Eq.(1). Specifically, according to the bellman-equation, only updating from the goal to the start point could make the reward propagates to other state, such as dynamic programming. Normally, the states around the goal are updated first. It's useless to update the states around the start point in the beginning of learning.

Consider 2 episode sampling results in a simple case where there is only one way to reach the goal. In order to simplify the presentation, we represent the state of quad (s, a, s', r) abbreviated as bigram (s, s'). The traditional sampling result is as below.

$$\begin{aligned} Episode1: &\quad (s_0, s_1), (s_1, s_2), ..., (s_{n-1}, s_n).\\ Episode2: &\quad (s_0, s_1), (s_1, s_2), ..., (s_{n-1}, s_n). \end{aligned}$$

where n is the length of path, s_0 means the start point and s_n the goal. Each episode ends once the agent reaches the goal.

While the sampling result of our proposed method is as below.

$$\begin{aligned} Episode1: \quad & (s_0, s_1), (s_1, s_2), ..., (s_{n-1}, s_n), \\ & (s_{n-1}, s_n), (s_{n-2}, s_{n-1}), ..., (s_0, s_1). \end{aligned}$$

where samples in the inverse order have been flipped, e.g. the second (s_{n-1}, s_n) in the sampling result. The sampling order would not change as the updating order changes.

As we know, in this case, the expectation of value function could be written as

$$\begin{aligned} E[V^t(s_i)] &= max_a(Q(s_i,a)) = Q(s_i,a) \\ &= P^t(s_i) * (r_i + E[V^{t-1}(s_{i+1})]) + (1 - P^t(s_i)) * E[V^{t-1}(s_i)] \end{aligned} \quad (4)$$

where $P^t(s_i)$ means the probability of updating sample (s_i, s_{i+1}) in time t. The n times' updating order of s_i is $(s_{n-1}, s_{n-2}, ..., s_3, s_2, s_1, s_0)$. The initializations of the function V are all 0. And the rewards are all 0 except r_{n-1}. After expand $E[V]$ repeatedly, we only consider the product of probabilities $\hat{P}$ below.

$$\begin{aligned} E[V^t(s_0)] &= P^t(s_0) * P^{t-1}(s_1) * P^{t-2}(s_2) * ... * P^1(s_{n-1}) * r_{n-1} \\ &= \prod_{t=1}^{n} P^t(s_i) * r_{n-1} = \hat{P} * r_{n-1} \quad s.t. \quad i = n - t \end{aligned} \quad (5)$$

Then we analyze our proposed method and the traditional sampling method.

The previous $n-1$ steps are the same for the two methods. The reward in the goal cannot be propagated from the goal in these $n-1$ steps. There are n samples $(s_{n-1}, s_{n-2}, ..., s_1, s_0)$ need to be updated, and the two methods have $2n-(n-1) = n+1$ times to update. Therefore they have $n+1-n = 1$ time to make mistake. In other word, they have one chance to update samples without following the order of $(s_{n-1}, s_{n-2}, ..., s_3, s_2, s_1, s_0)$.

Take $m = 0, n = 6$ as an example, where m is the number of the irrelevant samples in experience pool before Episode 1 starts. Table 1 shows the probability of picking up given samples from the experience pool without normalization. The first column in the table means the two methods make mistake in the ith update. The numbers in the rest $n+1$ columns mean the number of samples the two methods could use to update.

Look at first row of the two methods in Table 1, i.e. $j = 1$. For our method, it should not update s_5 at time $t = 1$, hence the number of samples the method could use to update is $6-1 = 5$. To fulfil the target, at time $t = 2$, it should update s_5. Since there are two samples s_5 in the experience pool, the number of samples the method could use to update is 2. Similarly, for traditional sampling method, it should not update s_5 at time $t = 1$, hence the number is $6-1 = 5$. Then it should update s_5 at time $t = 2$. Since at this time, Episode 2 starts. There is only one sample s_5 in the experience pool, hence the number of samples the method could use to update is 1. The probability of our method for $j = 1$ is product of $\frac{5}{6}$, $\frac{2}{7}$,..., $\frac{2}{12}$. Generally, the product of denominators in each row is same and equals to $(m+n)(m+n+1)(m+n+2)...(m+2n) = (m+2n)!/(m+n-1)!$. The probability of the whole method is the sum of the product of each row .

Table 1. Probability of picking up the given sample without normalization.

The time j making mistake	Our method							The time j making mistake	Traditional sampling method						
1	5	2	2	2	2	2	2	1	5	1	1	1	2	2	2
2	1	6	2	2	2	2	2	2	1	6	1	1	2	2	2
3	1	1	7	2	2	2	2	3	1	1	7	1	2	2	2
4	1	1	1	8	2	2	2	4	1	1	1	7	2	2	2
5	1	1	1	1	9	2	2	5	1	1	1	2	8	2	2
6	1	1	1	1	1	10	2	6	1	1	1	2	2	9	2
7	1	1	1	1	1	1	12	7	1	1	1	2	2	2	12

The product of probabilities of our method is equal to

$$\begin{aligned}
\hat{P}_{ours} &= \frac{(m+n-1)!}{(m+2n)!}[(m+n-1)2^n + (m+n)2^{n-1} + ... \\
&\quad + (m+2n-2)2^0 + (m+2n-1) + 1] \\
&= \frac{(m+n-1)!}{(m+2n)!}[(m+n-1)2^{n+1} + 2^n + 2^{n-1} + ... \\
&\quad + 2 - (m+n-1+n) + 1] \\
&= \frac{(m+n-1)!}{(m+2n)!}[(m+n-1)2^{n+1} + 2^{n+1} - 1 - (m+2n-1)] \\
&= \frac{(m+n-1)!}{(m+2n)!}[(m+n)2^{n+1} - (m+2n)]
\end{aligned} \tag{6}$$

And the product of probabilities of the traditional sampling method is equal to

$$\begin{aligned}
\hat{P}_{trad} &= \frac{(m+n-1)!}{(m+2n)!}2^{\frac{n}{2}}[(m+n-1) + (m+n) + ... + (m+n+\frac{n}{2}-2) \\
&\quad + (m+n+\frac{n}{2}-2) + ... + (m+2n-3) + (m+2n)] \\
&= \frac{(m+n-1)!}{(m+2n)!}2^{\frac{n}{2}}[(m+\frac{5}{4}n-\frac{3}{2})\frac{n}{2} + (m+\frac{7}{4}n-\frac{5}{2})\frac{n}{2} + (m+2n)] \\
&= \frac{(m+n-1)!}{(m+2n)!}2^{\frac{n}{2}}[(m+\frac{3}{2}n-2)n + (m+2n)]
\end{aligned} \tag{7}$$

Dividing Eq. (6) by Eq. (7), we can obtain

$$Y = \frac{(m+n)2^{n+1} - (m+2n)}{2^{\frac{n}{2}}[(m+\frac{3}{2}n)n + m]} \tag{8}$$

We only need to consider whether the last equation, i.e. Y, is larger than 1 or not. The partial derivative of Y with respect to m is

$$\frac{\partial Y}{\partial m} = \frac{(\frac{1}{2}n - 1)n2^{n+1} + \frac{1}{2}n^2 + 2n}{2^{n/2}[(m+\frac{3}{2}n)n + m]^2} \tag{9}$$

The denominator of Eq. (9) is larger than or equal to 0. When $n \geq 2$, $(\frac{1}{2}n - 1) \geq 0$, hence the numerator of Eq. (9) is larger than 0. That is

$$Eq.(9) > 0 \quad s.t. \quad n \geq 2, \quad m \geq 0 \tag{10}$$

Since Y has exponential items, the partial derivative of Y with respect to n is hard to compute the analytic solution. And Y can be written as

$$\begin{aligned} Y &= \frac{(m+n)2^n + (m+n)2^n - (m+2n)}{2^{\frac{n}{2}}[(m+\frac{3}{2}n)n + m]} \\ &> \frac{(m+n)2^n}{2^{\frac{n}{2}}[(m+\frac{3}{2}n)n + m]} = \frac{(m+n)2^n}{2^{\frac{n}{2}}[m(n+1) + \frac{3}{2}n^2]} \\ &= \frac{m2^{\frac{n}{2}} + n2^{\frac{n}{2}}}{m(n+1) + \frac{3}{2}n^2} \quad s.t. \quad n \geq 2, \quad m \geq 0 \end{aligned} \tag{11}$$

When $n \geq 7$, $2^{\frac{n}{2}} > (n+1)$ and $2^{\frac{n}{2}} > \frac{3}{2}n$. Then Eq. (11) is larger than 1. When $n = [2, 3, 4, 5, 6]$, Y is still larger than 1.

The value of the function Y is shown in Fig. 1. It can be observed that when $m \geq 0$ and $n \geq 2$, the function Y in Eq. (8) is larger than or equal to 1. Therefore our method is better than the traditional sampling method when $n \geq 2$. In the special case when $n = 1$, the two methods are the same.

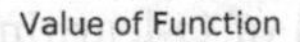

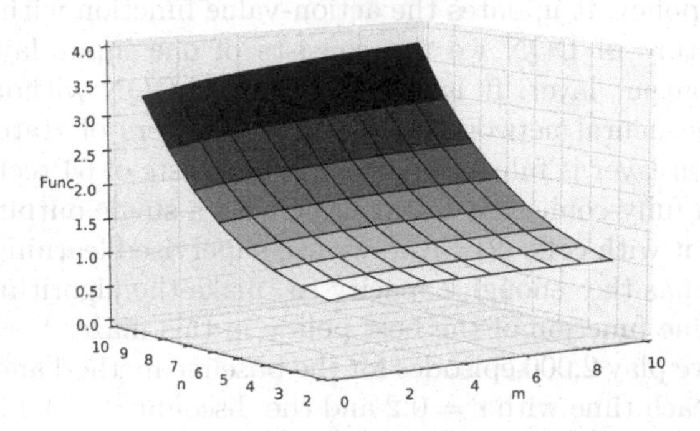

Fig. 1. The value of the function Y in Eq. (8). The larger the function Y value is, the deeper the corresponding gray scale is.

5 Experiments

5.1 Experimental Setting

In our experiment, we use the Gridworld as the environment. The Gridworld is from UC Berkeley CS188 [19]. It is a maze as shown in Fig. 2. The blue point in the maze represents the agent. All dark grids divided into four triangles are free to move. The gray grids not divided represent the obstacles. The grids with number 1 or -1 mean the terminal states. The agent could only get rewards from the terminal states.

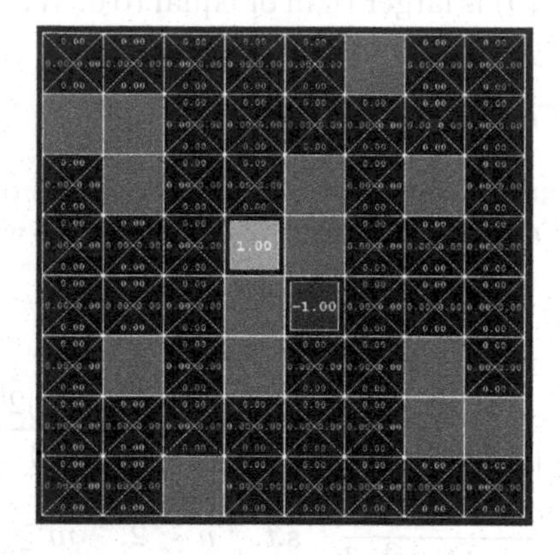

Fig. 2. The map of the maze is an 8×8 Gridworld. (Color figure online)

We use a policy search algorithm, i.e. Pegasus [20], to evaluate the performance of our method. Pegasus is an algorithm to evaluate policy performance by setting the environment from stochastic MDP to deterministic MDP. Then we can obtain episodes with the start points randomly generated in each episode.

We test our method on Deep Q-learning Network (DQN) [2] and use the new sampling method. DQN is a kind of Q-learning method with model-free, sample backup, and off-policy. It updates the action-value function with batch method.

The architecture of DQN we use consists of one input layer, one hidden layer, and one output layer. It is a lite version of DQN without CNN layers. The input to the neural network consists of 2 integers of state, e.g. (1, 1) or (3, 2). The hidden layer is fully-connected and consists of 64 rectifier units. The output layer is a fully-connected linear layer with a single output for each valid action. We train it with caffe [21]. And we use supervised learning to ensure that the architecture has the enough capacity to make the algorithm converge and represent the value function of the best policy in this maze.

For testing, we play 2,000 episodes for the baseline method and 1,000 episodes for our method each time with $\epsilon = 0.2$ and the discount $\gamma = 0.9$ in one game. To ensure the agent can reach the goal, the agent goes 5,000 steps under random policy or 200 steps under optimal policy. Then the agent finishes the rest episodes following its own policy, such as Greedy, Epsilon, and Softmax. The experimental steps and returns are the average of 32 times games. And the experiments' results are the average steps (the steps before the inverse phase of our method) and returns of 32 times for each action selection strategy. The curves of results are affected by stochastic fluctuation. Therefore, we present the result with the cumulative value function to observe the trend of results.

5.2 Result Analysis

DQN with and Without Noise. Figure 3 shows the mean steps and returns of DQN without noise on the maze of 8×8 Gridworld. The larger the returns are, the better the result is. Otherwise, the smaller the number of steps is, the better the result is.

It can be observed from Fig. 3 that our method with different action selection strategies performs better than the traditional sampling method (baseline) with DQN. Among the action selection policies, the greedy strategy performs best, and $\epsilon - greedy$ strategy performs better than softmax strategy. All these strategies perform better than the baseline method.

As we mentioned earlier, we give the agent a coarse model without considering the noise. This coarse model would be different from the real environment. Then we need to test whether the model can work with noise or not. The noise is defined as that the action the agent executes is different with the probability of 20% from the action selected by the strategy. Figure 4 shows the mean steps and returns of DQN with noise.

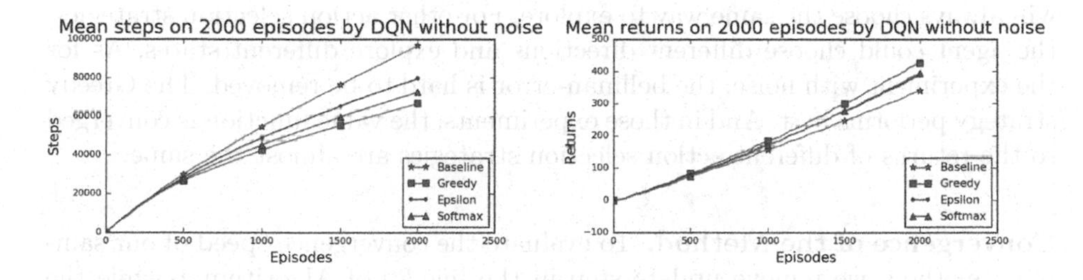

Fig. 3. Mean steps and returns of DQN without noise on 8×8 Gridworld.

It can be observed from Fig. 4 that our method can work with noise. Specifically, the tendency of the curves of the return in Fig. 4 is similar to that of Fig. 3. The reason lies that we use the same random seed of starting points in one experiment.

It can be observed from Table 2 that our method performs better than the traditional sampling method. The larger the returns are, the better the result is.

In the process of inverse sampling, we found that Greedy strategy performs better than other action selection strategies. It suggests that we should always select the direction with largest bellman-error to explore and adjust.

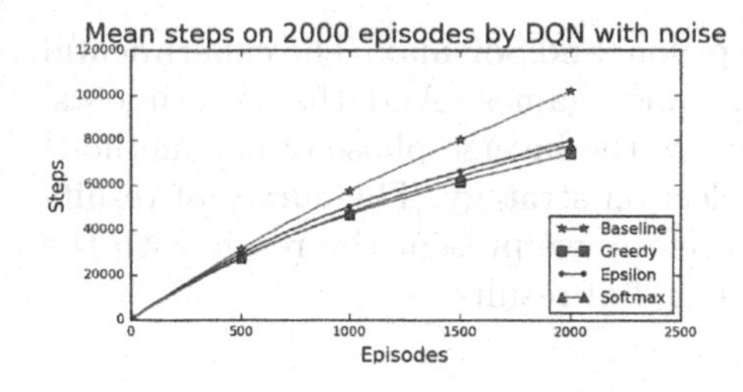

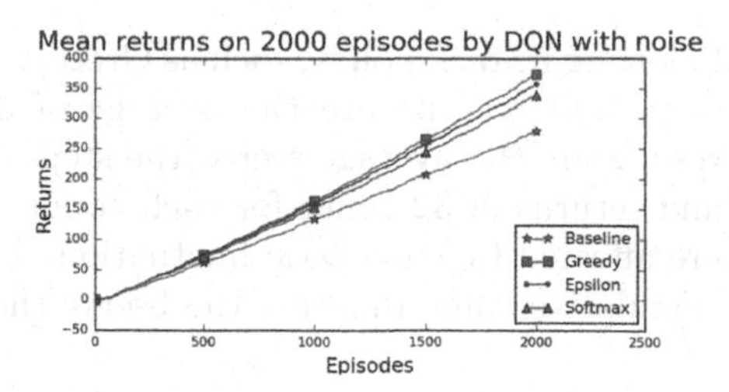

Fig. 4. Mean steps and returns of DQN with noise on 8×8 Gridworld.

Table 2. Mean returns of DQN

Strategy	w/o noise	Noise
Baseline	0.170	0.140
Greedy	**0.213**	**0.187**
Epsilon	0.204	0.179
Softmax	0.207	0.170

Table 3. Mean returns of look-up table

Strategy	w/o noise	Noise
Baseline	0.459	0.365
Greedy	0.458	**0.367**
Epsilon	0.460	0.366
Softmax	**0.461**	0.366

Look-Up Table with and Without Noise. We also test our method on a simple Q-learning with look-up table. It updates the action-value function with incremental method, and converges faster than DQN in a small set of states. The result is tabulated in Table 3.

It can be observed from Table 3 that Greedy strategy fails to perform better. The reason is that the look-up table updates the values relatively fast. The bellman-error near terminal states towards to 0 in short time, then the agent will always choose the same way to explore. For other action selection strategies, the agent could choose different directions and explore different states. As for the experiment with noise, the bellman-error is hard to be removed. The Greedy strategy performs best. And in those experiments, the value function is converged so the returns of different action selection strategies are almost the same.

Convergence of the Method. To evaluate the convergence speed of our sampling method, we remove update step in the line 20 of Algorithm 1 while the agent is in the inverse phase. So the times of updating in this method is the same as that of the baseline method. It means that in the methods, the agent both updates the value function when it searches the path towards the goal, and does not update with bellman-error searching. It is worth noting that the agent keeps sampling throughout the process. Experimental result is shown in Fig. 5. It can be observed from Fig. 5 that our sampling method still converges faster than traditional method without additional updating. It means our method collect the better samples for updating.

Experiment on Policy of Initializing the Experience Pool. In order to test whether the method can perform better on different policies of initializing

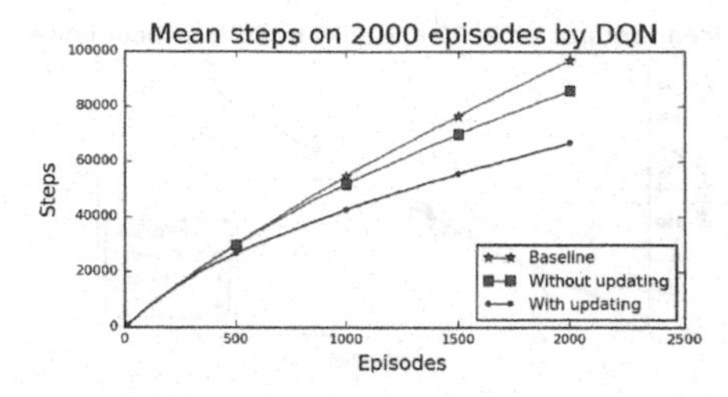

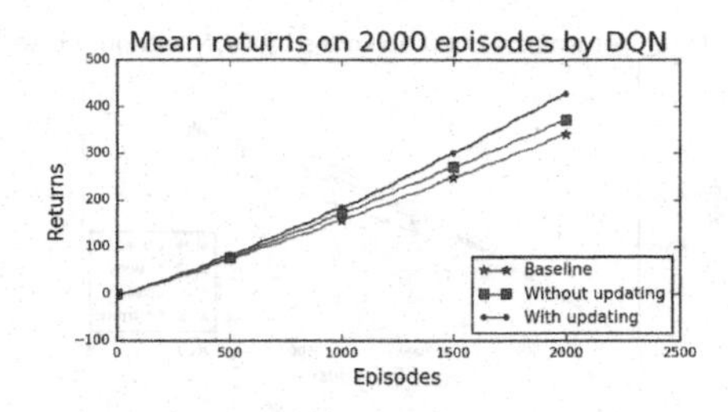

Fig. 5. Mean steps and returns of DQN without noise for convergence on 8×8 Grid-world.

the experience pool or not, we perform the experiment on the random and the optimal policies.

The random policy makes the agent randomly go in the first 5,000 steps. On the other hand, the optimal policy is the policy reach the goal with fewest steps. This policy is labeled by human. The agent goes under the optimal policy in the first 200 steps. Then the agent finishes the rest episodes following its own policy. The process that the agent goes with the optimal policy is similar to imitation learning. Experimental results are shown in Fig. 6.

It can be observed from Fig. 6 that our method preforms better than traditional sampling method. It can also be observed that 3 kinds of action selection strategies have different performance under different initialization policies. Specifically, softmax strategy performs better than greedy strategy under optimal policy to initialize the experience pool. The reason lies that the action selection strategy can only look forward one grid without model of the environment. When the values of the grids around one grid are converged to a fixed value, greedy strategy will only go to the grid in one direction, while softmax strategy can explore more neighborly grids.

On the other hand, greedy strategy performs better than softmax strategy under random policy to initialize the experience pool. When the values of the grids around one grid are not converged to a fixed value, it still has bellman error and makes greedy strategy find the better path than softmax strategy.

In addition, it can be observed from Fig. 6 that the first 50 episodes cost few steps, and get the higher returns. It is the beginning phase that the agent goes under the optimal policy. The optimal policy to initialize the experience pool makes the agent go to the goals directly. The number of samples in this path is less than that by the random policy. And it makes the agent's updating harder than that by the random policy. The results of our method are less affected by less samples than traditional sampling method.

Experiment on Size of the Maps. To test whether the method can perform better on maps with different sizes, we perform the experiment on maps with sizes of 6×6, 8×8, and 12×12.

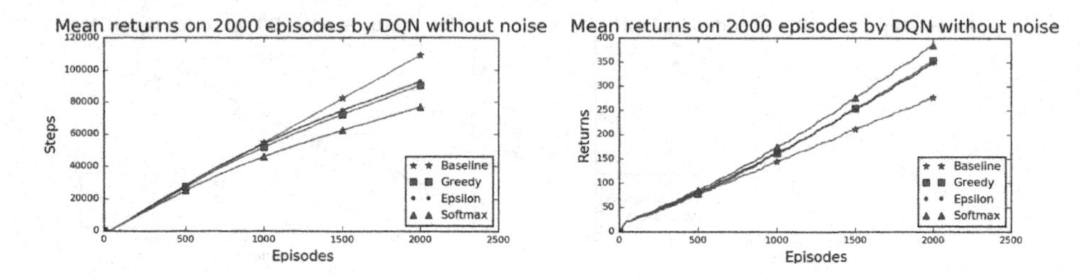

Fig. 6. Mean steps and returns of DQN under optimal policy without noise on 8×8 Gridworld.

The number of steps that random policy needed increases exponentially with the increase of size of the map. We select the optimal policy as the initialization policy for 12×12 map.

We compare the steps and returns on the 3 kinds of maps. The results are tabulated in Table 4. It can be observed that our sampling method performs better than traditional method no matter on what kind of maps.

The results of our sampling method with 3 kinds of action selection methods have nearly same performance on random initialization on map with size of 6×6. The map is small in this case. Those methods converge quickly in a small number of steps. Therefore the performances are mainly affected by the random initialization. And for map with sizes of 8×8 and 12×12, our method performs better than traditional method. Also greedy strategy performs best on random initialization in map with size of 8×8. And softmax strategy performs best on optimal initialization in all maps. The reason is the same as optimal policy in 8×8 grid as stated in Sect. 5.2. If you have known the optimal policy, you can train the agent with softmax strategy to select action. On the other hand, we normally do not know the optimal policy. Therefore, We suggest using $\epsilon - greedy$ strategy for training.

Table 4. Mean returns of DQN with different initialization policies on maps with different sizes.

Strategy	6×6		8×8		12×12
	Random	Opt.	Random	Opt.	Opt.
Baseline	0.392	0.324	0.170	0.139	0.125
Greedy	0.394	0.351	**0.213**	0.177	0.167
Epsilon	**0.396**	0.343	0.204	0.174	0.166
Softmax	0.394	**0.371**	0.207	**0.193**	**0.178**

6 Conclusions

In this paper, we propose a method that keeps the agent moving instead of stopping on the terminal state and exploring the spaces with high bellman-error which are rarely visited. This method makes the algorithm focus on the state around the goal at the beginning of updating the value function. So the expectation of reward would be propagated from the goal to other states early. In other words, this method makes the distribution more suitable for updating the value function of Q-learning during training phase. By exploring experience, it helps the Q-learning algorithm converge more effectively and efficiently. Our proposed method is useful for Q-learning with both incremental updating and batch updating. Experimental results demonstrate that our method is efficient.

Acknowledgments. This work was supported in part by National Natural Science Foundation of China (No.81373555) and Shanghai Committee of Science and Technology (14JC1402200).

References

1. Sutton, R., Barto, A.: Reinforcement Learning: An Introduction. MIT Press, Cambridge (1998)
2. Mnih, V., et al.: Human-level control through deep reinforcement learning. Nature **518**(7540), 529–533 (2015)
3. Watkins, C.J.C.H., Dayan, P.: Q-learning. Machine learning **8**(3–4), 279–292 (1992)
4. Hasselt, H.V., Guez, A., Silver, D.: Deep reinforcement learning with double Q-learning. In: Computer Science (2015)
5. Hasselt, H.V.: Double Q-learning, pp. 2613–2621. Mit Press, Cambridge (2010)
6. Schaul, T., Quan, J., Antonoglou, I., Silver, D.: Prioritized Experience Replay. arXiv preprint arXiv:1511.05952 (2015)
7. Mnih, V., et al.: Asynchronous methods for deep reinforcement learning. arXiv preprint arXiv:1602.01783 (2016)
8. Parisotto, E., Ba, J., Salakhutdinov, R.: Actor-Mimic: Deep Multitask and Transfer Reinforcement Learning. arXiv preprint arXiv:1511.06342 (2015)
9. Wang, Z., Freitas, N., Lanctot, M.: Dueling Network Architectures for Deep Reinforcement Learning. arXiv preprint arXiv:1511.06581 (2015)
10. Silver, D., et al.: Mastering the game of go with deep neural networks and tree search. Nature **529**(7587), 484–489 (2015)
11. Silver, D., Lever, G., Heess, N., Degris, T., Wierstra, D., Riedmiller, M.: Deterministic policy gradient algorithms. In: International Conference on Machine Learning, pp. 387–395 (2014)
12. Lillicrap, T.P., et al.: Continuous control with deep reinforcement learning. arXiv preprint arXiv:1509.02971 (2015)
13. MacAlpine, P., Depinet, M., Stone, P.: UT Austin villa 2014: RoboCup 3D simulation league champion via overlapping layered learning. In: Proceedings of the Twenty-Ninth AAAI Conference on Artificial Intelligence, pp. 2842–2848 (2015)
14. Yu, C., Zhang, M., Ren, F., Tan, G.: Emotional multiagent reinforcement learning in spatial social dilemmas. IEEE Trans. Neural Netw. Learn. Syst. **26**(12), 3083–3096 (2015)
15. Zhao, D., Zhu, Y.: MEC–A near-optimal online reinforcement learning algorithm for continuous deterministic systems. IEEE Trans. Neural Netw. Learn. Syst. **26**(2), 346–356 (2015)

16. Kusy, M., Zajdel, R.: Application of reinforcement learning algorithms for the adaptive computation of the smoothing parameter for probabilistic neural network. IEEE Trans. Neural Netw. Learn. Syst. **26**(9), 2163–2175 (2015)
17. Teng, T.H., Tan, A.H., Zurada, J.M.: Self-organizing neural networks integrating domain knowledge and reinforcement learning. IEEE Trans. Neural Netw. Learn. Syst. **26**(5), 889–902 (2015)
18. Deng, Y., Bao, F., Kong, Y., Ren, Z., Dai, Q.: Deep direct reinforcement learning for financial signal representation and trading. IEEE Trans. Neural Netw. Learn. Syst. **28**(3), 653–664 (2017)
19. DeNero, J., Klein, D.: Pacman Project (2012). http://ai.berkeley.edu/reinforcement.html
20. Ng, A.Y., Jordan, M.: PEGASUS: A policy search method for large MDPs and POMDPs. In: Proceedings of the Sixteenth Conference on Uncertainty in Artificial Intelligence. Morgan Kaufmann Publishers Inc., pp. 406–415 (2014)
21. Jia, Y., et al.: Caffe: Convolutional Architecture for Fast Feature Embedding. arXiv preprint arXiv:1408.5093 (2014)

Overlapping Community Detection Based on Community Connection Similarity of Maximum Clique

Xiaodong Qian[1(✉)], Lei Yang[2], and Jinhao Fang[3]

[1] School of Economics and Business Administration, Lanzhou Jiaotong University, Lanzhou 730070, China
qiaoxd@mail.lzjtu.cn
[2] School of Electronic and Information Engineering, Lanzhou Jiaotong University, Lanzhou 730070, China
[3] School of Institute of Electrical and Automation Engineering, Lanzhou Jiaotong University, Lanzhou 730070, China

Abstract. There are many problems in the community detection algorithm based on big data sets that can't effectively find the overlapping community and the rationality of the community structure is not high. This paper proposes overlapping community detection based on the connection similarity of maximum clique. The algorithm introduces an idea of maximum clique to initialize the network structure, quantifying analysis the community connection similarity, which is based on the sharing neighbor nodes and the connection between communities. On this basis, all cliques are merged to get a rational structure of overlapping community. The rationality of the proposed algorithm is tested on four real network dataset through comparing with CPM algorithm. The results show that the proposed algorithm has superior performance in term of accuracy, coverage and modularity on mining community structure. Thus, this algorithm is a kind of effective overlapping community detection algorithm.

Keywords: Big data · Community detection · Maximum clique Community connection similarity

1 Introduction

With the rapid development of communication technology and IT technology, the continuous expansion of the network scale and the gradual complexity of its structure, massive information data generated, namely: Big Data. The emergence of big data makes the human science and technology transit from the information age to big data era. Network big data has infiltrated every corner of human life, especially in the social, economic, scientific research and other fields have a significant role at the same time, it provides ample information for human being to cognize and understand the material world more thoroughly [1, 14]. The great value hidden in big data has aroused widespread concern in academia at home and abroad. Although big data has brought precious opportunities to the development of human society and the advancement of science and technology, the characteristics of big data such as complexity, diversity and

Q. Zhou et al. (Eds.): ICPCSEE 2018, CCIS 901, pp. 241–252, 2018.
https://doi.org/10.1007/978-981-13-2203-7_18

heterogeneity pose severe challenges to the development of big data [10]. Unlike simple networks, these complex networks often contain many individuals, and the relationship between individuals is also more complicated. How to mine useful information from these complex networks and apply these information to specific areas is the hot topic in the field of data mining at present [13].

Complex network is the main structure of big data; especially big data of the network, the methods based on complex network theory is one of the main methods to deal with big data mining and application. With the research on the structural characteristics of complex networks, people find that many real networks not only have the basic characteristics such as small world and scale-free, but also have topological feature-Community Structure [17], That is, the nodes in the same community are closely connected and the connections between different communities are sparse. Therefore, in order to study the big data of the network preferably, understand the function of the big data of the network and predict the behaviour of the big data of network, it is necessary to divide the network structure into communities. In 2002, Girven Etc. [11], proposed the concept of community mining, the representative algorithms include hierarchical clustering algorithm [4], module-based optimization partitioning algorithm [12, 16] and spectrum bisecting algorithm [3]. The above algorithms divide the network into several independent communities. That is, each node in the network belongs to only one community. However, communities are not completely independent in many real networks, Some nodes in the network can belong to multiple communities at the same time, Therefore, the study of structure overlapping community is more practical [15]. In order to discover structure of overlapping community in the network, Palla Etc [6] proposed The Clique Percolation Method (CPM algorithm), which identifies overlapping nodes in communities in 2005, but the community results divided by this algorithm are often affected by the selection of k values. Lancichinetti Etc [5] proposed LFM community detection algorithm, the algorithm randomly selects the seed node, through the calculation of fitness function, the network community structure is divided, and however, seed nodes selected randomly will affect the accuracy of the algorithm.

In summary, this paper proposes a big data set community detection algorithm based on maximum clique connection similarity. Through the introduction of maximum clique to initialize the community structure of the network, and the close connection of community was analysed and quantified, According to the quantified connection similarity, the closely connected communities in the network are merged to obtain structure of overlapping community.

2 Core Ideology

2.1 Introduction of Maximum Clique

In the given undirected graph G = (V, E), V is the assemblage of node in G, E is the assemble of edges in G. The clique of G represents the sub graph in which every node (in the same node's assemblage) can be connected by edges, it is also to say that the clique means a completed sub graph of the undirected graph G. If there is any completed sub graph which is not contained in another bigger completed sub graph of G, it is to say that this completed sub graph is not any other clique's proper subset of G, we call "maximum clique" of graph G.

The maximum clique represents a group of dense nodes in the given network, which possesses the best connexity, contains assemble information; based on above characters, the maximum clique is used in the research of community detection; it shows strongly cognition, accessibility, robustness etc. Architectural features, which are expected by the cohesive sub graph. Therefore we can conclude that all the nodes of maximum clique are under the same community.

In the view of situation that most layering cluster algorithm regards each node in the network as an independent community for prolongation; in this article we introduce the theory of maximum clique when initializing the structure of community network, in order to enhance the accuracy and the rationality of the algorithm.

2.2 Improving the Formula of Connection Semblance

Currently, most algorithms of community detection adopt concept we called "semblance", among which the most widely used is the semblance pointed at node. While, when we focus on the computation method in similarity of connection, the majority methods just take account in the situation of partaken nodes in the network, losing the sight of bridging connection. If there are more bridging connections between two communities, it means the two communities are closely connected, also, we can say the two communities are more likely to be the member of the same community. Therefore, if we neglect the bridging connection among the communities, it may lead the mistakes as follows; in the Fig. 1, we may merger community (a) and community (b), and the same way to community (c) and (d). Actually, after observing, it's more rational to merger community (c) and (d).

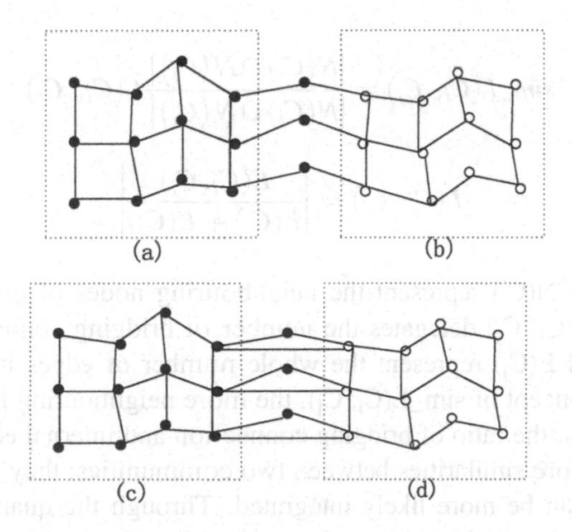

Fig. 1. Community Structure in case of Ignoring Bridging Edges

In the Fig. 2, If we don't take neighbouring nodes into account, there will be the same amount of bridging connections between community (a) and (b), community (c) and (d); besides, if we only focus on the connectivity of community, it will occurs that community (a) and (b), community (c) and (d) are merged, shown in Fig. 2. In comparison, it's more rational to merge community (c) and (d).

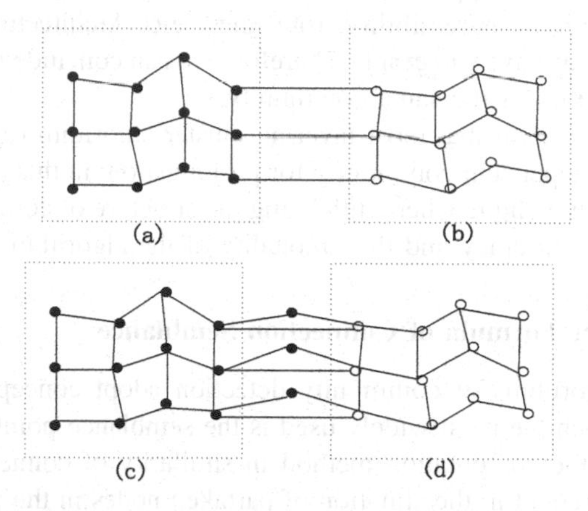

Fig. 2. Community Structure in case of Ignoring Neighboring nodes

Generally analyzing the advantage of neighboring nodes and bridging connections, this text proposes a computation method which considers about both the bridging connection in different kind of communities and the neighboring nodes owned by all communities, integrating the above-mentioned situation, will help us divide the community more rationally. We define the similarity of connection between communities as sim_L:

$$sim\_L(C_i, C_j) = \frac{|N(C_i) \cap N(C_j)|}{|N(C_i) \cup N(C_j)|} L(C_i, C_j) \tag{1}$$

$$L(C_i, C_j) = \left| \frac{E(C_i, C_j)}{E(C_i) + E(C_j)} \right| \tag{2}$$

In it, $N(C_i)$ and $N(C_j)$ represent the neighbouring nodes of two communities we called C_i and C_j; $E(C_i, C_j)$ delegates the number of bridging connection in two communities; $E(C_i)$ and $E(C_j)$ represent the whole number of edges in each community. According to the concept of $sim\_L(C_i, C_j)$, the more neighbouring nodes which shared by two communities, the ratio of bridging connection and internal edge will get bigger, besides, there are more similarities between two communities; they' are also connected more closely that can be more likely integrated. Through the quantitative analysis of how closely that communities are connected, we can improve the accuracy and rationality of the algorithm mentioned in the text.

3 Algorithm Introduction

This paper proposes an overlapping community detection algorithm based on the connection similarity of maximum clique. There are two main stages: the discovery stage of the maximum clique and the clustering stage of the maximum clique. In this paper, the algorithm based on the structure of maximum clique in the network to quantify the similarity of the connection between communities, Hierarchical clustering is carried out based on community structure initialized by a maximum clique, through the implementation of the two phases, and we get a structure of overlapping community ultimately.

3.1 Maximum Clique

Maximum Clique Detection Algorithm. In order to prevent duplication of selected nodes, each node is marked with a serial number firstly.

Step 1: Select the initial node with serial number 0 from the network and obtain the node's neighbour nodes;

Step 2: Select the node with the smallest serial number from the node's neighbour nodes and obtain the current clique; Get the set of nodes that connect with the current clique. That is, neighbour nodes of the current clique;

Step 3: Select the node with the smallest serial number from neighbour nodes of the current clique to expand the current clique, get neighbour nodes of expanded clique;

Step 4: Repeat the third step, until the neighbour nodes of the clique are empty, the maximum clique with the initial node 0 as the vertex is obtained;

Step 5: Select a node with the smallest serial number from the neighbour nodes of initial node which not included the selected node, repeat the second step until all the neighbour nodes have traversed, then you can get all the maximum cliques with the initial node as the vertex;

Step 6: select a node with the smallest serial number from the neighbour nodes of initial node which not included the selected node and repeat the first step until all the nodes are traversed.

Sample Heading (Third Level)

The analysis of overlap maximum clique is as follows: taking Fig. 3 as an example, the network shown in Fig. 3 includes two maximum cliques. When starting from node 0, a maximum clique {0, 1, 2, 3} is obtained; when starting from node 2, get the maximum clique {2, 3, 4, 5, 6}. The maximum clique obtained from node 0 and the maximum clique obtained from node 2 both contain node 2 and node 3, that is, there have overlapping nodes in the two maximum cliques. it can be inferred that the communities discovered based on these overlapping maximum cliques are likely to overlap with each other. So communities that discovered based on these maximum cliques also overlap with each other.

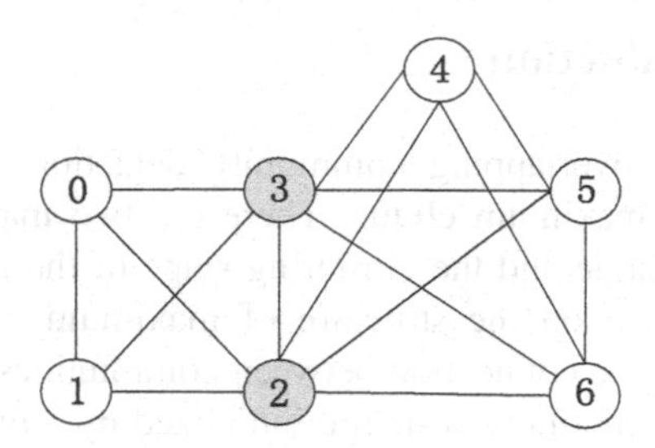

Fig. 3. Maximum Clique

3.2 Describe the Specific Algorithm

Overlapping community detection based on community connection similarity of maximum clique (where the parameter α equals 0.2):

Step 1: Initialization. Think of every discovered maximum clique of network G as a individual community C_i, the initial community structure of the network $C_S = \{C_0, C_1, \ldots\ldots, C_n\}$;

Step 2: According to the formula (1), we calculate the connection similarity of communities which connected with edge in C_S and put it into the queue C_{sim};

Step 3: Determine whether each two communities merged in C_S. If there is a value greater than parameter α in C_{sim}, Then merge the two related communities, update C_{sim} and C_S. Continue to operate the second step;

Otherwise operate the fourth step;

Step 4: Output overlapping community structure C_S.

When initializing the network community structure, some isolated nodes do not belong to any maximum clique, and these nodes tend to extend the algorithm running time, so this article does not consider these isolated nodes. In this paper, the algorithm finds out all the maximum cliques of a given network to initialize the network community structure. Each maximum clique is regarded as a separate community. Then, the improved inter-community connection similarity formula is used to calculate the inter-community closeness, The greater similarity value, the higher probability that two communities belong to the same community, merge the communities whose similarity value is larger than the parameters; then update the community structure and recalculate inter-community connection similarity, Until connection similarity of any two communities in the network less than the parameters, and finally output overlapping community structure of the network.

3.3 Time Complexity Analysis

In this paper, the time complexity of the algorithm in calculating the maximum clique is $O(c)$(where c is the number of initial communities), and because the algorithm in this paper uses the closeness of the two communities to determine whether the two communities are merged, The time complexity in calculating inter-community connection similarity is $O(s)$(where s is the number of neighbour communities), and the time complexity of merging communities repeatedly is $O(s^2(c-1))$. So the time complexity of the algorithm is $O(s^2c)$.

4 Experimental Analysis

In order to validate the rationality of overlapping community structure according to the algorithm mentioned in the text, we do experiments on four real networks in order to compare CPM (clique percolation method) with our algorithm.

4.1 Experimental Data

The detailed information we used in four real networks can be found in tabulation 1. The structure of network N1 and N2 is already known, while, it's unknown to in network N3 and N4.

In the text, the Karate network [9] is supposed to be a classical data set in social network analysing. In early nineteen seventies, the sociologist Zachary spent two years on observing the social relationship of 34 members in a karate club in a American university. Based on internal and external relationship among thus members, he constructed a social relationship network, including 34 nodes and 78 connections; two members are regarded as close friend at least if they're nodes are connected together.

Dolphins networks is proposed by Lusseau [2, 7] together with his colleagues, they had observed the living habit of 62 dolphins in New Zealand, getting conclusion that the engagement model of these dolphins emerged a special model, besides, they constructed a social network contains 62 nodes and 159 connections. Two nodes would be connected together if two corresponding dolphins have a frequent contact.

P2p-08 network is a file sharing network from the year 2002 August, collecting a huge number of p2p files, including 6301 nodes and 20777 connections, in it, nodes represent hosts, if there is connection between two nodes that means two hosts have shared some files.

Ca-HepPh network [8], it contains 12008 nodes, every node represents an author. Each connection indicates that two authors may have published a discourse together (Table 1).

Table 1. Real network dataset.

Structure of communities	Name		Node	Connection
Known	N1	Karate	34	78
	N2	Dophins	62	159
Unknown	N3	p2p-08	6301	20777
	N4	ca-HepPh	12008	237010

4.2 Evaluating Indicator of Community Detection

Accuracy and Coverage Rate. If the community's structure is known, we will adopt accuracy and coverage rate as the evaluating indicator which are used to judge the result of our algorithm. Supposing that C is the structure of community constructed by algorithm, C' is the structure already known;

Accuracy: the percentage of nodes in C that belong to the same community both in C and C';

Coverage rate: the percentage of nodes in C' that belongs to the same communities both in C' and C;

The larger two evaluation indexes, a better the detection effect of the algorithm is.

Modularity Qov. In the study of complex network, descriptive concepts can't be applied directly. While Newman and Girvan defined a kind of function we called modularity, we can make quantitative analysis on rationality of the excavated structure. Normally, Qov is used to describe the modularity of community's structure. A bigger Qov shows that the structure we have explored is better closes to the structure of real network.
Assuming that community's structure has been divided, we use formula as follows to indicate modularity:

$$Qov = \frac{1}{2M}\sum_{i=1}^{n}\sum_{j=1}^{m}\left[(a_{ij} - \frac{k_i k_j}{2M})\delta(\sigma_i,\ \sigma_j)\right] \quad (3)$$

In the formula, a_{ij} is element in the order matrix of network, if v_i and v_j are connected, a_{ij} equals to 1, else, $a_{ij} = 0$; $M = 0.5\sum a_{ij}$ is the number of connections in network; k_i is degree of node v_i, k_j is degree of v_j; in the network, if the structure is fixed, edges are connected randomly, the probability that v_i and v_j are connected together will be $\frac{k_i k_j}{2M}$;

δ is membership function, if v_i and v_j belong to the same community ($\sigma_i = \sigma_j$), we'll conclude that $\delta(\sigma_i,\ \sigma_j) = 1$, else, $\delta(\sigma_i,\ \sigma_j) = 0$.

4.3 Detection Result and Analysis

Result of known community's structure. As the known network like karate club and dolphin social network, we use accuracy and coverage rate to estimate the result of community detection, in Fig. 4 (a) and (b) have given the experimental result. In our algorithm, we set $\alpha = 0.2$, in clique percolation algorithm, $K = 3$.
We use both CPM and the textual algorithm to conduct the experiment of community detection, after comparing the accuracy and coverage rate, we show the result in Fig. 4. Comparing (a) with (b) in Fig. 4, In the karate club database, the distribution of nodes possess property of concentration, in the community structure of C', there are relatively fewer nodes belong to the same community, therefore at the step 2 and step 3 in algorithm, less communities will be merged. So in the final community detection according to the algorithm, the number of nodes that belong to the same community is much closed to such nodes in structure known communities. So in Karate club network, the accuracy shows an excellent performance, for example, in Fig. 4(a), accuracy equals to 1. We get higher accuracy and coverage through our algorithm than that through CMP; in the Dolphins network, we get the same conclusion. In comparison, algorithm mentioned in the text has advantage of dividing community on accuracy and coverage.

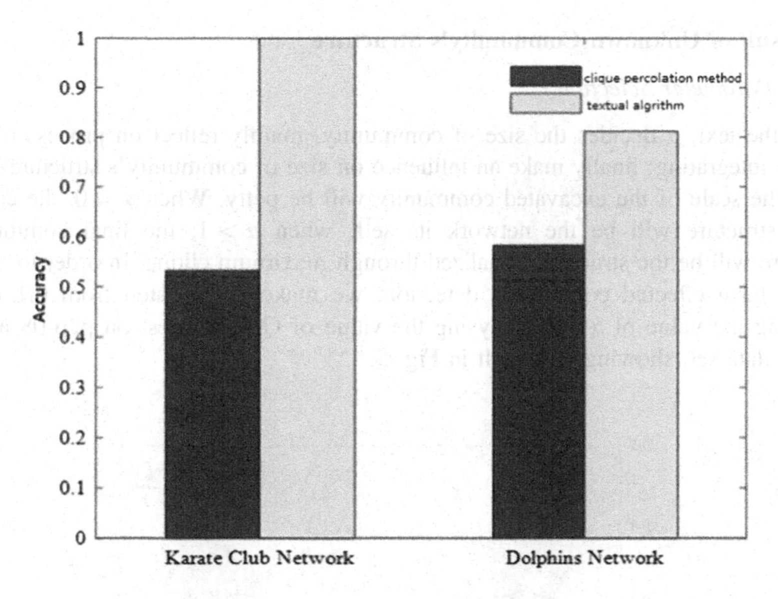

(a) Comparison of the accuracy of the two algorithms

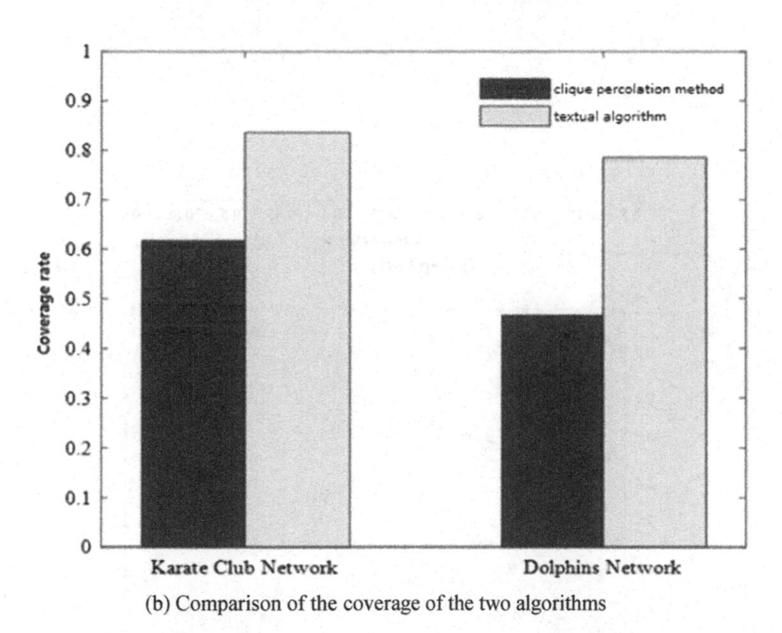

(b) Comparison of the coverage of the two algorithms

Fig. 4. Comparison of the accuracy and coverage of the two algorithms

Result of Unknown Community's Structure

(1) *α Parameter Selection*

In the text,. α decides the size of community, mainly reflect on process of community integrating; finally make an influence on size of community's structure. If α is large, the scale of the excavated community will be petty. When $\alpha = 0$, the community's structure will be the network its self; when $\alpha > 1$, the final community's structure will be the structure initialized through maximum clique. In order to analyse how α have effected community detection, we make 0.1 as step from 0.1 to 0.6, adjusting the value of α and analysing the value of Qov, we test on p2p-08 and ca-HepPh data set, showing the result in Fig. 5.

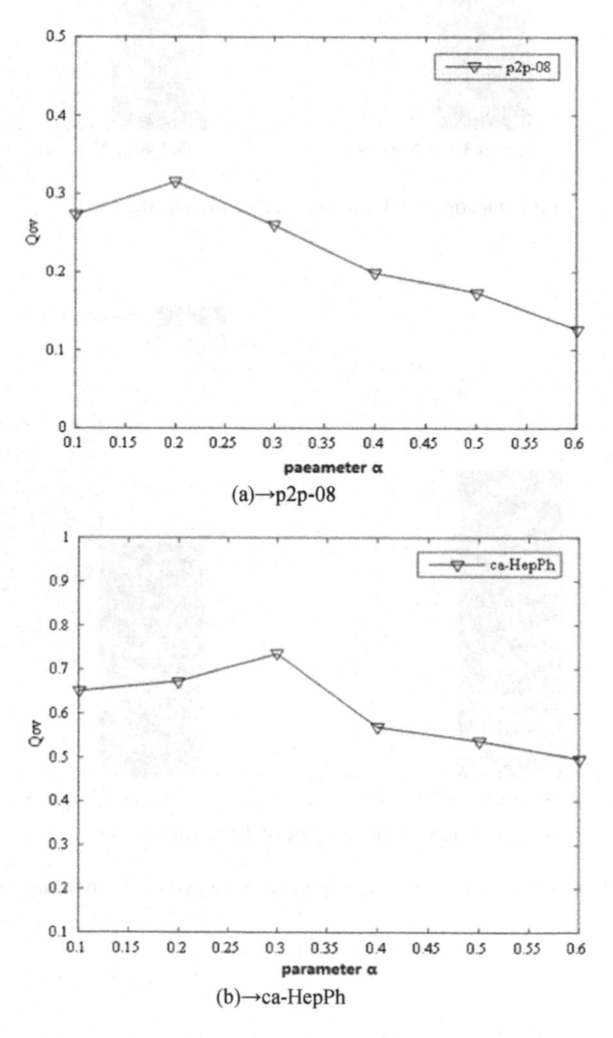

Fig. 5. The influence of α on the Qov in different data

Normally, it's supposed to be rational if Qov is larger than 0.3. Of course, the larger Qov is, the more reasonable detection we will get. From Fig. 5, to p2p-08 network, the result of community detection according to textual algorithm will get better along with the decrease of α, when α is 0.2, the Qov will not only be the highest, but also will be the most reasonable; to the ca-HepPh network, the result of community detection according to our algorithm will get better along with the decrease of α, when $\alpha = 0.3$, Qov will be the highest, also to be the most reasonable. In a word, α needs to choose corresponding value in the light of different network's structure.

(2) *Analysis of experimental Results*

In the text, we compare the best community detection with CPM. In the Fig. 2, we show the experimental results on real network dataset through the algorithm in mentioned in the text and CPM, in clique percolation method, we regard $K = 3$.

Table 2. Experiment results on real network dataset.

Network name	Textual algorithm		CPM	
	α	*Qov*	*K*	*Qov*
N3(p2p-08)	0.2	0.3157	3	0.2494
N4(ca-HePh)	0.3	0.7355	3	0.5772

In Table 2, we operated two algorithms on the real network dataset (p2p-08, ca-HepPh) and the results are shown in Table 2. It's clear that in the large-scale real network, textual algorithm has a better performance on modularity of overlapping community than CPM has. In summary, the textual algorithm enhances the rationality, and also conforms to the physical truth.

5 Conclusion

In the text, we composed a new overlapping community algorithm based on the similarity of maximum clique. Through the similarity calculation, we can merge to networks together if they're connected closely. Besides, we can get the structure of the overlapping community. Based on four experiment on real networks dataset, we can conclude: the textual algorithm could not only deal with the large scale dataset networks, but also has a better effect on dividing communities than CPM has, in the end, we get a rational community detection, once more illustrating that our algorithm has an excellent effect on community detection.

In the days to come, we can continue the study of social networks in the following aspects: (1) Continue the experiment on more real large-scale networks, and explore the relationship between the different scale networks and the similarities, selecting a reasonable parameter for the different networks; (2) taking a comprehensive consideration of community discovery algorithm on weighted networks in large dataset networks.

Acknowledgement. This article is supported by National Natural Science Foundation of China under grant No. 71461017.

References

1. Chopade, P., Zhan, J.: Community detection in large-scale big data networks. In: ASE International Conference, pp. 1–7 (2014)
2. Gaigai, G., Yuhua, Q., Xiaoqin, Z., et al.: Algorithm of detecting community in bipartite network with autonomous determination of the number of communities. Pattern Recognit. Artif. Intell. **28**(11), 969–975 (2015)
3. Liwei, H., Deyi, L., Yutao, M., et al.: A meta path-based link prediction model for heterogeneous information networks. Chin. J. Comput. **37**(4), 848–858 (2014)
4. Ruiyang, H., WuQi, Z.Y., et al.: Community detection in dynamic heterogeneous network with joint nonnegative matrix factorization. Appl. Res. Comput. **34**(10), 2989–2992 (2017)
5. Hu, J., Wang, B., Lee, D.: Research on centrality of node importance in scale-free complex networks. In: Proceedings of the 31st Chinese Control Conference, China, pp. 1073–1077 (2012)
6. Kitsak, M., Havlin, S., Paul, G., et al.: Betweenness centrality of fractal and nonfractal scale-free model networks and tests on real networks. Phys. Rev. E **75**(5), 056115 (2007)
7. Lancichinetti, A., Fortunato, S., Kertész, J.: Detecting the overlapping and hierarchical community structure of complex networks. New J. Phys. **11**(3), 19–44 (2018)
8. Xiaojia, L., Peng, Z., Zengru, D., et al.: Community structure in complex networks. Complex Systems and Complexity Science **5**(3), 19–42 (2008)
9. Scott, J., Hughes, M., Mackenzie, J.: The anatomy of scottish capital: Scottish companies and scottish capital, 1900–79. Econ. Hist. Rev. **33**(4) (1980)
10. Sun, Y., Norick, B., Han, J., et al.: Integrating meta-path selection with user-guided object clustering in heterogeneous information networks. In: ACM SIGKDD International Conference on Knowledge Discovery and Data Mining, pp. 1348–1356. ACM (2012)
11. Tang, L., Liu, H.: Community detection and mining in social media. Morgan & Claypool, San Rafael (2010). 137
12. Qi, W., Fucai, C., Ruiyang, H., et al.: Community detection in heterogeneous network with semantic paths. Acta Electron. Sinica **44**(6), 1465–1471 (2016)
13. Yajing, W., Zengru, D., Ying, F.: A Clustering algorithm for bipartite network based on distribution matrix of resources. J. Beijing Norm. Univ. Nat. Sci. **5**, 643–646 (2010)
14. Yuanzhuo, W., Xiaolong, J., Xueqi, C.: Network big data; present and future. Chin. J. Comput. **36**(6), 1125–1138 (2013)
15. Wei, D., Li, Y., Zhang, Y., et al.: Degree centrality based on the weighted network. In: 2012 24th Chinese Control and Decision Conference (CCDC), China, pp. 3976–3979. IEEE (2012)
16. Tao, W., Yang, L., Yaoyi, X., et al.: Identifying community in bipartite networks using graph regularized-based non-negative matrix factorization. J. Electron. Inf. Technol. **37**(9), 2238–2245 (2015)
17. Zhang, P., Wang, J., Li, X., et al.: Clustering coefficient and community structure of bipartite networks. Phys. A Stat. Mech. Appl. **387**(27), 6869–6875 (2007)

Heterogeneous Network Community Detection Algorithm Based on Maximum Bipartite Clique

Xiaodong Qian[1(✉)], Lei Yang[2], and Jinhao Fang[3]

[1] School of Economics and Business Administration, Lanzhou Jiaotong University, Lanzhou 730070, China
qiaoxd@mail.lzjtu.cn

[2] School of Electronic and Information Engineering, Lanzhou Jiaotong University, Lanzhou 730070, China

[3] School of Institute of Electrical and Automation Engineering, Lanzhou Jiaotong University, Lanzhou 730070, China

Abstract. The community mining of heterogeneous networks is a hot issue to study the big data of the network, and the original structure of the heterogeneous network and its information can fully exploit the community structure in the network. However, the existing algorithms mainly analysis one type of objects in the heterogeneous network, and the algorithms about the heterogeneous nodes which constitute the community structure is rarely studied. Therefore, this paper introduces the theory about maximum bipartite clique: Firstly, regarding the largest maximum bipartite clique that the key node belongs to as initial community. Then, the community is expanded based on the similarity between the neighbor node of the community and the initial community in quantitative. Finally, a reasonable community structure is mined and the simulation experiments are carried out on artificial networks and real heterogeneous networks. The experimental results show that the algorithm has relatively high community accuracy and modularity in community detection, which proves the rationality and validity of the algorithm.

Keywords: Heterogeneous network · Community detection
Maximum bipartite clique · Similarity

1 Foreword

With the rapid development of communication technology and IT technology, the scale of networks have gradually expanded, its structures have become more intricate, meanwhile, producing a huge number of information data, what we called: Big Data. The emergence of Big Data promotes human being's technology transmit from Information Age to Data Age. Big Data in internet have penetrated into every corner of human's life, especially in such fields: society, economy, and scientific research; at the same time, it provides enough information bases for people to cognize, understand the material world [15]. The huge wealth concealed in Big Data has attracted a lot of attention from scholars inside and outside the country.

Q. Zhou et al. (Eds.): ICPCSEE 2018, CCIS 901, pp. 253–268, 2018.
https://doi.org/10.1007/978-981-13-2203-7_19

Although Big Data have brought value opportunities to the development of society and improvement of technology, the features Big Data such as complexity, diversity and heterogeneity also bring severe challenges to the development of Big Data [1]. In real networks, heterogeneous networks with more complex structures are becoming more and more abundant. Compared with the homogeneous networks with single structure, types of nodes in heterogeneous networks put up difference, also the type of edges which connect nodes together shows diversity [10]. It's hard to analyzing the complex heterogeneous networks with the traditional social network analysis method, so the research on heterogeneous networks is on the upsurge, and is also becoming the very important research direction in the field of Data mining. For different social networks, whether it is homogeneous or heterogeneous, shares the same feature community structure, that is, in the internal connections of a community are closely, while the connections between two communities are sparse.

At present, the research on community detection algorithm in heterogeneous networks has attracted more and more people's attention. Based on the original two diagrams, Wu et al. [14] proposed the clustering method according to resource distribution. Zhang et al. [17] proposed the community detection method, this algorithm is based on the convergence factor of edges, successively removing the edges which have the minimum convergence factor, until the network is divided into the number established. Tang et al. [11] based on the spectral theory proposing the joint clustering algorithm, they divide the singular value in relational matrix which has been normalized, gaining the community division. While we need to set up the number of communities in advance through thus algorithms as above, generally, it's uncertain that how many communities should be divided will be the best suited. Huang et al. [5] proposed the community detection in dynamic heterogeneous network with joint non-negative matrix factorization, the algorithm first uses a multi-relational similarity matrix to describe heterogeneous relationships in dynamic heterogeneous networks, combining topological similarity with non-negative matrix decomposition method, realize community detection of dynamic heterogeneous networks; Wu et al. [13] proposed heterogeneous network community detection algorithm based on semantic path, calculating similarity information between different types of nodes in a heterogeneous network Based on semantic paths, use this as a prior condition for community detection to guide the community division of heterogeneous networks; Wang et al. [16] proposed a community detection algorithm in bipartite networks using graph regularized-based non-negative matrix factorization, through graph regularization achieve internal connection between the user subspace and the target subspace, use inter-class and intra-class information to improve the accuracy and stability of matrix decomposition, to improve the performance of the bipartite networks community detection. The above algorithms use some different methods to analyze the nodes of a target type in a heterogeneous network basically, use similarity information between different types of nodes as a priori condition to divide the community in the network constituted by the target type nodes, so above algorithms don not consider the situation some hetero-junctions belong to the same community. However, there are such cases that heterogeneous nodes together constitute a network in actual heterogeneous networks, the above algorithms cannot classify this heterogeneous network accurately. At present, there are few researches on community discovery algorithms for heterogeneous networks which have communities constituted by heterogeneous nodes.

In a comprehensive analysis, we propose a community detection algorithm for the heterogeneous networks based on two maximal groups, for the analysis and investigation of heterogeneous. Firstly, the most import two largest cliques are regarded as the initial communities, and then extend the communities based on the similarity of communities and neighbor nodes of quantified community, so that we can rationally divide the community structures of heterogeneous networks.

2 Correlation Analysis

2.1 Indirect Mapping

Nowadays, the research about heterogeneous networks has attracted widely attention. In the study on community structure of heterogeneous networks, the mainly work is about mapping and transformation of the heterogeneous; and then, using the existing community detection algorithm of homogeneous networks to divide the networks. However, in the process of mapping the heterogeneous networks to homogeneous networks, the information contained in heterogeneous networks may be lost or changed.

(1) *Unweighted Mapping*

As shown in Fig. 1, heterogeneous networks contain digital and alphabetic nodes. In (a), digital nodes 1, 2, 3, 4 are all connected with alphabetic nodes *a*, so digital nodes in (a) can be mapped into the network shown in (c); digital nodes 1, 2, 3, 4 in (b) are connected with all alphabetic node *a*, so the digital nodes in (b) also can be mapped into the network in (c). However, through the observation, we can find that digital nodes 2, 3, 4 in (a) are also connected with alphabetic node *c*, in (a), nodes 2, 3, 4 are connected more closely, while nodes 2, 3, 4 in (b) are relatively estranged, but (a) and (b) are mapped into the same homogeneous network. So while studying heterogeneous networks, neglecting the information of nodes and edges may lead the loss of network primitive semantics, and may have effect on the final divided communities.

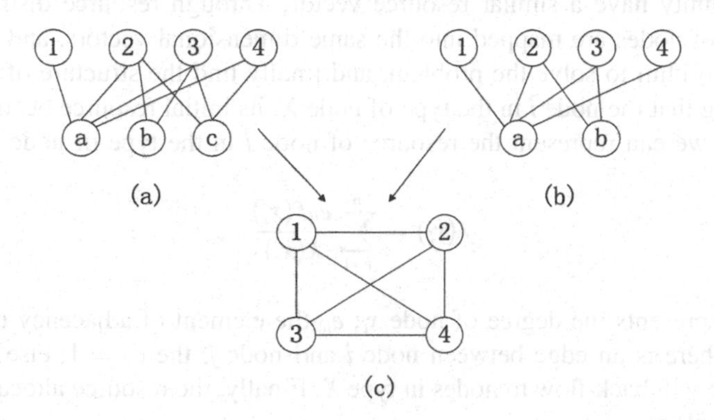

Fig. 1. Heterogeneous network mapping diagram

(2) *Weighted Mapping*

Weighted mapping uses the information of heterogeneous nodes and edges to project, such as community detection of heterogeneous networks method based on semantic path [3, 13]. The basis method of the algorithm: firstly determine the type of the target objects (that is to determine the type of object needed in the community detection), through the semantic path (semantic path is a series of combinations of two nodes in the heterogeneous networks. If the semantic path of two nodes is different, means the relationship of the two nodes is distinct.), we can evaluate the heterogeneous information's similarity of the different type of objects, and through constructing the reliability matrix as a regularization constraint of semi- supervised non-negative matrix factorization to realize the community detection of heterogeneous network.

In heterogeneous networks, objects are connected with multiple semantic paths, which represent a variety of relationships between objects. Although the method considers about the information of heterogeneous nodes and edges in community detection, there are still some deficiencies: firstly, the algorithm considers about only one relationship corresponding to a semantic path [13], while ignoring the relationship corresponding to other semantic path; The algorithm uses semantic paths to calculate similarity information between different types of nodes in a heterogeneous network, use this as a prior condition for community detection to divide target object type into communities, so this algorithm does not consider the situation where heterogeneous nodes belong to the same community, but for heterogeneous networks which have communities constituted by heterogeneous nodes, this algorithm unable to divide the community accurately.

2.2 Direct Analysis

The clustering method based on resource distribution [14] is a kind of clustering algorithm based on the original bipartite clique, which combines the resource distribution with community detection of the bipartite clique. The process of resource distribution is as follows: the node distributes resource to all the nodes connected with itself, through such process, all the nodes in network can be represented by three dimensional or multidimensional vectors. The algorithm considers that, nodes in the same community have a similar resource vector. Through resource distribution, different types of nodes are mapped into the same dimensional vectors, and then we use K-means algorithm to solve the problem, and finally find the structure of community.

Assuming that the node i in the type of node X, its initial resource occupancy value $f(x_i) \geq 0$, so we can represent the resource of node l in the type of node Y:

$$f(y_l) = \sum_{i=1}^{n} \frac{e_{il} f(x_i)}{k(x_i)} \tag{1}$$

In it, $k(x_i)$ represents the degree of node x_i; e_{il} the element of adjacency matrix in the network, if there is an edge between node i and node j, the $e_{il} = 1$, else $e_{il} = 0$; and then resource will back-flow to nodes in type X. Finally, the resource allocation of node i in type X will be:

$$f(x_i) = \sum_{l=1}^{m} \frac{e_{il} f(y_l)}{k(y_l)} = \sum_{l=1}^{m} \frac{e_{il}}{k(y_l)} \sum_{j=1}^{n} \frac{e_{jl} f(x_j)}{k(x_j)} = \sum_{j=1}^{n} \omega_{ij} f(x_j) \tag{2}$$

Actually, when we use K-means algorithm to divide the community, the initialization settings are random, the result of the final division is uncertain; Besides, we need to set the number of communities. When we are facing the unknown structure networks, it's hard to estimate the number of communities, and is also hard to divide the community structure accurately.

To solve the above problems, we propose a community detection algorithm based on maximum bipartite clique, which not only preserves the original information of heterogeneous network, but also considers about the situation of heterogeneous nodes belonging to the same community.

3 Detailed Introduction

In order to avoid matters mentioned above influence the heterogeneous network community detection algorithm, the paper optimizes and improves heterogeneous network community detection algorithm from the following aspects.

3.1 Key Nodes

Usually, there are unbalanced topology structures in communities, that is to say, there are a few key nodes, which play a dominant role to other non-critical nodes. In the real life, unbalance within community can be seen everywhere. For example, in the research cooperation network, a key member in a research team is more likely to be the academic leader than others. He may be the soul of the team and guide the other non-critical members in scientific research. Therefore, considering about the influence and representation of the key nodes in a community, we can use the character of "key" in nodes to discover the key nodes in the community, which can quickly locate the dense areas in the network as a key part of the community structure.

Generally, the higher key is, the more important the node is. Method of mining key nodes are usually based on the degree index [12], the interpreting index [6], the feature vector [4], and so on. In the networks topology, the degree index means the more important the node is, and the smaller the node's degree is, the lower importance of the node is. Therefore, the paper uses the node degree method to find the key nodes.

Starting from the right key nodes helps us to find community structure quickly and accurately, but if we start from the wrong key nodes, it may increase the complexity of the algorithm and affect the quality of community detection. The community detection algorithm based on key nodes has the following advantages: firstly, starting expanding the community from the key nodes will effectively avoid the restrictions and impacts on community detection by nodes with low influence or at the edge; secondly, when we get the key nodes, we can also get the key clique, through the expanding of the key clique (not the expanding of the key node) we will determine the key node belongs to which community accurately; besides, we also determine all the location of all the key nodes and key clique accurately.

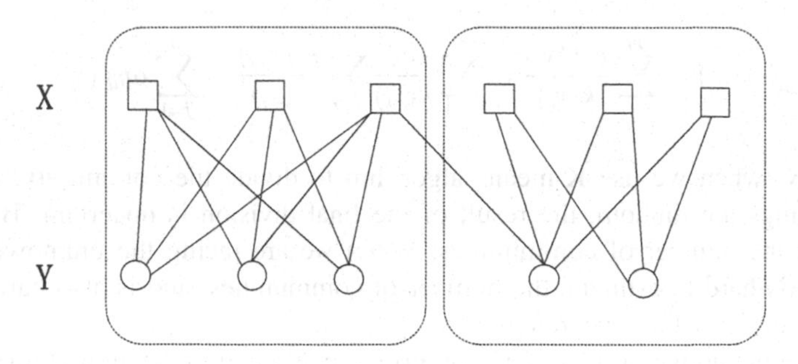

Fig. 2. Network of maximum bipartite clique

3.2 Maximum Bipartite Clique

Big data set networks generally show complexity, diversity, and heterogeneity. The bipartite clique is an important heterogeneous network structure which accord with the character of big data. The maximum bipartite clique is composed of two types of nodes and edges between two kinds of nodes, there is no connection between the same kind of nodes.

In the undirected bipartite clique $G = ((X, Y), E)$, given subgraph G_s, $X(G_s) \subseteq X(G), Y(G_s) \subseteq Y(G)$, for the arbitrary node $x_i \in X(G_s)$ x, $y_j \in Y(G_s)$, there will be the edge $e(x_i, y_j) \in E(G_s)$ e, then we call the G_s bipartite clique. Generally, it's required that $|X(G_s)| \geq 2$, $|Y(G_s)| \geq 2$. If there are no other bipartite clique G_s' exist which makes $X(G_s) \subset X(G_s')$, $Y(G_s) \subset Y(G_s')$, then G_s will be maximum bipartite clique. It is also called complete bipartite graph shown in Fig. 2.

The maximum bipartite clique is a set of nodes which connected closely in the heterogeneous network. it has the best connectivity, and is also regarded as the core part of dense substructure in heterogeneous networks, contains concentrated knowledge, so when it is applied to the community detection, it shows the structural features of awareness, accessibility, robustness which are expected possessing in coagulation of a subgroup. Therefore, we can conclude that all nodes in maximum bipartite clique belong to the same community.

3.3 Introduction of Similarity Formula

Similarity is the judgment indicator of expanding the key community of algorithm. Usually the similarity of nodes is set for the nodes in homogeneity network, hardly can be find in heterogeneous networks. In heterogeneous network $G = ((X, Y), E)$, in process of community detection, we define the set $N(v_i)$, it contains all the neighbor nodes which connected with node v_i. So the similarity between community and neighbor node is defined as follows:

$$sim(v, C) = \begin{cases} \left|\frac{N(v) \cap Y(C)}{N(v) \cup V(C)}\right| & (v \in X) \\ \left|\frac{N(v) \cap X(C)}{N(v) \cup V(C)}\right| & (v \in Y) \end{cases} \tag{3}$$

In it, $V(C)$ represents set of all the nodes in local community C, $X(C)$ represents the set of nodes belong to X type in local community C, $Y(C)$ represents the set of nodes belong to the type Y in local community C. The definition indicated that the ratio of the number of nodes which shared by the set of V_i neighbor node and the number of nodes in community C with the nodes in two sets, the larger ratio is, the closer the two part is connected, also they're more likely belonged to the same community.

4 Introduction of the Algorithm

In the paper, we have introduced the maximum bipartite clique theory: firstly, the largest scale maximum bipartite clique which contains the key nodes will be the initial community; and then, expanding the community based on the similarity of the community and the neighbor nodes in quantitative community; finally, dividing a reasonable community.

4.1 Key Nodes

Algorithm 1: Key nodes seeking algorithm

Input: initial network $G = ((X, Y), E)$;

Output: key nodes ;

1. select node v in the network randomly;
2. Repeat
3. acquire v's neighbor nodes, calculate the degree of v and its neighbor nodes;
4. if v has the largest degree
5. v will be the key node;
6. else
7. node with the largest degree will be the initial node v;
8. end
9. Until the initial node v has the largest degree;

4.2 Maximum Bipartite Clique Discovery Algorithm

Algorithm 2: maximum bipartite clique detection algorithm

Input: the selected node: v; the set of v's neighbor nodes: N(v);the selected: clique; the selected heterogeneous clique: bclique;

Output: the set of maximum bipartite cliques: Bi_Clique

```
01. clique ← v
2.  for each u ∈ N(v) do
3.    bclique ← u;
4.    repeat
5.      temp = ∩N(v_i), v_i ∈ clique ;
6.      temp = temp − bclique;
7.        if temp ≠ ∅
8.          clique ← temp;
9.          temp = clique;
10.         clique = bclique;
11.         bclique = temp;
12.       temp = ∅;
13.         break to repeat
14.       else
15.         if the node in clique and bclique isn't larger than 1
16.           break ;
17.         else
18.           Bi_Clique ← clique and bclique;
19.         end
20.       end
21.   Until Bi_Clique ← clique and bclique;
22. end for
```

4.3 Introduction of Community Discovery Algorithm

Input : network structure: $G = ((X, Y), E)$
Output: the final community division of the network: K

1. **execute algorithm 1, seek the key nodes;**
2. **execute algorithm 2, seek the maximum bipartite clique Bi_Clique that contains key nodes**
3. **calculate the largest number of nodes in Bi_Clique, and it will be the initial community C**
4. **acquire the set of neighbor nodes N(C) in community C**
5. **for each** $v \in N(C)$
6. **calculate** $sim(v, C)$ **based on formula (1)**
7. **if** $sim(v, C) > \alpha$
8. $temp \leftarrow temp \cup \{v\}$
9. **end**
10. **end for**
11. $C \leftarrow C \cup temp$
12. $K \leftarrow C$
13. $G = G - K$
14. **taking G as the given network, repeat the step (1) until all the nodes in the networks are divided into the corresponding communities;**

4.4 Analysis of the Time Complexity

The time complexity of the algorithm is mainly focus on the selection of key nodes, the detection of maximum bipartite clique, community expansion. If the number of communities in a network is k, then the selection of key nodes and the time complexity is O (kr), r means the shortest path from randomly selected nodes to the key nodes; O(k) is the time complexity in finding the maximum bipartite clique; the complexity time of expanding initial community is O(kc), in it, c is the number of neighbor nodes of each initial community; so the algorithm's time complexity is O(kr + k + kc).

5 Experimental Analysis

In order to verify the performance of the textual algorithm, we make experiments on artificial data set and three real networks, so as to estimate the rationality and the accuracy of the community division structure based on the algorithm.

5.1 Artificial Network

Artificial heterogeneous network contains two types of nodes, it is constructed as follows: the network contains 128 nodes, each type of node contains 64 nodes, divided into 4 equal communities, there are 32 nodes in each community, each type of node contains 16 nodes, set degree of each node is 16, which is $k = 16$, $k = k_{in} + k_{out}$, k_{in} means the edges of one node connected with other nodes in the community, k_{out} means the edges of one node connected with other nodes out of the community. Following the structure features of the bipartite clique, there is no connection between the same types of nodes. To test the performance of the algorithm, we select an artificial network $k_{in} = 16, 15, 14, \ldots, 9$ (shown in Fig. 3) to do the experiment of community division.

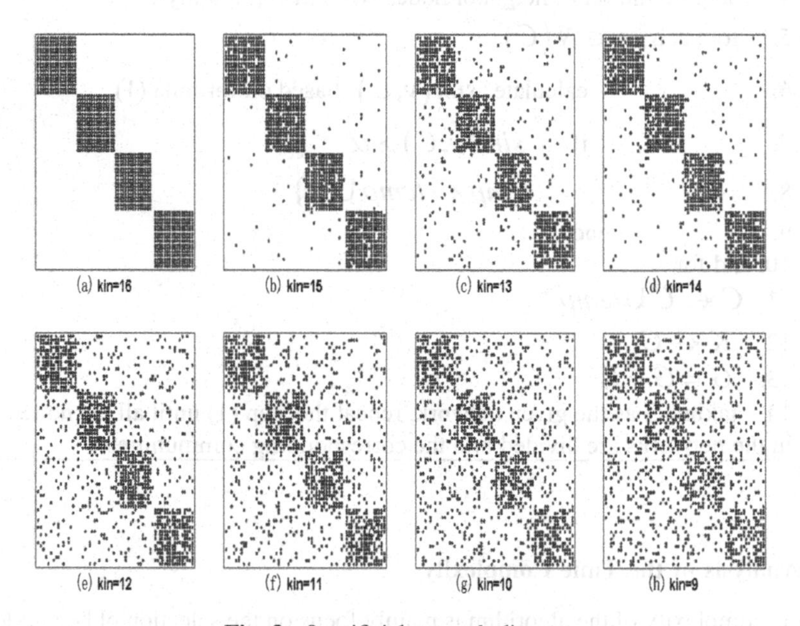

Fig. 3. 8 artificial network diagram

5.2 The Real Network

In this paper, we choose several real network data like: CEO Club Data [9] ,Southern Women Data [7] and Graph Products Data [2] to test the algorithm.

CEO Club Data is a heterogeneous network composed a set of CEO collected by Galaskiewicz, if a CEO is a member of the club, there is a connection between the CEO and interior of the heterogeneous network, otherwise, the connection will not exist.

Southern Women Data is a heterogeneous network composed of 18 women and 14 activities. If a woman takes part in an activity, there is a connection between the woman and the activity in this heterogeneous network, else, there is no connection. Graph Products Data comes from a book written by Klavzar S. in 1999. It is a heterogeneous network composed by 314 authors and 360 articles. If an author is the author of an article, there will be an edge connect the author and an article, otherwise, the edge is not exist.

5.3 Evaluation Index of Community Division Result

Correct Partition Rate

The correct division rate [11] is usually applied on the network analysis of known community structure. Through comparing the difference between the community structure and the real community structure to judge the performance of the algorithm. The larger the correct partition rate is, a better quality of the community is.

$$AC = \frac{\sum_{i=1}^{n} \delta(s_i, map(c_i))}{n} \tag{4}$$

In it, n represents the whole number of nodes in the network, s_i and c_i mean the community label and the true label of node v_i, besides we define the function $\delta(i,j)$ as follow: if $i = j$, we will get $\delta(i,j) = 1$; otherwise, $\delta(i,j) = 0$; $map(c_i)$ is a mapping function, it means the discovered community label map to the true value label.

Modularity

In the study of complex network community structure, the descriptive definition can not be directly used. So Newman and Girvan [8] define a modularity function, in order to quantitatively analyze the rationality of the detected community structure. Generally, we use Qov function to qualify the modularity of the community structure. The larger number Qov is, the community structure detected by the algorithm is closer to the situation in real network.

Assuming that the community structure of the heterogeneous network is well divided, the community module degree in the bipartite network is represented by the formula (5):

$$Qov = \frac{1}{M}\sum_{i=1}^{n}\sum_{j=1}^{m}\left(a_{ij} - \frac{k_{x_i}k_{y_j}}{M}\right)\delta(x_i, y_j) \tag{5}$$

In it, $M = 0.5\sum a_{ij}$ is the number of edges in the network; k_{x_i} is the degree of the node x_i, k_{y_j} is the degree of the node y_j; δ is the membership function, when node x_i and node y_j belong to the same community, there will be $\delta(x_i, y_j) = 1$, else, $\delta(x_i, y_j) = 0$.

5.4 Experimental Results and Analysis

Experimental Results of Artificial Network

We do the experiment of community division on both he textual algorithm and cluster method [14] based on the resource distribution, and compared the correct partition rate, the result is shown in Fig. 4. It can be seen from Fig. 4, when the number of edges of node is connected to the community is larger than 13 (that is the network community structure is obvious), the result of textual algorithm and resource distribution method are almost coincident, but with reducing the number of edges that connected with the nodes in community (that is the community structure is getting worse obvious), although the correct rate of community division based on both the textual algorithm and resource distribution method remain decline, the correct rate of the textual algorithm is still higher than the resource distribution method. In comparison, the community structure, divided by the algorithm, still has a great advantage in the accuracy.

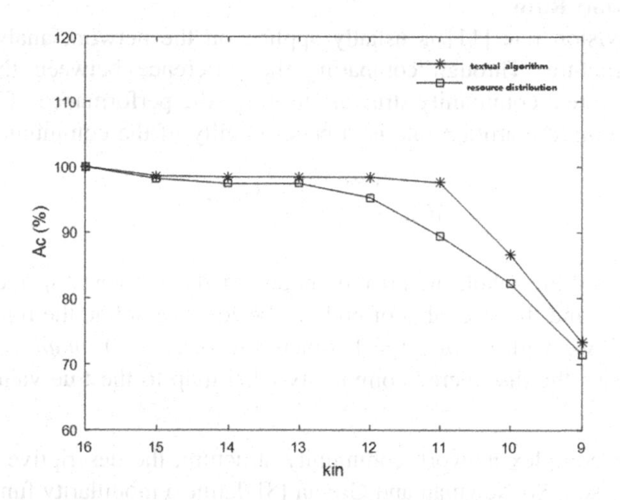

Fig. 4. Comparison of two method

Experimental Results of Real Network

(1) Analysis of parameter α

In the textual algorithm, we control the size of the community through parameter α its effect is manifested on the algorithm's process of merging the communities which are connected closely, so it will have certain influence on the size of the final discovered community structure. When the α is larger, the size of community based on the textual algorithm will be small; if $\alpha = 0$, the detected community structure will be the network itself; when $\alpha \geq 1$, we will find that the final community structure will be the discovered maximum bipartite clique after initialized. In order to analyze what the influence will have on the quality of community structure when use different α, we set 0.1 as the step from 0.1 to 0.9, adjusting parameter value, using *Qov* function as evaluation index; we test on three types of networks(CEO Club Data, Southern Women Data, Graph Products Data), results are shown in Fig. 5.

Generally, the number of *Qov* is larger than 0.3 means the community structure is rational, the larger Qov is, the better community is. Shown in Fig. 5(a), when $\alpha = 0.4$, the community division has the best performance; from Fig. 5(b), when $\alpha = 0.4$, community division has the best performance, from Fig. 5(c), when $\alpha = 0.2$, community division has the best performance. With the increasing of α, the value of *Qov* is getting smaller, until $\alpha \geq 0.5$, the value of *Qov* will not change. According to the experimental analysis, α has different value in networks with different structure.

(2) Analysis of Experimental Results

In Fig. 6, we show the comparison of *Qov* through the textual algorithm and resource distribution clustering method in three different data sets. We compared the

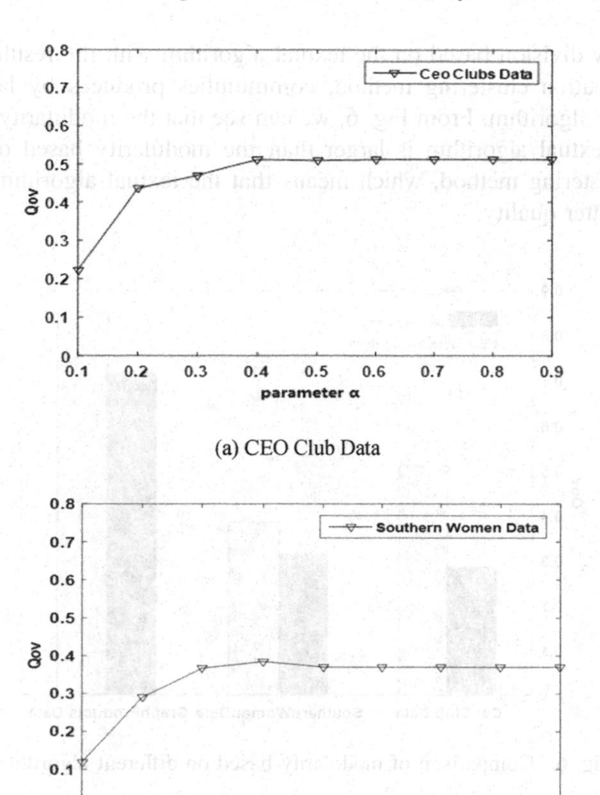

(a) CEO Club Data

(b) Southern Woman Data

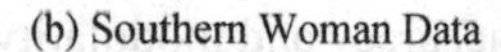

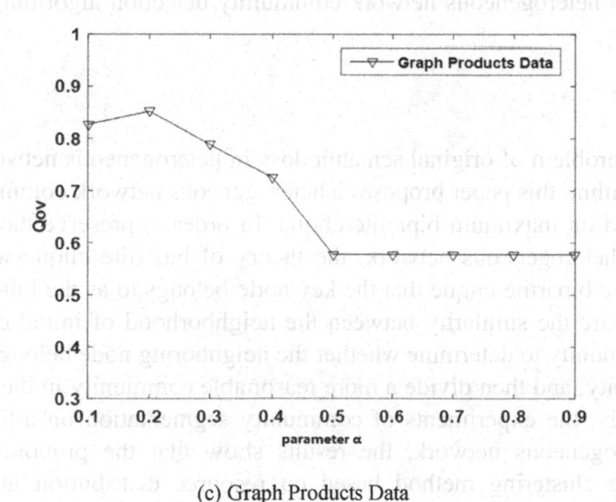

(c) Graph Products Data

Fig. 5. The influence of parameter α on the modularity in different data set best community division based on the textual algorithm with the results based on the resource distribution clustering method, communities produces by latter method is equal to former algorithm. From Fig. 6, we can see that the modularity of community based on the textual algorithm is larger than the modularity based on the resource distribution clustering method, which means that the textual algorithm can get community with better quality.

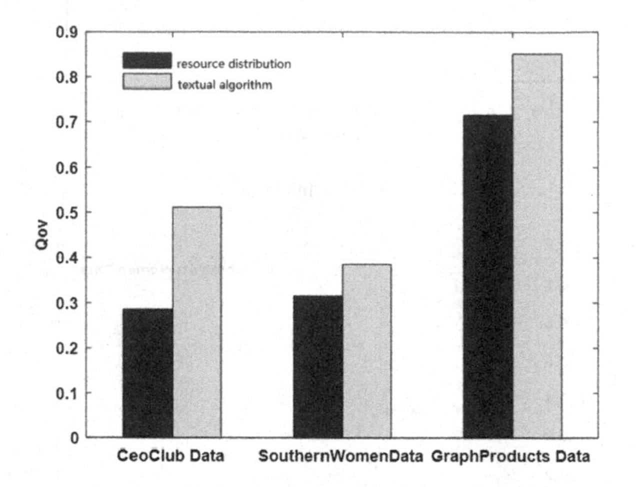

Fig. 6. Comparison of modularity based on different algorithms

Through the experimental results obtained from artificial networks and the real networks, we can see that the algorithm is superior to the resource distribution based on clustering algorithm in terms of accuracy and modularity. In summary, this algorithm is a more realistic heterogeneous network community detection algorithm.

6 Epilogue

Aiming at the problem of original semantic loss in heterogeneous network community detection algorithm, this paper proposes a heterogeneous network community detection algorithm based on maximum bipartite clique. In order to preserve the original information in the heterogeneous network, the theory of bipartite clique was introduced. First, we use the bipartite clique that the key node belongs to as the initial community. Then we compare the similarity between the neighborhood of initial community and the initial community to determine whether the neighboring node belongs to the current initial community, and then divide a more reasonable community in the heterogeneous network. Finally, the experiments of community segmentation on artificial networks and real heterogeneous network, the results show that the proposed algorithm is superior to the clustering method based on resource distribution in accuracy and modularity, so the proposed algorithm can find more realistic community. This proves the rationality and effectiveness of the proposed algorithm.

In order to improve and enhance the effectiveness and robustness of the proposed algorithm, the proposed algorithm will be further studied in the follow-up work in the following aspects: to continue experiments on more practical large-scale networks, to explore the relationship between different network sizes and Similarity parameter, to select more reasonable parameter values for different scales of networks; considering the community detection algorithm in the weighted network in the big data set network; and further analyzing the community detection algorithm of the network with n types of node.

Acknowledgement. This article is supported by National Natural Science Foundation of China under grant No. 71461017.

References

1. Chopade, P., Zhan, J.: Community detection in large-scale big data networks. In: ASE International Conference, pp. 1–7 (2014)
2. Guo, G., Qian, Y., Zhang, X., et al.: Algorithm of detecting community in bipartite network with autonomous determination of the number of communities. Pattern Recogn. Artif. Intell. **28**(11), 969–975 (2015)
3. Liwei, H., Deyi, L., Yutao, M., et al.: A meta path-based link prediction model for heterogeneous information networks. Chin. J. Comput. **37**(4), 848–858 (2014)
4. Hu, J., Wang, B., Lee, D.: Research on centrality of node importance in scale-free complex networks. In: Proceedings of the 31st Chinese Control Conference, China, pp. 1073–1077 (2012)
5. Huang, R., Wu, Q., Zhu, Y., et al.: Community detection in dynamic heterogeneous network with joint nonnegative matrix factorization. Appl. Res. Comput. **34**(10), 2989–2992 (2017)
6. Kitsak, M., Havlin, S., Paul, G., et al.: Betweenness centrality of fractal and nonfractal scale-free model networks and tests on real networks. Phys. Rev. E **75**(5), 056115 (2007)
7. Lancichinetti, A., Fortunato, S., Kertész, J.: Detecting the overlapping and hierarchical community structure of complex networks. New J. Phys. **11**(3), 19–44 (2018)
8. Li, X., Zhang, P., Di, Z., et al.: Community structure in complex networks. Complex Syst. Complex. Sci. **5**(3), 19–42 (2008)
9. Scott, J., Hughes, M., Mackenzie, J.: The anatomy of Scottish capital: Scottish companies and Scottish capital, 1900–1979. Econ. Hist. Rev. **33**(4) (1980)
10. Sun, Y., Norick, B., Han, J., et al.: Integrating meta-path selection with user-guided object clustering in heterogeneous information networks. In: ACM SIGKDD International Conference on Knowledge Discovery and Data Mining, pp. 1348–1356. ACM (2012)
11. Tang, L., Liu, H.: Community Detection and Mining in Social Media, p. 137. Morgan & Claypool (2010)
12. Wei, D., Li, Y., Zhang, Y., et al.: Degree centrality based on the weighted network. In: 2012 24th Chinese Control and Decision Conference (CCDC), pp. 3976–3979. IEEE, China (2012)
13. Wu, Q., Chen, F., Huang, R., et al.: Community detection in heterogeneous network with semantic paths. Acta Electronica Sinica **44**(6), 1465–1471 (2016)
14. Wu, Y., Di, Z., Fan, Y.: A clustering algorithm for bipartite network based on distribution matrix of resources. J. Beijing Normal Univ. Nat. Sci. **5**, 643–646 (2010)

15. Wang, Y., Jin, X., Cheng, X.: Network big data; present and future. Chin. J. Comput. **36**(6), 1125–1138 (2013)
16. Wang, T., Liu, Y., Xi, Y., et al.: Identifying community in bipartite networks using graph regularized-based non-negative matrix factorization. J. Electron. Inf. Technol. **37**(9), 2238–2245 (2015)
17. Zhang, P., Wang, J., Li, X., et al.: Clustering coefficient and community structure of bipartite networks. Phys. A: Stat. Mech. Appl. **387**(27), 6869–6875 (2007)

Novel Algorithm for Mining Frequent Patterns of Moving Objects Based on Dictionary Tree Improvement

Yi Chen, Yulan Dong, and Dechang Pi(✉)

College of Computer Science and Technology, Nanjing University of Aeronautics and Astronautics, 29 Jiangjun Road, Nanjing 211106, Jiangsu, China
1039032746@qq.com, 734823175@qq.com, dc.pi@nuaa.edu.cn

Abstract. Frequent pattern mining is one of the hotspots in the research of moving object data mining. In recent years, with the rapid development of various wireless communication technologies, such as Bluetooth, Wi-Fi, GPRS, 3G, and so on, more and more mobile devices have been used in various applications. In the face of so many data, the following problem is how to deal with and apply the massive data stored in the database, and the theory and technology of data mining emerge as the times require. Therefore, frequent pattern mining for moving objects is very necessary, meaningful and valuable. This paper draws on the existing algorithms for mining frequent patterns of moving objects, and puts forward some innovations. Aiming at the defect of low time efficiency for common frequent mode Apriori algorithm, an improved algorithm named IAA-DT based on dictionary tree is proposed in this paper. The algorithm first traverses all the trajectories in the database and adds it to the dictionary tree. At the same time, it also needs to maintain the linked list to facilitate the pruning of invalid entries. Because of the high degree of compressibility of the dictionary tree, it can be counted directly on the dictionary tree, which greatly improves the time efficiency. The IAA-DT algorithm and the existing improved Apriori-like algorithm based on SQL are compared and analyzed by using the open real data of taxi trajectories in San Francisco. The experimental results with different minimum support levels demonstrate the effectiveness and efficiency of our IAA-DT method.

Keywords: Frequent pattern · Moving trajectory · Apriori algorithm Data mining

1 Introduction

With the continuous development of sensing technology and global positioning system, people can use moving terminals to send their own locations to the server at anytime and anywhere, providing their own moving trajectories. Under this trend, a large number of moving data will be generated. At the same time, the technology of moving object database is also booming, among which the mining of frequent patterns of moving objects has become a hot spot of research.

Q. Zhou et al. (Eds.): ICPCSEE 2018, CCIS 901, pp. 269–278, 2018.
https://doi.org/10.1007/978-981-13-2203-7_20

The traditional algorithm is Apriori algorithm. Because of its low time efficiency, this paper proposes an improved Apriori algorithm based on dictionary tree. The improved algorithm is compared with the existing improved Apriori algorithm based on SQL, and the conclusion is drawn that the improved algorithm based on dictionary tree has better performance and higher time efficiency.

Section 2 of this paper introduces the related research work; Sect. 3 introduces the related knowledge of frequent pattern mining. In Sect. 4, the implementation process of the improved Apriori algorithm based on dictionary tree is introduced in detail. The experiment and result analysis are in Sect. 5. Section 6 summarizes the full text and looks forward to the future research work.

2 Related Work

Rakesh Agrawal and Ramakrishnan Srikant [1] first proposed Apriori algorithm based on association rules to solve frequent pattern mining problems. The purpose of association rules is to find out the relationship between items in a data set, also known as "Market basket Analysis". Apriori algorithm scans the database multiple times and uses candidate frequent sets to generate frequent item sets each time. The algorithm has good expansibility and can be used in parallel computing and other fields. Because Apriori algorithm needs to scan the database multiple times before generating frequent pattern complete sets and generates a large number of candidate frequent sets at the same time, this makes the time and space complexity of Apriori algorithm larger.

In 2000, Han [2] and others proposed FP-growth algorithm, which adopted the following divide-and-conquer strategy: the database providing frequent item sets is compressed into a frequent pattern tree (FP-tree), but the association information of item sets are still preserved. FP-tree is a special prefix tree, which consists of frequent item header table and item prefixes. It sorts each transaction data item in the transaction data table according to the degree of support in descending order, then inserts it into a tree with NULL as the root node successively, and records the support degree of the node at each node.

Both Apriori algorithm [3] and FP-growth algorithm [4] exploit the idea of searching to explore the candidates of frequent patterns and verify them by calculating the support degree, but the characteristics of moving objects determine that their frequent patterns are moving trajectories, which has the topology that the normal frequent patterns do not have. Therefore, Apriori algorithm and FP-growth algorithm have a lot of redundancy in time and space when dealing with moving object data, which is inefficient.

3 An Overview of Frequent Pattern Mining Algorithms

This section discusses the main definition of frequent pattern mining and the basic idea of algorithm.

3.1 Preparatory Knowledge

Definition 1. Item set: Suppose $I = \{x1, \cdots xn\}$, where the element *xi* is the item, *I* represents the collection of all *xi*. If there is a set $X{:}X \subseteq I$, and $X = \{x1, \ldots, xk\}$, *then X* is called a k-item set.

Definition 2. Data set: Suppose that data *D* is a collection of database transactions, where each transaction *T* is a non-empty item set, such that $T \subseteq I$. Each transaction has an identifier, that is called *TID*. Suppose *A* is an item set, transaction *T* includes *A*, if and only if $A \subseteq T$.

Definition 3. Frequent patterns: The patterns that frequently appear in the data set, such as item sets, subsequences and substructures. It has two basic measurements, namely, support degree and confidence degree.

Definition 4. Association rules: It is the implication of the form like $A \Rightarrow$ B, among which $A \subset I, B \subset I, A \neq \emptyset, B \neq \emptyset, A \cap B = \emptyset$.

Definition 5. Support degree: An association rule $A \Rightarrow$ B is established in a transaction set *D* and has a degree of support, that is, transaction *D* includes the percentage of $A \cup B$, which is expressed as: $support(\mathbf{A} \Rightarrow \mathbf{B}) = P(\mathbf{A} \cup \mathbf{B})$.

Definition 6. Confidence degree: An association rule $A \Rightarrow$ B has a degree of confidence in transaction *D*, that means *D* includes the percentage of transactions *A* and *B* at the same time, which is expressed as:$confidence(\mathbf{A} \Rightarrow \mathbf{B}) = P(\mathbf{B}|\mathbf{A})$.

Definition 7. Closed frequent item sets: If the item set *X* does not have its true super set *Y*, so that the support count of *Y* is the same as *X* in the data set *D*, then the item set *X* is closed in the data set *D*. And if the item set *X* is closed and frequent in the data set *D*, then the item set *X* is a closed frequent item set in the data set *D*.

Definition 8. Maximal frequent item sets (Maximal item sets): If the item set *X* is frequent, there is no super item set *Y* such that $X \subset Y$ and *Y* is frequent in the data set *D*, then the item set *X* is a maximal frequent item set or a maximal item set in the data set *D*.

According to the above definitions, the problem of mining association rules can be attributed to mining frequent item sets, the steps are as follows: (1) find out all frequent item sets; (2) generate strong association rules by frequent item sets.

Obviously, the time and space cost of the first step is much larger than that of the second step, that is, the performance bottleneck of the algorithm is mainly determined by the first step, so the solution to the frequent pattern mining problem should start with the first step.

3.2 The Basic Idea of the Algorithm

The Apriori algorithm uses an iterative method of layer-by-layer search, where k-item sets are used to search (k + 1)-item sets. First, scan the database, accumulate the counts of each item, and collect items that meet the minimum support. Find the set of frequent 1-item sets, denoted as L_1; Then, use L_1 to find the set of frequent 2-item sets, denoted as L_2; and then use L_2 to find out L_3, and so on until the frequent k-item sets can no longer be found. Finding out every frequent set L_k requires a complete scan of the database.

Figure 1 shows the pseudo-code of Apriori algorithm, first find the set of frequent 1-item sets L_1 (step 01). At steps 02-10, for $k \geq 2$, L_{k-1} is used to generate candidates C_k so that to find out L_k. The process *apriori_gen* generates the candidates, then those candidates with infrequent subsets are deleted using the A priori property (step 03). Once all candidates have been generated, the database is scanned (step 04). For each transaction, use the function *subset* to find all subsets in the transaction that are candidates (step 05), and add up a count of each such candidate (step 06 and 07). Finally, all candidates satisfying the minimum support degree (step 09) form a set of frequent item sets L (step 11).

```
Input:
• D:Tansaction Database
• min_sup:minimum support threshold
Output: frequent item sets in L,D
Steps:
(01)  L1 = find_frequent_1_itemsets(D);
(02)  for(k = 2; Lk-1 ≠ ∅; k++)
(03)  {
(04)    Ck = apriori_gen(Lk-1);
(05)    for each transaction t ∈ D  // scan D and count
(06)     {
(07)        Ct = subset(Ck, t);          // get the subsets of t,they are the candidates
(08)           for each candidate c ∈ Ct
(09)              c.count++;
(10)     }
(11)    Lk = {c(Ck | c.count ≥ min_sup}
(12)  }
(13)  return L = ∪k Lk;
```

Fig. 1. Pseudo-code of Apriori algorithm

After searching all frequent sets of the same depth, the Apriori algorithm extends to the next depth. It requires frequent access to the database and a large number of I/O operations, which greatly reduces the performance. Moreover, when the mode length is long, a large number of unqualified candidates will be produced and it's hard to exclude them, which makes it difficult to improve the performance of the algorithm [5].

4 An Improved Algorithm IAA-DT

4.1 Dictionary Tree

Dictionary tree is a kind of data structure which takes advantage of space for time and compresses the public prefix of a string to reduce the time overhead of query strings and to improve efficiency. Dictionary trees can also greatly avoid pointless comparisons between strings, far more efficient than hash tables. Based on these properties of the dictionary tree, multiple strings can be stored in the same tree to improve the efficiency of the query operation.

Given an example of a dictionary tree, as shown in Fig. 2.

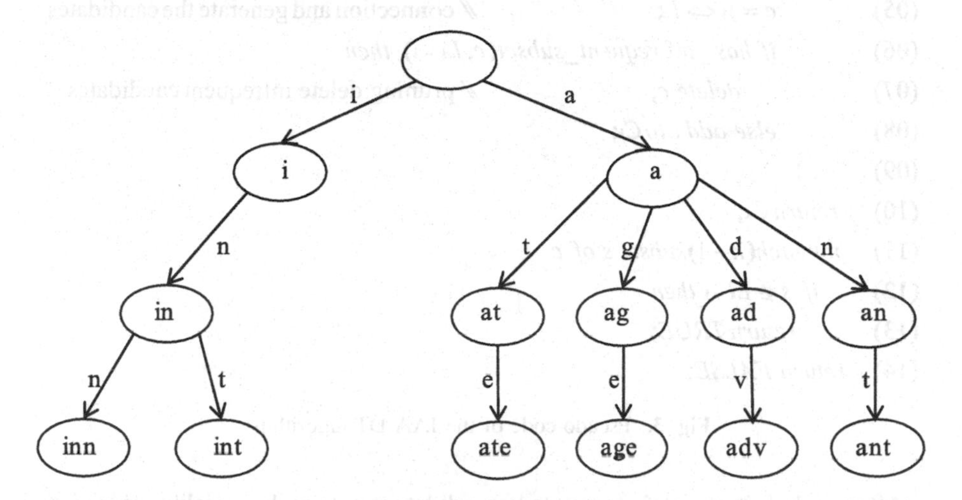

Fig. 2. An example of a dictionary tree

4.2 The Core Idea of the IAA-DT Algorithm

The IAA-DT algorithm generates candidate frequent item sets through association rules and validates them. The key improvements are pruning of the candidate item sets generated by the dictionary tree and optimizing the algorithm for the counting phase.

Figure 3 shows the pseudo-code of the improved algorithm IAA-DT. First, we traverse all the moving trajectories in the database and add them to the dictionary tree one by one. Due to the topological nature of motion trajectories of moving objects, each motion trajectory in the database is regarded as a string, which is read and stored in the dictionary tree of memory, thus solving the problem that requires a large number of I/O operations [6].

While building the dictionary tree, a lot of chain tables need to be maintained for the convenience of the pruning operation. These linked lists are composed of several List linked lists and one Head linked list. Lists store all the locations of an item corresponding to the linked list in the dictionary tree, while the Head stores the position of the List header corresponding to each item.

Input:

- D:Transaction Database
- min_sup: minimum support threshold

Output: Frequent item sets in L,D

Steps:

(01) $for\ each\ item\ set\ l_1 \in L_{k-1}$

(02) $for\ each\ item\ set\ l_2 \in L_{k-1}$

(03) $if(l_1[1] = l_2[1]) \wedge \ldots \wedge (l_1[k-2] = l_2[k-2]) \wedge (l_1[k-1] < l_2[k-2])\ then$

(04) {

(05) $c = l_1 \Leftrightarrow l_2;$ // connection and generate the candidates

(06) $if\ has\_\inf requent\_subset(c, L_{k-1})\ then$

(07) $delete\ c;$ // pruning:delete infrequent candidates

(08) $else\ add\ c\ to\ C_k;$

(09) }

(10) $return\ C_k;$

(11) $for\ each(k-1) subset\ s\ of\ c$

(12) $if\ s \notin L_{k-1}\ then$

(13) $return\ TRUE;$

(14) $return\ FALSE;$

Fig. 3. Pseudo-code of the IAA-DT algorithm

After each k-item set is generated, candidate sets can be quickly pruned and counted through the maintained List linked lists and the Head linked list (steps 03 to 07). When counting each candidate set, first find the first item in the candidate set and all the nodes appearing in the dictionary tree in its corresponding List linked lists, then match and count the candidate sets from these nodes downward. Because the dictionary tree has a high degree of compression, it can be counted directly on the dictionary tree. If the current node does not meet the sub nodes of the candidate (step 11), then quickly pruning, which can greatly improve the time performance of Apriori algorithm.

5 Experimental Results and Analysis

5.1 Experiment 1: Apriori Algorithm and IAA-DT Algorithm

In order to verify whether the IAA-DT algorithm proposed in this paper has better time performance than the traditional Apriori algorithm, experiments are carried out. The two algorithms are run on the real data and then compared and analyzed to verify which one has higher time efficiency and better performance.

The experimental data come from the GPS positioning data of taxi drivers in San Francisco in 2008 and the raw data are preprocessed as the data used in this experiment [7]. In order to facilitate the analysis of experimental results, we make eight data sets. The numbers 1 to 8 represent that with the number of taxi drivers increasing from 1 to 8, the number of moving trajectories also increases accordingly. The amount of time consumed by these eight data sets on two different algorithms is taken as a criterion, as shown in Fig. 4.

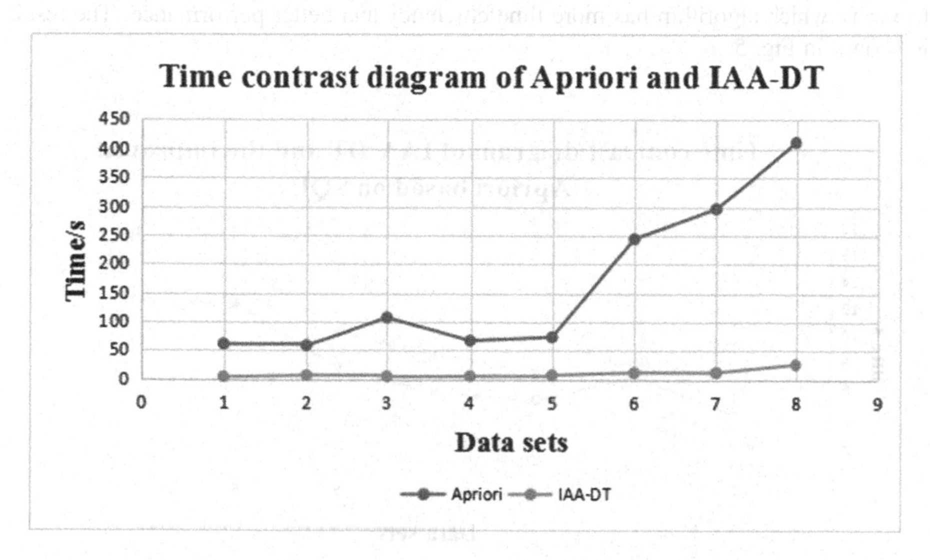

Fig. 4. Time contrast diagram of Apriori and IAA-DT

Figure 4 shows a comparison of time consumption between Apriori algorithm and IAA-DT algorithm. As can be seen from the diagram, the traditional Apriori algorithm has almost no change when the data sets are between 1 and 2; Time shows an increasing trend when data sets range from 2 to 3; Time reduces when data sets are between 3 and 4; Time growth is small when data sets range from 4 to 5; When the data sets are from 5 to 8, the time increase is very large, while the running time of the IAA-DT algorithm is almost the same as the data sets from 1 to 5, and the increase from 5 to 6 is also very small.

Comparing the time consumed by two algorithms, it is obvious that IAA-DT algorithm always takes less time than the traditional Apriori algorithm and the overall trend is that as data sets increases, the differences in time consumption between these two algorithms are getting bigger and bigger [8]. This proves that the proposed algorithm is indeed more efficient and has better performance than the traditional Apriori algorithm.

5.2 Experiment 2: Improved Apriori Algorithm Based on SQL and IAA-DT Algorithm Based on Dictionary Tree

Studies have shown that the improved Apriori algorithm based on SQL also takes less time than the traditional Apriori algorithm. In order to compare the time performance of the proposed IAA-DT algorithm based on dictionary tree and the existing improved algorithm based on SQL, the second experiment is now carried out. The two algorithms are run on the real data (the same data as in Experiment 1). Then compare and analyze to verify which algorithm has more time efficiency and better performance. The result is shown in Fig. 5.

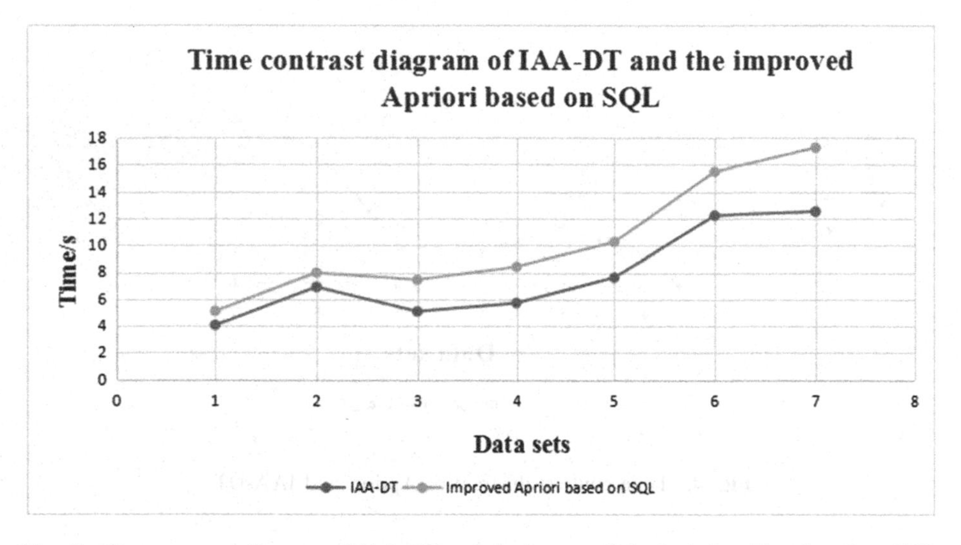

Fig. 5. Time contrast diagram of IAA-DT and the improved Apriori algorithm based on SQL

Figure 5 is a comparison diagram of the time consumption between the IAA-DT algorithm proposed in this paper and the existing improved Apriori algorithm based on SQL. It can be seen from the diagram that as data sets increase, the time of both algorithms shows an overall trend of growth [9], but the improved algorithm proposed in this paper always takes less time than the improved Apriori algorithm based on SQL, which shows that the proposed IAA-DT algorithm has higher time efficiency and better performance.

5.3 Experiment 3: Compare the Execution Time of the Three Algorithms Under Different Levels of Support

The traditional Apriori algorithm, the improved Apriori algorithm based on SQL and the IAA-DT algorithm proposed in this paper are combined to compare their execution time under different minimum support levels. Run three algorithms under different minimum support levels, measure the execution time of each algorithm, and then compare and analyze to verify the performance of these three algorithms. The result is shown in Fig. 6.

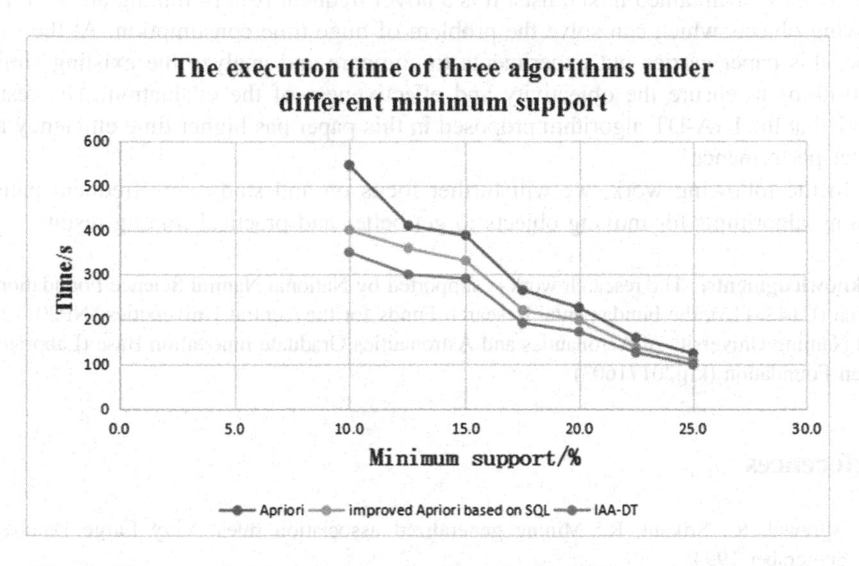

Fig. 6. The execution time of three algorithms under different minimum support

Figure 6 shows the execution time of three algorithms under different minimum support [10], it can be seen from the diagram that the execution time of these three algorithms decreases gradually with the increase of the minimum support degree. When the minimum support degree is small, the differences of the execution time of three algorithms are large, while the differences between them are gradually reduced as the minimum support degree increases. It can also be found from the graph that no matter how large is the minimum support degree, the execution time of the IAA-DT algorithm proposed in this paper is shorter than that of the traditional Apriori algorithm, which is because the Apriori algorithm contains the time overhead of connection and pruning while the proposed algorithm eliminates this overhead. In addition, comparing the proposed IAA-DT algorithm with the existing Apriori algorithm based on SQL, it can be found that the execution time of the IAA-DT algorithm proposed in this paper is always shorter than that of the existing improved Apriori algorithm based on SQL under different minimum support. This further proves that the proposed improved Apriori algorithm based on dictionary tree has better time performance and higher efficiency.

6 Conclusion

In this paper, the problem of frequent pattern mining for moving objects is studied. In order to solve the problem of low time efficiency of traditional Apriori algorithm, the IAA-DT algorithm based on dictionary tree is proposed. The main idea of this algorithm is to add all the trajectories to the dictionary tree and maintain all the linked lists, then the pruning and counting operation can be done directly on the dictionary tree by these maintained linked lists. It is a novel frequent pattern mining algorithm for moving objects which can solve the problem of huge time consumption. At the same time, this paper carries out experiments to compare and analyze the existing similar algorithms to ensure the objectivity and effectiveness of the evaluation. The results show that the IAA-DT algorithm proposed in this paper has higher time efficiency and better performance.

In the following work, we will further focus on and study new frequent pattern mining algorithms for moving objects to get better and practical mining results.

Acknowledgments. The research work is supported by National Natural Science Foundation of China (U1433116), the Fundamental Research Funds for the Central Universities (NP2017208) and Nanjing University of Aeronautics and Astronautics Graduate Innovation Base (Laboratory) Open Foundation (kfjj20171603).

References

1. Agrawal, R., Srikant, R.: Mining generalized association rules. Very Large Databases, September 1994
2. Han, J., Cheng, H., Xin, D., et al.: Frequent pattern mining: current status and future directions. Data Min. Knowl. Disc. **15**(1), 55–86 (2007)
3. Han, J., Kamber, M., Pei, J.: Concept and Technology of Data Mining, 3rd edn, p. 55. Mechanical Industry Press, Beijing (2012)
4. Qin, L., Shi, Z.: SFP-max–maximum frequent pattern mining algorithm based on ranking FP-tree. Comput. Res. Dev. **42**(2), 217–223 (2005)
5. Garg, K., Kumar, D.: Comparing the performance of frequent pattern mining algorithms. Int. J. Comput. Appl. **69**(25), 21–28 (2014)
6. Chen, Z., Shen, H.T., Zhou, X.: Discovering popular routes from trajectories. In: 2011 IEEE 27th International Conference on Data Engineering (ICDE), pp. 900–911. IEEE (2011)
7. Aggarwal, S., Singal, V.: A survey on frequent pattern mining algorithms. Int. J. Eng. Res. Technol. **3**(4), 2605–2608 (2014)
8. Chen, G., Chen, B., Yu, Y.: Mining frequent trajectory patterns from GPS tracks. In: International Conference on Computational Intelligence and Software Engineering, pp. 1–6. IEEE (2010)
9. Qiao, S., Han, N., Zhu, W., et al.: TraPlan: an effective three-in-one trajectory-prediction model in transportation networks. IEEE Trans. Intell. Transp. Syst. **16**(3), 1188–1198 (2015)
10. Aydin, B., Akkineni, V., Angryk, R.: Mining spatiotemporal co-occurrence patterns in non-relational databases. Geoinformatica **4**, 810–828 (2016)

MalCommunity: A Graph-Based Evaluation Model for Malware Family Clustering

Yihang Chen, Fudong Liu(✉), Zheng Shan, and Guanghui Liang

State Key of Laboratory of Mathematical Engineering and Advanced Computing, Zhengzhou 450001, Henan, China
lwfydy@126.com

Abstract. Malware clustering analysis plays an important role in large-scale malware homology analysis. However, the generation approach of the ground truth data is usually ignored. The Labels from Anti-virus(AV) engines are most commonly used but some of them are inaccurate or inconsistent. To overcome the drawback, many researchers make ground truth data based on voting mechanism such as *AVclass*, but this method is difficult to evaluate different-granularity clustering results. Graph-based method like *VAMO* is more robust but it needs to maintain a large-size database. In this paper, we propose a novel evaluation model named *MalCommunity* based on the graph named *Malware Relation Graph*. Different from *VAMO*, the construction of the graph is free from a large-size database and just needs the AV label information of the samples in the test set. We introduce community detection algorithm *Fast Newman* to divide the sample set and use modularity parameter to measure the target clustering results. The experiment results indicate that our model has the ability of noise immunity of malware family classification inconsistency and granularity inconsistency from AV labels. Our model is also convenient to evaluate different-granularity clustering methods with different heights.

Keywords: Malware family clustering · Evaluation model
Anti-virus label · Graph theory · Modularity

1 Introduction

In recent years, serious malware attacks emerged in an endless stream, and led to huge losses to the world. Ukrainian power, mining and railway systems was seriously infected by *KillDisk* and *BlackEnergy* [1]. The botnet *Mirai* led to network paralysis in America [2]. The series APT tools from *ATP28* and *APT29* may even affected the presidential election of the US [3] but so far there is still no assertive evidence to find out the criminal groups. The most famous ransomware *WanaCry* infected hundreds of thousands of computers all over the world [4]. Millions of polymorphous and complex malware samples are made every year

Q. Zhou et al. (Eds.): ICPCSEE 2018, CCIS 901, pp. 279–297, 2018.
https://doi.org/10.1007/978-981-13-2203-7_21

and existing anti-virus(AV) engines are hard to resist. Some reports from AV company showed that there were 324 million malware samples detected in 2014 [5] and the average number of infected computers per day was between 2–5 million [6].

To reduce the workload of malware analysis, analysts attempt to introduce clustering algorithms in malware automatic classification. Malware clustering analysis can also help analysts to discover the relations of malware samples, which can provide assertive evidences for tracing to the source, and even help to infer the criminal hacking groups. In recent years, malware family clustering and classification are research hotspots. Many researchers clustered malware samples by their extracted features such as behavior information collected by dynamic sandboxes, or metadata and disassemble information with static analysis.

However, the evaluation standard of malware clustering algorithm is still hard to reach a consensus. The production process of valid ground truth data is often ignored because it is not the kernel content in many clustering researches. At present, majority of researchers use labels of anti-virus(AV) products, namely AV labels, to make ground truth. However, the label mechanisms of different AV products are usually inconsistent. When in production process, most researchers used majority voting-based mechanism to decide which samples should be divided into the same families. An open-source tool *AVclass* [26], which was released in 2016, are still based on this mechanism. Li et al. [31] also pointed out that the samples with inconsistent AV labels may be abandoned when making the test set, but this samples may be the difficult samples for the clustering model to be evaluated. Different AV engines may have different malware family definition standards, which will seriously affect the validity of voting mechanism. Another approach detects the relations of samples by constructing graph of AV labels like *VAMO* [7], but *VAMO* needs to maintain a large-size database to guarantee the accuracy of the weights of edges in the graph.

In this paper, we propose a novel malware clustering evaluation model called *MalCommunity*. The model can be separated into two parts. The basic part is constructing a topological graph named *Malware Relation Graph*. This part of work is to reduce the effect of label inconsistency noise from raw data, which is the main problem of the effective evaluation model, and build a solid data foundation. The second part is the evaluation method, in which we compute a evaluation score for the target clustering results based on *Malware Relation Graph*. In *Malware Relation Graph*, each single sample in the dataset will be present as a node, and the weights of edges indicate the relation degrees of the two connected samples, namely the probability of whether they belong to the same family. The weights are also decided by AV label information, but different from *VAMO*, they are free from sample information outside the tested sample set and we needn't to maintain a large database. We also introduce a community detection algorithm, *Fast Newman*, to provide a serial of best divisions on *Malware Relation Graph* for reference. When evaluating other clustering results, we use modularity parameter to measure the division but not comparing the clusters. Because there may exist several best divisions and the marginal samples are difficult to categorize. These samples may lead to uncertainty results for the evaluation method of comparing the clustering sets. In addition, *Fast Newman* is hierarchical, therefore the model can be used to evaluate the clustering systems with different granularities.

We summarize the contributions of this paper as followed:

- We proposed a novel malware family clustering evaluation model *MalCommunity*. This model is based on *Malware Relation Graph* which relies on AV label information. We also improve the computational efficiency of this model.
- We introduce community detection algorithm *Fast Newman* and modularity parameter into our model. When evaluating, we measure the target clustering results by modularity but not precision or recall, which is computed by comparing the clustering sets.
- Our model is hierarchical so that it is suitable to evaluate different-granularity malware clustering system.

The rest of the paper is organized as followed. In Sect. 2 we will introduce the related works in malware analysis and evaluation areas. In Sect. 3, we discuss the methodology of our model in detail, including the construction of *Malware Relation Graph* and *Fast Newman* division algorithm. Then we present our experimental results in Sect. 4. Finally, we make a brief conclusion of our work.

2 Related Works

Malware analysis is a persistent and hot research area all along. Various analyzing techniques were proposed to detect and classify malware samples and applied in AV software, which leads to different detection mechanisms of AV engines. The extracted features of malware samples based on these different techniques are used to malware family clustering to discover the relations of malware samples. But the evaluation model of malware clustering results are often ignored. We focus on studying this kind of researches and propose our own model.

Static analysis is a traditional techniques to analyze malware samples, which needt extract features of samples in run time. Tamersoy et al. [29] introduced a large-scale malware detection method by mining file-relation graphs based on file similarity analysis. Drew et al. [9] used local sensitive hash for file summarization and analyzed the similarity among files at the physical level but not semantics. But obfuscation techniques are obstacle for the development of static analysis. In recent years, dynamic analysis is a hot topic in malware detection research. Xu et al. [11] implemented a prototype system, GOLDENEYE, with high success rate, high speed and little memory to dynamically trace malwares' behaviors in right environments. Spensky et al. [13] developed a low-level framework for performing automated binary analysis using hardware-based instrumentation. Dynamic analysis is good at bypassing obfuscation and package but the analysis is time-consuming. Now the researchers start to turn their attention in content-agnostic analysis, namely tracing the malware samples in the wild but not running them in the sandbox. Invernizzi et al. [17] and Taylor et al. [18]

checked and traced HTTP flow then detect the HTTP attack chains and find out malicious downloads and exploit kits. Kwon et al. [20] introduced two new concepts called downloader graph and influence graph to describe the features of downloaders. Content-agnostic analysis is usually for the malware samples that have frequent network behavior.

Malware clustering analysis is used to automatically classify large-scale malware sample set to multiple clusters which contain similar samples. This technique can sharply reduce the workload of malware analysis and provide assertive evidences for tracing to the source and even find out the criminal groups. Kirat et al. [8] proposed SigMal, which detected malware variants based on signal processing techniques. They considered that variants retained some similarity at the binary level. In another work, Kirat et al. [28] developed a leverages algorithms MALGENE, borrowed form bioinformatics, to locate evasive behavior in system call sequences. Hu et al. [12] built a system named DUET to trace malwares' run-time behaviors by extracting their system or API calls, then associated with some static features and clustered the malware set. Saxe et al. [15] used deep neural network to reach high true positive and low false positive rate. These approaches are based on different malware features and used different clustering models, which may lead to different results. The evaluation standard of malware clustering is difficult to reach a consistent level.

We focus on the evaluation method of malware clustering for a long time. Bayer et al. [30] chose consistent samples to make up of the test set, which are consistently voted into the malware families by all AV engines. But only very little size of sample set match the condition and Li et al. [31] also pointed out that the chosen samples may be easy to cluster and this method cannot construct a valid test set closed to the real world. Perdisci et al. [7] proposed *VAMO* system to evaluate malware clustering systems. They first construct a huge AV Label Graph to compute the distances of different AV labels, then calculate the relation degrees between every two samples in the test set. They also introduced hierarchical clustering algorithm to divide the test set and produce a reference division for evaluation. *VAMO* is an important inspiration source of our model. But *VAMO* needs to maintain a huge database and the accuracy of relation degrees among samples is decided by the database they maintained. Our method is free from a huge database and the relation degrees among samples is solid. Based on *VAMO* and majority voting mechanism, Sebastian et al. [26] developed an open-source tool *AVclass*, which can detect AV label alias and provide a single family name for each sample. However, this approach can't provide multiple reference divisions for different clustering mechanisms. Wei et al. [32] concludes the previous evaluation models and proposed a new approach based on the behavioral clustering system which developed by their research team. But this method rely on the feature extraction method they used and it is lack of objectivity.

An objective and valid malware clustering evaluation model should be proposed. After summarizing the previous works, we define the concept of *Malware Relation Graph* to describe the relations among samples in the target test set. We also introduce non-parameter division algorithm to generate reference division and we use modularity to evaluate target results instead of simply comparing the clusters between the target results and the reference results.

3 Methodology

In this section, we discuss our evaluation model *MalCommunity* in detail, which is based on AV labels and community detection algorithm. Firstly, we will introduce how to gather family name labels by AV software programs based on an open-source tool *AVclass* [26]. Then we will propose the concept of *Malware Relation Graph* and introduce the implement and improvement. Next we will describe the community detection algorithm, *Fast Newman*, to divide the *Malware Relation Graph* into MalCommunities. Finally, we will show the processing step to measure the performance of target clustering results.

3.1 Family Name Generation

After fetching all data back from VirusTotal, in this section, we introduce how to extract family names from the name information provided by AV software. It is thoughtless to use all AV information directly. Firstly, some of AV software vendors have more than one products in the AV's list of VirusTotal [24]. These software use the same AV engine as the backend so their scanning result are similar and it's pointless to record all results. Secondly, some AV engines have poor performance with low accuracy. Third, there is plagiarism in AV software. The detecting ability of some AV engines is weak, so their vendors copy the information from other AV software to improve their performance. We select 15 representative AV software from the AV's list of VirusTotal, mainly referencing some reports [25] of Anti-Virus Comparative, an authentic third-party testing institution for AV software. The selected AV software are listed in Table 1.

Table 1. The list of 15 selected AV software programs.

ID	AV software	ID	AV software	ID	AV software
1	AVG	6	F-Secure	11	GData
2	Avast	7	Emsisoft	12	Symantec
3	Avira	8	Microsoft	13	Panda
4	BitDefender	9	McAfee	14	K7AntiVirus
5	ESET-NOD32	10	Kaspersky	15	Qihoo 360

Every AV software provides their scanning information when detecting the malicious file. The AV information contains a malware name, the AV name and the scanning date. In our research, the scanning date is not necessary any more.

The name commonly could be separated into three parts: malicious type, family name and variant name. Take the name *Trojan.Spy.Delf.NNO* for example, *Trojan* and *Spy* is malicious type, while *Delf* is the family name and *NNO* is the variant name. As the named rules of numerous AV software are quite different, every samples have 15 different names. And what we need is their family name. The extraction work of family name is not cushy. The main problem is that various AV vendors defined their own name frameworks and the name fields are up-and-down depending on their detection scheme. Fortunately, the open-source tool named *AVclass* [26] to overcome this problem. *AVclass* takes malware names as the input, identifies and filters the irrelevant information, then extract the family names. In the second step, *AVclass* selects a possible family name by a voting mechanism. But we think the second step lose too much useful information, such as the relationship of malwares, which we will discuss in the next section. We modify *AVclass* to just extract family names and take them as output.

3.2 Malware Relation Graph

Malware Relation Graph is a undirected topological graph. *Malware Relation Graph*, containing all malware samples in the sample set, just presents the malware relation by undirected edges without coordinate. In the *Malware Relation Graph*, a node represents a malware while an edge connecting two different nodes represents that the connected nodes are in the same family. The weight value of the edge is the confidence level of the connection. Just as the above statement, single AV engine performs poorly on malware detection and family classification while the combination of all selected AV engines greatly improves these abilities. We use a voting mechanism by various AV engines to find out the connection and evaluate its confidence level.

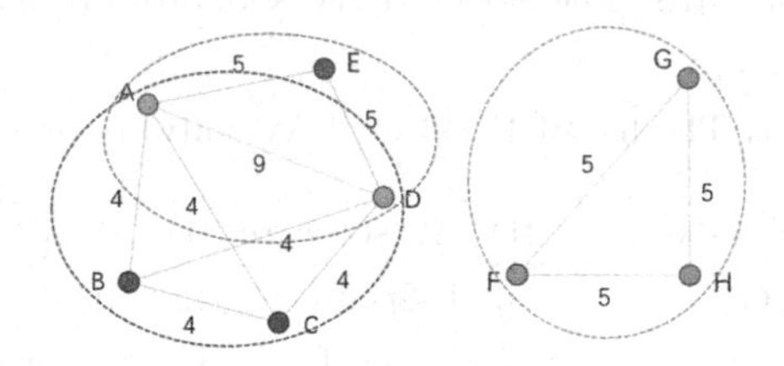

Fig. 1. Example of a small-size *Malware Relation Graph* (uninitialized).

Figure 1 illustrates a small-size example of *Malware Relation Graph*, containing eight samples and three families. Nodes A to H represent the malware samples. The 15 AV engines vote whether two different nodes are in the same family or not by the family names we extracted. If a single AV software marks two nodes with the same family name, the weight value of the edge connecting these two nodes increases 1. For example, assume that all edges are initialized to 0, then Avast labels node A and B with the same family name *Delf*, so the weight of edge AB increases to be 1. When *AVG* labels them with *Duch*, the edge AB is modified to be 2. Therefore we can see, *Malware Relation Graph* is independent of name contents and just retains relationship information from various AV information.

Under the mechanism of edge connection we introduce above, a malware family is a fully connected subgraph. The dotted circle we labeled in Fig. 1 are three full-connected families. Take family $\{F, G, H\}$ for example. The weights of edges FG, FH, GH are all 5, which means that there are 5 AV software which simultaneously classify samples F, G, H into the same family (thought they use different family names to label). But in the real *Malware Relation Graph*, cases like $\{A, B, C, D, E\}$ are much more common. Subgraph $\{A, B, C, D, E\}$, on the left of Fig. 1, contains two related families, $\{A, B, C, D\}$ and $\{A, D, E\}$, while nodes A and D are in the both families. It means, there are 4 AV engines that classify A, B, C and D into the same family while the other 5 AV engines classify A, D, E into another family. The AV engines which have not classified E into family $\{A, B, C, D\}$ may not detect E as malicious one or consider that E should be in another family. Briefly, the intersection of these two families is not empty, which indicates that they are related but not isolated. The relationship is observed and uncovered from different angles by the combination of various AV detecting mechanisms.

To regular the *Malware Relation Graph* model, we normalize the weight values of edges and set the confidence threshold. The normalization depends on the number of AV engines we select. As shown in Table 1, the aggregate number N_{AV} is 15. The formula or normalization is obtained as

$$\omega_{ij} = \frac{\omega_{ij}'}{N_{AV}}, \tag{1}$$

where ω_{ij}' is the weight of edge connecting node i and j and ω_{ij} is the normalization result of ω_{ij}'.

Now we provide the definition for the malware map abstraction. Malware map G can be defined as $G = (V, E, W)$, where the meaning of each symbol is as follows:

- V denotes the set of all samples in the malware map G.
- $E \subseteq V \times V$ denotes a set of undirected edges. Note that there is no self loop in E and the adjacency matrix of E is a symmetric matrix.
- W denotes a set of weight defined on the edges of the malware map. Notes that ω_{ij} in W is between 0 and 1.

3.3 Algorithm Improvement

When processing large-scale sample set, the *Malware Relation Graph* is difficult to construct because of its large scale. While generating the *Malware Relation Graph*, we should maintain a large sparse adjacency matrix in the memory,

Table 2. An example of malware family table.

Malware family	Member nodes in the malware family
Avira:Deil	G,F,H
AVG:Redir	G,F,H
McAfee:Netag	G,F,H
Mircosoft:Kryptik	G,F,H,A
ESET-NODE32:Eldorado	G,F,H,B

which is an unbearable consumption. Even though we change the data struct of the adjacency matrix to a chain table, it is poor in performance owing to a great number of edges. An equivalent but high-performance optimization is needed, which can separate the adjacency matrix into small parts so that the maintenance is no longer needed. Such an algorithm is designed by us. We first build and maintain a new table to store the information of every malware family labeled by every single AV engine. A simple example is shown in Table 2. Every malware family has a member node list in the table. The scale of the table is much smaller than the adjacency matrix. The query of the table can be implemented by hashing algorithm, whose complexity of time is $O(1)$.

We introduce a kind of virtual nodes to the graph. The virtual nodes represent the malware families shown in Table 2. Malware node is no longer connected to another one but connected to the family nodes it belongs to. For example, G is connected to 5 family virtual nodes: *Avira:Deil*, *AVG:Redir*, *McAfee:Netag*, *Mircosoft:Kryptik*, *ESET-NOD32:Eldorado*. At the same time, nodes F and H are connected to these five families. However, node A is only connected to *Mircosoft:Kryptik* and B is connected to *ESET-NOD32:Eldorado*.

Now consider which nodes may be in the same family with node i. In the virtual graph, as shown in Fig. 2, every node that could be reached in two hops from node i is considered to be *is* neighbor. Thus, a two-layer breadth first traversal is needed. Node j will be connected to i if there are more than N_{AV} different two-hop paths from j to i. The weight of edge ij is the number of two-hop paths between i and j. Take node G for example, F and H will be connected to G while A has only one two-hop path to G, and B, either.

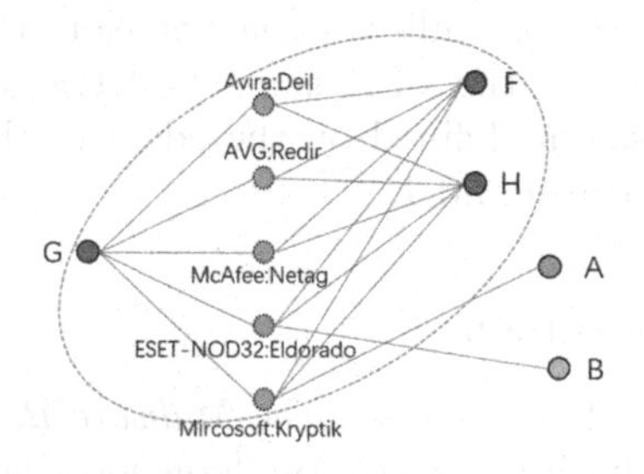

Fig. 2. An example of a virtual graph.

Using this methodology, the state of edges needn't to be maintained in the memory throughout the process. Only a portion of edges will be operated when finding the neighbors of one node then the weights of edges can be moved out of memory and are no longer needed in the construction. Figure 3 presents the *Malware Relation Graph* of a sample set we collect from *VXheaven* [23].

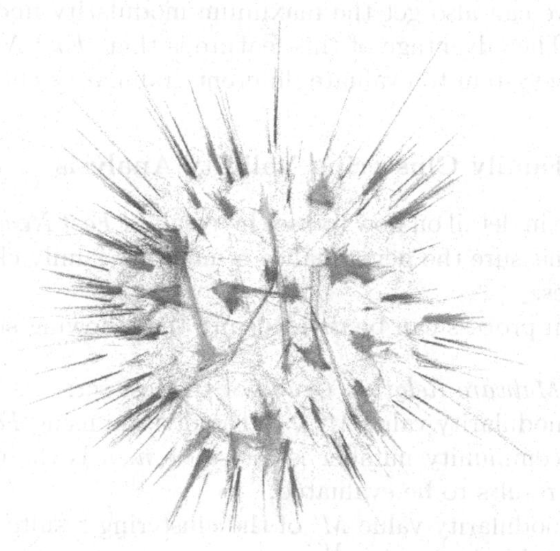

Fig. 3. The *Malware Relation Graph* of the ground true dataset.

3.4 MalCommunity

The method we introduced above has constructed an AV-label-based *Malware Relation Graph* of the sample set. In this section, we will divide the graph into multiple tight clusters by community detection algorithm, *Fast Newman* [33]. This algorithm is non-parameter and hierarchical, which is measured by modularity.

Suppose that there are n nodes and m edges in our graph. For a particular division of the graph, $\delta(c_i,c_j)$ represents the relation of the nodes i and j. If the two nodes are in the same group, $\delta(c_i,c_j)$ is 1 and otherwise $\delta(c_i,c_j)$ is 0. Therefore, the modularity of the division is as

$$Q = \frac{1}{2m}\Sigma_{ij}[A_{ij} - \frac{k_i k_j}{2m}]\delta(c_i, c_j) \tag{2}$$

where A_{ij} is the weight of the edge between node i and node j. When there is no edge between node i and node j, A_{ij} is 0. k_i and k_j denote the degree of node i and node j. The larger modularity denotes that the better the division result is.

Fast Newman algorithm is to find out a division to reach the maximal modularity value in a given graph. The algorithm can be separated into two steps. At the beginning, the graph is initialized into n different communities which contain only one node for each. In the first step, named modularity optimization, the nodes are divided into the neighbor community to increase the modularity value of the whole graph. Then in the second step, which is named community aggregation, replace the communities in the first step result by nodes to construct a new graph, and repeat the first step for further polymerization until the modularity value is maximum.

In addition, the *Fast Newman* algorithm is also hierarchical. In the polymerization process, we can also get the maximum modularity under different community number. The advantage of this feature is that, *Fast Newman* algorithm can be a reference system to evaluate different-granularity clustering results.

3.5 Malware Family Clustering Validity Analysis

We now introduce in detail on how to use the result of *Fast Newman* algorithm as a benchmark to measure the performance of malware family clustering methods in other researches.

The evaluation process can be divided into the following steps:

1. construct the *Malware Relation Graph* of the test set;
2. calculate the modularity value M_{FN} of the division using *Fast Newman* algorithm, whose community number is set to be n. n is the cluster number of the clustering results to be evaluated;
3. calculate the modularity value M' of the clustering results to be evaluated;
4. calculate the evaluation score $\frac{M'}{M_{FN}}$.

Fast Newman algorithm is a kind of non-parameter algorithm. Namely it can get the best community division result without setting any parameter. The result is one of best divisions that have the maximum modularity in the whole *Malware Relation Graph. Fast Newman* algorithm is also one kind of hierarchical algorithm. The community (or cluster) number can be given then get the corresponding result. This result is the best division with a given community number, whose modularity is always smaller than the maximum modularity in the whole graph.

In the second step, M_{FN} is the maximum modularity value of the n-community division where n is the cluster number of the clustering method to be evaluated. We choose n-community division as reference but not the best division result of the whole graph. That is because different clustering methods have different granularities. The best division result of *Fast Newman* is usually coarse grained and unreasonable to evaluate different-granularity clustering results. The best n-community division result of *Fast Newman* has the same granularity with the clustering result to be evaluated and the validity analysis is more reasonable.

In addition, evaluating the clustering results by calculating the modularity is more stable. Many previous evaluation models like *VAMO* used the division results of hierarchical clustering algorithm as reference. But hierarchical clustering algorithm may be disturbed by marginal samples and there may exist multiple best division results at the same time, which will lead to uncertainty when evaluating. *Fast Newman* algorithm may also exist multiple best division results. But different from *VAMO*, we measure the performance of the clustering results by modularity. The multiple best division results of *Fast Newman* algorithm have the same modularity so that the disturbance from marginal samples doesn't exist anymore.

4 Experimental Results

In this section we evaluate our model by comparing the validity and noise immunity of *MalCommunity* with some previous evaluation models *VAMO* and *AVclass*. The experiments can be separated into three groups. The first two experiments discuss the performance of *MalCommunity* under the noise of label inconsistency through the single factor experiment designs. The third experiment is an application of *MalCommunity*.

In the first experiment, we artificially construct a sample set then gradually increase the noise to simulate the malware family classification inconsistency of AV engines, and evaluate the noise immunity of our model. The second experiment also changes the single factor based on the artificial dataset but this time we control the granularity of malware family labels. This experiment is designed to evaluate our model under the noise of family granularity inconsistency among different AV engines. The third experiment is a real application, namely an evaluation process for an existing malware family clustering results.

In addition, the controlled models we used in this paper is *AVclass* and *VAMO*. *AVclass* is an open-source tool which can be downloaded from *GitHub* [27]. The methodology of *VAMO* was first presented in [7] but we can't find the source code of the tool, so we reimplement *VAMO* through [7] to complete the two controlled experiments.

The validity indexes we used in this paper include F1 score, Rand Statistic (RS), Jaccard Coefficient(JC) and Folkes and Mallows Index(FM). The derivation method of these four indexes is as follow.

Assume that in the sample set, there are s reference clustering $R_C = \{R(C_1), R(C_2), \ldots, R(C_s)\}$ and the malware clustering result is $C = \{C_1, C_2, \ldots, C_n\}$. In this setting, the indexes precision, recall and F1 score are defined as:

- **Precision.** $Prec = \frac{1}{n}\Sigma_{j=1}^{n} max_{k=1,\ldots,s}(C_j \cap R_{C_k})$
- **Recall.** $Rec = \frac{1}{s}\Sigma_{k=1}^{s} max_{k=1,\ldots,n}(C_j \cap R_{C_k})$
- **F1 score.** $F1 = \frac{Prec \cdot Rec}{Prec + Rec}$

Given a pair of samples (m_1,m_2) where $m_1, m_2 \in$, we can compute the following quantities:

- a is the number of (m_1, m_2) that m_1, m_2 are both in the reference cluster R_{C_k} and in the same cluster C_j as well.
- b is the number of (m_1, m_2) that m_1, m_2 are in the same reference cluster R_{C_k}, but they are separated into two different clusters.

- c is the number of (m_1, m_2) that m_1, m_2 are in the same cluster C_j but are assigned to different reference clusters R_{C_k} and R_{C_h}.
- d is the number of (m_1, m_2) that belong to two different reference clusters and separated into two different clusters, either.

Based on these four definitions, we can compute the cluster validity indexes RS, JC and FM as [34]:

- **Rand Statistic.** $RS = \frac{a+d}{a+b+c+d} = \frac{a+d}{|M|}$
- **Jaccard Coefficient.** $JC = \frac{a}{a+b+c}$
- **Foldes and Mallows Index.** $FM = \frac{a}{\sqrt{(a+b)(a+c)}}$

The four validity indexes F1 score, RS, JC and FM are all in [0, 1], and higher values indicate that the clustering result is closer similarity to the ground truth.

4.1 Malware Family Classification Inconsistency

Malware family classification inconsistency among different AV engines is caused by different AV detection mechanisms. For example, AV engine A detects sample a and b and categorizes them into malware family Fam_1, then categorizes the sample c into Fam_2. But AV engine B recognizes that sample a is in Fam_1 while b and c are in Fam_2 through its own detection method. This phenomenon is called malware family classification inconsistency.

The experiment in this section is to evaluate the validity of our model under the classification inconsistency noise. We first construct an artificial sample set, then simulate the inconsistent division methods of different AV engines in the real world by changing the AV labels of the samples and increasing the inconsistency degrees. *MalCommunity*, *VAMO* and *AVclass* are evaluated repeatedly under different inconsistency degrees of AV labels.

Controlled Datasets. The artificial sample set that we constructed includes 3000 malware samples, which can be divided into 15 different malware families, namely 200 samples per family. The labels were provided by four different AV engines, namely that each sample has four different family labels.

At initialization, the samples in the 15 malware families all have high-consistency AV family labels, namely the divisions for the sample set by the four AV engines is identical. For instance, malware family Fam_1 contains 200 samples, each AV engines will use a family label to mark these 200 samples and mark other samples by other family names. It is worth noting that there may exist family name alias, namely different AV engines may use different family label to indicate Fam_1. Therefore, for sample a, there is a set of labels $L_a = \{l_{1a}, l_{2a}, l_{3a}, l_{4a}\}$. And for sample b, there is $L_b = \{l_{1b}, l_{2b}, l_{3b}, l_{4b}\}$. If sample b belongs to the same family with a, $l_{ib} = l_{ia}$ where $i = 1, 2, 3, 4$; otherwise $l_{ib} \neq l_{ia}$. Since there are 15 different families in the ground truth, the labels provided by each AV engines have 15 kinds, namely for sample a, $l_{ia} \in \{F_{i1}, F_{i2}, \ldots, F_{is}\}$ where $s = 15$ and $i = 1, 2, 3, 4$.

In the next step, we increase the inconsistency noise in the dataset through controlling the malware family labels for each samples. At the beginning, the divisions of sample set by the four AV engines are identical. But when the noise is introduced, the division is no longer consistent. The method of controlling variables is as followed. When the noise index is N, namely the labels of N samples in the dataset are with noise, we randomly choose N samples and flip some of their labels. Take sample a as an example. The label set of sample a is $L_a = \{l_{1a}, l_{2a}, l_{3a}, l_{4a}\}$ when initializing, we randomly change some of its labels and get the result as $L'_a = \{l'_{1a}, l_{2a}, l'_{3a}, l_{4a}\}$, where $l'_{1a} \in \{F_{11}, F_{12}, ..., F_{1s}\}/\{l_{1a}\}$, $l'_{3a} \in \{F_{31}, F_{32}, ..., F_{3s}\}/\{l_{3a}\}$ and $s = 15$.

Performance Evaluation. We use previous works *AVclass* and *VAMO* as control groups to evaluate the amending ability of our malware classification evaluation model *MalCommunity*. *AVclass* is a majority-voting-based evaluation model, namely deciding which malware family the target samples should belong to by joint voting. Since the research of *AVclass* is later than *VAMO*, *AVclass* draws lesson from the principles of *VAMO* to detect family name alias to improve the performance of traditional majority-voting-based evaluation models. Therefore, *AVclass* is better than previous majority-voting-based evaluation models. *VAMO* is proposed by Perdisci et al. [7] and in this paper we reimplemented this tool through the related paper. The principles of *VAMO* is similar to *MalCommunity*, which both construct topological graph based on AV labels to search the best family division. But the graph that *VAMO* constructs is AV Label Graph whose nodes are family labels, and the graph of *MalCommunity* is *Malware Relation Graph* whose nodes indicate the malware samples.

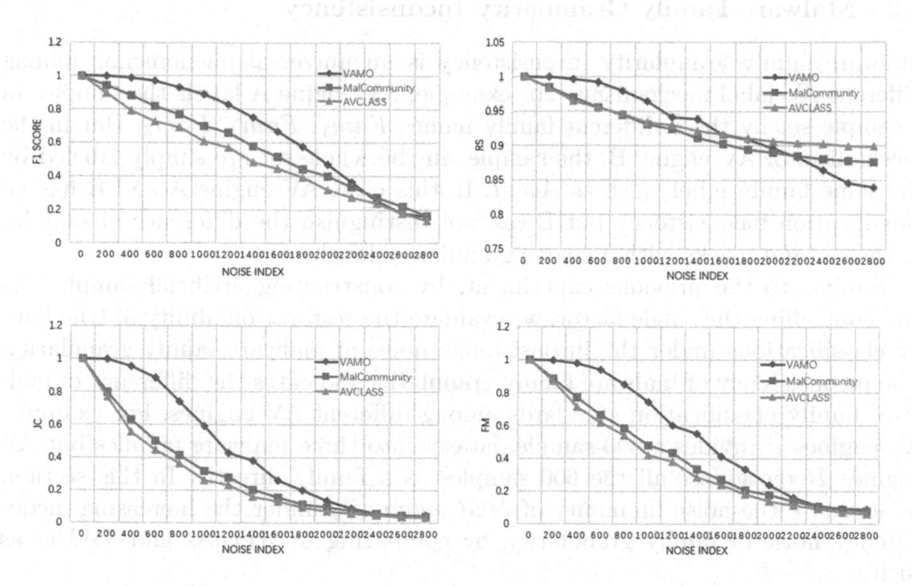

Fig. 4. The F1 score, RS, JC and FM indexes of *VAMO*, *MalCommunity* and *AVclass* under different noise indexes.

Figure 4 indicates that the overall performance of *MalCommunity* is slightly better than *AVclass*. *VAMO* performs the best under classification inconsistency noise. Especially when the noise index is less than half of the dataset, *VAMO* and *MalCommunity* is relatively stable. However, because *VAMO* needs to compute the average distance values between every two samples in the dataset for hierarchical clustering, it is time-consuming though Floyd Algorithm is introduced into the model. The time-consumming curves are shown in Fig. 5. Therefore, *VAMO* is undesirable for the evaluation of large-scale clustering results. Generally, *MalCommunity* is better performance and less time consumed, which indicates that *MalCommunity* is suitable for large-scale clustering evaluations.

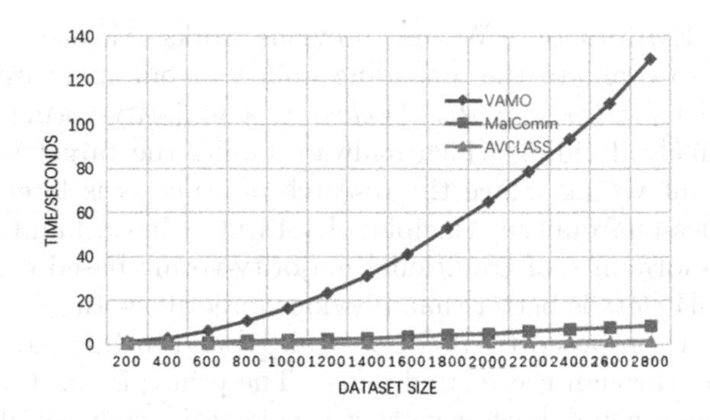

Fig. 5. The time-consuming curves of *VAMO*, *MalCommunity* and *AVclass* with different sizes of datasets.

4.2 Malware Family Granularity Inconsistency

Malware family granularity inconsistency is an universal phenomenon among different AV label mechanisms. For example, AV engine A labels the samples in a sample set by three different family names Fam_1, Fam_2, Fam_3. But in the mechanism of AV engine B, the samples in the whole set are simply labeled by the same family label, such as *Agent*. In this case, AV engine A and B has no classification inconsistency but B can not distinguish the difference of samples in the set and marks with a coarse granularity label.

Similar to the previous experiment, by constructing artificial sample sets and controlling the single factor, we evaluate the restoration ability of true family classifications under the inconsistency noise of malware family granularity. The inconsistency of malware family granularity indicates the difference of malware family classification standards among different AV engines. For example, AV engines *A* divides a 600-sample dataset into three malware families but AV engines *B* recognizes all the 600 samples as a family directly. In this section, we evaluate the noise immunity of *MalCommunity* under the increasing inconsistency noise of family granularity, by comparing to *AVclass* and *VAMO* as such.

Controlled Datasets. In this experiment, we initialed a 3000-sample dataset, whose samples are all labeled by four different AV engines. Similarly, the dataset can be divided into 15 malware families, namely 200 samples per family.

This time, we keep the consistency of malware classification but control the inconsistency degrees of the family granularities. The consistency of malware classification means that if AV engine *A* labels sample *a* and *b* by the same family name, AV engine *B* will not them into two different families. But coarse-grained family labels may category multiple malware families into a large family.

Therefore, we expand the family labels of a certain AV engine to control the granularities of family labels. Take AV engine *A* as an example. We expand the family labels of *A* from 200 samples per family to 400 samples per family or more. Relatively, we increase the number of AV engines with coarse-grained labels from one to two or three. Under different granularity degrees and different number of AV engines with coarse-grained labels, we evaluate the restoration ability of the ground-truth divisions of our model.

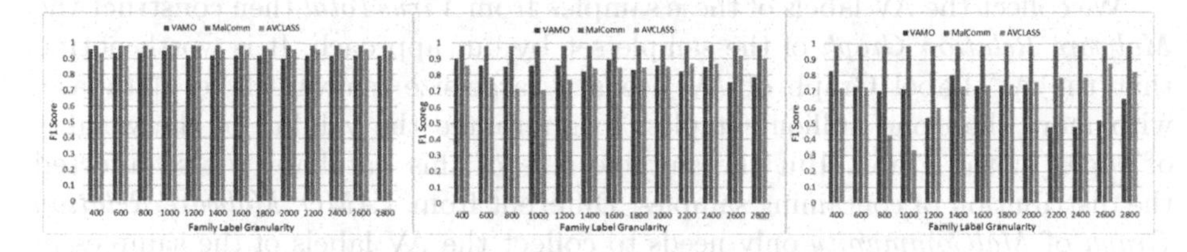

Fig. 6. (a) The F1 score of the three models under different granularities by controlling only one AV engines; (b) The F1 score of the three models under different granularities by controlling two AV engines; (c) The F1 score of the three models under different granularities by controlling three AV engines.

Performance Evaluation. As Fig. 6 shown, (a) indicates that only control the family label granularities of a certain AV engines from 200 per family to 400 per family, 600 per family and etc., but keep the labels of the other three AV engines and evaluate the performance of the three models. And (b) is the F1 results when expanding the family labels of two AV engines at the same time. (c) is the results when changing the labels of three AV engines. It is worth noting that the cluster number should be given for *VAMO* before hierarchical clustering but this parameter is unknown for *VAMO*. Therefore, we can only use the average number of different family label types among the four AV engines to be the cluster number parameter.

The experiment results suggest that the division of *MalCommunity* will not be affected no matter how many AV engines change their labels or what the granularities of the labels are. The F1 scores of *MalCommunity* is always nearly 1, which means that the division results of *MalCommunity* are very close to ground truth. Relatively, with the increase of the number of coarse-grained AV engines, *VAMO* and *AVclass* is seriously affected and their F1 scores sharply decrease. When the granularity degree is five times larger than ground truth, namely 1000 samples per family, *AVclass* reaches the lowest performance; *VAMO* performs worst when the granularity degree is nearly 13 times larger than ground truth. Therefore, *MalCommunity* is robust to the inconsistency noise of family label granularity.

4.3 Real-World Application

In this section, we discuss how to apply our model *MalCommunity* to evaluate other malware family clustering results in the real world. We take a self-developed malware clustering system as an example to introduce the specific process of *MalCommunity* to evaluate the target clustering system. This system clusters the sample set based on the disassemble information of binary samples. We first choose a real sample set, including 93 binary malware samples, which are all from the revealed hacking group *APT28*. The samples in this sample set are a serial Trojan tools that *APT28* used to attack and permeate their targets. These tools have various variants so that there are multiple malware families.

We collect the AV labels of these samples from *VirusTotal* then construct the *Malware Relation Graph* of the sample set by our approach. It is worth noting that the AV Label Graph of AV needs a large-size malware label database, with more than one million samples, to guarantee the validity of the weights of edges in the graph. But the suitable size of this database is also affected the distribution of containing samples. Different from *VAMO*, *Malware Relation Graph* of *MalCommunity* only needs to collect the AV labels of the samples in the dataset and the weights of edges in the graph are firm and certain.

We can get a serial of division results on the *Malware Relation Graph* of *APT28* using the community detection algorithm Fast Newman. As shown in Table 3, the 93 samples of *APT28* can be categorized into 15 clusters by *Fast Newman* with the highest modularity value 0.413. With the increase of cluster

Table 3. Top 20 division results of *Fast Newman* on the *Malware Relation Graph* of *APT28*.

Clustering number	Modularity	Clustering number	Modularity
15	0.413254	25	0.375044
16	0.409168	26	0.370511
17	0.407947	27	0.367446
18	0.404186	28	0.365708
19	0.400329	29	0.360979
20	0.392758	30	0.357377
21	0.390023	31	0.354132
22	0.381991	32	0.350845
23	0.379825	33	0.346329
24	0.376522	34	0.342941

numbers, the modularity will decrease. Table X lists the cluster number and their corresponding modularity values of the Top 20 division of *Fast Newman*. The disassembly-based clustering system divides this sample set into 26 clusters. Therefore, the reference modularity value for this system is 0.371 in this granularity degree. On account, the modularity value of this clustering result on the *Malware Relation Graph* is 0.352. Thus, The evaluation score for this clustering method is $score = 0.352/0.371 = 0.949$.

5 Conclusion

In this paper, we propose a novel evaluation model named *MalCommunity* to automatically evaluate the target malware clustering results. We use two groups of experiments in controlled setting environment to verify the validity and noise immunity against malware family classification inconsistency and granularity inconsistency of AV labels. We also make a real-world application to show the evaluation process of target malware clustering models.

In previous works, many evaluation models are based on majority voting mechanism such as *AVclass*. But voting mechanism is lack of the ability to evaluate the clustering methods with different granularity. Some of models also used label graph to analyze the relation of samples such as *VAMO*. But *VAMO* needs to maintain a huge database and its division approach may lead to uncertain results if there are marginal samples. *MalCommunity* uses community detection algorithm based on *Malware Relation Graph* and evaluate other clustering models through modularity parameter. It is free from huge database and the uncertainty of marginal samples. It is also able to evaluate different-granularity clustering models with different heights.

References

1. http://www.securityweek.com/blackenergy-killdisk-infect-ukrainian-mining-railway-systems
2. Biggs, J.: Hackers release source code for a powerful DDoS app called Mirai. TechCrunch, 10 October 2016. Accessed 19 Oct 2016
3. https://www.fireeye.com/blog/threat-research/2014/10/apt28-a-window-into-russias-cyber-espionage-operations.html
4. https://www.fireeye.com/blog/threat-research/2017/05/smb-exploited-wannacry-use-of-eternalblue.html
5. Internet Security Center Qihoo 2015. 2014 Internet Security Research Report in China. http://zt.360.cn/report/
6. Kingsoft: 2015–2016 Internet Security Research Report in China (2016). http://cn.cmcm.com/news/media/2016-01-14/60.html
7. Perdisci, R., Manchon, U.: VAMO: towards a fully automated malware clustering validity analysis. In: Computer Security Applications Conference, pp. 329–338 (2012)
8. Kirat, D., Nataraj, L., Vigna, G., et al.: Sigmal: A static signal processing based malware triage. In: Proceedings of the 29th Annual Computer Security Applications Conference, pp. 89–98. ACM (2013)

9. Drew, J., Moore, T., Hahsler, M.: Polymorphic malware detection using sequence classification methods. In: 2016 IEEE Security and Privacy Workshops (SPW), pp. 81–87. IEEE (2016)
10. Yakdan, K., Dechand, S., Gerhards-Padilla, E., et al.: Helping Johnny to analyze malware: a usability-optimized decompiler and malware analysis user study. In: 2016 IEEE Symposium on Security and Privacy (SP), pp. 158–177. IEEE (2016)
11. Xu, Z., Zhang, J., Gu, G., Lin, Z.: GoldenEye: efficiently and effectively unveiling malware's targeted environment. In: Stavrou, A., Bos, H., Portokalidis, G. (eds.) RAID 2014. LNCS, vol. 8688, pp. 22–45. Springer, Cham (2014). https://doi.org/10.1007/978-3-319-11379-1_2
12. Hu, X., Kang, G.S.: DUET: integration of dynamic and static analyses for malware clustering with cluster ensembles. In: Computer Security Applications Conference, pp. 79–88 (2013)
13. Spensky, C., Hu, H., Leach, K.: LO-PHI: low-observable physical host instrumentation for malware analysis. In: Network and Distributed System Security Symposium (2016)
14. Kittel, T., Vogl, S., Kirsch, J., Eckert, C.: Counteracting data-only malware with code pointer examination. In: Bos, H., Monrose, F., Blanc, G. (eds.) RAID 2015. LNCS, vol. 9404, pp. 177–197. Springer, Cham (2015). https://doi.org/10.1007/978-3-319-26362-5_9
15. Saxe, J., Berlin, K.: Deep neural network based malware detection using two dimensional binary program features. In: 2015 10th International Conference on Malicious and Unwanted Software (MALWARE), pp. 11–20. IEEE (2015)
16. Rajab, M.A., Ballard, L., Lutz, N., et al.: CAMP: content-agnostic malware protection. In: Network and Distributed System Security Symposium (2013)
17. Invernizzi, L., Miskovic, S., Torres, R., et al.: Nazca: detecting malware distribution in large-scale networks. In: Network and Distributed System Security Symposium (2014)
18. Taylor, T., Snow, K.Z., Otterness, N., Monrose, F.: Cache, trigger, impersonate: enabling context-senstive honeyclient analysis on-the-wire. In: Network and Distributed System Security Symposium (2016)
19. Li, Z., Alrwais, S., Xie, Y., et al.: Finding the linchpins of the dark web: a study on topologically dedicated hosts on malicious web infrastructures. In: 2013 IEEE Symposium on Security and Privacy (SP), pp. 112–126. IEEE (2013)
20. Kwon, B.J., Mondal, J., Jang, J., et al.: The dropper effect: insights into malware distribution with downloader graph analytics. In: ACM SIGSAC Conference on Computer and Communications Security, pp. 1118–1129. ACM (2015)
21. Plohmann, D., Yakdan, K., Klatt, M., et al.: A comprehensive measurement study of domain generating malware. In: 25th USENIX Security Symposium (USENIX Security 16), pp. 263–278. USENIX Association (2016)
22. Le Blond, S., Gilbert, C., Upadhyay, U., Gomez-Rodriguez, M., Choffnes, D.R.: A broad view of the ecosystem of socially engineered exploit documents. In: Network and Distributed System Security Symposium (2017)
23. http://vxheaven.org/
24. https://www.virustotal.com/
25. https://www.av-comparatives.org/
26. Sebastián, M., Rivera, R., Kotzias, P., Caballero, J.: AVclass: a tool for massive malware labeling. In: Monrose, F., Dacier, M., Blanc, G., Garcia-Alfaro, J. (eds.) RAID 2016. LNCS, vol. 9854, pp. 230–253. Springer, Cham (2016). https://doi.org/10.1007/978-3-319-45719-2_11

27. https://github.com/malicialab/avclass
28. Kirat, D., Vigna, G.: MalGene: automatic extraction of malware analysis evasion signature. In: Proceedings of the 22nd ACM SIGSAC Conference on Computer and Communications Security, pp. 769–780. ACM (2015)
29. Tamersoy, A., Roundy, K., Chau, D.H.: Guilt by association: large scale malware detection by mining file-relation graphs. In: Proceedings of the 20th ACM SIGKDD International Conference on Knowledge Discovery and Data Mining, pp. 1524–1533. ACM (2014)
30. Bayer, U., Comparetti, P.M., Hlauschek, C., et al.: Scalable, behavior-based malware clustering. In: Network and Distributed System Security Symposium, NDSS 2009, San Diego, California, USA, February. DBLP (2009)
31. Li, P., Liu, L., Gao, D., Reiter, M.K.: On challenges in evaluating malware clustering. In: Jha, S., Sommer, R., Kreibich, C. (eds.) RAID 2010. LNCS, vol. 6307, pp. 238–255. Springer, Heidelberg (2010). https://doi.org/10.1007/978-3-642-15512-3_13
32. Wei, F., Li, Y., Roy, S., Ou, X., Zhou, W.: Deep ground truth analysis of current android malware. In: Polychronakis, M., Meier, M. (eds.) DIMVA 2017. LNCS, vol. 10327, pp. 252–276. Springer, Cham (2017). https://doi.org/10.1007/978-3-319-60876-1_12
33. Newman, M.: Modularity and community structure in networks. Proc. Nat. Acad. Sci. U.S.A. **103**(23), 8577–8582 (2006)
34. Halkidi, M., Batistakis, Y., Vazirgiannis, M.: On clustering validation techniques. J. Intell. Inf. Syst. **17**(2), 107–145 (2001)

Negative Influence Maximization in Social Networks

Jinghua Zhu[1,2](✉), Bochong Li[1], Yuekai Zhang[1], and Yaqiong Li[1]

[1] School of Computer Science and Technology, Heilongjiang University, Harbin 150001, China
zhujinghua@hlju.edu.cn

[2] Key Laboratory of Database and Parallel Computing of Heilongjiang Province, Harbin, China

Abstract. Influence maximization is one of the key research problems in social networks due to its wide applications like spread of ideas and viral marketing of products. Most of the existing work focus on social networks containing only positive relationships (e.g. friend or trust) between users, but in reality social networks containing both positive and negative relationships (e.g. foe or distrust). Ignoring the negative relations may lead to over-estimation of positive influence in practical applications. Thus, in this paper, we study influence maximization problem with negative effects (NIM). To address the NIM problem, we use the polarity Independent Cascade (IC-P) diffusion model which extends the standard Independent Cascade (IC) model with negative opinions. Then we propose the positive influence maximization algorithm (PIM) and negative influence maximization (NIM) problem. We prove that influence function of the NIM problem is monotonic and submodular, so we propose a CELF based algorithm (CELF_NIM) to solve it. Experiments results show that our algorithm has matching influence spread compared with greedy algorithm and achieves several orders of magnitude time improvement.

Keywords: Social networks · Influence maximization · Negative effects

1 Introduction

The rapid growth of large-scale social networks such as Facebook, WeChat and Weibo has enabled people around the world to share information and share their hobby without leaving home. The definition of influence maximization problem is proposed by Kempe *et al.* for the first time [1]. Given a social network graph G and the number of k, based on a certain propagation model, influence maximization algorithm will find the k nodes which can influence as many nodes as possible. The propagation models are mainly divided into two types: the independent cascade model (IC) and the linear threshold model (LT). The diffusion procedure is as follows: The seeds try to activate their neighbors with a certain activity probability. And then in the later periods, all the active nodes continue to try to activate their neighbors until no more nodes can be activated. The aim of the influence maximization problem is to select k nodes as seeds and make the propagation as wide as possible from the seed nodes.

Q. Zhou et al. (Eds.): ICPCSEE 2018, CCIS 901, pp. 298–307, 2018.
https://doi.org/10.1007/978-981-13-2203-7_22

However, most of the existing research work [2–4] focus on social networks only containing positive relationships (e.g. friend or trust) between users, but in reality social networks containing both positive relationships and negative relationships (e.g. foe or distrust) between users. Ignoring the negative relations may lead to over-estimation of positive influence in practical applications. The negative influence maximization problem in signed social network is a key research problem that has not been studied and it is the focus of this paper. Although Li *et al.* has studied the influence maximization problem in signed social network [7], they only give a greedy solution for the problem. The high time complexity hinder the application of their solution to a large scale social network.

In this paper, we extend the IC model by adding a quality factor q and propose a new propagation model IC-P. Based on this model, we divide the influence maximization problem into positive influence maximization problem PIM and negative influence maximization problem NIM. Since the greedy simulations are expensive, we present CELF_NIM and MIA_NIM algorithm to reduce the running time. CELF_NIM optimization utilizes submodularity such that in each round the incremental influence spread of a large number of nodes do not need to be re-evaluated because their values in the previous round are already less than that of some other nodes evaluated in the current round. In a large scale social network, the running time of CELF_NIM is still very long. Thus, we propose MIA_NIM algorithm which borrows the idea of Maximum Influence Arborescence (MIA) heuristic algorithm to solve the negative influence maximization problem. MIA_NIM algorithm is proved to be highly efficient and scalable.

The contributions of this paper are as the following:

- We extend the IC model to IC-P model in signed social network by introducing quality factor q.
- We propose CELF_NIM and MIA_NIM algorithms to effectively solve the negative influence maximization problem in signed social network.
- We conduct a series of experiments on real social network dataset to verify the superiority of our algorithms.

2 Related Work

Kempe et al. formalized the maximization problem as a discrete optimization problem [1]. They gave the proof that the problem was NP-hard. Leskovec et al. proposed "Lazy Forward" optimization strategy, and proposed CELF (Cost Effective Lazy Forward) and CELF++ algorithm, which can significantly improve the efficiency of the greedy algorithm [5]. The experimental results show that the CELF algorithm is similar to the greedy algorithm in terms of the influence range, but the CELF algorithm is up to 700 times faster than greedy algorithm.

Chen *et al.* proposed NewGreedy algorithm and MixGreedy algorithm which are used to solve the deficiencies of greedy algorithm [6]. The NewGreedy algorithm eliminates the edges of the network which are not useful for influence propagation, in this way, it can make an efficient propagation on the network. The MixGreedy algorithm combines the NewGreedy algorithm and the CELF algorithm together. In addition, Chen et al. also proposed an improved algorithm based on maximum degree, which is called DegreeDiscount algorithm. Li *et al.* put forward the problem of maximizing the impact in signed social networks for the first time [7]. They introduced the concept of maximizing positive and negative effects in signed social networks and pointed out that the relationships among people in social networks are not always friendly and positive. Negative relationships also exist in real situation. Thus people's reviews of products usually have both good and bad effects.

IC (Independent Cascade Model) and LT (Linear Threshold) model are the two most popular influence propagation model in social network. The independent cascade model is proposed by Glodenberg [8] and the linear threshold model is firstly introduced by Granovetter [9].

3 Propagation Model

In signed social network, there exists both positive influence and negative influence. Traditional propagation model only considers the positive influence. In order to incorporate the negative influence, we extend the IC model to Polarity-related Independent Cascade model which is called IC-P model in this paper.

In the IC-P model, when a node is activated, q determines the probability of the node becoming a positive node. The model works as follows. In network graph G, the initial set S contains activated nodes, nodes in set S become positive nodes with probability q, and a negative node with probability $1 - q$. At each step of information propagation, the node that is being activated in the previous step will deactivate all its neighbor nodes that have not been activated in this phase with the probability p (edge activation probability). If the activation is successful, then the activated node will become a positive node with probability q and a negative node with probability $1 - q$. Meanwhile, if a node is negatively activated in the previous step, it will also be deactivated with a probability p in this phase. If all its inactive nodes are activated, then the active node changes to a negative node with a probability of one. In the process of model propagation, if there are more than one node to activate a node at the same time, then the probability of activation is random.

4 CELF_NIM and MIA_NIM Algorithm

4.1 CELF_NIM Algorithm

The aim of maximizing the impact is to find the initial set S of nodes of size k, which maximizes the number of nodes that can be affected in the network graph G. The polarity-related influence maximization problem can be divided into two parts: positive influence maximization (PIM) and negative influence maximization (NIM). Given a network G = (V, E, P), the initial set S of size k, the quality factor q, the problem is defined as: $S \in \arg\max_{s \in v, |s|=k} \sigma_G(S, q)$.

Next we will discuss $\sigma_G(S, q)$ and use the properties. Given a network graph $G = (V, E, P)$, initial set S, quality factor q, we define positive activation probability as $pap_G(v, S, q)$ negative activation probability as $nap_G(v, S, q)$. According to linear expect, we calculate:

$$\sigma_{+G}(S,q) = \sum\nolimits_{v \in V} pap_G(v,S,q) \tag{1}$$

$$\sigma_{-G}(S,q) = \sum\nolimits_{v \in V} nap_G(v,S,q) \tag{2}$$

The distance from the set S to the node v in graph G is defined as $d_G(S,v)$, and $d_G(S,v)$ is the shortest distance between nodes in set S and the set v, and if there is no path to node v in the set S, then $d_G(S,v) = +\infty$. If the distance is infinite, we agree $q^{+\infty} = 0 (0 \leq q \leq 1)$. In social network G, a set of nodes that i steps away from the initial set S is defined as $a_G(S,i) = |\{v | d_G(S,v) = i\}|$. The following Lemma 4.1 gives the proof of formulas (1) and (2).

Lemma 4.1. Given a network graph G = (V, E, P), assume that the edge probability p (e) = 1, for $e \in E$, $v \in V$:

$$pap_G(v,S,q) = q^{d_G(S,v)+1}$$

$$nap_G(v,S,q) = (1-q)(1+q+q^2+\ldots+q^{d_G(S,v)+1}) = 1 - q^{d_G(S,v)+1}$$

thus:

$$\sigma_{+G}(s,q) = \sum_{i=0}^{n-1} a_G(S,i) q^{i+1}$$

$$\sigma_{-G}(s,q) = \sum_{i=0}^{n-1} a_G(S,i)(1-q^{i+1})$$

For a network graph G, we simulate all random events on the basis of the probability of edge propagation, and we obtain a subgraph $G' = (V', E', p')$, where $V' = V$, $E' \in E$, for each e in E', $p'(e) = 1$, G' with probability $\Pr(G') = \prod_{e \in E'} p(e) \cdot \prod_{e' \in E \setminus E'} (1 - p(e'))$ calculated. Defining all such sets of subgraphs G' as δ_G, this edge must be active if one can be found in subgraph G'.

For calculation, we can first figure out the subgraph G', and then calculate the subgraph on the impact of communication problems, so the problem becomes simpler. In G', when multiple neighbor nodes of node v try to activate them at the same time, we do not need to randomly sort these neighbor nodes anymore. Because the probability of edge is one, the activate result is the same. Therefore, after the calculation, we simply choose one of these neighbors in the node set.

Theorem 4.1. Given a network graph $G = (V, E, p)$, the initial node set $S \in V$, we have:

$$\begin{aligned} \sigma_{+G}(S,q) &= E_{G' \leftarrow \delta_G}[\sigma_{G'}(S,q)] = \sum\nolimits_{G' \in \delta_G} \Pr_G(G') \cdot \sigma_{G'}(S,q) \\ &= \sum\nolimits_{G' \in \delta_G} \Pr_G(G') \sum_{i=0}^{n-1} a_{G'}(S,i) q^{i+1} \end{aligned} \tag{3}$$

$$\sigma_{-G}(S,q) = \sum\nolimits_{G' \leftarrow \delta_G} [\sigma_{G'}(S,q)] = \sum\nolimits_{G' \in \delta_G} \Pr_G(G') \cdot \sigma_{G'}(S,q)$$
$$= \sum\nolimits_{G' \in \delta_G} \Pr_G(G') \sum_{i=0}^{n-1} a_G(S,i)(1-q^{i+1}) \tag{4}$$

Theorem 4.2. For any network $G = (V, E, P)$, the influence function is monotonic with submodel characteristic when the initial set of nodes S is fixed.

Without loss of generality, we take the positive impact function as an example and prove as follows:

$$\sigma_G(S,q) = \sum\nolimits_{G' \in \delta_G} \Pr_G(G') \sum\nolimits_{u \in V} q^{d_{G'}(S,v)+1} \tag{5}$$

Since the CELF algorithm adopts the submodel property, with the initial S increases, the marginal benefit brought by new seed node decreases. It does not need to calculate the marginal revenue of all nodes in the same way as the traditional greedy algorithm dose. If the marginal revenue of influence value of node u in the previous round is less than the marginal revenue of influence value of node v in the current round, the marginal revenue of the influence value of node u in the current round must be smaller than that of node v. Therefore, u cannot be the node with the largest marginal revenue in the current round. CELF algorithm greatly reduces the number of marginal revenue to be calculated for each iteration, so the time efficiency is increased nearly 700 times compared with the traditional greedy algorithm.

CELF_NIM (negative influence maximization) aims to find a set of initial nodes so that the negative impact propagation is maximized, which can be calculated by the following formula:

$$S^{-} = \arg \max_{S \subseteq V; |S| = k} \sigma_{-}(s,q) \tag{6}$$

The specific algorithm is as follows:

CELF_NIM(*G*, *k*, *S*)

```
1:  initialize S =∅
2:  For i =1 to k Do
3:      While pq not empty Do ( pq is priority queue).
4:          (inc_u,u) ← pq.pop()
5:          (ninc_u) ← σ(S ∪ {u}) − σ{s}
6:          (inc_w,W) ← pq.pop()
7:          If ninc_u ≥ inc_w Then
8:              S ← S ∪ {u}
9:          Else pq.push(ninc_u,u)
10: Return S
```

4.2 MIA _NIM Algorithm

In the large scale social networks, CELF_NIM algorithm will take a long time to solve the impact maximization problem, therefore its efficiency is hard to satisfy user's demand. In this paper, we borrow the idea of Maximum Influence Arborescence (MIA) heuristic algorithm to solve the negative influence maximization problem and propose MIA_NIM algorithm. MIA_NIM algorithm is proved to be highly efficient and has good scalability. The model is constructed by the largest impact path, and its global influence value is approximated by the influence of the node in its periphery. MIA_-NIM algorithm calculates the local influence of a node by constructing a local tree structure graph to approximate the global influence. Local trees are divided into two types: one type is the in-neighbor tree, all the edges point to the root node; the other type is the out-neighbor tree, all edges away from the root node.

We use $A = (V, E, p)$ to represent a local tree. Given an initial node set S, quality factor q, influence propagation spread on local tree A is defined as $\sigma_A(S,q)$. $\sigma_{+A}(S,q)$ and $\sigma_{-A}(S,q)$ represent the positive influence and negative influence respectively.

For any node $u \in V$, the probability of u being successfully activated is defined as $ap(u)$, and $\sigma_A(S,q) = \sum_{u \in V} ap(u)$. In out-neighbor tree, path(u) represents the path from node s to node u in node set S, $|path(u)|$ is the length of path.

$$ap(u) = \begin{Bmatrix} \prod_{e \in E(path(u))} p(e) \cdot q^{|path(u)|+1} & \textit{if } path(u) \neq NULL \\ 0 & \textit{else} \end{Bmatrix} \tag{7}$$

In the in-neighbor tree, when the parameter $q = 1$, $pap(u)$ can be formulated as $ap(u) = 1 - \prod_{w \in N_u^{in}} (1 - ap(w)p(w,u))$.

When $q < 1$, the situation becomes complicated because the neighbor nodes of node u have both positive and negative nodes. If the negative neighbor succeeds in activating u, u will become a negative node, and the positive node succeeds in activating u, u will become a negative node or a positive node, so the state of u depends on the order in which the neighboring nodes activate it. This sequence depends on two factors: one is the sequence of time steps in which u is activated; the other is the sequence of random activated at the same time step. To calculate $pap(u)$, we adopt the method of dynamic programming. First we define $ap(u,t)$ as the probability of node u being activated at time step t, then $ap(u) \sum_{t \geq 0} ap(u,t)$.

When $t > 0 \wedge u \notin s$

$$ap(u,t) = \prod_{w \in N_u^{in}} [1 - \sum_{i=0}^{t-2} ap(w,i)p(w,u)] - \prod_{w \in N_u^{in}} [1 - \sum_{i=0}^{t-1} ap(w,i)p(w,u)] \tag{8}$$

The set of in-neighbor tree nodes for node u is $N^{in}(u)$, and for each node $w \in N^{in}(u)$, $ap(w,i)p(w,u)$ represents the probability that w is activated at time t. If the u node state is inactive, it may be activated by the w node at time t + 1 and this event is mutually exclusive for w nodes that are activated at different time step t. Thus,

$1 - \sum_{i=0}^{t-2} ap(w,i)p(w,u)$ is the probability that node u was not activated by node w at time step $t-1$ or before time step $t-1$.

So $\prod_{w \in N_u^{in}} [1 - \sum_{i=0}^{t-2} ap(w,i)p(w,u)]$ is the probability that node u is not activated by its in-neighbor node at time step $t-1$ or before time step $t-1$. Thus $\prod_{w \in N_u^{in}} [1 - \sum_{i=0}^{t-1} ap(w,i)p(w,u)]$ is the probability that node u is not activated by its in-neighbor node at before time step t.

5 Experiment Evaluation

5.1 Data Set

Two real data sets were used for testing. We extracted a considerable number of data sets from the datasets and conducted simulation experiments on them. In the dataset, the appropriate deletion was performed, that is, nodes that were too scattered to have edge connections with other node. The specific size of the datasets are shown in Table 1 below:

Table 1. Social site data set statistics

Dataset	Nodes	Edge	Average nodes
Slashdot	16K	55K	6.8K
Digg	12K	43K	7.3K

5.2 Algorithm Performance Analysis

To verify the effectiveness of our algorithm, the accuracy of CELF_NIM algorithm and MIA_NIM algorithm based on IC-P model are verified on the three datasets. The effect of different seed set size and different quality factor q on the influence spread are evaluated too.

Because the impact probability of propagation cannot be calculated from these data sets, we get the propagation probability on the edge with $p_{u,v} = 1/d(v)$, where $d(v)$ is the in-neighbor of node v. The algorithms test the dataset for 10,000 Monte Carlo simulations.

Traditional greedy algorithms and other impact maximization algorithms do not consider negative effects. In signed social networks, there exist positive and negative influences, ignoring the positive influence will decrease the accuracy of results. We evaluate the number of the seeds and the value of q on the performance of various algorithms. We vary the size of seeds set from 10 to 60, and the value of quality factor q from 0.5 to 0.9.

As shown in Fig. 1, x-axis is the seed set size and y-axis is the influence spread. When the seed set size is the same, CELF and Greedy have similar propagation spread. CELF-NIM is the worse because it has positive influence and is low in value, thus reflecting the accuracy of the CELF algorithm based on IC-P model.

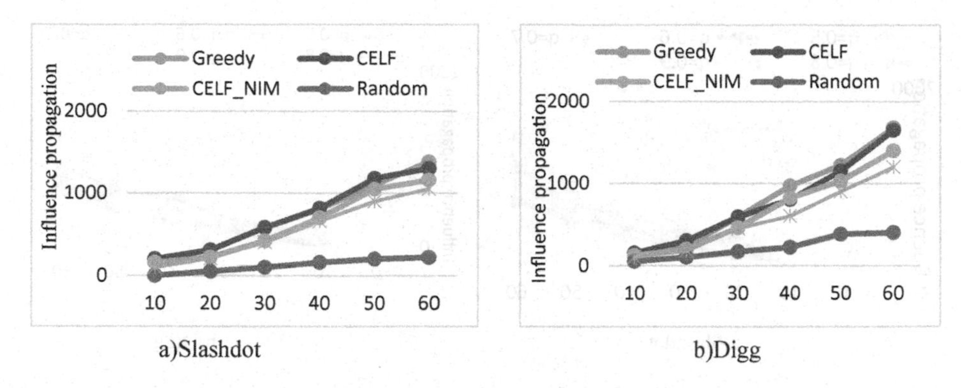

Fig. 1. Influence spread of different algorithms with various seed size

As shown in Fig. 2, the Random is far better than the other algorithms in terms of time efficiency. The running time of the greedy algorithm is the longest and it cannot scale to deal with the large amount of data.

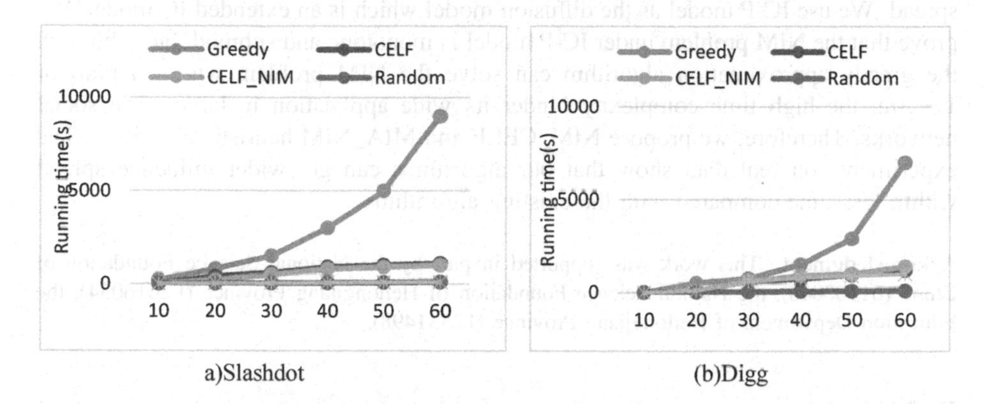

Fig. 2. Running time(s) of different algorithms

As shown in Fig. 3, as the quality factor increases, the size of the positive influence propagation also increases linearly. When the size of q increases to 1, there is no difference between the impact propagation size calculated by the traditional greedy algorithm. It can be seen from the figure in the "viral marketing" type of product promotion. The better the product quality, the easier it is to promote.

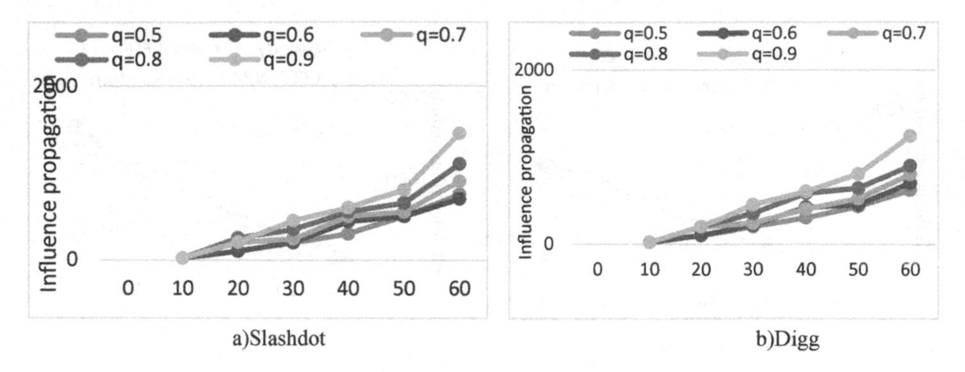

a)Slashdot b)Digg

Fig. 3. Influence propagation spread

6 Conclusion

We study the problem of negative influence maximization problem in signed social networks which aims to find the nodes set that with maximum negative influence spread. We use IC-P model as the diffusion model which is an extended IC model. We prove that the NIM problem under IC-P model is monotone and submodular. Although the greedy approximation algorithm can solve the NIM problem within a ratio of $1 - 1/e$, the high time complexity hinder its wide application in large scale social networks. Therefore, we propose NIM_CELF and MIA_NIM heuristic algorithm. The experiments on real data show that our algorithms can get wider influence spread within less time compared with the existing algorithms.

Acknowledgment. This work was supported in part by the National Science Foundation of China (61100048), the Natural Science Foundation of Heilongjiang Province (F2016034), the Education Department of Heilongjiang Province (12531498).

References

1. Kempe, D., Kleinberg, J.M., Tardos, E.: Maximing the spread of influence through a social network. In: The 9th ACM SIGKDD Conference on Knowledge Discovery and Data Mining, pp. 137–146. ACM, Washington, DC (2003)
2. Anagnostopoulos, A., Kumar, R., Mahdian, M.: Influence and correlation in social networks. In: Proceedings of the 14th ACM SIGKDD International Conference on Knowledge Discovery and Data Mining, pp. 7–15. ACM, Las Vegas (2008)
3. Tianyu, C., Xindong, W., Song, W., Xiaohua, H.: OASNET: an optimal allocation approach to influence maximization in modular social networks. In: SAC, pp. 1088–1094 (2010)
4. Zhu, J., Yin, X., Wang, Y., Li, J., Zhong, Y., Li, Y.: Structural holes theory-based influence maximization in social network. In: Ma, L., Khreishah, A., Zhang, Y., Yan, M. (eds.) WASA 2017. LNCS, vol. 10251, pp. 860–864. Springer, Cham (2017). https://doi.org/10.1007/978-3-319-60033-8_73

5. Leskovec, J., Krause, A., Guestrin, C.: Cost-effective outbreak detection in networks. In: The 13th ACM SIGKDD International Conference on Knowledge Discovery and Data Mining, pp. 420–429. ACM, San Jose (2007)
6. Chen, W., Wang, Y., Yang, S.: Efficient influence maximization in social networks. In: Proceedings of the 15th ACM SIGKDD International Conference, pp. 199–208. ACM, Paris (2009)
7. Li, D., Xu, Z.-M., Chakraborty, N.: Polarity related influence maximization in signed social networks. PLoS One **9**(7), 1–10 (2014)
8. Goldenberg, J., Libai, B., Muller, E.: Talk of the network: complex systems look at the underlying process of word-of-mouth. Mark. Lett. **12**(3), 209–211 (2001)
9. Granovetter, M.: Threshold models of collective behavior. Am. J. Sociol. **83**(6), 1420–1443 (1978)

Mining Correlation Relationship of Users from Trajectory Data

Zi Yang and Bo Ning(✉)

Dalian Maritime University, Dalian 116026, China
1162568728@qq.com, ningbo@dlmu.edu.cn

Abstract. With the development of location based applications, more and more personal trajectory is recorded by the application providers, and it brings opportunities and challenges for mining potential information from trajectory data. In this paper, the correlation between users is explored from the trajectory data. Firstly, we preprocess original trajectory by sub-trajectory matching. Secondly, we analyze the factors in trajectory data that reflect the users' close relationship, and quantify these factors to measure the correlation relationship between users. Then we propose mining algorithm by using these factors to mining the similar trajectories, and we also improve the algorithm efficiency by using a formula filtering method. Moreover, we use sigmoid function to express the intimacy degree between users so that it is more sensitive when the correlation is relatively small. Finally, the performance of our algorithm is evaluated by the experiment and the validity of our algorithm is verified.

Keywords: Trajectory data mining · User correction
Influencing factor · Similarity sub-trajectory

1 Introduction

With the rapid development of mobile internet, more and more location based services and applications emerge. The location function of these applications make it possible that the users' trajectories can be collected by the application providers, and the potential information can be mined from the large mount of trajectory data. This makes people's lives very convenient, which is of great help to urban computing, intelligent transportation and many other fields. In recent years, more and more applications make use of the user's trajectory information for interest recommendation or path planning [1,2]. Users can find more interesting knowledge and potential friends from the recommendation. Many works are based on user's online transaction records to recommend products, books, food and music through correlation among users.

In this paper, we focus on how to get the users' correlation of privacy based on their trajectories. The main purpose of trajectory mining in recent years is to use it in recommendation systems [2]. In this paper we preprocess the original trajectory to find the sub-trajectory which is reflect the correlation relationship

Q. Zhou et al. (Eds.): ICPCSEE 2018, CCIS 901, pp. 308–321, 2018.
https://doi.org/10.1007/978-981-13-2203-7_23

between users. We analyze the factors of relationships in trajectory data, and the factors are considered from three aspects, including POI collection similarity and time similarity, sequence similarity. And then, in mining relationship phase, we together these factors and define a comprehensive formula to measure the correlation between users. And the formula filter can be used to improve the efficiency of calculation. To improve the sensitivity of the intimacy degree, we use sigmoid function to amplify the intimacy degree when it is small. So that there is a small difference in the degree of closeness among the users with strong related relationships, and there is a big difference in the degree of closeness value between users with no relation.

The contribution of this paper are listed as follows:

(1) We preprocess the trajectories by sub-trajectory matching. We select the part that is helpful to the relationship mining from the original trajectory.
(2) We propose the influencing factors to measure the correlation between users, which concern about many aspects of trajectory information. We also quantify these factors to mining relationship.
(3) In mining relationship phase, we calculate the influencing factors synthetically. In the process of calculation, we use formula filtering to improve the efficiency. we also use sigmoid function, it makes the intimacy degree more sensitive when the correlation is relatively small.
(4) We evaluate our algorithm on the real data set, and the result shows that our algorithm is effective and efficient.

The rest of this paper is organized as follows: Sect. 2 introduces the related work and the relevant definitions. In Sect. 3, we present sub-trajectory matching to preprocess trajectories. In Sect. 4, we introduce and quantify the factors that reflect the relationship between users, and propose algorithms to mining the correlation between users. Section 5 shows our experimental results, finally conclusions are included in Sect. 6.

2 Related Work and Problem Definition

There are many works of personal mining location data are studied, including important position detection of the user, predicting the user's movement among these locations, identifying individual specific activities on each location [3–6]. These works only concern about the mining from individual's location information.

There are also many works about mining similar trajectories from large amount of trajectory data. In [7,8], the trajectory patterns of a set of individual that share the property of visiting the similar place sequence at the similar time is mined. [9,10] mines the user's behavior pattern by clustering, and gets the similarity according to the number of users in the same cluster. Our work is to explore the correlation between users according to the trajectories,

In recommendation systems, trajectory mining is also be employed for interest recommendation or site recommendation. Recommendation systems use opinions of a community to help individuals in that community, and more effectively identify content of interest from a potentially overwhelming set of choices [11,12]. Many recommendation systems take into account the current geographic location of a specific user when recommending content to the user. The recommendation system tries to recommend stores or predict the destination to users by analyzing their past location histories, according to their personal preferences and needs. When attempting to predict the destination about a particular user, a well-known technique used in such systems is called collaborative filtering. The basic idea of collaborative filtering is that similar users vote similarly on similar items. If similarity is determined between the user, we can make a potential prediction for the destination. The focus of these systems is to use the real-time location of customers as a constraint to provide information to customers, while our purpose is to mine multi-user trajectories and explore correlation between individuals.

Many works [13] predict driver's destination through extracting from a historical trajectories of drivers. In [14], the mobility can be predict, such as walking or driving, etc. In [15], by the semantics, the destination of users can be predicted. In [16], they propose a deep-learning-based approach, called $ST-ResNet$, to collectively forecast the inflow and outflow of crowds in each and every region of a city.

Most of the related works only study the similarity of the trajectory, however, we further study the correlation of users by their trajectory. The user's closeness relationship is extracted from several aspects of trajectory data, including POI collection similarity, time similarity, the sequence similarity. Furthermore, our goal of mining users' correlation is inclined to discover the privacy relationship among users.

Definition 1 *(Trajectory). The trajectory is defined as follows:* $tr_i = p_1^i(t_{1.ar}^i \sim t_{1.le}^i) \rightarrow p_2^i(t_{2.ar}^i \sim t_{2.le}^i) \rightarrow \ldots \rightarrow p_z^i(t_{z.ar}^i \sim t_{z.le}^i)$

The location point p *contains the latitude and longitude of the location. The* z *represents the total number of locations in the user* i*'s trajectory. The* $t_{k.ar}^i$ *represents the time to arrive of user* i *in the* k *th location. Similarly, the* $t_{k.le}^i$ *represents the time to leave of user* i *in the* k *th location.*

Definition 2 *(Trajectory Set). Trajectory data set* $Tr=[tr_1,tr_2, \ldots,tr_n]$, *the* tr_i *defines the original trajectory of the user* i, *where the* n *represents the number of users.*

Definition 3 *(Correlation Relationships). The degree of correlation between* tr_i *and* tr_j *about user* i *and user* j *is represented by* $S(sim_{ij})$. *If* $S(sim_{ij}) \geq S_{th}$, *we define* tr_i *and* tr_j *have correlation relationship.* S_{th} *is an predefined closeness value threshold, and if this threshold is reached, we call they are relevant, and vice versa.*

3 Sub-Trajectory Matching

In this section, we preprocess the original trajectory by sub-trajectory matching. We only concerned about the sub-trajectory that reflects the correlation relationship between users. Firstly, we use the method that filter the head and tail of trajectory sequence to reduce the computational complexity. Secondly, we use similarity sub-trajectory matching to get max-length similarity sub-trajectory set. When we are mining user relationships, we are concerned about the similar sub-trajectories of user trajectories, because other dissimilar parts are considered meaningless in this paper.

3.1 Filter Head and Tail in Trajectory Sequence

Suppose that the two trajectories are represented by tr_i and tr_j, respectively. $tr_i = p_1^i(t_{1.ar}^i \sim t_{1.le}^i) \rightarrow p_2^i(t_{2.ar}^i \sim t_{2.le}^i) \rightarrow ... \rightarrow p_z^i(t_{z.ar}^i \sim t_{z.le}^i)$ $tr_j = p_1^j(t_{1.ar}^j \sim t_{1.le}^j) \rightarrow p_2^j(t_{2.ar}^j \sim t_{2.le}^j) \rightarrow ... \rightarrow p_z^j(t_{z'.ar}^j \sim t_{z'.le}^j)$

Factor ①: POI collection similarity(We also call text similarity): $\exists\ p_u^i\ and\ p_v^j, p_u^i \in tr_i, p_v^j \in tr_j, p_u^i = p_v^j$;

Factor ① describes that user i and j went to the same location.

Factor ②: time similarity

$\text{cov}_{uv} = \min\{t_{u.le}^i, t_{v.le}^j\} - \max\{t_{u.ar}^i, t_{v.ar}^j\} > 0$;

Factor ② describes that the time of user i and j overlaps. The cov_{uv} represents how long time overlaps.

As far as we know, sometimes we don't have to calculate all the locations of the entire trajectory, and the operation is to filter heads and tails of the trajectory sequences. From the front and back of the two trajectories, our method to find the locations to meet the factors ①&②, only the remaining parts of the two trajectories are considered, which reduces the workload and improves the efficiency of calculation.

For example, Fig. 1 shows the trajectories of tr_1 and tr_2 respectively.

tr_1:

$P(6:00 \sim 6:30) \rightarrow A(7:00 \sim 8:00) \rightarrow B(9:00 \sim 10:00) \rightarrow$
$C(11:00 \sim 12:00) \rightarrow D(13:00 \sim 14:00) \rightarrow E(15:00 \sim 16:00) \rightarrow$
$F(17:00 \sim 18:00) \rightarrow G(19:00 \sim 21:00) \rightarrow M(21:30 \sim 22:00)$

tr_2:

$A(6:00 \sim 7:30) \rightarrow C(8:00 \sim 11:30) \rightarrow B(12:00 \sim 13:00) \rightarrow$
$D(13:30 \sim 14:00) \rightarrow E(15:00 \sim 16:00) \rightarrow F(17:30 \sim 18:00) \rightarrow$
$G(19:00 \sim 21:00) \rightarrow N(21:30 \sim 22:00)$

First, the algorithm scans the heads of the two trajectories and finds that the second location of tr_1 and the first location of tr_2 match factors text similarity and time similarity(Time overlaps in the same location). The first location P of tr_1 does not need to be calculated. After scanning from the tail of the two trajectories, it is found that the penultimate location G of tr_1 and tr_2 match factors text similarity and time similarity. Then the last location M of tr_1 and the last location N of tr_2 can be ignored.

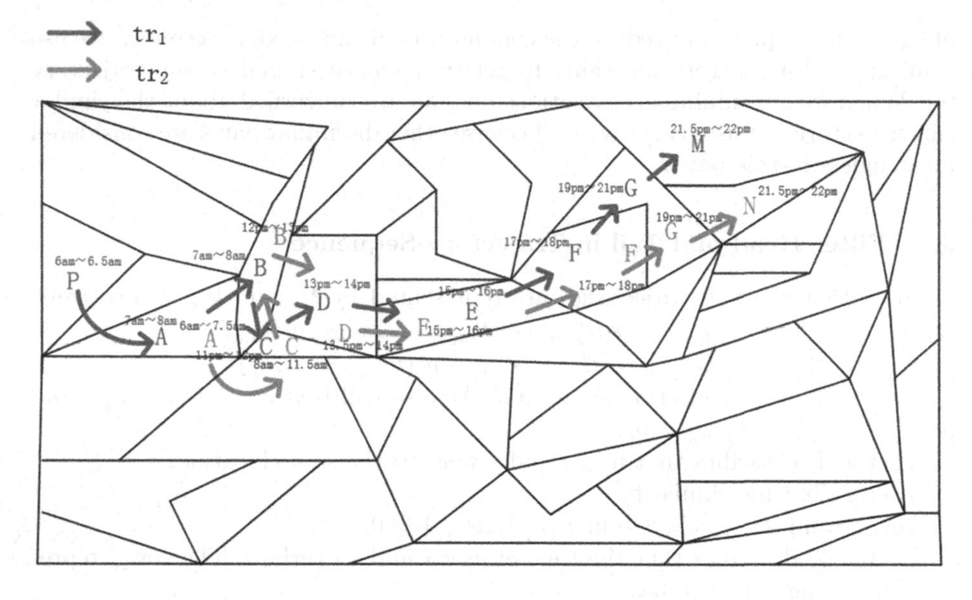

Fig. 1. Trajectories of tr_1 and tr_2 in geographical position

3.2 Sub-Trajectory Matching

Firstly, we find some location that reflect of relationship between users from the trajectories. Then, we use expansion and pruning operations to get trajectory sequence.

Given the two trajectories, we only consider the parts that reflect of relationship between users. we use the following conditions to find location.

Condition (1): Constraint factors ①&②;

It means that user i and j stay in the same location at the same time.

Condition (2): Sequence similarity

$\Delta t_u^i = |t_{u.ar}^i - t_{u'.le}^i|, \Delta t_v^j = |t_{v.ar}^j - t_{v'.le}^j|$

The u' and v' denote the former locations of u and v respectively. The Δt_u^i represents the time spent on the road when user i travel from u' to u.

$\exists\, u, v, 1 \leq u \leq z, 1 \leq v \leq z', |\Delta t_u^i - \Delta t_v^j| \leq t_{th}$

The t_{th} is a threshold of time interval between two locations, which indicates that two users have similar travelling time between in the same paths. It means that if condition (1) and (2) are reached, the location can be described below:

$p_k^i(\max\{t_{u.ar}^i, t_{v.ar}^j\}, \min\{t_{u.le}^i, t_{v.le}^j\})$

Here, we define the l-length similar sub-trajectories, if the length of a similar sub-trajectory is l, we define it is a l-length similar sub-trajectory. We also define max-length similar sub-trajectory, if it not included in other similar sub-trajectories.

The sub-trajectory matching procedure consists of the following two phases. Based on the locations reaching condition (1) and (2), we use expansion phase and pruning phase to get similar sub-trajectories.

Expansion Phase: The l-length similar sub-trajectories are extended to the $(l+1)$-length similar sub-trajectories. For example, we find the 1-length similar sub-trajectories, then search for 2-length similar sub-trajectories based on 1-length similar sub-trajectories. The specific operation: In trajectories about user i and j, for the next location of the last location in the $(l-1)$-length sub-trajectories, if it satisfy the sequence similarity, we extended the next location from the $(l-1)$-length to l-length sub-trajectories.

Pruning Phase: The pruning process is to prune the $(l-1)$-length similar sub-trajectories contained in the l-length similar sub-trajectories. Because in the expansion phase, there are many sub-trajectory sequences in the l-length and the parts of $(l+1)$-length are the same, we remove the shorter l-length similar sub-trajectories from the collection.

The result of similar sub-trajectories is describe below: $TRsub_{ij} = p_1^i(\max\{t_{u.ar}^i, t_{v.ar}^j\}, \min\{t_{u.le}^i, t_{v.le}^j\}) \rightarrow ... \rightarrow p_z^i(\max\{t_{u.ar}^i, t_{v.ar}^j\}, \min\{t_{u.le}^i, t_{v.le}^j\})$

The Similar Sub-Trajectories Matching Algorithm includes two processes of expansion phase and pruning phase(Algorithm 1 line 1–7). The expansion phase is to extend the l-length to $(l+1)$-length (Algorithm 1 line 8–16). The extension phase condition is that the time from the last position of the l-length to the next position be expanded is less than t_{th}, and the extended position point needs to satisfy the condition (1) and (2) (Algorithm 1 line 11–14). During he pruning process, l-length sub-trajectory needs to be removed because it has been contains $(l+1)$- length similar sub-trajectories (Algorithm 1 line 17–24). Finally, the *max*-length similar sub-trajectories are obtained, and the counts may be more than one. Each one belongs to the set is not contained with each other.

For example, given the time threshold t_{th} is as 2, according to the similarity of text and time, all 1-length trajectories of the two trajectories are found.

1-length:

$A\,(7:00 \sim 7:30)\,C\,(11:00 \sim 11:30)\,D\,(13:30 \sim 14:00)$
$E\,(15:00 \sim 16:00)\,F\,(17:30 \sim 18:00)\,G\,(19:00 \sim 21:00)$

According to the expansion operation, we get 2-length similar sub-trajectories. For 1-length, from the next location of the last location in the 1-length sub-trajectories, we perform expansion operations and get to the 2-length sub-trajectories.

2-length:

$$AC = |(11-8)-(8-7.5)| = |3-0.5| = 2.5 > t_{th},$$

AC can not be put in collection, because the time is lager than t_{th}.

$$CD = |1-2| = 1 < 2, DE = 0 < 2, EF = 0.5 < 2, FG = 0 < 2$$

After expansion phase, we can get 2-length sub-trajectories: CD, DE, EF and FG.

Pruning operation: The 2-length sub-trajectories contain 1-length sub-trajectories that is C, D, E, F, G, so we remove C, D, E, F, G from collection of sub-trajectories, and only A remains.

Algorithm 1. $Simlar\ Sub-Trajectories\ Matching(tr_i, tr_j)$

```
   Data: Trajectory: tr_i, tr_j; Time threshold: t_th.
   Result: Set of max-length sub-trajectory of tr_i and tr_j: MaxSimiSequence_ij.
 1 begin
 2   | According factors ① and ② adding 1 − length trajectories into
     | MaxSimiSequence_ij;
 3   | extendSeq = ExtendSequence(seq, t_th);
 4   | Add extendSeq into MaxSimiSequence_ij;
 5   | PruneSequence(MaxSimiSequence_ij);
 6   | return MaxSimiSequenceij;
 7 end
 8 Function ExtendSequence(seq, t_th)
 9 begin
10   | foreach l − length sub − trajectories ∈ MaxSimiSequence_ij do
11   |   | for the next location of the last location in the l − length
     |   | sub − trajectories
12   |   | while (|Δt^i_u − Δt^j_v| ≤ t_th and constraint ①&②) do
13   |   |   | l − length sub − trajectories is extended (l + 1) − length;
14   |   |   | Adding (l + 1) − length into MaxSimiSequence_ij;
15   | return MaxSimiSequence_ij
16 end
17 Function PruneSequence(MaxSimiSequence_ij)
18 begin
19   | foreach (l + 1) − length sub − trajectories do
20   |   | while l − length sub − trajectories ∈ MaxSimiSequence_ij &&
     |   | l − length sub − trajectories ⊂ (l + 1) − length do
21   |   |   | removing l − length sub − trajectories that satify above formula
     |   |   | from MaxSimiSequence_ij;
22   | return MaxSimiSequence_ij
23 end
```

similarly, we can get 3-length, 4-length, 5-length as follows.

3-length: CDE, DEF and EFG. Then 2-length sub-trajectories are all removed because they are all contained in 3-length sub-trajectories.

4-length: CDEF and DEFG. Removing all in 3-length sub-trajectories.

5- length: CDEFG. Then all 4-length sub-trajectories are removed.

Finally, two *max*-length similar sub-trajectories can be get.

$STsub_{ij} = A(7:00 \sim 7:30)$;

$STsub_{ij} = C(11:00 \sim 11:30) \rightarrow D(13:30 \sim 14:00) \rightarrow E(15:00 \sim 16:00) \rightarrow F(17:30 \sim 18:00) \rightarrow G(19:00 \sim 21:00)$.

4 Mining Correlation Relationship Between Users

The text similarity, time similarity and sequence similarity are discussed above, and in this section, we discuss the quantification of factors, and sensitive locations and sensitive time similarity are also taken into account.

①POI collection similarity: The proportion of the location count at the same time to the total location count in two users' trajectory sequences. The m is the total location count in two trajectories. The formula is shown below.

$$\forall 1 \leq u \leq z, \forall 1 \leq v \leq z', \mathrm{f} = \begin{cases} 1, p_u^i = p_v^j \\ 0, p_u^i \neq p_v^j \end{cases} sim(①) = \frac{\sum_{u=1}^{u=z} \mathrm{f}}{\max\{z, z'\}} \tag{1}$$

②Time similarity: The proportion of the overlapping time interval at the same place to the statical time interval. The formula is as follows.

$$\forall q, 1 \leq q \leq z, if \, \mathrm{cov}_{uv} > 0, sim(②) = \frac{\sum_{q=1}^{q=z} \mathrm{cov}_{uv}}{T.end - T.beg} \tag{2}$$

The *T.end-T.beg* indicates that the data collected from the beginning time *T.beg* to the end time *T.end*. The time similarity reflects how long the two users spend together in a day.

We comprehensively considers the influence of these factors on the user's closeness value. First of all, for sequence similarity, we've taken it into account when we get similar sub-trajectories above. Secondly, for the POI collection similarity and time similarity, because the two users appear in the same place at the same time, factors ① and ② are juxtaposed, we consider factors ① and ② together and use the sum of the factors ②, two factors to indicate that they are juxtaposed. For the According to the above discussions, we present a formula for calculating the closeness value of similar sub-trajectories of l-length, as shown below.

$$sim = sim(①) + \mathrm{sim}(②) \tag{3}$$

The quantitative calculation formula for two users' closeness values is shown as follows, and h indicates the count of h *max*-length similar sub-trajectories, where k is the k th *max*-length similar sub-trajectory.

$$sim(ij) = \sum_{k=1}^{k=h} sim_k \tag{4}$$

During the process of calculating the user's closeness value, we find that the value of the user's closeness mostly fall into a small region and its distribution is nonuniform. To make the closeness quota more sensitive, we employ the sigmoid function to measure users' closeness degree, so that the distribution of closeness values can be more uniform and the difference of closeness values with tiny correlations is easy to be distinguished. The formula is shown below:

$$S(sim(ij)) = \frac{1}{1 + e^{-\{w \times (sim(ij)) + d\}}} \tag{5}$$

The w and d are parameter values about the range of sigmoid function. We also evaluate the best w and d in the later experiments.

The value range of function value is (0,1), the closer it is to 1, the closer the relationship is. When the independent variable is small, the variation range of function value is larger, but when the independent variable is larger, the range of function value is smaller. Moreover, the formula of closeness value is normalized.

Formula Filter. In this step, we study how to improve the computational efficiency of the *sim* function, and propose a filter method for the *sim* function. When factors ① is calculated for the trajectories of both users, if the closeness threshold is met, factors ② is not need to be calculated, similarly, in the calculation of ②, if the S_{th} threshold is met in the calculation, the time of the remaining locations do not need to be calculated. The calculation efficiency is improved greatly.

Algorithm 2. *Computing similarity according formula* ($MaxSimiSequence_{ij}$)

Data: *max*-length sub-trajectory set between tr_i and tr_j: $\text{MaxSimiSequence}_{ij}$
Closeness value threshold: S_{th}
Result: If or not tr_i and tr_j is correlation.

```
1  begin
2  |  foreach max − length sub − trajectory set of tr_i and tr_j : sub_ij ∈
   |  MaxSimiSequence_ij do
3  |  S(①)= sim(①) = Computing similarity about① (TRsub_ij);
4  |  if S(①) > S_th then
5  |  |  return tr_i and tr_j is correlation;
6  |  |  continue;
7  |  else
8  |  |  S(① + ②_part) = (sim(①) + sim(②_part)) = Computing similarity
   |  |  about ① and the parts of② (TRsub_ij);
9  |  |  if S(① + ②_part) > S_th then
10 |  |  |  return tr_i and tr_j is correlation;
11 |  |  |  continue;
12 |  |  else
13 |  |  |  Calculate other max − length similar sub − trajectory set , added
   |  |  |  to the previous result : S(①+②);
14 |  |  |  repeat the above steps ;
15 |  Return correCollection.
16 end
```

Algorithm 2 quantifies the *max*-length similar sub-trajectory set of tr_i and tr_j obtained by Algorithm 1 to get the degree of correlation between the users (Algorithm 2 line 1–16). For each element in set, we consider the factors ① and ② step by step. First, we calculate the correlation between factors ① according to the formula (Algorithm 2 line 3). If the threshold S_{th} is reached, the algorithm stop (Algorithm 2 line 4–6); if not, the algorithm continue to calculate factors ① and the parts of ② (Algorithm 2 line 8–14), the *sim*(②*_part*) represents that we only calculate the parts of ② so that result met the threshold. Then the following steps are similar. Finally, we can get the correlation between users.

An example of the quantitative calculation process is as follows: we define the closeness value threshold S_{th} is set as 0.84, w is set as 8, d is set as -4;

$STsub_{ij} = A(7:00 \sim 7:30)$,

$sim_1 = sim(①)+\text{sim}(②)=\frac{1}{9}+\frac{0.5}{16}=0.14 < S_{th}$;

$STsub_{ij} = C(11:00 \sim 11:30) \rightarrow D(13:30 \sim 14:00) \rightarrow E(15:00 \sim 16:00) \rightarrow F(17:30 \sim 18:00) \rightarrow G(19:00 \sim 21:00)$,

$sim_2 = \frac{5}{9} + \frac{4.5}{16} = 2.227$

$sim(ij) = sim_1 + sim_2 = 2.367$

$S(sim(ij)) = \frac{1}{1+e^{-\{8\times(sim(ij))-4\}}} = \frac{1}{1+e^{-\{8\times 2.367)-4\}}} \geq S_{th}$, thus, user i and user j are related.

If the formula filter is used to calculate the text similarity and time similarity of sim_2, $sim_2 = \frac{5}{9} + \frac{4.5}{16} = 0.995$, $sim(ij) = 0.995$, $S(sim(ij)) = \frac{1}{1+e^{-\{8\times 0.995)-4\}}} = 0.9 \geq S_{th}$, the threshold of $sim(ij)$ has been reached and no need to calculate the next steps. From the example, we can see that the computation complexity is reduced.

Algorithm 3. *Computing similarity about Tr*

```
Data: Trajectory data set:Tr.
Result: A collection of two users with related relationships, named
        correCollection
begin
    for int i = 1; i < |Tr| ; i++ do
        for int j = i + 1; j < |Tr| ; j++ do
            (tr_i sub, tr_j sub) = From front to back to filter and from back to
            front to filter (tr_i, tr_j);
            MaxSimiSequence_ij = Algorithm_simlar Sub − Trajectories
            Matching( tr_i sub, tr_j sub);
            ifCorrection = Computing similarity according formula(
            MaxSimiSequence_ij);
            if ifCorrection==true then
                add [tr_i , tr_j] into correCollection;
    return correCollection.
end
```

The Algorithm 3 shows the pseudo code for computing the closeness value between user trajectories that is the whole process of finding twins users that have a correlation in trajectories set (Algorithm 3 line 1–10). The first step is to filter the locations of heads and tails of the trajectories (Algorithm 3 line 4), and the second step is to perform similar sub-trajectories matching to get the *max*-length similar sub-trajectory set (Algorithm 3 line 5). The third step is to calculate the correlation of the trajectories according to the formula (Algorithm 3 line 6). Finally, we get twins users who are related to each other in trajectory set (Algorithm 3 line 7–8).

5 Experiments

5.1 Experimental Setup

All the algorithms implement in JDK 1.7. Real-world data include two groups data_1 and data_2. data_1 is the trajectory data set in BeiJing, it has 1042 trajectories of 20 users. data_2 is a total of 2389 trajectories data of 48 users. Attributes include user name, date, location name, latitude and longitude, time of arrival and leave etc. And there are also correlation relationship between users. We perform three groups of experiments, it includes that finding the best w and d, and the time efficiency of algorithms.

5.2 Performance Study

Influence of w and d

In this paper, we test the value of w and d of the closeness function. The closeness function refers to formula (5). We choose the variable that makes the highest slope, because it makes the closeness value more sensitive.

In Fig. 2, we evaluate the $w's$ best value. The purpose of this paper is to set the closeness function in that the closeness relationship of the user has a smaller difference, while the user with the less close relationship has a larger difference. As shown in Fig. 3, the slope is the largest when we set w value is 8, so in this experiment, the value of w is 8. Similarly, in Fig. 3, we can see that when the slope is the largest, the value of d is set to 4. Therefore, in the following experiment, we set equation as $S(sim(ij)) = \frac{1}{1+e^{-\{8\times(sim(ij))-4\}}}$.

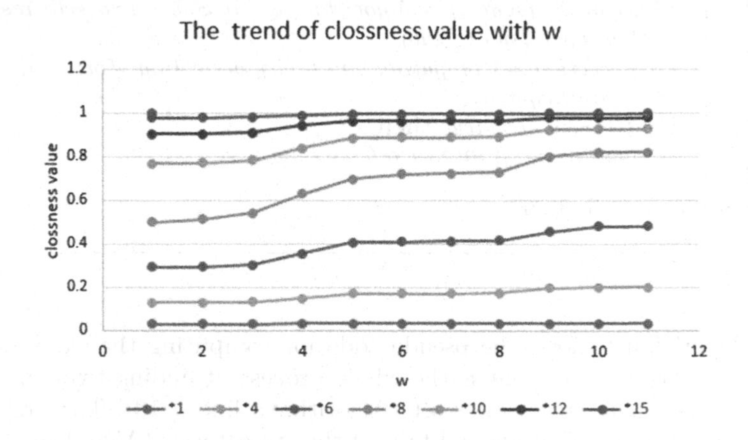

Fig. 2. The trends of closeness value with w

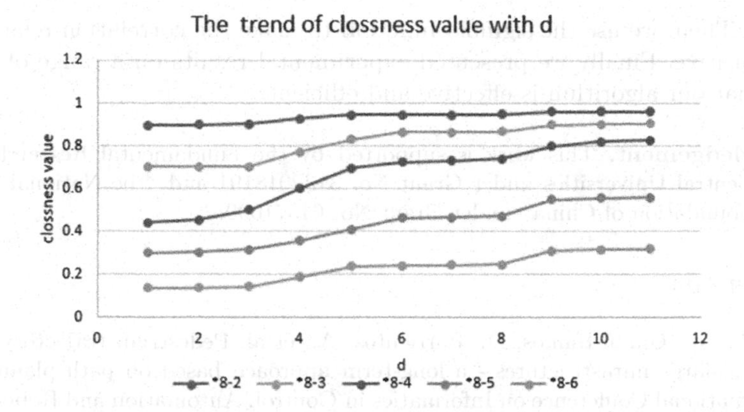

Fig. 3. The trends of closeness value with d

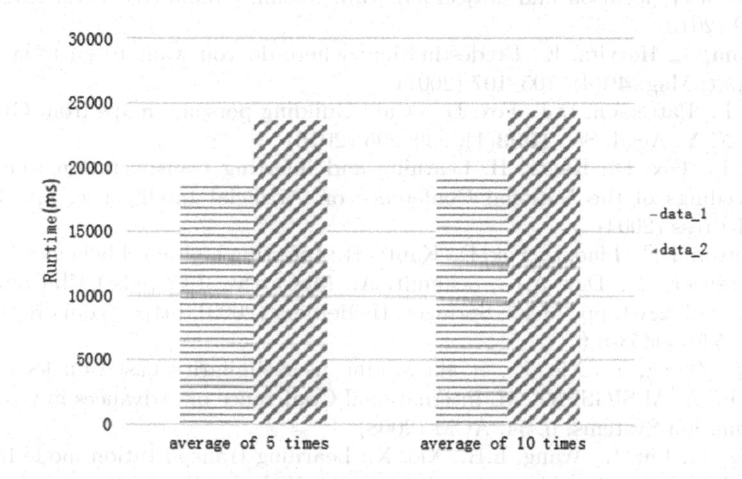

Fig. 4. The trend of time efficiency with on k

Time Efficiency. We tested the runtime of the two sets of data. Each set was performed 5 times and 10 times. The number of trajectories and the correlation relationships in data_2 is large than data_1, that is why the runtime of date_1 is shorter than data_2.

6 Conclusions

In this paper, we proposed the algorithm which can get correlation relationship between users. Firstly, we preprocessed the trajectories by filtering the heads and tails of trajectories and matching the sub-trajectories. Secondly, we measured the correlation relationship of users from many factors, including sequence similarity, POI collection similarity and time similarity. Then we proposed a composite formula. Then, we use the sigmoid function to make the correlation relationship more sensitive. Finally we presented experimental results on a range of real. It shows that our algorithm is effective and efficient.

Acknowledgement. This work is supported by the Fundamental Research Funds for the Central Universities under Grant No. 3132018191 and "the National Natural Science Foundation of China" under Grant No. 61371090.

References

1. Garzn, M., Garzn-Ramos, D., Barrientos, A., et al. Pedestrian trajectory prediction in large infrastructures - a long-term approach based on path planning. In: International Conference on Informatics in Control, Automation and Robotics, pp. 381–389 (2016)
2. Zheng, Y., Xie, X., Ma, W.Y.: GeoLife: a collaborative social networking service among user, location and trajectory. Bull. Techn. Committee Data Eng. **33**(2), 32–39 (2010)
3. Krumm, J., Horvitz, E.: Predestination: where do you want to go today? IEEE Comput. Mag. **40**(4), 105–107 (2007)
4. Liao, L., Patterson, D.J., Fox, D., et al.: Building personal maps from GPS data. Ann. N. Y. Acad. Sci. **1093**(1), 249–265 (2010)
5. Liao, L., Fox, D., Kautz, H. Learning and inferring transportation routines. In Proceedings of the National Conference on Artificial Intelligence, pp. 348–353. ACM Press (2004)
6. Patterson, D.J., Liao, L., Fox, D., Kautz, H.: Inferring high-level behavior from low-level sensors. In: Dey, A.K., Schmidt, A., McCarthy, J.F. (eds.) UbiComp 2003. LNCS, vol. 2864, pp. 73–89. Springer, Heidelberg (2003). https://doi.org/10.1007/978-3-540-39653-6_6
7. Li, Q., Zheng, Y., Xie, X., et al.: Mining user similarity based on location history. In: ACM SIGSPATIAL International Conference on Advances in Geographic Information Systems, p. 34. ACM (2008)
8. Zheng, Y., Liu, L., Wang, L.H., Xie, X.: Learning transportation mode from raw GPS data for geographic applications on the Web. In: Proceedings of the WWW 2008, pp. 247–256. ACM Press (2008)
9. Higgs, B., Abbas, M.: Segmentation and clustering of car-following behavior: recognition of driving patterns. IEEE Trans. Intell. Transp. Syst. **16**(1), 81–90 (2015)
10. Wang, Y., Qin, K., Chen, Y.: Detecting anomalous trajectories and behavior patterns using hierarchical clustering from taxi GPS data. Int. J. Geo-Inf. **7**(1), 25 (2018)
11. Adomavicius, G., Tuzhhilin, A.: Toward the next generation of recommender systems: a survey of the state-of-the-art and possible extensions. IEEE Trans. Knowl. Data Eng. **17**(6), 734–749 (2006)
12. Sarwar, B., Karypis, G., Konstan, J., Riedl, J.: Application of dimensionality reduction recommender system - a case study, In: ACM WebKDD Workshop (2000)
13. Xue, A.Y., Zhang, R., Zheng, Y.: DesTeller: a system for destination prediction based on trajectories with privacy protection. Proc. VLDB Endow. **6**(12), 1198–1201 (2013)
14. Krumm, J., Horvitz, E.: Predestination: inferring destinations from partial trajectories. In: Dourish, P., Friday, A. (eds.) UbiComp 2006. LNCS, vol. 4206, pp. 243–260. Springer, Heidelberg (2006). https://doi.org/10.1007/11853565_15

15. Huang, C.M., Ying, J.C., Tseng, V.S., et al.: Location semantics prediction for living analytics by mining smartphone data. In: International Conference on Data Science and Advanced Analytics, pp. 527–533. IEEE (2015)
16. Zhang, J., Zheng, Y., Qi, D.: Predicting citywide crowd flows using deep spatio-temporal residual networks. Artif. Intell. **259**, 147–166 (2018)

Context-Aware Network Embedding via Variation Autoencoders for Link Prediction

Long Tian[1], Dejun Zhang[1(✉)], Fei Han[1], Mingbo Hong[1], Xiang Huang[1], Yilin Chen[2], and Yiqi Wu[3]

[1] Sichuan Agricultural University, Yaan 625014, China
djz@sicau.edu.cn
[2] Wuhan University, Wuhan 430072, China
[3] China University of Geosciences, Wuhan 430074, China

Abstract. Networks Embedding (NE) plays a very important role in network analysis in the era of big data. Most of the current Network Representation Learning (NRL) models only consider the structure information, and have static embeddings. However, the identical vertex can exhibit different characters when interacting with different vertices. In this paper, we propose a context-aware text-embedding model which seamlessly integrates the structure information and the text information of the vertex. We employ the Variational AutoEncoder (VAE) to statically obtain the textual information of each vertex and use mutual attention mechanism to dynamically assign the embeddings to a vertex according to different neighbors it interacts with. Comprehensive experiments were conducted on two publicly available link prediction datasets. Experimental results demonstrate that our model performs superior compared to baselines.

Keywords: Networks Embedding · Variational AutoEncoder Attention mechanism · Link prediction · Data mining

1 Introduction

Network Representation Learning (NRL) is an important research area in data mining issue because it is the basis of many applications, such as link prediction in citation networks. The goal of network embedding is learning a vector which can represent all information in the network, has attracted interest in recent years. Although there are not a few recent work proposed to study the issue [1,16], however it is still far from satisfactory. Because most network vertices have abundant external information (e.g. text). But the traditional NRL based models [12] mainly rely on network topology information to realize link prediction, which overlook the external information.

Inspired by the above observations, we propose context-aware network embedding via variation autoencoder model, which takes full account of vertex structure information and text information. Thus, context-aware embedding

Q. Zhou et al. (Eds.): ICPCSEE 2018, CCIS 901, pp. 322–331, 2018.
https://doi.org/10.1007/978-981-13-2203-7_24

can significantly improve the quality of the network representation, which further enhance the accuracy of the network analysis tasks. Autoencoder is used in our model to efficiently extract feature information from the vertex. In order to get highly non-linear structure of large-scale networks [17], we introduce an approach based on Variation AutoEncoder (VAE) [7]. At the same time, the mutual attention mechanism is adopted, which is expected to guide VAE models to emphasize those words that are focused by its neighbor vertices and eventually obtain context-aware embeddings.

Our work has two main contributions: (1) We propose a model which combines the structure and context of vertices. Experimental results show that our model is effective. (2) The combination of VAE and attention mechanism in our model improve the accuracy of link prediction. And the attention mechanism make our model more realistic. We report results on the two different datasets to show that our model achieves highly accuracy in vertices link prediction.

The remainder of this paper is organized as follows. In Sect. 2, we present the recent related work. In Sects. 3 and 4, we show the problem formulation, and introduce the goal of the network embedding. Meanwhile we present our context-aware embeddings for details. In Sect. 5, we present link prediction study and compare our method with baseline results of datasets. In Sect. 6, the conclusion and the future plan are demonstrated.

2 Related Work

We briefly introduce existing NRL methods. Recently, neural network-based methods have been proposed for constructing vertex representation in large-scale graphs. DeepWalk [13] presents a two-phase algorithm for graph representation learning. In the first phase, DeepWalk samples sequences of neighboring vertices of each node by random walking on the graph. Then, the vertex representation is learned by training a skip-gram model [11] through the random walks in the second phase.

Several methods have been proposed which extend this idea. LINE [15] learns graph embeddings which preserves both the 1-order and 2-order proximities in a graph, meanwhile LINE optimizes the joint and conditional probabilities of edges in large-scale networks to learn vertex representation. Node2vec [4] combines DFS-like and BFS-like exploration within the random walk frameworks. And then, matrix factorization methods [19] and deep neural networks have also been proposed as alternatives to the skip-gram model for learning the latent representations. The above NRL methods focus on network topology.

Most of these Networks Embedding (NE) models [4,13] only encode the structural information into vertex embedding, without considering heterogeneous information accompanied with vertices in real-world social networks. Researchers also explore algorithms to incorporate meta information such as text information into NRL. TADW [19] incorporates text features of nodes into embedding learning. MMDW [18] learns semi-supervised network embeddings with max-margin constraints between vertices from different labels.

Both structure and text information are taken to consideration for vertex representation in our approach. We postulate that a vertex has different embeddings according to which the vertex it interacts with, constraining our model to learn context-aware embeddings.

3 The Method

3.1 Problem Formulation

Let $G = (V, E, T)$ denote a given network, where V is the set of vertices, $E \subseteq V \times V$ are edges between vertices, $e_{i,j} \in E$ denotes the relationship between vertices(i, j). There is also a weight coefficient $\omega_{i,j}$ that denotes the relationship between vertices (i, j). T denotes the text information of vertices. The objective of the network embedding is that allocate a real-valued vector representation base on structure and text information.

3.2 The Overall Framework

As mentioned above, the effect of structure information and text information on study of vertex link prediction are fully considered. We introduce important ingredient of the model separately, in following parts.

Without loss of generality, given an edges $e_{i,j}$, we can obtain the context-aware embeddings of vertices with their structure embeddings and context-aware text embeddings as $\boldsymbol{i}_{(j)} = \boldsymbol{i}^s \oplus \boldsymbol{i}^t_j$, where $\oplus$ indicates the concatenation operation. Note that, $\boldsymbol{i}^s$ denotes structure-based embedding which encodes network structure information, while $\boldsymbol{i}^t_j$ denotes text-based embedding which captures the textual meanings lying in the associated text information. More detail of text-based embedding will be introduced in Sect. 4.2.

All vertices context-aware embeddings can be obtained by the same operation. We can achieve link prediction with these embeddings. The target of the model is to maximize the overall objective of edges and minimize the VAE loss. The loss function is defined as follows:

$$L = \sum_{e \in E} \left(L_s(e) + L_t(e) - loss_{VAE} \right), \tag{1}$$

where $L_s(e)$ denotes structure-based objective, $L_t(e)$ denotes the text-based objective and $loss_{VAE}$ denotes VAE loss. In Sects. 4.1, 4.2 and 4.4, we describe three objective functions in detail.

4 Three Sub-objectives

4.1 Structure-Based Objective

The structure-based objective aims to measure the log-likelihood of a directed edge using the structure-based embeddings as

$$L_s(e) = \omega_{i,j} \cdot \log p(\boldsymbol{j}^s | \boldsymbol{i}^s). \tag{2}$$

Besides, we follow LINE [15] to define the conditional probability of i generated by j in Eq. (3) as

$$p(\boldsymbol{j}^s|\boldsymbol{i}^s) = \frac{\exp(\boldsymbol{i}^s \cdot \boldsymbol{j}^s)}{\sum_{z \in V} \exp(\boldsymbol{i}^s \cdot \boldsymbol{z}^s)}. \quad (3)$$

4.2 Text-Based Objective

Vertices usually contain a lot of text information in real social networks. In conventional NE models, each vertex is represented by a static embedding vector that means the embeddings are fixed, this may be incomplete. Because one vertex probably plays different roles when interacting with different neighbors. Our model is dynamic which means it assign the different text-embeddings to a vertex according to different neighbors it interacts.

We propose the text-based objective to take advantage of these text information, as well as learn text-based embeddings for vertices. In order to fully consider the impact of structural information on the expression of textual information, and make $L_t(e)$ compatible with $L_s(e)$, we define $L_t(e)$ as follows:

$$L_t(e) = a_1 \cdot L_{tt}(e) + a_2 \cdot L_{ts}(e) + a_3 \cdot L_{st}, \quad (4)$$

where a_1, a_2 and a_3 denote three different hyper-parameters, the loss of three parts as follows:

$$L_{tt}(e) = \omega_{i,j} \cdot \log p(\boldsymbol{i}^t|\boldsymbol{j}^t), \quad (5)$$

$$L_{ts}(e) = \omega_{i,j} \cdot \log p(\boldsymbol{i}^t|\boldsymbol{j}^s), \quad (6)$$

$$L_{st}(e) = \omega_{i,j} \cdot \log p(\boldsymbol{i}^s|\boldsymbol{j}^t). \quad (7)$$

On the one hand, the objective of our model aims to maximize the conditional probabilities of the two vertices on the edge. On the other hand, we expect that the structure and text representation vectors of the same vertex are consistent. So, the conditional probabilities in Eqs. (5), (6) and (7) are defined to map the two types of vertex embeddings into the same representation space.

4.3 Context-Aware Text-Emebedding

The architecture of context-aware text embeddings is shown in Fig. 1. The textual matrix $\mathbf{Z}_i \in R^{d \times m}$ and $\mathbf{Z}_j \in R^{d \times n}$ are obtained by the VAE. By introducing an attentive matrix $\mathbf{A} \in R^{d \times d}$, the correlation matrix $\mathbf{F} \in R^{m \times n}$ can be calculated as follows:

$$\mathbf{F} = \tanh \mathbf{Z}_i^T \mathbf{A} \mathbf{Z}_j. \quad (8)$$

Note that, each element $\mathbf{F}_{ij}$ in $\mathbf{F}$ represents the correlation score between two corresponding vectors. After that, we conduct pooling operations along rows

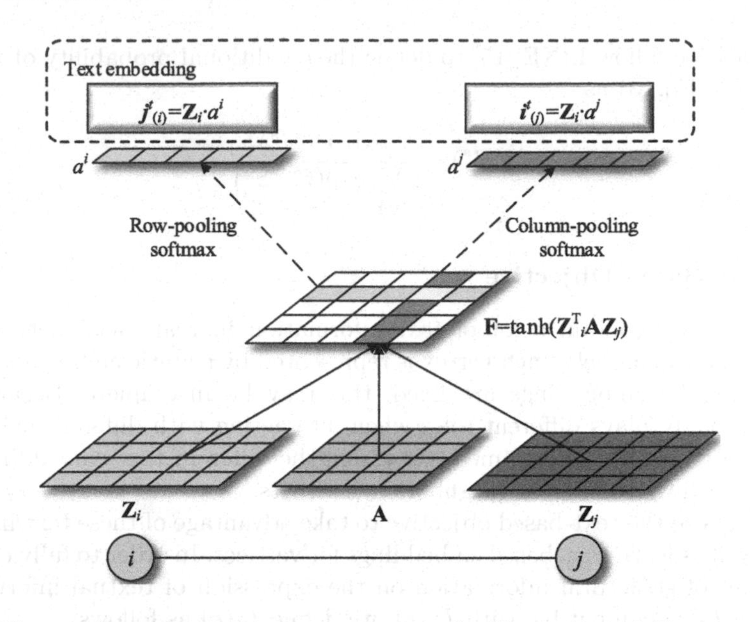

Fig. 1. The architecture of context-aware text embeddings. $\mathbf{Z}_i$ and $\mathbf{Z}_j$ are two representation textual matrix learning by the VAE module, $\mathbf{A}$ is attentive matrix.

and columns of $\mathbf{F}$, named as row-pooling and column-pooling. The mean-pooling operations are computed as follows:

$$g_i^p = \mathbf{mean}(\mathbf{F}_{i,1}, \ldots, \mathbf{F}_{i,n}), \tag{9}$$

$$g_i^q = \mathbf{mean}(\mathbf{F}_{1,i}, \ldots, \mathbf{F}_{m,i}). \tag{10}$$

The important vectors of $\mathbf{Z}_i$ and $\mathbf{Z}_j$ are obtained as $\boldsymbol{g}^p = [g_1^p, \ldots, g_m^p]^T$, $\boldsymbol{g}^q = [g_1^q, \ldots, g_n^q]^T$. The softmax function is employed to transform importance vectors $\boldsymbol{g}^p$ and $\boldsymbol{g}^q$ to attention vectors $\boldsymbol{a}^p$ and $\boldsymbol{a}^q$, individually. For instance, the *i-th* element of $\boldsymbol{a}^p$ is formalized as follows:

$$a_i^p = \frac{exp(g_i^p)}{\sum_{j \in [1,m]} exp(g_j^p)}. \tag{11}$$

Then, the context-aware text embeddings of i and j are computed as:

$$\boldsymbol{i}_{(j)}^t = \mathbf{Z}_i \mathbf{a}^p, \tag{12}$$

$$\boldsymbol{j}_{(i)}^t = \mathbf{Z}_j \mathbf{a}^q. \tag{13}$$

Finally, we can obtain the context-aware embeddings of vertices with their structure embeddings and context-aware text embeddings as

$$\boldsymbol{i}_{(j)}^t = \boldsymbol{i}^s \oplus \boldsymbol{i}_{(j)}^t, \tag{14}$$

$$\boldsymbol{j}_{(i)}^t = \boldsymbol{j}^s \oplus \boldsymbol{j}_{(i)}^t. \tag{15}$$

4.4 Variation Autoencoder Objective

There are a large number of algorithms available for text embedding, e.g. Convolution Neural Network (CNN) [6], Recurrent Neural Network (RNN) [8] bidirectional RNN [14], GRU [3], autoencoder and VAE. To accommodate the characteristics of our model, we finally select VAE. In this section, we introduce the three different parts of VAE separately, and VAE loss function is introduced at the end of this part.

The VAE is employed for converting the input into an embedding and transforming it back into an approximation of the input. The encoding part aims to find the representation **Z** of a given data **X**, and the decoding part is reflection of the encoder used to reconstruct the original input **X**. The illustration of VAE [7] is shown in Fig. 2 in which imposes a prior distribution on the hidden layer and re-parameterizes the network according to the parameters of the prior distribution. Through the parameterization process, the means and variance values of the input data can be learned.

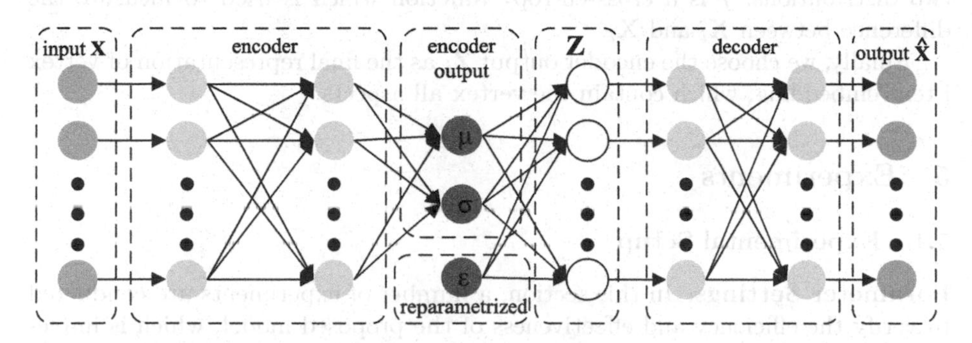

Fig. 2. Through the looking-up transforms each word into its corresponding word embeddings, as the input of VAE, the encoder and decoder are stack full-connected layers, μ and σ are the mean and variance of the distribution of the content data. ε is the sample data from the Gaussian distributions.

Encoder. The word embeddings are feed into the encoder which consists of several dense layers that formed a multiple non-linear mapping function. Thus, we can map input data to a highly non-linear latent space. Given the input $\mathbf{X}_i$, the output h^k layer is shown as follow:

$$h^1 = \psi(W^1\mathbf{X}_i + b^1), \tag{16}$$

$$h^k = \psi(W^k h^{k-1} + b^k), k = 1, 2, \ldots, K \tag{17}$$

where ψ is the nonlinear activation function of each layer, and the value of K varies with the data. Finally, the mean μ and variance σ of the distribution of text information can be learned from the encoder.

Sample. We sample a values ε from previous distribution (e.g. Gaussian distribution). The re-parameterized y_i can be obtained for vertex i. The text information can be representation by $\mathbf{Z}_i$ through y_i. Consequently, the gradient descent method can be applied in optimization. The operations can be expressed as follows:

$$y_i = f(\mu_i, \sigma_i, \varepsilon_i) \tag{18}$$

Decoder. The decoding phase is a reflection of the encoder. According to the text matrix $\mathbf{Z}_i$, the decoder output is $\hat{\mathbf{X}}_i$, which should approximate input $\mathbf{X}_i$. Specifically, the original information is restored as much as possible by the decoder layer.

The loss function of VAE should be minimized as follows:

$$loss_{VAE} = -KL(q(\mathbf{Z}_i|\mathbf{X}_i)||p(\mathbf{Z}_i) + f(\mathbf{X}_i, \hat{\mathbf{X}}_i), \tag{19}$$

where KL is the KL divergence which is used to measure of the difference two distributions, f is a cross-entropy function which is used to measure the difference between $\mathbf{X}_i$ and $\hat{\mathbf{X}}_i$.

Finally, we choose the encoder output $\mathbf{Z}_i$ as the final representation of vertex i text-embedding, which contain the vertex all aspects.

5 Experiments

5.1 Experimental Setup

Parameter Settings. In this section, a number of experiments are conducted to verify the efficiency and effectiveness of the proposed model, which is implemented using on Windows 7/Inter(R) core(TM) i7-4470 (3.4 GHz)/8.00 GB of memory with TensorFlow.

The performance of different methods with varying dimensions has been evaluated. For a fair comparison, we set the embedding dimension as 200 for all baseline models. In LINE [15], we set the number of negative samples as 5. In node2vec [4], we employ gird search and select the best-performed hyper-parameters for training. Grid search is also adopted in our model, it has best performance when $a_1 = 0.7, a_2 = 1.0, a_3 = 0.1$.

Dataset. We conduct experiments of link prediction on two real-world datasets (Cora and HepTh). Cora is a typical paper citation network constructed by [10]. Some of them are without text information. We single out 2277 papers in this network. HepTh (High Energy Physics Theory) is another citation network from arXiv released by [9]. We filter out papers without abstract information and retain 1038 papers at last.

We randomly divide the edges into two parts according to a certain proportion, one for training and the other for testing. Meanwhile, we randomly initialize the word embeddings. The detailed statistics are listed in Table 1.

Table 1. Statistics of datasets.

Datasets	Cora	HepTh
Vertices	2,277	1,038
Edges	5,214	1,990

5.2 Baseline

MMB [2] extends block models for relational data to one which capture mixed membership latent relational structure, and providing an object-specific low-dimensional representation. DeepWalk [13] performs random walks over networks and employ skip-gram model [11] to learn vertex embeddings. LINE [15] learns vertex embeddings in large-scale networks using first-order and second-order proximities. Node2vec [4] proposes a biased random walk algorithm based on DeepWalk to explore neighborhood architecture more efficiently. TADW [19] is text-based DeepWalk, which incorporates text information into network structure by matrix factorization. MMB, DeepWalk, LINE and Node2vec only consider the structure information of the vertices, while TADW combine the structure information and text information.

5.3 The Accuracy of Link Prediction

As shown in Tables 2 and 3, we evaluate the AUC [5] values while removing different ratios of edges on Cora and HepTh respectively. Due to random walks can explore the sparse network structure well even with limited edges, the DeepWalk-based methods (LINE, Node2vec and TADW) perform much better under small training ratios. However, when the training ratio rises, their performance is not as good as our model because its simplicity and the limitation of bag-of-words assumption.

According to the comparison of experiments, our model effectively improves the accuracy for link prediction in different data set and different training ratios. The result of this experiment can be explained by the fact that our model can seamlessly integrates the structure information and the text information of the vertex.

Table 2. AUC values on Cora.

Training edges	15%	25%	35%	45%	55%	65%	75%	85%	95%
MMB [2]	54.7	57.1	59.5	61.9	64.9	67.8	71.1	72.6	75.9
DeepWalk [13]	56.0	63.0	70.2	75.5	80.1	85.2	85.3	87.8	90.3
LINE [15]	55.0	58.6	66.4	73.0	77.6	82.8	85.6	88.4	89.3
Node2vec [4]	55.9	62.4	66.1	75.0	78.7	81.6	85.9	87.3	88.2
TADW [19]	86.6	88.2	90.2	90.8	90.0	93.0	91.0	93.4	92.7
OUR	**66.3**	**73.6**	**79.6**	**86.8**	**87.9**	**90.7**	**91.9**	**94.5**	**94.9**

Table 3. AUC values on HepTh.

Training edges	15%	25%	35%	45%	55%	65%	75%	85%	95%
MMB [2]	54.6	57.9	57.3	61.6	66.2	68.4	73.6	76.0	80.3
DeepWalk [13]	55.2	66.0	70.0	75.7	81.3	83.3	87.6	88.9	88.0
LINE [15]	53.7	60.4	66.5	73.9	78.5	83.8	87.5	87.7	87.6
Node2vec [4]	57.1	63.6	69.9	76.2	84.3	87.3	88.4	89.2	89.2
TADW [19]	87.0	89.5	91.8	90.8	91.1	92.6	93.5	91.9	91.7
OUR	**71.4**	**80.8**	**86.8**	**88.9**	**93.3**	**95.0**	**95.1**	**97.0**	**96.7**

Moreover, this experiments demonstrates that the attention mechanism can extract different text-embedding according to different neighbor vertices. To sum up, all the above observations demonstrate that our model not only can learn high-quality context-aware embeddings, but also has stability and robustness.

6 Conclusion and Future Work

In this paper, the vertex structure and text information are adopted to improve performance of vertices representation. We add the variation autoencoder and attention mechanism in our model for assign dynamic context-aware embeddings according to its neighbors. Experimental results of link prediction demonstrate that our model is effective for mining the relationship between vertices.

In real life, vertices information will change with time. It needs to consider the influence of time when link with other vertices. In the future, we will consider the effect of time on the text-based embeddings.

Acknowledgments. This paper was supported by the National Science Foundation of China (Grant No. 61702350).

References

1. Belkin, M., Niyogi, P.: Laplacian Eigenmaps for Dimensionality Reduction and Data Representation. MIT Press, Cambridge (2003)
2. Blei, D.M., Airoldi, E.M., Fienberg, S.E., Xing, E.P.: Mixed membership stochastic blockmodels. J. Mach. Learn. Res. **9**(5), 1981–2014 (2008)
3. Cho, K., et al.: Learning phrase representations using RNN encoder-decoder for statistical machine translation. In: International Conference on Empirical Methods Natural Language Process, pp. 1724–1734 (2014)
4. Grover, A., Leskovec, J.: node2vec: scalable feature learning for networks, p. 855 (2016)
5. Hanley, J.A., Mcneil, B.J.: The meaning and use of the area under a receiver operating characteristic (ROC) curve. Radiology **143**(1), 29 (1982)
6. Kalchbrenner, N., Grefenstette, E., Blunsom, P.: A convolutional neural network for modelling sentences. Eprint Arxiv 1 (2014)

7. Kingma, D.P., Welling, M.: Auto-encoding variational bayes. In: Conference Proceedings: Papers Accepted to the International Conference on Learning Representations (ICLR) (2014)
8. Kiros, R., et al.: Skip-thought vectors. In: International Conference on Neural Information Processing Systems, pp. 3294–3302 (2015)
9. Leskovec, J., Kleinberg, J., Faloutsos, C.: Graphs over time: densification laws, shrinking diameters and possible explanations. In: Eleventh ACM SIGKDD International Conference on Knowledge Discovery in Data Mining, pp. 177–187 (2005)
10. Mccallum, A.K., Nigam, K., Rennie, J., Seymore, K.: Automating the construction of internet portals with machine learning. Inf. Retr. **3**(2), 127–163 (2000)
11. Mikolov, T., Chen, K., Corrado, G., Dean, J.: Efficient estimation of word representations in vector space. ICLR Workshop (2013)
12. Pan, S., Wu, J., Zhu, X., Zhang, C., Wang, Y.: Tri-party deep network representation. In: International Joint Conference on Artificial Intelligence, pp. 1895–1901 (2016)
13. Perozzi, B., Al-Rfou, R., Skiena, S.: Deepwalk: online learning of social representations, pp. 701–710 (2014)
14. Schuster, M., Paliwal, K.K.: Bidirectional recurrent neural networks. IEEE Trans. Sig. Process. **45**(11), 2673–2681 (1997)
15. Tang, J., Qu, M., Wang, M., Zhang, M., Yan, J., Mei, Q.: Line: large-scale information network embedding, vol. 2, no. 2, pp. 1067–1077 (2015)
16. Tenenbaum, J.B., Silva, V.D., Langford, J.C.: A global geometric framework for nonlinear dimensionality reduction. Science **290**(5500), 2319 (2000)
17. Tian, F., Gao, B., Cui, Q., Chen, E., Liu, T.Y.: Learning deep representations for graph clustering. In: Twenty-Eighth AAAI Conference on Artificial Intelligence, pp. 1293–1299 (2014)
18. Tu, C., Zhang, W., Liu, Z., Sun, M.: Max-margin deepwalk: discriminative learning of network representation. In: International Joint Conference on Artificial Intelligence, pp. 3889–3895 (2016)
19. Yang C, Zhao D, Zhao D, et al.: Network representation learning with rich text information. In: International Conference on Artificial Intelligence, pp. 2111–2117 (2015)

SFSC: Segment Feature Sampling Classifier for Time Series Classification

Fanshan Meng, Tianbai Yue, Hongzhi Wang(✉), Hong Gao, and Yaping Li

Massive Data Computing Research Center, Harbin Institute of Technology, Xidazhijie. 92, Harbin, China
wangzh@hit.edu.cn

Abstract. Time series classification research is important in data mining. However, the existing methods are not fast and accurate enough. Common classification algorithms cannot satisfy the requirements of time series classification. Since the dimensions of each time series sample may be different. Based on this character, we propose a new approach combining the feature sampling algorithm and the random forest classifier (SFSC). To test the efficiency and effectiveness of SFSC, UCR time series datasets are used for our experiments. We also discuss the performance of Dynamic Time Warping (DTW) compared with our approach. We conclude the suitable situations in which our method behaves better. Besides, We discuss the method to determine the best parameters for feature sampling method. The experiment results shows that in most cases, the SFSC algorithm behaves better than DTW and can be used for real-time query and large dataset.

Keywords: Time series · Ensemble learning · Sampling
Cross validation · Random forest · Parameters determination

1 Introduction

The format of time series is a list contains continuous data points. It is recorded by machines in equal time interval. The time series research contains many aspects in data mining, such as business analysis [6], prediction of stock, the heart disease prediction, motion capture [2] and so on. In the aspects of the time series research, time series classification is an important field. It is expected to give the correct class for the given time series. Due to the importance of time series classification, some methods have been proposed to provide relatively accurate prediction, such as Dynamic Time Warping (DTW) [11], Long Short-term Memory (LSTM) [14], and some methods based on machine learning.

Compared with the common classification datasets, the dimensions of the time series may be different. The time series with different dimensions are widely used in some aspects. For example, in the field of mobile communications, customer loss is becoming a major problem. According to the consumption behavior

Q. Zhou et al. (Eds.): ICPCSEE 2018, CCIS 901, pp. 332–346, 2018.
https://doi.org/10.1007/978-981-13-2203-7_25

of the lost users, building a model to analyse the loss of the users is becoming a major concern for companies. The time series can be used to represent the consuming behavior of the users. In this situation, the length of the time series are different.

The common classification algorithms, such as the Support Vector Machine (SVM) [3], Logistic Regression (LR) are not suitable for time series classification. These algorithms require the input with the equal dimensions and the same meaning in each dimension. Thus transform the time series with equal dimensions and combine the machine learning method is a problem worth researching.

Time series representation varies according to the application scenarios. According to the application scenarios, different algorithms are proposed for certain situation such as Single Value Decomposition (SVD) [13]. Besides, Discrete Wavelet Transform (DWT) [12] is also a popular choice. Symbolic Aggregate Approximation (SAX) [10] is based on the Piecewise Aggregate Approximation (PAA) method [4].

Therefore, in order to get efficient and accurate time series classification results, it is important to transform the time series [8] to the training dataset with the same dimensions for each sample. Besides, the classifier also influences the efficiency and accuracy of the experiments, algorithms such as boosting sampling and random sampling could be a better choice.

In the last ten years, the main researches about the time series classification focus on getting the nearest time series under different metrics, such as Euclidean Distance (ED) [5] and DTW. DTW is a approach firstly used to recognize the phonetic sequence. Berndt and Clifford [1] applied it for data mining. It is more robust than ED. Many methods have been proposed based on it. For example, in order to speed up the DTW, index method is proposed. Besides, an improved edition for DTW is widely used named FastDTW [7,15].

Recent years, transforming the time series to the format suitable for common machine learning method is becoming popular. For example, the Ye proposed a method based on shapelets [9] for time series transformation. Lines improved it and apply it to the time series classification. But this method has many shortcomings, shapelets ignores the logical combination relationship of the time series, this will reduce the accuracy and the time costs for the computation of the shapelets is also large.

To support efficient analysis on time series classification, we have proposed an optimization algorithm for time series transformation and classification. In order to get more accuracy, the main feature of the time series need to be collected. To achieve this goal, we develop segment feature sampling algorithm to transform the time series. We use segment feature sampling algorithm to get more information from the original time series. It changes the time series with different dimensions to the dataset with the same dimensions. We use random forest to train the model with the transformed dataset. The experiment results show that the segment feature sampling algorithm can get main features about the dataset. The algorithm behaves better than DTW in most cases.

In summary, this paper makes the following contributions.

1. First, We design a time series transformation method. This method can deal with large time series dataset and the time cost is very little.
2. Using the ensemble models for time series classification is the second contribution. As the time series data have different length and have the topological similarity character, we combine the random forest with the segment feature sampling algorithm.
3. Finally, we use Python to implement the algorithm. The experiment results shows that the algorithm can improve the effect of classification and fast enough for real-time and offline query.

The organization structure of the paper is as follows. We first give some definitions about our method in Sect. 2. We introduce the detail design about the time series classification algorithm in Sect. 3. The experiment results and analysis are presented in Sect. 4. Finally, we conclude the experiments and discuss the future work in Sect. 5.

2 Definition

In this paper, we research the classification algorithm for time series. In this section, we will give some basic definitions and some basic operations about our algorithm.

Definition 1. *S is defined as a time series with values recorded in time order, i.e., $S = [(t_1, s_1), (t_2, s_2), \ldots, (t_n, s_n)]$, in which t_i stands for the time index of s_i and $t_i < t_j$ when $i < j$. s_i presents the value of the time index t_i.*

Definition 2. *T is defined as a sub-time series which contains values recorded in time order from S, i.e., $T = [(t_m, s_m), (t_{m+1}, s_{m+2}), \ldots, (t_{m+l}, s_{m+l})]$, in which t_i stands for the time index of s_i, and $t_i < t_j$ when $i < j$. s_i presents the value of the time index t_i, m is the sub-series starting location, l is the length of T.*

Definition 3. *A class specified time series dataset D^c is described as a list contains all the time series with same class index c. For dataset D, $D.classcount$ presents the class count of the dataset D, D_i presents the ith sample in the dataset D.*

Definition 4. *Some basic operations in our model. $length(S)$ presents the length of time series S. S_i presents the ith dimension of time series S. $S.add(i)$ presents adding i at the end of the time series S. $S.class$ presents the class index of the time series S.*

3 SFSC: Segment Feature Sampling Classifier for the Time Series Classification

In this section, we introduce our method, <u>S</u>egment <u>F</u>eature <u>S</u>ampling <u>C</u>lassifier and the detailed procedures. Firstly, we introduce our three kind sampling methods, we will compare the sampling methods with experiment results. Next, we present methods to determine the optimal parameters. Finally, we use the random forest combining the segment feature sampling algorithm to train the classification model. We conclude the suitable situation for our algorithm and analysis the experiment results. The details of the model are shown as follow.

Subsection 3.1 presents an overview of entire approach workflow. Subsection 3.2 presents an overview of feature sampling algorithms. Subsection 3.3 presents the random forest classifier. Subsection 3.4 presents the method to determine the best parameters for feature sampling algorithm.

3.1 Model Overview

Figure 1 presents overview of our algorithm. The step of workflow is as follows:

1. For the training dataset, use the z-zero normalization to normalize it.
2. Divide the training dataset for parameter determination algorithm. Run the parameter determination to get the best parameters.
3. Run the Segment Feature Sampling Algorithm with the best parameters. Segment Feature Sampling Algorithm will transform original time series into the new training dataset with the same feature length.
4. Use the Random Forest classifier with the new training dataset as input to train the classifier model.
5. For the test dataset, use the z-zero normalization to normalize it.
6. Run the Segment Feature Sampling Algorithm with the best parameters to transform the test dataset.
7. Use the trained model to predict the results and evaluate it.

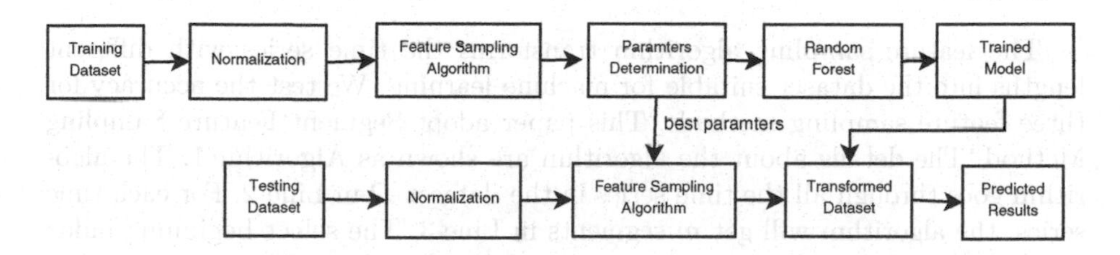

Fig. 1. The structure of SFSC algorithm.

3.2 Sampling Algorithm

Feature sampling algorithm is proposed to handle unequal length of time series. The feature sampling algorithm accepts the time series dataset as input and output the training samples with the same dimensions. For one time series, the feature sampling algorithm will transform it to several training sample according to the given parameter. In this paper, we propose three kind feature sampling algorithms. These sampling algorithms aim to apply the machine learning methods to the time series. They are described as follows.

1. **Random Feature Sampling Method**
   Random Feature Sampling Methods contains three parameters, randomly sampling features count l, sampling times m and sampling starting location b. The random feature sampling method firstly random generate a candidate set s containing m numbers. Then it will repeatedly sample m times. At each time, the algorithm picks one number from s as b. Then it starts to sample at the location b, randomly select l features in time order to generate a new training sample. When the selecting index exceed the max length of the time series, it will select feature from the beginning. After the sampling, it will change one time series to m training samples.
2. **Equal Interval Feature Sampling Method**
   Equal Interval Feature Sampling Methods contains three parameters, sampling time interval g, sampling feature count m, sampling starting location b. The algorithm starts to sample at the location b, it will select m features in time interval g to generate a new training sample. When the selecting index exceed the max length of the time series, it will select features from the beginning. After the sampling, it will change one time series to one training sample.
3. **Segment Feature Sampling Method**
   Segment Feature Sampling Method contains three parameters, segment length l, segment count m, the interval between segments g. The segment feature sampling method will repeatedly sample m times. At the sth time, the algorithm selects l continuous features from the location $g * s$. When the selecting index exceed the max length of the time series, it will select features from the beginning. After the sampling, it will change one time series to m training samples.

The feature sampling algorithm transforms the time series with different lengths into the dataset suitable for machine learning. We test the accuracy for three feature sampling methods. This paper adopt Segment Feature Sampling Method. The details about the algorithm are shown as Algorithm 1. The algorithm goes through all the time series in the dataset D in Line 2. For each time series, the algorithm will get m segments in Line 3. The select beginning index will be changed according to the segment index in Lines 4–6. Then the algorithm begins to sample the features in order to generate the new training sample in Lines 9–18. Figure 2 presents an overview of the segment feature sampling algorithm.

3.3 Random Forest Classifier

In last section, we discuss three kind feature sampling algorithms. The feature sampling algorithm transforms the time series so that the machine learning algorithms can be applied. In this section, We will discuss how to choose the machine learning algorithms and the reasons.

The choice of the classifiers should be on the basis of dataset features. The time series dataset have segment similarity and strict time sequence features. In

Algorithm 1. SegementFeatureSampling

```
Input:
   D: time series dataset
   l: segment length
   m: segment count
   g: interval between segments
Output: D': transformed time series dataset
1  D' ← null;
2  for every time series S of D do
3      for i = 0 to m do
4          b ← i * g;
5          if b ≥ len then
6              b ← b%len;
7          end
8          S' ← null;
9          for j = 0 to l do
10             len ← length(S);
11             if b + j ≤ len then
12                 S' ← S'.add(S_{b+j});
13             end
14             else
15                 index ← (b + j)%len;
16                 S' ← S'.add(S_{index});
17             end
18         end
19         D' ← D'.add(S');
20     end
21 end
22 return D';
```

this paper, we choose the random forest as the basic classifier, it contains many advantages suitable for time series. Firstly, it contains many decision trees, the final result is determined by all the trees, this will make the result more accurate. Secondly, it select the features randomly as the final input from the original input, this satisfies the time series segment similarity feature. Finally, in the training process, the random forest can be realized concurrently, this can deal with large datasets. In the experiment section, the results shows that the random forest can get better results in most cases.

In the time complexity aspect, the segment feature sampling algorithm complexity is $O(log(m*l))$. In the experiments, we set m to $S_{classcount}+1$. Therefore the time complexity in the segment feature sampling algorithm is $O(logn)$. In the model training, the training size is $(c+1)*n$ and the feature length is l. We choose the random forest as the classifier and the time complexity is $O(nlogn)$. In the predicting procedure, the time cost is $O(C+1)$. Therefore, the SFSC can satisfy the real-time query.

3.4 Feature Sampling Algorithm Parameters Determination

Segment Feature Sampling Methods contains three parameters shown in Table 1 These three parameters have an important effect on the selected features. The parameters determination will improve the classification accuracy. We take the cross validation to determine the best parameters. Compared with the original

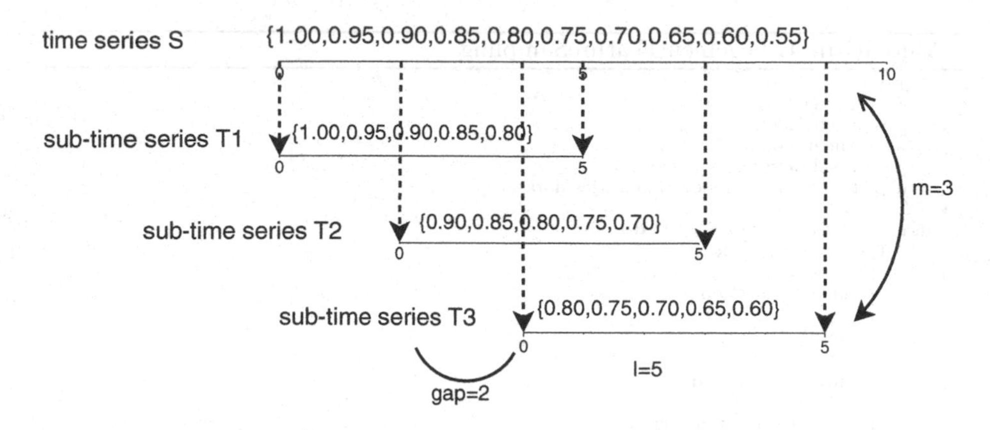

Fig. 2. The procedure of segment feature sampling.

method, the cross validation not only considers the training error, but also generation error. We use the K-fold cross validation method to make full use of the dataset. The K is set according to the size of the training dataset. If the size of the dataset is lower than 300, we set K to 2, else we set K to 3. We split the training dataset into K parts randomly. We use one part as test dataset, the left as the training dataset. In each experiment we use different test dataset, and get the min error l and g. We choose the parameters which behave best in the K experiments.

Table 1. Segment feature sampling method parameter table

Parameters	Meaning
l	The selecting features count
m	The segments count
g	The interval between segments

We set the parameter m a fixed value $m = D.classcount + 1$ because it will influence the final class of the time series. After the feature sampling algorithm, one test time series S will get m test sample. In order to make the result unique, the class of time series S will be voted by m results predicted by m test sample.

We make some changes based on cross validation. In the procedure of dividing training dataset, randomly division will make training dataset and test dataset different distribution. We except that for each class, training dataset and test dataset will proportionally assigned after division. Thus we divide training dataset according to K in each class. We divide each class into average K part randomly and pick one as test dataset each time. This will ensure training dataset and test dataset same distribution. The detailed algorithm is shown as Algorithm 2.

Algorithm 2. ParametersDetermination

```
Input:
    D: training time series dataset
    K: training dataset division count
Output:
    l: segment length
    g: interval between segments
1  for every time series S of D do
2  |   c ← S.class; D^c ← D^c.add(S);
3  end
4  l ← 0;
5  m ← D.classcount + 1;
6  g ← 0;
7  bestaccuracy ← 0;
8  for j = 0 to k do
9      for g = 1 to 10 do
10         for l = 1 to 10 do
11             trainingDataset ← null;
12             testDataset ← null;
13             for every class c in D do
14                 len ← len(D^c);
15                 klen ← len/k;
16                 for i = 0 to len do
17                     D' ← SegementFeatureSampling(D^c_i, l, m, g);
18                     if i ≥ klen * j and i<klen * (j + 1) then
19                     |   testDataset.add(D');
20                     end
21                     else
22                     |   trainingDataset.add(D');
23                     end
24                 end
25             end
26             model ← RandomForestClassifier(trainingDataset);
27             accuracy ← model.predict(testDataSet);
28             if accuracy ≥ bestaccuracy then
29             |   update(l, g);
30             end
31         end
32     end
33 end
34 return l, g, m;
```

4 Evaluation

We carry out the experiments on UCR time series datasets and evaluate our algorithm. The main points we concern about are as follows:

1. Segment feature sampling algorithm time cost.
2. The entire model building and training time cost.
3. The accuracy comparison between SFSC and DTW.
4. The existing problems in our algorithm and future work.

To prove effectiveness and accuracy of our algorithm, we test various UCR datasets. Our experiments are carried out on a PC with intel i5, 2.4 GHz CPU and 8 GB memory. We use python 3.5 to implement the algorithm. The operating system is macOS Sierra. The runtime of program and the error-rate are used to evaluate our algorithm. Subsection 4.1 presents l and g impact on accuracy.

Subsection 4.2 presents the feature sampling time cost and give an empiric value for l and g. Subsection 4.3 presents the runtime comparison between SFSC and DTW. Subsection 4.4 presents the effectiveness comparison between SFSC and DTW. Subsection 4.5 presents the experiment results analysis and the problem existed in our algorithm.

4.1 Parameters Impact on Accuracy

Parameters l and g have an import impact on Accuracy. We test the impacts using three datasets, BeetleFly, Lighting2 and ProximalPhalanxOutlineCorrect (PPOCorrect). Figure 3 presents l impact on accuracy. l_rate presents the ratio of sample length to the maximum time series length. From the Fig. 3 we can observe that when the l_rate set to 0.93, the experiment can get well results. The sample length influence the model building. When the sample length is too long, there will be much redundant information. When the sample length is too short, training sample will contain little features of original time series. Figure 4 presents g impact on accuracy. It presents the interval between sampling procedures. The g influences the time relationship among samples. It will break time relationship when it is long. From the Fig. 4, we set g to 5 and the result shows it works. These empirical values will be used in next experiments.

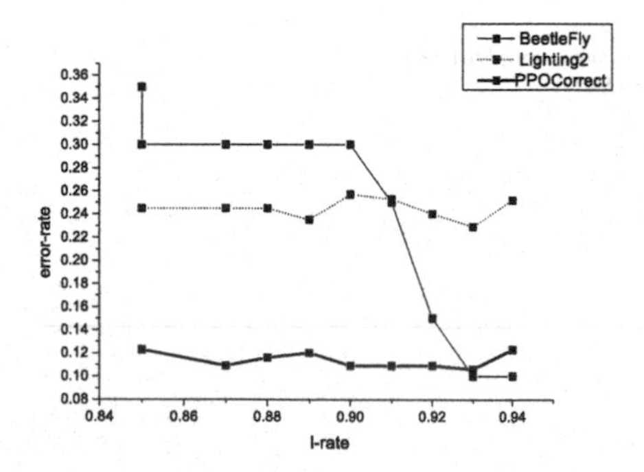

Fig. 3. The l impact on accuracy

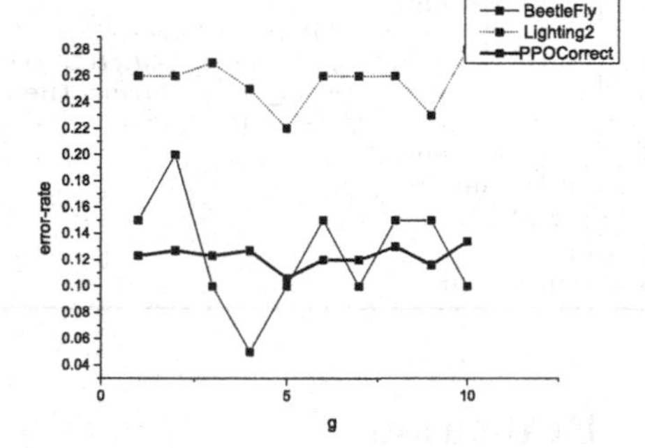

Fig. 4. The g impact on accuracy

4.2 Preprocessing Time

Before using random forest classifier to train the model, we need to determine best parameters l and g, and the original time series dataset need to be transformed, whose time complexity is $O(n)$. As for the parameters determination, after we carry out most experiments, we find an empiric value for l and g. Assume the length of the longest time series S in dataset D is len, we suggest that l is set to $0.93 * len$ and g is set to 5 which will get an accurate result in most cases. The empirical value saves a lot of time costed by the parameters determination algorithms. We use the empiric value as the best parameters in the following experiments. The real datasets are used to test segment feature sampling algorithm runtime.

In the UCR time series datasets, the datasets can be split by the representative meaning. We use ECGFiveDays (EFD), MoteStrain (MS) and ElectricDevices (ED). The details of datasets, the set parameters l, g and the preprocessing time are described in Table 2.

In Table 2, we can conclude that the transformation costs little time. The major cost is to get the best parameters for SFSC. But the empiric value solve it and save the time. Thus in the following experiments, the time costs is ignored by the runtime the training process costs.

Table 2. Segment feature sampling methods runtime

Dataset	#query	#series	l	g	Runtime
EFD	861	23	126	5	117
MS	1252	20	78	5	97
ED	7711	8926	89	5	3713

4.3 Comparisons of Efficiency

In this subsection, we compare the efficiency among DTW, ED and SFSC to show the fast speed of our approach. We test some optimization techniques. For DTW, we used FastDTW with multi-layer filtering technique.

We use StarLightCurves (SLC), FacesUCR (FU) and ElectricDevices (ED) to test the efficiency. The details about the datasets, the parameters l, g are described in Table 2. The predicting results runtime of three approaches are shown in Table 3.

Table 3. Runtime of different approaches

Dataset	SFSC	ED	FastDTW
EFD	30	57	45
MS	75	345	111
ED	198	3091	535

In Table 3, we can conclude that our algorithm is fastest. In addition, our algorithm can save the trained model and make the classification real-time using the trained model.

Table 4. Error rate of different approaches in UCR archive

Name	Train-size	Test-size	#Classes	ED	DTW	SFSC
50words	450	455	50	0.369	0.242	0.481
Adiac	390	391	37	0.389	0.391	0.514
Beef	30	30	5	0.333	0.333	0.233
BeetleFly	20	20	2	0.250	0.300	0.100
BirdChicken	20	20	2	0.450	0.300	0.200
CBF	30	900	3	0.148	0.004	0.000
ChlorineConcentration	467	3840	3	0.350	0.350	0.272
CinC_ECG_torso	40	1380	4	0.103	0.070	0.213
Coffee	28	28	2	0.000	0.000	0.035
Computers	250	250	2	0.424	0.380	0.372
Cricket_X	390	390	12	0.423	0.228	0.500
Cricket_Y	390	390	12	0.433	0.238	0.476
Cricket_Z	390	390	12	0.413	0.254	0.497
DiatomSizeReduction	16	306	4	0.065	0.065	0.088
DistalPhalanxOutlineAgeGroup	139	400	3	0.218	0.228	0.160
DistalPhalanxOutlineCorrect	276	600	2	0.248	0.232	0.198
DistalPhalanxTW	139	400	6	0.273	0.272	0.240
Earthquakes	139	322	2	0.326	0.258	0.223
ECG200	100	100	2	0.120	0.120	0.190
ECG5000	500	4500	5	0.075	0.075	0.066
ECGFiveDays	23	861	2	0.203	0.203	0.291
ElectricDevices	8926	7711	7	0.450	0.376	343
FaceFour	24	88	4	0.216	0.114	0.079
FacesUCR	200	2050	14	0.231	0.088	0.306
FISH	175	175	7	0.217	0.154	0.097
FordA	1320	3601	2	0.341	0.341	0.182
FordB	810	3636	2	0.442	0.414	0.353
Gun_Point	50	150	2	0.087	0.087	0.073
Ham	109	105	2	0.400	0.400	0.323
HandOutlines	370	1000	2	0.199	0.197	0.213
Haptics	155	308	5	0.630	0.588	0.512
Herring	64	64	2	0.484	0.469	0.359
InlineSkate	100	550	7	0.658	0.613	0.676
ItalyPowerDemand	67	1029	2	0.045	0.045	0.042
LargeKitchenAppliances	375	375	3	0.507	0.205	0.405
Lighting2	60	61	2	0.246	0.131	0.229
Lighting7	70	73	7	0.425	0.288	0.301
MALLAT	55	2345	8	0.086	0.086	0.043
Meat	60	60	3	0.067	0.067	0.100
MedicalImages	381	760	10	0.316	0.253	0.306

(continued)

Table 4. *(continued)*

Name	Train-size	Test-size	#Classes	ED	DTW	SFSC
MiddlePhalanxOutlineAgeGroup	154	400	3	0.260	0.253	0.222
MiddlePhalanxOutlineCorrect	291	600	2	0.247	0.318	0.275
MiddlePhalanxTW	154	399	6	0.439	0.419	0.385
MoteStrain	20	1252	2	0.121	0.134	0.156
NonInvasiveFatalECG_Thorax1	1800	1965	42	0.171	0.185	0.251
NonInvasiveFatalECG_Thorax2	1800	1965	42	0.120	0.129	0.232
OliveOil	30	30	4	0.133	0.133	0.133
OSULeaf	200	242	6	0.479	0.388	0.367
PhalangesOutlinesCorrect	1800	858	2	0.239	0.239	0.168
Phoneme	214	1896	39	0.891	0.773	0.849
ProximalPhalanxOutlineAgeGroup	400	205	3	0.215	0.215	0.131
ProximalPhalanxOutlineCorrect	600	291	2	0.192	0.210	0.106
ProximalPhalanxTW	205	400	6	0.292	0.263	0.205
RefrigerationDevices	375	375	3	0.605	0.560	0.464
ScreenType	375	375	3	0.640	0.589	0.584
ShapeletSim	20	180	2	0.461	0.300	0.455
ShapesAll	600	600	60	0.248	0.198	0.438
SmallKitchenAppliances	375	375	3	0.659	0.328	0.272
SonyAIBORobotSurface	20	601	2	0.305	0.305	0.339
SonyAIBORobotSurfaceII	27	953	2	0.141	0.141	0.211
StarLightCurves	1000	8236	3	0.151	0.095	0.05
Strawberry	370	613	2	0.062	0.062	0.044
Symbols	25	995	6	0.100	0.062	0.208
synthetic_control	300	300	6	0.120	0.017	0.036
ToeSegmentation1	40	228	2	0.320	0.250	0.350
ToeSegmentation2	36	130	2	0.192	0.092	0.361
Trace	100	100	4	0.240	0.010	0.050
Two_Patterns	1,000	4,000	4	0.090	0.002	0.005
TwoLeadECG	23	1139	2	0.253	0.132	0.187
uWaveGestureLibrary_X	896	3582	8	0.261	0.227	0.195
uWaveGestureLibrary_Y	896	3582	8	0.338	0.301	0.289
uWaveGestureLibrary_Z	896	3582	8	0.350	0.322	0.256
UWaveGestureLibraryAll	896	3582	8	0.052	0.034	0.043
wafer	1,000	6,174	2	0.005	0.005	0.010
Wine	57	54	2	0.389	0.389	0.240
WordsSynonyms	267	638	25	0.382	0.252	0.536
Worms	77	181	5	0.635	0.586	0.546
WormsTwoClass	77	181	2	0.414	0.414	0.370
yoga	300	3000	2	0.170	0.155	0.163

4.4 Comparisons of Accuracy

In this subsection, we compare the efficiency among DTW, ED and SFSC. The UCR datasets are used to test our algorithm. In the datasets, each datasets contains TRAIN and TEST with labels for each sample. We use the error-rate to evaluate our algorithm. The error-rate is defined as follow.

$$errorRate = \frac{inaccurate\ time\ series\ count}{size\ of\ TEST}$$

Table 4 shows the comparasons among three algorithms. It contains all the accuracy comparison results on UCR time series datasets. From the results we observe that in most cases, the ED behaves worst in these three algorithms. As for SFSC and DTW, we can observe that SFSC outperforms DTW in 50% of the whole datasets. Besides, for the convenience of our discussion, we split the whole datasets into two parts, one part named $PARTONE$ contains all the datasets with more than 10 class counts, the other part named $PARTTWO$ contains the left datasets. We can observe that for datasets in $PARTONE$, the DTW outperforms SFSC in most cases, but for datasets in $PARTTWO$, SFSC outperforms DTW in 70% datasets. That is to say, SFSC behaves better when the class count is less than 10.

4.5 Analysis

In the segment feature sampling procedure, we select m samples representing the original time series S. Each training sample contains partly l features from the original time series. But this m training samples are selected from different start location b and will contains all the features in S. In the model training, we use m samples with same class as the original time series. The model will learn the features without loss of original information.

From the results we observe that when the class count of the dataset is less tan 10, SFSC behaves better. In the feature sampling procedure, assume that the class count of the dataset D is c, the time series count is n, after the segment feature sampling, the training sample size is $(c+1)*n$. The new dataset becomes larger than before and the samples with same class increase. This will make the classifier learn more about the difference among the classes. But when the dataset has many classes, the size of the samples with same class is very few. Though the feature sampling will increase the samples number, the classifier will not learn well about feature. How to deal with the dataset with many classes is worth researching.

5 Conclusion

In this paper, we propose a time series classification algorithm. We design the algorithm with feature sampling algorithm for selecting features from time series. We propose three feature sampling methods and choose the random forest as the basic classifier. After feature sampling algorithms, we discuss the way to determine the best parameters for SFSC. In the experiments, we test the runtime of the feature sampling algorithm and make the comparison between DTW and SFSC in efficiency and effectiveness. We implement our model with python and make the experiments in real time series datasets. From the results we can see SFSC improve the accuracy. Finally, we analysis on the results and conclude the suitable datasets for SFSC.

In future, we plan to improve our algorithm for time series classification in two aspects. Firstly, the transformation of time series is the crux of the matter. We propose three methods for the transformation in feature sampling. But the parameters determination is still a problem. Secondly, the design about model and the model designed for the multiple class datasets need to be improved. In this paper, we adopt the single model random forest. But the multiple models combination maybe works better than single model and SFSC works not very well for the multiple class dataset. Therefore, the next step for our work mainly contains two aspects. Firstly, we will find the way to determine the parameters with little time costs. Secondly, we will carry out the experiments with multiple models and design algorithms for multiple class. Besides, we will also consider the application of the algorithm for the industrial data and parallelize it.

Acknowledgments. This paper was partially supported by NSFC grant U1509216, The National Key Research and Development Program of China 2016YFB1000703, NSFC grant 61472099,61602129, National Sci-Tech Support Plan 2015BAH10F01, the Scientific Research Foundation for the Returned Overseas Chinese Scholars of Heilongjiang Provience LC2016026.

References

1. Berndt, D.J., Clifford, J.: Using dynamic time warping to find patterns in time series. In: KDD Workshop (1994)
2. Fox, E.B., Hughes, M.C., Sudderth, E.B., Jordan, M.I.: Joint modeling of multiple time series via the beta process with application to motion capture segmentation. Ann. Appl. Stat. **8**(3), 1281–1313 (2013)
3. Ghorbani, M.A., Khatibi, R., Goel, A., Fazelifard, M.H., Azani, A.: Modeling river discharge time series using support vector machine and artificial neural networks. Environ. Earth Sci. **75**(8), 685 (2016)
4. Guo, C., Li, H., Pan, D.: An improved piecewise aggregate approximation based on statistical features for time series mining. In: Bi, Y., Williams, M.-A. (eds.) KSEM 2010. LNCS (LNAI), vol. 6291, pp. 234–244. Springer, Heidelberg (2010). https://doi.org/10.1007/978-3-642-15280-1_23
5. Keogh, E., Wei, L., Xi, X., Vlachos, M., Lee, S.H., Protopapas, P.: Supporting exact indexing of arbitrarily rotated shapes and periodic time series under euclidean and warping distance measures. VLDB J. Int. J. Very Large Data Bases **18**(3), 611–630 (2009)
6. Lai, R.K., Fan, C.Y., Huang, W.H., Chang, P.C.: Evolving and clustering fuzzy decision tree for financial time series data forecasting. Expert Syst. Appl. Int. J. **36**(2), 3761–3773 (2009)

7. Lohrer, J., Lienkamp, M.: Building representative velocity profiles using FastDTW and spectral clustering. In: International Conference on ITS Telecommunications, pp. 45–49 (2016)
8. Martnez, B., Gilabert, M.A.: Vegetation dynamics from NDVI time series analysis using the wavelet transform. Remote Sens. Environ. **113**(9), 1823–1842 (2009)
9. Rakthanmanon, T., Keogh, E.: Fast shapelets: a scalable algorithm for discovering time series shapelets (2013)
10. Sun, Y., Li, J., Liu, J., Sun, B., Chow, C.: An improvement of symbolic aggregate approximation distance measure for time series. Neurocomputing **138**(11), 189–198 (2014)
11. Tormene, P., Giorgino, T., Quaglini, S., Stefanelli, M.: Matching incomplete time series with dynamic time warping: an algorithm and an application to post-stroke rehabilitation. Artif. Intell. Med. **45**(1), 11–34 (2009)
12. Wang, H.F.: Clustering of hydrological time series based on discrete wavelet transform. Phys. Procedia **25**, 1966–1972 (2012)
13. Weng, X., Shen, J.: Classification of multivariate time series using two-dimensional singular value decomposition. Knowl.-Based Syst. **21**(7), 535–539 (2008)
14. Wollmer, M., Eyben, F., Keshet, J., Graves, A., Schuller, B., Rigoll, G.: Robust discriminative keyword spotting for emotionally colored spontaneous speech using bidirectional LSTM networks. In: IEEE International Conference on Acoustics, Speech and Signal Processing, pp. 3949–3952 (2009)
15. Xu, H., Xing, Y., Shen, F., Zhao, J.: An online incremental learning algorithm for time series. In: International Joint Conference on Neural Networks, pp. 1–5 (2015)

Fuzzy C-Mean Clustering Based: LEO Satellite Handover

Syed Umer Bukhari[1], Liwei Yu[1,2], Xiao qiang Di[1,2(✉)], Chunyi Chen[1], and Xu Liu[1,2]

[1] School of Computer Science and Technology, Changchun University of Science and Technology, Changchun, China
umershah81@hotmail.com, dixiaoqiang@cust.edu.cn
[2] Jilin Province Key Laboratory of Network and Information Security, Changchun, China

Abstract. Satellite networks are better alternatives to terrestrial networks because of their global coverage. Different satellite networks like GEO, MEO, and LEO can be used for the purpose of global communication. Especially, LEO satellites are more suitable for communication because of their lower propagation delays and fewer power consumptions. However, the high speed of LEO satellites is a big issue which leads to frequent cell and satellite handovers. Many techniques have been proposed to deal with satellite handover, but most of proposed techniques did not consider the situation when "the required channels are more than available channels" which causes more delays, terminations, and blocking. To handle this situation, we proposed a fuzzy C-mean cluster-based handover technique to improve the QoS of the network during handover. In the proposed technique, users (waiting for handover in a queue) are divided into clusters based on their geographic locations. Cluster heads are selected from each cluster and they act as the relay between their members and the satellite. During handover, only cluster heads reserve the channels in upcoming satellite and perform handover with their members. The experimental results prove that the proposed technique helps to reduce the handover failure, terminations, and number of waiting users, as well as it also improves the utility of network.

Keywords: LEO satellites · Fuzzy clustering · Handover · Elevation angle MMCK queue

1 Introduction

The satellite infrastructure can play an important role in the field of data sciences and big data (as an analytical architecture of satellite networks is given [1] for big data), when we need to send the data or communicate to remote locations. The satellite infrastructure provides global coverage and less complex paths for remote users because it can cover large areas. For communication purpose, the infrastructure of LEO satellites [2] is the best choice because of low altitude, less propagation delay, energy consumption, installation cost and relatively better communication performance than GEO and MEO satellites. Inevitably, the high velocity and short live time (usually 7–8 min for iridium constellation) of LEO satellites lead to high number of handovers. High number of

Q. Zhou et al. (Eds.): ICPCSEE 2018, CCIS 901, pp. 347–358, 2018.
https://doi.org/10.1007/978-981-13-2203-7_26

handovers introduces more delay, termination, and data loss etc. There are plenty of LEO satellite constellations such as Globalstar, Teledesic, Ellipso, and Iridium. The Iridium constellation offers a good experience to study the performance and other issues of LEO satellite networks. In this paper, we will consider the Iridium constellation. Satellite and cell handovers are very common in satellite networks shown in Fig. 1. Many solutions are given in to improve the QoS of LEO satellite networks like combined satellite handover algorithm [3], channel reservation techniques [4, 5], dynamic Doppler priority-based channel reservation [6–8] schemes. The elevation angle is frequently used in previous literature to determine the handover time. In literature, [9] it is assumed that the GPS enabled users can determine the service time of visible satellites by using elevation angle and then select the satellite which provides largest service time. Satellite handover is well studied in existing literature, but the issue "when available channels are less than the required channels for handover while many users are waiting in the queue" is not well studied.

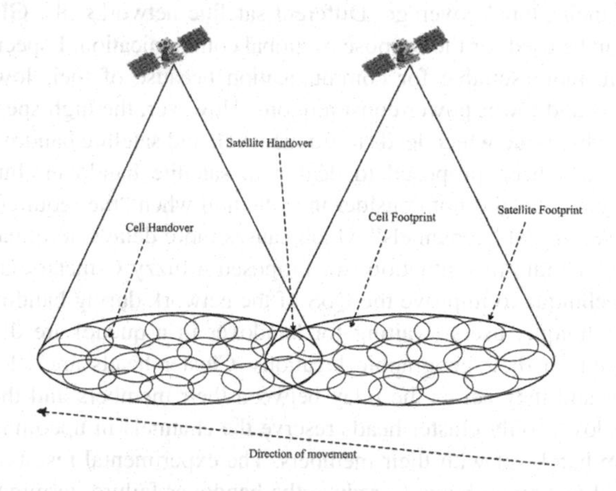

Fig. 1. Satellite and cell handover

In this paper, we develop and simulate a simple multi-server queuing model (MMCK) [10] based satellite handover technique and then to solve the issue of "less available channels than the required channels" we enhance the MMCK technique with Fuzzy C-Mean clustering algorithm (FCM) and proposes a Fuzzy C-mean clustering based: LEO satellite handover technique. In FCM based handover technique, users wait in the queue when there is no free channel in upcoming satellite and if the waiting time of users exceeds from specified threshold time then the users are terminated. If the queue size exceeds from specified threshold then system divides the queued users into clusters using FCM clustering algorithm. After clustering, the cluster heads, selected from each cluster act as a relay between cluster members and the satellite. Finally, the cluster heads switch to the upcoming satellite with their members using reserved channels. So that one cluster consumes only one channel during having handover. The number of clusters is based on the reserved channels. Simulation work is carried out on

MATLAB, and the results show that the FCM based handover technique improves the utility and QoS of the network by minimizing the individual handovers.

2 Related Work

Satellite handover is well studied in existing literature. Many strategies are introduced to improve the handover performance of LEO satellite networks. Different existing satellite handover schemes like hard/soft handover, satellite handover, cell handover, some queuing and priority-based schemes are briefly discussed and compared in [8]. In terms of throughput and latency, the terrestrial networks are better than satellite networks. A technique for hybrid satellite-terrestrial networks is presented in [11] to optimize the satellite signaling using resource manager. In this technique, it is suggested that, use the terrestrial network for communication when it is available otherwise satellite network can be used. To make handover process more smooth and efficient, different authors introduced many channel reservation and management techniques. In time-based channel reservation techniques [4, 8] the satellite handover time is determined by utilizing the previous (cell visiting time) history of the user and the channels are reserved in advance in next upcoming satellite. In literature [7] the channel reservation is based on Dynamic Doppler prioritizing technique (DDBHP) [12]. Authors assume that, the users are in common area between different satellites and they can select any of them. Different satellite selection schemes are also defined using some parameter i.e. maximum service time, the maximum number of free channels and maximum distance. In Graph-based satellite handover framework [13] the simplified perturbations model (SPG4) is utilized to predicted the visibility of forthcoming satellites and the graph for future handovers is drawn, user selects the shortest path for future handovers which helps to reduce the unnecessary handovers.

In addition, the ongoing calls are more important than new calls in terms of QoS of the network. A novel call admission scheme based on user's location is given in [14]. By using GPS functionality, the system can determine whether to block the new call or not to avoid the unnecessary handover failure of ongoing calls. A real-time handover management approach [9] is a very simple and efficient. In this approach, it is assumed that the users are GPS enabled and they can determine their trace angle, using trace angle users can calculate the service time of all currently visible satellite and the satellite with largest service time can be selected for handover. Software-defined network (SDN) is a newly emerging field. It separates the data plan from control plan so that the routing, load balancing, and security policies can be updated remotely in satellite networks. A seamless handover scheme using SDN is presented in [15] and an extra module called "location server" is introduced, which keeps track of satellite movement. When a user needs handover, it sends a request to the location server and location server notify the controller to update the flow tables after updating the flow tables the user starts handover. Combined satellite handover algorithm given in literature [3], in this algorithm the handover time is determined by using signal strength and the location of a user. The movement of users from one network to another network at the same time causes congestion and complexity in the network due to simultaneous handovers. In such conditions, it is very difficult to manage simultaneous handovers, the group handover management approach [16] is presented to solve this problem. In this approach, the bandwidth adaptation policy and dynamic bandwidth reservation policy is used to manage the group handover in mobile femto-cellular networks. Satellite handover is well studied in existing literature but as in our knowledge, the situation "when required resources are more than the available resources" is not well considered. Our proposed FCM clustering based satellite handover technique can handle this situation more efficiently and it helps to reduce the delays, terminations and blocking and improves the utility of the network.

3 Elevation Angle

The elevation angle is frequently used in existing literature to determine the handover time. By using GPS functionality user can calculate its elevation angle; if the elevation angle of user is less than the threshold then it sends the channel reservation request to next upcoming satellite and start handover process. The final formulation for calculating elevation angle [9] is given in Eq. (1).

$$\theta_a = cos^{-1}\left(\frac{\sqrt{x_u^2 + y_u^2}}{\sqrt{x_s^2 + y_s^2 + z_s^2}}\right) \tag{1}$$

In Eq. (1) x and y denote the longitude and latitude of user and satellite respectively, the altitude of a satellite is denoted by z. The scenario of elevation angle and minimum threshold elevation angle is shown in Fig. 2.

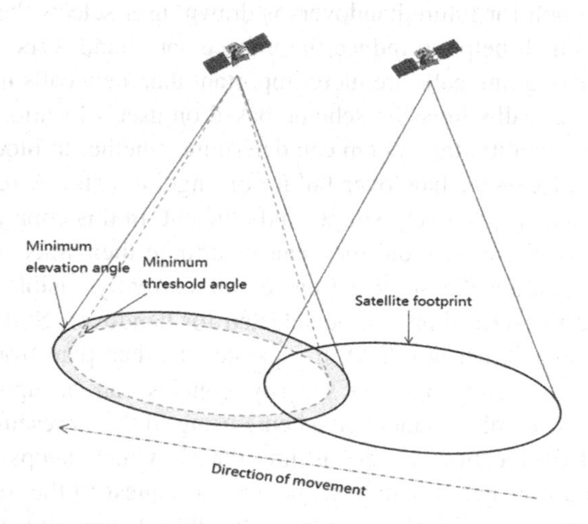

Fig. 2. Threshold and minimum elevation angles

It is necessary to check that if the satellite is approaching to the user or going away, it can be decided by comparing the current time instant of the user and the time stamp sent by satellite.

4 Fuzzy C-Mean Clustering Based Satellite Handover

During satellite handover, users may have to wait in queue for free channels (because of the limited resources) which lead to more delay, blocking, and sometimes call termination. The FCM clustering based handover technique is a good solution to resolve this limited resources issue and it also enables us to utilize the network resources more efficiently.

4.1 Fuzzy C-Mean Clustering

In our proposed fuzzy c-mean clustering handover technique, we took the advantage of Fuzzy C-Mean clustering algorithm [17–19] which helps us to perform the fuzzy distribution of the given data based on their similarity and dissimilarity. It divides the given data set $S = (x_1, x_2, x_3, x_4, \ldots\ldots\ldots, x_n)$ into C clusters $(1 < C < N)$. To get the optimal value of an objective function (j_m) given in Eq. (2), the membership function u_{ij} (3) and the cluster center function c_j (4) are gradually updated. Cluster center function (c_j) gives the most centered user of the cluster; here we consider it as a head of that cluster. It determines its member users by checking their received signal power. If the received signal power of the user is less than the specified threshold then this user will not be the member of a cluster.

$$j_m = \sum_{i=1}^{N} \sum_{j=1}^{C} u_{ij}^m \left\| x_i - c_j \right\|^{-2} \tag{2}$$

$$u_{ij} = \frac{1}{\sum_{k=1}^{C} \left(\frac{\|x_i - x_j\|}{\|x_i - c_k\|} \right)^{\frac{2}{m-1}}} \tag{3}$$

$$c_j = \frac{\sum_{i=1}^{N} u_{ij}^m * x_i}{\sum_{j=1}^{N} u_{ij}^m} \tag{4}$$

N denotes the number of data set (users), C denotes the number of clusters, x_i is the ith data of set S and m represents the degree of fuzziness of the function and it belongs to $m \in [1 - \infty]$.

4.2 Utility

Three utility functions are defined to verify the performance of our work, in steady state (when available channels are more than the required channels) Eq. (5) is used to calculate the utility and if the system is not in steady state (when the required channels are more than available channels) Eq. (6) is used. System performs the FCM clustering when queued users exceed from the specified threshold and the utility is calculated by Eq. (7).

$$u_1 = 1 + \sum\nolimits_{j=1}^{length(hl)} hl - (\text{cp} \ + \text{hc} \ ) \quad if(ch > 0) \tag{5}$$

$$u_2 = \frac{1 + \sum\nolimits_{j=1}^{length(hl)} hl - (\text{cp} \ + \text{hc} \ )}{1 + \sum\nolimits_{j=1}^{length(tl)} tl} \quad if(ch < 1) \tag{6}$$

$$u_3 = \frac{1 + \sum\nolimits_{j=1}^{length(tlc)} \text{tlc} - (hc)}{1 + \sum\nolimits_{j=1}^{length(blu)} \text{blu} \ * (cp)} \quad if(clustering) \tag{7}$$

In Eqs. (5) and (6), hl, cp and hc denote the total handover load, the computational cost of FCM clustering algorithm and handover cost respectively, tl is the total blocked when there are no available channels. In Eq. (7), tlc and blu represent the total handover load and a load of total blocked users (while having FCM clustering) respectively.

4.3 FCM Based LEO Satellite Handover

1. Users are coming by poison process and the average call duration time is 180 s.
2. User can determine that if it needs handover or not by checking its elevation angle (given in Eq. 1).
3. If the elevation angle is less than the specified threshold then user checks if there is a free channel in an upcoming satellite.
4. If there is a free channel in upcoming satellite, then user will send a channel reservation request to upcoming satellite.
5. Otherwise, the user will wait in queue. "The length of the queue is unlimited but there is a specific waiting time. If the waiting time of user in queue is more than the specified threshold then user is terminated".
6. In the case of users waiting in queue exceeds from the maximum queued users threshold: Then

   - System checks if the clustering time of previously formed clusters is expired. If so, the FCM clustering is performed on waiting users. Membership function u_{ij} and the cluster center function C_j are gradually updated to get the optimal value of objective function j_m. FCM functions are given in Eqs. (2), (3) and (4) to calculate the optimal objective values, memberships and cluster centers respectively:
   - Cluster heads are selected from each cluster by Eq. (4).
   - Cluster heads determine their member users by checking their received signal strength.
   - Cluster heads relay the traffic of their members to the satellite.

7. Finally, each cluster head reserves one channel in upcoming satellite and performs handover with its members.

The satellites diversity is shown in Fig. 3 in case of FCM clustering. We can see, the users in common area (between two satellites) are divided into clusters based on their locations. The cluster heads are chosen from each cluster and they select their members by checking their received signal strength.

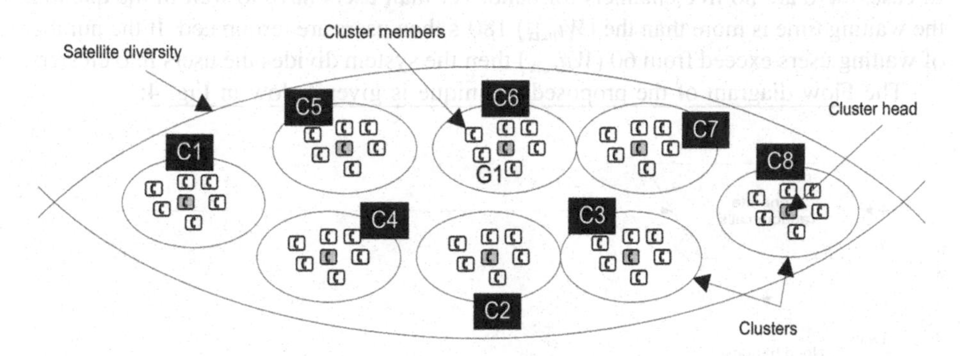

Fig. 3. Clustering in satellite diversity

5 Simulation and Results

The major contribution of our work is; to handover more users using less channels and to handover the users in clusters unlike the individual handovers during busy times. In order to evaluate the performance of proposed technique, we simulate and compare results of MMCK based and FCM clustering based handover techniques. In our simulation, the handover users are coming by poison process (handover time can be determined by checking the elevation angle [9] between satellite and user). Parameters used in our simulation are given in Table 1.

Table 1. Simulation parameters

Handover cost	hc	0.0025
Computational cost	cp	0.2254
Maximum communication time of cluster	cc_t	180 s
Arrival rat	λ	1.0
Average call duration	Cd	180 s
Channels	C	20
Reserved channels for clustering	r_{ch}	5.0
Maximum waiting time in queue	Wt_{max}	180 s
Waiting users threshold for clustering	Wu_{max}	60.0
Total simulated users	U_t	17734
Received signal strength	th_{rss}	0.3131

We assume that each satellite have 20 channels (C) for normal handover and 5 reserved channels (r_{ch}) for clustering handover. Handover cost (hc) and computational cost (cp) coefficients are used to define the computation cost of given techniques. The maximum communication time of the cluster (cc_t) defines that, after this time the previous clusters are terminated and new clusters can be formed for new queued users. In case, there are no free channels for handover than users have to wait in the queue if the waiting time is more than the (Wt_{max}) 180 s then users are terminated. If the number of waiting users exceed from 60 (Wu_{max}) then the system divides the users into clusters.

The Flow diagram of the proposed technique is given below in Fig. 4:

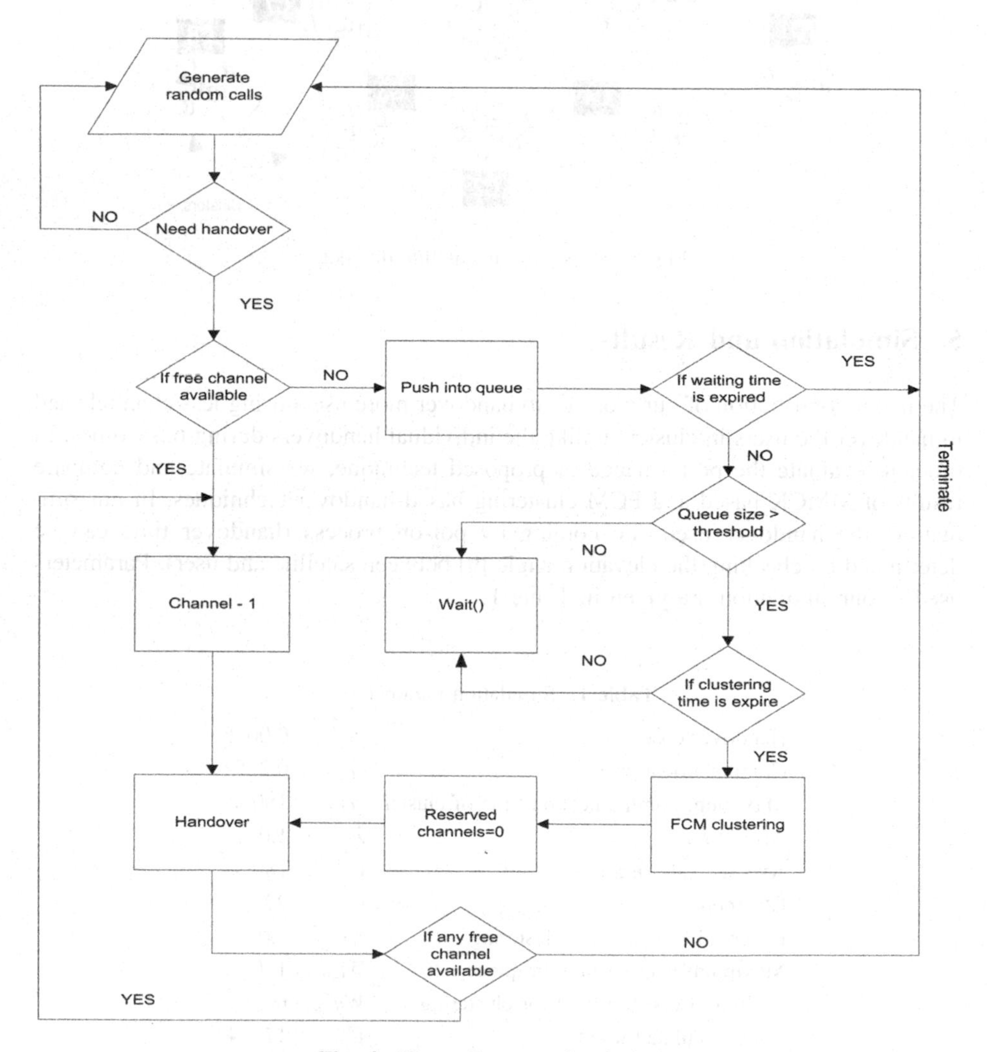

Fig. 4. Flow of proposed technique

The total numbers of simulated users are 11734. Here we scale the received signal index between 0 and 1, and the threshold (th_{rss}) for getting a member of the cluster is 0.3131, by changing the threshold value the selection criteria of cluster members can be changed. Figure 5 shows that the waiting users in queue are less by using FCM technique as compared to MMCK based handover technique, indicating that less queued users, less handover delay. As shown in Fig. 6 the numbers of successful handovers are significantly increased by using FCM clustering handover technique.

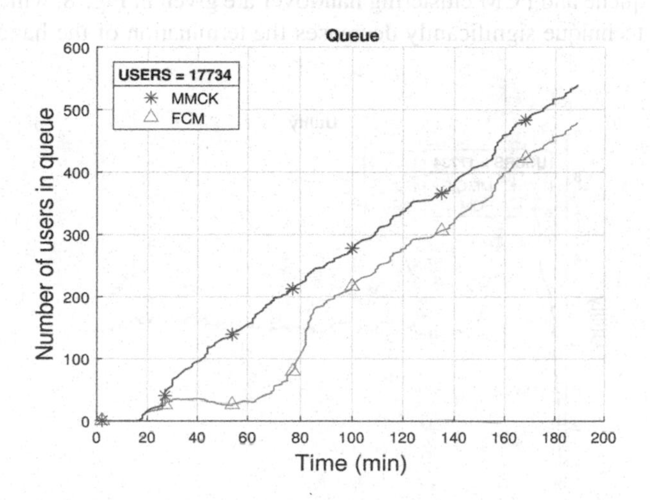

Fig. 5. The number of users waiting in queue for handover

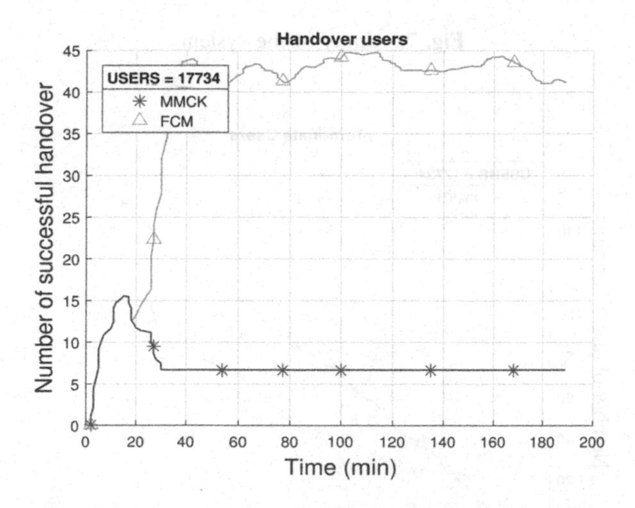

Fig. 6. Number of handover while using FCM + MMCK and MMCK techniques

According to Fig. 6, if the rate of handover requests exceeds from specific amount; then the success rate of the proposed technique become constant between 40 and 45 while the success rate of MMCK based handover remains at 7.

The utility of the whole network is calculated by using three utility functions given in Eqs. (5), (6) and (7) for both MMCK based handover and the FCM cluster-based handover. The utility of the system is given in Fig. 7. If channels >0 the utility is calculated by Eq. (5), if channels <1 then the utility is calculated by Eq. (6) and in clustering mood, utility is calculated by Eq. (7). The number of terminated users using both MMCK queue and FCM clustering handover are given in Fig. 8, which shows that the proposed technique significantly decreases the termination of the handover users.

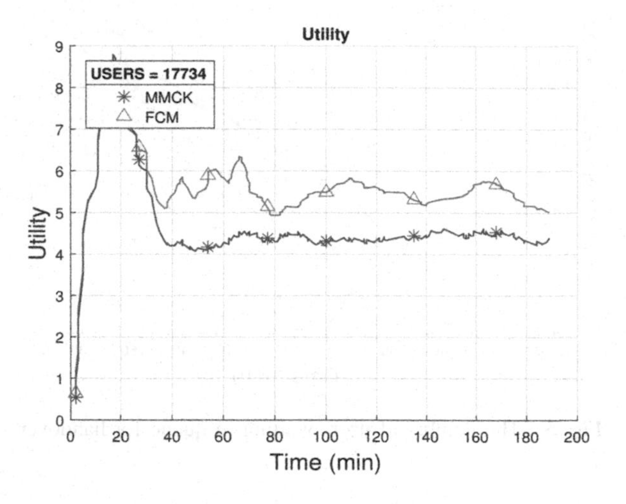

Fig. 7. Utility of the system

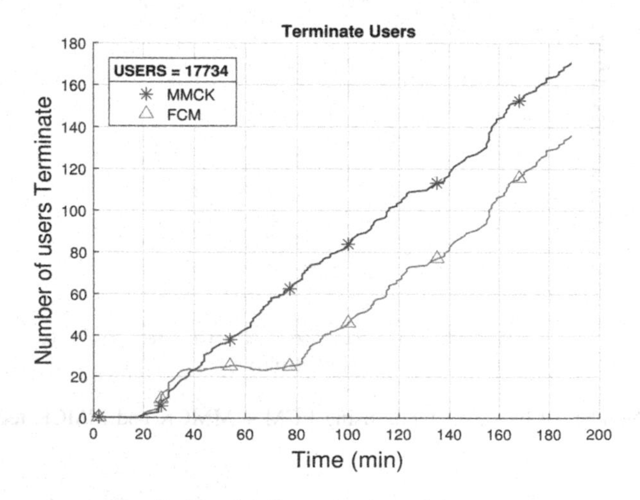

Fig. 8. Total call termination of the system

According to the simulation results, it is concluded that the FCM clustering handover technique always outperforms then MMCK based handover technique. Moreover, it increases the number of successful handovers; network utility while reduces the handover failure and number of queued users.

6 Conclusions

In this paper, we have studied the issue of satellite handover in LEO satellite networks when the required resources are less than available resources and proposed a "Fuzzy C-mean Clustering Based LEO Satellite Handover" technique to resolve this issue. Based on assumption that every user needs a channel to continue its ongoing call, we consider scenario that if multiple users need handover at the same time when the numbers of required channels are more than the available channels, then users will wait in a queue, if the numbers of waiting users in the queue are more than the specified threshold, then the FCM function will perform the clustering and divide the waiting users into the cluster. After clustering, the system will select cluster heads from each group, and the cluster head will act as a relay between member users and the satellite. Finally, we carried out the simulation work using MATLAB and evaluate the performance of FCM clustering based handover technique with respect to call termination, number of handovers, number of waiting users in queue and utility. Through the simulation results, we conclude that FCM clustering based handover technique provides less termination, better utility and less queued users than the MMCK queue based handover technique.

7 Future Work

In future, we will extend our work by adding the iridium satellite mobility models and we will use software-defined network (SDN) to make the handover more efficient. Using SDN we will get the current traffic load information of the satellites and will select the satellite accordingly to avoid the congestion in the network.

Acknowledgements. This work was partially supported by the international cooperative project (20180414024GH) of Jilin Province of China and the Development Program of Science and Technology of Jilin Province (20180519012JH). In the end, authors would like to thank the reviewers who helped to improve the quality of this paper.

References

1. Rathore, M.M.U., et al.: Real-time big data analytical. architecture for remote sensing application. IEEE J. Sel. Top. Appl. Earth Obs. Remote Sens. **8**(10), 4610–4621 (2015)
2. Earthobservatory. http://earthobservatory.nasa.gov/Features/OrbitsCatalog/
3. Zhao, W., Tafazolli, R., Evans, B.G.: Combined handover algorithm for dynamic satellite constellations. Electron. Lett. **32**(7), 622–624 (1996)

4. Boukhatem, L., et al.: TCRA: a time-based channel reservation scheme for handover requests in LEO satellite systems. Int. J. Satell. Commun. Netw. **21**(3), 227–240 (2003)
5. Wang, X., Wang, X.: The research of channel reservation strategy in LEO satellite network. In: 2013 IEEE 11th International Conference on Dependable, Autonomic and Secure Computing (DASC). IEEE (2013)
6. Papapetrou, E., et al.: Satellite handover techniques for LEO networks. Int. J. Satell. Commun. Netw. **22**(2), 231–245 (2004)
7. Papapetrou, E., et al.: Satellite Handover Techniques in LEO Systems for Multimedia Services
8. Chowdhury, P.K., Atiquzzaman, M., Ivancic, W.: Handover schemes in satellite networks: state-of-the-art and future research directions. IEEE Commun. Surv. Tutor. **8**(4), 2–14 (2006)
9. Wu, Z., et al.: A simple real-time handover management in the mobile satellite communication networks. In: 17th Asia-Pacific Network Operations and Management Symposium (APNOMS) 2015. IEEE (2015)
10. Adan, I., Resing, J.: Queueing systems. Department of Mathematics and Computing Science. Eindhoven University of Technology (2015)
11. Dhaou, R., et al.: Optimized handover and resource management: an 802.21-based scheme to optimize handover and resource management in hybrid satellite-terrestrial networks. Int. J. Satell. Commun. Netw. **32**(1), 1–23 (2014)
12. Papapetrou, E., Stathopoulou, E., Pavlidou, F.-N.: Supporting QoS over handovers in LEO satellite systems. In: Mobile & Wireless Telecommunications Summit 2002, 17–19 June 2002, Thessaloniki – Greece (2002)
13. Wu, Z., et al.: A graph-based satellite handover framework for LEO satellite communication networks. IEEE Commun. Lett. **20**(8), 1547–1550 (2016)
14. Pan, C., et al.: Leo satellite communication system handover technology and channel allocation strategy. Int. J. Innov. Comput. Inf. Control **9**(11), 4595–4602 (2013)
15. Yang, B., et al.: Seamless handover in software-defined satellite networking. IEEE Commun. Lett. **20**(9), 1768–1771 (2016)
16. Chowdhury, M.Z., Sung, H.C., Yeong, M.J.: Group handover management in mobile femtocellular network deployment. In: 2012 Fourth International Conference on Ubiquitous and Future Networks (ICUFN). IEEE (2012)
17. Torra, V.: On the selection of m for Fuzzy c-Means. In: IFSA-EUSFLAT (2015)
18. Cannon, R.L., Dave, J.V., Bezdek, J.C.: Efficient implementation of the fuzzy c-means clustering algorithms. IEEE Trans. Pattern Anal. Mach. Intell. **2**, 248–255 (1986)
19. Elena, M.: Fuzzy C means clustering in matlab. In: The 7th International Days of Statistics and Economics, Prague, pp. 905–914 (2013)

An Improved *Apriori* Algorithm Based on Matrix and Double Correlation Profit Constraint

Yuan Liu(✉), Ya Li, Jian Yang, Yan Ren, Guoqiang Sun, and Quansheng Li

Yellow River Engineering Consulting Co., Ltd., Zhengzhou 450000, China
liuyhhuc@163.com, 248123833@qq.com, 12776573@qq.com,
renhyan@163.com, yrccsgq@163.com, lqsyrcc@163.com

Abstract. In view of the performance bottlenecks of Apriori algorithm association rules, this paper puts forward a new type of profit constraint extensional association rules mining algorithm named *MDC_Apriori*. The algorithm constructs a 0-1 transaction matrix through scanning the database once, considering establishing the AE arrays to weight each column in matrix and calculating the completely-weighted support for frequent itemsets. Frequent itemsets and negative itemsets are pruned by the minimum support and the strategy of double correlation, finally using the correlation coefficient to mine the profit constraint extensional association rules and determine the strength degree of the correlation. Experiments show that the method has a good effect of pruning, effectively reduce the number of generated redundant frequent itemsets and association rules, meanwhile improving the operational efficiency of the algorithm.

Keywords: Constraint association rules
MDC_apriori algorithm double correlation · Weighted support

1 Introduction

Data mining is the process of mining useful information from huge amounts of data. Agrawal et al. first put forward the concept of association rules, it is a key branch of data mining [1]. *Apriori* algorithm is the most classic no weighted correlation pattern mining algorithm in the field of association rules algorithm. And the interaction between users and the system simply sets the minimum support threshold and the minimum confidence threshold to wait for the mining output, finally the users get a large number of redundant association rules. On the basis of the algorithm, scholars from different angles and methods have done a lot of improvement work in the field of constraint association rules, such as Narmadha and others put forward a new pruning strategy to improve the efficiency of algorithm. Jean- Francois put forward the concept of 'negative border' [2–5], monotone constraint in combined with the anti-monotone constraint at the same time, proved that using the anti-monotone constraint in the *Apriori* algorithm is effective, demonstrating the *free* set is a type of the anti-monotone constraint. Pei et al. put forwards a mining algorithm of frequent itemsets (FICM, FICA) [6], whose basic principle is to sort the itemsets in a certain order, satisfies the constraint anti-monotonicity or monotonicity. But these improved algorithms still have some deficiencies in the generation of

Q. Zhou et al. (Eds.): ICPCSEE 2018, CCIS 901, pp. 359–370, 2018.
https://doi.org/10.1007/978-981-13-2203-7_27

candidate itemsets, calculation of the support of candidate itemsets and pruning of itemsets. At the same time, there is no consideration for the importance in different items and the condition of different weight of items in the transaction database, resulting in a large number of invalid, redundant and uninteresting association patterns. Furthermore, in a large number of transaction data, there may be more mutexes that are not found, which we called negative association rules [7]. Due to the algorithm usually focus in the study of the mining of mutually reinforcing information, lack of the negative ones, but negative association rules not only can use the contradiction between the positive and negative association rules to remove invalid association rules, but also can dig up new negative association rules, so the negative association rules take an important significance for mining association rules [7–9]. And the traditional algorithms don't think about the value of the mined rules, we refer to the traditional association rules and the negative association rules that are mutually exclusive as extended association rules. In order to solve above problems, this paper sets a new calculation method of completely-weighted support and pruning strategy, combining with the factor of correlation coefficient at the same time, build a new type profit constraint extended association rule mining *MDC_Apriori* algorithm, take the item weight change depends on the transaction situation into consideration, prune for the frequent and negative itemsets by setting the double correlation standards for digging out the interesting weighted frequent and negative itemsets, then taking a simple calculation of itemsets weight, excavated the effective profit constraint type extension association rules. The experiments prove that it is more suitable for large databases with small support, so as to realize effective mining and improve the efficiency of association rules.

2 The Analysis of the *Apriori* Algorithm

Apriori algorithm is a classic algorithm in the field of association rules mining, its main idea is based on the breadth-first search strategy, using the priori principle to make low-dimensional frequent itemsets to high-dimensional frequent itemsets by the way of iteration step by step. Pruning the candidate itemsets C_i by the *minimum support* threshold, then rotativing low-order frequent itemsets step by step, thus dugging up all the frequent itemsets.

First step scanning database, take out all the items to constitute candidate *1*-itemsets C_1, after scanning the database again for calculating each item support in the candidate *1*-itemsets C_i, compared the candidate *1*-itemsets support $sup(C_1)$ with the *minimum support* threshold, then generating the frequent *1*-itemsets F_1. Second combinating F_1 in pairs to generate the candidate *2*-itemsets C_2, scanning the database, calculating the support of candidate *2*-itemsets, and compared the support with *minimum support* threshold, generating the frequent *2*-itemsets. Repeating the above steps again and again, until $(k-1)$-itemsets just only one which can't generate *k*-candidate itemsets. It is necessary to scan the database when computing support C_i to generate F_i.

3 The Improved Algorithm Based on the Matrix and the Double Correlation Profit Constraints

In traditional *Apriori* algorithm, due to frequently scanning database, bring about the big cost of algorithm and heavy I/O burden, at the same time large amount of calculation of generating *k*-itemsets will also seriously reduce the efficiency of *Apriori* algorithm. If users can guide the mining process really, give weight to the database in the items to reflect the importance of the different items, which only excavating what we want and filtering the meaningless rules for users, it would make the results more close to the actual situation, so as to improve the execution efficiency of the algorithm greatly. In general, we call the data whose weight changes with the transaction record as the completely-weighted data [12, 13]. The so-called users' guidance based on the weight is profit constraint [14]. Extensional association rules include positive and negative association rules, and the distribution range of positive and negative itemsets support is also different. If there is only a single correlation, algorithm will miss a lot of negative itemsets who has a low correlation degree when a relatively high correlation threshold happened, and will produce a lot of redundant frequent itemsets when the correlation of threshold has been set too low. In order to reduce the I/O overhead cost and reflect the importance of the different items and transactions, digging up more valuable rules, we propose a *MDC_Apriori* algorithm based on the matrix and a profit constraint algorithm combining the double correlation degree threshold of frequent and negative itemsets in this paper.

3.1 The Related Concept of the Algorithm

Definition 1.

(1) Assuming the transaction database *D* contains *n* transactions and *m* transaction items, we can construct a *n-row* and *m-column* 0-1 matrix as follows:

$$M = \begin{bmatrix} a_{11} & a_{12} & \ldots & a_{1m} \\ a_{21} & a_{22} & \ldots & a_{2m} \\ \ldots & \ldots & \ldots & \ldots \\ a_{n1} & a_{n2} & \ldots & a_{nm} \end{bmatrix}$$

In the formula, $a_{ij} = \begin{cases} 1, & a_{ij} \in T_i \\ 0, & a_{ij} \notin T_i \end{cases}$ $(i = 1, 2, \ldots, n,\ j = 1, 2, \ldots, m)$, The rows of matrix *M* above are corresponded with the transactions in database, and the columns are corresponded with the items which included in the transaction.

(2) Defines the vector of *k*-itemsets $\{I_1, I_2 \ldots I_k\}$ as $M_{12\ldots k} = M_1 \wedge M_2 \wedge \ldots \wedge M_k = (M_1 \wedge M_2 \wedge \ldots \wedge M_{k-1}) \wedge M_k$, so the support _count of the *k*-itemsets $\{I_1, I_2 \ldots I_k\}$ is $\sup\_count\{I_1, I_2 \ldots I_k\} = \sum_{k=1}^{n} \{(M_{n1} \wedge M_{n2} \wedge \ldots \wedge M_{n(k-1)}) \wedge M_{nk}\}$.

Definition 2. Take AND operation bit-by-bit for the elements in the same line in any column of the matrix M, the so-called k- itemsets support is the value of the number of "1" in the calculation result divided by the total number of itemsets.

Definition 3. $I = \{I_1, I_2, \ldots, I_m\}$ express the itemsets in the transaction database D, and m expresses the number of the items. The occurred frequency of I_j in the transaction database is $P(I_j)$, whose calculating as formula (1), and $w(I_j)$ represents the weight of I_j, whose calculating as formula (2).

$$P(I_j) = l/n \tag{1}$$

$$w(I_j) = 1/P(I_j) \tag{2}$$

In above formula, l expresses the number of I_j occurred in the transaction sets, which is the totally number of 1 in j_{st} column and n expresses the number of transaction records.

Definition 4. Transaction T_k expresses the k_{st} record in the transaction data sets. Whose weight named $wt(T_k)$, refers to the average weight of the items contained in the transactions, means averaging for all of $w(I_j)$ when $a_{kj} = 1$, among them, $j = 1, 2, \ldots, m$, and calculate as formula (3).

$$wt(T_k) = \sum_{j=1}^{n \in T_k, I_j \in T_k} w(I_j)/|T_k| \tag{3}$$

In above formula, $|T_k|$ expresses the number of items contained in the transaction T_k.

Definition 5. Denotes the completely-weighted support of items as $cwS(I)$, the proportion of the weight of weighted itemsets I in the total weighted transaction database as the calculation method of the support of itemsets I, and calculate as formula (4).

$$cwS(I) = 1/(n \times m) \times \sum_{i=1}^{n} \sum_{i_j \in I} w[r_i][i_j] = \sum_{k=1}^{k=n} wt(T_k)/(n \times m) \tag{4}$$

Definition 6. We called itemsets I completely-weighted frequent itemsets when existing a completely-weighted itemsets I and $cwS(I) \geq cs$, and we called itemsets (I_1, I_2) completely-weighted negative itemsets when itemset I_1 and I_2 both are completely-weighted itemsets and $cwS(I_1, I_2) < cs$, among them, cs is the minimum support threshold.

Definition 7. If existing a completely-weighted frequent itemsets $FI = (i_1, i_2, \ldots, i_m)(m > 1)$ and its sub-itemsets are $\{I_1, I_2, \ldots, I_q | I_1 \subset FI, I_2 \subset FI, \ldots, I_q \subset FI, q > 1\}$, we call the frequent itemsets FI occured conditional probability as the association degree of FI when the largest support subsets of the completely-weighted frequent itemsets occurred. The computational formula of association degree *cwFIR*(*FI*) between subsets of the completely-weighted frequent itemsets *FI* as formula (5).

$$cwFIR(FI) = \frac{cwS(FI)}{\max\{cwS(I_k)|\forall k \in \{1, 2, 3, \ldots, q\}|\}} \tag{5}$$

Definition 8. If existing a completely-weighted negative itemsets $NI = \{i_1, i_2, \ldots, i_r\}(r > 1)$ and its sub-itemsets are $\left\{I_1, I_2, \ldots I_p \middle| I_1 \subset NI, I_2 \subset NI, \ldots, I_p \subset NI, p > 1\right\}$, we call the negative itemsets *NI* occured conditional probability as the association degree of *NI* when the largest support subsets of the completely-weighted negative itemsets not occurred. The computational formula of association degree *cwNIR*(*NI*) between subsets of the completely-weighted negative itemsets *NI* as formula (6).

$$cwNIR(NI) = \frac{cwS(NI)}{1 - \max\{cwS(I_k)|\forall k \in \{1, 2, 3, \ldots, p\}\}} \tag{6}$$

3.2 Mining Steps of the *MDC_Apriori* Algorithm

This algorithm is proposed by scanning the database once to get the frequent itemsets L_k, structured a matrix and then taking AND operation bit-by-bit to obtain the support of frequent itemsets, remove redundant column vectors by means of compression matrix to generate frequent itemsets. At the same time, pruning the generated completely-weighted frequent itemsets and negative itemsets by setting the minimum correlation threshold and the *min_support* threshold, delete the itemsets with weak connection strength and not close enough relationship between itemsets. The first step is to scan the completely-weighted database *D* and construct the 0-1 matrix, and set up an AE array to store the number of columns that the latter columns which are completely identical to the current column. Scan the matrix only once and calculate the support count of all transactions. If the support count of the items is less than the minimum support count, delete the row vector. The results are multiplied by the corresponding weight values of each column in the AE array, and the column vectors whose value is less than the minimum support count should be deleted. The completely-weighted support is also should be calculated. Based on comparing completely-weighted support with the *minimum support* threshold, calculating the correlation degree of completely-weighted frequent itemsets and completely-weighted negative itemsets. Respectively compared with the positive and negative correlation threshold again, excavated completely-weighted frequent *1*-itemsets and completely- weighted *1*-negative itemsets. Then starting from 2-itemsets, linking the *k*-itemsets frequent itemsets $L_{k-1}(k \geq 2)$ by *Apriori* to obtain completely-weighted candidate *k*-itemsets, then calculating the weight of each transaction and item, and completely-weighted support at the same time. Digging out the completely-weighted frequent *k*-itemsets and negative *k*-itemsets whose correlation strong enough, until emptied algorithm to jump to the association rules generation phase.

Achieving frequent itemsets and negative itemsets through the double correlation pruning strategy. And the specific pruning methods of double correlation shown as below:

$cwS(I) \geq cs$:	if $cwFIR(FI) \geq cFr$, achieve an effective completely-weighted frequent itemsets FI.
	if $cwFIR(FI) < cFr$, delete the itemsets FI.
$cwS(I) < cs$:	if $cwNIR(NI) \geq cNr$, achieve an effective completely-weighted negative itemsets NI.
	if $cwNIR(NI) < cNr$, delete the itemsets NI.

In order to describe the algorithm whose itemsets pruning strategy based on minimum support and double correlation more vividly, following give the flow chart to describe the process of generating frequent itemsets and negative itemsets from the completely-weighted database (see Fig. 1).

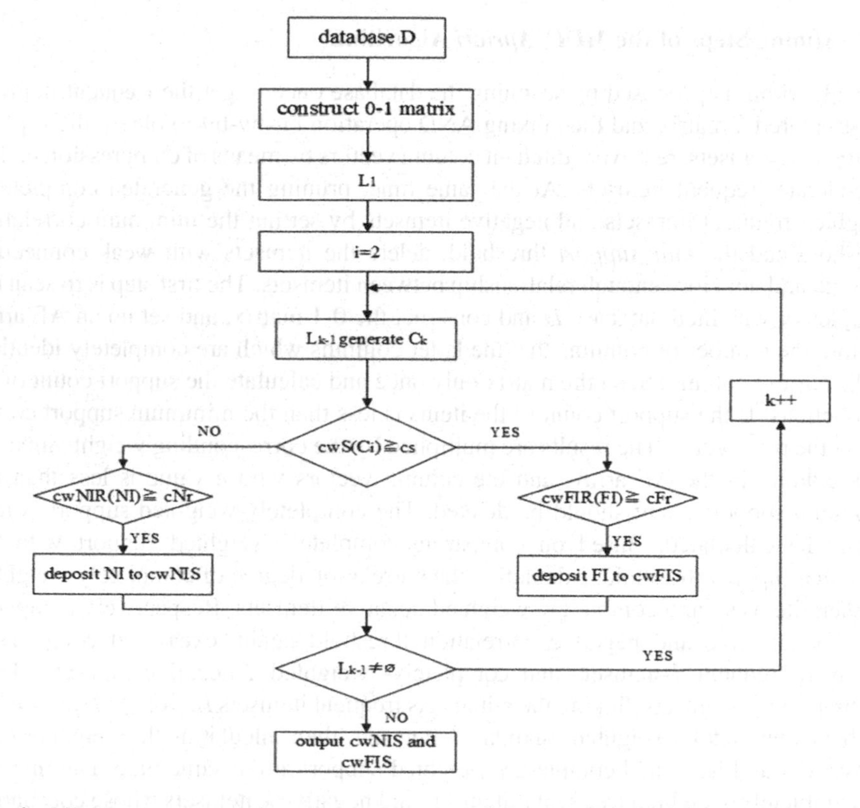

Fig. 1. Mining process meeting the requirements of the correlation of frequent itemsets and negative itemsets

3.3 The Association Rules Generation of *MDC_Apriori* Algorithm

The algorithm uses correlation coefficient instead of the traditional correlation, not only can distinguish the difference between the positive and negative association rules, but also can measure the related degree of association rules. The calculation formula of correlation coefficient $cwPCC(I_1, I_2)$ of the completely-weighted itemsets (I_1, I_2) as formula (8). Among them, $cwS(*) > 0$, $cwS(*) \neq 1$.

$$cwPCC(I_1, I_2) = \frac{cwS(I_1, I_2) - cwS(I_1) \times cwS(I_2)}{\sqrt{cwS(I_1) \times cwS(I_2) \times (1 - cwS(I_1))(1 - cwS(I_2))}} \tag{7}$$

In the field of completely-weighted data, on the basis of the nature of the traditional correlation coefficient, we can deduce the correlation coefficient $cwPCC(I_1, I_2)$ with the following properties:

Property 1. When $cwPCC(I_1, I_2) > 0$, completely-weighted items I_1 and I_2 were positively correlated; I_1 and $\neg I_2$ are negatively correlated; $\neg I_1$ and I_2 are negatively correlated; $\neg I_1$ and $\neg I_2$ were positively correlated.

Property 2. When $cwPCC(I_1, I_2) < 0$, completely-weighted items I_1 and I_2 are negatively correlated; I_1 and $\neg I_2$ are positively correlated; $\neg I_1$ and I_2 are positively correlated; $\neg I_1$ and $\neg I_2$ were negatively correlated.

Proof:

$$\because cwS(*) > 0,\ cwS(*) \neq 1 \Rightarrow \sqrt{cwS(I_1)cwS(I_2)(1 - cwS(I_1))(1 - cwS(I_2))} > 0 \tag{8}$$

$$cwPCC(I_1, I_2) > 0 \text{ 和式 } (9) \Rightarrow cwS(I_1, I_2) - cwS(I_1)cwS(I_2) > 0$$

$$\Rightarrow cwS(I_1, I_2) > cwS(I_1)cwS(I_2) \Rightarrow \frac{cwS(I_1, I_2)}{cwS(I_1)cwS(I_2)} > 1 \tag{9}$$

It can be seen from (9) that the completely-weighted itemsets are positively correlated. The proof of the other propositions in Properties 1 and 2 are similar to the above process and no longer to describe. We can inferred from the above properties that the algorithm can mine completely-weighted positively association rule $I_1 \Rightarrow I_2$ and negative completely-weighted association rule $\neg I_1 \Rightarrow \neg I_2$ when $cwPCC(I_1, I_2) > 0$, and mine completely-weighted negative association rule $I_1 \Rightarrow \neg I_2$ and $\neg I_1 \Rightarrow I_2$ when $cwPCC(I_1, I_2) < 0$.

3.4 Instance for *MDC_Apriori* Algorithm

Take transaction data in Table 1 as an example, the data table has *8-rows* and *10-columns*, assuming the minimum support is 0.25.

Table 1. Data of transactions.

TID	Transaction
1	A, B, C, F, G, H
2	A, E, F, G, H
3	A, B, C, E, H, I, J
4	A, B, C, F, G, H, J
5	B, C, D, I
6	C, E
7	D
8	B, D

① The process of mining by *MDC_Apriori* algorithm is as follows:

Structure a 0-1 transaction matrix M with *8-rows* and *8-columns* according to the transaction database as follows:

$$AE = \begin{bmatrix} 2 & 1 & 1 & 1 & 1 & 2 & 1 & 1 \end{bmatrix}$$

$$M = \begin{bmatrix} 1 & 1 & 1 & 0 & 0 & 1 & 0 & 0 \\ 1 & 0 & 0 & 0 & 1 & 1 & 0 & 0 \\ 1 & 1 & 1 & 0 & 1 & 0 & 1 & 1 \\ 1 & 1 & 1 & 0 & 0 & 1 & 0 & 1 \\ 0 & 1 & 1 & 1 & 0 & 0 & 1 & 0 \\ 0 & 0 & 1 & 0 & 1 & 0 & 0 & 0 \\ 0 & 0 & 0 & 1 & 0 & 0 & 0 & 0 \\ 0 & 1 & 0 & 1 & 0 & 0 & 0 & 0 \end{bmatrix}$$

② According to the formulas (1) and (2), P(A) = 1/2, w(A) = 2.00, and calculate the weight of other items as shown in Table 2 in the same way.

Table 2. The weight of items

I_i	$W(I_i)$
A	2.00
B	1.60
C	1.60
D	2.67
E	2.67
F	2.67
G	4.00
H	4.00

③ According to the formula (3), take the first record in the database as an example, the transaction contains A, B, C, F, $wt(T_1) = (2.00 + 1.60 + 1.60 + 2.67)/4 = 1.97$, and we can calculate the weight of other affairs in the same way as Table 3.

Table 3. Weight of transactions

TID	$wt(T_k)$
1	1.97
2	2.45
3	2.65
4	2.37
5	2.47
6	2.14
7	2.67
8	2.67

Compute the completely-weighted support according to the formula (4) is $(1.97 + 2.45 + 2.65 + 2.37 + 2.47 + 2.14 + 2.67 + 2.67)/8 \times 8 = 0.303$.

④ Due to the calculated completely-weighted support is greater than the minimum support threshold, take an AND operation bit-by-bit of each row in the transaction matrix. After deleting the row vectors less than the minimum support, the matrix with frequent 2-itemsets is shown as follows.

$$AE = \begin{bmatrix} 2 & 1 & 1 & 1 & 1 & 2 & 1 & 1 \end{bmatrix}$$

$$M_2 = \begin{bmatrix} 1 & 0 & 0 & 0 & 0 & 1 & 0 & 0 \\ 1 & 1 & 1 & 0 & 0 & 0 & 0 & 0 \\ 1 & 1 & 1 & 0 & 0 & 1 & 0 & 0 \\ 0 & 1 & 1 & 0 & 0 & 0 & 0 & 0 \\ 1 & 0 & 0 & 0 & 1 & 0 & 0 & 0 \\ 1 & 0 & 0 & 0 & 0 & 1 & 0 & 0 \\ 1 & 1 & 1 & 0 & 0 & 0 & 0 & 1 \\ 0 & 1 & 1 & 0 & 0 & 0 & 1 & 0 \\ 0 & 0 & 1 & 0 & 1 & 0 & 0 & 0 \\ 0 & 1 & 1 & 0 & 0 & 0 & 0 & 0 \\ 0 & 1 & 0 & 1 & 0 & 0 & 0 & 0 \end{bmatrix}$$

If the sum of the columns is less than the minimum support count, delete the column. According to the actual situation to calculate the correlation degree of completely-weighted frequent itemsets, compared with the threshold set in advance, to dig up the completely-weighted frequent itemsets.

4 The Experiment and the Analysis of Algorithms

In order to verify the effectiveness of the improved algorithm *MDC_Apriori*, select the representative completely-weighted extensional association rule mining algorithm *(WPNARM-SCCI)* [6] as the comparison algorithm, analyzing the number of mined positive and negative frequent itemsets and the running time of the algorithms. PC (Intel

Core i5, 2.67 GHz CPU, memory, 2 GB) as the Experimental platform, using SQL Server 2008 database and Java language, test data sets selected from UCI public Car Evaluation data set. The data set with a total of 1728 data attribute and each attribute has 3–4 value. In order to better experiment, this paper take different value of each expanded data instance as a transaction, formed a data set contains 19 columns to deposit in the database.

4.1 Compare the Number of Generated Frequent Itemsets

The number of generated positive and negative frequent itemsets reflect the merits of the double correlation standard. By setting the minimum correlation thresholds *cFr* of positive itemsets and minimum correlation thresholds *cNr* of negative itemsets, analyzing the numbers of graduated positive and negative frequent itemsets and the decreasing amplitude compared with the *WPNARM_SCCI* algorithm to contrast the pruning performance. Setting *cs* = *0.027*, the experimental results as follows in Figs. 2 and 3.

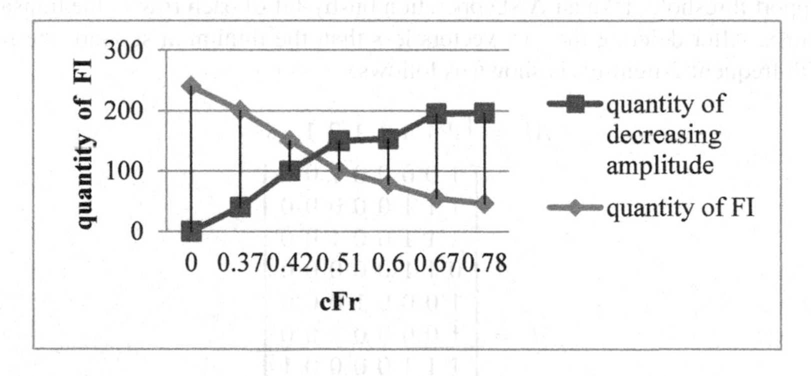

Fig. 2. The number of generated frequent itemsets when *cFr* changing

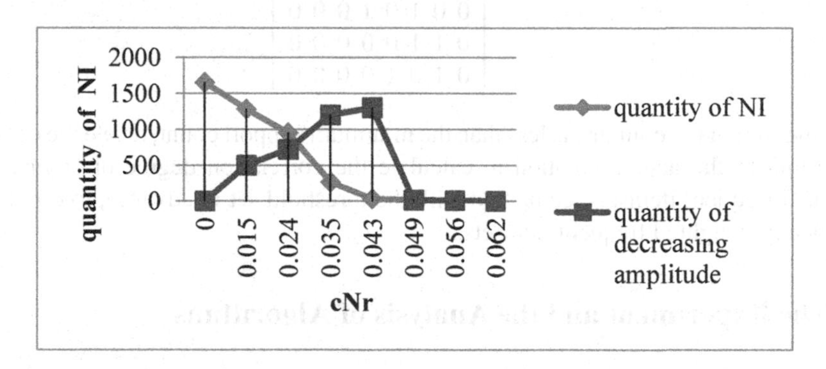

Fig. 3. The number of generated negative itemsets when *cNr* changing

From Figs. 2 and 3, we can see that when the minimum correlation threshold *cFr* of frequent itemsets and *cNr* of negative itemsets increasing, the number of generated frequent itemsets and negative itemsets of *MDC_Apriori* algorithm are gradually reduced, certifying the effect of pruning is obvious. Especially, when *cNr = 0.043* it has the biggest decreasing amplitude compared with algorithm *(WPNARM-SCCI)* also reflect the advantage of *MDC_Apriori* algorithm.

4.2 Comparison of Algorithm Mining Time

The running time of algorithm reflects the running efficiency, the shorter time reflects the higher computational efficiency. The experimental results are as follows:

From Fig. 4, we can get the mining time of *MDC_Apriori* algorithm is less than *WPNARM-SCCI* algorithm obviously when the *minimum support* especially small. With the increase of *minimum support*, the execution time of the two algorithm both are declining on the whole, performance is approaching.

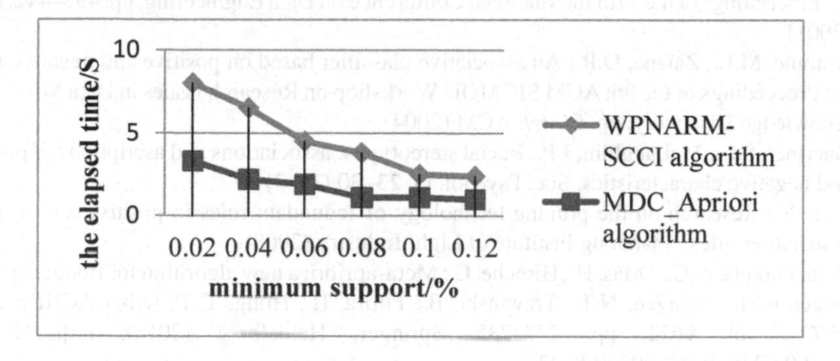

Fig. 4. Comparison of algorithm mining time

5 Conclusion

This paper aimed at optimizating the lack of association rule mining algorithm currently, proposed an improved algorithm *MDC_Apriori*, and put forward to take correlation coefficient to determine the relevance of the generated association rules, just scanning the transaction database once and constructing a 0-1 matrix, considering setting up an AE array for each column weight of the matrix to obtain the frequent itemsets. Through calculating the weighted support of transaction and the completely-weighted support of itemsets, and through the pruning strategy based on double correlation, excavated the positive and negative association rules with low redundancy profit constraints. Through theoretical analysis and experiments the proposed *MDC_Apriori* algorithm has a better effect of pruning, the number of generated positive and negative itemsets meeting the correlation requirements and mining time of association rules is significantly reduced, greatly reduces the number of ineffective and boring frequent itemsets and association rules. It has a good scalability meanwhile.

References

1. Han, J., Kamber, M., Pei, J.: Data Mining: Concepts and Techniques. Elsevier, New York City (2011)
2. Jeudy, B., Boulicaut, J.F.: Optimization of association rule mining queries. Intell. Data Anal. **6**(4), 341–357 (2002)
3. Lakshmanan, L.V.S., Ng, R., Han, J., et al.: Optimization of constrained frequent set queries with 2-variable constraints. In: ACM SIGMOD Record, vol. 28, no. 2, pp. 157–168. ACM (1999)
4. Mannila, H., Toivonen, H.: Levelwise search and borders of theories in knowledge discovery. Data Min. Knowl. Discov. **1**(3), 241–258 (1997)
5. Srikant, R., Vu, Q., Agrawal, R.: Mining association rules with item constraints. In: KDD, vol. 97, p. 67 (1997)
6. Pei, J., Han, J., Lakshmanan, L.V.S.: Mining frequent itemsets with convertible constraints. In: Proceedings of the 17th International Conference on Data Engineering, pp. 433–442. IEEE (2001)
7. Antonie, M.L., Zaïane, O.R.: An associative classifier based on positive and negative rules. In: Proceedings of the 9th ACM SIGMOD Workshop on Research Issues in Data Mining and Knowledge Discovery, pp. 64–69. ACM (2004)
8. Gaertner, S.L., McLaughlin, J.P.: Racial stereotypes: associations and ascriptions of positive and negative characteristics. Soc. Psychol. Q. 23–30 (1983)
9. Yue, Y.: Research on the pruning technology of redundant rules in positive and negative association rules. Shandong Institute of Light Industry (2008)
10. Benhamouda, N.C., Drias, H., Hirèche, C.: Meta-apriori: a new algorithm for frequent pattern detection. In: Nguyen, N.T., Trawiński, B., Fujita, H., Hong, T.-P. (eds.) ACIIDS 2016. LNCS, vol. 9622, pp. 277–285. Springer, Heidelberg (2016). https://doi.org/10.1007/978-3-662-49390-8_27
11. Zaki, M.J.: Scalable algorithms for association mining. IEEE Trans. Knowl. Data Eng. **12**(3), 372–390 (2000)
12. Cai, C.H., Fu, A.W.C., Cheng, C.H., et al.: Mining association rules with weighted items. In: Proceedings of the IDEAS 1998 International Database Engineering and Applications Symposium, pp. 68–77. IEEE (1998)
13. Pears, R., Koh, Y.S.: Weighted association rule mining using particle swarm optimization. In: Cao, L., Huang, J.Z., Bailey, J., Koh, Y.S., Luo, J. (eds.) PAKDD 2011. LNCS, vol. 7104, pp. 327–338. Springer, Heidelberg (2012). https://doi.org/10.1007/978-3-642-28320-8_28
14. Zhu, Y., Sun, Z., Zhang, Z.: An effective constraint association rule mining algorithm. Comput. Eng. **28**(2), 29–31 (2002)
15. Huang, M., Huang, F., Yan, X.W., et al.: Weighted positive and negative association rules mining based on changing of item weight and SCCI framework. Control Decis. (10), 1729–1741 (2015)

Mining and Ranking Important Nodes in Complex Network by K-Shell and Degree Difference

Jianpei Zhang[1], Hui Xu[1,2](✉), Jing Yang[1], and Lijun Lun[3]

[1] College of Computer Science and Technology, Harbin Engineering University, Harbin, China
hzytsg2009@163.com

[2] Library, Heilongjiang University of Chinese Medicine, Harbin, China

[3] College of Computer Science and Information Engineering, Harbin Normal University, Harbin, China

Abstract. Identifying important nodes in complex networks can help us effectively design protection strategies, improve the security and protection capabilities of network hub nodes, and enhance the network survivability and structural stability. In view of nodes partition being too coarse by the k-shell decomposition method, this paper proposes a new index named k-shell and degree difference, which considers the network node location, the local characteristics of the node and its neighbors and the impact of multi-level nodes on it. In this paper, the network efficiency index is used to quantify the impact of the node removal on the network structure and function, and the destruction-resistance experiment is carried out in four actual networks. Experimental results show that the method proposed in this paper is more accurately to assess the importance of nodes than other four methods.

Keywords: Complex network · Important nodes · Network efficiency
K-shell and degree difference

1 Introduction

Research of node important measurement is an important branch of the study of the structural characteristics of complex networks [1], and study for robustness and vulnerability has important theoretical significance and wide practical application of complex networks [2, 3]. Identifying the important nodes accurately can improve the reliability and survivability of the whole network by protecting them, such as effectively inhibiting the spread of virus [4], effectively avoiding cascading failures of electric power network [5], and restraining the spread of gossip in the society [6]. Common algorithms for evaluating the importance of complex network nodes are degree centrality [6], betweenness centrality [7], closeness centrality [8] and Pagerank [9] and so on. Most of these methods can effectively measure the importance of nodes, but with the further research of complex networks, more and more researchers realize that node important assessment algorithm that integrates more information can more reasonably evaluate nodes in complex networks. However, the fusion of more information means to select more information sources, and the algorithm has a higher complexity, which reduces the

Q. Zhou et al. (Eds.): ICPCSEE 2018, CCIS 901, pp. 371–381, 2018.
https://doi.org/10.1007/978-981-13-2203-7_28

practicability of the algorithm. Therefore, the study of the node importance of complex networks needs to find the relative equilibrium between the effectiveness and the complexity of the algorithm. In complex network research, especially in real network research, it takes a long time to get the overall structural characteristics of complex networks. In some cases, we cannot obtain the overall structural characteristics of complex networks in time. Therefore, the localization of the overall network structure and the lack of some network structures restrict the timeliness of our research and exploration of complex networks. For example, in the incompletely connected network, the betweenness centrality of the complex network nodes is basically invalid, that is, it is impossible to evaluate the importance of nodes in complex networks by using betweenness centrality and the shortest path based on the shortest path between nodes. Therefore, researchers need to study and explore the structural characteristics of complex networks through local information under some special constraints.

This paper proposes a new node important evaluation index named k-shell and degree difference, which fully considers the local characteristics of nodes and their neighbors, and also considers their network locations. The invulnerability experiments were performed on four real networks. The experimental results show that the method proposed in this paper more accurately assess the importance of the node, compared with the k-shell decomposition method, the extended neighborhood coreness, the extended gravity method and the semi-local centrality method. The remainder of this paper is structured as follows. In Sect. 2, we discuss some related work. We briefly review several typical centrality indices for subsequent comparative analysis, and then propose and describe our method in Sect. 3. In Sect. 4, we conduct related experiments and evaluate the experiment results. In Sect. 5, we summarize the full text and look forward to future research directions.

2 Related Works

With the migration of time, the structure of large-scale complex networks tends to change. It has limitations in obtaining network global information for evaluating the importance of nodes. Researchers had more and more attention to taking advantage of the local characteristics to mine and rank important nodes.

Ruan et al. [10] propose a node important evaluation algorithm that defines the similarity between nodes by quantifying the coincidence degree of the node's local network topology. Kitsak et al. [12] proposed the k-shell decomposition method (labeled as k-shell) to determine the location of the core nodes in the network by iterative pruning. The k-shell decomposition method has a low computational complexity and is widely used in the mining and analysis of the core nodes of various complex networks. However, the ranking results are too coarse-grained, and the difference of node importance is not very significant. Liu et al. [13, 14] found that because of the existence of small groups with very close local connection in the network, the core found by the k-shell decomposition is the false core. Based on the definition of information entropy, they put forward the connection entropy and discuss the diversity of different k-shell layers. Shan et al. [15] proposed a ranking algorithm based on neighborhood coreness, which overcomes the defects of the non-comprehensive ranking process and the high algorithm complexity. Lalou et al. [16] summarized recent advances and results obtained from the critical node detection problem and proved new complexity results and induce some solving algorithms through relationships established between different variants. Liu et al. [17] proposed a weight degree centrality which measures the spreading ability of nodes based on their degree and their ability of spreading out using a tuning weight parameter. Adebayo et al. [18] proposed the network response to structural characteristics theory participation factor and the degree centrality based on the network response to structure characteristic indices can be used for voltage stability assessment and identification of important nodes.

Chen et al. [11] used a complex network locality to measure the structural characteristics of complex networks and proposed a semi-local centrality index. The index limitedly expanded the coverage of the node's domain and has a good balance between algorithm accuracy and time complexity (labeled as SL). Bae and Kim [19] evaluated the importance of nodes in the network by considering the k-shell values of the neighborhood. This method is highly effective, but it requires global information about the network and may be infeasible in some networks such as scale-free networks (labeled as C_{nc+}). Ma et al. [20] proposed an important node identification method based on the gravity formula. The method uses the k-shell value of each node as its mass and the shortest distance between two nodes as the gravity center of their distance (labeled as G_+). Motivated by authors in these previous studies, we propose k-shell and degree difference index in view of the nodes' local attributes and compare our approach with the k-shell decomposition, SL, C_{nc+} and G_+.

3 Methods

Assume that an undirected network G = (N, M) consists of N nodes and M edges. Here we briefly review the definitions of four comparative indices in Table 1.

Table 1. List of comparative indices in the experiment.

Indices	Descriptions	Comments
k-shell	The degree of any node i must satisfy $k_i \geq ks$	1. k_i is the degree of node i 2. $\Gamma(i)$ is the nearest neighbors of node i 3. ks_i is the k-shell value of node i 4. d_{ij} is the shortest path distance between node i and node j 5. $\Psi(i)$ is the number of the nearest neighbor of node i
SL	$Q(s) = \sum_{j \in \Gamma(s)} \Psi(j)$ $SL(i) = \sum_{s \in \Gamma(i)} Q(s)$	
C_{nc+}	$C_{nc}(i) = \sum_{j \in \Gamma(i)} ks_j$ $C_{nc+}(i) = \sum_{j \in \Gamma(i)} C_{nc}(j)$	
G_+	$G(i) = \sum_{j \in \Gamma(i)} ks_i ks_j / d_{ij}^2,$ $G_+(i) = \sum_{j \in \Gamma(i)} G(j)$	

Each above algorithm has a good effect in identifying important nodes in some networks, but there are still some limitations. These algorithms either only consider the important nodes through computing nodes' network locations, or determine the importance of the nodes by calculating the degree of the nodes and their neighboring nodes, ignoring the correlation between them. Based on this, we propose a novel node importance evaluation index named k-shell and degree difference (labeled as KSD index), which takes into account the location of the node in the network and the diversity of outward links, and the impact of the multi-level nodes on this node, including the k-shell values and the shortest distances to it. KSD index is defined as follows:

$$KSD_i^{(n)} = \sum_{j \in \Gamma_i^{(n)}} \frac{\left|(ks_i - k_i) * (ks_j - k_j)\right|}{(d_{ij})^2} \tag{1}$$

$$KSD_i = \sum_{n \in \{1,2 \ldots m\}} KSD_i^{(1)} + KSD_i^{(2)} + \cdots + KSD_i^{(n)} \tag{2}$$

where k_i is the degree of node i, ks_i is the k-shell value of node i, and d_{ij} is the shortest distance between node i and node j. $\Gamma_i^{(1)}$ represents the set of node i's neighbor nodes with the shortest distance of being one, which is called the first-order neighbor set of node i. $\Gamma_i^{(2)}$ represents the set of node i's neighbor nodes with the shortest distance being two, which is called the second-order neighbor set of node i. And so on, $\Gamma_i^{(n)}$ represents the set of node i's neighbor nodes with the shortest distance being n, which is called the n-order neighbor set of node i. In this method, the value of m is between one and the average path length. The greater the KSD value is, the more important the node is.

As showed in Fig. 1, we calculate the first-order KSD index of the four nodes in the kernel as an example. Node A's KSD value is twelve, node B's KSD value is zero, node C and node D's KSD values are both four. It is not difficult to see that Node B is in the same core as the other three nodes, but node B lacks outbound links, that resulting from its importance in a significant reduction.

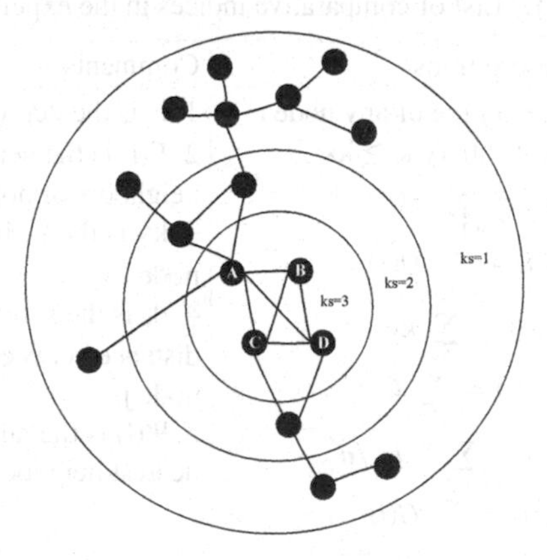

Fig. 1. K-shell decomposition.

4 Experimental Studies

In this section, we compare KSD index with the k-shell decomposition, SL, C_{nc+} and G_+ in four real networks to verify the validity of KSD index.

4.1 Evaluation Criteria

This paper evaluates the ranking results based on the robustness of the network and uses the network efficiency to quantify the impact on the network structure and function to evaluate the structural importance of the nodes after removing the nodes. Network efficiency [21, 22] is used to evaluate the connectivity of the network. The worse the connectivity of the network is, the lower the network efficiency is. Removing the nodes and all their corresponding edges in the network will cause the paths between them to be interrupted and make the network connectivity worse. The network efficiency is expressed as:

$$\varphi = \frac{1}{N(N-1)} \sum_{i \neq j \in G} \varphi_{ij} \tag{3}$$

where $\varphi_{ij} = 1/d_{ij}$, the value of φ is between zero and one. If φ is one, it indicates that the network connectivity is best. If φ is zero, it indicates that the network is consisted of isolated nodes. φ is normalized to its possible largest value N(N – 1), for totally connected graph having N(N – 1)/2 edges. In this paper, we will delete some nodes at a certain proportion in order to carry out the deliberate attack simulation experiment. The descending ratio of network efficiency is used to quantitatively describe the importance of each node before and after the network attack. The descending ratio of the network efficiency is expressed as $f = 1 - \varphi/\varphi_0$. Assuming that φ_0 is the initial network efficiency without being subjected to network attack. Then suffering attacks make the network efficiency become φ. f is in the range between zero and one. When f equals one, it indicates the network efficiency is down to zero after attacks. Namely the network is composed of isolated nodes. When f equals zero, it means that deleting nodes don't affect the network efficiency after attacks. In Eq. (3), it can be seen that the greater f is, the worse the network connectivity gets after attacks and the more accurate this method measure nodes importance.

4.2 Datasets

Considering that different types of complex networks represent different characteristics of network topology, we chose four real networks data sets to analyze and compare, such as the dolphin social network [23], American college football network [24], books about US politics and neural network [25–27]. Table 2 lists the structural properties of these four networks.

Table 2. Structural properties studied in this work. N and E are the number of nodes and edges, respectively, <k> is the average degree, <d> is average path length and φ_0 is network efficiency of the initial network.

Network	N	E	<k>	<d>	φ_0
Dolphin social network	62	159	5.1290	3.2504	0.3792
Books about US politics	105	441	8.4000	3.0207	0.3971
American college football	115	613	10.6609	2.4649	0.4504
Neural network	297	2359	14.4646	2.4388	0.4448

4.3 Simulation Experiments

Complex networks with scale-free characteristic are highly resistant to random attacks, but they are vulnerable when they are deliberately attacked. The failure of the top 5%–10% important nodes can cripple the entire network [28]. We implement the five indices of k-shell, C_{nc+}, G_+, SL and KSD to four networks to rank important nodes. Table 3 lists the ranking results of the top 10% important nodes in each network.

Table 3. The ranking results of the top 10% important nodes.

Networks	Indices	Ranking results
Dolphin social network	k-shell	7, 10, 11, 17, 19, 22
	SL	37, 41, 19, 8, 29, 11
	C_{nc+}	37, 8, 9, 1, 41, 53
	G_+	15, 46, 21, 38, 41, 34
	KSD	15, 38, 46, 34, 52, 30
Books about US politics	k-shell	3, 9, 10, 11, 12, 13, 14, 15, 21, 24
	SL	59, 10, 20, 54, 13, 9, 31, 8, 73, 5
	Cnc+	59, 73, 10, 32, 5, 20, 21, 13, 52, 31
	G+	9, 13, 73, 85, 31, 67, 74, 48, 41, 10
	KSD	13, 9, 85, 4, 74, 73, 31, 67, 12, 75
American college football	k-shell	1, 2, 3, 4, 5, 6, 7, 8, 9, 10, 11
	SL	107, 89, 14, 26, 59, 25, 81, 17, 64, 115, 111
	C_{nc+}	78, 89, 107, 25, 52, 44, 8, 26, 69, 14, 3
	G_+	1, 2, 3, 4, 6, 7, 8, 16, 54, 68, 89
	KSD	89, 16, 7, 54, 4, 8, 68, 105, 3, 1, 2
Neural network	k-shell	2, 3, 4, 5, 7, 8, 9, 10, 13, 14, 15, 17, 32, 43, 45, 48, 49, 60, 64, 68, 73, 74, 75, 76, 78, 85, 87, 88, 91
	SL	61, 113, 30, 36, 37, 47, 76, 24, 49, 74, 41, 44, 25, 62, 35, 210, 18, 46, 51, 57, 211, 78, 197, 65, 105, 251, 48, 75, 215
	C_{nc+}	61, 76, 49, 74, 113, 36, 37, 47, 78, 44, 35, 24, 30, 210, 105, 211, 75, 18, 25, 41, 65, 45, 48, 57, 62, 215, 46, 197, 198
	G_+	45, 13, 3, 87, 85, 5, 119, 4, 118, 126, 173, 138, 49, 143, 74, 7, 192, 48, 205, 76, 208, 75, 115, 131, 196, 78, 2, 14, 133
	KSD	45, 13, 3, 85, 87, 173, 126, 5, 119, 4, 118, 138, 143, 205, 120, 99, 131, 208, 115, 192, 188, 7, 227, 196, 133, 178, 14, 112, 2

If each evaluation index has higher resolution, the differences between nodes can be more easier distinguished. Monotonicity M of ranking list L [19] is used to quantitatively measure the resolution of each index by selecting a certain percentage of the top important nodes.

$$M(L) = \left[1 - \frac{\sum_{u \in I} N_u(N_u - 1)}{N_e(N_e - 1)} \right]^2 \tag{4}$$

Where N_e is the number of selecting nodes at a certain proportion e, and N_u is the number of nodes that have the same index value u. If M(L) = 1, this means that the ranking method is completely monotonic, and the values of each node are different. Otherwise, all nodes are of the same rank when M(L) = 0. Table 4 lists the M values of SL, C_{nc+}, G_+ and KSD, respectively, when e is approximately equal to 25%. M values of k-shell are all zero in the four networks, which indicate that the k-shell index is too coarse-grained to distinguish important nodes. From Table 4, it is easy to see that the M values of the KSD index are relatively higher than the M values of the other three indices, indicating that the KSD index can better distinguish the differences between nodes.

Table 4. The monotonicity of these four indices.

Network	M(SL)	M(C_{nc+})	M(G_+)	M(KSD)
Dolphin social network	0.9623	0.9253	1.0000	1.0000
Books about US politics	0.9877	0.9634	1.0000	1.0000
American college football	0.9894	0.8234	0.9121	0.9789
Neural network	0.9633	1.0000	0.9985	1.0000

In order to further verify the effectiveness of the KSD index for identifying important nodes, we select the top 25% important nodes ranked by k-shell, SL, C_{nc+}, G_+ and KSD, and perform the simulation contrasting experiments respectively. The experiment results are shown in Fig. 2(a)–(d). If the selected nodes have the same value such as k-shell, we carry out N random experiments, which N represents the number of nodes that are repeated. In Fig. 2(a), the two curves of KSD and G_+ are close to each other and act better than the other three indices. There is a sudden rise in k-shell and C_{nc+} when p = 17.74%, that maybe they identify top important nodes. In Fig. 2(b), KSD and G_+ act better than the other three indices. When some nodes are deleted at p = 24.76%, the f of KSD and G_+ are 86.02% and 86.00% respectively. K-shell and C performances are relatively poor in this network. In Fig. 2(c), the five curves are close to each other, indicating that the five indices have relatively consistent recognition capabilities in the network. In Fig. 2(d), the fluctuation of the SL curve is relatively large and C_{nc+} performs worse. The reason for these phenomena is that they only consider a single attribute, so they do not apply to various types of network structures to mine important nodes. Figure 2(a)–(d) show that in different scale and structural properties of the networks, KSD is more stable and more accurate than other four indices.

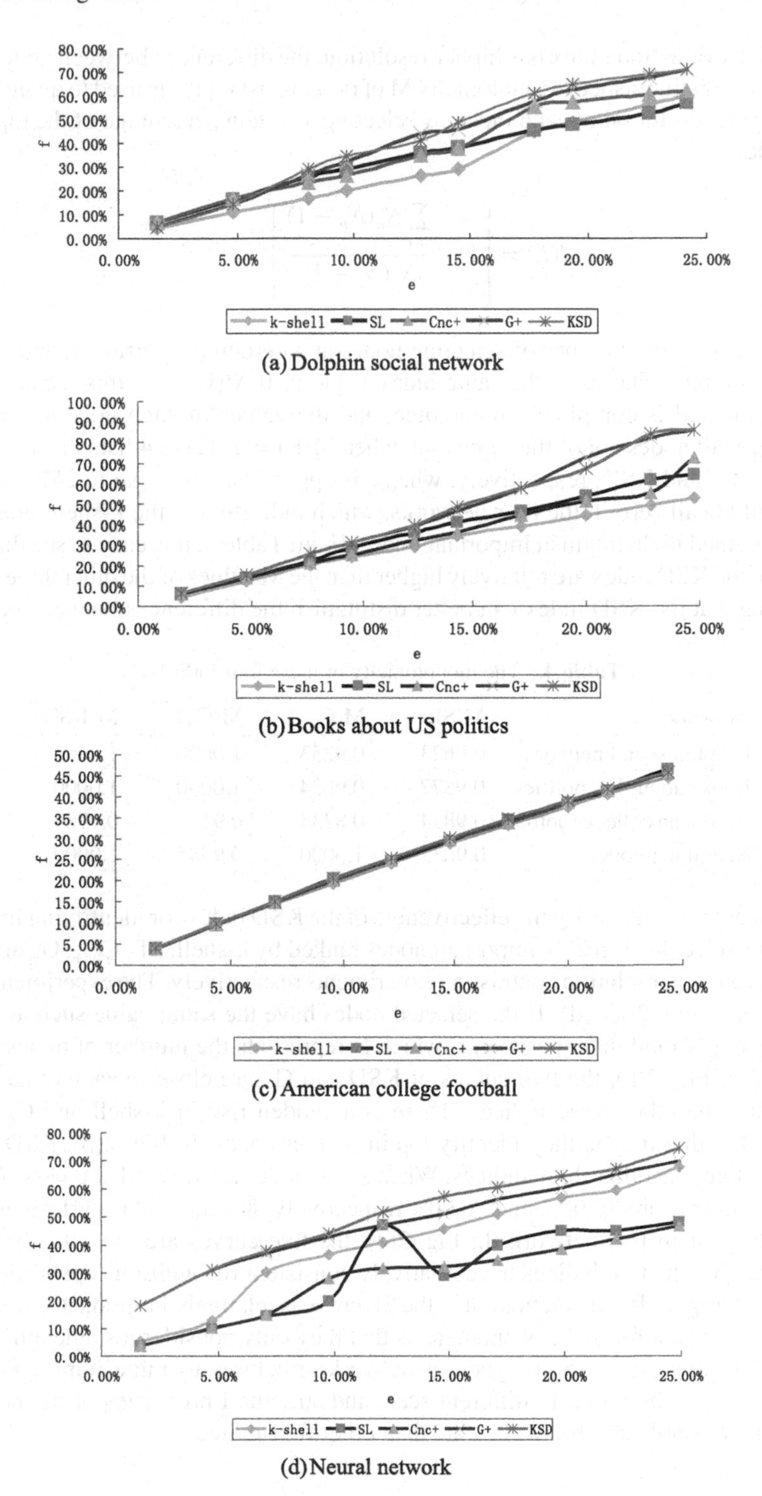

Fig. 2. (a)–(d) The f after removing nodes at different e.

Table 5 shows the average increasing ratio of f by KSD compared with G_+, C_{nc+}, SL and k-shell after removing the top 25% important nodes. In Dolphin social network, the average ability of identifying the important nodes from strong to weak is KSD, G_+, C_{nc+}, SL and k-shell. In Books about US politics network, the average ability to identify important nodes from strong to weak is KSD, G_+, SL, C_{nc+} and k-shell. In American college football network, the average ability to identify important nodes from strong to weak is KSD, SL, G_+, C_{nc+} and k-shell. In this network, it is not difficult to see that the effect of these five indices to identify important nodes is similar, as showed in Fig. 2(c). In Neural network, the average ability to identify important nodes from strong to weak is KSD, G_+, k-shell, C_{nc+} and SL. In this network, C_{nc+} and SL are extremely unstable in the performance of identifying important nodes in these networks. It is easy to see that KSD has better performance to mining and ranking the important nodes.

Table 5. Average increasing ratios of f by KSD compared to k-shell, C_{nc+}, G_+ and SL.

Networks	Compared to k-shell/%	Compared to C_{nc+}/%	Compared to G_+/%	Compared to SL/%
Dolphin social network	42.28	11.55	2.45	16.52
Books about US politics	42.53	25.03	0.19	17.84
American college football	3.63	2.37	1.08	0.81
Neural network	52.55	120.82	5.04	121.13

5 Conclusions

Accurately identifying important nodes in a complex network is playing an important role in improving the robustness of the actual system and designing efficient system architecture. The method proposed in this paper involves not only the local characteristics of nodes and neighbors, but also their network locations. It makes up for the defects of the coarse-grained partition of k-shell decomposition method in theory, and has practical significance in describing the survivability and structural reliability of large-scale networks. It is proved that the KSD method is better than k-shell, C_{nc+}, G_+ and SL by comparing the experimental results of the M and f in the four real networks. In this paper, the importance of nodes in a single-layer network is analyzed from the point of view of structure. Next, it will focus on the study of the importance of nodes based on the dynamic characteristics and the network structure.

Acknowledgments. The work was supported by the National Natural Science Foundation of China under Grant Nos. 61370083, 61672179, 61402126, the Heilongjiang Province Natural Science Foundation of China under Grant No. F2015030, the Youth Science Foundation of Heilongjiang Province of China under Grant No. QC2016083, and the Postdoctoral Support of Heilongjiang Province of China under Grant No. LBH-Z14071.

References

1. Opsahl, T., Agneessens, F., Skvoretz, J.: Node centrality in weighted networks: generalizing degree and shortest paths. Soc. Netw. **32**(3), 245–251 (2010)
2. Ren, X.L., Lü, L.Y.: Review of ranking nodes in complex networks. Chin. Sci. Bull. (Chin. Ver.) **59**(59), 1175–1197 (2014). (in Chinese)
3. Ren, Z.M.: Analysis of the spreading influence of the nodes with minimum K-shell value in complex networks. Acta Phys. Sin. **62**(10), 956–959 (2013). (in Chinese)
4. Balthrop, J., Forrest, S., Newman, M.E., et al.: Computer science. Technological networks and the spread of computer viruses. Science **304**(5670), 527–529 (2004)
5. Kinney, R., Crucitti, P., Albert, R., et al.: Modeling cascading failures in the North American power grid. Eur. Phys. J. B – Condens. Matter Complex Syst. **46**(1), 101–107 (2005)
6. Moreno, Y., Nekovee, M., Pacheco, A.F.: Dynamics of rumor spreading in complex networks. Phys. Rev. E Stat. Nonlinear Soft Matter Phys. **69**(6 Pt 2), 066130 (2004)
7. Sabidussi, G.: The centrality index of a graph. Psychometrika **31**(4), 581–603 (1966)
8. Goh, K.I., Oh, E., Kahng, B., et al.: Betweenness centrality correlation in social networks. Phys. Rev. E Stat. Nonlinear Soft Matter Phys. **67**(1 Pt 2), 017101 (2016)
9. Brin, S., Page, L.: Anatomy of a large-scale hypertextual web search engine. J. Comput. Netw. ISDN Syst. **30**, 107–117 (1998)
10. Ruan, Y.R., Lao, S.Y., Wang, J.D., et al.: Node importance measurement based on neighborhood similarity in complex network. Acta Phys. Sin. **66**(3), 038902 (2017). (in Chinese)
11. Chen, D., Lü, L., Shang, M.S., et al.: Identifying influential nodes in complex networks. Phys. A Stat. Mech. Appl. **391**(4), 1777–1787 (2012)
12. Kitsak, M., Gallos, L.K., Havlin, S., et al.: Identifying influential spreaders in complex networks. Nat. Phys. **6**(11), 888–893 (2010)
13. Liu, Y., Tang, M., Zhou, T., et al.: Core-like groups result in invalidation of identifying super-spreader by k-shell decomposition. Sci. Rep. **5**, 9602 (2015)
14. Liu, Y., Tang, M., Zhou, T., et al.: Improving the accuracy of the k-shell method by removing redundant links: from a perspective of spreading dynamics. Sci. Rep. **5**, 13172 (2015)
15. Shan, B., Tao, F.: Design change control of complex products based on important nodes. Comput. Eng. Appl. **54**(6), 222–227 (2018)
16. Lalou, M., Tahraoui, M.A., Kheddouci, H.: The critical node detection problem in networks: a survey. Comput. Sci. Rev. **28**, 92–117 (2018)
17. Liu, Y., Wei, B., Du, Y., et al.: Identifying influential spreaders by weight degree centrality in complex networks. Chaos Solitons Fractals Interdiscip. J. Nonlinear Sci. Nonequilib. Complex Phenom. **86**, 1–7 (2016)
18. Adebayo, I., Jimoh, A.A., Yusuff, A.: Voltage stability assessment and identification of important nodes in power transmission network through network response structural characteristics. IET Gener. Trans. Distrib. **11**(6), 1398–1408 (2017)
19. Bae, J., Kim, S.: Identifying and ranking influential spreaders in complex networks by neighborhood coreness. Phys. A Stat. Mech. Appl. **395**(4), 549–559 (2014)
20. Ma, L.L., Ma, C., Zhang, H.F., et al.: Identifying influential spreaders in complex networks based on gravity formula. Phys. A Stat. Mech. Appl. **451**, 205–212 (2016)
21. Vragović, I., Louis, E., Díaz-Guilera, A.: Efficiency of informational transfer in regular and complex networks. Phys. Rev. E **71**(3 Pt 2A), 036122 (2005)
22. Latora, V., Marchiori, M.: A measure of centrality based on network efficiency. New J. Phys. **9**(6), 188 (2007)

23. Lusseau, D., Schneider, K., Boisseau, O.J., et al.: The bottlenose dolphin community of Doubtful Sound features a large proportion of long-lasting associations. Behav. Ecol. Sociobiol. **54**(4), 396–405 (2003)
24. Girvan, M., Newman, M.E.J.: Community structure in social and biological networks. Proc. Natl. Acad. Sci. U. S. A. **99**(12), 7821–7826 (2001)
25. Watts, D.J., Strogatz, S.H.: Collective dynamics of 'small-world' networks. Nature **393**(6684), 440–442 (1998)
26. White, J.G., Southgate, E., Thomson, J.N., et al.: The structure of the nervous system of the nematode Caenorhabditis elegans. Philos. Trans. Roy. Soc. B Biol. Sci. **314**(1165), 1–340 (1986)
27. Zhou, T., Lü, L., Zhang, Y.C.: Predicting missing links via local information. Eur. Phys. J. B **71**(4), 623–630 (2009)
28. Lai, Y.C., Motter, A.E., Nishikawa, T.: Attacks and cascades in complex networks. In: Ben-Naim, E., Frauenfelder, H., Toroczkai, Z. (eds.) Complex Networks. LNP, vol. 650, pp. 299–310. Springer, Heidelberg (2004). https://doi.org/10.1007/978-3-540-44485-5_14

Representation Learning for Knowledge Graph with Dynamic Step

Yongfang Li, Liang Chang(✉), Guanjun Rao, Phatpicha Yochum, Yiqin Luo, and Tianlong Gu

Guangxi Key Laboratory of Trusted Software, Guilin University of Electronic Technology, Guilin, China
changl@guet.edu.cn

Abstract. Representation learning aims to represent the entities and relations in a knowledge graph as dense, low-dimensional and real-valued vectors by machine learning. The translation-based model is a typical representation learning method and has shown good predictive performance in large-scale knowledge graph. However, when modeling complex relations such as 1-N, N-1 and N-N, these models are not very effective. To solve the limitation of traditional learning model in modeling complex relations, a representation learning method based on dynamic step is proposed. Defining a dynamic step according to the different types of relations can significantly improve the efficiency of learning. The algorithm is used to solve the problem of single optimization goal, and the experimental results show that the dynamic step method can mainly improve the performance in the link prediction task.

Keywords: Representation learning · Knowledge graph · Dynamic step Link prediction

1 Introduction

With the development of the times, there is more abundant information, and how to extract effective information from mass data has become the focus of attention. Knowledge graph is an ideal technique for extracting structured knowledge from a large amount of texts and images. We can focus and conduct semantic search, an intelligent question and answer and document understanding through the knowledge graph. At the same time, because the knowledge graph usually contains tens of thousands of nodes and edges, any of their reasoning and calculation may not be easy.

For instance, when conduct experiments on the link prediction task on Freebase, we need to handle 68 million nodes and billions of edges. Furthermore, knowledge graphs are generally represented by symbols and logic, which is insufficient for intensive numerical calculations. In recent years, representation learning [1] has emerged in the field of artificial intelligence, which attracted the attention of many researchers. Representation learning aims to represent the entities and relations in a knowledge graph as dense, low-dimensional and real-valued vectors by machine learning, which is to achieve distributed representation of entities and relations (Distributed Representation).

Q. Zhou et al. (Eds.): ICPCSEE 2018, CCIS 901, pp. 382–393, 2018.
https://doi.org/10.1007/978-981-13-2203-7_29

Not only this method can improve the computational efficiency significantly, alleviate data sparseness effectively, but also achieve heterogeneous information fusion [2].

Knowledge graph was introduced by Google Inc. in May 2012 for improving the quality of search results, which launched the research project and application of large-scale knowledge graphs. Different from traditional search engine, knowledge graph can effectively find the complex associated information, understand the user's intention from the semantic meaning, and improve the query quality. For instance, if you enter Durant in Google's search box, Durant's related information will appear on the right side of the page, such as date of birth, family background, life experiences, careers, technical characteristics, etc.

Knowledge representation is an important foundation for knowledge acquisition and application, so knowledge representation learning has become the focus on the whole process of building and applying knowledge graph. As a new method of knowledge representation, knowledge graph belongs to a part of the semantic web which is a data structure used to store knowledge. The purpose of knowledge graph is to describe the various entities and concepts that exist in the real world, as well as the association between these entities and concepts, to capture and present the semantic relationships between domain concepts. Knowledge graph represents knowledge in the form of triple (entity 1, relationship, entity 2). Entities represent real world objects, such as, things, places, people and abstract concepts (genres, religions, professions). Relations represent a graph-based data model where relationships are first-class. At present, this representation is used in all common knowledge graphs, such as the Resource Description Framework (RDF) technical standard published by the World Wide Web (W3C).

As knowledge graphs (KGs) still have some problems, such as incomplete facts or errors, we need to reason about completely the missing knowledge in the reality [3]. Knowledge graph completion has been proposed to find missing relation of knowledge graph.

The knowledge graph completion task can be divided into two sub-tasks: -(1) entity prediction and (2) relationship prediction. The entity prediction task takes a partial triplet (h, r, ?) or (?, r, t) as input (where "?" represents t or h) and produces a ranked list of candidate entities as output. Other task, the relationship prediction aims to find a ranked list of relationships that connect a head-entity with a tail-entity <h, ?, t>. In this paper, we focus on the entity prediction task.

A distributed representation of KG representation learning is mainly applied in tasks, such as similarity calculation, KG completion and other applications, such as relation extraction, automatic question and answer and entity link. It means that representation learning technology can efficiently calculate the corresponding semantic between entities and relationships in low-dimensional vector space, effectively alleviate data sparseness, achieve heterogeneous information fusion, establish a unified space and realize knowledge transfer. It is of great significance for the construction, reasoning and the application of the knowledge graph, which is worthy of our in-depth exploration.

This paper is organized as follows. Section 2 introduces related work. In Sect. 3, representation learning for knowledge graph with dynamic step is presented. The experiments and analysis are shown in Sect. 4. The last Section provides conclusion and future work.

2 Related Work

2.1 Traditional Representation Learning Model

In recent years, many researchers have successively proposed multiple representation learning models to represent entities and relations in a knowledge graph. There are several representative models, such as structural models, single-layer neural network model, energy model, bilinear model, tensor neural network model, matrix analysis model and translation model. Structural embedding (SE) [4] is a distance model that uses knowledge representation obtained from learning to conduct prediction link. That is, to find the relation matrix that has the closest distance between two entities as the relationship between them through calculation. The single layer model (SLM [4]) uses the nonlinear operation in a single-layer neural network to solve the problem that SE can not jointly represent the semantic relationship between entities and relations. Semantic matching energy (SME) [5] embeds a more complex approach to obtain semantic relevance between entities and relations. A latent variable model (LFM) is a bilinear model which uses a bilinear transformation based on relation to represent the second-order relationship. The neural tensor network (NTN) [4] utilizes bilinear tensors to associate the head and tail entity vectors in different dimensions. The RESACL model is a representative method of the matrix factorization model, which obtained the low-dimensional vector representation by matrix factorization. After Mikolov et al. proposed the word2vec model for learning vector representation of word [6] in 2013, representation learning has attracted much attention in the field of natural language processing, while the word2vec also was found that there is an imbalance in the word vector space. Inspired by this phenomenon, Bordes et al. proposed the TransE model (Translating Embeddings for Modeling Multi-relational Data) [7] in 2013, embedding entities and relations into low-dimensional vector space and expressing the relationship as a translation operation among entities in a low-dimensional embedding space, so TransE is also called a translation model.

So from then on, most of presentation learning models have been extended on the basis of TransE. TransH model [8] was proposed by Wang et al., it is shown that this model solves the problems encountered in TransE when dealing with complex relationships by setting up the relation hyperplane, which makes that entities under different relations have different representations. In TransH model, entities and relations are still in the same semantic space, and it is unreasonable. In 2015, TransR model [9] was projected each triplet into the corresponding relation space, and built a translation between the head and tail entities in the relation space. At the same year, Ji et al. proposed the TransD model [10], setting up two projection matrices respectively to project the head and tail entities into the relation space, which made the projection matrix relevant to both entities and relations, and it can also solve the problem of too many parameters in TransR. TranSparse model [11] was used in sparse matrices instead of dense matrices in TransR to solve the problem of heterogeneity and imbalance of entities and relations. In 2015, Xiao et al. proposed TransA model [12], which used

Euclidean distance to calculate the loss value and weighted the different dimensions of the vector to predict entities and relations more accurately and effectively.

2.2 Problem Statement

As described in the related work section, TransE is a representative model for knowledge representation learning or called an energy-based model (also known as a translation-based model). Figure 1 depicts the TransE model. It shows the vector coordinates of entities and relationships. The basic idea of TransE is to embed all entities and relations in knowledge graph into the same low-dimensional vector space. Using the gradient descent algorithm to minimize the loss function can reduce the value of the scoring function, so as to achieve the purpose $\boldsymbol{h} + \boldsymbol{r} \approx \boldsymbol{t}$. We define the scoring function for training embedded vectors as:

$$f_r(h, t) = \|\boldsymbol{h} + \boldsymbol{r} - \boldsymbol{t}\|_{L_{1/2}}^2 \tag{1}$$

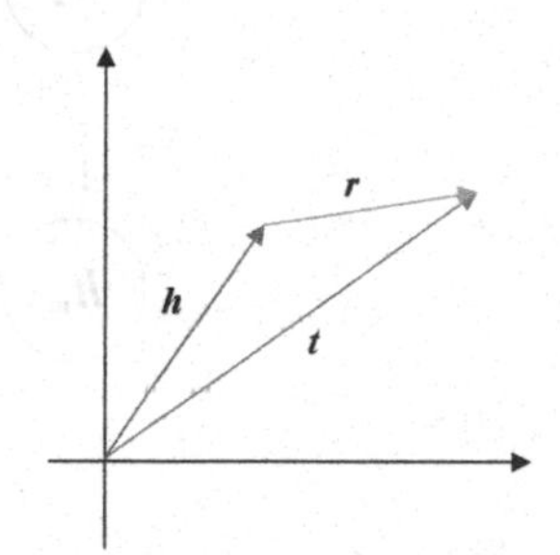

Fig. 1. TransE model illustration

TransE has a stable significant predictive performance in the large scale knowledge graph, but it seems like this model is not so good when dealing with different types of complex relations. Figure 2 shows the relation types of 1-1, 1-N, N-1 and N-N. From the figure, we can clearly see that the number of entities corresponding with different types of relation is distinct, and the complexity is also different. TransE does not discriminate the complexity of entities and relations in modeling complex relations such as 1-N, N-1 and N-N, which leads to a poor performance.

In order to solve the problem of single object optimization in the existing methods, this paper proposes a new representation learning method based on dynamic step. For relations 1-1, 1-N, N-1, and N-N [5], a dynamic step is defined to train the entities and relations in a knowledge graph. We will explain it in detail in the next section.

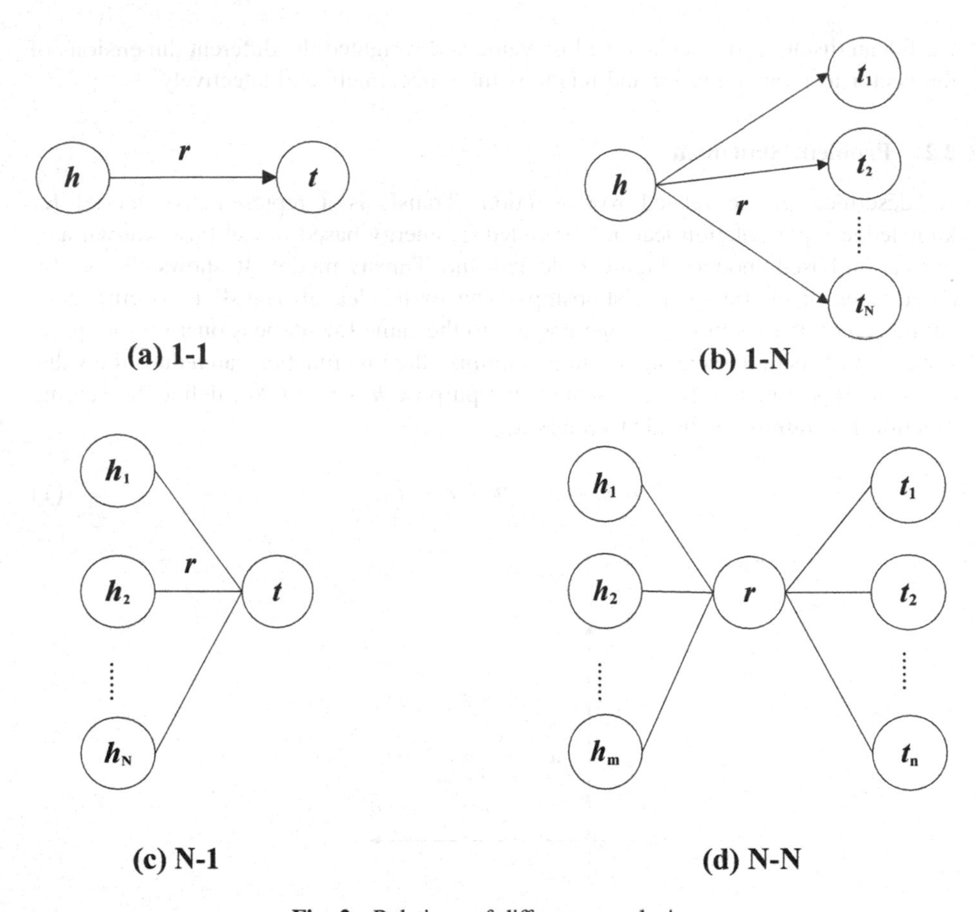

Fig. 2. Relations of different complexity

3 Representation Learning for Knowledge Graph with Dynamic Step

3.1 Motivation

Traditional representation learning methods have shown good performance for modeling a knowledge graph, but it is still challenging for modeling complex relations and the fusion of multi-source information because the objects (entities and relations) in the knowledge graph are heterogeneous and unbalanced. Specifically, (1) some relations connected many entity pairs are called complex relations while others are called simple relations. (2) Some relations can connect many head (tail) entities and less tail (head) entities, for instance, gender relations can connect many head names and only contact men or women in the tail.

According to statistics, the complexity of different relations varies greatly, and the imbalanced relation occupies a large proportion in knowledge graphs. If we use a step with the same length for training during the entire iteration process, we cannot distinguish the relations between different complexities. For instance, the same step has a greater influence on complex relations and less influence on simple relations. In all previous work in Trans (E, H, R, and D) area did not consider both of these issues and all of them used the same way when modeling the relation. Heterogeneity can lead to over-fitting of simple relations or under-fitting of complex relations. At the same time, the imbalance between head and tail shows that it is unreasonable for treating them equally.

In this viewpoint, we propose an algorithm based on Dynamic Step (DS) combined with TransE to form a representation learning model based on dynamic step (TransE-DS). According to different complexity of the relation, we set the dynamic step for knowledge representation learning. The link prediction task is conducted on two data sets: FB15K and WN11, and the results of experiments show that the performance of the model has been significantly improved.

3.2 Method Based on Dynamic Step

According to the number of entity pairs corresponding to the relation, we set up the dynamic step with a piecewise function to adjust step size of the entity and relation with different complexity, details as follows:

$$\lambda = \begin{cases} 0.01, & x \leq 10 \\ \frac{1}{10x}, & 10 < x \leq 1000 \\ 0.0001, & x > 1000 \end{cases} \tag{2}$$

Where λ is the step size and x represents the number of entity pairs which are connected with the relation. Specifically, when the number of entity pairs connect with a relation is less than or equal to 10, the step size is set to be 0.01. As the number of entity pairs connected with the relation increasing, the relation becomes more complex and the step size has a greater influence on it. Therefore, when the number of entity pairs connected with the relation is larger than 10 or less than or equal to 1000, we set an inverse proportional function that is the step size becomes smaller as the number of entity pairs connected with the relation increases. However, after increasing to a certain extent, the number of entity pairs connected with the relation is larger than 1000, the step length remains unchanged at 0.0001. The reason why the inverse proportional function adjusts the relation in this interval [10, 1000] is because the number of relations existing in this range is relatively large and the complexity is moderate. So, we make the step size in this range be dynamic, if the relation is too simple or too complex, the step size is 0.01 or 0.0001. In addition, for triples have not seen in the training set, we fix the step size of them with 0.001.

This paper, we use the same score function in TransE:

$$f_r(h, t) = \|\boldsymbol{h} + \boldsymbol{r} - \boldsymbol{t}\|_{L_1/L_2} \tag{3}$$

Where $r \in R^m$, a lower score of the positive triple and higher score of the negative triple indicate that the performance of the model is better.

We denote the ith triplet in a knowledge graph as (h_i, r_i, t_i) $(i = 1, 2, \ldots)$, and every triple has a label to indicate the triple is correct $(y_i = 1)$ or wrong $(y_i = 0)$. Then the positive triplets are represented by $\Delta = \{(h_i, r_i, t_i)|y_i = 1\}$, and the negative triplets are represented by $\Delta' = \{(h_i, r_i, t_i)|y_i = 0\}$. The negative triplet set is generated by replacing the head entity or tail entity in the positive triple set, then it can be expressed as $\Delta' = \{(h_i', r_i, t_i)|h_i' \neq h_i \wedge y_i = 1 \cup (h_i, r_i, t_i')|t_i' \neq t_i \wedge y_i = 1\}$.

We use (h, r, t) to represent a correct triple and select an error triplet (h', r, t') for each correct triple (replace head or tail) and define margin-based loss function as the optimization target for training:

$$L = \sum_{(h,r,t) \in S_{(h,r,t)}} \sum_{(h',r,t') \in S'_{(h,r,t)}} [f(h, r, t) + \gamma - f(h', r, t')]_+ \tag{5}$$

Where, $[f(h, r, t) + \gamma - f(h', r, t')]_+ = max(0, f(h, r, t) + \gamma - f(h', r, t'))$, γ is the margin between positive triples and negative triples. S (h, r, t) represents the set of all correct triples in a knowledge graph. The error triples are not generated randomly but replaced randomly by other entity or relation with each triple in the positive triple set S, which resulting in a negative triple set S′, so $S' = \{(h', r, t),(h, r, t')\}$. $f_r(h, r, t)$ represents the score function, which is used to represent some kind of connection between entities and relations.

Algorithm 1 Learning TransE-DS

input Training set S={(h, r, t)}, entities ad rel. sets E and R, margin γ,embedding dim. k,step λ,thenumber of entity pairs with a relation to.x

1: **initialize**$r\leftarrow$ uniform$\left(-\frac{6}{\sqrt{k}},\frac{6}{\sqrt{k}}\right)$ for each $r\in R$

2: $r\leftarrow r/\|r\|$ for each r$\in$R

3: $e\leftarrow$ uniform$\left(-\frac{6}{\sqrt{k}},\frac{6}{\sqrt{k}}\right)$ for each entity $e\in E$

4: **loop**

5: $e\leftarrow e/\|e\|$ for each entity $e\in E$

6: $S_{batch}\leftarrow$ sample (S, b) // sample a minibatch of size b

7: $T_{batch}\leftarrow\varnothing$ // initialize the set of pairs of triplets

8: **for** (h, r, t) $\in S_{batch}$ **do**

9: (h', r, t') $\leftarrow$ sample ($S'_{(h,r,t)}$) // sample a corrupted triplet

10: $T_{batch}\leftarrow T_{batch}\cup$ { ((h, r, t) ,(h', r, t')) }

11: **if** ($10<x\leqq1000$)

12: $\lambda\leftarrow0.01$;

13: **else if**($x\leqq10$)

14 : $\lambda\leftarrow\frac{1}{10x}$;

15: **else if**($x>1000$)

16: $\lambda\leftarrow0.0001$;

17: **else if**(triples not seen in the training set)

18: $\lambda\leftarrow0.001$;

19: **end for**

20: Update embedding h.r.t. $L=\sum_{(h,r,t)\in S_{(h,r,t)}}\sum_{(h',r,t')\in S'_{(h,r,t)}}[f(h,r,t)+\gamma-f(h',r,t')]_+$

21: **end loop**

The algorithm for learning TransE-DS is illustrated in Algorithm 1, which describes the detailed optimization process. Firstly, we initialize all embedded vectors of entities and relations. During main iteration of this algorithm, the entity's embedded vectors are first normalized. Next, we set different steps for triples according to the number of entity pairs connect with a relation. After that, a small group of triples are taken from the training set as mini-batch train triplets. Moreover, for each triple, we sample a corrupt triple. The gradient descent method is used to update the parameters and the algorithm won't stop until optimal results are obtained.

4 Experiments and Analysis

4.1 Datasets and Experiment Settings

We conduct the link prediction task on two typical knowledge graphs, WordNet and Freebase. In this paper, we choose a subset of Wordnet: WN18 [5] and a subset of Freebase: FB15K [5], which is a relatively dense subgraph in Freebase, and all entities are in the Wikipedia database. The statistics for these data sets are illustrated in Table 1.

As Table 1, WN18 contains 18 relationship types, 40,493 entities, 141442 triples in training set and 5000 triplets in both of valid set and test set. The same applies to FB15K.

Table 1. Statistics of the data sets

Dataset	#Rel	#Ent	#Train	#Valid	#Test
WN18	18	40,493	141,442	5,000	5,000
FB15 K	1,345	14,951	483,142	50,000	59,071

The parameters and complexity analysis are illustrated in Table 2. We observe that the complexity of almost all the models introduced in the Sect. 2 and compare with the complexity of our model. N_e and N_r denote the number of entities and relations respectively, N_t denotes the number of triples in a knowledge graph, m represents the dimension of the entity's embedding space, n denotes the dimension of the relation embedding space, and d denotes the number of clusters for a relation, k denotes the number of implied nodes in a neural network, s denotes the number of slices of the tensor and $\hat{\theta}\left(0 < < \hat{\theta} \leq 1\right)$ denotes the average sparseness of all translation matrices. It can be seen from the table that the complexity of TransE-DS is smaller than TransH and almost the same as TransE, which means that our model doesn't increase in complexity and even less than others.

Table 2. Parameters and complexity analysis

Model	#Parameters	#Operations (Timecomplexity)
SLM [4]	$O(N_e m + N_r(2k + 2nk))(m = n)$	$O((2mk + k)N_t)$
NTN [4]	$O(N_r m + N_r(n^2 s + 2ns + 2s))(m = n)$	$O((((m^2 + m))s + 2mk + k)N_t)$
TransE [7]	$O(N_e m + 2N_r n)(m = n)$	$O(N_t)$
TransH [8]	$O(N_e m + 2N_r n)(m = n)$	$O(2mN_t)$
TransR [9]	$O(N_e m + N_r(m + 1)n)$	$O(2mnN_t)$
CTransR [9]	$O(N_e m + N_r(m + d)n)$	$O(2mnN_t)$
TransD [11]	$O(2N_e m + 2N_r n)$	$O(2nN_t)$
TransE-DS	$O(N_e m + 2N_r n)(m = n)$	$O(N_t)$

4.2 Link Prediction

Link prediction is intended to predict a missing entity *h*(or *t*) in a fact triple (*h*, *r*, *t*). This task is principally focused on the ranking of candidate entity sets obtained from the knowledge graph, rather than looking for the best answer. In this method, we use two data sets: WN18, a subset of Wordnet and FB15K, a relatively dense subgraph in Freebase, and all entities exist in Wikipedia database.

For each triple needs to be test (*h*, *r*, *t*), we will first replace the head or tail entity by all of the entities in the knowledge graph, and calculate the similarity scores of these entities by the score function f_r. Then rank these entities in descending order according to their similarity score. Similar to TransE, we use two measure methods as our evaluation criteria: (1) the correct entity's average ranking (represented by Mean Rank) and (2) the top 10 ratio (represented by Hit@10). Obviously, a good model should obtain a low average ranking and a high top 10 ratio.

It should be noted that for a triple (*h*, *r*, *t*), its negative triplet may also exist in a knowledge graph and be treated as a correct triple. However, the above assessment may rank these incorrect triples in front of the correct triples and then lead to underestimating its performance. Therefore, we should filter out these triples from training set, validation set and test set before ranking. We denote this evaluation criteria setting as "Flit" and the original evaluation criteria setting as "Raw".

Table 3. TranE-DS parameter statistics

Dataset	Model	γ	n, m	B	D.S
WN18	TransE-DS	2	20	1,440	L_1
FB15 K	TransE-DS	2.5	100	4,800	L_1

Table 3 shows the training parameters used in our experiment for TransE-DS. The parameter in this table is the optimal parameter value. With these parameter values, our experiment achieves the best results. λ is the learning rate, γ is the margin, n and m are embedding dimensions for entity and the relation, B is the mini-batch size and D. S is the dissimilarity measure in the score function. The iteration number for Stochastic Gradient Descent (SGD) is 3000.

The experiment results of the link prediction are shown in Table 4. The upper part of the results comes from the literature directly because all the methods discussed in this paper use the same testing data sets. Since there are no DS's experiment results on link prediction in the literature, we won't discuss it in this table.

Table 4. Results for link prediction

Dataset	WN18				FB15 K			
Metric	Mean rank		Hits@10		Mean rank		Hits@10	
	Raw	Filt	Raw	Filt	Raw	Filt	Raw	Filt
RESCAL	1,180	1,163	37.2	52.8	828	683	28.4	44.1
SE	1,011	985	68.5	80.5	273	162	28.8	39.8
SME	542	533	65.1	74.1	274	154	30.7	40.8
LFM	469	456	71.4	81.6	283	164	26.0	33.1
TransE	263	251	75.4	89.2	243	125	34.9	47.1
TransH	318	303	75.4	86.7	211	84	42.5	58.5
TransR	232	219	78.3	91.7	226	78	43.8	65.5
C-TransR	243	230	78.9	92.3	233	82	44.0	66.3
TransD	242	229	79.2	92.5	211	67	49.4	74.2
TransE-DS	261	249	76.2	90.5	**18152**		48.270.1	

Result Analysis: From the Table 4, we can draw the following conclusions: (1) TrasnE-DS's results are better than TrasnE, TransH, and previous presentation learning models in all of the link prediction tasks, especially on FB15K where the Mean Rank is 73 higher than TrasnE and HIT@10 is increased by 23% under the Filt settings. Compared with TransH, the test results on WN11 of link predict, Mean Rank are increased by 102 under the Raw setting, and HIT@10 on FB15K is increased by 11.6% under the Filt setting. On the FB15K data sets, results of Mean Rank are better than all previous models including TrasnR, C-TrasnR and TransD, and the Mean Rank under the Flit settings are increased by 30 than the most effective model TransD in Table 4. It shows that the method based on dynamic step combined with TransE can significantly improve the performance of this model and better represent the data in a knowledge graph.

5 Conclusion and Future Work

In this paper, a knowledge graph representation learning model based on dynamic step is proposed. This model mainly focuses on the different complexity of relations and entities, and defines a piecewise function to sets up the step size of the relations with different complexity. By setting up the dynamic step size, different types of relations can be effectively separated, so the optimization goal of the triple is more clearly in training process and the prediction results are more accurate and effective without increasing the complexity. The method is simple and effective. Not only it can effectively solve the heterogeneity problem and imbalance in knowledge graph, but also can be combined with other models and improve the accuracy of them.

In the future work, in addition to distinguishing different types of the relation, when dealing with the computational efficiency and data sparse problems in a knowledge graph, many researchers have focused on multi-information fusion and reasoning of complex relation paths. We will deeply study the fusion of textual information and structured information, infer complex relation paths for reasoning and completing of knowledge graphs and apply them to large-scale knowledge graphs effectively.

Acknowledgements. This work is supported by the National Natural Science Foundation of China (Nos. 61572146, U1501252, U1711263), and the Natural Science Foundation of Guangxi Province (Nos. 2015GXNSFAAI39285).

References

1. Liu, Z., Sun, M., et al.: Knowledge representation learning: a review. J. Comput. Res. Dev. **53**(2), 247–261 (2016)
2. Needlakantan, A., Roth, B., et al.: Compositional vector space models for knowledge base completion. In: Proceedings of ACL, pp. 156–166. ACL, Stroudsburg (2015)
3. Dong, X., Gabrilovich, E., et al.: Knowledge vault: a web-scale approach to probabilistic knowledge fusion. In: Proceedings of ACM SIGKDD, New York, pp. 601–610 (2014)
4. Socher, R., Chen, D., Manning, C.D., et al.: Reasoning with neural tensor networks for knowledge base completion. In: Proceedings of NIPS, pp. 926–934. MIT Press, Cambridge (2013)
5. Bordes, A., Glorot, X., Weston, J., et al.: A semantic matching energy function for learning with multi-relational data. Mach. Learn. **94**(2), 233–259 (2014)
6. Mikolov, T., Chen, K., Corrado, G., et al.: Efficient estimation of word representations in vector space. Proceedings of ICLR (2013). arXiv:1301.3781
7. Bordes, A., Usunier, N., et al.: Translating embeddings for modeling multi-relational data. In: Proceedings of NIPS, pp. 2787–2795. MIT Press, Cambridge (2013)
8. Wang, Z., Zhang, J., Feng, J., et al.: Knowledge graph embedding by translating on hyperplanes. In: Proceedings of AAAI, pp. 1112–1119. AAAI, Menlo Park (2014)
9. Lin, Y., Liu, Z., Sun, M., et al.: Learning entity and relation embeddings for knowledge graph completion. In: Proceedings of AAAI. AAAI, Menlo Park (2015)
10. Ji, G., He, S., Xu, L., et al.: Knowledge graph embedding via dynamic mapping matrix. In: Proceedings of ACL, pp. 687–696. ACL, Stroudsburg (2015)
11. Ji, G., Liu, K., He, S., Zhao, J.: Knowledge graph completion with adaptive sparse transfer matrix. In: Proceedings of AAAI, pp. 985–991. AAAI, Phoenix (2016)
12. Xiao, H., Huang, M., et al.: TransA: an adaptive approach for knowledge graph embedding. arXiv preprint arXiv:1509.05490 (2015)

An Improved K-Means Parallel Algorithm Based on Cloud Computing

Xiaofeng Li[1(✉)] and Dong Li[2]

[1] Department of Information Engineering, Heilongjiang International University, Harbin 150080, China
mberse@126.com
[2] School of Computer Science and Technology, Harbin Institute of Technology, Harbin 150001, China

Abstract. Through deeply analyzing of the problem in K-Means algorithm, this topic proposed an improved scheme based on Hadoop distributed platform. Using the proposed clustering analysis system to configure the experimental environment, the algorithm is optimized from three aspects: parallel random sampling, parallelization of sample distance computation and parallelization of data clustering process. At the same time, the improved K-Means parallel algorithm flow was described in detail. The experimental result shows that the cluster analysis system based on Hadoop distributed cloud computing platform can provide efficient, stable and configurable clustering analysis service. Improved K-Means parallel clustering algorithm can quickly deal with large scale calculation of cluster analysis.

Keywords: Cloud computing · Hadoop · K-Means · Clustering analysis

1 Introduction

As one of the oldest clustering algorithms, the K-Means algorithm has been invented for half a century. Due to its relatively simple and time-consuming features, the K-Means algorithm has been favored by many researchers [1, 2]. Up till now, the K-Means algorithm has also been active in the field of data mining. For K-Means algorithm, there are many factors affecting its clustering accuracy. However, the most intuitive and significant impact is its input parameter K, which refers to the number of final cluster centers specified by the user, that is, it is divided into several types of data [3]. The change of this value directly determines the accuracy of the algorithm's final clustering result. Therefore, there are many researches on how to initialize the cluster center by users [4, 5]. Among many clustering algorithms, K-Means algorithm is one of the most widely used algorithms, but the algorithm itself still has many problems. In this paper, the traditional serial K-Means algorithm will be used as a starting point to fully study the algorithm flow and characteristics, and conduct parallel optimization based on Hadoop cloud computing platform to solve the problem of its efficiency in the face of large-scale data sets.

Q. Zhou et al. (Eds.): ICPCSEE 2018, CCIS 901, pp. 394–402, 2018.
https://doi.org/10.1007/978-981-13-2203-7_30

2 Traditional K-Means Algorithm

The idea of the K-Means algorithm: First, the user needs to determine the number of clusters of the final clustering result, and then randomly selects the initial cluster center of number K in the original data set. Then, iteratively iterative process requires calculating the spacing of the full amount of data objects to the center of each cluster and merging them into their respective clusters according to the spacing. After all the data points are categorized, the average spacing of the objects in each cluster is calculated, and the original center is replaced with the new cluster center. This iterative process continues until the objective function converges. The convergence of the objective function is that after the end of a classification, the recalculation of the new cluster center does not change, and the algorithm ends.

2.1 Algorithm Equation

It is convenient to describe the improved algorithm of K-Means. This paper introduces the symbol $X = \{x_i \in R^n, i = 1, 2, \ldots, n\}$ to represent the original dataset, $M_1, M_2, \ldots, M_k$ represents the center of K class cluster, $L_1, L_2, \ldots, L_k$ represents a different class of K. The Euclidean distance Equation between two arbitrary data objects is in Eq. (1).

$$d(x_i, x_j)^2 = \sqrt{(x_{i1}, x_{j1})^2 + (x_{i2}, x_{j2})^2 + \ldots + (x_{in}, x_{jn})^2} \quad (1)$$

In the Eq. (1), x_i and x_j are data objects of dimension n. Define the center points of the same class cluster as shown in Eq. (2).

$$M_j = \frac{1}{n_j} \sum_{x \in w_j} x \quad (2)$$

In the Eq. (2), n_j is the number of data objects in the same class cluster. The definition of convergence is shown in Eq. (3).

$$J = \sum_{i=1}^{k} \sum_{j=1}^{n_i} d(x_j, z_j) \quad (3)$$

The target function requires the user to enter the specified parameters, R and z, when the number of data objects contained in the spherical cluster with radius R exceeds the value z, the current region is considered as a high-density area, whereas the other is the low-density region.

2.2 Problems Existing in K-Means Algorithm

1. The traditional K-Means algorithm is a stand-alone operation algorithm, which is limited by the hardware of a single machine, and the algorithm cannot adapt to the growing clustering of massive data.

2. The traditional K-Means algorithm uses a completely random selection strategy to initialize the clustering center point, which not only affects the accuracy of the algorithm, but also reduces the efficiency of the algorithm.
3. In order to ensure the accuracy of the replacement cluster center operation, the traditional K-Means algorithm uses the global sequence to replace the cluster center, but such coarse-grained operations increase the time complexity of the algorithm and thus affect the execution efficiency.

3 Improved Scheme of K-Means Algorithm

3.1 Parallel Random Sampling

The calculation of the traditional K-Means algorithm uses a full amount of data objects, which is very inefficient in the face of very large-scale data sets. In order to reduce the time consumption of the algorithm, this paper designs a preprocessing operation for initializing the clustering center, i.e., pre-sampling processing. In order to improve the efficiency of K-Means algorithm, a parallel random sampling process based on Top K processing is designed. And the parallel process is based on Hadoop distributed system. The parallel process algorithm is based on Hadoop distributed system is as follow:

Input: the random number range H, the sample data capacity N, and the number R_n of Reducer.

Output: N sample data samples.

1. In the Map phase, the total amount of data object is assigned, the value range is H, and the random value is key, the data value is value, and the key value is output.
2. The output results are sorted internally, each Reducer outputs a sorted previous N/R_n data.
3. The sample is preprocessed to get the initialization cluster center point. The preconditioning Eq. (4) is defined as follows:

$$V_j = \sum_{i=1}^{n}((\sum_{i=1}^{n} d_{i1}) - d_{ij}), j = 1, 2, \ldots, n \tag{4}$$

3.2 Parallelization of Sample Distance Calculation

The K-Means parallel algorithm is based on the independence of elements. The traditional K-Means algorithm calculates the distance of a full data object in a circular manner. Therefore, the distance calculation process is parallelized. In the Map Reduce parallel computing framework, Map plays a major role in mapping. Therefore, this paper considers the use of the mapping function of the Map stage to map the full amount of data in the form of <key, value> to different Reducers for parallel clustering calculations, and the Reducer in this case is different K clusters so as to make full use of the independence of the original data objects and parallel cluster analysis [6]. After parallelization at the Map stage, multiple nodes can simultaneously calculate the sample distance and speed up the algorithm operation efficiency.

3.3 Clustering and Parallelization of Data Object

After the mapping of the Mapper function, the data objects are mapped to the respective cluster Reducer according to the distance. Because each cluster corresponds to its own Reducer, the reducer parallelism is set to K. At the Reducer stage, it is necessary to iteratively calculate the center point of clusters, replacing the initial center point that was originally calculated based on parallel random sampling. The calculation rule here is the sum of the squares of the Euclidean distances of the full data objects in the cluster, and the minimum point is chosen as the new center point.

At the first stage of execution, each cluster corresponds to its own Reducer, and the parallel data strategy is performed sequentially on the entire data object in the cluster. First, all data objects are taken as input data sets, and then any data object is selected as the center point of the temporary clusters. The sum of the squares of the Euclidean distances from other elements in the class to the current center point is calculated, and the least squared point and the numerical minimum point are selected as the new center point.

In this paper, the minimum Euclidean distance is calculated, and the characteristics of kv structure are optimized by using the Map Reduce distributed computing framework. In the key value pair, key implements the compareTo() method of interface Writable Comparable. compareTo() can compare the numeric size between elements, so that it can be sorted [7]. Therefore, the iterative calculation of the comparison process of cluster center point steps can be realized by using the distributed sorting function of kv structure.

4 Implementation of Improved K-Means Parallel Algorithm

Through in-depth analysis of the characteristics of the traditional K-Means algorithm, this paper studies and implements an improved K-Means parallel algorithm based on the Hadoop clustering analysis system. The algorithm is optimized from the three directions: parallel random sampling, parallelization of Mapper, and Parallelization of Reducer. At the Mapper stage, the parallelization of the sample distance calculation is improved, and the data object clustering process and the Euclidean distance sorting are improved at the Reducer stage. The specific execution process of the algorithm is as follows:

1. The user enters the original data set with the final cluster number K and data size n. The output condition is that the objective function converges, i.e., the Euclidean distance at the center of each cluster is less than the threshold.
2. The original data set is processed by the Top K-based parallel random sampling. After the sample is preconditioned by Eq. (4), the center point of the cluster is initialized.
3. The data serial number is used as the key, and the distance calculation Eq. (1) is used to calculate the Euclidean distance for each data point.
4. Map the entire data object to its own classifier Reducer using the parallel mapping at the Map stage. This process requires the intermediate file storage of the HDFS distributed file system.

5. In the Reducer, the sum of the squares of the distances of each cluster is calculated in parallel to calculate the new cluster center.
6. Determine whether the Euclidean distance of the current cluster center is greater than the threshold. If yes, Replace the center point of the original cluster with the center point of the current cluster and return to step 3 to recalculate, otherwise the algorithm ends.

The full data set first undergoes parallel random sampling and preprocessing before performing clustering calculations. The sample distance calculation and data classification of cluster analysis are performed by MapReduce. Compared with the traditional K-Means serial algorithm, this improved algorithm parallelizes the cluster analysis process, which makes the efficiency of the algorithm greatly improved when running large-scale data.

5 Experimental Analysis and Results

In the environment of cluster analysis system, the design experiment of the improved k-means parallel algorithm is combined with the experimental results. Firstly, the experimental environment and data preparation of cluster analysis system are introduced. Then, the traditional k-means algorithm and k-means parallel algorithm are compared experimentally from the four directions of convergence speed, accuracy, initial sampling rate and acceleration ratio in the cluster environment. Finally, the improved algorithm is analyzed and summarized.

5.1 Experimental Environment and Data of Cluster Analysis

In order to simulate the distributed cluster environment in real situation, six PC computers were used in cluster analysis system experiment environment. The operating system is Cent OS6.4. Software Java_1.7.0_79, Zookeeper-3.4.5, Hadoop2.6.0 and HBase 0.96.2 were installed respectively.

Because the experiment needs the accuracy of the test and the speedup in the cluster environment, two data are prepared. One is the Iris open source dataset commonly used for cluster analysis, and the other is a large-scale dataset generated. There are three classic Iris datasets, each with a data capacity of 50, and each data object contains four different attributes. Due to the small capacity of the Iris dataset, it is impossible to test the improvement effect of the algorithm in large-scale clustering. Therefore, the attribute dimensions and the capacity of the Iris dataset are increased, and a large-scale random dataset is constructed with code. In this experiment, five sets of data sets with different sizes are generated. Each set of data is divided into three clusters, and the number of elements in each cluster is the same.

Table 1. Generated random data set

Data set file	Data set size	Total number of data elements	Data dimension	Cluster center point number
File A	0.2 M	8000	5	3
File B	150 M	8000000	5	3
File C	450 M	24000000	5	3
File D	1.3 G	64000000	5	3
File E	2.2 G	150000000	5	3

5.2 The Convergence Speed Comparison

The convergence speed comparison experiment is to compare the number of iterations required for running the algorithm when computing the same data set in a stand-alone environment, comparing the traditional K-Means algorithm and the improved K-Means parallel algorithm.

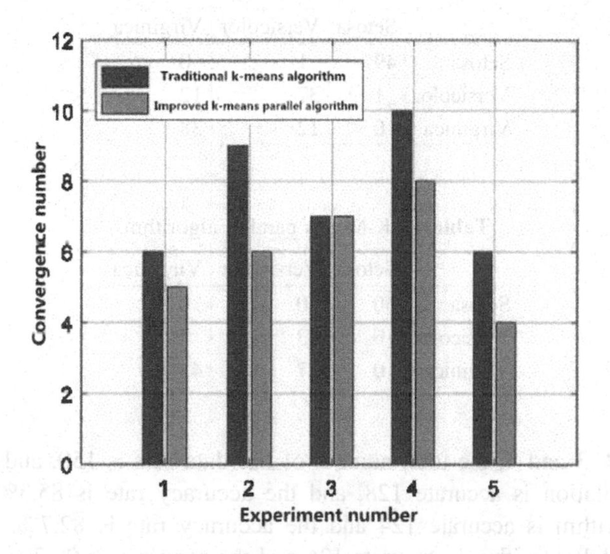

Fig. 1. Convergence performance comparison

In order to eliminate the interference of parallel computing, the traditional K-Means algorithm runs in the common stand-alone environment, and the improved K-Means parallel algorithm runs in the pseudo-distributed mode. The above two algorithms are run on 5 machine nodes with File A as the original data set. The test data is shown in Fig. 1.

The experimental results show that the improved k-means parallel algorithm has fewer average iteration times in the single-machine pseudo-distributed mode, so it has better convergence. The reason that the algorithm converges faster is that the pre-treatment process makes the initial class cluster center more accurate than the traditional algorithm.

5.3 Accuracy Comparison

The purpose of the accuracy comparison experiment is to test the accuracy of traditional K-Means, mahout K-Means algorithm and improved K-Means parallel algorithm for standard Iris data clustering, where, the mahout K-Means algorithm is the K-Means parallel algorithm implemented by Hadoop platform. The clustering effects of the three algorithms are shown in Tables 2, 3 and 4.

Table 2. Traditional K-Means

	Setosa	Versicolor	Virginica
Setosa	50	0	0
Versicolor	0	39	11
Virginica	0	11	39

Table 3. Traditional K-Means

	Setosa	Versicolor	Virginica
Setosa	49	1	0
Versicolor	1	37	12
Virginica	0	12	38

Table 4. K-Means parallel algorithm

	Setosa	Versicolor	Virginica
Setosa	50	0	0
Versicolor	0	43	7
Virginica	0	7	43

In Tables 2, 3 and 4, the total number of Iris data sets is 150, and the traditional K-Means calculation is accurate 128, and the accuracy rate is 85.3%. The mahout K-Means algorithm is accurate 124 and the accuracy rate is 82.7%. The improved K-Means parallel algorithm is accurate 136 and the accuracy is 90.7%. Therefore, the experimental results show that the improved k-means parallel algorithm has better accuracy. After analysis, this result is caused.

5.4 Initial Sampling Rate Comparison

This experiment is to compare the operation efficiency of several different random sampling methods. The sampling methods of the comparison are sequential traversal, byte offset, and parallel random sampling based on Top K improvement. The k-means parallel algorithm runs on 6 nodes. In this experiment, the File B File is the original data set, and the timeout period is 1 h. The sampling time of each method is shown in Table 5.

Table 5. Time comparison of different sampling methods

	The number of elements in the sample data set			
	90	900	900000	9000000
Line-by-line through	461.2 s	2605.1 s	Timeout	Timeout
Byte offset	1 s	9.6 s	624.1 s	Timeout
The parallel sampling	32.4 s	32.5 s	43.1 s	52.1 s

5.5 Cluster Environment Speedup Over Validation

Due to improved algorithm is a kind of parallel algorithm design in this paper, the speed ratio is a parallel algorithm is one of the most intuitive indicator of fine performance, so the improved k-means algorithm is used to speedup ratio the experiment. The speedup ratio is the ratio of the same task running in a different number of processors. The formula is defined as follows:

$$S_p = \frac{T_s}{T_p} \tag{5}$$

In Table 1, there are 5 orders of magnitude of different artificial data sets, including File A, File B, File C, File D and File E as the original input data The speedup ratio of parallel algorithm is calculated by using 1, 2, 3, 4 and 5 computing nodes respectively. The speedup ratio is shown in Fig. 2.

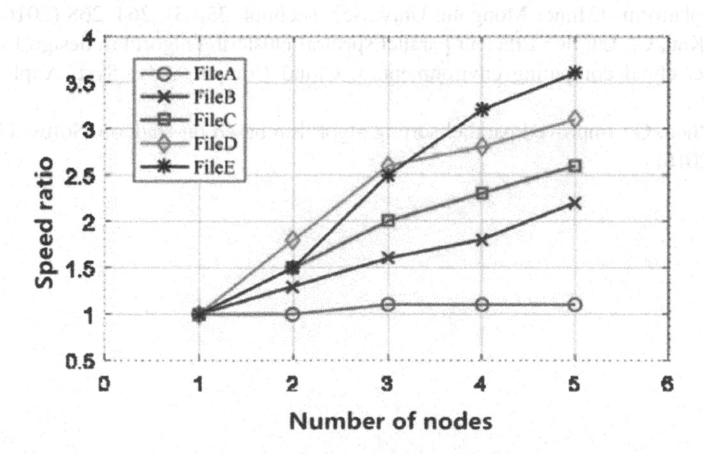

Fig. 2. Speedup ratio test

From the experiment result shows that all data set speedup ratio increases with the increase of computing nodes, the speedup increases with the increase of data amount. It shows that the improved K-Means algorithm in the distributed cluster parallel environment can significantly improve the operation efficiency, and can adapt to large-scale data set of cluster computing.

6 Conclusion

This paper first analyzes the process and existing problems of k-means algorithm. Then, the improvement scheme of k-means algorithm is studied, which mainly includes parallel random sampling, sample distance computation parallelization and data object clustering process. In this paper, the improved k-means parallel algorithm is tested in four directions from the convergence speed, accuracy, the initial sampling rate and the clustering environment speedup ratio. The experimental results show that the clustering analysis system was designed and implemented in this paper can efficient and stable distributed clustering services, improvement of K-Means parallel algorithm has good convergence and accuracy, initialization, sampling rate and the speedup ratio of the cluster environment.

References

1. Deng, Q., Yang, Y.: Research on improved parallel K-means algorithm based on Spark framework. Intell. Comput. Appl. **8**(01), 76–78 (2018)
2. Li, X., Yu, L., Lei, H., Tang, X.: A parallel implementation and application of K-means improved algorithm. J. Univ. Electron. Sci. Technol. China **46**(01), 61–68 (2017)
3. Li, H.: Improved K-means clustering method and its application, pp. 15–17. Northeast Agricultural University (2014)
4. Li, G.B., Han Qing, J.: An improved K-means clustering algorithm for MapReduce parallelization. Digit. Technol. Appl. (12), 134–136 (2016)
5. Lu, S., Wang, J., Zhang, X., Gao, J.: Optimization of K-means clustering algorithm based on Hadoop platform. J. Inner Mongolia Univ. Sci. Technol. **35**(03), 264–268 (2016)
6. Ran, J., Kou, C., Liu, R.: Efficient parallel spectral clustering algorithm design for large data sets under cloud computing environment. J. Cloud Comput. Adv. Syst. Appl. **2**(1), 1–10 (2013)
7. Fu, C., Zhou, G.: Improved parallel sorting algorithm based on Hadoop. Softw. Guide **15**(4), 68–70 (2016)

Statistical Learning-Based Prediction of Execution Time of Data-Intensive Program Under Hadoop2.0

Haoran Zhang, Jianzhong Li, and Hongzhi Wang(✉)

Massive Data Computing Research Center, Harbin Institute of Technology, Xidazhijie. 92, Harbin, China
wangzh@hit.edu.cn

Abstract. This paper is mainly to predict the running time of data-intensive MapReduce program under Hadoop2.0 environment. Although MapReduce programs are diverse, they can be divided into data-intensive and computationally intensive, depending on the time complexity and the nature of the program. The prediction of computationally intensive programs has always been difficult, and Hadoop has exhibited certain database attributes that are basically data-intensive. Moreover, the relationship between data-intensive programs and the amount of data is more closely related and shows certain statistical characteristics. So the method of statistical learning is applied to predict the execution time. This paper first generates training data and test data according to requirements, and then selects the appropriate features through the analysis of the logs. The prediction was first performed using the KCCA algorithm. However, the deficiencies were found. Then based on the characteristics of the kernel function, a prediction method based on deep learning was proposed, and the result was significant.

Keywords: Training data · Feature extraction · KCCA
Deep learning

1 Introduction

1.1 Background

We generally believe that we are facing an era of information prosperity. The amount of data has become very large, and traditional management and processing methods are no longer effective. In this case, MapReduce has proven to be an effective way to deal with "big data". It is a programming model that is scalable and has a fault-tolerant operating environment. Due to its simplicity, versatility and maturity, it has become the most popular platform for data analysis. Hadoop is an open source implementation of MapReduce. Many Internet companies deploy many Hadoop clusters to provide core services such as log analysis, data mining, feature extraction, and so on. After studying the performance of some Hadoop clusters, you can find some interesting issues as follows:

Q. Zhou et al. (Eds.): ICPCSEE 2018, CCIS 901, pp. 403–414, 2018.
https://doi.org/10.1007/978-981-13-2203-7_31

- How to design an efficient scheduling strategy? Predecessors did a lot of work on Hadoop scheduling strategies. However, some advanced strategy algorithms need to perform high-level estimation of job performance in advance to determine the scheduling strategy;
- How to optimize job performance? More and more companies are concerned about the return on investment They not only need to solve problems, but also optimize job performance, thereby reducing time and resources to solve more problems;
- How to evaluate the performance of the job? Even if all possible optimization strategies are made, this is not the optimal strategy. So a standard is needed to judge the performance of the job;
- How to balance cost and performance? Some users prefer to use on-demand services instead of deploying their own. In this case, they need to determine exactly how long they are and how many nodes they use;

If the cluster's performance improvement does not reach its ultimate goal, resources may be wasted. Even in a production environment, performance tuning is a tedious task. To properly solve these problems, the focus is on estimating job performance in advance. So it's particularly critical for the prediction of time.

Although MapReduce programs are diverse, they can be divided into data-intensive and computationally intensive, depending on the time complexity and the nature of the program [1]. The prediction of computationally intensive programs has always been more difficult because the results are more uncertain. For example, for the different accuracy requirements, the execution result of pi will also be different, and the natural execution time will also be very different. In contrast, data-intensive programs are relatively friendly. The execution time and data volume of this type of program are strongly related and have statistical characteristics. In addition, Hadoop embodies strong database operation features such as counting, querying, etc. These are all data-intensive programs.

1.2 Related Work

There have been ideas for the prediction of MapReduce programs. For example, some research work through detailed analysis of various stages of the MapReduce program execution, and then the detailed modeling of each stage to achieve the purpose of prediction [2,3]. They generally give more complex modeling. Others model the entire process and achieve prediction by learning parameter coefficients [4]. However, the process of MapReduce execution is influenced by many factors. When the amount of data increases, is it still a linear model? Others use sampling to create a simulator to predict execution time [1]. However, there are few statistical learning methods to predict. There has been research of the prediction of hive [5] and achieved good results. However the prediction of the MapReduce program was only conjecture.

1.3 Main Research Content

This paper aims to predict the execution time of data-intensive programs through statistical learning methods. Firstly we generate the required training data according to certain rules. Then, we pick out features that fit the model. Finally, according to the results, improvements have been made.

The contributions of this paper are listed as follows:

- In the absence of enterprise-level data, based on the actual situation, we generate suitable training data;
- We propose an improved method based on the KCCA regression problem;
- We have designed a data distribution that can find out the inherent laws of the MapReduce model.

We begin our research with data generation and feature extraction in Sect. 2. Then we apply KCCA algorithm [6,7] in Sect. 3. We propose our optimization in Sect. 4. Finally, the results are shown in Sect. 5.

2 Data Generation and Feature Extraction

2.1 Data Generation

In previous research results [1,5], they give us ideas on how to get training data. The best way is to directly use the company's data. If we do not have corporate data, we can generate data that meets the requirements ourselves. So according to the actual situation, we chose to generate training data. After considering the overall factors, we summed up the following requirements for generating data:

- Have enough coverage
- In the implementation of the example program, we need to consider the performance of the cluster.

Because statistical learning methods are more based on the probability of summing up a certain law, the predicted results are actually the spatial division of the results of the training set, especially for regression. Therefore, we have to cover all possible situations as far as possible. Secondly, the execution time is closely related to the performance of the cluster. We do not want to change the environment in which the program is executing. For example, when we run a program on a large data set, we will have to use a lot of resources in the cluster. However, at the same time when other people submit an assignment during this process, the environment in which our program runs will change. This involves Hadoop's scheduling allocation mechanism. Not only will it affect other people, but more importantly, our results will be affected.

Therefore, according to the above conditions, according to the situation of the laboratory cluster, it is proposed to adopt the following scheme:

(1) The data set size is within 1M, multiplied by 64k, eg 64k, 128k, 194k,..., 1024k

(2) The data set is more than 1M, multiplied by 1M, for example, 1M, 2M,..., 1024M.

According to the performance of the integrated cluster and the execution time of the cluster, it is appropriate that the maximum data volume is 1G. Because in the previous literature it has been discussed that the running time of datasets within 64k of the Shuffle and Sort stage executed by MapReduce is negligible. Then design a data set with a step size increase to 1M. This can achieve sufficient coverage.

This paper generates training datasets through the teragen method under the example package that comes with the Hadoop tool. By observing data, this method uses pseudo-random numbers to follow the output string of the line cycle. One benefit of such data set is that the resulting data will not be skewed. In this way, there will be no situation that the data distribution affects the execution of the program. Through the experiment, 100 lines were measured to be 40B, so that the corresponding data set can be calculated by calculation. We generate the corresponding execution command through the program, and finally execute the script to generate the data set. Then, we need to pick a program as a benchmark. We chose the wordcount program in the example folder of the Hadoop tool.

2.2 Log Extraction

This paper analyzes the possible influencing factors including [8,9] of program execution time and proposes to select features through the following three aspects:

(1) System hardware configuration
(2) Hadoop configuration information [10]
(3) Execution process of Hadoop.

Our goal is to predict the execution time of the program before running the MapReduce program. Then the process of extracting features becomes the behavior of extracting valid information before executing MapReduce program. We can analyze how to extract reasonable features by splitting the execution of the MapReduce program.

When the MapReduce program starts, the system does not immediately start executing the Map program. The system will initialize the execution environment first, then this overhead can be defined as the initialization overhead when the Map task starts execution. We define this time as **MapSysCost**.

After the Map program is initialized, Hadoop begins using a circular buffer to read the data specified on the HDFS. Then, in the process of reading information, the disk read speed and cache ratio are the key to this process. The read speed of the disk belongs to the hardware configuration of the system cluster, which has a key influence on the execution of any job, especially critical for applications such as MapReduce that handles large data sets. On the other hand, the ratio of buffer thresholds is also more critical in this issue. The Map process has been completed when the data is in the buffer and when the data continuously fills the buffer to a certain threshold, the split phenomenon will occur. Therefore, the buffer ratio will affect the rate of sorting.

Meanwhile, a MapReduce program will have a shuffle between Map and Reduce program execution. At the same time, when the buffer area is fixed, the proportion becomes more important. The Hadoop system has a mechanism to save time. When the Map program is about to finish, Reduce will start beforehand. This is a reason why the reduce will start when the progress of the map execution does not reach 100.

We can summarize the selected features in a table as shown in Table 1:

Table 1. Feature we select

NO.	Symbol	Description
1	SeqRead	Hard disk read speed
2	MapSysCost	The overhead of the Map task
3	mapreduce.task.io.sort.mb	Output memory buffer size in Map
4	mapreduce.reduce.shuffle.parallelcopies	The number of concurrent threads to copy data in Reduce
5	mapreduce.map.sort.spill.percent	Buffer overflow threshold in Map
6	mapreduce.reduce.shuffle.input.buffer.percent	Buffer overflow threshold ratio in shuffle

Of course, the most important thing is the size of the input file. However, we may have some deviations from the experimental data set size calculated by the number of rows, and after the task is executed, it is difficult for us to distinguish the programs executed by the job by the name of the job. Therefore, we use the log information to get the size of the execution file.

3 KCCA Model

Kernel Canonical Correlation Analysis (KCCA), is a variant of CCA that captures similarity using *kernel function*. The correlation analysis is on pairwise distance, not the raw data itself. This approach provides much more expressiveness in capturing similarity and its correlations can then be used to quantify performance similarity of various workloads. KCCA is the statistical machine learning technique we use in this dissertation.

The projection resulting from KCCA provides two key properties:

- The dimensionality of the raw datasets is reduced based on the number of useful correlation dimensions.
- Corresponding datapoints in both projections are collocated. Thus, there is a clustering effect that preserves neighborhoods across projections.

We next give the description of the KCCA algorithm and application methods.

3.1 KCCA Overview

Using the kernel-cca algorithm, we try to obtain a standard eigenproblem for the kernel mapping of the text and image kernels. The objective function is as follows:

$$\rho = \max_{\alpha,\beta} \frac{\alpha' K_x K_y \beta}{\sqrt{\alpha' K_x^2 \alpha \beta' K_y^2 \beta}} \tag{1}$$

The corresponding Lagrange is:

$$L(\lambda_\alpha, \lambda_\beta, \alpha, \beta) = \alpha' K_x K_y \beta - \frac{\lambda_\alpha}{2}(\alpha' K_x^2 \alpha - 1) - \frac{\lambda_\beta}{2}(\beta' K_y^2 \beta - 1) \tag{2}$$

Taking derivatives in respect to α and *beta* we obtain

$$\frac{\partial f}{\partial \alpha} = K_x K_y \beta - \lambda_\alpha K_x^2 \alpha \tag{3}$$

$$\frac{\partial f}{\partial \beta} = K_y K_x \alpha - \lambda_\beta K_y^2 \beta \tag{4}$$

Subtracting β' times the second equation from α' times the first we have

$$\alpha' K_x K_y \beta - \lambda_\alpha \alpha' K_x^2 \alpha - \beta' K_y K_x \alpha + \lambda_\beta \beta' K_y^2 = 0 \tag{5}$$

Which together with the constraints implies that

$$\lambda_\alpha - \lambda_\beta = 0 \tag{6}$$

So considering the case:

$$\lambda = \lambda_\alpha = \lambda_\beta \tag{7}$$

We eventually get the following matrix form:

$$\begin{bmatrix} 0 & K_x K_y \\ K_y K_x & 0 \end{bmatrix} \begin{bmatrix} A \\ B \end{bmatrix} = \lambda \begin{bmatrix} K_x K_x & 0 \\ 0 & K_y K_y \end{bmatrix} \begin{bmatrix} A \\ B \end{bmatrix} \tag{8}$$

This procedure finds subspaces in the linear space spanned by the eigenfunctions of the kernel functions such that projections onto these subspaces are maximally correlated. Operationally, KCCA produces a matrix A consisting of the basis vectors of a subspace onto which K_x may be projected, giving $K_x \times A$, and a matrix B consisting of basis vectors of a subspace onto which K_y may be projected, such that $K_x \times A$ and $K_y \times B$ are maximally correlated.

3.2 Prediction Method

KCCA uses kernel functions to compute distance metrics between all pairs of workload vectors and pairs of performance vectors. A kernel function allows us to transform non-scalar data into scalar vectors, allowing us to use algorithms that require input vectors in the form of scalar data. Since our features and performance results contain categorical and non-numeric data, we create custom kernel functions to transform our datasets.

Given N instances, we form an $N \times N$ matrix K_x whose $(i, j)th$ entry is the kernel evaluation $k_x(\mathbf{x}_i, \mathbf{x}_j)$. We also form an $N \times N$ matrix K_y whose $(i, j)th$ is the kernel evaluation $k_y(\mathbf{y}_i, \mathbf{y}_j)$. Since $k_x(\mathbf{x}_i, \mathbf{x}_j)$ represents similarity between $\mathbf{x}_i$ and $\mathbf{x}_j$, and similarity for $k_y(\mathbf{y}_i, \mathbf{y}_j)$, the kernel matrix K_x and K_y are symmetric and their diagonals are equal to one. Our similarity metric is constructed from Gaussian kernel functions.

We then project K_x and K_y onto subspaces α and β. For this projection step, KCCA calculates the projection matrices A and B, respectively consisting of the basis vectors of subspaces α and β. In particular, the matrices A and B are calculated using the generalized eigenvector problem formulation in (8) such that the projections $K_x \times A$ and $K_y \times B$ are collocated on subspaces α and β. Once we build KCCA model, performance prediction is described as follows. For the new instance $\mathbf{x}_{new}$, we calculate $u_{new} = K_{xnew}A$. The K_{new} is different from K_x. It is $1 \times N$, and its every dimension is $K(\mathbf{x}_i, x_new)$. Then we infer the jobs coordinates on the performance projection subspace β by using its 3 nearest neighbors in the job projection. In the original sample, we used the weighted average method to complete the forecast.

In the question of weighted average, we consider an idea that when the distance between two points is infinitely close to 0, then the weight of this point should be infinitely close to 1. Conversely, if two points are far enough, then the weight of this point is infinitely close to zero. Thus we divide into two steps to get the distance-weighted weights.

We have chosen the *tanh* function, in which $x \in \mathbf{R}$, $y \in [0, 1]$. The specific form of the function is as follows:

$$tanh(x) = \frac{1 - e^{-2x}}{1 + e^{-2x}} \tag{9}$$

The function image is:

We denote the distance between the new point u_new and the nearest three points as d_1, d_2, d_3 and tanh function can be transform into:

$$tanh(d) = \frac{1 - exp(-2/d)}{1 + exp(-2/d)} \tag{10}$$

Thus the function of the tanh function is to map the distance of points to an interval of [0,1]. Then let

$$D = tanh(d_1) + tanh(d_2) + tanh(d_3) \tag{11}$$

Then the weights are $\frac{tanh(d_1)}{D}$, $\frac{tanhd_2}{D}$, $\frac{tanh(d_3)}{D}$ respectively. Finally, according to the characteristics of the kernel function, we can calculate the distance between two points which are in the high-dimensional space in the current feature space as follows (Fig. 1):

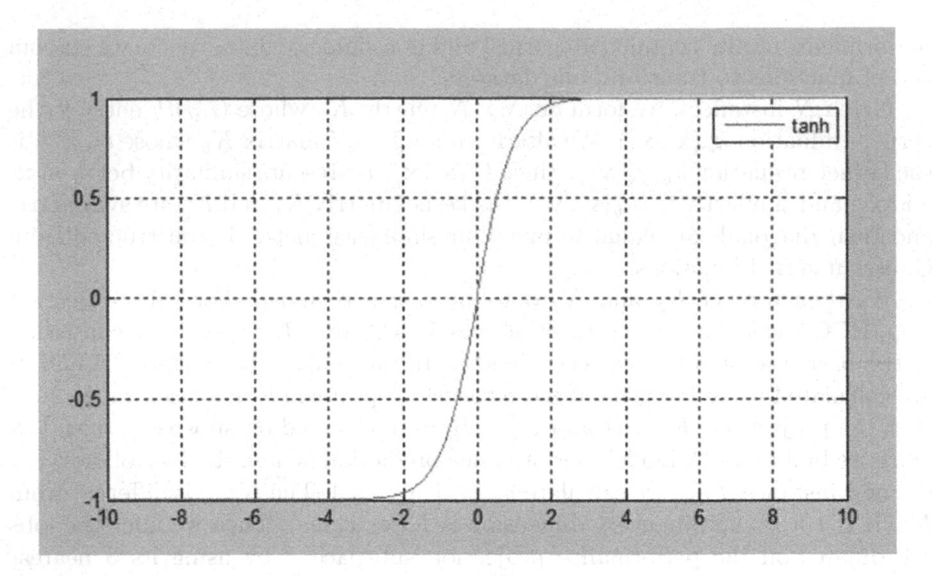

Fig. 1. The function image of tanh.

$$\begin{aligned}\|\phi(x)-\phi(x')\|^2 &= (\phi(x)-\phi(x'))^T(\phi(x)-\phi(x')) \\ &= \phi(x)^T\phi(x) - 2\phi(x)\phi(x') + \phi(x')^T\phi(x') \\ &= \langle\phi(x),\phi(x)\rangle - 2\langle\phi(x),\phi(x')\rangle + \langle\phi(x'),\phi(x')\rangle \\ &= \kappa(x,x) - 2\kappa(x,x') + \kappa(x',x')\end{aligned} \tag{12}$$

4 Deep Learning Based Model

Before the advent of deep learning, people generally compensated for the lack of computing power in high-dimensional space through the use of kernel functions. Designing specific mapping functions in a specific area has a huge effect on improving accuracy. It has been proved that as long as there is a mapping function, there will be a corresponding kernel function. After the theory of deep learning has been perfected, it is generally believed that the deep learning model is actually learning mapping functions. Therefore, for each type of task, the deep learning model can learn specific mapping functions. Meanwhile, the current trend is to make training predictions through a single model so as to achieve the goal of becoming less desirable [11]. We designed a two-layer model based on deep learning.

Data Analysis. Before we select our model, we should analyze the distribution of data intuitively to find out best model. We use word count program, which is in Hadoop example package, to run on the prepared data set. We just pick up some important relationships between time and the size of data set.

First, we look at the relationship between average map time and the input size in Fig. 2.

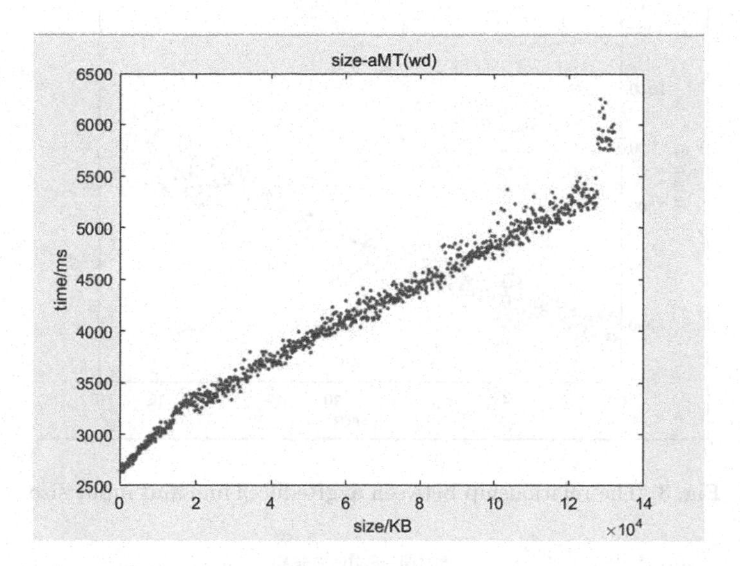

Fig. 2. The relationship between avgReduceTime and input size.

We can clearly find that the correlation is almost equal to 1, and it shows good function characteristics. The same phenomenon occurs in the reduce stage, which we can see in Fig. 3.

The point of supplemental explanation is that the input size of reduce is the same as the output of map, therefore we can easily acquire through history server. It shows a great relationship between time and size of data set because of $o(n)$ time complexity. The last but not least, we show the relationship between the sum of every phase and size in Fig. 4, and we surprisingly find out the number of input split has a great influence on the final result, because the breakpoints occur where data jumps occur.

Model Design. After analyzing the issue we studied in this paper, a simple deep learning neural network was finally selected. Since we need more to explore a functional relationship between selected feature and performance. The number of hidden neurons is generally determined based on experience. Since we do not have many features and the amount of data is not large, we determine that we have two hidden layers and each hidden layer is three times the number of neurons in the input layer. Meanwhile, as the input to the underlying model, we need to determine a suitable output for the model. Looking through the logs, it is found that the following three times are closely related to the total execution time: avgMapTime, avgShuffleTime, avgMergeTime respectively denotes average Map time, average shuffle time and average merge time. We also find an interesting phenomenon from the log that the average reduce time is the opposite number of shuffle time. Thus we only select one of them.

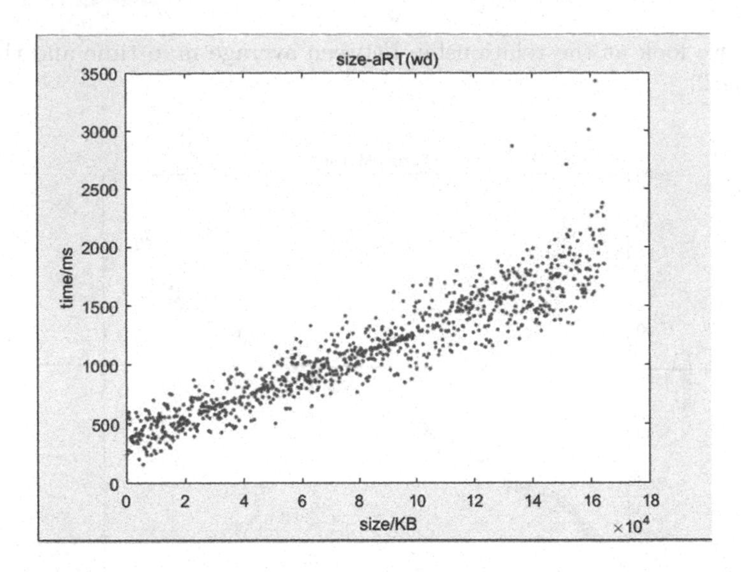

Fig. 3. The relationship between avgReduceTime and input size.

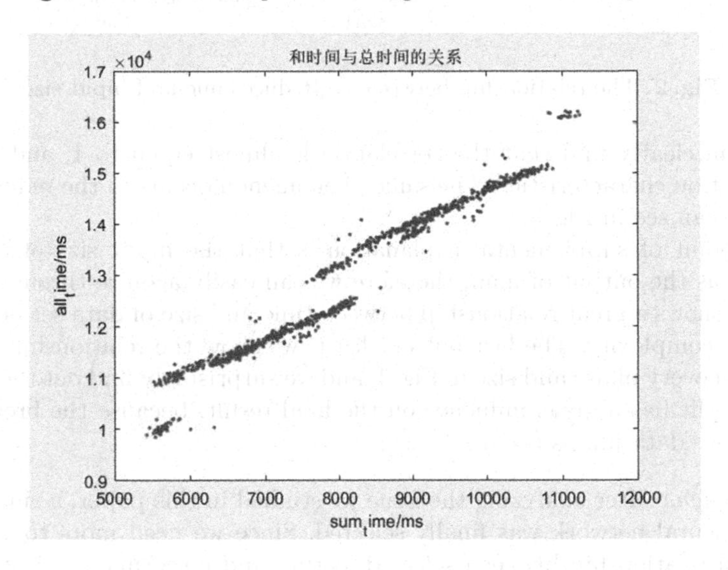

Fig. 4. The relationship between avgReduceTime and input size.

Then we need to determine the underlying model. The inspiration comes from our previous prediction model. Our final result is a weighted average of the original values. Previous research has used tree structures to solve this regression problem [12]. Random forest will have a good performance on the general data set. However, when we analyze the specific problem, bagging is actually more appropriate. Since strictly speaking, under the condition that we control the environmental conditions, the uniform task will run the same time on the same data set. What we are looking for is a better fit to the data. Thus we select Gradient Descent Regression Tree. It is a compound tree and each of its new tree learns the previous residuals under the square loss function. Therefore, under the conditions of our strategy for generating training data, we will achieve a very accurate result. Of course the final output is the total implementation time.

5 Experiment

We use a non-replacement method to collect the training set, with a ratio of 10%. In KCCA model, we need data for preprocessing. The preprocessing process needs to make the average of the data zero. Taking into account the characteristics of the training data, that is, the data span is too large, thus the variance of the Gaussian kernel function is set to 0.1. In this experiment, we chose the average error rate and the result is as follows (Fig. 5):

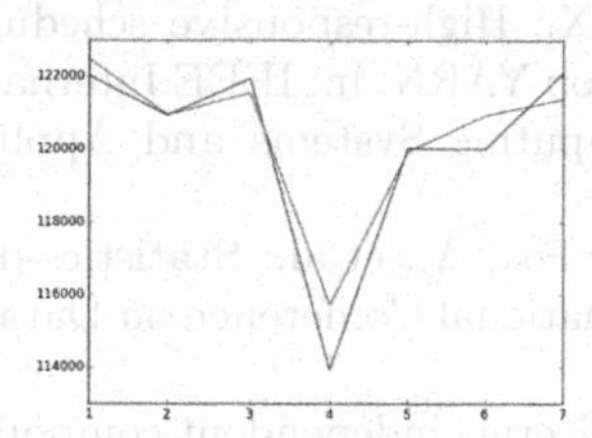

Fig. 5. The result of KCCA model.

The abscissa is the sample number, and the ordinate is time.

The average error rate is 18.5%. However, the most important thing is that the time it takes for the calculations is unbearable for us, even if the amount of data is not so big. Then we try to use the second model which is combined with deep learning model and gradient descent regression tree. We use the Sigmod function as an activation function, because of the regression problem. Then we set the maximum depth of the tree to no more than 3 to guarantee a certain degree of generalization performance. We use random sampling for verification. We also use the average error rate and the result image is as follows (Fig. 6):

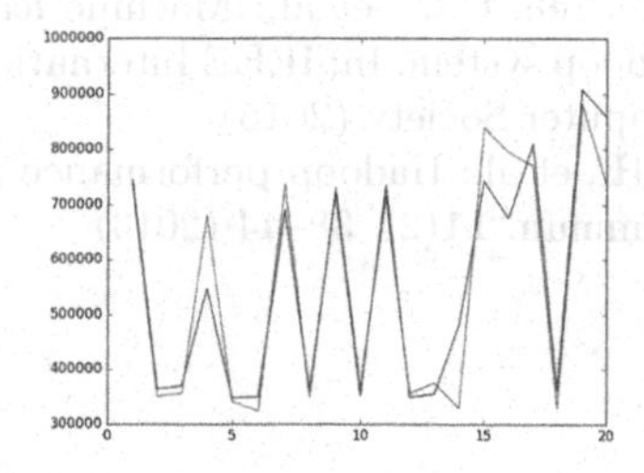

Fig. 6. The result of mixing model.

Similarly, the abscissa is the sample number, and the ordinate is time.
The average error rate is 0.2%. Therefore, this result proves that as long as we select the right features and have enough data, we can make a very accurate prediction of a task.

References

1. Song, G., Meng, Z., Huet, F., et al.: A hadoop mapreduce performance prediction method. In: IEEE International Conference on High Performance Computing and Communications and 2013 IEEE International Conference on Embedded and Ubiquitous Computing, pp. 820–825. IEEE (2013)
2. Lin, X., Meng, Z., Xu, C., et al.: A practical performance model for hadoop mapreduce. In: IEEE International Conference on CLUSTER Computing Workshops, pp. 231–239. IEEE (2012)
3. Khan, M., Jin, Y., Li, M.: Hadoop performance modeling for job estimation and resource provisioning. IEEE Trans. Parallel Distrib. Syst. **27**(2), 441–454 (2016)
4. Liu, Y., Zeng, Y., Piao, X.: High-responsive scheduling with mapreduce performance prediction on hadoop YARN. In: IEEE International Conference on Embedded and Real-Time Computing Systems and Applications, pp. 238–247. IEEE (2016)
5. Ganapathi, A., Chen, Y., Fox, A., et al.: Statistics-driven workload modeling for the cloud. In: IEEE International Conference on Data Engineering Workshops, pp. 87–92. IEEE (2010)
6. Bach, F.R., Jordan, M.I.: Kernel independent component analysis. J. Mach. Learn. Res. **3**(1), 1–48 (2002)
7. Hardoon, D.R., Szedmak, S., Shawe-Taylor, J.: Canonical correlation analysis: an overview with application to learning methods. Neural Comput. **16**(12), 2639–2664 (2014)
8. Malekimajd, M., Ardagna, D., Ciavotta, M.: Optimal map reduce job capacity allocation in cloud systems. ACM Sigmetrics Perform. Eval. Rev. **42**(4), 51–61 (2015)
9. Verma, A., Cherkasova, L., Campbell, R.H.: ARIA: automatic resource inference and allocation for mapreduce environments. In: International Conference on Autonomic Computing, ICAC 2011, Karlsruhe, Germany, June 2011, pp. 235–244. DBLP (2011)
10. Mathiya, B.J., Desai, V.L.: Apache hadoop yarn parameter configuration challenges and optimization. In: International Conference on Soft-Computing and Networks Security, pp. 1–6. IEEE (2015)
11. Chen, C.O., Zhuo, Y.Q., Yeh, C.C., et al.: Machine learning-based configuration parameter tuning on hadoop system. In: IEEE International Congress on Big Data, pp. 386–392. IEEE Computer Society (2015)
12. Bei, Z., Yu, Z., Zhang, H., et al.: Hadoop performance prediction model based on random forest. ZTE Commun. **11**(2), 38–44 (2013)

Scheme of Cloud Desktop Based on Citrix

Xia Liu[1(✉)], Xu-lun Huo[1(✉)], Zhao Qiu[2], and Ming-rui Chen[2]

[1] Sanya Aviation and Tourism College, Sanya 572000, Hainan, China
paolo_lx@qq.com, 284627151@qq.com

[2] Information and Technology School, Hainan University, Haikou 570228, Hainan, China
5454734@qq.com, 1607885098@qq.com

Abstract. As the increasing number of applications in Office Mobile, it's hard for the traditional IT network infrastructure to carry out remote management and maintenance. The Citrix-based Cloud Desktop can manage and maintain users' terminal equipment, reduce the operation and maintenance costs and fulfils the enterprise's centralized management. Meanwhile, Cloud Desktop provides users with a personalized desktop and a variety of access methods which improve service levels and business continuity needs. Compared to the traditional IT infrastructure, the Cloud Desktop provides users with faster & higher predictability, higher cost performance and richer applications. Moreover, it is particularly prominent in energy conservation and cost savings.

Keywords: Cloud computing · Citrix · Desktop virtualization
Network architecture · Centralized management

1 Introduction

Cloud Computing has been always the hot topic in IT industry since 2007. Its definition has been described in detail by IBM technical white book about "Cloud Computer" [1], Professor Liu [2] and Wikipedia [3]. About the desktop cloud, it was once defined by IBM Smart Business Desktop Cloud as "the application and the entire client's desktop through the client's terminal or any other equipment connected with internet" [4].

2 Advantages of Cloud Desktop and Citric-Based Network Structure

2.1 Safer System

As the hard disk is prohibited, it can prevent virus infection from the network internally, data leakage from the internal network, result in invisible local disk on chip, prevent the operator from damaging the client's equipment due to misoperation, guarantee the manageability of the user's authority and safety of system data and application software, enable the administrator to supervise illegal users to stop his/her dangerous operation in time; besides, the transmission safety can be guaranteed because only information will be transmitted when input or output via screen, keyboard

Q. Zhou et al. (Eds.): ICPCSEE 2018, CCIS 901, pp. 415–425, 2018.
https://doi.org/10.1007/978-981-13-2203-7_32

and mouse. The application programs will operation completely based on server. In this sense, illegal shutdown will not be concerned because the terminal itself uses DOM card whose performance is far stronger than the hard disk. Therefore, the illegal shutdown will not damage the disk's sectors; in addition, the terminal adopts the remote control mode with important documents stored in the server. Consequently, the illegal shutdown will not result in the data loss.

2.2 More Convenient Management

The administrator can release the applications to the users in time. There's no need of setting the user terminal. The administrator may implement remote control the terminal uses to avoid any irrational use. When the users demand any assistance, the system administrator may carry out interactive operation by remote control. The system setting, applications and data at the terminal end can also be fully disposed by the administrator to avoid losses resulting from the user's misoperation as much as possible [5].

2.3 Reduction of Upgrade Fees and Upgrade Workload

At present, the computer is usually served for three to five years with upgrading afterwards when the terminal mode doesn't have to be upgraded but expanding the general server. If hundreds of work stations have to be installed, it will not only consume a great amount of manpower and material resources, but also postpone the engineering progress. Thus WIN terminal adopts WINCE embedded system. In this way, the users can work upon resetting and getting reconnected with the backend server when modifying the terminal mode.

2.4 Speed Increase

The current applications are mostly troubled by slow speed; but the terminal can increase speed greatly in that the terminal only transmits screen and the data are exchanged between PC server and minicomputer. In this way, the speed will not encounter the bottleneck.

In addition, the cloud desktop is also characterized by energy conservation, lower cost and short response time to IT service, etc.

2.5 Citrix-Based IT Network Structure and Advantages

At present, the primary cloud computing service providers have their respective cloud desktops. Citrix, VMware, MED-V and SUN, etc. are mostly applied on the market. Among others, the Citrix enterprise-level network solution based on the server involving the computer users in diverse industries including large and medium-sized enterprises and public institutions and the emerging application service providers (ASP) have provided abundant applications to enormous clients which have been universally applied in a wider area and scope by virtue of higher efficiency, favorable predictability and satisfying cost performance [6]. Such solution, in comparison with

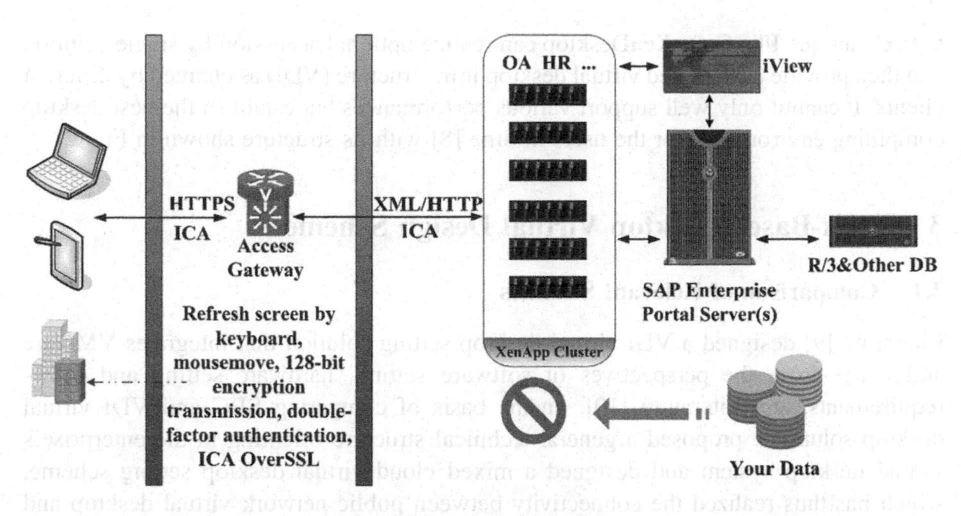

Fig. 1. Citrix-based IT structure

Table 1. Comparison between Citrix and non-Citrix based solution

Non-Citrix based solution	Citrix-based solution
Shall be installed on the desktop	Shall be separated from the operation system
One image to each desktop	Only store and maintain an OS image
Each desktop has to be administered independently	Operate independently with dynamics released to the users
Still a large amount of maintenance and storage only transferred from the front end to back end	Manage the user files independently in a logic and concentrated manner

non-Citrix based structure highlight unique advantage. Its advantages are shown in Table 1 [7] and its structure is shown in Fig. 1.

Citrix adopts uniform receiver at the user terminal so that users may visit their own desktop applications via any computer. With Citrix HDX technology, XenDesktop may guarantee the users to have sound experience via whatever equipment. Upon integration

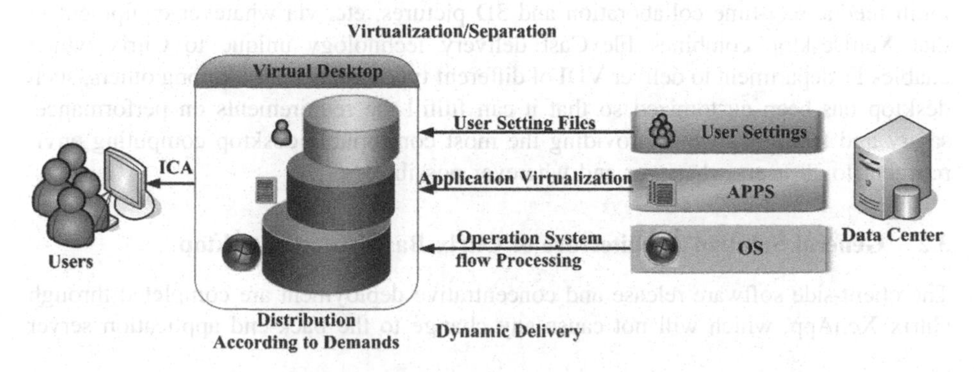

Fig. 2. The optimal VDI solution provided by Citrix

Citrix's unique FlexCast, XenDesktop can realize optional accession by single solution and then provide customized virtual desktop infrastructure (VDI) as claimed by different clients. It cannot only well support various performances but establish the best desktop computing environment for the users in time [8] with its structure shown in Fig. 2.

3 Citrix-Based Desktop Virtual Design Scheme

3.1 Comparison of Relevant Schemes

Literature [9] designed a VDI virtual desktop setting solution that integrates VMware and Citrix from the perspectives of software setting, hardware setting and safety requirements, etc. Literature [10], on the basis of comparing SBC and VDI virtual desktop solutions, proposed a general technical structure orienting to the enterprise's virtual desktop system and designed a mixed cloud virtual desktop setting scheme, which has thus realized the connectivity between public network virtual desktop and Intranet system. Literature [11] proposed a VMware View-based desktop visual solution and analyzed its deficiencies. Literature [12] proposed the basic framework, deployment units and application mode of a desktop cloud teaching platform based on Citrix virtualization technology. Literature [13] proposed a virtualization-based cloud desktop technical solution, designed the general structure scheme of the desktop at the business hall based on the cloud desktop technical principles, and further designed the software, application structure and network, etc. in a detailed manner. Literature [14], upon comparison of such virtual solutions as VMware, vSphere, Hyper-V, Citrix Xen, Oracle VirtualBox and Ret Hat KVM, etc. on the basis of a typical cloud desktop system structure, proposed the selection of cloud desktop in terms of availability, easy use, expansibility and cost performance, etc. Literature [15] proposed the new mode how the cloud desktop would manage the multi-media classroom, designed a cloud administration structure based on the terminal mode of "concentration, dispersion and isomerism" for coupling of the applications and the operation system, which has created an environment of virtual operation for the applications.

In general, with Citrix's receiver as a uniform user terminal, XenDesktop allows the users to visit their own desktop and the enterprise's applications via any PC, isomerism, thin client and smartphones, etc. XenDesktop adopts Citrix HDX technology to guarantee all the users may have excellent HD experience especially when involving multi-media, real-time collaboration and 3D pictures, etc. via whatever equipment in that XenDesktop combines FlexCast delivery technology unique to Citrix which enables IT department to deliver VDI of different types to all users. Among others, such desktop has been customized so that it can fulfill the requirements on performance, safety and flexibility while providing the most comfortable desktop computing environment to all users whenever and wherever possible.

3.2 General Solution Architecture of Citrix-Based Cloud Desktop

The client-side software release and concentrative deployment are completed through Citrix XenApp, which will not cause any change to the back-end application server architecture so that all office software and applications as released may visit at the client side by XenApp. Its overall framework is shown in Fig. 3 as follows:

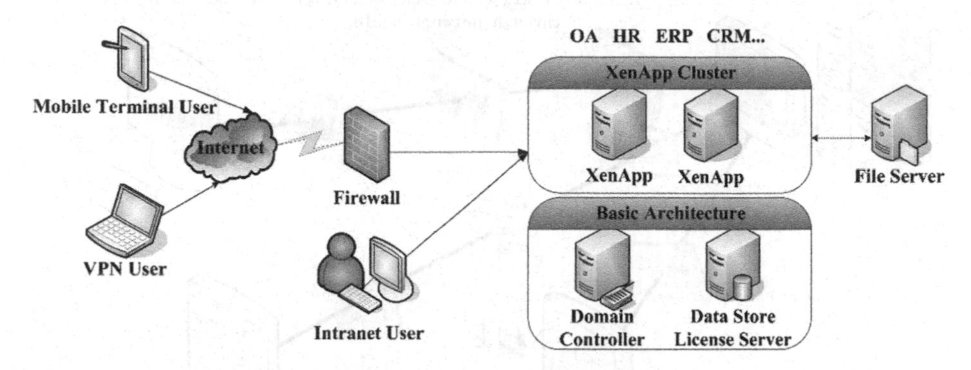

Fig. 3. Overall structure

3.3 Realization of Citrix Cloud Desktop Functions

3.3.1 Centralized Management

Allocate applications within the data center by virtualized manner [16]. As the server and the client-side are set within the same intranet, its safety and applicability can be greatly upgraded; therefore, the users may visit the data center at any network and terminal rapidly. In the course, the enterprises just have to manage and maintain the data center within the intranet, which has greatly simplified the operation and maintenance [17]. Besides, the installation and configuration of applications can be carried out at the server; therefore, it is convenient to deploy and configure the office environment.

XenApp provides a server-based computing to the users, which supports the release of virtualized applications [18]. ICA agreement, as the core technology, connects the remote client-side equipment on its server and the application process to relocate various input and output data associated with application process on the I/O equipment at the remote client side through up to 32 ICA virtual channels including keyboard, mouse, port and printing, etc. In comparison with the installation and operation of the client-side software at the client side, although no such client-side software operates on the client's equipment, the users will not feel any change in operation. The release principle of XenApp virtualized application is shown in Fig. 4 as follows:

As a highly efficient data exchange agreement, information from mouse, keyboard and screen refresh is transmitted upon encryption and compression between ICA remote terminal equipment and the central server. At this moment, the single connection takes up less than 20 KB bandwidth [19].

The utilization of the said mode enables the enterprises to deploy their overall framework in a more efficient manner, realizes the transition from the distributed to centralized management, and thus greatly improves the application visit and safety performance, etc.

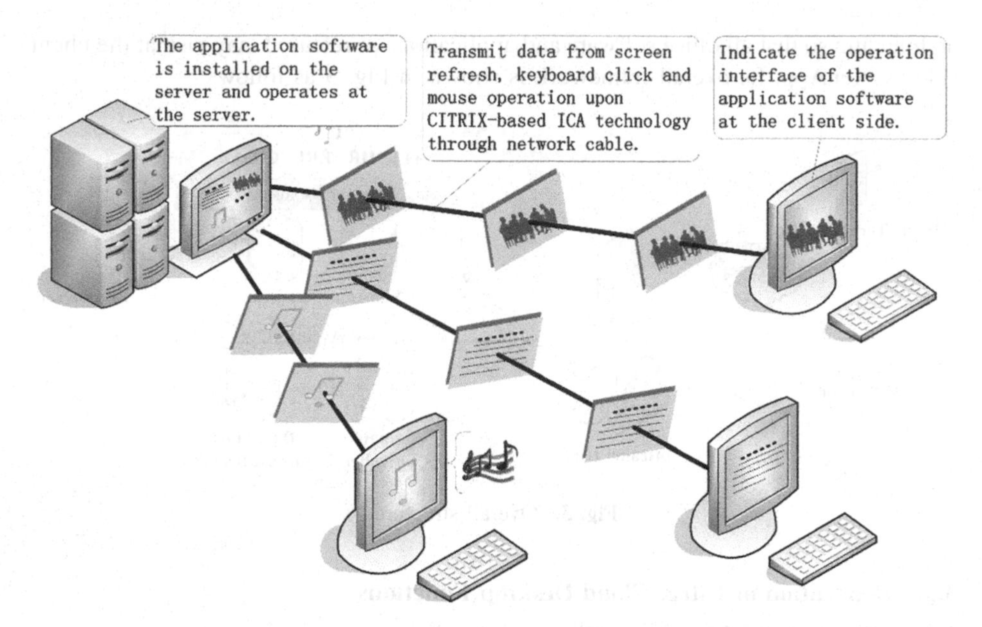

Fig. 4. Release principles of XenApp virtualization application

3.3.2 Store and Separation

The combined action of Windows Server users' profile mechanism and NTFS file system provides individual private space to the users at the server; moreover, the server stores the users' AD, file data as stored and personal space under safe management environment through NTFS special file permission mode [9], which can not only grant the authority of acquiring and visiting the users' data to the specific administrator but also manage the behavior of interleaving access to other users' data.

Meanwhile, the reorientation by configuration of Windows Server's files can guarantee the users to visit personal data by logging on any server.

3.3.3 Remote Access and Visit Control

Citrix provides uniform safe access to the users. Common access flow is shown as follows [9]:

Firstly, identify the user's identity. XenApp integrates diverse ways of identity authentication including double factor authentication, provision of the user's name, command and passcode by the visiting users.

Secondly, after authentication, the user will obtain an encrypted link; by this way, when the user logs on his/her own portal, he/she can use various applications.

Thirdly, manage with the virtualization server and command provided by Citrix to call and log on system or software rapidly and automatically; thus the users may experience as if native applications.

Fourthly, the administrator may configure the users group permitted by each application very conveniently and record the uses' use of each application completely. The users' visit of applications is strictly controlled by Citrix application delivery platform, which provides diverse access strategies to manage the user's access requirements under different situations. And their operations in the entire course and duration of visiting the enterprise's resources are controllable, even printing, copy and paste, save to local place and terminal, etc. which are controllable operations and resources.

3.4 Realization of Citrix Cloud Desktop Technology

3.4.1 Horizontal Expansion

The sever embedded with Citrix highlights cluster function. With such configuration, each cluster is a Farm and the users only have to add the server to the Server Farm for horizontal expansion; in this way, the scale of office system can be magnified.

3.4.2 Load Balancing

As shown in Fig. 5, on the Server Farm of Citrix, the exchange with servers and the collection of dynamic information in respect of CPU and the internal memory, etc. are all completed by the load balancing dispatch server of "Data Collector", which will also be responsible for arranging the application requests to the server of the minimum load [11].

The Server Farm guarantees the server's high availability. When the single server malfunctions, the user will be forwarded to other server with smaller load so as to guarantee his/her normal use and reduce the single malfunction as well.

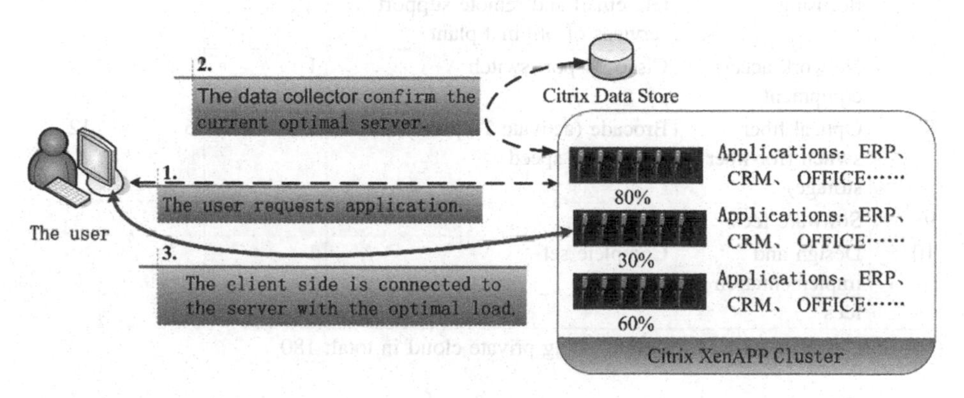

Fig. 5. Load balancing

4 Price Analysis of Desktop Virtualization

The one-time investment in establishing private cloud is relatively higher than that by traditional pure PC method; but the operation cost upon establishment is much lower than the latter, for example, a set of traditional computer host is about 200 W, which consumes 0.2 degree of power per hour while the power of a cloud terminal is about 10 W; the service cycle of a computer is about five years with 8–10 years for the cloud terminal; besides, the cloud terminal basically does not demand hardware maintenance, etc. In comparison with 400 traditional personal computers and 400 cloud terminals within five years, it is not hard to find that the total cost of cloud desktop is much lower than that of the former. See Table 2 for the price budget of desktop virtualization (about RMB1.6 million only for application virtualization).

Table 2. Price budget (unit: RMB10,000)

No.	Equipment type	Equipment description	Quantity	Unit price	Total
1	Desktop server	Ultramicro 2-way, 6-core, CPU98G	7	4	28
2	Server chassis	Ultramicro 7U	1	12.5	12.5
3	Access server	Dell 2-way 12G internal memory	1	1.5	1.5
4	Virtual machine storage	hp P6000 double control, 24 pieces of 600 15K FC hard disks, 24 pieces of 2 TB SATA hard disks	1	15	15
5	Thin client terminal	Chinavdi cloud base thin client	400	0.15	60
6	Virtual desktop software licensing	Free license version for one year, upgrading and one-year tel., email and remote support services of original plant	300	0.1665	49.95
7	Network access equipment	Cisco 24-port switch	1		
8	Optical fiber switch (for fiber storage)	Brocade (activate 16 ports); 8 GB port speed	2	6	12
9	Software fees	Windows			
10	Design and implementation fees	Complete set	1		

One-time investment fees for establishing private cloud in total: 180

Table 3 compares the Total Cost of Ownership (TCO) of desktop virtualization and PC. Assuming the investment of RMB1.8 million in 400 sets of cloud desktop, the service life of PC machine is 5 years while that of the cloud desktop is 8 years. The maintenance and management cost of PC is RMB30/month. The power consumption of cloud desktop is 10 W/set/h with 200 W/set/h for PC machine as per the unit price of RMB0.59/degree, power on for 8 h a day and 198 days a year (9 months * 22 days/month):

Table 3. TCO analysis of cloud desktop and PC in five years (unit: Yuan)

Category	Traditional PC			Cloud desktop		
	Cost per unit	Quantity	Category	Cost per unit	Quantity	Category
Display	1000	400	Display	1000	400	Display
Depreciation of mainframe hardware cost	2500	400	Depreciation of mainframe hardware cost	2500	400	Depreciation of mainframe hardware cost
Maintenance & management expenses	360/set/year	400	Maintenance & management expenses	360/set/year	400	Maintenance & management expenses
Terminal energy charges	187/set/year	400	Terminal energy charges	187/set/year	400	Terminal energy charges
Air-conditioner energy charges	PUE = about 2400 persons		260,000	Air-conditioner energy charges	PUE = about 2400 persons	
Total			Total			Total

5 Risk and Countermeasures

5.1 Risk Description

The store of all the terminal user operations and data relies on the network terminal server. If the network service performance is insufficient or shutdown, the operations of all users as loaded on such terminal server shall be affected.

The store of all the terminal user operations and data relies on the network transmission. In case of problem of the network transmission or interruption of the network transmission connection point (Switch/Hub), all the users through such network transmission connection point shall be affected and negative effect of inappropriate operation will be caused to a large amount of users.

Deployment implementation. Online system switch will be implemented in the course of later implementation, which may exert effect on the system continuity and availability to a certain extent.

User experience. As the mode of cloud desktop differs from the use mode of physical PC to a certain extent, it differs in the extent to which the users accept the mode; therefore, the users may misunderstand or contradict IT services and management.

System risk. The cloud desktop provides services to the users through network on the basis of server cluster and uniform store, therefore, the malfunction of network, server and store will exert significant effect on the users.

5.2 Countermeasures

The server shall adopt the load balancing scheme and be deployed with hardware of sufficient performance.

Enhance the stability of network transmission, upgrade the core switch and double backup the network lines within various regions, and add UPS protection of the network nodes, etc.

In terms of system design, carry out perfect planning of the entire implementation plan. The system switch shall avoid the office time to reduce effect on the business system and regular working.

The preliminary survey and test of user scenario shall fully understand the client's demand of desktop application and fulfill the user's demands in the course of implementation as much as possible.

The system deployment shall take full consideration of the high reliability and availability of server, store and network, reduce the possibility of system malfunction and mitigate the effect on cloud desktop users.

6 Conclusion

The Citrix cloud desktop solution provides higher speed, higher predictability and more superior cost performance to the computer users from all walks of life while providing more abundant applications to more visitors. XenDesktop may fulfill the requirements on performance, safety and flexibility, provide the most suitable desktop computing environment to all users whenever and wherever. In meanwhile, the users may visit diverse enterprise applications and office tools centrally released by XenApp at the client side; in this sense, the client equipment can be managed and maintained in a uniform manner. Although this solution has several ways of safety ID authentication, which only aim at the user's ID verification; therefore, efforts are still devoted to soling the platform integrity verification of both parties involved in the communication and the ID verification of the sending party.

Acknowledgements. We acknowledge the support received from A cooperative science project between colleges and local government in Sanya (2013YD64). In addition, thanks to Lecturer Xu-lun Huo and Professor Zhao Qiu, correspondents of this paper.

References

1. IBM White Paper (2007). http://download.boulder.ibm.com/ibmdl/pub/softeware/dw/wes/hipods/cloud_computing_wp_final_8oct.pdf. Accessed 13 July 2016
2. Liu, P.: Cloud Computing, 2nd edn. Publishing House of Electronics Industry, Beijing (2011)
3. Wikipedia. http://zh.wikipedia.org/wiki/cloud. Accessed 10 December 2016
4. Xia, L.: Research and applications of cloud desktop. Comput. Appl. Syst. **23**(7), 12–16 (2014)

5. Yin, J.: Master of surfing the internet in the new century—one of technical lectures of windows 2000 access to internet. Softw. World (08), 154–156 (1999)
6. Yang, B., Ren, X.: Brief introduction to virtualized application platform. Silicon Valley (02),33(2011)
7. Baidu Library. http://wenku.baidu.com. Accessed 31 October 2017
8. Qing, S.: Let the user's experience to decide the success or failure of desktop virtualization. Comput. World (04), 42 (2010)
9. Yang, Z.: Virtual desktop technology and research. Inf. Safety Technol. (12), 75–77 + 91 (2011)
10. Wang, F., Lei, B.: Research on enterprise oriented virtual desktop system. Telecommun. Netw. Technol. (2), 1–6 (2012)
11. Xu, Y.: Research on application of VMware view desktop virtualization technology to practical training machine room. China Educ. Inf. (14), 81–83 (2014)
12. Li, B.: Research on construction and application of citrix-based virtual cloud desktop education platform. Comput. Disk Softw. Appl. **12**, 184–185 (2013)
13. Cheng, J.: Research and design of virtualization-based cloud desktop technical solution. Guangdong Commun. Technol. **6**, 36–39 (2011)
14. Zhong, Y.: Probe into application of library cloud desktop. Libr. Forum **6**, 106–111 (2014)
15. Lin, F., Lin, X.: Application of cloud desktop to teaching management. Libr. Res. Explor. (32), 336–338 + 343 (2013)
16. Li, M.: Probe into the design and implementation of concentrated application of the release platform at our plant. Sci. Technol. Inf. (25), 12 (2012)
17. Chen, J., Ye, M.: Construction of elastic and scalable smart office cloud platform. Cable TV Technol. (10), 74–77 (2012)
18. Liu, Y.: Application of citrix application virtualization to telecom IT supporting system. Guangdong Commun. Technol. **32**(3), 75–78 (2012)
19. Chen, X., Ding, B., He, G., et al.: Probe into construction of uniform access platform among mobile operators. Guangdong Sci. Technol. (8), 89–90 (2009)

A Constraint-Based Model for Virtual Machine Data Access Control in Cloud Platform

Zhixin Li[1,2](✉), Lei Liu[1], and Xin Wang[2]

[1] College of Computer Science and Technology, Jilin University, Changchun 130012, China
52868081@qq.com

[2] School of Computer Technology and Engineering, Changchun Institute of Technology, Changchun 130012, China

Abstract. Virtual machine (VM) data access control provides a cloud-computing platform with guaranteed safety. Given that the cloud platform environment is dynamically variable, static VM data access operational authorization is different from the dynamic cloud platform environment in state determination. This difference affects the safety and performance of VMs in the entire cloud platform. A constraint-based VM data access control model was proposed in this study to evaluate the influence of dynamic environmental change in a cloud platform on VM data access control operation. The state information of the dynamic cloud environment was considered a constraint evaluation function. The model realized organic integration of static Bell–LaPadula model safety level and dynamic cloud platform environmental information. A safety policy of VM data access control operation was established, and the capability of the constraint-based access control model to improve the safety of VMs was verified. A model implementation framework and the main functions in combination with the proposed model were realized. The effectiveness and performance of the constraint-based VM data access control model were also evaluated. Results showed that the performance loss was within 7% when the constrained VM data access control model was used for operations, such as VM management. The test of communication intensive workload of a VM indicated that the operating time of the model was increased by approximately 4%. The constraint-based VM data access control model in cloud platform could adapt to the complex dynamic cloud platform environment and improve the safety of VMs. This study provided technical and theoretical bases for VM data access control in cloud platform.

Keywords: Data access control · Virtual machine security · Cloud security

1 Introduction

Virtualization technology has been rapidly developed and widely applied in cloud platforms [1]. The safety problem [2, 3] of virtual machines in cloud platforms has become a crucial factor restricting its development. As an important type of safety mechanism, security data access control technology of virtual machines (VMs) is an effective measure of guaranteeing the safety of cloud platforms. VM data access control realizes protection of system information and resources mainly through access permission

Q. Zhou et al. (Eds.): ICPCSEE 2018, CCIS 901, pp. 426–443, 2018.
https://doi.org/10.1007/978-981-13-2203-7_33

management. Considering that cloud platform environment is under highly dynamical variations [4], such as operations like creation, deletion, migration, and restoration of VMs, the safety need and safety state of resources are constantly changing. Moreover, illegal traffic access may be generated among VMs, and traditional data access control model [5] under traditional calculation pattern does not fully consider the environmental influencing factors of cloud computation. Therefore, how to establish a VM data access control model that combines the dynamically variable cloud environment, restricts operations of VMs and their access range, and guarantees the safe use and authorized access of the VMs are the key problems that need solutions.

VM data access control models in cloud platforms emerge under such a circumstance. Based on traditional data access control model, a VM data access control model is constructed with VM resources as subject and object of access control. In terms of the present development of VM data access control models, most access control models have not handled problems in VM data access control operation efficiently; they combine the static security attribute authorization of VMs and information in the dynamic cloud platform environment. In the aspect of static security attribute authorization problem, Bell et al. [6] proposed the Bell–LaPadula (BLP) model and implemented "read down and write up" through multilevel access control (MAC) of resources, which could effectively protect information resources and prevent illegal visit. To apply the BLP model to VM access control in cloud platforms, Weng et al. [7] introduced the BLP-model-based VM access control mechanism so that the safety of VM was improved. However, when these models are applied to VM systems in cloud platforms and frequent operations like communication and migration occur among VMs, the influence of VM operation on the safety of access control cannot be guaranteed. Qian et al. [8, 9] proposed the Virt-BLP model and a virtual medium access control framework related to multilevel safety in VM systems, which solved the safety communication and access control problems of VMs but lacked consideration of the influences of such factors as the dynamic environmental state changes of subject and object, physical environment, and system state on VM access control.

In this study, based on the BLP model, VM data access control operation was implemented by introducing the state information of the dynamic cloud environment of VMs into the access control model. Static security attribute authorization and dynamic cloud platform information were combined. The influencing problem of the environmental change of VMs in cloud platforms on the security attribute of the VMs was solved to a certain degree; therefore, the multilevel safety needs in VM systems of cloud platforms were satisfied. Moreover, the safety of VMs in cloud platforms was effectively improved under the circumstance that the influence on the performance of VM platforms was minor.

2 State-of-the-Art Approaches

Traditional VM data access control models cannot effectively solve the VM safety problem due to the dynamics and complexity of the cloud platform environment [10]. Therefore, numerous studies have been conducted on VM data access control models in recent years. Virtual resource control and management [11] through an access control model are important to guaranteeing the safety and usability of virtual cloud platforms [12, 13]. VM data access control models in present cloud platforms mainly concentrate on two aspects [14]; one is VM-isolated access control, and the other is access control that controls the safety of resources shared by VMs. Most of these models realize safety access control of VMs in cloud platforms using traditional access control models. Among VM-isolated access control models, Shi et al. [15] formulated access control policies of information flow among domains in cloud platforms and verified the effectiveness of isolation mechanism. Brewer et al. [16] proposed Chinese wall policy, in which VMs in the same conflict set could not operate in the same VM monitor (VMM), and this policy guaranteed isolation among VMs, controlled information flow among VMs, and improved the safety of cloud platforms. However, these models lacked safety grading of resources while guaranteeing isolation among VMs, and illegal visit would be generated easily among VMs. In the aspect of protection of the safety of resources shared by VMs, MAC of VM resources ensures the safety of the VMs. Boebert et al. [17] established simple-type enforcement policy and conducted access control through type labels. Only when the subject VM owned an object VM label could the subject gain the access to the object to control resource sharing among different VMs. Venelle et al. [18] conducted cascade protection of all information flows in a Java VM (JVM) through application of MAC model and trans-platform monitor to JVM, and results showed high effectiveness and efficiency. These MAC models lacked restrictions from dynamic environmental constraints. During operations of VMs, such as migration or deletion, static MAC attribute authorization could not effectively adapt to the dynamic environment of cloud computation. Therefore, Fan et al. [19] proposed a time-limit multilevel safety model, and the problem of safety level time limit of the safety model was solved by introducing time limit functions. Su et al. [20] introduced a behavior-based access control model and used behavioral mapping function to realize access of the subject to object within specific range by combining BLP model. The above two models solved the access control problems of subject–object time limit and specific range. However, in cloud platforms, the frequency operations of subject–object VMs and the environmental factors of dynamic cloud computation could not guarantee the influence of change in the cloud platform environment on VM access control.

A cloud platform is a dynamic distributed system; hence, VMs usually operate on different servers in a distributed pattern, and different servers have different service capabilities and time. Most present VM access control models have not considered the dynamic environmental factors of cloud platforms. Therefore, the VM data access control model in this study not only judged the safety level of subject–object VMs, but also took the dynamic consideration of state change of subject and object, the physical environment where they are located, and the system state during the judgment process. The environment where VMs were located in a cloud platform was mapped into the environmental state set through an environment-checking function to realize the demands of VM access control in environmental constraints of the cloud platform, compensate for the influence of dynamic environmental change of the cloud platform on VM operations, and improve the safety of VM data access control operations in the cloud platform.

The remainder of this paper is organized as follows. Section 3 expounds the composition of a VM system in a cloud platform, presents the formalized abstract description of a VM system in a cloud platform, and discusses the development of a model based on constrained VM access control. Section 4 consists of the experiment and result analysis. Section 5 concludes.

3 Methodology

3.1 VM System Constitution in a Cloud Platform

Cloud platform system is a dynamic distributed system mainly consisting of two parts:

Infrastructure layer: This layer consists of computing server, storage center, and network switch, and it is abstracted into the environmental constraint of the cloud platform during modeling analysis process.

Software application layer: This layer mainly refers to the VM system consisting of software, such as VMM installed on infrastructure layer, VMs, VM template, and mirror image.

A VM system generally consists of VMM and privileged VM (*domain*0). *domain*0 is a management system on the host computer, which is used to manage physical resources and VMs of the host computer, as shown in Fig. 1.

The above abstract models include infrastructure and application software in cloud platform and can basically depict functional implementation of cloud platform and constraint of operation process with favorable completeness.

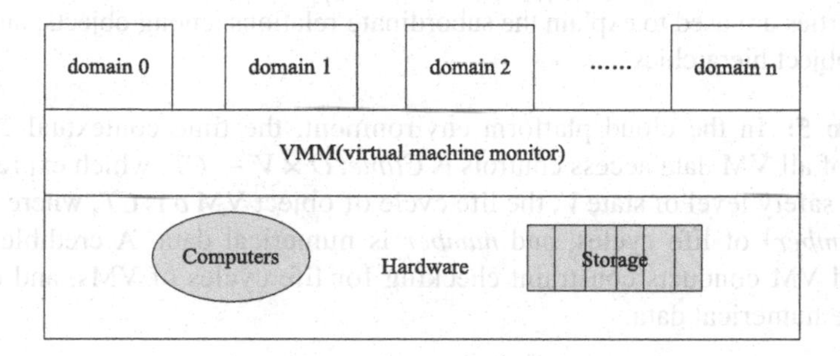

Fig. 1. VM system diagram

3.2 Formalized Abstract Description of a VM System in a Cloud Platform

A VM system in a cloud platform can be formally described with reference to BLP model.

Definition 1: Subject refers to the activity entity of active access to resources in a cloud platform system, and it is mainly VM in this paper. Subject set is expressed by S, where subject set element is expressed by s.

Definition 2: Object refers to the activity entity of passive access to a cloud platform system, and object set in this paper refers to VM or server. Object set is expressed by *O*, where object set element is expressed by *o*.

Definition 3: Operation set refers to the action sequence set *OP* of VMs in a cloud platform system, and it expresses the operations among VMs.

$OP = \{VM\_create, VM\_del, VM\_migrate, VM\_res, VM\_clone, VM\_com\}$, where *VM_create* is to create VM, *VM_del* is to delete VM, *VM_migrate* is to migrate VM, *VM_res* is to restore VM, *VM_clone* is to clone VM, and *VM_com* means VM data communication and access. Operation set *OP* includes all motions from creation of VM to resource release, and it refers to the complete life cycle of the VM; thus, it can reflect all operations of a virtualized cloud platform.

Definition 4: The state set of a cloud platform system is $V = \{v_1 \ldots v_i\}$, and the state element in the set is $V = (B, M, F, H)$, where B is the present access set, and element $b \in P(S \times O \times A)$ expresses the access of the present subject to the object in all access attributes. $A = (r, w, a)$ is the set of access attributes, where r means read only, w is read and write, and a is write only. $P(\bullet)$ is the power set. M is a set of access matrixes. One element in M_{ij} represents the access control set of subject s_i to object o_j. F is a safety level function set $\{f_s(s), f_c(s), f_o(o)\}$, where $f_s(s)$ and $f_c(s)$ respectively express the safety level function of subject and the present safety level function of subject, and $f_s(s) \geq f_c(s)$. $f_o(o)$ is the safety level function of object. H is a set $H(o)$ of object hierarchies, and it has the two following properties: (1) when $o_i \neq o_j$, $H(o_i) \cap H(o_j) = \emptyset$; (2) no set $(o_1, o_2, \ldots o_k) \subseteq O$ that results in $o_{i+1} \in H(o_i)$ and $o_{k+1} = o_1$ for $i = 1, 2, 3 \ldots k$ exists. The two properties are used to explain the subordinate relations among objects, and no ring exists in object hierarchies.

Definition 5: In the cloud platform environment, the time contextual constraint function of all VM data access controls is $Ctime: O \times V \rightarrow CT$, which expresses that under the safety level of state V, the life cycle of object VM o is CT, where CT is set $CT = \{number\}$ of life cycles, and *number* is numerical data. A credible-subject-privileged VM conducts constraint checking for life cycles of VMs, and checking results are numerical data.

Definition 6: Environmental constraints of cloud platform refer to objective factors under the cloud environmental platform where subject and object VMs are located, such as external information related to VM access control, including hardware platform server and network position. Cloud platform can realize access motions among VMs using environmental restrictions related to safety. The environmental contextual constraint function of all VM data access controls is $Evaluate: O \times V \rightarrow CE$, which represents the environmental constraint CE of object VM o under state V. Subject VM checks the environmental constraints of object VM, and checking results decide whether this operation is allowed under state V. The state of all resources in cloud platform is known through this function. Here, the quantity of resources is defined as an entity to make the quantity concept of resources universal.

Definition 7: One cloud platform system is a quintuple $(S, OP, O, Ctime, Evaluate)$. Subject and object refer to software VM entities, which can be recognized with certain functions. Operation refers to interactions used to establish VMs. Time and environmental constraints dominate these interactive VM entities.

Through Definition 7, Definition 3 stipulates cloud platform system and system state, and cloud platform system operates among VMs through constraints of infrastructure layer. The basic access attribute *A* in system state is read and write, and operation *OP* realizes access control of cloud platform through basic attributes. Change in system state denotes the change in the resource state of one VM. Accordingly, state transition function should be established to describe and record the resource allocation, resource utilization, and resource consumption of VMs.

Definition 8: The state transition function $\rho{:}OP \times v_i \rightarrow \mathrm{J} \times v_j$ describes the result of operation *OP* under the state. J is the judgment set $\mathrm{J} = \{yes, no, error, ?\}$, namely, response to the operation. *yes* means request enabled, *no* is request rejected, *error* is error, and ? means illegal request. For example, $\rho(op, v_i) \rightarrow (J, v_j)$ expresses that operation *OP* transfers system state from v_i to v_j and judges whether it can be executed.

Definition 9: Access rules express the request types of operation *OP* for basic access attributes. GR is a set of request types of operation *OP* for access attributes, namely, $\mathrm{GR} = \{g, r\}$, where *g* is (get) request, and *r* is (release) request. $\mathrm{R} = \{\mathrm{R}^1, \mathrm{R}^2, \mathrm{R}^3\}$ is a set of VM accesses, where $\mathrm{R}^1 = \mathrm{GR} \times S \times O \times A$ expresses that the subject VM requests or releases access to the object through access attribute set *A*, $\mathrm{R}^2 = \mathrm{GR} \times S \times O \times L$ means that the subject VM creates an object VM or changes the safety level of the object VM, and $\mathrm{R}^3 = S \times O$ means that the subject VM requests to delete the object VM.

3.3 Constraint-Based VM Data Access Control Modeling

VM data access control is the restriction of resource utilization and decides whether the subject can execute one operation for the object. Therefore, the objective of a VM data access control model in a cloud platform is proposed as follows: to realize access control among VMs in cloud platform by establishing a model; therefore, it can support multi-level safety. On this basis, the interaction between VM and cloud platform environment is modeled, and the cloud environmental constraint model is used to control the access control between different subject and object VMs. The interaction process of access control operation among VMs is depicted using access control rules, which can reflect the relationships among VMs in cloud platform. Moreover, they can effectively improve the operating safety of VMs in a cloud platform when used.

3.3.1 Constraint-Based Data Access Control Model

BLP model is a finite-state machine model that defines a complete set of safety model elements, safety axioms, and state transition rules, where safety model elements and state transition have already been defined in Sect. 3.2, and safety axiom is the foundation of the BLP model. Axioms 1 and 2 are compulsory access control policies, and axiom 3 is a discretionary access control policy.

Axiom 1 (ss-characteristic)**:** a state $v = (b, M, f, H)$ satisfies safety characteristic when and only when $s \in S$, $o \in b(s{:}r, w) \rightarrow f_s(s) \geq f_o(o)$. Subject safety level dominates object safety level, and the subject can read and write object operations.
Axiom 2 (*-characteristic)**:** a state satisfies *-characteristic when and only when for $s \in S$, the following are satisfied:

$$o \in b(s{:}r) \rightarrow f_c(s) \geq f_o(o),\ o \in b(s{:}a) \rightarrow f_c(s) \leq f_o(o),\ o \in b(s{:}w) \rightarrow f_c(s) = f_o(o).$$

If the present safety level of the subject dominates the object safety level, then the subject can read only (r) the object. If the object safety level dominates the present safety level of the subject, then the subject can write only (a) the object. If the object safety level is equal to the present safety level of the subject, then the subject can read and write (w) the object.
Axiom 3 (ds-characteristic)**:** a state $v = (b, M, f, H)$ satisfies discretionary safety characteristic when and only when for each $s \in S$, $(s_i, o_j, x) \in b \rightarrow x \in M_{ij}$, and it should be ensured that each of the present access will be allowed by access matrix.

Definition 10: Under the VM system state of a cloud platform, when and only when a state v simultaneously exhibit ss-, *- and ds- characteristics, it is called safety state. If each input state of state transition function ρ is safe and the state transition after operation *OP* remains safe, then the VM system of the cloud platform can be regarded as safe.

In a virtualized cloud platform, a VM *domain*0 is higher than the limits of authorization of other VMs, and it can be used to manage the VMs of other clients; thus, it is expressed by monitor management function *dom*(). The function is introduced to indicate that all operations for the object in the system must pass monitor function.

When the following conditions are satisfied, a VM system in a cloud platform implements constraint-based VM data access control policy.

(1) In VMM management function *dom*(), after operation *op* is executed, state v_i is transited into state v_j only by depending on object state, which can be observed by the subject in *dom*(*op*), as follows:

$$K_1{:}v_i \overset{dom(op)}{\rightarrow} v_j \Rightarrow [\forall o_i \in read(dom(o_i, v_j))].$$

(2) If operation *dom*(*op*) changes the environmental attribute value of object o_i, then the environmental attribute of the object can be read only using the read function through *dom*() under state v_j, as follows:

$$K_2{:}(v_i \overset{dom(op)}{\rightarrow} v_j)\ and\ (Evlauate(o_i, v_i) \neq Evlauate(o_i, v_j)) \Rightarrow [\forall o_i \in read(dom(Evaluate(o_i, v_j)))].$$

(3) If *dom*(*op*) provides the subject executing *op* with authority to change the object and the cloud platform environment allows the subject's operation for the object, then operation *op* changes the environmental attribute value of the object, as follows:

$$K_3{:}[\forall s_i \in write(dom(op))] \Rightarrow (Evlauate(o_i, v_i) \rightarrow Evlauate(o_i, v_j)) \cup (v_i \overset{dom(op)}{\rightarrow} v_j).$$

(4) Constraint check is conducted for the life cycle of a VM. If the object VM remains in the life cycle at present, then it satisfies the contextual attribute, and *dom*(*op*) can execute subject operation *op*, as follows:

$$K_4 : [Ctime(o_i, v_i) = \infty \text{ or } Ctime(o_i, v_i) = CT] \Rightarrow (v_i \overset{dom(op)}{\rightarrow} v_j).$$

The definition of safety state implies that the system satisfying simple safety characteristic, *- characteristic, and discretionary safety characteristic is a safe system. To sum up the above four conditions, when and only when initial state is safety state and each state transition satisfies the environmental constraint and life cycle check of VMs, then the system remains safe.

Theorem 1: If the initial state of a cloud platform system is safe, then the constraint-based access control policy of VMM management function *dom*() is used for operations of VMs, and the VM remains under safety state.

Proof: Only proving that the constraint-based access control policy still satisfies the three axioms of BLP model is necessary.

ss-characteristic combines four conditions of four VMMs. When and only when $s \in S, o \in b(s{:}r, w) \rightarrow (f_s(s) \geq f_o(o)) \cap (K_1, K_2, K_3, K_4)$. If the subject safety level dominates the object safety level and satisfies the conditions of the VMM management function, then the subject can read and write object operations.

Proof by Contradiction: The above conditions are assumed to be met. The subject cannot read or write the object, namely, $(f_s(s) \geq f_o(o)) \cap (K_1, K_2, K_3, K_4)$ is satisfied. Under this assumption, only constraints are added when the subject reads and writes the object; hence, constraint conditions satisfy $(K_1, K_2, K_3, K_4) \equiv 1$. $f_s(s) \geq f_o(o)$ does not hold, which is contradictory to simple safety characteristic. Therefore, when four constraint conditions of privileged VMs for VMs are satisfied, subject and object safety levels are not changed, and the subject can read and write object operations.

In a similar way, after constraint conditions are met, axioms 2 and 3 are still satisfied. The VM remains under safety state according to Definition 10.

Therefore, when privileged VM (*domain*0) is the subject, it has all access attributes to the object VM. When four constraint conditions of privileged VM for VM are met, this operation will not affect object VMs in the VM system in satisfying simple safety, *-, and discretionary safety characteristics. Hence, this practice not only keeps constraints for object VM, but also adapts to the new application scene of the VM system.

3.3.2 Operation Description of a Constraint-Based VM System in a Cloud Platform

The VM operation sets defined in Sect. 3.2 are first modeled, and each operation can be regarded as main constituent part of cloud platform management.

(1) VM (*VM_create*) is created, and subject s_i requests to create object VM o_j. The creation of a new VM is allowed only when the subject passes *dom*() constraint check. Once created, the new VM must have safety level and should be controlled by *domain*0. *VM_create* is described in the VMM management function as $dom(VM\_create) = \{(g, s_i, o_j, f_o(o_j)) \in R^2\}$, and the operation description is as follows:

$$VM\_create = \begin{Bmatrix} (?, v) & \text{if } R^2 \notin dom\{VM\_creat\} \\ (\text{yes}, (O + o_j), f_s(s_i) \geq f_o(o_j), v_i \rightarrow v_j) & \text{if } R^2 \in dom\{VM\_creat\} \\ & \&R^2 \in (\mathrm{K}_1, \mathrm{K}_2, \mathrm{K}_3) \\ (no, v) & \text{other} \end{Bmatrix}.$$

If *VM_creat* does not pass *dom*() function, then it will be an illegal request. The subject safety level dominates the object safety level, and function *dom*() and conditions $(\mathrm{K}_1, \mathrm{K}_2, \mathrm{K}_3)$ are satisfied. The request is successful, and object VM o_j is created. Related elements are added to safety level and object hierarchies, and state change is $[v_i = (b, M, f, H)] \rightarrow [v_j = (b, M, f_o \cup f_o(o_j), H \cup H(o_j))]$, where $f_o \cup f_o(o_j)$ expresses that the safety level of object VM is added to f_o and then to object hierarchy H.

Lemma 1: If the initial system state is safe, then it will still be under safety state after the creation operation of VM.

Proof: *VM_create* input state is a safety state. After the creation of VMs, state v_j shows that present access set b is not changed, and an initial safety level is only given to the newly created VM o_j. $f_s(s_i) \geq f_o(o_j)$ indicates that axiom 1 is satisfied. v_j is obviously related to the fact that S satisfies axioms 2 and 3. The constraint conditions that should be met during the creation of VMs imply that the created VM is safe.

(2) VM (*VM_del*) is deleted, and subject s_i requests to delete object VM o_j. The subject usually refers to *domain*0 VM only, which can delete VM. *VM_del* is described in the VMM management function as $dom(VM\_del) = \{(r, s_i, o_j) \in R^3\}$, and the operation description is as follows:

$$VM\_del = \begin{Bmatrix} (?, v) & \text{if } R^3 \notin dom\{VM\_del\} \\ (\text{yes}, (O - o_j), f_s(s_i) \geq f_o(o_j), v_i \rightarrow v_j) & \text{if } R^3 \in dom\{VM\_del\} \\ & \&R^3 \in (\mathrm{K}_1, \mathrm{K}_2, \mathrm{K}_3, \mathrm{K}_4) \\ (no, v) & \text{other} \end{Bmatrix}.$$

If *VM_del* does not pass *dom*() function, then it will be an illegal request. The subject safety level dominates the object safety level, and function *dom*() and conditions $(\mathrm{K}_1, \mathrm{K}_2, \mathrm{K}_3, \mathrm{K}_4)$ are met. The request is successful, object VM o_j is deleted $(O - o_j)$, and the state change is $[v_i = (b, M, f, H)] \rightarrow [v_j = ((b - b_{ij}), M \cup (M_{ij} * 0), f_o, H - H(o_j))]$. $(b - b_{ij})$ expresses that all accesses implemented in present access set b to object VM o_j are removed. $M_{ij} * 0$ means that the access attribute of subject to object in the access control matrix is set as 0. Objective VMs are deleted in object hierarchies.

Lemma 2: If the initial system state is safe, then it will still be safe after the deletion operation of VMs.

Proof: After the transition of state v_i, all accesses to the deleted VM o_j are removed. However, for the remaining VMs, their safety level relations remain unchanged, and constraint conditions (K_1, K_2, K_3, K_4) are still met. Therefore, state v_j remains safe.

(3) VM (*VM_migrate*) is migrated, and the subject s_i requests to migrate object VM o_j. The subject usually refers to *domain*0 VM only, which can migrate object VM. *VM_migrate* is described in the VMM management function as $dom(VM\_migrate) = \{(r, s_i, o_j) \cup (g, s_i, o_u, f_o(o_u))) \in (R^2 and\, R^3)\}$, and the operation description is as follows:

$$VM\_migrate = \begin{cases} (?, v) & \text{if } (R^2 \text{ and } R^3) \notin dom\{VM\_migrate\} \\ (\text{yes}, (o_j = o_u) \cap (O - o_j), & \text{if } (R^2 \text{ and } R^3) \in dom\{VM\_migrate\} \\ (f_s(s_i) \geq f_o(o_j)) \cap (f_o(o_j) = f_o(o_u)), v_i \rightarrow v_j) & \&(R^2 \text{ and } R^3) \in (K_1, K_2, K_3, K_4) \\ (no, v) & \text{other} \end{cases}.$$

If operation *VM_migrate* does not pass function *dom*(), then it will be an illegal request. The subject safety level dominates the object safety level, and function *dom*() and conditions (K_1, K_2, K_3, K_4) are met. The request is successful, and object VM o_j is replicated $(o_j = o_u)$. Object VM o_j is deleted $(O - o_j)$, and state change is

$$[v_i = (b, M, f, H)] \rightarrow [v_j = ((b - b_{ij}) \cup (b_{ij} = b_{iu}), M \cup (M_{ij} = M_{iu}), f_o, H].$$

$(b - b_{ij}) \cup (b_{ij} = b_{iu})$ means that all accesses in present access set b to object VM o_j are migrated to VM o_u, the access attributes of subject to object in $M \cup (M_{ij} = M_{iu})$ access control matrix are also migrated, and the original attributes of object safety level and object hierarchies of object VMs are retained.

Lemma 3: If the initial system state is safe, then it will still be safe after the migration operation of VM.

Proof: After the transition of state v_i, all accesses to the deleted VM o_j and access control matrix are migrated to VM o_u. Nevertheless, for the remaining VMs, their safety level relations remain unchanged, and constraint conditions (K_1, K_2, K_3, K_4) are still met. State v_j is thus still safe.

(4) VM (*VM_res*) is restored, and subject s_i requests to restore object VM under temporary storage o_j. The subject usually refers to *domain*0 VM only, which can restore object VM. *VM_res* is described in the VMM management function as follows: $dom(VM\_res) = \{(g, s_i, o_j, A) \in R^1\}$.

The operation description is as follows:

$$VM\_res = \begin{cases} (?, v) & \text{if } R^1 \notin dom\{VM\_res\} \\ (\text{yes}, (O \cup o_j), f_s(s_i) \geq f_o(o_j), v_i \rightarrow v_j) & \text{if } R^1 \in dom\{VM\_res\} \\ & \& R^1 \in (K_1, K_2, K_3, K_4) \\ (no, v) & \text{other} \end{cases}.$$

If operation *VM_res* does not pass function *dom*(), then it will be an illegal request. The subject safety level dominates the object safety level, and function *dom*() and conditions (K_1, K_2, K_3, K_4) are met. The request is successful; $O \cup o_j$. The application of object VM o_j is restored, and related elements are added to safety level and object hierarchies. The state change is $[v_i = (b, M, f, H)] \rightarrow [v_j = (b, M, f_o \cup f_o(o_j), H \cup H(o_j))]$.

The restore operation of VM is mainly to add the safety level of VM o_j under temporary storage to f_o and to object hierarchies, the original attributes of object safety level and VM object hierarchies are retained, and access attributes remain unchanged.

Lemma 4: If the initial system state is safe, then it will still be safe after the restore operation of VM.

Proof: The system state change is the same as that in Lemma 1, and therefore it is still safe.

(5) Clone VM (*VM_clone*) mainly refers to the replication operation of VM through VM mirror image or template, and the subject s_i requests to clone object VM o_j. The subject mainly refers to *domain*0 VM only, which can clone object VM. *VM_clone* is described in the VMM management function as $dom(VM\_clone) = \{(g, s_i, o_j, o_u, f_o(o_u)) \in R^2\}$.

The operation description is as follows:

$$VM\_clone = \begin{cases} (?, v) & \text{if } R^2 \notin dom\{VM\_clone\} \\ (\text{yes}, (o_j = o_u) \cap (O + o_u), & \text{if } R^2 \in dom\{VM\_clone\} \\ (f_s(s_i) \geq f_o(o_j)) \cap (f_s(s_i) \geq f_o(o_u)), v_i \rightarrow v_j) & \& R^2 \in (K_1, K_2, K_3, K_4) \\ (no, v) & \text{other} \end{cases}.$$

If operation *VM_clone* does not pass *dom*() function, then it will be an illegal request. The subject safety level dominates the object safety level, and function *dom*() and conditions (K_1, K_2, K_3, K_4) are met. The request is successful; $(o_j = o_u) \cap (O + o_u)$, namely, object VM o_j is cloned to o_u, related elements are added to safety level and object hierarchies, and the state change is $[v_i = (b, M, f, H)] \rightarrow [v_j = (b, M, f_o \cup f_o(o_u), H \cup H(o_u))]$..

$f_o \cup f_o(o_u)$ means that the safety level of object VM is added to f_o and to object hierarchy H.

Lemma 5: If the initial system state is safe, then it will still be safe after the clone operation of VM.

Proof: The system state change is the same as that in lemma 1, and thus it is still safe.

(6) In VM data communication and access (*VM_com*), the subject VM s_i requests data communication access to object VM, and the subject is allowed to implement communication access to object VM only after passing *dom*() constraint check. Access attribute A is determined according to the safety levels of object and subject VMs. The safety labels of subject and object VMs should be controlled by *domain*0. Once the safety level is determined, VM data communication is implemented according to axioms and access control conditions. *VM_com* is described in the VMM management function as $dom(VM\_com) = \{(g, s_i, o_j, A) \in R^1\}$, and the operation description is as follows:

$$VM\_com = \begin{cases} (?, v) & \text{if } R^1 \notin dom\{VM\_com\} \\ (\text{yes},(s_i, o_j, A),\ (f_c(s_i) \succ f_o(o_j)), v_i \to v_j) & \text{if } R^1 \in dom\{VM\_com\} \\ & \& R^1 \in (\mathrm{K}_1, \mathrm{K}_2, \mathrm{K}_3, \mathrm{K}_4) \\ (no, v) & \text{other} \end{cases}.$$

If operation *VM_com* does not pass *dom*() function, then it will be an illegal request. $f_c(s_i) \succ f_o(o_j)$. The present safety level of subject VM is compared with that of object VM, and the results are assessed according to axiom 2 attribute. *dom*() function and conditions $(\mathrm{K}_1, \mathrm{K}_2, \mathrm{K}_3, \mathrm{K}_4)$ are met, and the request is successful. The state change is $[v_i = (b, M, f, H)] \to [v_j = (b + b_{ij}, M + M_{ij}, f_o, H)]$.

$b + b_{ij}$ means that b_{ij} of access attribute A is added to *b*, $M + M_{ij}$ means that b_{ij} of access attribute A is added to *M*, and addition principles are determined according to the comparison results of $f_c(s_i) \succ f_o(o_j)$.

Lemma 6: If the initial system state is safe, then it will still be safe after the communication access operation of VM.

Proof: If the input state v_i of the system is safe, then proving that the output state v_j is also safe is necessary. Assuming $b_{ij} \in b$, state $v_i \to v_j$ does not change, and it is still safe. If $b_{ij} \notin b$, then $b + b_{ij}$ will be implemented, and addition principles are determined according to the comparison results of $f_c(s_i) \succ f_o(o_j)$. According to axiom 2, the addition of access attribute b_{ij} does not damage *-characteristic. Moreover, $b_{ij} \in M_{ij}$ is still satisfied according to axiom 3 (ds-characteristic). Therefore, Lemma 6 can be proven.

To sum up, the constraint-based VM data access control model is used to describe the access control process of a cloud platform. This model conforms to the multilevel safety model, and the VM data access control operation is safe.

3.4 Model Implementation Framework and Main Functions

The framework of the constraint-based VM data access control model is mainly implemented in Xen, as shown in Fig. 2. VM data access control operations are implemented through the policy management module in *domain*0. The implementation of access control policies is defined through the description in Sect. 3.3, including the definition of safety level and safety-level-based access control rules. When subject VM implements the access control operations of object VM, access control module captures this

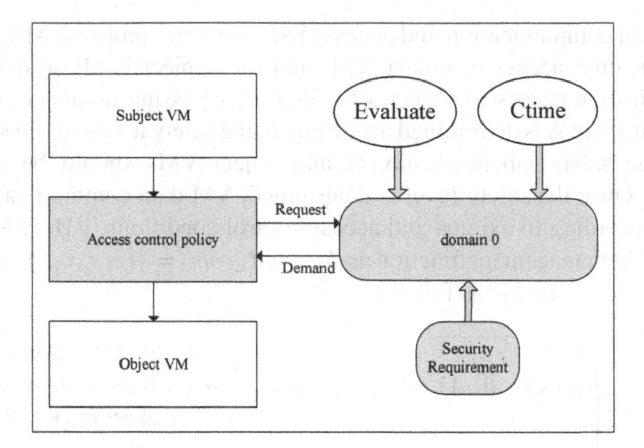

Fig. 2. Implementation framework diagram of the constraint-based VM access control model

```
struct acm_ssid_domain{
...
void *ssid;
struct domain *subject;
domid_t domainid;
...
}//Define the virtual machine ssid structure
manage_domain(domain0 domain *d)
{
read virtual machine configuration information from the file / vm;
read virtual machine running information from the file /local/domain;
Evaluate(struct domain *d);
Ctime(struct domain *d);//Constraint evaluation function
dom(OP);//determine the operating status of the virtual machine
static void acm_domain_create(struct domain *d);
static void acm_domain_destroy(struct domain *d);
static void acm_domain_migrate(struct domain *d);
static void acm_domain_restore(struct domain *d);
static void acm_domain_clone(struct domain *d);
static void acm_domain_communication(struct domain *d,struct domain *d,);
}
```

Fig. 3. Main algorithm diagram of the constraint-based VM access control model

operation request using Hooks function and sends the request to *domain*0. Whether this operation is allowed is judged according to the constrained VM data access control model. Related information of subject VM constraints are evaluated according to Evaluate and Ctime functions, judgment is made in accordance with the description of operating rules, and preparation is made for safety state transition. Whether the operation is allowed is decided according to the judgment results (Fig. 3).

The implementation of this model in Xen framework mainly includes two parts: the initialization and VM operation parts. Initialization mainly includes (1) the distribution of a memory space to VM objects, (2) the evaluation of the context of the operating environment of VMs, and (3) the configuration of the safety level information for VMs. The VM safety operation part mainly includes (1) the creation and destroying of VM, (2) the migration of VM, (3) the restoration of the execution of VM, (4) the cloning of VM, and (5) VM data communication and access. *domain*0 controls and manages other VM domain through the control interface provided by Xen, and *domain*0 can create and delete VMs according to safety level, control domain operation, suspend, restore, and migrate. The communication access among VMs is managed through *domain*0.

4 Experiment and Result Analysis

This section presents the analysis of the effectiveness and performance of the constraint-based VM data access control model. A related confirmatory experiment is conducted according to VM safety level, and an operating functional test of VMs under different safety levels is carried out. A performance test of the constraint-based VM access control model is also conducted.

Two Dawning A840-G10 servers are configured for cloud service platform, with AMD 6376 CPU, 16-core 2.3 GHz × 4 ea, 256 GB memory, and 1,000 Mbps card. Xen3.1 virtual platform is deployed in the above servers to test access control function. *domain*0 provides management service in cloud platform, and it has a high performance requirement. The configuration is shown in Table 1.

Table 1. Xen configuration

Xen version	Xen3.1
domain 0	8VCPU/8 GB memory
domain u	1VCPU/1 GB memory
os	Ubuntu Linux 2.6.38
Network	Bridging via domain 0/100 Mb

4.1 Validity Test

In the experiment, after safety levels are set, the safety level judgment of access requests, such as creation, deletion, migration, restore, cloning, and communication of VMs, is carried out to control VM operations. In the Xen virtual platform, *domain*0 has the highest safety level. In the VM data access control model in this paper, two tasks are completed through *domain*0; one is to determine the environmental constraints of object VM, and the other is to compare safety levels of subject and object VMs. In the experiment, *domain*0 is a credible subject with the highest safety level. Two VMs, *domain*1 and *domain*2, are created, and their safety levels are sequenced as *domain*0 > *domain*1 > *domain*2.

Step 1: When the subject VM operates the object VM, whether the operation is allowed by present rules will be evaluated according to the description of operation set (Sect. 3.3.2) after checking of constraint function (*Ctime*, *Evaluate*). If allowed, then authorization will be made through *domain*0, and VM operation will be implemented. On the contrary, operation request will not be allowed.

Step 2: Under the highest safety level of *domain*0, creation, deletion, migration, restore, cloning, and communication operations will be implemented for *domain u*.

Test 1: *domain*0 > *domain*1 > *domain*2, and *domain*1 and *domain*2 can be separately operated through *domain*0, as shown in Table 2. Test 2: *domain*0 > *domain*2 > *domain*1, and operation results through *domain*0 are shown in Table 2.

Table 2. Xen operation test analysis and results.

VM operation sets	Test case 1	Result	Test case 2	Result
dom(*VM_create*)	0 → 1	SO	0 → 2	SO
dom(*VM_del*)	0 → 1	SO	0 → 2	SO
dom(*VM_migrate*)	0 → 1	SO	0 → 2	SO
dom(*VM_res*)	0 → 1	SO	0 → 2	SO
dom(*VM_clone*)	0 → 1	SO	0 → 2	SO
dom(*VM_com*)	1 → 2	SO	2 → 1	FO

Experimental result 1 shows that VM *domain*1 can authorize through *domain*0 and implements data access control operations for VM *domain*2, but *domain*2 cannot. *domain*0 can implement the above data access control operations for the VMs through environmental constraints. Experiment result 2 is on the contrary. In Table 2, 0 represents *domain*0, and 1 is *domain*1. → means operation, SO means successful operation, and FO is failed operation.

4.2 Performance Test

The addition of the constrained data access control model in the Xen cloud platform system will unavoidably cause performance loss to *domain*0 and VM. Therefore, the influences of VM data access control operations and performance loss due to parallel load to VM is tested in this section under the model.

Under the above experimental environment, 20 VMs are virtualized. Reaction time is measured for virtual platform before and after constraint-based access control. Reaction time to creation, deletion, cloning, restoration, migration, and communication of VMs is measured.

The measurement results are shown in Fig. 4 (the average value of 20 times is taken). The experimental results imply that for VM management operations, the caused time delay percentage is within 7%, which is caused by the interposition of environment constrained evaluation function during the VM operation process. The delay size is closely related to corresponding operations and complexity. The average reaction time to migration, cloning and communication of VM is slower than the unconstrained situation by approximately 3% to 7%, mainly because more environmental constraint conditions and time delay of network communication should be analyzed, and some information are located on physical server. The average reaction time delay to creation, deletion, and restoration of VMs is slower by approximately 2%, mainly due to few network communication operations. The experiment shows that the proposed constraint-based access control model has a minor influence on the operating performance of VM, and its influence on system deployment is also restricted.

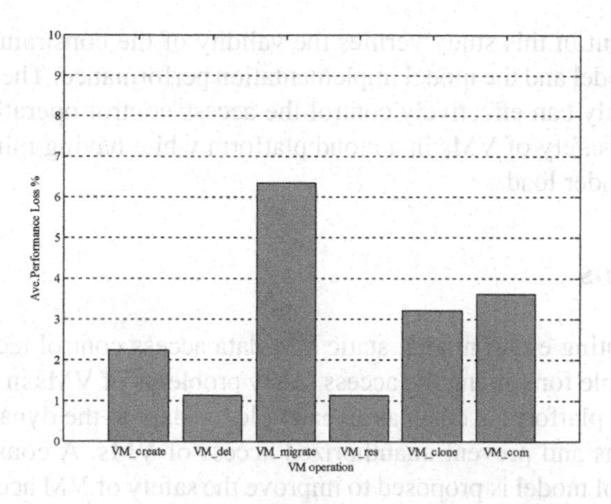

Fig. 4. Performance loss diagram of the constraint-based access control model

The influence of the constraint-based data access control model on the data communication among VMs is realized through three testing programs—scalar penta-diagonal (SP), lower upper triangular (LU), and block tri-diagonal (BT)—in a communication-intensive test example, namely, NAS Parallel Benchmark (NPB). Two VMs are used for parallel program test, and their configurations are shown in Table 1. In the experiment, the VMs can be read and written only when only two VMs have the same safety level, and only in this way can three testing programs operate normally. The test results are shown in Fig. 5.

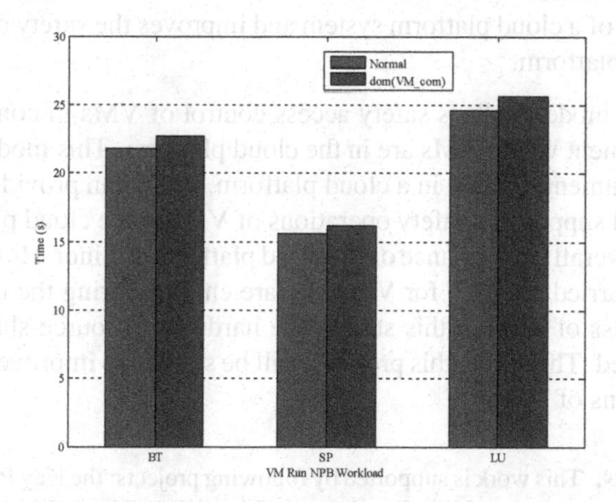

Fig. 5. Comparison diagram of NPB loads under the operation of VMs

Figure 5 compares the operation time of three loading programs (BT, SP, and LU) under normal circumstance and in constraint-based VM data access control model. The experimental results show that the operation time of the three programs increases by approximately 4% in this model and have a minor influence on the performance of VM platform.

The experiment in this study verifies the validity of the constraint-based VM data access control model and the model implementation performance. The results show that this model not only can effectively control the access control operations of VMs, but also improve the safety of VMs in a cloud platform while having minor influences on VM operations under load.

5 Conclusions

In a cloud-computing environment, static VM data access control technology will not be certainly suitable for solving the access safety problems of VMs in a cloud platform. Xen virtual cloud platform is taken as an example to adapt to the dynamic environment of cloud platforms and prevent unauthorized access of VMs. A constraint-based VM data access control model is proposed to improve the safety of VM access control in the dynamic environment. The following research conclusions are drawn:

(1) The constraint-based VM data access control model can be used to solve the safety problems of VM access operations in the dynamic cloud environment and complete such operations as creation, deletion, migration, restoration, cloning, and communication of VMs in the cloud platform.
(2) The constraint functional model can accurately reflect the dynamic environment of a cloud platform, solve the influence of dynamic environmental change in the cloud platform on the safe operations of VMs, and prevent unauthorized access of VMs due to the change in the cloud environment.
(3) The constraint-based VM data access control model exerts a minor influence on the performance of a cloud platform system and improves the safety of VM operations in the cloud platform.

The proposed model realizes safety access control of VMs in consideration of the dynamic environment where VMs are in the cloud platform. This model adapts to such a dynamic environment of VMs in a cloud platform, and it can provide convenient and accurate technical support for safety operations of VMs in the cloud platform when the influence on the overall performance of the cloud platform is minor. However, multilevel safety access is carried out only for VM software entities during the modeling process of the safety access of VMs in this study. The hardware-resource-sharing problem of VMs is disregarded. Therefore, this problem will be studied to improve the applicability of safety operations of VMs.

Acknowledgements. This work is supported by following projects: the Key Program for Science and Technology Development of Jilin Province of China (Grant No. 20130206052GX). Project supported by the National Natural Science Foundation of China (Nos. 61602057), the China Postdoctoral Science Foundation (No. 2017M611301), the Science and Technology Department of Jilin Province, China (No. 20170520059JH).

References

1. Min, L., Yu, T., Wu, X., et al.: C-DEVA: detection, evaluation, visualization and annotation of clusters from biological networks. Biosystems **150**, 78–86 (2016)
2. Reuben, J.S.: A survey on virtual machine security. Helsinki University of Technology, vol. 2, no. 36 (2007)
3. Xu, Z., Zhipeng, C., Jianzhong, L., et al.: Location-privacy-aware review publication mechanism for local business service systems. In: 36th Annual IEEE International Conference on Computer Communications, pp. 1–9. IEEE, Istanbul (2017)
4. Chen, K., Zheng, W.M.: Cloud computing: system instances and current research. J. Softw. **20**(5), 1337–1348 (2009)
5. Feng-hua, L., Mang, S., Guo-zhen, S., Jian-feng, M.: Research status and development trends of access control model. Acta Electronica Sin. **40**(4), 805–813 (2012)
6. Bell, D.E., LaPadula, L.J.: Secure Computer Systems: Mathematical Foundations. MITRE Corp, Bedford (1973)
7. Weng, C., Luo, Y., Li, M., Lu, X.: A BLP-based access control mechanism for the virtual machine system. In: 9th Proceedings of the Young Computer Scientists, pp. 2278–2282. IEEE, Hunan (2008)
8. Qian, L., Guanhai, W., Chuliang, W., et al.: A mandatory access control framework in virtual machine system with respect to multi-level security I: theory. China Commun. **7**(4), 137–143 (2010)
9. Qian, L., Guanhai, W., Chuliang, W., et al.: A mandatory access control framework in virtual machine system with respect to multi-level security II: implementation. China Commun. **8**(2), 86–94 (2011)
10. Wang, Y.D., Yang, J.H., Xu, C., Ling, X., Yang, Y.: Survey on access control technologies for cloud computing. J. Softw. **26**(5), 1129–1150 (2015)
11. Almutairi, A., Sarfraz, M., Ghafoor, A.: Risk-aware management of virtual resources in access controlled service-oriented cloud datacenters. IEEE Trans. Cloud Comput. **99**, 1–14 (2015)
12. Tourani, R., Misra, S., Mick, T., et al.: Security, privacy, and access control in information-centric networking: a survey. IEEE Commun. Surv. Tutor. **99**, 1–15 (2017)
13. Wang, J., Liu, J., Zhang, H.: Access control based resource allocation in cloud computing environment. IJ Netw. Secur. **19**(2), 236–243 (2017)
14. Bauman, E., Ayoade, G., Lin, Z.: A survey on hypervisor-based monitoring: approaches, applications, and evolutions. ACM Comput. Surv. **48**(1), 1–33 (2015)
15. Shi, Y., Guo, Y., Liu, J.Q., Han, Z., Ma, W., Chang, L.: Trusted cloud tenant separation mechanism supporting transparency. J. Softw. **27**(6), 1538–1548 (2016)
16. Brewer, D.F., Nash, M.J.: The Chinese wall security policy. In: Proceedings of the Security and Privacy, pp. 206–214. IEEE, Oakland (1989)
17. Boebert, W.E.: A practical alternative to hierarchical integrity policies. In: Proceedings of the 8th National Computer Security Conference, pp. 18–27. NIST, Gaithersburg (1985)
18. Venelle, B., Briffaut, J., Clévy, L., et al.: Security enhanced Java: mandatory access control for the Java virtual machine. In: Proceedings of the Object/Component/Service-Oriented Real-Time Distributed Computing (ISORC), pp. 1–7. IEEE, Paderborn (2013)
19. Fan, Y., Zhen, H., Cao, X., et al.: A multilevel security model based on time limit. J. Comput. Res. Dev. **47**(3), 508–514 (2010)
20. Su, M., Li, F., Shi, G.: Action-based multi-level access control model. J. Comput. Res. Dev. **51**(7), 1604–1613 (2014)

Improved DES on Heterogeneous Multi-core Architecture

Zhenshan Bao, Chong Chen(✉), and Wenbo Zhang(✉)

Faculty of Information Technology,
Beijing University of Technology, Beijing 100124, China
ccmoli@126.com, zhangwenbo@bjut.edu.cn

Abstract. DES (Data Encryption Standard) is one of the most classical algorithms of cryptography and its higher security makes it hard to be broke for a very long time. However, along with the constant development of computer technology, especially in the 21st century, DES cannot be applied widely because of its low efficiency. Recently, the novel heterogeneous multi-core architecture represented by APU (Accelerated Processing Unit), provides a new solution for the above problems. APU integrates CPU and GPU in a groundbreaking manner and makes the algorithm to make full use of the performance advantage of heterogeneous multi-core system by realizing the HSA (Heterogeneous System Architecture) standard. This paper realizes DES on the fresh APU processor. By analyzing the performance, two kinds of improved schemes are proposed. The experimental results show that the running efficiency of algorithm can be greatly improved by using APU with reasonable optimization. In the same way, the other DES-like algorithm would also be optimized on these heterogeneous multi-core architecture.

Keywords: Data Encryption Standard (DES) · APU
Heterogeneous multi-core architecture · Heterogeneous computing

1 Introduction

DES [1, 2] is an important algorithm in the field of cryptography. The algorithm was developed by IBM in America in 1972 and has been widely spread internationally. Till the early 21st century, along with the constant improvement of the performance of processor, DES is replaced gradually in part of the fields due to its key length and arithmetic speed.

However, so far, the optimization of DES has not been stopped, 3DES algorithm can make up the problem of short key of DES to some extent. As for the performance, there are research results in both software and hardware level [3–6]. The improvement of software mainly focuses on optimization of internal storage and data structure, but the effect is barely satisfactory. On the contrary, in the hardware level, using the heterogeneous system jointly constituted by FPGA (Field Programmable Gate Array), ASIC (Application Specific Integrated Circuit), GPU and CPU to run the algorithm has acquired excellent performance effects. So to speak, the emergence of heterogeneous platform provides a new opportunity for classic algorithm such as DES. APU [7] is the

Q. Zhou et al. (Eds.): ICPCSEE 2018, CCIS 901, pp. 444–451, 2018.
https://doi.org/10.1007/978-981-13-2203-7_34

"CPU-GPU" heterogeneous multi-core processor launched by AMD. From the fourth generation of Kaveri, by using HSA standard, APU can truly get the integration of software and hardware. In [8, 9], some related researches about HSA and the key technology we have done before are shown. This paper realizes DES algorithm on Kaveri based platform and proposes two different methods for improvement by analyzing the structure of processor. The application of APU optimization algorithm needs not only sufficient understanding of algorithm structure but also deep comprehension of system structure, which brings about certain challenges to the realization of this paper. The final experimental result shows that the application of APU can greatly improve the running efficiency of algorithm. But it is not real that those two methods can all realize favorable effects and one of them produces a contrary effect.

This paper is organized as follows. We give an overview of processor architecture in Sect. 2. In Sect. 3 we show the HSA-based implementation and two possible optimization methods, the iterations way and the S box way. Next, experiment results and analysis are presented in Sect. 4. Finally, Sect. 5 provides conclusions.

2 The Latest APU Architecture

APU, formerly known as AMD fusion, comes from the product concept of "The Future is Fusion". It applies the special design of placing CPU and GPU on the same chip and boasts the processing performance of high performance processor and the latest discrete graphics, which can greatly improve the operation efficiency of computers. The first generation APU Llano came out in 2011, and upgrade at the frequency of once per year. Early APU architectures, from the 1st Llano to the 3rd Kabini, do not support HSA although they designed as single chip already. In another word, the tremendous potential on both hardware and software level can not be used directly. To change this negative situation, the HSA Foundation were established in 2012 at the initiative of AMD, and they put forward HSA standard in the same year. Two years later, Kaveri, the products compliant HSA standard came out, all of these make it easier to use the advantages of fusion. See the rest of this section for more information about advanced APUs and HSA.

2.1 Kaveri and Carrizo

Kaveri, the 4th generation of APU, was launched by AMD in 2014 and was the first generation of APU supporting the HSA standard. Kaveri is a 28 nm process and has 4 processor cores of "Steamroller" microarchitecture as well as 8 Radeon R7 graphics core of GCN (Graphics Core Next) framework. Since it conforms to HSA standard, CPU and GPU in Kaveri can perform task on an equal footing, equivalent to 12 calculation cores. Carrizo, the 5th generation of APU, was launched in 2015 and can fully support HSA1.0 standard. It also applies the manufacturing technique of 28 nm, CPU core of "Excavator" microarchitecture and GPU core of GCN framework. Comparing with further APU, Kaveri and Carrizo are greatly improved in aspects such as tessellation performance, multimedia processing performance and power consumption control. Meanwhile, Kaveri and Carrizo also supports DirectX11 technology.

In a word, it is reasonable to regard Kaveri and Carrizo, two products with outstanding performance, as excellent products under the concept of "The Future is Fusion".

2.2 HSA

The HSA standard was proposed by HSA Foundation in 2012 and the HSA Technical Specification 1.0 was issued normally in January 2015. As shown in Fig. 1, HSA is devoted to make CPU and GPU undertake tasks jointly and allocate different types of loads to the most suitable computing units reasonably, thus to achieve true chip-level integration. In a system which complies with HSA standard, due to the two key technologies, hUMA (heterogeneous Unified Memory Access) and hQ (heterogeneous Queuing) [9, 10], data of any processor unit can be accessed by other processor unit. Furthermore, all processor units can access to the virtual memory accessibly. It improved the computing capability of a processor greatly. hUMA has subverted the mutual isolation mode of CPU and GPU, so that both can take uniform addressing. When the CPU has task and distribute it to the GPU, it can be completed only by transferring pointers without transferring large number of data. After GPU processing is completed, the CPU can check the results directly. It reduces unnecessary overhead significantly. hQ has changed the dominant position of CPU in a heterogeneous system, so that the GPU can be run independently and the CPU and GPU are in equal positions. Additionally, HSA also provides an intermediate programming language, HSAIL (Heterogeneous System Architecture Intermediate Language), so that the portability of program is improved greatly.

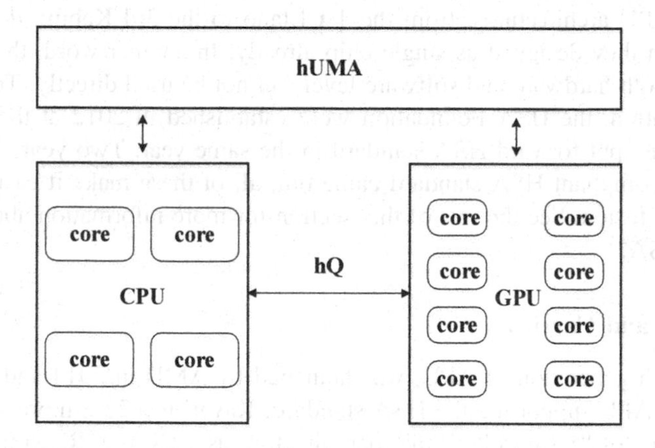

Fig. 1. HSA model

3 Implementation and Optimization

This section describes the DES, the implementation of how to change the algorithm in high-level language to HSA environment through SNACK (Simple No Api Compiled Kernels) compiler. Two possible optimization options are also given, the one is to run sixteen identical iterations on GPU while the other only run the s box. We give analysis on all of these implementations and validate later.

3.1 The Feature of DES and Improvement

The DES algorithm encrypts and deciphers the data in 64-bit blocks per time under the control of a 56-bit key. The 64-bit input plant text need a initial transposition depend on a certain transposition table, then the plant text handled with sixteen identical iterations under the control of sixteen different 48-bit subkeys. After the iterations, a final transposition is also needed before we get the cipher text. The algorithm is symmetric, so the decryption performs as the same of the encryption only with keys in reverse order. In each iteration, the 64-bit plant text is divided into left part and right part, the left Li+1 is only a copy of Ri, and the right Ri+1 is the XOR result of Li and F function. The 32-bit right part expanded to 48-bit then XOR with 48-bit subkey, the result is partitioned into eight groups of 6-bit each. Each group is operated separately by S-box and change back to 32-bit at last. Before the algorithm starts, the 64-bit original key also need to be handled to sixteen 48-bit subkeys to fit the iterations. Only 56-bit of the key participate in initial transposition, and the output is partitioned into two 28-bit blocks. In the next step, each 28-bit block is rotated left by one or two bits then follows a compress transposition.

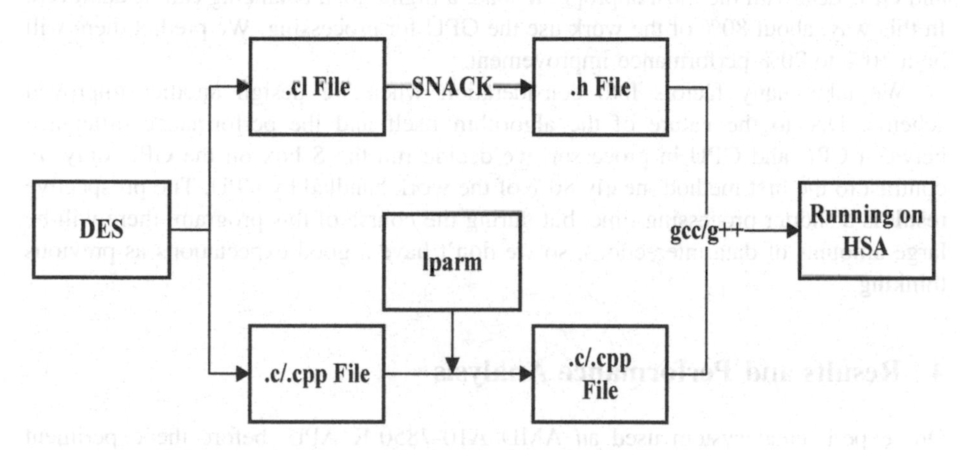

Fig. 2. HSA-based transformation

The implementation of change the DES into HSA environment can be easily reached with the help of CLOC (CL Offline Compiler) and SNACK compiler. Figure 2 shows a formal process of the transformation. Ahead of the modification, a small part in the original algorithm must be separated out using OpenCL mode. CL kernels can be compiled to HSAIL file by SNACK, then we can include the .h file and use the auto-generated API. It is more complicated to modify .c or .cpp file. Lparm, an additional argument is required to provide when calling a SNACK function, it provides the global and local run dimensions to execute the kernel, and it is a simple structure that can be initialized with the SNK_INIT_LPARM string macro provided in the generated header.

In our implementation, plain text are stored in global memory arrays and the global symbols are in demand, we set the lparm into one dimension. After all the above contents are complete, we can run the APU-based algorithm by using gcc or g++ as normal.

3.2 Optimization Strategy

In order to get better performance, it is necessary to decide which part is more suitable to running on experimental platform rather than offload all the encryption process to GPUs. For the current situation, when GPU in APU deal with tasks, CPU just wait until it get the results. To solve this problem, we divide the encryption process into two parts, one running on CPU while the other running on GPU. The processor in our platform, A10-7850 K, has 8 GCN cores and 512 stream processors. That is to say, the more the number of independent tasks close to the stream processors, the more tasks can be simultaneous processed, and it is also means less time consumption.

In order to better study and use the characteristics of the APU, we propose two optimization strategies, put different parts of the DES on the GPU, and examine the specific results. In the first improved scheme, we use CPU to generate subkeys, do the initial and final transposition. As soon as the transposition is done, global data will send to GPU to do sixteen identical iterations. We expect that by using this method, CPU and GPU deal with the most appropriate load, a higher load balancing can be achieved. In this way, about 80% of the work use the GPU for processing, We predict there will be a 10% to 20% performance improvement.

We take many factors into consideration when we design another improved scheme. Due to the nature of the algorithm itself and the performance difference between CPU and GPU in processor, we decide run the S box on the GPU only. In contrast to the first method, nearly 80% of the work handled by CPU. The prospective result is a shorter processing time, but during the course of this program, there will be large amounts of data interactions, so we don't have a good expectations as previous thinking.

4 Results and Performance Analysis

Our experimental system used an AMD A10-7850 K APU, before the experiment begin, several steps needed to be completed. We ran Ubuntu 15.04 (kernel version 3.19), installed HSA-Drivers, HSA-Runtime and CLOC in accordance with the order strictly and used the check file in HSA-Drivers to make a detection. Then, we installed AMD CodeXL, a comprehensive tool suite that enables developers to harness the benefits of APU. Additionally, the graphics card driver was forbidden.

After modifications as we proposed before, we ran the DES and 3DES respectively in three ways. And as a comparison task, an original version also ran on A10-7850 K CPU only. In each case, the size of data increased from 8 MB to 64 MB, and the execution time was monitored by using a Linux command "time". We aimed to observe the overall performance, through previous investigation and research, a better result were expected in the HSA-based implementation and optimization.

Figure 3 shows the results of workloads running on Kaveri with the size of data (MB) increasing from 8 to 64. As the size of data increases in linear, the execution time increases correspondingly. Figure 3(a) shows the results of DES, compared with CPU, the fully GPU-handled implementation reduces the time consumption by almost a half. And on this basis, the optimized design that GPU deal sixteen identical iterations gets a further 10% decrease. On the contrary, when GPU only run the S Box, the execution time increases over five times. Because of the too long time consumption, only a fraction of S Box are tested. Figure 3(b) is the 3DES one, it has the same changing trend with DES but a nearly three times longer expenditure. All of the specific value are given in Table 1(a) and (b).

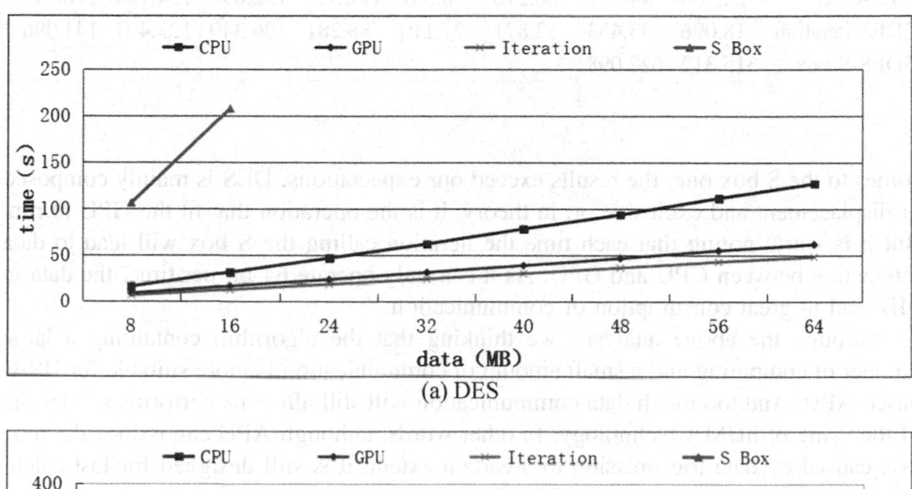

(a) DES

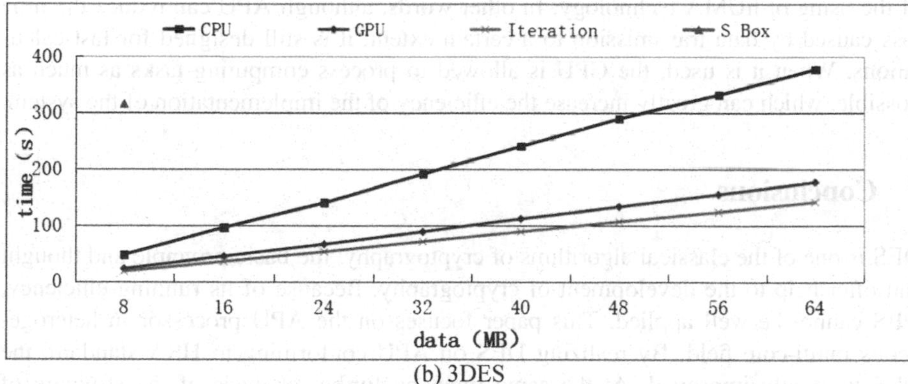

(b) 3DES

Fig. 3. The execution time of workloads

When the whole experiment are considered, the HSA-based implementation and the iteration one run as expected. In comparison, GPU is more suitable than CPU to deal with tasks in high degree of parallelism, so the HSA-based implementation costs little time than working on CPU.

Because of the CPU's meaningless waiting, the execution time are also reduced as the result of moving the transposition and key generation to CPU. Finally, when it

Table 1. The specific execution time

(a) DES								
Data(MB)	8	16	24	32	40	48	56	64
DES-CPU	15.754	31.286	46.561	61.849	78.277	93.460	111.375	127.138
DES-GPU	8.355	15.993	23.523	31.112	38.689	46.369	54.792	61.589
DES-iteration	6.232	12.220	18.132	24.044	30.018	36.074	42.645	48.054
DES-S box	106.098	207.674						
(b) 3DES								
Data(MB)	8	16	24	32	40	48	56	64
DES-CPU	47.322	95.731	139.584	191.001	240.378	289.462	332.053	377.390
DES-GPU	22.708	44.792	66.276	88.357	111.359	132.698	154.744	176.541
3DES-iteration	18.096	35.433	52.871	71.110	88.281	106.349	123.407	141.096
3DES-S box	315.312	622.098						

comes to the S box one, the results exceed our expectations. DES is mainly composed of displacement and exclusive or, in theory, it is the operation that fit the GPU to run. But it is worth noting that each time the iteration calling the S box will lead to data interaction between CPU and GPU. As it can only operate 64-bit per time, the data in MB lead to great consumption of communication.

Through the above analysis, we thinking that the algorithm containing a large number of computing and a small amount of communication is more suitable for HSA-based APU. And too much data communication will still affect its performance despite of the using of hUMA technology. In other words, although APU can reduce the time loss caused by data transmission to a certain extent, it is still designed for fast calculations. When it is used, the GPU is allowed to process computing tasks as much as possible, which can greatly increase the efficiency of the implementation of the system.

5 Conclusions

DES is one of the classical algorithms of cryptography, the basic principle and thought can offer help to the development of cryptography. Because of its running efficiency, DES cannot be well applied. This paper focuses on the APU processor in heterogeneous multi-core field. By realizing DES on APU conforming to HSA standard, the DES is greatly improved. At the same time, by further analysis of the structure of processor, this paper also provides two possible improvement schemes and conducts verification respectively. When the improved scheme is more suitable for the structure of processor, the corresponding performance improvement can be acquired. On the contrary, the performance can be influenced and could be far from the common realization. It is fully showed that the heterogeneous multi-core processor of “CPU-GPU” structure has become a new way to solve the performance bottleneck. The DES-similar classical algorithm will still be worthy of learning.

Acknowledgement. This work was supported by the Foundation of Beijing key laboratory on Trusted Computing (Project No. BZ005), the significant special project for Core electronic devices, high-end general chips and basic software products (2012ZX01039-004), and also supported by Beijing Key Laboratory on Integration and Analysis of Large Scale Stream Data.

References

1. Diffie, W., Hellman, M.E.: Exhaustive cryptanalysis of the NBS data encryption standard. Computer **10**(6), 74–84 (1997)
2. Smid, M.E., Branstad, D.K.: The data encryption standard - past and future. Proc. IEEE **76** (5), 550–559 (1988)
3. Shepherd, S.J.: A high-speed software implementation of the data encryption standard. Comput. Secur. **14**(4), 349–357 (1995)
4. Han, S.J., Oh, H.S., Park, J.: The improved data encryption standard (DES) algorithm. In: IEEE 4th International Symposium on Spread Spectrum Techniques and Applications (IEEE ISSSTA 1996), Mainz, Germany, vol. 3, pp. 1310–1314 (1996)
5. Arich, T., Eleuldj, M.: Hardware implementations of the data encryption standard. In: 14th International Conference on Microelectronics, Beirut, Lebanon, pp. 100–103 (2002)
6. Lupescu, G., Gheorghe, L., Tapus, N.: Commodity hardware performance in AES processing. In: 13th IEEE International Symposium on Parallel and Distributed Computing (ISPDC), France, pp. 82–86 (2014)
7. APU. https://www.amd.com/en/products/desktop-processors-7th-gen-am4
8. Zhang, W.B., Chen, C., Liu, F., et al.: Linux kernel driver support to heterogeneous system architecture. In: The 2015 International Conference on Computer Science (CCSE 2015), Guangzhou, China (2015)
9. Bao, Z.S., Chen, C., Zhang, W.B., et al.: Study on heterogeneous queuing. In: International Conference on Information Engineering and Communications Technology (IECT 2016), Shanghai, China (2016)
10. HSA Foundation. http://www.hsafoundation.com/

Task Scheduling of Data-Parallel Applications on HSA Platform

Zhenshan Bao, Chong Chen(✉), and Wenbo Zhang(✉)

Faculty of Information Technology, Beijing University of Technology, Beijing 100124, China
ccmoli@126.com, zhangwenbo@bjut.edu.cn

Abstract. As CPU processing speed has slowed down year-on-year, heterogeneous "CPU-GPU" architectures combining multi-core CPU and GPU accelerators have become increasingly attractive. Under this backdrop, the Heterogeneous System Architecture (HSA) standard was released in 2012. New Accelerated Processing Unit (APU) architectures – AMD Kaveri and Carrizo – were released in 2014 and 2015 respectively, and are compliant with HSA. These architectures incorporate two technologies central to HSA, hUMA (heterogeneous Unified Memory Access) and hQ (heterogeneous Queuing). This paper realizes radix sort and matrix-vector multiplication – two data-parallel applications on Kaveri platform. By analyzing the performance, a dynamic task scheduling stratgy is proposed. The experimental results show that the running efficiency of algorithm can be greatly improved by using APU with reasonable task scheduling. In the same way, the other data-parallel algorithm would also be optimized on these heterogeneous multi-core architecture.

Keywords: Data-parallel application · HSA · APU · Heterogeneous computing

1 Introduction

In recent years, as a result of slowing CPU performance, GPU acceleration has become more mainstream. Compared with CPUs, GPUs have shown their ability to provide better performance in many applications such as image processing and floating point arithmetic. As a result, heterogeneous multi-core "CPU-GPU" architectures are becoming an increasingly attractive platform, bringing increased performance and reduced energy consumption. This has brought about several novel avenues of research for academia and industry.

In 2012, a non-profit organization called the HSA Foundation was established under the advocacy of AMD, and they proposed HSA standard [1, 2]. This aims to reduce communication latency between CPUs, GPUs and other compute devices, making them more compatible from the programmer's perspective, by making the task of planning the moving of data between devices' disjoint memories more transparent.

As shown in Fig. 1, HSA covers both hardware and software, and provides users a unified model based on shared storage, aiming at decreasing the heterogeneous computing programmability barrier. At the beginning of 2014, the APU Kaveri [3] was the first generation hardware to implement HSA. In 2015, the APU Carrizo [4, 5] was

Q. Zhou et al. (Eds.): ICPCSEE 2018, CCIS 901, pp. 452–461, 2018.
https://doi.org/10.1007/978-981-13-2203-7_35

released and fully supported the HSA standard. hUMA and hQ [6], the two core technologies of HSA, were both implemented in Kaveri and Carrizo. However, research into the actual vs. perceived benefits of these technologies have not been fully determined. In this paper, we designed radix sort and matrix-vector multiplication on Kaveri platform. By analyzing hQ, we then did some research on task scheduling of these two data-parallel applications. Experimental results show that the application designed scientifically running on APU can greatly improve the efficiency of algorithm.

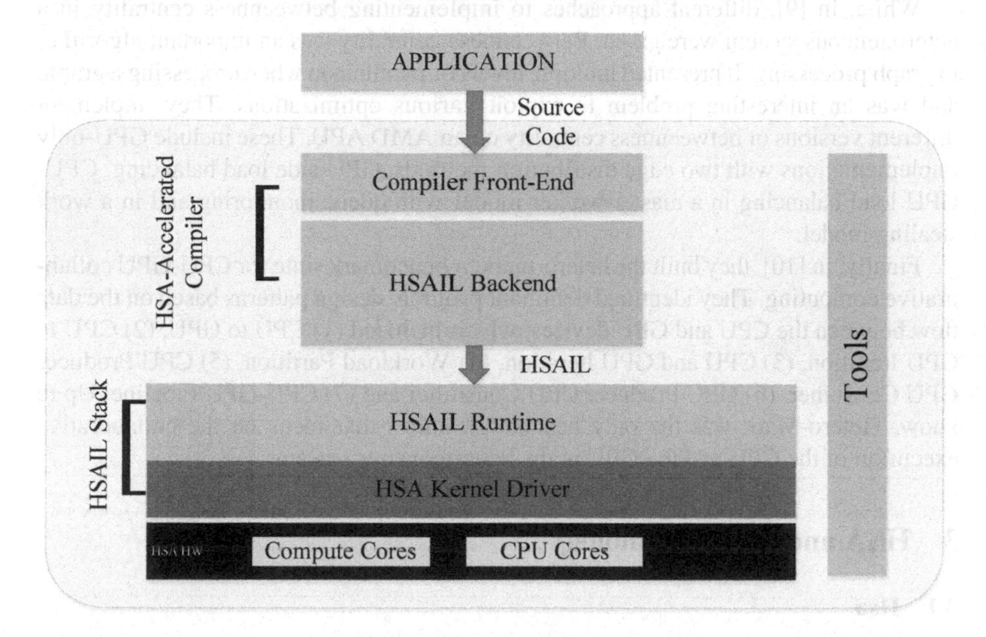

Fig. 1. The architecture of HSA

This paper was organized as follows. Some related works were given in Sect. 2. And we gave an overview of HSA in Sect. 3, along with a description of the APU architecture. In Sect. 4, we showed an overall about algorithm design and the task scheduling strategy. Next, experiment results and analysis were presented in Sect. 5. Finally, Sect. 6 provided conclusions.

2 Related Work

Recently, some researchers did some experiences to compare the fused CPU-GPU chips with discrete CPU-GPU system.

In [7], the researchers acquired the performance and energy consumption on discrete and fused HSA for different FFT implementations with different input sizes. With the growing input data size, the energy efficiency of these HSA increases due to better utilization of the data-parallel resources on the GPUs. They also concluded that the power consumption increases with the number of OpenCL kernel calls and increased use of the GPU fetch unit.

In [8], they designed an OpenCL-Based Multi-swarm PSO Algorithm, the implementation of each of the kernels in this algorithm and their optimization were both given, especially the data layout, the author employ a lookup table called an Estimated Time to Complete (ETC) matrix to prevent redundant makespan calculations. The result show that the best-performing algorithm get approximately a 29% improvement.

While, in [9], different approaches to implementing betweenness centrality in a heterogeneous system were given. Betweenness centrality was an important algorithm in graph processing. It presented multiple levels of parallelism when processing a graph, and was an interesting problem to exploit various optimizations. They implement different versions of betweenness centrality on an AMD APU. These include GPU-only implementations with two edge distribution methods, GPU-side load balancing, CPU-GPU load balancing in a master-worker model with queue monitoring and in a work stealing model.

Finally, in [10], they built the hetero-mark, a benchmark suite for CPU-GPU collaborative computing. They identified dominant program design patterns based on the data flow between the CPU and GPU devices, which included (1) CPU to GPU, (2) CPU to GPU Iteration, (3) CPU and GPU Iteration, (4) Workload Partition, (5) CPU Producer GPU Consumer, (6) GPU Producer CPU Consumer and (7) CPU-GPU Pipeline. Up to know, Hetero-Mark was the only benchmark suites that focus on the collaborative execution of the CPU and the GPU in the heterogeneous systems.

3 HSA and APU Technology

3.1 Hsa

The HSA standard was proposed by the HSA Foundation in 2012 and the HSA Technical Specification 1.0 was issued formally in January 2015. HSA aims to make CPU and GPU integration more seamless, allocating appropriate loads to the most suitable computing units to achieve true chip-level integration. In a system which complies with HSA standard, due to the two key technologies, hUMA and hQ, data of any processor unit can be accessed by other processing unit. Furthermore, all processing units can access to the virtual memory significantly improving the computing capability of the system. hUMA has subverted the mutual isolation mode of CPU and GPU, so that both can take uniform addressing. When the CPU distributes a task to the GPU, it only by transfers pointers, avoiding the transfer of the large amounts of associated data. After the GPU processing is completed, the CPU can check directly access the results, significantly reducing unnecessary overhead. hQ has changed the dominant position of the CPU in a heterogeneous system, so that the GPU can be run independently, making the CPU and GPU more equivalent than in the traditional CPU-GPU relationship.

3.2 Apu

AMD Accelerated Processing Unit (APU) [11], formerly known as AMD fusion, places CPU and GPU on the same chip, harnessing the processing performance of high performance processors and the latest discrete graphics processing technologies. Since pre-APUs did not support HSA, we do not consider them here. Kaveri, the 5th generation of APU, was launched by AMD in 2014 and was the first generation of APU supporting the HSA standard. Kaveri is a 28 nm process and has 4 "Steamroller" microarchitecture processor cores and 8 Radeon R7 graphics cores of GCN (Graphics Core Next) framework.

4 Task Scheduling of Data-Parallel Applications

4.1 Matrix-Vector Multiplication

Matrix-vector multiplication is an efficient algorithm. Matrix is one of the basic concepts in linear algebra. An m by n matrix is just a number of m by n rows in m row n columns. Because it compacts a lot of data together, it is sometimes easy to express complex models. matrix-vector multiplication looks strange, but it's actually very useful and extensively applied. Matrix-vector multiplication is a typical application in parallel computation. If there are two matrices A and B, which are M * N and N * P, and if C = A * B, then C is M * P. The sequential process of the matrix vector multiplication algorithm is a three-layer cycle, and its time complexity is approximately O (M * P * N). When M, N and P are very large, the calculation will be very time-consumlng. On the HSA platform, we can optimize it. In this paper, the matrix is divided into submatrices, which can be realized by block matrix multiplication, which can reduce the frequency of individual elements from the cache and improve the operation efficiency. The block size of BLOCK_SIZE * BLOCK_SIZE is used to block A and B matrices. The piece here just corresponds to the working group of the HSA programming model, a single element to the work item. In this way, the number of individual elements in and out of the memory is reduced and the efficiency is greatly improved.

4.2 Radix Sort

The radix sort belongs to the distributive sort, also known as the bucket sorting method, as the name implies, it is a sorting algorithm based the information through the key value, which is assigned to some bucket to complete the sort process. Radix sort is a stable sort algorithm. At some point, the radix sort method is more efficient than other stability sorting methods. The invention of radix sort dates back to 1887. The implementation detail is as follows. Unify the values that need to be compared to the same digital length; the digits with shorter length are zero-padded on the left. Then, starting from the lowest order, do the sorting one at a time. So, after visiting from the lowest order to the highest order, the sequence becomes an ordered sequence.

Set the number element key to be sorted as m-bit d-nary integer (zero pad to m-bit), and set d buckets, numbered 0, 1, 2, 3 …, d − 1. First, put each data element in the corresponding bucket according to the lowest value of the key. Then, the data elements in each bucket are collected according to the sequence of data elements from large to small and into buckets. In this way, a new permutation of the data element collection is formed. The sequence of data elements obtained from a base sequence is placed in the corresponding bucket according to the value of the key order; then the data elements in the buckets are collected in the order of the bucket number from small to large. This process is repeated, and when the m order is completed, the sequence of data elements is sorted.

4.3 Task Scheduling

In the heterogeneous APU platform, the main task of CPU is to allocate task, while the main task of GPU is to compute processing with a clear division. How to maximize the functions of these to processors is the key issue of the heterogeneous GPU platform in the perspective of the development of software. This kind of task allocation model is also called load balancing. In a broad sense, the load balancing method of computing task could be divided into two categories, static task scheduling and dynamic task scheduling.

Static scheduling is to set the task allocation according to the expected uptime before executing the load. This method needs no task synchronization and communication overhead is small. However, it is inconvenient to be flexibly applied to various computing tasks, and the problem of uneven load may still be serious. Dynamic scheduling is to determine the load allocation in the process of task execution according to the CPU and GPU performance dynamically. Although the cost of dynamic scheduling is larger than static scheduling, the prediction is more accurate.

We consider the static model first, increase the size of the load data in turn, and get the speed line. The HSA heterogeneous computing platform involved in this paper, has 4 CPU cores and 8 GPU cores. However, in the HSA heterogeneous platform, the number of cores in the GPU cannot be involved in the user's level. In the HSA programming model, the number of cores controlling task allocation in CPU is one. When GPU is computing, the core of CPU has to wait. In this way, there are three idle cores in CPU.

Next, we will think about the dynamic model. A task could be conducted on a variety of devices. For example, matrix-vector multiplication can be done both on the CPU and on the GPU. In this case, we will allocate tasks in queue. The task will be divided in to N equal parts, stored in the task queue. M (the number of CPU cores) threads will be opened at the same time, assigning this N equal share tasks. A variable n would be set on behalf of the next task to complete the label. This variable n would be locked, only one thread at a time can change this variable. In this case, the task is non-preemptive, that is, once a task is allocated, other threads cannot obtain the task.

In HSA heterogeneous computing platform, because the GPU cannot initiate communication to the CPU, it means that the GPU cannot initiate a load request to a non-local work pool. Therefore, a CPU thread is used to assign tasks to the GPU, called as a dedicated thread for that GPU. When the GPU dedicated thread detects that GPU is idle, the dedicated thread will assign tasks to the GPU. In addition, we also open up

3 CPU threads, so that 12 computing cores in HSA platform are all involved in the operation.

5 Experiments on Kaveri with Analysis

5.1 Experimental Environment

APU experimental environment hardware construction method is the same as the general desktop computer, but due to the conflict between the HSA and the discrete graphics card, it is necessary to pay attention to use the HSA function can not use the discrete graphics card. After installing the hardware, install any version of Linux that includes the KFD driver. This paper uses Ubuntu 15.04 with kernel version 3.19, AMD provided all HSA-related software in [12]. In order to set up the experimental environment, it is necessary to download and install the HSA-Drivers and HSA-Runtime in sequence. In the Ubuntu environment, these two parts use the regular dpkg - i command. After installation, add the environment variable LD_LIBRARY_PATH. At this point, the vector_copy payload in the HSA-Runtime is run, and if passed, the HSA environment has been successfully built.

In order to better use the HSA environment, you also need to install the CL Offline Compiler development tool. The integrated SNACK allows researchers to program in the HSA environment. After the installation is complete, the amd_kernel_code.h is copied to the system. In the include folder, use the apt-get install command to install the three system packages libbsd-dev, libtinfo-dev, and libdw-dev. Finally, use the dpkg - i routine to install the CodeXL performance analysis tool and execute the vector_copy payload in SNACK. If the GPU Performance Counters in CodeXL can monitor GPU performance data, then the overall environment is set up.

5.2 Basic Performance of Matrix-Vector Multiplication

Shown in Fig. 2, in the experiment, the size of matrix increasing from 5000 to 29000 was selected for testing and in the same way, a separate CPU running test was performed in the above two groups of tests. With the increase of data size, the executing time of matrix-vector multiplication increased rapidly because the time complexity of matrix-vector multiplication algorithm itself is O(n^2) and the amount of computation increased exponentially with the vector length. In general, it shows that GPU has better scalability for such algorithm and is better suitable for running these compute-intensive loads while it is quite difficult for CPU to run loads with large computation.

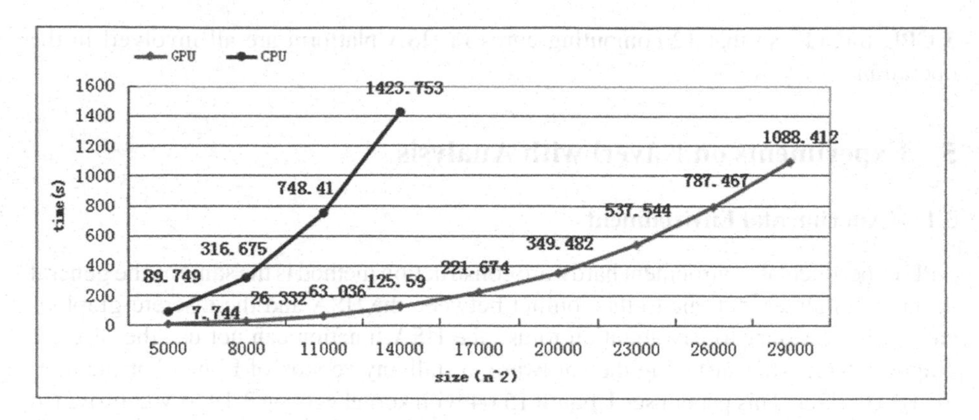

Fig. 2. Basic performance of matrix-vector multiplication on CPU and GPU

5.3 Basic Performance of Radix Sort

We can find in Fig. 3, the number of elements selected by radix sorting is increasing from 81920000 to 819200000 respectively for testing. The computing memory access of the load was relatively low, mostly the instructions of memory access and comparison and vector computing operation was not common. Therefore, it is not necessarily suitable for GPU. We can see from the above test results that CPU could also complete the task rapidly when running these programs.

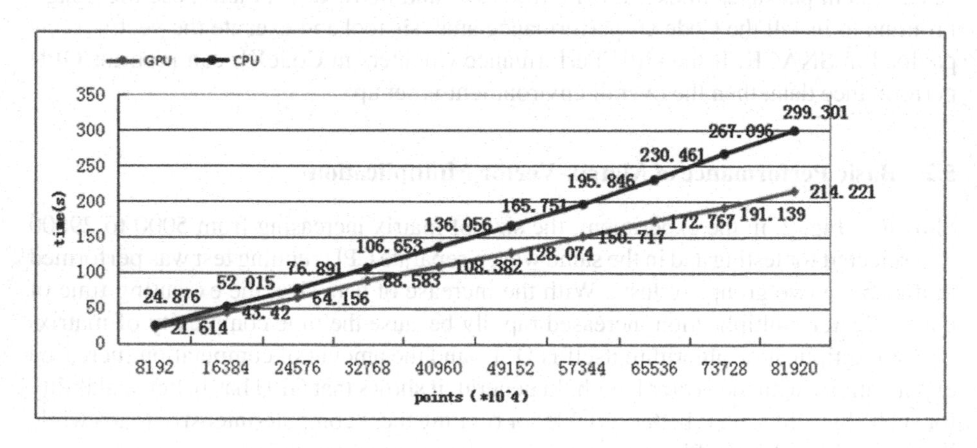

Fig. 3. Basic performance of radix sort on CPU and GPU

5.4 Experiments of Task Scheduling

In the experiment of task scheduling, we tested each load at 0, 0.2, 0.4, 0.6, 0.8 and 1.0 respectively, based on the ratio of CPU running load, and tested the dynamic task allocation loads at the same time. Among them, CPU running load ratio of 0 means that the load runs completely on the GPU, 1.0 means running completely on the CPU, 0.2 to 0.8

means that the CPU and GPU were executed at the same time according to the corresponding ratio. When running dynamically, CPU and GPU were running freely and competitively. Data processing state was maintained through a shared array. After completing their tasks, CPU and GPU queried unprocessed data to continue the rest of the operations.

The data size for matrix multiplication was the matrix of 20000 * 20000 with a block size of 30 and floating point data. It can be seen that it took much less time for matrix-vector multiplication to run entirely on the GPU than on the CPU and CPU had difficulty in running such algorithm. Even though CPU is not suitable for running matrix-vector multiplication operations, allocating certain tasks to run on the CPU can also have some effects on acceleration. As shown in the Fig. 4, the overall efficiency of task running has been improved when statically allocating about 10% of the load to run on the CPU. The performance of CPU and GPU could be better played when programs are automatically filling in the CPU and GPU running queues, which allows DYN allocation achieve the fastest execution time.

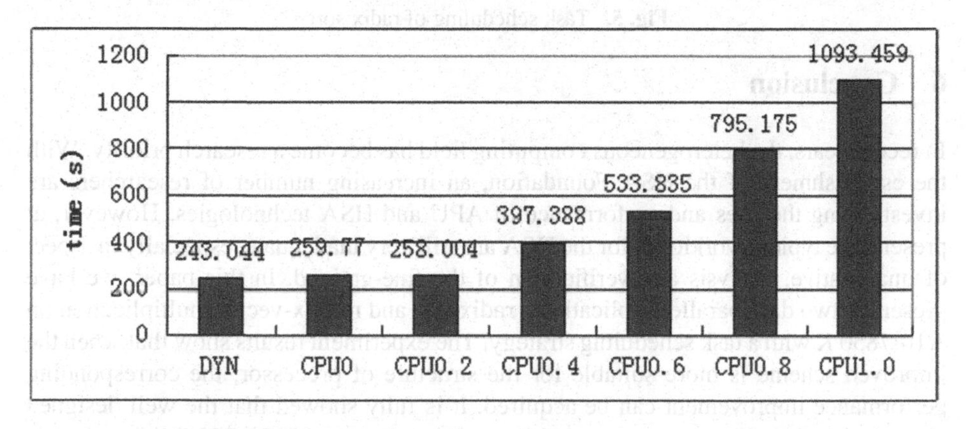

Fig. 4. Task scheduling of matrix-vector multiplication

While in Fig. 5, the data of radix sorting was floating point type with the size of 491520000. As the performance of CPU and GPU was equivalent in processing such loads, the best effects can be achieved when the static allocation ratio approached to 0.4. This was also showed in the method of dynamic scheduling.

From the above experimental results and analysis, we can see that the scheduling method proposed in this paper is suitable for GPU load, and can effectively find the optimal allocation ratio. Based on this, the proposed dynamic scheduling method has higher adaptability for data parallel applications. In practical use, the load's execution characteristics and the matching degree of the underlying hardware resources should be fully considered. Reasonable utilization of the task scheduling method proposed in this paper can indeed accelerate application execution and further highlight the performance advantages of the HSA heterogeneous platform.

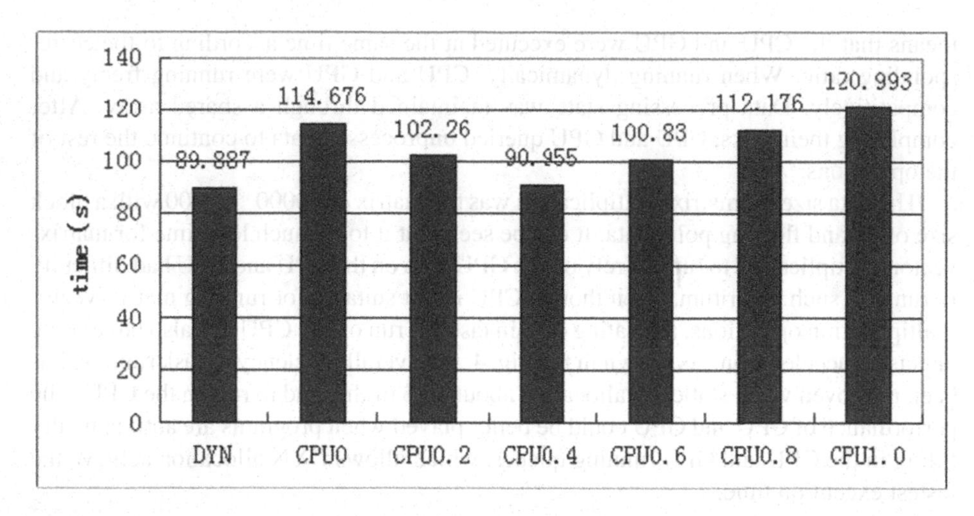

Fig. 5. Task scheduling of radix sort

6 Conclusion

In recent years, the heterogeneous computing field has become a research priority. With the establishment of the HSA Foundation, an increasing number of researchers are investigating the uses and performance of APU and HSA technologies. However, at present, the typical workloads for the HSA are still very inadequate, especially in aspect of quantitative, analysis and verification of the fine-grained. In this paper, we have presented two data-parallel applications, radix sort and matrix-vector multiplication on A10-7850 K with a task scheduling strategy. The experiment results show that when the improved scheme is more suitable for the structure of processor, the corresponding performance improvement can be acquired. It is fully showed that the well designed algorithm and the heterogeneous multi-core processor of "CPU-GPU" structure has become a new way to solve the performance bottleneck.

Acknowledgement. This work was supported by the significant special project for Core electronic devices, high-end general chips and basic software products (2012ZX01039-004), and also supported by Beijing Key Laboratory on Integration and Analysis of Large Scale Stream Data.

References

1. Rogers, P.: Heterogeneous system architecture overview. In: 25th IEEE Hot Chips Symposium, HCS 2013, pp. 7–48. IEEE, New York (2016)
2. Heterogeneous System Architecture: A Technical Review. http://developer.amd.com/wordpress/media/2012/10/hsa10.pdf
3. Bouvier, D., Sander, B.: Applying AMD's Kaveri APU for heterogeneous computing. In: 2014 IEEE Hot Chips 26 Symposium, vol. 30, no. 4, pp. 1–42 (2014)

4. Krishnan, G., Bouvier, D., Zhang, L., et al.: Energy efficient graphics and multimedia in 28 nm Carrizo APU. In: 2015 IEEE Hot Chips 27 Symposium, HCS 2015, pp. 1–34. IEEE, New York (2015)
5. Krishnan, G., Bouvier, D., Naffziger, S.: Energy-efficient graphics and multimedia in 28-nm carrizo accelerated processing unit. IEEE Micro **36**(2), 22–33 (2016)
6. Bao, Z.S., Chen, C., Zhang, W.B., et al.: Study on heterogeneous queuing. In: International Conference on Information Engineering and Communications Technology (IECT2016), Shanghai, China (2016)
7. Ukidave, Y., Ziabari, A.K., Mistry, P., Schirner, G., Kaeli, D.: Quantifying the energy efficiency of FFT on heterogeneous platforms. In: Proceedings of the 2013 IEEE International Symposium on Performance Analysis of Systems and Software (ISPASS 2013), pp. 235–244 (2013)
8. Franz, W., Thulasiraman, P., Thulasiram, R.K.: Optimization of an OpenCL-based multi-swarm PSO algorithm on an APU. In: Wyrzykowski, R., Dongarra, J., Karczewski, K., Waśniewski, J. (eds.) PPAM 2013. LNCS, vol. 8385, pp. 140–150. Springer, Heidelberg (2014). https://doi.org/10.1007/978-3-642-55195-6_13
9. Che, S., Orr, M., Rodgers, G., et al.: Betweenness centrality in an HSA-enabled system. In: Proceedings of the ACM Workshop on High Performance Graph Processing, Co-located with HPDC 2016, pp. 35–38. ACM, New York (2016)
10. Sun, Y.F., Gong, X., Ziabari, A.K., et al.: Hetero-mark, a benchmark suite for CPU-GPU collaborative computing. In: Proceedings of the 2016 IEEE International Symposium on Workload Characterization, pp. 13–22. IEEE, New York (2016)
11. Calandra, H., Dolbeau, R., Fortin, P., Lamotte, J.-L., Said, I.: Evaluation of successive CPUs/APUs/GPUs based on an OpenCL finite difference stencil. In: 2013 21st Euromicro International Conference on Parallel, Distributed, and Network-Based Processing (PDP 2013), Belfast, United kingdom (2013)
12. AMD: CLOC. https://github.com/HSAFoundation

Dual-Scheme Block Management to Trade Off Storage Overhead, Performance and Reliability

Ruini Xue, Zhongyang Guan(✉), Zhibin Dong, and Wei Su

University of Electronic Science and Technology of China, Chengdu, China
xueruini@gmail.com, zhongyangguan@gmail.com, developerdong@gmail.com, suwei779@gmail.com

Abstract. Distributed storage systems usually adopt replication for reliability and fast access. However, as the data volume grows, many large-scale storage systems are tending to employ *erasure coding* to reduce the storage overhead of replication while deliver the same reliability. Unfortunately, erasure coding could result in performance degradation due to less data locality and degraded reads. To trade off among reliability, performance and storage overhead at the same time, we propose FlexBM, a flexible dual-scheme block management approach. FlexBM supports both replication and erasure coding simultaneously, and applies them dynamically according to the recent data temperature. Erase coding is for cold data to reduce storage, while replication is for hot files so that applications can leverage data locality. To guarantee the same reliability as replication with fewer replicas, FlexBM models block placement with bipartite graphs. The prototype of FlexBM is implemented based on HDFS. The experimental results show that FlexBM succeeds in reducing the storage overhead even for a scenario with many small files without reliability compromising, and meanwhile, providing better data locality for frequently accessed datasets.

Keywords: Hadoop · Erasure coding · Distributed file system
Task scheduling · Data-locality

1 Introduction

Distributed storage systems provide reliable access to data over unreliable commodity hardware by replication, typically three copies of everything. However, replicating the entire data footprint becomes infeasible when the amount of data reaches petabytes scale even if storage resource is relatively cheap. Therefore, many large-scale distributed storage systems intend to use *Erasure Coding* (EC) to reduce storage overhead while providing equivalent reliability. For example,

This work is supported by the National Natural Science Foundation of China (No. 61272528) and the Fundamental Research Funds for the Central Universities (No. ZYGX2016J088).

Q. Zhou et al. (Eds.): ICPCSEE 2018, CCIS 901, pp. 462–476, 2018.
https://doi.org/10.1007/978-981-13-2203-7_36

Facebook developed a RAID layer [2] on top of HDFS as early as 2010 and in 2015 they applied erasure coding in their warm BLOB storage system [15]; A novel encoding strategy, *Local Reconstruction Codes* (LRC), based on the observation of failure locality, has been proposed by Microsoft and deployed in Windows Azure [10]; Google also declared that they adopted $RS(6,3)$ in ColossusFS [4].

The basic idea is to choose a group of data blocks, apply an encoding algorithm to them and maintain these blocks as well as their parity blocks. Then, the system would be able to tolerate a certain number of block error within the group. However, applications now would suffer from reduced data locality and *degraded read* problem. Computing frameworks like MapReduce [5,6] can not create adequate local tasks due to less replication factor. Besides, reconstructing an unavailable block in EC requires fetching multiple blocks, resulting in increased read latency as well as communication overhead compared with replication schemes. Many erasure coding implementations in Hadoop ecosystem are based on HDFS-RAID [2], which is inspired by DiskReduce [7]. For example, HDFS-Xorbas [18], a module that replaces Reed-Solomon codes with LRCs in HDFS-RAID, and HAFCS [22], an extension to HDFS-RAID that adapts to workload change by using two different coding families. These systems not only suffer from performance degradation due to reduced data locality and degraded reads, but also can not maintain the same reliability as replication without carefully placing data blocks.

To address these issues, we propose FlexBM, a flexible dual-scheme block management approach, which focuses on the block management trade-off for storage overhead, performance as well as reliability. Implementing an adaptable physical block storage scheme, FlexBMsupports storing a block with replication and erasure coding at the same time. To the best of our knowledge, FlexBMis the first attempt to support the dual schemes simultaneously. By refering to the term "dual-scheme", FlexBM achieves the following: 1. translation between the two schemes is bidirectional; 2. data may be stored with both schemes simultaneously. This is different from current systems, in which some data is stored by replications while the rest by erasure coding. Besides, FlexBM supports both in-file and cross-file coding groups to avoid unnecessary padding for small files. Moreover, we devise a novel metric *file temperature* that allows FlexBM to adjust replication factor adaptively so as to achieve a balance between high performance and low storage overhead. FlexBM models block placement as bipartite graphs to survive similar failures and the algorithm can reduce IO cost during block relocating greatly. The prototype of FlexBM is implemented based on HDFS [19]. Experimental results show that it succeeds in reducing the storage overhead even when there are large number of small files, and meanwhile, providing better data locality for frequently accessed datasets.

2 Motivation

2.1 EC in Distributed Storage Systems

An EC codec takes a piece of data as input, divides it into a number of uniformly sized data shards, and then outputs a number of redundant parity shards for fault

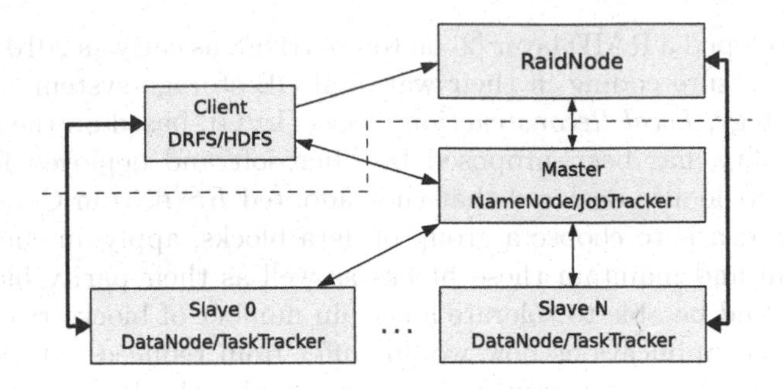

Fig. 1. The architecture of HDFS-RAID.

tolerance. The simplest example of EC is XOR. Suppose we have a tiny binary file whose content is `01011101`, then divide it into `0101` and `1101`, apply XOR operation and get `1000`. Now we can put these three shards into three containers and tolerate any single failure of these containers due to $X \oplus (X \oplus Y) = Y$. However, 4 extra bits are required to achieve the same reliability by replication because we at least need to duplicate the file. The typical model of distributed storage systems is similar to the above example. In these systems, data is usually divided into fixed-size blocks, which are very suitable to be taken as the input of EC codecs, and the nodes in a cluster just act like the shards containers.

In production environments, it is far insufficient to tolerate only one failure, thus other forms of EC have been introduced to address the limitation of XOR. One of the most famous erasure codes is *Reed-Solomon* (RS). An RS codec can be described as a function $RS(k, m)$, which accepts k input data streams, and outputs m parity streams. These $k + m$ streams form a coding group and the system can tolerate any m corruptions. In practice, a data stream denotes a sequence of input shards and is typically formed by a fixed-size data block. There are also cases called *striped layout*, where each single data block is splitted into several input streams.

2.2 HDFS-RAID Architecture

HDFS-RAID [21] is a classic implementation of EC in HDFS. Data reliability guarantees are maintained by creating parity files through an EC algorithm. It's an open-source project and has been wildly used in EC-related researches, such as HDFS-Xorbas [18], HACFS [22], Degraded-first Scheduling [12], etc.

Figure 1 illustrates the architecture of HDFS-RAID. `RaidNode` consists of *block integrity monitor*, *purge monitor*, *placement monitor* and some other components. Block integrity monitor scans the specified directory periodically, encodes the suitable files and maintains the efficiency of parity data. It's also responsible for recovery of missing blocks. Purge monitor cleans the parity files after deletions of corresponding data files. Placement monitor guarantees reliability by avoiding co-locating blocks of the same erasure group. This is done by directly communicating with DataNodes via underlying data transfer protocol.

Distributed Raid File System (DRFS) on the client side is a wrapper of the native implementation that provides transparency to upper-level applications.

The input streams of an EC codec under HDFS-RAID are divided within a file. If the file size divided by the stream size has remainder, the file will be padded at the end in order to fit the stream size. The output parity data is stored in as a companion file to the original one. There are two main reasons for such design. First, by this way the inode tree can be directly used as the index of parity data for maintenance; Second, HDFS-RAID is designed to be an transition layer on top of HDFS and thus it can only manipulate data at file level.

2.3 Summary

The philosophy of HDFS-RAID is simple. However, issues arise due to its limitations in interacting with the HDFS core. First, the storage overhead will get worse if there are too many small files, thanks to the padding strategy. Though directory-level encoding may ease such situation [21], things would become complex on directory changes. Besides, the placement monitor in HDFS-RAID is inefficient and becomes a bottleneck for large-scale data transfer. Furthermore, the placement policy may also be overriden by HDFS core since HDFS is unaware of the RAID layer. Finally, parity blocks under the HDFS-RAID are stored as regular files, which are user-visible and prone to misoperations.

These observations indicate that EC schemes should be embedded inside a storage system rather than implemented as extensions. Besides, replication is so critical for performance that it should not be abandoned. These findings inspired the design of FlexBM.

3 Design

3.1 Basic Decisions

Block Layout. In distributed storage systems, a file is usually divided into fixed-size byte sequences called `blocks`. The dividing method can be straightforward, as shown in Fig. 2a, which is termed a contiguous block layout and has been widely used in many systems because of its simplicity. By contrast, a striped block layout, as Fig. 2b shows, breaks the raw data into much smaller splits and writes these repeated stripes of splits across a set of blocks in a round robin way. Since these splits are small enough to be cached in memory, clients can directly apply an EC codec to each group of splits during writing [24]. FlexBM uses the contiguous instead of stripped block layout for several considerations:

1. Most of the existing systems use contiguous layout and the cost of migration to striped layout is massive.
2. Striped layout abandons data locality and heavily relies on faster networks to deliver the data, which is not friendly to typical applications like MapReduce.

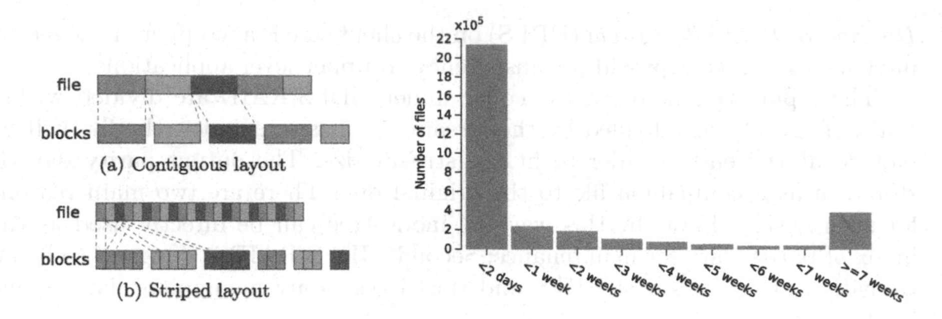

Fig. 2. Two kinds of block layout. **Fig. 3.** Histogram of the file lifespan.

When to Encode. Typically, EC can be performed online, which means to encode on writing, or offline, where data is encoded on demand after its creation. Distributed storage systems usually use a large block size, e.g., 64 MB in HDFS by default, making the total size of a group of blocks too large to be buffered for online EC. However, FLEXBM adopts offline EC not only because of the contiguous block layout, but also the following two facts:

1. Most data access happens within a short time after its creation: more than 99% of data access happens within the first hour of a data block's life [7].
2. Most files are ephemeral: 90% of deletions target files that are less than 22.27 min old for the *PROD* workload [1].

The histogram of the file lifespan in one of Yahoo cluster shown in Fig. 3 also indicates the fact. Thus, by using offline EC, FLEXBM can avoid unnecessary encoding. Besides, storing new files with replication scheme in the beginning also helps to provide better data locality.

Grouping Blocks. Most of EC implementations generate blocks from the same file for simplicity. The *in-file* group of blocks have the same lifespan, and therefore parities can be safely purged after file deletion. However, if data blocks in a group are from different files, which is then called the *cross-file* group, more efforts will be involved during deletion of one of the files since the parity data is still necessary for recovery of other blocks.

FLEXBM supports both of them for different cases: it first tries best to select blocks according to the in-file policy, and the remaining blocks of different files will be tracked and grouped together later. This simple combination solves the storage overhead problem introduced by small files [21]. In FLEXBM, most files will be grouped with in-file policy, and only small files or remaining blocks of huge files will be grouped with cross-file policy. Since most files are ephemeral, FLEXBM can further reduce cross-file groups by *delayed-encoding* strategy.

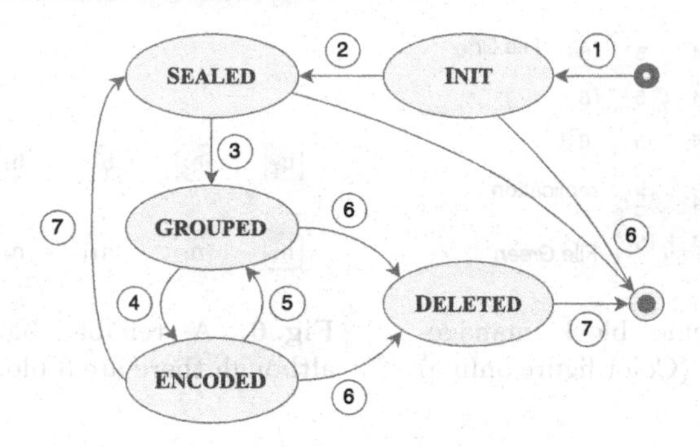

Fig. 4. State transition diagram of data block.

3.2 Block Management Model

To support the dual scheme block management model, where a block can be replicated, encoded or both at the same time, FLEXBM focuses on the physical storage of blocks rather than logical block groups described by file.

Finite State Machine. FLEXBM defines the following states to tracking the erasure coding process of blocks:

Init. All newly created blocks are replicated in the INIT state.
Sealed. A sealed block is *immutable*, where the content cannot be modified. This is common in many existing systems such as HDFS and Windows Azure.
Grouped. A GROUPED block has been assigned to an EC group. But its replication factor cannot be reduced since some blocks in the group may be missing.
Encoded. If all blocks in a group are present and are safely distributed in the cluster, they become ENCODED.
Deleted. A block is DELETED if its corresponding file has been deleted. These blocks are still preserved and may act as parity blocks if it's in a cross-file group.

The transition between these states is shown in Fig. 4. An immutable block starts with SEALED. If a SEALED block is not deleted after a period of time, FLEXBM will assign it to an group if there are enough other SEALED blocks. The blocks are selected according to the policy described in Sect. 3.1. FLEXBM then appends the newly created group into a queue and encode it later. If and only if the parity blocks are persisted and the placements of all blocks in the group are reliable, all blocks in the group move to ENCODED. FLEXBM scans encoded blocks and check their placement periodically in the background. Once a missing block is found or its placement is detected to be *unhealthy*, the state of the block is degraded to GROUPED. In the meanwhile, FLEXBM will attempt to fix the problem so that the block can return to ENCODED again. If a file is deleted, the encoded blocks of this file turn into DELETED, and will serve as special parities

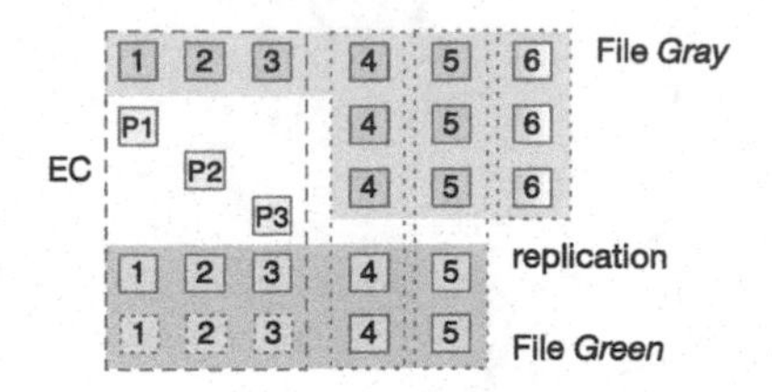

Fig. 5. Dual-scheme block management in FLEXBM. (Color figure online)

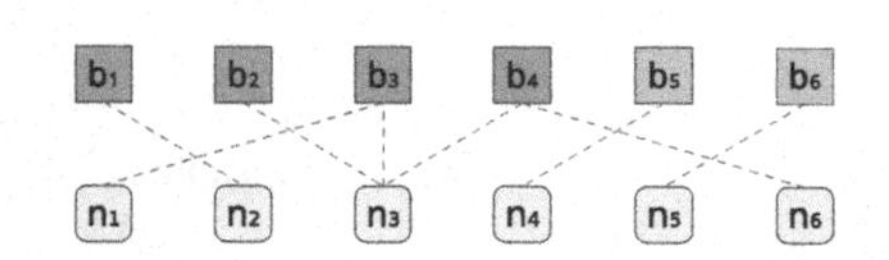

Fig. 6. A reliable block distribution although there are 3 blocks on 3rd node.

and only be used for repairing other *grouped* blocks in the same group. Only after coding groups are disbanded can these blocks be purged. Disbandment of a coding group occurs when the number of *deleted* blocks achieves a threshold. Before disbanding a group, FLEXBM also checks the states of remaining blocks to guarantee data durability. These remaining blocks will become SEALED and wait to be grouped again.

Dual Schemes. Putting all together comes to Fig. 5, the dual scheme block management model in FLEXBM. There are two files `Gray` and `Green`, whose blocks are denoted as gray and green rectangles respectively. Block 1, 2, 3 of `Gray` and Block 1, 2, 3 of `Green` are encoded with parity blocks P1, P2 and P3. Because `Green` is at high temperature, its first 3 blocks are replicated as they are encoded in the same time. These replicas will be erased when its temperature falls. Block 4, 5 of `Gray` and Block 4, 5 of `Green` are sealed but not grouped since there are not enough number of SEALED blocks. They will remain replicated until grouped and encoded. The unfinished `Gray` block 6 is not sealed, and thus it is in INIT state and is replicated. Apparently, `Gray` and `Green` blocks are stored in dual schemes: some of them are encoded, some of them are replicated, and some of them are both replicated and encoded, which is totally different from existing systems.

3.3 Block Relocation Algorithm

Problem Description. Blocks in the same coding group are relevant and not supposed to be allocated on the same node. In HDFS-RAID, the placement monitor is designed to address the problem. The monitor daemon periodically fetches and scans the topology of the live nodes, and once it finds there are too many blocks on the same node, a number of *block mover* will be launched to disperse these blocks. However, such mechanism is subject to the following issues:

1. Block movers are synchronous and the number of running movers is limited by the pool size, leading to a bottleneck as the data grows.
2. Count-based relocation trigger would be inaccurate when EC and replication co-exist. For example, Fig. 6 shows a reliable block distribution where block movement is not required, but movers still get launched in such scenario.

Algorithm 1. Block Relocation Algorithm. km refers to a *Kuhn-Munkres* algorithm implementation [14].

```
1: matching := km(G')
2: for all edge in matching do
3:     if edge.weight = 0 then
4:         move edge.block to edge.node
5:     end if
6: end for
```

3. HDFS-RAID placement policy is transparent to HDFS core and thus it may be overriden by HDFS itself or other utilities like balancers.

Bipartite Model. To address these problems, FLEXBM models block placement with the bipartite graph. The bipartite graph G is defined as Eq. (1).

$$G = (V, E) \quad \text{and} \quad V = B \cup N \tag{1}$$

where B is the set of blocks in the group and N is a set of living nodes where these blocks are stored. E describes location relationship between B and N and can be defined as in Eq. (2).

$$E = \{(b, n) \mid b \in B, n \in N, b \text{ is stored in } n\} \tag{2}$$

Hungarian algorithm can find the maximum matching in $O(V \cdot E)$. In fact, the size of B is usually small and depends on the EC codec, and the replication factor is also bounded. Therefore, its time complexity can be seen as constant at runtime. Given M, a subset of E, the matching found in the previous step, the reliability of current block distribution can be simply determined by $|M|$. If $|M| = |B|$, the group is *healthy*, otherwise some blocks in B should be relocated. In addition, M also presents the minimum number of blocks which should be relocated to fix the reliability problem. To relocate, first add $|B| - |M|$ candidate nodes to the node set which contains the matched node. Let N' be the extended node set and $G' = (V', E')$, where $V' = B \cup N'$ and $E' = \{(b_i, n_j) \mid b_i \in B, n_j \in N'\}$. Then, assign the weight for each edge e in E' to 1 if $e \in E$, otherwise 0. Finally, apply the relocation algorithm presented in Algorithm 1. In other words, a block should be moved to the candidate node.

3.4 Adaptive Replication Mechanism

For some circumstances, EC codecs are only used to store "colder" data that is less often accessed. For example, f4 [15] is designed mainly for storing warm BLOB, and HACFS [22] uses compact code only when it's *write-cold*. Different from these systems, FLEXBM adjusts the replication factor per file dynamically varying with the temperature of file.

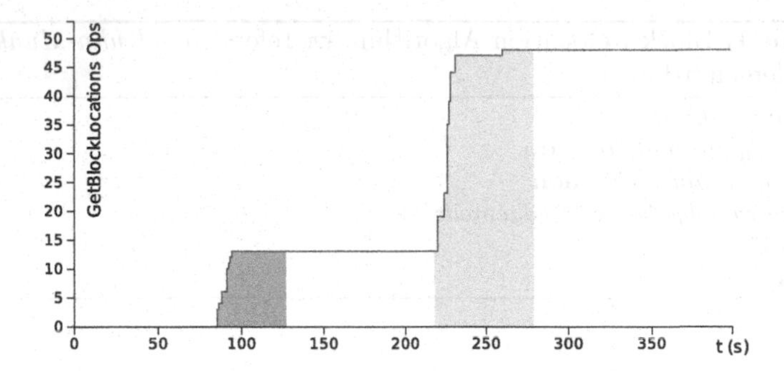

Fig. 7. Run the same job (a wiki dumps parser) against 2 different sized inputs, the red one is smaller (159 MB) and the green one is larger (748 MB). (Color figure online)

Data Temperature. HACFS uses *read frequency* to determine the temperature of data. However, there are two facts keeping us from using it. First, it is expensive to maintain read frequency on each file access. Besides, read frequency is usually related to file size in a distributed system. Figure 7 illustrates the cumulative distribution function of `getBlockLocations` invocations for a MapReduce job accessing 2 files of different sizes. The figure shows smaller files only contribute about 15 file accesses, while larger ones take up 35. Hence, the larger files would have higher read frequency, even if the job is launched at the same rate. *Access time* is also inappropriate for temperature measurement because: 1. a single *atime* doesn't provide enough information to evaluate the accurate data temperature; 2. most distributed systems only maintain coarse-grained access times or even strike access time support [3] for the sake of performance.

To cope with the problem, FlexBM proposes an RRA-based (Round Robin Archive) access recorder with following advantages: 1. it's in-memory and cost-effective, requiring only 128 bits for each target data without demands for data persistence; 2. it accurately reflects the recent access pattern of data, which is necessary for a real-time replica adjuster.

The idea is inspired by RRDtool [16], a famous toolkit for handling time-series data. As shown in Fig. 8, multiple *primary data points* (PDP) are consolidated into one *consolidation data point* (CDP) by *consolidation function* (CF), allowing the history data to be stored in a RRA with relatively small size. FlexBM uses a primitive of 64-bit long as a RRA, where each bit indicates whether the data was accessed in the corresponding period. A demo of 8-bit RRA is presented in Fig. 9. Whenever the data is accessed, the RRA gets updated by setting the first bit of the variable to 1. Before performing an update, RRA should be shifted first if the *last time point* (LTP) is out-of-date. The value of RRA get changed only once even if the data is accessed multiple times in that time window, which matches the access pattern shown in Fig. 7.

Integration. To integrate the RRA-based access recorder, FlexBM introduces three replication factors:

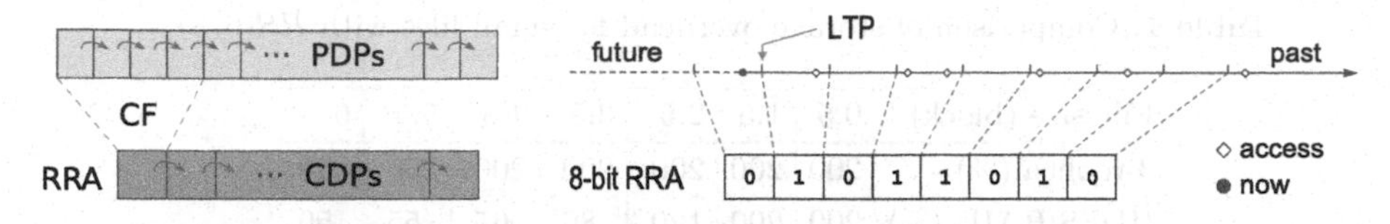

Fig. 8. RRA in RRDtool. **Fig. 9.** RRA in FLEXBM.

safeRep. The minimum number of replicas required to provide basic fault tolerance. For example, 3 for newly-created blocks and 1 encoded ones.

extraRep. The extra number of replicas for better data locality and load balance. This value is evaluated from the RRA. FLEXBM offers following built-in implementation, and users can customize *MaxExtras* in Eq. (3).

$$extraRep = \frac{bitcount(RRA) \times (MaxExtras + 1)}{sizeof(RRA) + 1} \tag{3}$$

realRep. The real number of replicas that should present in the system. FLEXBM adjusts the replication of each block according to Eq. (4). The *realRep* of a file's blocks might be different, since the blocks might be in different block schemes as is presented in Fig. 5.

$$realRep = safeRep + extraRep \tag{4}$$

In addition, FLEXBM doesn't reduce replicas immediately as *realRep* decreases in case that the data might become hot soon.

4 Implementation

The prototype of FLEXBM is built based on HDFS. The state of a block is mainly maintained in `BlockInfo`. Instead of adding more state fields, FLEXBM prefers evaluating the state by method invocations whenever necessary, so as to avoid consistency problems and excessive memory footprint.

A singleton `ECGroupManager` has been developed to maintain the index of all groups. It also scans groups periodically to check the state of each block and performs corresponding actions when the state changes. Group metadata, consisting of a *group id*, a list of *block ids*, and an `ECType` field, is persisted in *fsimage* and *editlog*, is similar to what `INode` does in native HDFS, making it possible for high-availability strategies for inodes to be applied to group metadata as well.

The last unfilled block of each file should never be treated as a SEALED block because it may be appended later. For efficiency, FLEXBM does not scan SEALED blocks, but marks them every time of block reporting.

An encoding or recovery task is performed by one of the DataNodes instead of a MapReduce job, considering a storage-layer system oughtn't to be coupled with upper applications. Selection of the worker node follows a simple principle: FLEXBM always chooses an idle node that has the most blocks in the specified

Table 1. Comparison of storage overhead for small files with $RS(6,3)$.

File size (block)	0.5	1.5	2.5	3.5	4.5	5.5	6
3-replica (%)	200	200	200	200	200	200	200
HDFS-RAID (%)	200	200	120	86	67	55	50
FLEXBM (%)	200	100	80	71	67	64	50

group in order to save network traffic. Degraded reads are supported by adding an RPC method to the `ClientProtocol`, which returns locations of other blocks in the group, allowing the client to fetch them for data recovery.

Replicas of blocks are adjusted in block level, which differs from native HDFS. That is, blocks of the same file may have a different number of replicas. The time window size of an RRA bit is currently a global setting for simplicity.

5 Evaluation

Experiments are conducted on a cluster consisting of 11 nodes. The master node has a 2.9 GHz Intel i5 CPU and 8 GB RAM, and the other 10 are slaves with 1.40 GHz Intel Celeron processors, 4 GB of memory, and 500 GB hard disk. All nodes are connected by 1 Gb/s Ethernet network. $RS(6,3)$ is used in the evaluation. Repairing time is not evaluated because FLEXBM is mainly concerned about block management rather than codec optimization. Additionally, we didn't compare with other real-world systems because most of these systems are not open-sourced and thus it is hard for us to carry out parallel comparison.

5.1 Storage Overhead

One of the most important assumptions in HDFS design is to handle big files. In such a premise, both HDFS-RAID and FLEXBM can guarantee almost minimum storage overhead. That is, with $RS(6,3)$ scheme, the systems can reduce storage overhead by approximately 50% compared to replication. However, there are cases where HDFS-RAID may fail to reduce storage overhead due to small files. For instance, with $RS(6,3)$, a file with only a single data block would still end up writing three parity blocks, accounting for a storage overhead of 300%, which is even worse than 3-way replication. Table 1 presents the storage overhead of different systems, where file-size is measured in the number of blocks. FLEXBM performs better for small files.

To verify this observation, several small datasets are stored in the cluster and a wide variety of MapReduce jobs from Cloud9 [13] are launched to produce more files. Figure 10 shows the final disk usage by different sized files. Files between 1 and 2 blocks account for 58.1% disk usage, and all files less than 8 blocks account for 90%. This indicates that small files are still common under some workloads, and even so FLEXBM can still reduce the storage overhead to a reasonable level — 87.1%. This is because more than 75% of the disk space is occupied by SEALED blocks, which can be later grouped properly for EC codec.

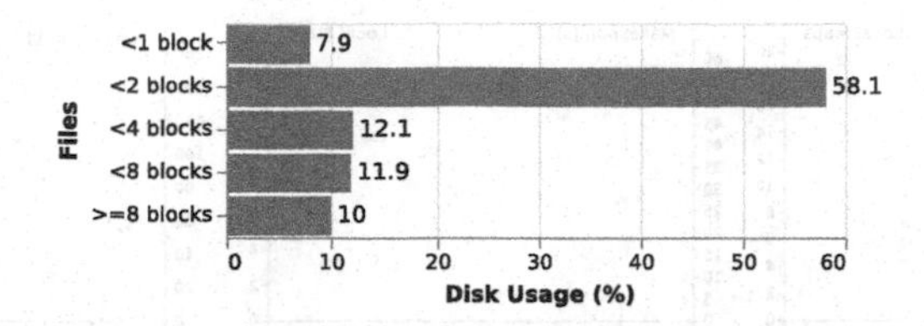

Fig. 10. Disk usage of different sized files.

Table 2. The distribution of the number of file accesses. [23]

Access times	File count	Disk usage (TB)
0	8673686	387.57
(0, 10]	23257527	515.40
(10, 20]	3684566	214.99
(20, 40]	3228428	202.81
(40, 60]	886926	124.80
(60, 100]	548699	159.54
(100, +∞)	943820	626.01

Table 3. Throughput and IO Rate for each map slot in TestDFSIO.

Temperature	Throughput (MB/s)	Average IO rate (MB/s)
Cold	4.36	11.40
Warm	7.13	22.17
Hot	8.83	25.60

In a production environment, the replication factor of a file may be increased by FLEXBM according to its access pattern. However, as shown in Table 2, files accessed more than 100 times only account for no more than 30% of disk usage. Besides, they rarely get accessed frequently at the same time. Thus, the storage overhead caused by extra replicas would be very limited.

5.2 I/O Performance

To evaluate I/O performance of FLEXBM, *TestDFSIO*, an official benchmark bundled with Hadoop distribution, is executed on same data at three temperature levels: cold, warm and hot. Each level is defined by a different *extraRep*, which are 0, $\lfloor 0.7\, MaxExtras \rfloor$ and $MaxExtras$ for cold, warm and hot blocks, respectively. In our test case, $MaxExtras$ is set to 3. *Throughput* and *Average IO rate* are used to evaluate FLEXBM, both of which are based on the file size read by individual map tasks and the elapsed time. They are defined as Eq. (5), where N is the number of map tasks.

$$Throughput = \frac{\sum_{i=1}^{N} filesize_i}{\sum_{i=1}^{N} time_i} \quad \text{and} \quad Average\, IO\, rate = \frac{1}{N}\sum_{i=1}^{N} \frac{filesize_i}{time_i} \qquad (5)$$

The results in Table 3 imply that data locality is important for IO bound applications and replication is necessary for the performance guarantee, especially for those datasets with high temperature.

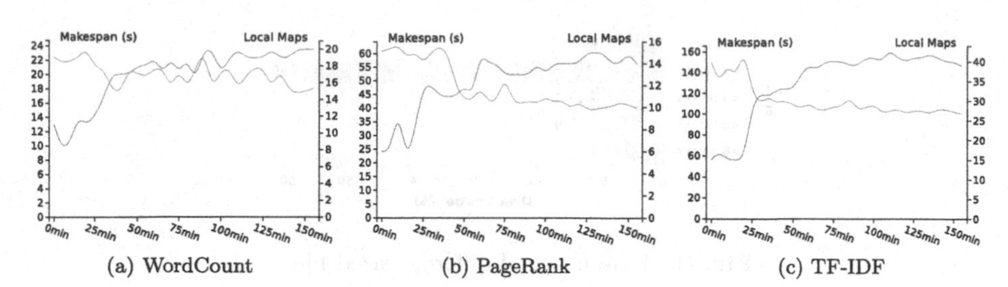

(a) WordCount (b) PageRank (c) TF-IDF

Fig. 11. Adaptive replication can reduce makespans by increasing replicas dynamically.

5.3 Effect of Adaptive Replication

Since FlexBM provides flexible balance between storage overhead and performance with adaptive replication as described in Sect. 3.4, three typical MapReduce jobs are executed to further evaluate how adaptive replication can increase data locality and reduce execution time (makespan) in turn. There are *WordCount*, *PageRank* and *TF-IDF*.

The input datasets are the `bible` and `shakes` from Cloud9 [13], the preprocessed wiki-links file from Henry's work [9] and the first five subsets of latest enwiki article pages [20]. To trigger the increase of replication factor, all jobs were committed periodically to get the input data warmer gradually. The scissors curves in Fig. 11 depict how makespan reduces as the adaptive replication increases replicas, which is measured by *local map tasks* in the figure.

6 Related Work

Although there are lots of efforts focusing on erasure coding theories, this paper concentrates on how to incorporate existing EC algorithms into distributed storage systems instead of proposing new codecs, so we will mainly relate current EC-based system implementations, especially for HDFS, in this section.

QFS [17] is an efficient alternative to HDFS and is compatible with Hadoop MapReduce. It uses striped block layout and online EC. For a $RS(k, m)$ codec, QFS client collects data stripes, usually 64 KB each, into k 1 MB buffers. When these buffers fill, the client calculates m parities and sends all $k+m$ data shards to different chunk servers. These servers continually write the 1 MB data to blocks, until the blocks reach 64 MB. By such a way, QFS can achieve almost minimum storage overhead regardless of the file size. Besides, read and write operations are performed on $k+m$ nodes simultaneously, thus the throughput can be greatly improved if the network bandwidth is adequate. Also, the HDFS-EC [24] project, whose design is very similar to that of QFS, is conducted by the Hadoop community to build native EC support inside HDFS in order to alleviate high storage overhead.

Because of stripped layout, both QFS and HDFS-EC lack data locality and heavily rely on high bandwidth networks to guarantee performance. Besides,

HDFS hosts a wide variety of workloads: from batch-oriented MapReduce jobs to interactive, latency-sensitive queries in Impala [11] and HBase [8]. Many applications will produce temporary files, which are ephemeral and with high temperatures. Contiguous block layout and offline EC are more suitable for such use cases.

7 Conclusion and Future Work

Cloud computing relies on reliable, cost-efficient and high performance storage systems. However, most current solutions fail to satisfy all these needs. FlexBM bridges the gap with flexible dual-scheme block management by incorporating replication and erasure coding simultaneously: archival data is encoded via EC to reduce storage overhead, while hot data is replicated to guarantee performance. FlexBM proposes a light-weight RRA-based temperature monitoring approach, and devises an adaptive replication mechanism to dynamically adjust the number of replicas. To ensure reliability, FlexBM adopts a bipartite graph model to resolve block placement. The prototype of FlexBM is built on HDFS and experimental results indicate that it succeeds in reducing the storage overhead, while providing better data locality for hot data.

The storage overhead of FlexBM is not optimal because of the existence of un-Sealed blocks. The trade-off to encode these blocks will be investigated in the future. Besides, a few erasure coding families like LRC require special placement policies, which hasn't been discussed in the current implementation.

References

1. Abad, C.L., Roberts, N., Lu, Y., Campbell, R.H.: A storage-centric analysis of Mapreduce workloads: file popularity, temporal locality and arrival patterns. In: 2012 IEEE International Symposium on Workload Characterization (IISWC), pp. 100–109. IEEE (2012)
2. Borthakur, D., Schmidt, R., Vadali, R., Chen, S., Kling, P.: HDFS RAID. In: Hadoop User Group Meeting (2010)
3. Borthakur, D.: Access times of HDFS files, May 2013. https://issues.apache.org/jira/browse/HADOOP-1869
4. Corbett, J.C.: Spanner: google's globally distributed database. ACM Trans. Comput. Syst. (TOCS) **31**(3), 8 (2013)
5. Dean, J., Ghemawat, S.: Mapreduce: Simplified data processing on large clusters. In: Proceedings of the 6th Conference on Symposium on Operating Systems Design & Implementation, OSDI 2004, vol. 6, p. 10. USENIX Association, Berkeley, CA, USA (2004). http://dl.acm.org/citation.cfm?id=1251254.1251264
6. Dean, J., Ghemawat, S.: Mapreduce: simplified data processing on large clusters. Commun. ACM **51**(1), 107–113 (2008)
7. Fan, B., Tantisiriroj, W., Xiao, L., Gibson, G.: DiskReduce: raid for data-intensive scalable computing. In: Proceedings of the 4th Annual Workshop on Petascale Data Storage, pp. 6–10. ACM (2009)
8. George, L.: HBase: The Definitive Guide, 1st edn. O'Reilly Media, Sebastopol (2011)

9. Haselgrove, H.: Using the wikipedia page-to-page link database (2010). http://users.on.net/~henry/home/wikipedia.htm
10. Huang, C., et al.: Erasure coding in windows azure storage. In: Usenix Annual Technical Conference, pp. 15–26. Boston, MA (2012)
11. Kornacker, M., et al.: Impala: a modern, open-source SQL engine for Hadoop. In: Proceedings of the Conference on Innovative Data Systems Research (CIDR 2015) (2015)
12. Li, R., Lee, P.P., Hu, Y.: Degraded-first scheduling for Mapreduce in erasure-coded storage clusters. In: 2014 44th Annual IEEE/IFIP International Conference on Dependable Systems and Networks (DSN), pp. 419–430. IEEE (2014)
13. Lin, J., Dyer, C.: Cloud9: A hadoop toolkit for working with big data (2010). http://cloud9lib.org/
14. Munkres, J.: Algorithms for the assignment and transportation problems. J. Soc. Ind. Appl. Math. **5**(1), 32–38 (1957)
15. Muralidhar, S., et al.: F4: Facebook's warm blob storage system. In: Proceedings of the 11th USENIX Conference on Operating Systems Design and Implementation, OSDI (2014)
16. Oetiker, T.: Rrdtool: round robin database tool (2005). http://oss.oetiker.ch/rrdtool/
17. Ovsiannikov, M., Rus, S., Reeves, D., Sutter, P., Rao, S., Kelly, J.: The quantcast file system. Proc. VLDB Endow. **6**(11), 1092–1101 (2013)
18. Sathiamoorthy, M., et al.: XORing elephants: novel erasure codes for big data. In: Proceedings of the VLDB Endowment. vol. 6, pp. 325–336. VLDB Endowment (2013)
19. Shvachko, K., Kuang, H., Radia, S., Chansler, R.: The Hadoop distributed file system. In: Khatib, M.G., He, X., Factor, M. (eds.) IEEE 26th Symposium on Mass Storage Systems and Technologies, MSST 2012, Lake Tahoe, Nevada, USA, 3–7 May 2010, pp. 1–10. IEEE Computer Society (2010)
20. The Wikimedia Foundation: Wikipedia html data dumps (2016). http://dumps.wikimedia.org/enwiki/latest/
21. Wang, W., Kuang, H.: Saving capacity with HDFS RAID, June 2014. https://code.facebook.com/posts/536638663113101/saving-capacity-with-hdfs-raid/
22. Xia, M., Saxena, M., Blaum, M., Pease, D.A.: A tale of two erasure codes in HDFS. In: Proceedings of the 13th USENIX Conference on File and Storage Technologies, pp. 213–226. USENIX Association (2015)
23. Yahoo! Webscope Dataset: Yahoo! statistical information regarding files and access pattern to files in one of yahoo's clusters (2010). https://webscope.sandbox.yahoo.com/catalog.php?datatype=s&did=39
24. Zhang, Z., Wang, A., Zheng, K.: Introduction to HDFS erasure coding in apache Hadoop, September 2015. http://blog.cloudera.com/blog/2015/09/introduction-to-hdfs-erasure-coding-in-apache-hadoop/

Cooperation Mechanism Design in Cloud Manufacturing Under Information Asymmetry

Haidong Yu[1,2] and Qihua Tian[3](✉)

[1] Institute of Information Science and Technology, China Three Gorges University, Yichang 443002, Hubei, China
[2] Institute of System Engineering, Huazhong University of Science and Technology, Wuhan 430074, Hubei, China
[3] College of Mechanical and Power Engineering, China Three Gorges University, Yichang 443002, Hubei, China
hustyhd@163.com

Abstract. The paper proposed the overall optimization and innovation-driven strategy in industrial big data service system combined with the demand of wisdom cloud manufacturing chain. It provided a framework to systematically explore the relationships between enterprise as alliance leader and the union member in cloud manufacturing. It researched Bayesian Nash equilibrium under incomplete information condition which further implied that the incompleteness of information had an effect on the union member's asking price. Thus it revealed a methodology to analyze the stability of Bayesian Nash equilibrium and gave a detailed algorithm. Though Bayesian Nash implementation, teams in industrial big data service system can work best under information asymmetry.

Keywords: Industrial big data · Cloud manufacturing · Mechanism design

1 Introduction

Technical complexity and manufacturing chain demand complexity have brought serious challenges to the development of wisdom cloud manufacturing, so it is important to put forward industrial big data service system combined with the demand of wisdom cloud manufacturing chain. It is a significant way for enterprises to enhance core competence and to rapid respond to the data of market changing that integrating outside manufacturing resources based on the web technologies. With the growing trend of global economic integration, users propose continuously the diversified and individualized requirements for market products. Therefore, enterprises must have a high degree of flexibility and rapid response capability. As a result, organizational structure of modern enterprise has turned toward simplification and flatness and the virtual enterprise form emerges and develops rapidly through combining knowledge in R&D and other resources. The core concept of virtual enterprise is to share enterprise resources including knowledge sharing, which improves the agility, flexibility of manufacturing system and achieve the goal of multiparty win-win [1–4].

Virtual enterprise team is a new form of organization, which emerges accompanied by the development of information technology and the changes of user's demand. It is a

Q. Zhou et al. (Eds.): ICPCSEE 2018, CCIS 901, pp. 477–483, 2018.
https://doi.org/10.1007/978-981-13-2203-7_37

dynamic union set up by many subjects who are consistent with the target, which takes the principles of knowledge sharing and focusing on trust in production and business activities. On the one hand is to meet the needs of economic globalization while on the other hand advanced information and communication technologies lay out the technical foundation for the emergence of virtual enterprise team.

In recent years, many new network technologies are emerging, which provide the new background and breeding environment for virtual enterprise cooperation. A new service-oriented networked manufacturing model called cloud manufacturing was put forward to further promote the network, intelligent and service development of enterprise information construction. In addition, manufacturing cloud service can provide dynamic on-demand access to manufacturing resources or manufacturing capacity for virtual enterprise team, which provides support for the full life cycle of task collaboration. The above is the main channel of virtual enterprise team. Although its platform is highly open, enterprise can realize the free access to manufacturing resources and manufacturing capabilities through cloud&client with the form of cloud service, there exist some problems. For example, a lot of times the community doesn't have that potential. A bad product is taken to the market to take a chance, the enterprise allows good business opportunities to slip away from under their noses, or to perform ineffective competitive strategies. The same goes for virtual enterprise team: policy misjudgments often hurt thousands or even millions of member enterprise [5–9]. In group decision making is very easy to go astray, which is called "myth of the group". Experiments have found that a closed group discussion produced simplification and stereotyped thinking in virtual enterprise team. Specific to a team in the decision making process, because the members tend to make their views and group consensus, can make the decision making participants cannot be objective analysis, the result is group decision making lack of wisdom.

Moderate controlling the size of the virtual enterprise team in cloud manufacturing is necessary. This paper tries to build a virtual enterprise cooperation mechanism of industrial big data service system. When the mechanism can be implemented, there exists cooperation path between enterprises and the manufacturing partner, which automatically adjusts the size of the virtual enterprise team.

2 Cooperation Mechanism Design

Now investigate the condition of the enterprise (in terms of F) and the union member (with U) of virtual enterprise team on the formation issue of virtual collaboration. To simplify the analysis, it is assumed that the number of union members is certain. w_r means that a union member who does not collaborate with the enterprise can still obtain the income(reservation wage), while π represents the profit of the enterprise. Assuming that the real value of π is the private information of the enterprise, only the enterprise knows. The union member does not know its true value, but knows it will obey the uniform distribution on the interval. Therefore, profit π can be regarded as the type of enterprise. To simplify the analysis, let's assume $w_r = \pi_L = 0$.

Definition. Bayesian game. Bayesian game consists of the following:

(1) Player set $\Gamma = \{1, 2, \ldots, n\}$.
(2) Players' type space sets for each player are $T_1, \ldots, T_n$.
(3) The deduction of a player about other player's type $p_1(t_{-1}|t_1), \ldots, p_n(t_{-n}|t_n)$.
(4) Possible action set according to each player $A_1(t_1), \ldots, A_n(t_n)$.
(5) Player's payoff function is $u_i(a_1(t_1), a_2(t_2), \ldots, a_i(t_i); t_i)$.

Where t_i denotes player i's special type, T_i denotes player i's all types set(type space set), $t_i \in T_i$, $t = (t_1, \ldots, t_n)$ denotes a type profile of all players' types combination, $t_{-i} = (t_1, \ldots, t_{i-1}, t_{i+1}, \ldots, t_n)$ denotes a type profile of all players except player i, so $t = (t_i, t_{-i})$. $p_i(t_{-i}|t_i)$ denotes player i knows his own type and deduces other player type probability(conditional probability) is $p_i(t_{-i}|t_i) = \frac{p(t_{-i}, t_i)}{p(t_i)} = \frac{p(t_{-i}, t_i)}{\sum_{t_{-i} \in T_{-i}} p(t_{-i}, t_i)}$, $p(t_i)$ is marginal probability density function [10].

Assume that negotiations of income distribution continue for at least two periods. In the first period, U offers the asking price w_1, and the game ended if F accepts the asking price. At this point, the benefits of U and F are respectively w_1 and $\pi - w_1$. If the F refuses to charge, the game enters the second period, and U gives a second asking price w_2. If F accepts the asking price, the present value of the earnings of the U and F is respectively δw_2 and $\delta(\pi - w_2)$. δ reflects both the discounting factor and the reduction in the benefits resulting from the extension of the negotiations to a shorter period than the first phase. If the F rejects the U's second asking price, the game will end. Both sides' earnings are zero.

Figure 1 gives an extended description of the simplified game about negotiations in the formation of virtual collaboration. There are only two values of $\pi(\pi_L$ and $\pi_H)$ in the figure, and there are only two possibilities (w_1 and w_2) for the union's asking price. In this simplified game, U have three turn it information set of action, so its strategy also contains three price claims, namely the first issue of the price w_1, as well as two in the second phase of the price, w_2 after $w_1 = w_H$ being rejected or after $w_1 = w_L$ being rejected. These three actions are carried out on three non-single junction information sets, where the U's inferences are expressed as $(p, 1-p)$, $(q, 1-q)$ and $(r, 1-r)$ respectively. In the full game shown in Fig. 1, a strategy of U is the first issue of the price w_1 and function $w_2(w_1)$ in the second phase. The function shows w_2 in every condition that every possible price w_1 is rejected. These actions take place in the non-single section of the information set, and there is a second set of information for each of the different initial asking price rates that U may propose. There is a decision node for each possible value in the first and second consecutive information set. In each information set, U's inference is the probability distribution of these decisions. In the complete game, we use $\mu_1(\pi)$ to show the inference that U has in the first issue, while $\mu_2(\pi|w_1)$ in the second phase after the first asking price is rejected.

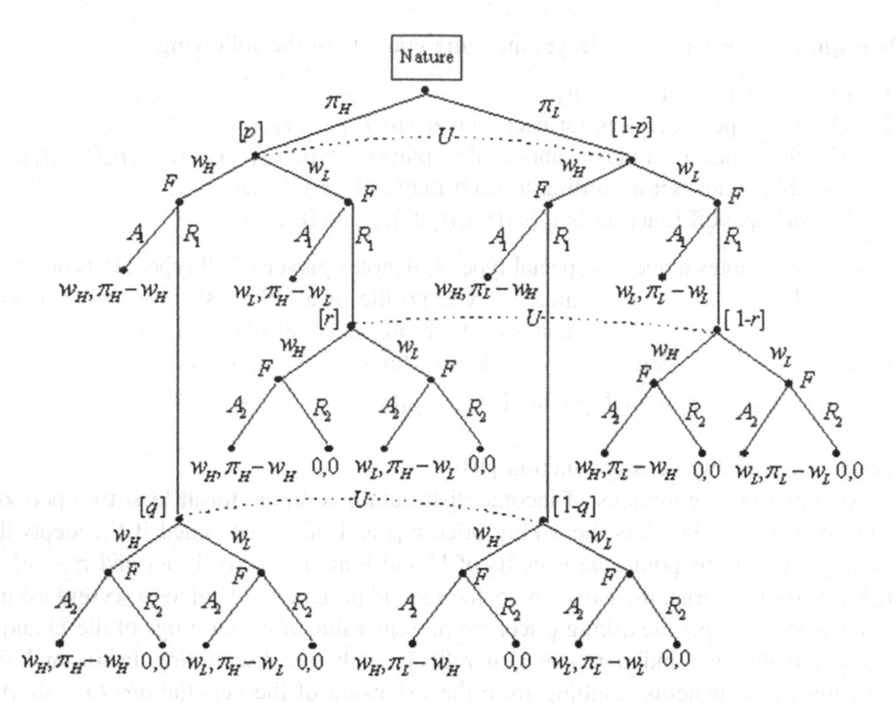

Fig. 1. Bargaining for incomplete information in the formation of virtual collaboration.

Theorem. For the above model, the game has refined Bayesian Nash equilibrium:

(1) the asking price of U in the first period

$$w_1^* = \frac{(2-\delta)^2}{2(4-3\delta)}\pi_H$$

$$\pi_1^* = \frac{2w_1^*}{2-\delta} = \frac{2-\delta}{4-3\delta}\pi_H$$

If $\pi \geq \pi^*$, U will accept w_1^*, otherwise refuse.

(2) If the asking price in the first period is rejected, U corrects its extrapolation of F's profit and considers the uniform distribution of π is $[0, \pi_1^*]$. The asking price of U in second stage is

$$w_2^* = \frac{\pi_1^*}{2} = \frac{2-\delta}{2(4-3\delta)}\pi_H < w_1^*$$

If $\pi \geq w_2$, U will accept w_2^*, otherwise refuse.

Proof: The decision problems faced by U can be expressed as

$$\underset{w}{Max}\,\pi_U = w \cdot \Pr ob\{\text{F accepts } w\} + 0 \cdot \Pr ob\{\text{F refuses } w\} \tag{1}$$

For $\forall w \in [0, \pi_1]$, $\Pr ob\{\text{F accepts } w\} = \frac{\pi_1 - w}{\pi_1}$, so $\pi_U = \frac{w(\pi_1 - w)}{\pi_1}$, thus

$$w^*(\pi_1) = \frac{\pi_1}{2} \tag{2}$$

Now let's go back to the negotiations of income distribution continue for at least two periods. The optimal strategy for the F is

$$A_1(w_1|\pi) = \begin{cases} 1 & \text{if } \pi \geq \max\{\pi^*(w_1, w_2), w_1\} \\ 0 & \text{if } \pi < \max\{\pi^*(w_1, w_2), w_1\} \end{cases} \tag{3}$$

For the type of the F obeys the uniform distribution on the interval $[0, \pi_1]$, so U's best asking price for the second issue must be

$$w^*(\pi_1) = \frac{\pi_1}{2} \tag{4}$$

For $\pi_1 = \max\{\pi^*(w_1, \pi_1/2), w_1\}$, so

$$\begin{cases} \pi_1(w_1) = \frac{2w_1}{2-\delta} \\ w_2(w_1) = \frac{w_1}{2-\delta} \end{cases} \tag{5}$$

Now the multi-period dynamic optimization problem becomes single-period optimization problem:

$$\underset{w_1}{Max}\,\pi_U = w_1 \cdot \Pr ob\{\text{F accepts } w_1\} + \delta w_2(w_1) \cdot \Pr ob\{\text{F refuses } w_1,\ \text{but accepts } w_2\}$$
$$+ \delta \cdot 0 \cdot \Pr ob\{\text{F refuses both } w_1 \text{ and } w_2\} \tag{6}$$

Because

$$\Pr ob\{\text{F accepts } w_1\} = \frac{\pi_H - \pi_1(w_1)}{\pi_H}$$

$$\begin{aligned} \Pr ob\{\text{F refuses } w_1 \text{ but accepts } w_2\} &= \Pr ob\{\text{F accepts } w_2 | \text{F refuses } w_1\} \cdot \\ &\quad \Pr ob\{\text{F refuses } w_1\} \\ &= \frac{\pi_1 - w_2}{\pi_1} \cdot \frac{\pi_1}{\pi_H} \\ &= \frac{w_1}{\pi_H(2-\delta)} \end{aligned}$$

so

$$\underset{w_1}{Max}\, \pi_U = w_1 \cdot (1 - \frac{2w_1}{(2-\delta)\pi_H}) + \delta \cdot \frac{w_1^2}{(2-\delta)^2 \pi_H} \tag{7}$$

thus

$$w_1^* = \frac{(2-\delta)^2}{2(4-3\delta)}\pi_H, \quad w_2^* = \frac{\pi_1^*}{2} = \frac{2-\delta}{2(4-3\delta)}\pi_H \tag{8}$$

3 Conclusion

The paper provided a framework to systematically explore the relationships between enterprise as alliance leader and the union member in cloud manufacturing, which could improve the ability to capture and process data in real time. Firstly, it studied the game strategy decisions of enterprise and members in collaboration. Secondly, it researched Bayesian Nash equilibrium under incomplete information condition which further implied that the incompleteness of information had effected on the union member's asking price. Lastly, it revealed a methodology to analyze the stability of Bayesian Nash equilibrium and gave a detailed algorithm. It showed that virtual enterprise alliance was an essential aspect of modern organizational work. What the mechanism ultimately does is it tries to discover what's meaningful for both enterprise and the union member and it correlates that to their data trading behavior. Though Bayesian Nash implementation, teams in industrial big data service system can work best under information asymmetry.

Acknowledgment. This work was supported by National Natural Science Foundation under Grant No. 51475265.

References

1. Xue, J.: Selection algorithm of virtual enterprise partner based on task-resource assignment graph. Int. J. Grid Distrib. Comput. **9**(9), 185–192 (2016)
2. Xu, X.: From cloud computing to cloud manufacturing. Robot. Comput. Integr. Manufact. **28**(1), 75–86 (2012)
3. Lyu, Y., Zhang, J.: Big-data-based technical framework of smart factory. Comput. Integr. Manufact. Syst. **22**(11), 2691–2697 (2016)
4. Filipe, F., Ahm, S., Americo, A., et al.: Virtual enterprise process monitoring: an approach towards predictive industrial maintenance. Adv. Intell. Syst. Comput. **330**(1), 285–291 (2015)
5. Benjamin, K., Michele, M., Klaus-Dieter, T.: Collaborative open innovation management in virtual manufacturing enterprises. Int. J. Comput. Integr. Manufact. **30**(1), 158–166 (2017)
6. Zhang, W., Zhu, Y., Zhao, Y.: Fuzzy cognitive map approach for trust-based partner selection in virtual enterprise. J. Comput. Theor. Nanosci. **13**(1), 349–360 (2016)

7. Kang, H., Lee, J., Choi, S., et al.: Smart manufacturing: past research, present findings, and future directions. Int. J. Precis. Eng. Manufact. Green Technol. **3**(1), 111–128 (2016)
8. Kusiak, A.: Smart manufacturing must embrace big data. Nature **544**(7648), 23–25 (2017)
9. Zandieh, M., Adibi, M.: Dynamic job shop scheduling using variable neighbourhood search. Int. J. Prod. Res. **48**(8), 2449–2458 (2010)
10. Luo, Y.: A Course in Game Theory. Tsinghua University Press, China (2007)

A Scheduling Algorithm Based on User Satisfaction Degree in Cloud Environment

Feng Ye[1,2(✉)], Yong Chen[2], and Qian Huang[1]

[1] College of Computer and Information, Hohai University, Nanjing 211100, China
yefeng1022@hhu.edu.cn

[2] Postdoctoral Centre, Nanjing Longyuan Micro-Electronic Company, Nanjing 211106, China

Abstract. Efficient task scheduling strategy in cloud environment plays a vital role. Because the size of computing tasks and the time of arrival to the cloud are uncertain, and users tend to have certain expectations in the respect of carrying out the tasks, how to allocate computing resources reasonably for task scheduling is an important problem while satisfying the users' expectations. Combining the idea of greedy algorithm, this paper presents a task scheduling algorithm named UTS. UTS adopts user satisfaction degree model as the evaluation criteria for task scheduling. Comparing with RR, max-min and min-min scheduling policies by simulation using CloudSim, experimental results show that UTS is a more effective task scheduling algorithm.

Keywords: Scheduling algorithm · Cloud computing · User satisfaction degree CloudSim

1 Introduction

Due to the critical characteristics of elasticity, quality of service (QoS) guaranteed and on-demand resource provisioning model, more and more service providers (SP) have adopted cloud computing [1, 2] to handle large-scale computing tasks of services consumers. For computing tasks, different users usually have diverse expectations or user satisfaction degree, such as task priority, execution time, etc. How to allocate computing resources reasonably for task scheduling while satisfying users' expectations is a challenge worth studying for a long time. The size of computing tasks and the time of arrival to the cloud are uncertain, however existing works [3–16] show that many scheduling algorithms fail to consider the dynamic characteristics of computational tasks, and they also ignore the use of multicore concurrent processing. Therefore, aim at above problems, a scheduling algorithm named UTS based on user satisfaction degree is presented.

The rest of the paper is organized as follows. The Sect. 2 describes some of the works related to our topics of interest. The model of user satisfaction degree is proposed and introduced in Sect. 3. In Sect. 4, the task scheduling algorithm is introduced. The Sect. 5 presents the result of simulation and performance evaluation by extending the CloudSim. At last, conclusion along with the direction for future research has been provided.

Q. Zhou et al. (Eds.): ICPCSEE 2018, CCIS 901, pp. 484–492, 2018.
https://doi.org/10.1007/978-981-13-2203-7_38

2 Related Works

For task scheduling in the cloud environment, many algorithms have been presented. From the point of view of reducing the completion time, the works in [3–9] improved the traditional algorithms, and there were varying degrees of optimization in the completion of the task. For example, Liu [3] proposed an improved min-min algorithm that not only based on the min-min algorithm but also the three constraints, including QoS, the dynamic priority model and the cost of service. Compared with the traditional min-min algorithm, the experimental results show it can make long tasks execute at reasonable time, increase resource utilization rate and meet users' requirements. Li, in [5], put forward a cloud task scheduling policy based on Load Balancing Ant Colony Optimization (LBACO) algorithm, which can balance the entire system load while trying to minimizing the makespan of a given tasks set. Shi [6] proposed a task scheduling algorithm based on dynamic programming model. Compared to max-min and min-min algorithms, the proposed algorithm had better performance in terms of task completing time and resource load than the classical algorithms. Considering the load balancing of cloud server, [10, 11] proposed the optimization scheme of task scheduling. Wang [10] introduced a multi-dimensional scheduling algorithm based on CPU and memory, which implements task scheduling according to task requirements and resource load conditions. In [11], the authors discussed a two levels task scheduling mechanism based on load balancing in cloud computing. This task scheduling mechanism can not only meet user's requirements, but also get high resource utilization for virtual machines. Although these algorithms can achieve better results than the traditional algorithms, the downside is that the dynamic nature of the tasks is ignored.

From the perspective of QoS or non-functional requirement, some researchers introduced their contributions such as [12–16]. Take Jung's work [14] for example. In [14], the author proposed a cloud-simulating system with QoS using CloudSim. Providing CloudSim with a priority queue is basic solution for offering QoS. Additionally, this paper introduced the implementation of priority queue in CloudSim. Proposed system was able to control cloudlets with priority and process them differentially. This advanced CloudSim showed faster complete time than default system in time-sharing policy. But there were still many shortcomings, including: the authors did not consider multiple batches of tasks to reach the cloud at irregular intervals; high priority tasks in the subsequent batch will affect low priority tasks in the first batch.

To sum up, different algorithms above have different emphases on task scheduling, but their implementation is static and they fail to consider the dynamic characteristics of computational tasks, and they also ignore the use of multicore concurrent processing. Therefore, we propose a task scheduling algorithm named UTS, which considers the characteristics of "multiple batches, irregular timing and different sizes".

3 The Model of User Satisfaction Degree

To simplify unnecessary complexity and build an efficient model, we make the assumptions as follows: (1) Different tasks are run independently, and there is no correlation between tasks; (2) High-priority tasks do not preempt low-priority tasks that are currently running.

In order to describe the model clearly, we make the following definition.

Definition 1. $task_{ij}$ represents the task *j* in the task sequence of the batch *i* to the cloud server. $Task_i$ represents the set of all tasks in the batch *i*. Both *i* and *j* are positive integer. L_{ij} represents instruction length of $task_{ij}$; the size of $task_{ij}$ is S_{ij}. The number of cores required for parallel processing is N_{ij}.

Definition 2. The *VM* represents the set of all the virtual machines used by a user; vm_k represents the virtual machine *k* in *VM* used by a user, and *k* is positive integer. $mips_k$ represents the number of messages that can be proceed by vm_k in per second. $mbps_k$ represents the bandwidth of vm_k. The number of CPU cores of vm_k is N_k, and $N_k > 0$.

When $Task_i$ arriving at the cloud server, task dispatcher component is responsible for scheduling tasks to virtual machine *VM*.

Definition 3. $Delay_{ij}$ represents the transmission time delay generated during the task $task_{ij}$ is scheduled to vm_k, and it can be expressed as formula (1):

$$Delay_{ij}=S_{ij}/mbps_k \tag{1}$$

Definition 4. T_{ij} represents the execution time that $task_{ij}$ is scheduled to the virtual machine vm_k. And, it can be expressed as formula (2):

$$T_{ij}=L_{ij}/mips_k, N_{ij} \leq N_k. \tag{2}$$

3.1 The Model of User Satisfaction Degree

Before the task is executed, users usually have an expected value for the final completion of the tasks. But the expected value for the final completion of the tasks doesn't have to be exactly the same as the actual execution time.

Definition 5. T_{exp_ij} represents the expected time of completion of the task $task_{ij}$, and T_{act_ij} represents the actual time of completion of the task $task_{ij}$.

Definition 6. In order to measure the user satisfaction degree, here we introduce the Sat_{ij}. It can be expressed as formula (3):

$$Sat_{ij}\begin{cases} = 1 - \frac{T_{act\_ij}-T_{\exp\_ij}}{T_{act\_ij}}, if\ T_{act\_ij} \geq T_{\exp\_ij} \\ = 1, if\ T_{act\_ij} < T_{\exp\_ij} \end{cases}. \tag{3}$$

From formula (3), we can see that when the completion time of a task is less than expected completion time, the user satisfaction degree Sat_{ij} is equal to 1; when the completion time of a task is more than expected completion time, the user satisfaction degree Sat_{ij} is gradually decreasing as the difference between the two gets bigger.

In the real cloud environment, the exact time of task execution is difficult to determine. Further, we construct a prediction mechanism for the completion time and user satisfaction degree of the task to provide the basis for the scheduling algorithm below.

Definition 7. $Ewait_{ij}^{k}$ represents predicted waiting time that the $task_{ij}$ is scheduled to the virtual machine vm_k. And, $Efinish_{ij}^{k}$ represents predicted completion time that the $task_{ij}$ is scheduled to the virtual machine vm_k. According to formulas (1) and (2), we can get the formula (4):

$$Efinish_{ij}^{k} = Delay_{ij} + Ewait_{ij}^{k} + T_{ij} \tag{4}$$

Because the execution time of each task is predictable, $Ewait_{ij}^{k}$ is predictable. According to formulas (3) and (4), we can get the formula (5) for predictable user satisfaction degree:

$$Esat_{ij}^{k} = \begin{cases} 1 & f_{ij}^{exp} \geq Efini_{ij}^{k} \\ 1 - \frac{Efini_{ij}^{k} - f_{ij}^{exp}}{Efini_{ij}^{k}} & f_{ij}^{exp} < Efini_{ij}^{k} \end{cases} . \tag{5}$$

3.2 The Model of Task Priority

Task priority represents a setting of the order of execution of the task submitted by the user. High-priority tasks are scheduled to a specific virtual machine, and their order of execution is preferred than low priority tasks. To avoid the low priority tasks to starve, it is necessary to set dynamic priority mechanism, namely: tasks with low priority gradually increase the priority over time so that they have a chance to be scheduled to execute.

Definition 8. $p_{ij}^{k,\Delta t}$ represents a priority of task $task_{ij}$ in waiting for execution status, among this: $\Delta t \geq 0$ represents the wait time in the queue and p_{ij} represents initial priority.

$$p_{ij}^{k,\Delta t} = \begin{cases} p_{ij} & 0 < \Delta t \leq f_{ij}^{exp} \\ p_{ij} - \left\lfloor \frac{\Delta t - f_{ij}^{exp}}{f_{ij}^{exp}} \right\rfloor - 1 & f_{ij}^{exp} < \Delta t \leq (p_{ij} - 1) \times f_{ij}^{exp} \\ 1 & \Delta t > (p_{ij} - 1) \times f_{ij}^{exp} \end{cases} . \tag{6}$$

From formula (6), we can see that when the wait time is not greater than the difference value between the expected completion time and the actual execution time in the virtual machine, the priority can stay the same. If the waiting time of the task is too long, the priority will be promoted to the highest priority "1".

4 The Task Scheduling Algorithm Proposed

When the user's tasks reach the cloud server, they will be added into task scheduling queue, and then the task dispatcher component performs the scheduling process. The specific scheduling process is as follows.

The virtual machine data collection module periodically collects task execution information for the user's virtual machine, including tasks that have been performed, tasks that are being performed, and task information that is in waiting status. In order to compute relevant data of the models above, the task dispatcher component creates a snapshot of task execution queue for each virtual machine. Using snapshot, the latency waiting time for a task in the virtual machine is obtained.

Assume that the task dispatcher component try to schedules a task $task_{ij}$ into the virtual machine vm_k at time T_1. Before $task_{ij}$, there are some tasks waiting for performing, and the transmission time of these tasks is *Time*. The evolution of snapshot of task execution queue is shown in Fig. 1 below. According to formula (1), assume that the transmission time of $task_{ij}$ in vm_k is TM_1, and vm_k is performing T_1, indicated in the figure by dotted lines. The possible assigned position is ①, ②, ③ and ④. In practice, because of the delay, when the task $task_{ij}$ goes into the queue, T_1 has finished and T_2 has reached ②. Therefore, without loss of generality, we can assume the task $task_{ij}$ is assigned to ③, and $Ewait_{ij}^{k} = TM_2 - TM_1$. And according to formula (2), we can obtain the value of $Efinish_{ij}^{k}$. Furthermore, the task $task_{ij}$ has waited *Time*, so $Efinish_{ij}^{k} = Delay_{ij} + Ewait_{ij}^{k} + T_{ij} + Time$.

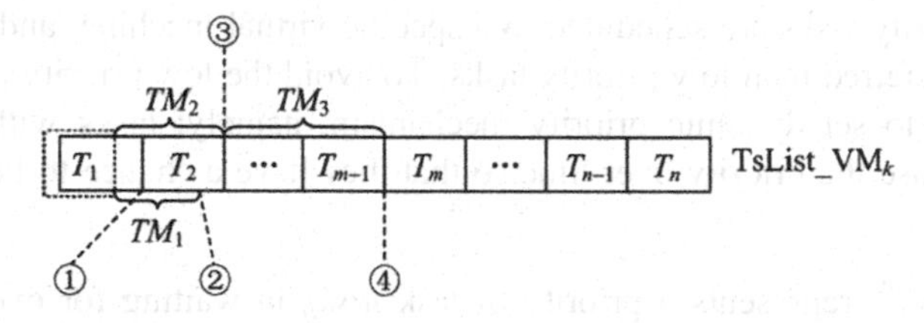

Fig. 1. The evolution of snapshot of task execution queue

Similarly, we can calculate the predicted completion time of task $task_{ij}$ in other virtual machines, and then find out the virtual machine with most user satisfaction degree as the final scheduling target.

Based on the introduction above, the UTS proposed is as follows.

Step 1: The task dispatcher component puts the tasks from high priority to low priority into the task scheduling queue Q_{tasks}.

Step 2: The virtual machine data collection module collects related information and sends them to task dispatcher component.

Step 3: The task dispatcher component gets the tasks with highest priority from Q_{tasks}, and add them into the set *taskCol*.

Step 4: If the set *taskCol* is empty, the algorithm jumps to step 7; If the set is not empty, the algorithm performs the looping execution with steps 5–6.

Step 5: For each task in *taskCol*, the algorithm calculates every predicted completion time in different virtual machine, and then calculates the corresponding predicted user satisfaction degree. At last, it picks out the best $Esat_{ij'}^{k'}$.

Step 6: According to the result of step 5, it finds out the virtual machine and sends the task to the queue of this virtual machine. Remove the task from *taskCol* and Q_{tasks}. At last, the algorithm jumps to step 4.

Step 7: If the set Q_{tasks} is empty, the algorithm continues to Step 8; otherwise, the algorithm jumps to step 3.

Step 8: The data collection module of the virtual machine is cleared to return the information to the task scheduling module; the task dispatcher component is completed, and the scheduler progress is withdrawn.

5 Experiments and Discussion

CloudSim [17, 18] is a generalized and extensible simulation framework that allows seamless modeling, simulation, and experimentation of emerging Cloud computing infrastructures and application services. By using CloudSim, researchers and industry-based developers can test the performance of a newly developed application service in a controlled and easy to set-up environment. Therefore, here we adopt and extend CloudSim for verifying the UTS algorithm.

At first, to simulate the untimed arrival of a task, we add the dynamic event of the task to the cloud in class DataCloudTags. To simulate the features of task priority, we add static and dynamic priority properties in class Cloudlet. In order to describe the expected completion time of the task and the time of arrival to cloud, we expend the class Cloudlet with these two attributes using user setting mechanism and clock of the cloudsim. In order to implement the task scheduling algorithm UTS,we expend the event that request the queue of virtual machine and task queue information in class DataCloudTags. At last, in order to verify UTS, RR, Max-min, Min-min are also implemented in CloudSim.

The simulation environment parameter setting constructed in the experiment is shown in Table 1.

Table 1. The setting list of basic experiment parameter

Names	Values
PE number of virtual machine	1 or 2 or 4
MIPS of virtual machine	500–1000
Bandwidth of virtual machine	500–1000 Mbps
Memory of virtual machine	1–4 GB

5.1 Verifying the Effectiveness of the Dynamic Priority

At first, we need to set the parameters of the task as follows.

(1) The size of the tasks is within 1000;
(2) The number of PE is 1 or 2;
(3) The task is within the range of 500–100,000 instructions.

We simulated users submit 20 batches of tasks within 10 h. Our target is to study the change of satisfaction degree as the number of virtual machines increases. The experimental result is shown in Fig. 2. The scheduling scheme with dynamic priority strategy is always better than the static priority scheduling method.

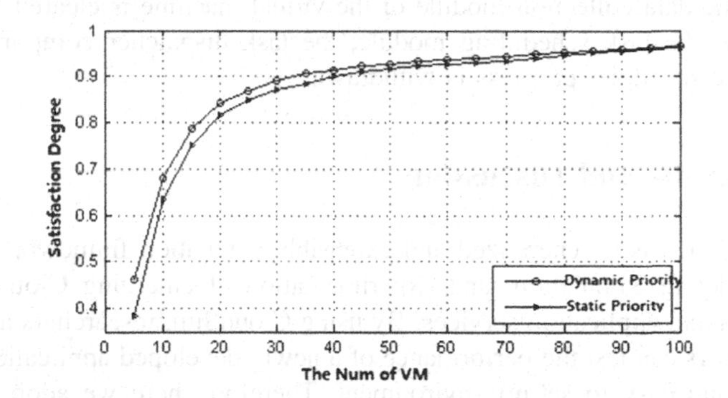

Fig. 2. The comparison of dynamic priority and static priority

5.2 Comparison of Different Scheduling Algorithms

We need to set the parameters of the task as follows.

(1) The number of each batch of tasks was randomly generated at the maximum of 50.
(2) The users submit 20 batches of tasks within 10 h.
(3) The scale of the task will increase in proportion to the initial value.
(4) When different algorithms are tested, the data is exactly the same.

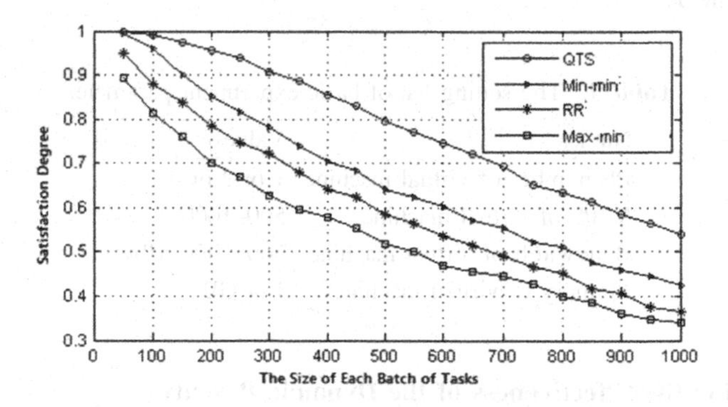

Fig. 3. The comparison of four scheduling schemes

As can be seen from Fig. 3, UTS has a better performance than other scheduling strategies in terms of satisfaction.

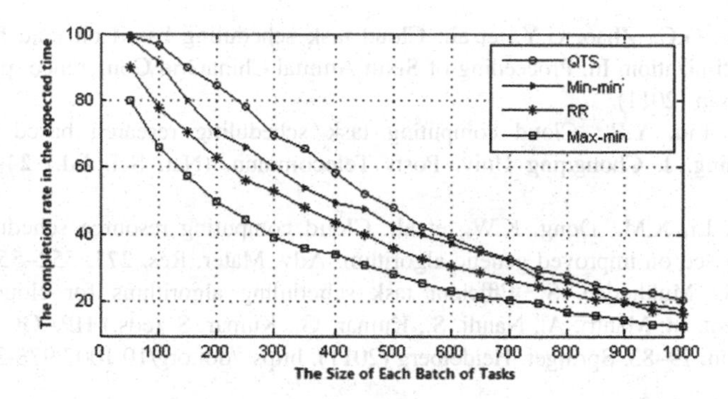

Fig. 4. User satisfaction degree comparison under the four scheduling schemes

From Fig. 4, using UTS, the number of tasks that can be completed in the expected time is higher than the other three scheduling strategies.

6 Summary

In the paper, we propose a task scheduling algorithm, which considers the user's completion time expected and task priority comprehensively. To better evaluate the performance of the scheduling algorithm, user satisfaction degree is introduced for quantification of task scheduling. Compared to other task scheduling algorithms, multi-core concurrency requirements for tasks are also considered. Experimental results show that UTS is a more effective task scheduling algorithm, such as RR, Max-min and Min-min.

Acknowledgment. This research was supported by the National Natural Science Foundation of China [grant No. 61300122]; the Fundamental Research Funds of China for the Central Universities [grant Numbers 2009B21614 and 2017B42214]; 2017 Jiangsu Province Postdoctoral Research Funding Project [grant number 1701020C]; Six Talent Peaks Endorsement Project of Jiangsu [grant number XYDXX-078].

References

1. Chen, K., Zheng, W.M.: Cloud computing: system instances and current research. J. Softw. **20**(5), 1348–1377 (2009)
2. Wang, L., Ranjan, R., Chen, J., et al.: Cloud Computing: Methodology, Systems and Applications. CRC Press, Boca Raton (2012)
3. Liu, G., Li, J., Xu, J.: An improved min-min algorithm in cloud computing. In: Du, Z. (ed.) Proceedings of the 2012 International Conference of MCSA. AISC, vol. 191, pp. 47–52. Springer, Heidelberg (2013). https://doi.org/10.1007/978-3-642-33030-8_8
4. Guo, L.Z., Zhao, S.G., Shen, S.G., et al.: Task scheduling optimization in cloud computing based on heuristic algorithm. J. Netw. **7**(3), 547–553 (2012)

5. Li, K., Xu, G.C., Zhao, G.Y., et al.: Cloud task scheduling based on load balancing ant colony optimization. In: Proceeding of Sixth Annual ChinaGrid Conference, pp. 3–9. IEEE Press, Dalian (2011)
6. Shi, S.F., Liu, Y.B.: Cloud computing task scheduling research based on dynamic programming. J. Chongqing Univ. Posts Telecommun. (Nat. Sci. Ed.) **24**(6), 687–692 (2012)
7. Cui, Y.F., Li, X.M., Dong, K.W., et al.: Cloud computing resource scheduling method research based on improved genetic algorithm. Adv. Mater. Res. **271**, 552–557 (2011)
8. Sindhu, S., Mukherjee, S.: Efficient task scheduling algorithms for cloud computing environment. In: Mantri, A., Nandi, S., Kumar, G., Kumar, S. (eds.) HPAGC 2011. CCIS, vol. 169, pp. 79–83. Springer, Heidelberg (2011). https://doi.org/10.1007/978-3-642-22577-2_11
9. Zhu, Z.B., Du, Z.J.: Improved GA-based task scheduling algorithm in cloud computing. Comput. Eng. Appl. **05**, 77–80 (2013)
10. Wang, L., Laszewski, G., Kunze, M., Tao, J.: Schedule distributed virtual machines in a service oriented environment. In: Proceedings of the 24th IEEE International Conference on Advanced Information Networking and Applications, pp. 230–236. IEEE Press, Perth (2010)
11. Fang, Y., Wang, F., Ge, J.: A task scheduling algorithm based on load balancing in cloud computing. In: Wang, F.L., Gong, Z., Luo, X., Lei, J. (eds.) WISM 2010. LNCS, vol. 6318, pp. 271–277. Springer, Heidelberg (2010). https://doi.org/10.1007/978-3-642-16515-3_34
12. Wang, J.P., Zhu, Y.L., Feng, H.Y.: A multi-task scheduling method based on ant colony algorithm. Adv. Inf. Sci. Serv. Sci. **4**(11), 185–192 (2012)
13. Rahman, M.M., Thulasiram, R., Graham, P.: Differential time-shared virtual machine multiplexing for handling QoS variation in clouds. In: Proceedings of the 1st ACM Multimedia International Workshop on Cloud-based Multimedia Applications and Services for E-Health, ACM, pp. 3–8. ACM Press, Nara (2012)
14. Jung, J.K., Kim, N.U., Jung, S.M., et al.: Improved cloudsim for simulating QoS-based cloud services. In: Han, Y.H., Park, D.S., Jia, W., Yeo, S.S. (eds.) Ubiquitous Information Technologies and Applications. LNEE, vol. 214, pp. 537–545. Springer, Dordrecht (2013). https://doi.org/10.1007/978-94-007-5857-5_58
15. Sun, R.F., Zhao, Z.W.: Resource scheduling strategy based on cloud computing. Aeronaut. Comput. Tech. **40**(3), 103–105 (2010)
16. Lin, W.W., Chen, L., James, Z., et al.: Bandwidth-aware divisible task scheduling for cloud. Comput. Softw. Pract. Exp. **44**(2), 163–174 (2014)
17. Buyya, R., et al.: Modeling and simulation of scalable cloud computing environments and the cloudsim toolkit: challenges and opportunities. In: Proceedings of High Performance Computing & Simulation, pp. 1–11. IEEE, Leipzig (2009)
18. Calheiros, R.N., Ranjan, R., Beloglazov, A., et al.: CloudSim: a toolkit for modeling and simulation of cloud computing environments and evaluation of resource provisioning algorithms. Softw. Pract. Exp. **41**(1), 23–50 (2011)

E-CAT: Evaluating Crowdsourced Android Testing

Hao Lian[1], Zemin Qin[1], Hangcheng Song[2], and Tieke He[1(✉)]

[1] National Key Laboratory for Novel Software Technology, Nanjing University, Nanjing 210093, China
hetieke@gmail.com

[2] CASIC Intelligence Industry Development Co., Ltd, Beijing 100039, China

Abstract. Everyday, millions of crowdsourcing tasks are accomplished in exchange for payments. Pricing acts as an important role in crowdsourcing campaigns, not only for the interest of requesters and workers, but also for the fair competition among the crowdsourcing markets, as well as its sustainable development. All the previous pricing strategies are based on the evaluation of results, however, in the scenario of crowdsourced android testing *(CAT)*, the testing process of a worker is a factor that we cannot overlook. In this paper, we propose a unified model that combines *E*valuation on both process and results of *CAT (E-CAT)*. And based on the proposed *E-CAT*, we can construct the pricing strategy for *CAT*. On one hand, *E-CAT* enables the requesters to investigate the testing process of a worker from both aspects of depth and width. On the other hand, it helps the requesters evaluate the coming-outs of each worker.

Keywords: Crowdsourcing · Auction · Android testing

1 Introduction

In addition, the proposed Evaluating Crowdsourced Android Testing (E-CAT) also helps to validate the test reports by reviewing the testing process of a worker.

Mobile apps market has changed dramatically in the way that people get, install and use apps [11]. The market is using a new supply model that make manufacturers can have the ability to deploy, modify and maintain applications rapidly. The new supply model uses a platform which can make it easier for large amount of users to get to different apps with a low cost. This allows small manufactures compete with big manufactures.

Google play is a typical case of android app market that use this kind of supply model. It grows rapidly in recent years. However it brings new problems. It is very hard for small organizations to have enough resources to sufficiently test their apps. Therefore it is very easy that apps with relatively more defects installed in users' devices. These defects will harm the usability of applications even security of users' devices. Numerous security attacks affect the apps [14].

Q. Zhou et al. (Eds.): ICPCSEE 2018, CCIS 901, pp. 493–504, 2018.
https://doi.org/10.1007/978-981-13-2203-7_39

The situation is becoming more serious as apps have becoming more and more complex.

Automated testing is helpful in finding specific bugs. But manual testing can not be fully replaced by automated testing. Because some defects need to be found in specific mobile phones, operating systems or network environments that automated testing can not stimulate all of them. So if a manufacture wants to test its apps, the best way is to give its apps to as many users as possible to use. Crowdsourced testing can achieve this goal.

Crowdsourcing engages large, distributed groups of people to complete subtasks or to generate information [10]. Crowdsourced testing is crowdsourced testing task to people. Quality assurance is a major challenge of crowdsourcing [8,9]. If a manufacturer wants to crowdsourcing its app testing work, many problems about payment should be solved. How much should workers be paid to test one application? Would workers testing the same applications should get same payment? Should only those who find bugs need to get payment? Answering these questions are very important. As the payment shall be highly related the quality of workers' work.

Even though finding bugs can be a very effective way to evaluate workers. But the nature of android platform makes finding bug a difficult task. Some may spend a lot of effort but finding no bug. Giving his work a proper evaluation is important to the development to the entire crowdsourcing platform.

A mature crowdsourcing market like Amazons' Mechanical Turk pose challenges for monitoring quality of employees' work. Workers may cheat on jobs as their work result are mixed with large number of results of other person. This is more natural to those subjective tasks or those with multiple outputs. Cheating rates can be as high as 30% sometimes [13]. Giving workers having low quality work even cheating is unfair to others. Even if workers do not cheat, their work quality can be highly different due to variability in their effort or skills [3].

There are many research efforts have been made to find and correct outcoming of low quality work in order to improve the quality. People propose different approaches to solve this problem. Some people use gold standards to post-hoc weighting based on worker agreement or reputation [5,7]. Most of these approaches depend on only one aspect of business process in human computation markets: the final output. With only the final output and some minimal reputation metrics about the employees involved. Employers must make difficult tradeoffs between quality and the cost. For example, some methods need multiple redundant judgements by workers. Some methods need standard answers such as labeled data. These are all a lot of extra cost need to insure quality.

But for the crowdsourcing of mobile apps only focusing on the final output, bugs found is unfair. We propose a model of crowdsourced testing task quality to evaluate the effort of a worker paid to find bugs. We present a new approach that record workers testing procedure by collecting the accessibility events that trigged by the system when users interact with the device along with the bug reports workers submitted. By analyzing these events along with the submitted bug report we can know the testing behavior of workers. All the reports will have labels that created by workers that describing whether they find a bug. But some of the reports will be inspected or replay by testers to verify the results of the submitted reports.

We will analyze these events and verified reports to determines We use collected events to build a model to describe how hard the workers work to find bug. In our model, even a worker works very hard, he may not find bug. We extract some features that may be high related with whether bugs found from the events and use these features to build the model.

We first use some knowledge about android apps and system to determine some features that are highly related to whether bug is found. Then we analyze collected events, and use linear regression to build model on these features. We also validate constructed features predict power. We believe that our model can be applied to different types of android apps crowdsourced testing platform to evaluate task results that workers submit.

Our contribution is as following:

- We first bring up android crowdsourced testing task evaluation;
- We first use android accessibility events to evaluate;
- We fist extract features from events to evaluate;
- We first use linear regression to build the evaluating model.

We will first introduce the framework of android, then the accessibility event. Then we will give an introduction to the feature we extract. Then we explain how we use linear regression to build model on the features.

2 Android Framework

Before giving a better description about our evaluating model, we need to first give an description about android. Android Apps developers can get access to development environment, the android Application Development Framework (ADF) to build apps, create GUIs and manipulate data by using APIs provided by ADF. In this paper we focus on GUIs as it represent user behaviors and functions of app. Android apps have four main components: Activities, Services, Broadcast Receivers, and Content Providers. An activity is the interface between the back and the front. It generate a screen to the user by manipulating some layout designs. Layouts contain view widgets which are the basic elements of GUI.

Activity is one of the most important component in the android system framework. Activities provide user interfaces. They are the visible parts of an android app [4]. It corresponds to different screens or windows in an app. Activities are launched with Intents. They return data to the component which invokes it.

An activity contain typical GUI elements such as pop-ups, scrolls text views and so on. When using app, users will navigate different activities by manipulating those GUI elements. An activity has an lifecycle. The lifecycle of an activity is presented as of events and states switching. The states are Running, Paused, and Stopped. After onCreate(), onStart() or onResume() event is fired by the system, the corresponding activity will reach Running state. The activity will be suspended if onPaused() is fired. onStop() make an activity stopped. OnResume() make an activity change from stopped to running. Services Broadcast Receivers and Content Providers are all components for back. We will not use them in our model.

3 Accessbility Event

In this section we will introduce the conception of accessibility event.It is different from the events we mentioned above. An accessibility event is sent by android system when something happens from the user interface, e.g., a button is clicked. An accessibility event is sent by an view populating the event with data for its state. The view will then request from its parent to send the event to corresponding parties. The accessibility event is sent by the top view of the view tree. Hence, an accessibility can access all attributes in an accessibility event to get more information about the event. We will use these information to differentiate events. The purpose of accessibility events are providing enough information for an Accessibility Service. Accessibility service gives meaningful feedback to users.

Table 1 is the form of a recorded accessibility event. There several attributes for one event. In the example, we just show 3 important attributes, we do not show other attributes. In the example, the event type is TYPE_VIEW_CLICKED. It means that the event indicates that a view is clicked. The ClassName is the source of the event. In this example the event comes from a button. It means that the button is clicked. The text means that the button has a text attribute whose content is "login".

Table 1. Example of an accessbility event

EventType: TYPE_VIEW_CLICKED	ClassName: android.widget.Button	Text:[login]

There are many types of AccessbilityEvents, we list some of the most often occurring types of events.

- TYPE_VIEW_CLICKED Represents the event of clicking on a View like Button, CompoundButton
- TYPE_VIEW_LONG_CLICKED: It represents long clicking on a View like a Button
- TYPE_VIEW_SCROLLED: It represents scrolling a view.
- TYPE_VIEW_SELECTED: It represents selecting an item.
- TYPE_VIEW_TEXT_SELECTION_CHANGED: It represents changing the selection in an EditText.
- TYPE_WINDOWS_CHANGED: It represents change in the windows shown on the screen. This often followed with the change of activity

Several events will form a path that represent the operation track of a user. When the user entering an activity, an event that the value of attribute ClassName is the activity will be fired. We call these events activity-involved event. Operation within this activity will generate some events whose ClassName are GUI elements. We call this non-activity-involved event. These non-activity-involved events will occurs until a new activity is triggered and a new activity-involved event is generated.

These activity transitions according to events form a path. In Fig. 1, we illustrate the above procedure. On top is the description of user action. In the middle is the screen shot about the activitity. Initially the app is in the main activity, when the user click the textbox, the the event is activated then the app transfer to activity, so the event related to activity is trigged.

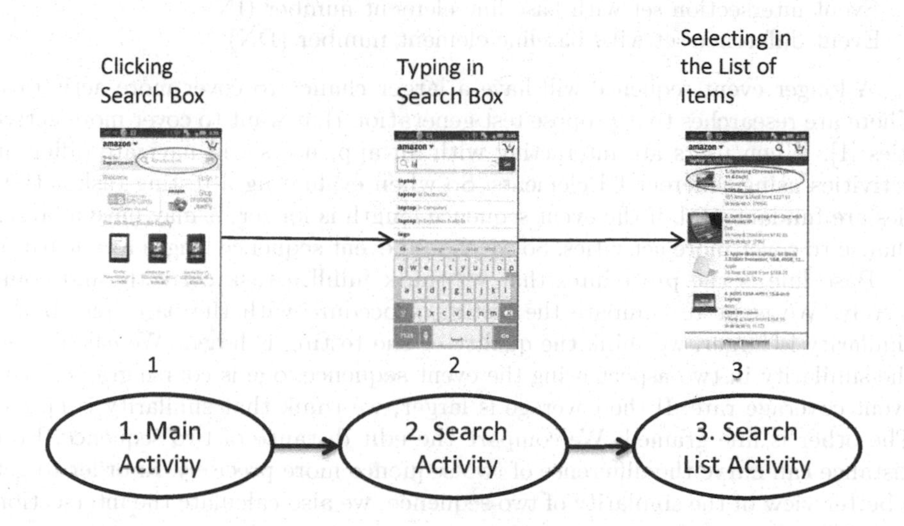

Fig. 1. Procedure

4 Work Flow

In this section, we describe our work flow in Fig. 2.

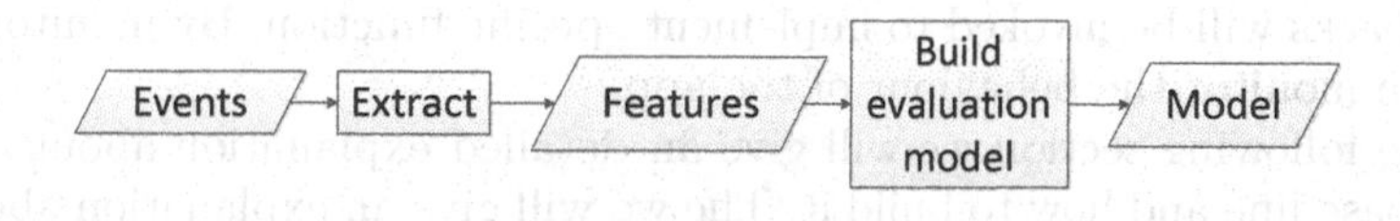

Fig. 2. Workflow

First we will extract some features from the event sequences. In order to extract the features, we need to create a base line and compare each event sequence with the base line. In order to create the base line, we have two ways.

One way is to create it manually. The other is using clustering. We will describe how to create it in detail in the section Extract Features. After The features are extracted, we will build a evaluation model. We will describe the model in detail in the section Build a evaluation model.

5 Extract Features

In this section we will first describe the features we extract. And then describe why we choose these features. Our model is based on following features:

- Event sequence length (SL)
- Event coverage rate to base line (BLC)
- Event sequence distance with base line (DBL)
- Event intersection set with base line element number (IN)
- Event difference set with baseline element number (DN)

A longer event sequence will have a larger chance to cover more activities. There are researches that propose test generation that want to cover more activities [1]. When users are interacting with an app, users will navigate different activities using different UI elements. So when evaluating a testing task activities are fundamental, if the event sequence length is longer, it may have a larger chance to cover more activities. So we choose event sequence length as a feature.

Base line is the procedures that we think fulfill a task correctly and completely. We want to compare the testing procedure with the base line. If the similarity is larger, we think the quality of the testing is larger. We calculating the similarity in two aspect using the event sequence, one is course grained, the event coverage rate. If the coverage is larger, we think that similarity is larger. The other is fine grained. We compare the edit distance of two sequence. Edit distance can curve the difference of two sequence more precisely. In order to get a better view of the similarity of two sequence, we also calculate the intersection and difference set of two sequence.

An android app consists of several separate screens named activities. An activity will define several tasks that can be put into several groups. Each group represent specific behaviors and is represented by several GUI elements, buttons or images. Developers can implement the activities by extending the android.app.Activity class. Android apps are GUI guided. It means the entrance of an app is several callbacks instead of main(). when an GUI events are fired the call backs will be invoked to implement specific function. By monitoring the events we monitor the behaviour of the app.

In the following section we will give an detailed explanation about the conception base line and how to build it. The we will give an explanation about how to get features according to base line.

5.1 Create Baseline

Base line is the procedures that we think fulfill a task correctly and completely. It is represents a sequence of event. We have two ways to generate a base line.

One is to let a professional tester manipulate the app according to the task description. As the test is very familiar with the app and task. So he can correctly and completely fulfill the task. Another way is to use clustering. The intuition is that the sequence that generate by most of people must be right. There are other crowdsourced task researchers that use the task that fulfilled by most of people as the standard answer. First we will cluster all the events. After clustering similar event sequence, event sequences will be grouped together. Then we will a heuristic to determine a sequence based on the clustering result. After clustering, similar event sequences will be grouped into one clustering. Different clustering has different number of event sequences. We choose from the cluster that has the largest number of sequences. In the cluster, we choose the longest event sequence as the baseline (Fig. 3).

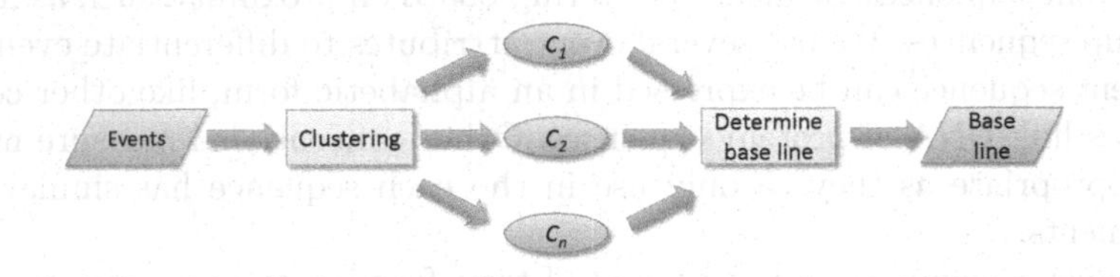

Fig. 3. Generate base line

5.2 Sequence Clustering

In this paper we cluster sequential data. Sequential data are sequences with various length and other characteristics, e.g. dynamic behavior. In this paper, the event sequence we collect represent the dynamic behaviour of users of app. and the events are collected in a span of time. In this paper we use cluster analysis to explore potential patterns hidden in the event sequences and there for build a standard event sequence for each task. There are three categories of sequential clustering Sequence similarity, Indirect Sequence Clustering and Statistical Sequence Clustering. In this paper we mainly focus on sequence similarity.

5.3 Sequence Similarity

Before measure the similarity of event sequence, we need to know how to differentiate events. There are may attributes for one event. We will choose some of the attributes to differentiate event. The attributes are ClassName, EventType and Text. If one of the attributes value is different, we say that the two events are different. There is an attribute showing the location of the source GUI elements of the event. It use the pixel of the screen as the X and Y axis of the GUI element. It seems that this can be also used to differ events. But different devices has different size of the screen. So using this is not precise. Table 2 shows an example of using these three attributes to differentiate 7 events e_1,..., e_7.

Table 2. Events differentiate

Event ID	EventType	ClassName	Text
e_1	TYPE_WINDOW_STATE_CHANGED	LoadingActivity	[cloud music]
e_2	TYPE_WINDOW_STATE_CHANGED	MainActivity	[cloud music]
e_3	TYPE_VIEW_SELECTED	Gallery	[topic]
e_4	TYPE_VIEW_SELECTED	Gallery	[action]
e_5	TYPE_VIEW_CLICKED	RelativeLayout	[local music, (4)]
e_6	TYPE_WINDOW_STATE_CHANGED	ScanMusicActivity	[local music]
e_7	TYPE_VIEW_CLICKED	LinearLayout	[playing, flying heart]

Sequence similarity is based on the measure of the distance between each pair of event sequences. Sequence clustering based on proximity, such as hierachy can group sequences. We use several event attributes to differentiate events such that event sequence can be expressed in an alphabetic form, like other common sequences link DNA or protein sequences. Other conventional measure methods are inappropriate as they ca only use in the each sequence has similar length and elements.

A sequence comparison is a process of transforming one sequence to another with a series of actions. The actions including substitution, insertion and deletion operation. The distance between two sequence is defined as the minimum number of required actions. The distance is known as edit distance or Levenshtein distance. These operators can be weighted according to some prior domain knowledge. In this paper, operations to some accessibility events that are more commonly used will be have a larger weight. By this means, the distance between two sequence is the minimum cost of completing the transformation. The similarity or distance between two sequences can also be formulated as optima alignment problem.

Two event sequences $Seq_1 = (e_1^1, e_2^1...e_i^1...e_N^1)$ and $Seq_2 = (e_1^2, e_2^2...e_j^2...e_M^2)$ The basic dynamic programming-based sequence alignment [6,12], Needleman-Wunsch algorithm can be represented as the following equation.

$$S(i,j) = max \begin{cases} S(i-1, j-1) + a(e_i^1, e_j^2) \\ S(i-1, j) + a(e_i^1, \phi) \\ S(i, j-1) + a(\phi, e_j^2) \end{cases} \tag{1}$$

in this equation S(i,j) is the best alignment score between sub event sequence $(e_1^1, e_2^1...e_i^1)$ of Seq_1 and $(e_1^2, e_2^2...e_j^2$ of Seq_2. $a(e_i^1, e_j^2)$,$a(e_i^1, \phi)$, $a(\phi, e_j^2)$ is the cost for aligning e_i^1 to e_j^2, aligning e_i^1 to a gap symbol (ϕ) or aligning e_j^2 to a gap symbol.

5.4 Hierarchical Clustering

Hierarchical clustering (HC) algorithms group event sequences into a hierarchical structure based on specific proximity matrix. Results are usually represented as binary tree or dendrogram. The root node represents the whole event sequences. Each leaf node is regarded as an event. The intermediate nodes describe that the leaf nodes of the subtree are proximal to each other. The hight of the dendrogram represent the distance between each pair of event sequences or clusters, or one event sequence and a cluster. We can cut the dendrogram at different levels to obtain the clustering result.

HC has two main classification, one is agglomerative method and the other is divisive method. Agglomerative clustering begins with N clusters. Each cluster is one single event sequence. Merge operations will lead some event sequences group to one group. Divisive clustering is opposite to agglomerative clustering. It treat all the event sequences as one cluster. The procedure divides iteratively until each event sequence becomes a single cluster For N event sequences, there are 2N-1-1 subsets of division method. It is very expensive in computation[Cluster Analysis]. So in this paper we mainly focus on agglomerative clustering.

The general agglomerative clustering procedure can be summerized as follows:

1. Calculate the distance matrix for the N cluster, N is the total number of event sequences.
2. Find the minimal distance between two event sequence cluster SC_i and SC_j which is shown in function 2. In function 2, $D(*, *)$ calculate distance between two event sequence cluster SC_m and SC_l. In the proximity matrix, sequence cluster SC_i,SC_j will be combined to form a new cluster.
3. Calaulte the distances between the new cluster and the other clusters.
4. Repeat steps 2-3 until all event sequences are in the same cluster.

$$MIND(SC_i, SC_j) = \max_{1 \leq m,l \leq N(m \neq l)} D(SC_m, SC_l) \tag{2}$$

There are different definitions for distance between two clusters. Different definition indicate agglomerative clustering methods. The most popular methods are single linkage [The application of computers to taxonomy] and complete linkage method [A method of establishing groups of equal amplitude in plant sociology based on similarity of species content and its application to analyzes of the vegetation on Danish commons]. The single linkage method determine the distance between two clusters using the two closet event sequences in these two cluster. It is also called nearest neighbor. The complete linkage determine inter-cluster distance using the farthest of the pair of event sequences in the two clusters. In this paper, we use single linkage method.

Table 3 shows an event sequences that make up of events in Table 2. In Fig. 4 is the clustering dendrogram. All the leaf nodes are event sequences. The upper nodes records the distances of two clusters. In this figure we use linkage distance. We can see that the distances between Seq3 and Seq 4 and Seq5 and Seq6 are all 1. The distance between cluster of Seq3, Seq4 and cluster of Seq5 and Seq5 are 2. If we cut at second level, Seq 3, 4, 5, 6 are grouped together and Seq 1

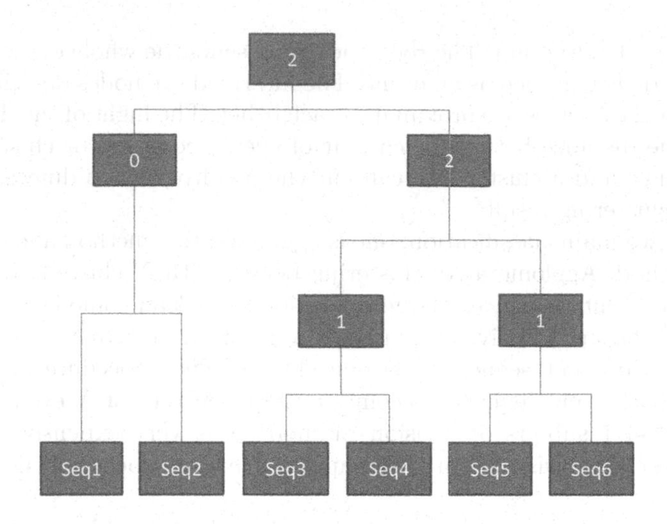

Fig. 4. Hierarchical clustering example

Table 3. Event sequences

Event seqence id	Events
Seq_1	e_1, e_2
Seq_2	e_1, e_2
Seq_3	e_1, e_2, e_3, e_5
Seq_4	e_1, e_2, e_4, e_5
Seq_5	e_1, e_2, e_3, e_5, e_6, e_7
Seq_6	e_1, e_2, e_3, e_5, e_6

and 2 are grouped together. As the definition of base line we choose Seq5 as the base line.

5.5 Example of Extracting Features

We use Seq5 as the base line, as depicted in Table 4.

Table 4. Baseline

Event seqence id	SL	CBL	DBL	IN	DN
Seq_1	2	2/6	4	2	0
Seq_2	2	2/6	4	2	0
Seq_3	4	4/6	2	4	0
Seq_4	4	3/6	3	3	1
Seq_5	6	1	0	6	0
Seq_6	5	5/6	1	5	0

6 Build the Evaluation Model

We want to produce an model of android crowd sourced tasks that reduce the cost of people to evaluate tasks manually.

6.1 Linear Regression

In statistics linear regression models the relationship between a scalar dependent variable y and several explanatory variables X. If the number of explanatory variable is one, it is called a simple linear regression. Other wise it is multiple linear regression. In this paper Xs are the features we extracted from event logs and Y is whether a bug is found.

In linear regression, the relationships between Xs and Y are modeled using linear predictor functions. The model parameters are estimated from the data which we extracted from accessibility events we collected. This model is called linear model.

Linear regression are used in many applications especially for software defect prediction [2]. Models depending linearly on their parameters are easier to fit than models whose parameters are non-linearly related. In this paper, the application of linear regression falls into the following two category.

1. Test the prediction ability of our variable. After a linear regression model has been developed. If X is given without y, the generated model can be used to predict the value of y. We want to see whether our extracted variable can have a good ability of predict that the bug.

2. In the condition that y and corresponding Xs is set. Linear regression analysis can be used to measure the strength of the relationship between Y and each variable. We can know which variable is closely related to Y.

6.2 Model Construction

According to discussion above, we choose the activity of interest from the test report and use these activities to filter the original text. Use sequence hierarchical clustering on activities. In each category, we choose the longest text as the base line. The feature is built according to the baseline with an event sequence length with five factors: Event coverage rate to baseline, Event sequence distance with base line, Event intersection set with baseline element number and Event difference set with baseline element number. When a new bug report is generated, we calculate these five factors and do the weighted average to get the score of a new test report. Then we get the result of the sample set by linear regression of the $X->y$ of the sample.

7 Conclusion

The testing process of a worker is a factor that we cannot overlook in the scenario of crowd sourced android testing (CAT). On this condition, we propose a unified model that combines Evaluation on both process and results of CAT (E-CAT) in this paper. And based on the proposed E-CAT, we can construct the pricing strategy for CAT. First, E-CAT enables the requesters to investigate the testing process of a worker from both aspects of depth and width. Secondly, it helps the requesters evaluate the coming-outs of each worker.

Acknowledgment. This work is supported in part by the National Key Research and Development Program of China (2016YFC0800805), and the National Key Technology Research and Development Program of China (2015BAJ04B00).

References

1. Azim, T., Neamtiu, I.: Targeted and depth-first exploration for systematic testing of android apps. In: Acm Sigplan Notices, vol. 48, pp. 641–660. ACM (2013)
2. Bibi, S., Tsoumakas, G., Stamelos, I., Vlahavas, I.P.: Software defect prediction using regression via classification. In: AICCSA, pp. 330–336 (2006)
3. Callison-Burch, C.: Fast, cheap, and creative: evaluating translation quality using amazon's mechanical turk. In: Proceedings of the 2009 Conference on Empirical Methods in Natural Language Processing, vol.1, pp. 286–295. Association for Computational Linguistics (2009)
4. Chin, E., Felt, A.P., Greenwood, K., Wagner, D.: Analyzing inter-application communication in android. In: Proceedings of the 9th International Conference on Mobile Systems, Applications, and Services, pp. 239–252. ACM (2011)
5. Dekel, O., Shamir, O.: Vox populi: collecting high-quality labels from a crowd. In: COLT (2009)
6. Durbin, R., Eddy, S.R., Krogh, A., Mitchison, G.: Biological sequence analysis: probabilistic models of proteins and nucleic acids. Cambridge University Press, Cambridge (1998)
7. Ipeirotis, P.G., Provost, F., Wang, J.: Quality management on amazon mechanical turk. In: Proceedings of the ACM SIGKDD Workshop on Human Computation, pp. 64–67. ACM (2010)
8. Kamps, J., Geva, S., Peters, C., Sakai, T., Trotman, A., Voorhees, E.: Report on the SIGIR 2009 workshop on the future of IR evaluation. In: ACM SIGIR Forum, vol. 43, pp. 13–23. ACM (2009)
9. Kazai, G., Milic-Frayling, N.: On the evaluation of the quality of relevance assessments collected through crowdsourcing. In: SIGIR 2009 Workshop on the Future of IR Evaluation. vol. 21 (2009)
10. Le, J., Edmonds, A., Hester, V., Biewald, L.: Ensuring quality in crowdsourced search relevance evaluation: the effects of training question distribution. In: SIGIR 2010 workshop on crowdsourcing for search evaluation, vol. 2126 (2010)
11. Mahmood, R., Mirzaei, N., Malek, S.: Evodroid: Segmented evolutionary testing of android apps. In: Proceedings of the 22nd ACM SIGSOFT International Symposium on Foundations of Software Engineering, pp. 599–609. ACM (2014)
12. Needleman, S.B., Wunsch, C.D.: A general method applicable to the search for similarities in the amino acid sequence of two proteins. J. Mol. Biol. **48**(3), 443–453 (1970)
13. Rzeszotarski, J.M., Kittur, A.: Instrumenting the crowd: using implicit behavioral measures to predict task performance. In: Proceedings of the 24th Annual ACM Symposium on User Interface Software and Technology, pp. 13–22. ACM (2011)
14. Shabtai, A., Fledel, Y., Kanonov, U., Elovici, Y., Dolev, S., Glezer, C.: Google android: a comprehensive security assessment. IEEE Secur. Priv. **2**, 35–44 (2010)

Dual-Issue CGRA for DAG Acceleration

Li Zhou(✉), Jianfeng Zhang, and Hengzhu Liu

Institute of Microprocessor and Microelectronics, School of Computer,
National University of Defense Technology, Changsha 410073, Hunan, China
zhouli06@nudt.edu.cn

Abstract. Coarse-grained Reconfigurable Array (CGRA) is suitable candidate hardware architecture for many computation-intensive applications due to its flexibility and efficiency. In current CGRA architecture, each Processing Element (PE) in CGRA performs one operation or transfers data onto neighbors per cycle. In this paper, a dual-issue scheme is proposed to execute Data Acyclic Graph (DAG). Two operations with mutual precedents can be executed at the same cycle in the PE in the proposed design to improve efficiency. Since some hardware is shared by operations, the overhead can be lowered. We also proposed an ant colony based algorithm for mapping DAG to dual-issue CGRA. Experimental results demonstrate that dual-issue CGRA consumes 5.24% less hardware resource while the performance of some DAG improved 2.6%.

Keywords: CGRA · Dual-issue · DAG mapping

1 Introduction

Various applications such as software-defined radio and media processing, require real-time and programmability features at the same time. In these situations, hardware platform needs to encourage high performance and flexibility both. CGRA is such an architecture which has been used in these stream fields for several years [1, 2]. CGRA is also suitable for machine learning [3]. Figure 1 shows a typical 4 × 4 CGRA structure. PE is the basic component of CGRA. ALU in PE can be set up to perform different operations of DAG at word level in contrast to the bit level reconfiguration logic in FPGA. Parallelism in the application can be better explored by the immense amount of PE connected with network. So, CGRA revivals ASIC in performance, meanwhile, CGRA maintains a certain kind of flexibility by loading different context from context cache [4].

According to the architecture of PE, only one operation can be performed on PE at a specific cycle. The operation gets its operands by selecting data from neighbors and registers within the PE. Thus, operation costs multiplex and data transfer link resources in addition to ALU. However, some operations in DAG have mutual precedents. It is possible to make these operations sharing some resources so that the hardware overhead can be reduced.

In this paper, we present a dual-issue scheme for CGRA. This architecture gets a better utilization of hardware resource without declining performance much. The rest of this paper is organized as follows. Section 2 gives a motivation example and related

Q. Zhou et al. (Eds.): ICPCSEE 2018, CCIS 901, pp. 505–511, 2018.
https://doi.org/10.1007/978-981-13-2203-7_40

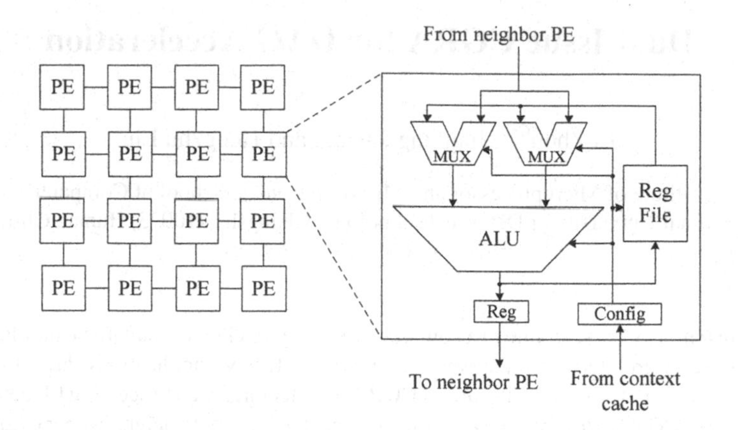

Fig. 1. Typical CGRA architecture.

work. Section 3 indicates the detailed dual-issue architecture and mapping algorithm. Section 4 presents the experimental evaluations, and Sect. 5 concludes the paper.

2 Motivation Example

In order to illustrate the benefit of the dual-issue scheme, we give a simplified CGRA and DAG example as shown in Fig. 2. There are 3 PEs connected to a straight line. In the given DAG, operation C, D, and E have the same precedents A and B. Assume operation A and B are mapped to PE 0 and PE 1 on the example CGRA at cycle 0, then at cycle 1, only operation C and D can be executed on PE 0 and PE 1. Since there are only 2 PEs neighboring the location of A and B both, operation E cannot be mapped in cycle 1. As showed in Fig. 2, DAG cost 3 cycles due to the limited data path resource, additional delay or data route cycle needs to be inserted, and the result of operation A will live one more cycle in the register file.

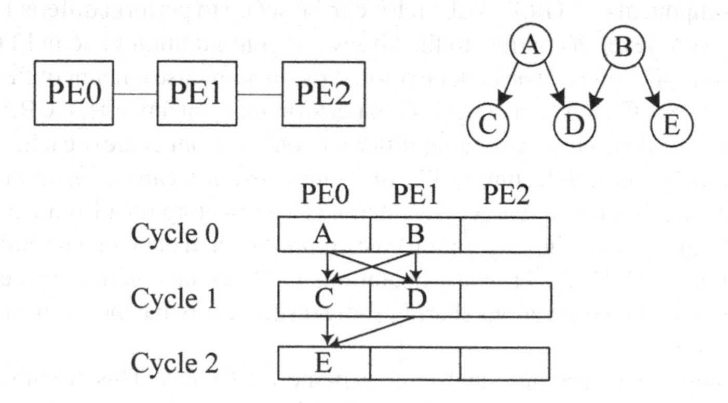

Fig. 2. Example CGRA and DAG mapping.

Considering the 3 operations have exactly the same operands. They can share the data path in the PE. So, we can add another ALU in the PE without any additional data path resource, then, one PE can issue two operations at the same cycle. In the example CGRA, 1 cycle can be saved if PE 1 supports dual-issue. This idea has been taken in solving branch instruction execution problems on CGRA [5, 6]. The dual-issue scheme enables each PE performing two distinct branches simultaneously. However, the architecture employed in [7] duplicates computation and data path resource both. Actually, this scheme can also be adopted without data path duplication if 2 operations share data path. Thus, CGRA hardware overhead can be lower.

3 Dual-Issue CGRA for DAG

3.1 Architecture

Figure 3 is detailed dual-issue CGRA architecture. It is composed of single-issue PE and dual-issue PE (DPE). In each DPE, function unit in ALU is doubled compared to single-issue PE. To support data path resource sharing, we add an extra ALU to carry out another operation on the same cycle without duplicate all multiplexers. Moreover, 2 ALU can be heterogeneous according to application feature for better efficiency concerns. The results of these two ALUs are selected through configuration. Thus only one of them is output to neighbors. The connection network of CGRA does not require modification. However, the input/output port and the size of register file need to increase to enable the other operation getting its operands and storing the result.

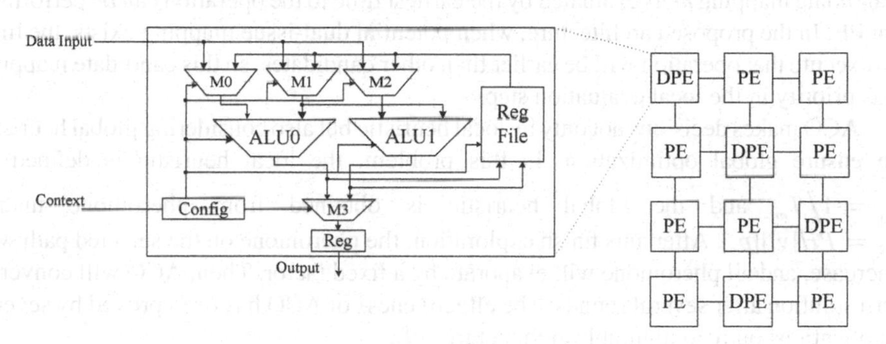

Fig. 3. Dual-issue CGRA architecture.

We note that not only operations with exactly the same precedents but also the operations share parts of precedents can be issued at the same cycle. So, there are 3 multiplexers in the DPE to support both full and partial data path share. A dual-issue PE has less operand multiplexer and connection logic with the identical effect when performing 2 operations compared to 2 single-issue PEs.

The ratio and location of the DPE in CGRA are some other important factor which determines efficiency. DPE's performance decreases when dealing with no precedent sharing operations due to data path shortage. For a fair comparison with the single CGRA in Fig. 1, 4 DPEs are distributed on the diagonal in the dual-issue CGRA topology as showed in Fig. 3, the total number of ALU in dual-issue architecture equals the CGRA in Fig. 1.

3.2 Mapping Algorithm

An application DAG can be denoted as a graph $G = <V, E>$, where vertices $v \in V$ represent operations, the edge $e = <v_1, v_2> \in E$ represents the fact that operation v_2 is data-dependent on v_1. Given a target CGRA, we can also use a directed graph $C = <P, L>$ to represent the computation resource it contains. The edge $l = <p_1, p_2> \in <P, L>$ represents that p_2 can use the result of p_1 directly next cycle either through connection network or register files. Mapping DAG to CGRA is to find a function $f: V \rightarrow P$ and make sure a path exists from $f(v_1)$ to $f(v_2)$ for each element in E. Ant colony optimization has been adopted for mapping DAG to CGRA [8]. We are required to modify the specified CGRA architecture representation for ACO algorithm in order to do mapping DAG.

In the proposed dual-issue architecture, a DPE should be denoted by 2 computation elements p_1, p_2 in CGRA graph C. Networks connected to DPE are viewed as edges connected to p_1, p_2 both. The p_1, p_2 also have self-connected and mutual-connected edges. Algorithm 1 is the pseudo code how ACO mapping DAG to dual-issue CGRA.

The algorithm runs *Nc* time exploration, every ant find a solution step by step in one iteration as depicted in line 3–11. At each step, ants find all operations v_x which can be executed at the current step and their potential mapping location p_x firstly; then each candidate mapping m_x is evaluated by the earliest time to the operation can be performed on PE. In the proposed architecture, when potential dual-issue mapping exists, the time to execute that operation will be earlier than other candidates, so this candidate mapping has priority in the local evaluation step.

ACO makes decisions not only by local heuristic but also considering global heuristic to ensure global optimization. In this problem, the local heuristic is defined as $\eta_x = 1/T_{m_x}$ and the global heuristic is obtained from pheromone matrix $\tau_x = PH[v_x][p_x]$. After ants finish exploration, the pheromone on the selected path will increase, and all pheromone will evaporate by a fixed factor. Then, ACO will converge to a solution after several rounds. The effectiveness of ACO has been proved by several applications on reconfigurable architectures [9].

Algorithm 1: ACO for CGRA mapping

Input: DAG graph $G=<V,E>$ and CGRA graph $C=<P,L>$

Output: mapping function $f:V\rightarrow P$

1: initial pheromone matrix *PH*;
2: **for** *n=1* to *Nc*
3: **for** each ant *i*
4: $f_i=\phi$;
5: **while** unmapped operation exist
6: find all candidate mapping $M=\{m_x=(v_x,p_i)|v_x\in V,p_x\in P\}$;
7: calculate local heuristic for each candidate $\eta_x=1/T_{m_x}$;
8: calculate the probability for each candidate $P_x=\frac{\tau_x^\alpha\eta_x^\beta}{\sum_M\tau_x^\alpha\eta_x^\beta}$;
9: select one candidate by round robin and add it to f_i;
10: **end while**
11: **end for**
12: update *PH* according to the mapping quality;
13: **end for**
14: output the best f_i;

4 Experimental Result

We implemented the typical CGRA in Fig. 1 and the dual-issue CGRA in Fig. 3 on a Xilinx Virtex-7 FPGA. Table 1 gives the detailed hardware consumption of these two architectures after synthesis.

Table 1. Synthesis result comparison for 2 CGRA.

Resource	Single-issue CGRA	Dual-issue CGRA	Reduction%
Slice LUTs	13792	13068	5.24
Slice registers	2576	2440	5.28
DSP	48	48	0

Compared to single-issue architecture, dual-issue topology not only reduces hardware consumption for computation data path, but also saves a few connection networks since there are less PE in the dual-issue scheme. When the computational resource is equivalent in these two architectures, 5.24%–5.28% hardware overhead is reduced due to some data path resource is shared and network is removed. The use of DSP does not change because the number of computation unit remains the same in these 2 CGRAs.

To evaluate the efficiency of dual-issue CGRA, 4 DAG extracted from the actual application are mapped by ACO algorithm. There are all from stream signal processing applications. Table 2 shows the execution cycle of performing each DAG.

Table 2. DAG mapping result comparison.

DAG	Single-issue CGRA	Dual-issue CGRA
fft	109	106
idct	237	239
sha	474	484
mac	226	220

In the *fft* and *mac* DAG case, the performance of dual-issue CGRA improved 2.75% and 2.65% due to the higher ratio of operations that share operands in the 2 DAG. Less cycle is required because of less data transfer is required. The more the DPE is utilized for computing, the more time will be saved in execution time. We also note that in the application *idct* and *sha*, the advantage of dual-issue CGRA failed to embody due to the lower possibility that 2 operations can be mapped in DPE compared to *fft* and *mac* DAG. Several operations in these 2 DAG share operands, but they often appeared in predecessor-successor relations so that cannot be performed simultaneously. Thus, dual-issue PE may be wasted in this situation. So, source code profiling is necessary in order to apply the dual-issue scheme. Special cases exist if the instruction level parallelism in the application is low due to data dependency. However, even in these cases, the performance loss is acceptable considering the hardware area reduced.

5 Conclusion

In the paper, dual-issue CGRA architecture is presented with an ant colony based mapping algorithm. This architecture is built by mixing single-issue and dual-issue PE. It saves hardware area because of the sharing of data path among operations which share operands, while its performance can be improved slightly if an application contains a certain amount of these operations. The dual-issue CGRA outperforms single issue CGRA architecture with 5.24% less hardware resource. The performance improved 2.65% when executing DAG which is appropriate for data path sharing and does not lose much performance when DAG has not many dual-issue operations exist.

Acknowledgement. This paper is supported by the National Natural Science Foundation of China (61602496).

References

1. Zainul, A., Svensson, B.: Evolution in architectures and programming methodologies of coarse-grained reconfigurable computing. Microprocess. Microsyst. **33**(3), 161–178 (2009)

2. Sangyun, O., Hongsik, L., Jongeun, L.: Efficient execution of stream graphs on coarse-grained reconfigurable architectures. IEEE Trans. Comput. Aided Des. Integr. Circuits Syst. **36**(12), 1978–1988 (2017)
3. Xitian, F., Di, W., Wei, C., Wayne, L., Lingli, W.: Stream processing dual-track CGRA for object inference. IEEE Trans. Very Larg. Scale Integr. (VLSI) Syst. **26**, 1–14 (2018)
4. Chattopadhyay, A.: Ingredients of adaptability: a survey of reconfigurable processors. VLSI Design **2013**, 1–18 (2013)
5. Hamzeh, M., Shrivastava, A., Sarma, B.K.: Branch-aware loop mapping on CGRAs. In: Proceedings of 2014 Design Automation Conference (DAC), pp. 1–6. IEEE, San Francisco (2014)
6. Tajas, R., Lukas, J., Dennis, W., Christian, H.: Scheduler for inhomogeneous and irregular CGRAs with support for complex control flow. In: Proceedings of 2016 IEEE International Parallel and Distributed Processing Symposium Workshops, pp. 198–207 (2016)
7. Radhika, S.H.R., Shrivastava, A., Hamzeh, M.: Path selection based acceleration of conditionals in CGRAs. In: Proceedings of 2015 Design, Automation & Test in Europe Conference & Exhibition (DATE), pp. 121–126. IEEE, Grenoble (2015)
8. Zhou, L., Liu, D., Zhang, B., Liu, H.: Ant colony optimization for application mapping in coarse-grained reconfigurable array. In: Brisk, P., de Figueiredo Coutinho, J.G., Diniz, P.C. (eds.) ARC 2013. LNCS, vol. 7806, p. 219. Springer, Heidelberg (2013). https://doi.org/10.1007/978-3-642-36812-7_22
9. Merkle, D., Middendorf, M.: Fast ant colony optimization on runtime reconfigurable processor arrays. Genet. Progr. Evol. Mach. **3**(4), 345–361 (2002)

Interruptible Load Management Strategy Based on Chamberlain Model

Zhaoyuan Xie[1(✉)], Xiujuan Li[2], Tao Xu[1], Minghao Li[3], Wendong Deng[1], and Bo Gu[1]

[1] State Grid Chongqing Jiangbei Power Supply Company, Chongqing, China
newlife201301@163.com

[2] State Grid Chongqing Power Company, Chongqing, China

[3] Power Supply Service Center of State Grid Chongqing Power Company, Chongqing, China

Abstract. As one of the important measures of power demand response, the interruptible load management has been widely used. The key to interruptible load management is the interruption compensation price and interruption capacity, which is related to the profits of power companies and users. To maximize the profits of both users and power companies, this article comprehensively considers the profits of power companies and users, and in view of the problem of interruption compensation price setting in interruptible load, it establishes a bargain model for power companies and users. In order to solve the limitations of single-user negotiation, the Chamberlain model considering product diversity was introduced to establish the multi-rounds bidding model of multi-users participation. At the same time, the neural network optimized by the genetic algorithm and particle swarm optimization was used to solve the problem of the initial price. According to the experiment, the model is effective and has superiority in mobilizing users' enthusiasm to participate in interruptible load management.

Keywords: Bargain model · Chamberlain model · Neural network model
Genetic algorithm · Particle swarm optimization

1 Introduction

With the continuous development of economy and improvement of economic system, the power consumption of users is increasing year by year, and the power load is also increased, which seriously aggravates the pressure of normal operation of the power system, and also exacerbates the supply and demand imbalance between the power companies and users. The power demand response effectively alleviates these problems [1, 2], and becomes an indispensable part of the construction of smart grid [3–5].

Interruptible load management is one of the important measures of DR, interruptible load management takes advantages of some users' electricity flexibility. When power system is in the peak or emergency, the power company and the user sign an interruptible load contract in advance. The contract stipulates the load capacity which the users can interrupt or reduce, and the power company will give the user some

Q. Zhou et al. (Eds.): ICPCSEE 2018, CCIS 901, pp. 512–524, 2018.
https://doi.org/10.1007/978-981-13-2203-7_41

compensation. Interruptible load management can effectively alleviate electricity peak period or emergency conditions, which can facilitate the normal operation of the power system and improve the service quality for users. Interruptible load contract should reflect the nature of interruptible load management, and should clearly specify the related expense, including early notice time of interruption, the contract's period of validity, the duration of interruption, the load capacity of interruption, and the load compensation price of interruption [6, 7]. The load capacity and load compensation price of interruption are the key content of the interruptible load contract, which is related to the interests of power companies and users. And users want to use their own interruption losses as little as possible to obtain greater interruption compensation, while the power companies want to use less interruption compensation cost to obtain more interruption load capacity.

In summary, there is a game relationship between the users and power companies that participate in interruptible load management. This paper comprehensively considers the interest of both power companies and users, and according to the setting of interruption compensation price in interruptible load, it established the bargaining model between power companies and users. At the same time, in order to allow more users to participate in the bidding, Chamberlain model which considers product differences is introduced to establish the multi-rounds bidding model of multi-users participation. In order to make the initial electricity price more reasonable, it uses the neural network model optimized by the genetic algorithm and particle swarm optimization to predict the initial interruption compensation price. The experiment shows the effectiveness of the model and the superiority in mobilizing users' enthusiasm in participating management.

2 Correlation Research

In the interruptible load management, there is a game relationship between power companies and users. Considering the interests of both users and power companies, many scholars used game theory in recent years to establish a reasonable incentive model to mobilize users' enthusiasm to participate in management. An Xuena et al. established the Cournot equilibrium model for the electricity wholesale market considering the interruptible load contract, and proposed that the interruptible load contract can effectively reduce and balance the market price and its volatility, and in order to reduce the market price and its volatility, the interruption threshold price in interruptible load contract needs to be reasonably selected [8]. Li et al. combined with Nash bargaining theory with a study of the bidding strategies of power generation companies and power supply companies under incomplete information, and obtained the optimal bidding strategy by solving Bayesian Nash equilibrium [9]. These documents fully prove the application of game theory in electricity market, but there is a limitation of single user participation in bidding. This article draws on the bargaining theory, and introduces the Chamberlain model which considers product differences to solve this limitation, so as to establish the multi-rounds bidding model of multi-users participation.

The key to establish a multi-rounds bidding model is how to set the initial interruption compensation price, which can mobilize the users' enthusiasm and protect the interests of power companies. According to the consumer psychology [10], users have obvious response to the price changes within a certain effective range, but when the price is not adjusted within this range, users will not have any response. Blindly setting the electricity price may not be able to mobilize users' enthusiasm, or cannot maximize the benefit of the power companies and users. Neural network algorithm is the most commonly used algorithm in current electricity price prediction. Therefore, reasonable interruption compensation price is significant to motivate users to participate in the interruption load. At present, the power supply capacity is an important factor that determines the users' participation in the interruption load. Accurate power load forecasting is essential to forecast the compensation price that users expected. Neural network algorithm is the most common algorithm in power load forecasting. He Yaoyao et al. aiming at the influence of temperature factors on the mid-term power load, proposed a mid-term power load probability density prediction method based on neural network quantile regression [11]. Considering that the neural network model has the disadvantages of slow convergence rate, local minimum and the influence of initial weights, many scholars have made optimization of neural network to compensate for these shortcomings. Shi Biao et al. combined the advantages and disadvantages of the traditional particle swarm optimization with radius vector neural networks, and proposed a particle swarm optimization with adaptive variable coefficients [12]. However, the optimization algorithm only uses the particle swarm optimization as the optimization of the initial weights and thresholds, and the gradient descent method is used to solve the weights and thresholds of each layer during the internal iteration of the neural network, which makes low convergence speed. According to the characteristics of interruptible load, this paper uses the neural network model optimized by the genetic algorithm and particle swarm optimization to predict the interruption load compensation price. Because the predicted electricity price does not take into account of the interests of users and power companies, which cannot maximize the interests of both parties, so it adjusts the multi-rounds bidding model which considers the interests of both parties to obtain a more reasonable interruption compensation price.

3 Neural Network Model Optimized by Hybrid Algorithm of Genetic Algorithm and Particle Swarm Optimization

3.1 Prediction Model of Standard BP Neural Network

This paper use a three-layer neural network model, including input layer, hidden layer and output layer. The number of nodes in input layer is m, and the number of nodes in output layer is n.

For the number of nodes in the hidden nodes, the estimated value of the nodes in hidden layer can be obtained according to the empirical formula (1), and then adjusted according to the values in the vicinity of the estimated values. Finally, select the number of nodes in the hidden layer according to the minimum error.

$$h = \sqrt{m+n} + a \tag{1}$$

Among them, h is the number of nodes in the hidden layer, m is the number of nodes in the input layer, n is the number of nodes in the output layer, and a is the adjusted integer between [1, 10].

3.2 Power Load Forecasting with Neural Network Model Optimized by Hybrid Algorithm of Genetic Algorithm and Particle Swarm Optimization

Genetic algorithm and particle swarm optimization are often used as the optimization of BP neural network model. Genetic algorithm has particularly prominent superiority in searching global optimality, but has slow convergence speed. Particle swarm optimization converges quickly, but it is easy to fall into local optimum. It combines their advantages and disadvantages to optimize the BP neural network. The optimized BP neural network model can predict the characteristics of power load more quickly and reasonably.

The key steps of improved neural network model optimized by the genetic algorithm and particle swarm optimization are as follows:

Determine the Neural Network Structure

According to the characteristics of interruptible demand response programs such as large time interval, short duration, few occurrences, etc., and considering the influence of seasons and other factors, select the data of interruptible demand response programs of power system in recent five years, and divide the data into training data sets and test data sets according to a certain proportion. Determine the topology of the BP neural network, including the number of layers of the hidden lay, and the number of nodes in the input layer, the output layer and the hidden layer.

Initialization

Initialize the initial position and velocity of M particles in the population, and set the particle's the effective range of position and velocity, initial inertia weight ω and effective rang, learning factors c1 and c2, the size of the population, the number of iterations, the crossover probability and mutation probability of genetic algorithm, and other parameters. The number of independent variables in the particle swarm is:

$$d = (m+1) * h + (h+1) * n \tag{2}$$

Among them, d is the number of independent variables of particle swarm, m is the number of input layer nodes of BP neural network, h is the number of hidden layer nodes of BP neural network, and n is the number of output lay nodes of BP neural network.

Determine the Optimal Value
According to the fitness function, calculate the fitness function value of each particle, and determine the global optimal value of all the particles and the historical optimal value of each particle. The optimal solution of initialization to the current iteration number is the historical optimal solution of the particle, and store the optimal position in P_h. For all the particles, compare their optimal value with their global optimal fitness value, and store P_h in P_e. The formula of fitness function is as follows:

$$J = \frac{1}{S}\sum_{i=1}^{S}\sum_{j=1}^{n}\left(y'_{i,j} - y_{i,j}\right)^2 \tag{3}$$

Among them, S is the total number of training samples, n is the number of output nodes, $y'_{i,j}$ is the actual output of the neural network, and $y_{i,j}$ is the expected output of the neural network.

By finding the optimal value, the optimal weight of the neural work is determined, the power load is predicted more accurately.

Update the Velocity and Position of the Particle
In a given range, the velocity $v_{i,j}(t+1)$ and position $x_{i,j}(t+1)$ of each particle are updated, in which t is the current number of iterations.

$$v_{i,j}(t+1) = \omega v_{i,j}(t) + c_1 r_1\left[p_{i,j} - x_{i,j}(t)\right] + c_2 r_2\left[p_{g,j} - x_{i,j}(t)\right] \tag{4}$$

$$x_{i,j}(t+1) = x_{i,j}(t) + v_{i,j}(t+1)\ i = 1,2,\cdots,n,\ j = 1,2,\cdots,d \tag{5}$$

Among them, r_1 and r_2 are random numbers between 0–1, c_1 and c_2 are learning factors, n is the number of particles, and d is the number of independent variables.

By updating the velocity and position of the particles, a better weight is found to ensure the optimal weight of the neural network.

Update the Weight Value
By using the characteristic that the inertia weight changes linearly with the number of iterations, update the inertia weight ω of the particle swarm optimization.

$$\omega = \omega_{\max} - \frac{t * (\omega_{\max} - \omega_{\min})}{t_{\max}} \tag{6}$$

Among them, $\omega_{\max}$ and $\omega_{\min}$ represent the maximum and minimum values in the range of limited inertia weight, t_{max} is the maximum number of iterations, and t is the current number of iterations.

Cross Operation
The new individual crossover function is used to cut the sequence of elements from two selected parents and exchange them to produce two new candidates. Cross operation is crucial to the success of predicting the model. This paper uses relative cross operation [13] instead of absolute cross operation, which can control the convergence and avoid local optimum. According to the fitness function value, N individuals are randomly selected, and the cross operation is performed on the configuration candidates with the

probability which is called the cross probability (P_c). This operation increases new individuals, expands the search scope, and makes the weights of the neural network more reasonable, thereby improving the accuracy of the prediction model.

Mutation Operation
The mutation operation is performed on all individuals, and select high value of the fitness function in $M + N$ to enter the next generation. The variation aims to avoid falling into a local optimum by randomly mutating elements which have a given probability. Instead of taking gene mutations in each generation, a random number r is generated for each individual. If r is bigger than the probability of mutation (P_m), the specific individual undergoes a mutation process. Otherwise, perform the mutation operation. In order to establish the prediction model as soon as possible, the range of numbers substituted of mutation is the range set in (3). This process prevents the premature convergence of the population, ensures the global optimal weight of the neural network, and improves the predicting speed of the prediction model.

Establish the Neural Network Model
Check whether the algorithm meets the condition to be finished, which means that it reaches the given number of iterations or meets the minimum error requirement). If it meets the condition, stop iterating and output the final weight and threshold of the neural network. Otherwise, go to step "Determine the optimal value".

Actual Application Prediction
Apply the BP neural network model which is already trained and optimized by the genetic algorithm and particle swarm optimization to the actual power load prediction.

When training the neural network model optimized by genetic algorithm and particle swarm optimization, the connection weights and thresholds between the layers are obtained by searching for the optimal value in each particle. The improved model has a simple iterative formula, and the calculation speed is faster than the gradient descent method. Moreover, the optimized genetic algorithm can effectively avoid local optimum and prevent the whole algorithm from premature convergence.

3.3 Interruptible Load Price Compensation Model Based on Load Characteristics

The interruptible load compensation price could motivate users to reduce their electricity load. Therefore, there is obviously linear relationships between the compensation price that users expected and the related indicators of load at that time such as load supply capacity, load gap. So the interruptible load compensation price could be estimated by linear regression method based on the load forecasting results got from Sect. 3.2.

Linear regression is a statistical analysis method used to determine the quantitative relationship between two or more variables, using the regression analysis in mathematical statistics.

In this paper, define the eigenvectors as $X = \{x_0, x_1, .., x_n\}$ based on the load forecasting results got from Sect. 3.2. Their characteristic coefficient vectors are defined as $\theta = \{\theta_0, \theta_1, .., \theta_n\}$, and the linear regression function is shown as:

$$h_\theta(X) = \sum\nolimits_{i=0}^{n} \theta_i x_i = \theta^T X \tag{7}$$

In order to obtain the characteristic coefficient vector, build the loss function $J(\theta)$ as follow:

$$J(\theta) = \frac{1}{2m} \sum\nolimits_{i=0}^{n} \left(h_\theta(x_i) - y_i\right)^2 \tag{8}$$

$h_\theta(x_i)$ represents the function that needs to learn, m represents the number of samples in training set, x_i represents the eigenvector of the i-th sample in training set, y_i represents the label of the i-th sample.

Because that $J(\theta)$ is a convex function of θ, the gradient descent would be used to solve the problem. Gradient descent algorithm is a search algorithm. The basic idea is to give θ an initial value, and update the value of θ to minimize the value of $J(\theta)$ based on iterative method.

$$\theta_j = \theta_j - \alpha \frac{\vartheta}{\vartheta \theta_j} J(\theta) \tag{9}$$

α represents the learning rate.

And on the foundation, adopt batch gradient descent method. The function of θ_j is shown as:

$$\theta_j = \theta_j - \alpha \sum\nolimits_{i=0}^{n} \left(h_\theta\left(x_0^{(i)}, x_1^{(i)}, \ldots, x_0^{(i)}\right) - y_i\right) x_j^{(i)} \tag{10}$$

When $J(\theta)$ satisfies the convergence conditions, calculate the θ. And the interruptible load price compensation model based on load characteristics could be constructed.

4 Multiple Rounds of Bidding Model

4.1 Application of Bargain Theory in Interruptible Load

In interruptible load management, there is a game relationship between users and power companies that maximizes their own interests. The two parties have a sequence of priorities when conducting price negotiations, and the information is asymmetric. For the user's interruption costs, the user knows, but the power company does not know. Similarly, for the power company's earnings, the power company knows, but the user does not know. According to its own load capacity requirements, the power company actively announces the interruption capacity and interruption compensation price to the user. According to the user's interruption cost and characteristics, they report their acceptable interruption compensation price and interruption capacity to the power company. According to the information reported by the user, the power company chooses to accept or reject it based on its own load capacity requirements and revenue conditions. If the power company chooses to accept, the negotiation is over, otherwise the power company continues to offer price until it reaches the final balance which is Nash equilibrium. Considering the actual situation, a benefit discount factor $\delta \in [0, 1]$ is introduced to ensure that the negotiation will not proceed indefinitely.

When performing the interruptible load management, the power company can relieve the load pressure and reduce the operating costs of the power system. Through the interruptible load management, the power company can receive the profit C which includes rotating reserve cost, operation and maintenance cost of power system, saving cost of capacity production, management cost, but not includes social profits.

$$C = C_r + C_o + C_c + C_m * P * t \tag{11}$$

Among them, C_r is the rotating reserve cost saved after the interruptible load management, C_o is the saving cost of the power system operation and maintenance, C_c is the saving cost of capacity production, C_m is the management cost, P is the load capacity of interruptible load management, t is the duration of interruption.

The power company's interruption economic compensation to users who participate in interruptible load management actually comes from the interruption profits of the power company. The power companies and users actually play the game against this interruption profit. When the interruption capacity and interruption duration are determined, the cost in the power company's profit is the internal data of the power company, so that the profits of the power company can be determined. According to the bargaining theory, we can see that after several rounds of bidding, the power companies and users will achieve Nash equilibrium, and the equilibrium point is the interruption compensation price which maximizes the profits of all the parties.

Assume that the total interruption capacity is P, the interruption duration is t, and the profit discount factor is δ. The specific process of the game between the power company and the user is as follows:

According to the information such as total interruption capacity, interruption duration, and initial electricity price of interruption announced by the power company, the user reports his own interruption compensation price and interruption capacity based on his own situation. The user's profit is B_1, then the power company's profit is $C - B_1$. If the power company accepts, the negotiation is over, otherwise the next round of bidding will be conducted. According to its own shortage capacity and profit situation, the power company reports the interruption compensation price and interruption capacity. At this time, the actual interruption profit of the power company is δB_2, and the actual interruption profit of the user is $\delta(C - B_2)$. The user chooses to accept or reject according to his own situation. If the user accepts, the bidding is completed. If the user does not accept, the next round of bidding will be conducted. At this time, he actual profit of the user and the company is $\delta^2 B_3$ and $\delta^2 (C - B_3)$. If the company does not accept is, the next round of bidding will be conducted. This cycle will be continued until both parties can accept it.

4.2 Multiple Rounds of Bidding Model Based on Chamberlain Model

The object of the bargaining theory is a single user. Taking the multi-user situation into account, the Chamberlain model, which considers the difference of product, is introduced. Chamberlain model believe that the difference in the sales volume of each manufacturer comes from the difference between products. The sales volume of each manufacturer is the result of three factors: the unit price of sales, the nature of the product, and the development of sales.

Product difference is introduced in participating in interruptible load management. For participating users, there is a significant difference in interruption capacity, interruption compensation, and enthusiasm of user's participation, which can divide users into different types. And different types of users can provide different interruptible load services. When selecting the user to participate in the interruption response, the power company is more willing to choose users with large interruption capacity, low interruption compensation price, high credit and participation enthusiasm, because such users are less likely to default participating in demand response management, so the power company will bear low risk. User enthusiasm can be expressed θ according to his type. The larger the type of user, the greater the enthusiasm of user participation. The interruptible costs of the company are as follows:

$$C_{LC} = \sum\nolimits_{i=1}^{n} [y_i + \alpha(1 - \theta_i)(1 - \gamma_i)]x_i \tag{12}$$

Among them, y_i is the interruption capacity that user i can provide, θ_i is the type of user i, γi is the credit of user i, x_i is the interruptible compensation price of user i, n is the number of users participating in interruptible load management, α is the adjustment factor, and $\alpha \in [0, 1]$.

The basic process of the multi-rounds bidding model which considers product difference is as follows:

(1) Based on its own load requirements, the power company announces the information such as interruption capacity, initial price of interruption (from Sect. 3.2), interruption duration to users. The users according to their own situation combines with the power company's interruption compensation price and interruption capacity, and reports their own electricity price and capacity.
(2) If the load capacity reported by users does not meet the demand of the power company, the power company will calculate its own power profit, raise the interruption compensation price, report to the users, and conduct the next round of bidding.
(3) If the load capacity reaches the demand of power company, it is ranked according to each user's interruptible cost from low to high. Users are screened according to the load capacity, and each user's winning result, interruption compensation price, and interruption capacity are fed back to the users.
(4) The user checks the feedback result of the power company, and decides whether to proceed to the next round of bidding according to the power company's clearing price and situation. When all users no longer change the price, the bid is finished.

5 Experimental Design and Result Analysis

5.1 Experimental Setup

According to the electricity price prediction model described above, the training data is related to the relevant power consumption data and the users' information of a provincial power company's interruptible load management in the past five years. Three layers of BP neural network are used in the experiment: input layer, hidden layer, and output lay. And the number of nodes is 4/7/1.

In multiple rounds of bidding, assuming that the total demand of system load is 800 MW and the interruption duration is 1.5 h, there are 100 users who can provide services.

5.2 Experimental Design and Result Analysis

Forecast Electricity Price Model

Using the selected data to train the model, the BP neural network was trained separately using gradient descent (GD), particle swarm optimization (PSO), genetic algorithm (GA), genetic algorithm and particle swarm optimization (GPSO). The training results are shown in Table 1. The parameters in the algorithm are set as follows: the population size is 30, c_1 is 2.4, c_2 is 1.1, ω gradually adjusts from 0.8 to 0.3 with the optimization process, the crossover probability is 0.7, and the mutation probability is 0.2.

Table 1. Comparison of training results of different algorithms.

Algorithm	Training time/s	Iterations number	Minimum error
GD	195.85	983	8.7×10^{-3}
PSO	26.14	63	3.2×10^{-4}
GA	310.73	587	1.4×10^{-4}
GPSO	16.47	29	8.3×10^{-5}

From the above table, we can see that in the process of predicting electricity price, compared with GD algorithm, PSO algorithm and GA algorithm, the algorithm proposed in this paper performs better in convergence speed and accuracy. The prediction model obtained by optimizing the neural network through this algorithm, provides a more reasonable and fast initial price for the multi-rounds bidding model, and accelerates the running time of the whole process.

Multi Wheel Bidding Model

According to the given data, experiments were performed on multi-rounds bidding model, and the number of iterations was 120. The three diagrams in Fig. 1 respectively show the curves of three parameters changing with the number of iterations, and the parameters are interruption compensation price, the power company's cost, and the number of users participating in interruption load management.

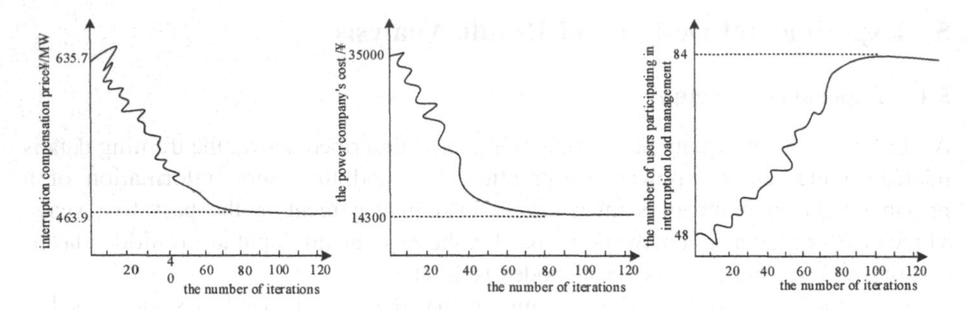

Fig. 1. The curves of three parameters changing with the number of iterations.

Through the observation of diagrams in the above Figure, in the process of using the multi-rounds bidding model, at first interruption compensation price, the power company's cost and the number of users participating in interruption load management are in the continuous fluctuations, but with the increase of the number of interactions, the fluctuations gradually flatten, and finally the interruption price converges to a lower price level, and the cost of the power companies tends to be stable. It can be seen that the multi-rounds bidding model can greatly mobilize the enthusiasm of users and ensure that the power companies can get higher profits.

Algorithm Parameters

In electricity price prediction model, the crossover operation and mutation operation are the core operations of the genetic algorithm. Through these two operations, we can guarantee individual diversity, expand the search scope, and avoid falling into local optimum. The key of the operations is the setting of crossover probability and mutation probability. In this paper, by changing the values of the crossover probability and mutation probability, we can study their influence on the completion time of the model.

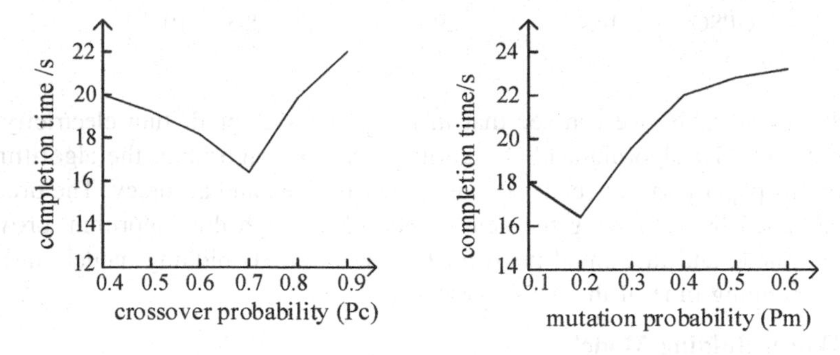

Fig. 2. The effect of completion time of training model changing with crossover and mutation probability.

The first diagram in Fig. 2 shows that the completion time of the model decreases with the increase of the crossover probability, and then increases with the increase of the probability, which shows that a very large crossover probability may lead to deterioration of the model completion time. Therefore, this article empirically sets the crossover probability to 0.7, which is the tradeoff between the search speed and the improvement of the model completion time. The second diagram in Fig. 2 shows that the adjusting mutation probability is similar to the crossover probability in the experiment. This article empirically sets the mutation probability to 0.2 without affecting the convergence in the experiment. The crossover probability and mutation probability set in this way can not only improve the completion time of the prediction model, but also avoid local optimum and get a more reasonable prediction price.

6 Conclusion

For the problem of setting the interruption load price, this paper analyzed the characteristics of the users who participated in the interruptible load management, and used the neural network model optimized by the genetic algorithm and particle swarm optimization to predict the interruption compensation price. Because the predicted electricity price did not take into account of the profit of users and power companies, it learned the bargaining theory of game theory. In order to solve the limitation of single-user, the Chamberlain model considering product difference was introduced. So learning from the above models, a multi-rounds bidding model was comprehensively established to obtain a more reasonable interruption compensation price. The experiments not only validated the rationality of the model, but also provided a more reasonable interruption compensation price and mobilized the enthusiasm of the users to participating in the interruptible load management.

References

1. Tan, H., Chen, S., Zhong, M., et al.: Research and design of demand side energy efficiency management and demand response system. Power Syst. Technol. **39**(1), 42–47 (2015). https://doi.org/10.13335/j.1000-3673.pst.2015.01.007
2. Walawalkar, R., Femands, S., Thakur, N., et al.: Evolution and current status of demand response (DR) in electricity markets: insights from PJM and NYISO. Energy **35**(4), 1553–1560 (2010). https://doi.org/10.1016/j.energy.2009.09.017
3. Zhang, Q., Wang, X., Fu, M., et al.: Smart grid from the perspective of demand response. Autom. Electr. Power Syst. **33**(17), 49–55 (2009)
4. Yan, H., Chen, S., Du, C., et al.: Research and design of power demand response standard system about smart grid. Power Syst. Technol. **39**(10), 2685–2689 (2015). https://doi.org/10.13335/j.1000-3673.pst.2015.10.001
5. Rahimi, F., Ipakchi, A.: Demand response as a market resource under the smart grid paradigm. IEEE Trans. Smart Grid **1**(1), 82–88 (2010). https://doi.org/10.1109/TSG.2010.2045906

6. Zhu, L., Zhou, X., Tang, L., et al.: Multi-objective optimal operation for microgrid considering interruptible loads. Power Syst. Technol. **41**(6), 1847–1854 (2017). https://doi.org/10.13335/j.1000-3673.pst.2016.2504
7. Dong, J., Zhang, J., Chen, X., et al.: Study on short-run congestion management model and incentive mechanism considering demand response. Power Syst. Prot. Control **38**(3), 24–28 (2010)
8. An, X., Zhang, S., Wang, X., et al.: Cournot equuilibrium analysis of electricity markets considering interruptible load contracts. Power Syst. Prot. Control **40**(16), 49–53 (2012)
9. Li, L., Tan, Z., Guan, Y.: Study on the optimal model of interruptible loads implemented by power generation and supply company. Electr. Power Sci. Eng. **24**(2), 16–19 (2008)
10. Luo, Y., Xing, L., Wang, Q., et al.: Least-squares estimation of parameters of customer response models for peak and valley time-of-use electricity price. East China Electr. Power **37**(1), 67–69 (2009)
11. He, Y., Wen, C., Xu, Q., et al.: A method to predict probability density of medium-term power load considering temperature factor. Power Syst. Technol. **39**(1), 176–181 (2015). https://doi.org/10.13335/j.1000-3673.pst.2015.01.027
12. Shi, B., Li, Y., Yu, X., et al.: Short-term electricity price forecast model based on adaptive variable coefficients particle swarm optimizer and radial basis function neural network hybrid algorithm. Power Syst. Technol. **34**(314(1)), 98–106 (2010). https://doi.org/10.13335/j.1000-3673.pst.2010.01.036
13. Deb, K., Pratap, A., Agarwal, S., Meyarivan, T.: A fast and elitist multi objective genetic algorithm: NSGA-II. IEEE Trans. Evol. Comput. **6**(2), 182–197 (2002). https://doi.org/10.1109/4235.996017

A Method to Identify Spark Important Parameters Based on Machine Learning

Tianyu Li, Shengfei Shi(✉), Jizhou Luo, and Hongzhi Wang

Harbin Institute of Technology, Xidazhi Str. 92, Harbin 150001, China
shengfei@hit.edu.cn

Abstract. Apache Spark is the most popular open-source framework today that uses an in-memory-oriented abstraction Resilient Distributed Dataset (RDD) to process large-scale data. Recently, research work on performance prediction and optimization for Spark platform continues to increase rapidly. However, selecting important configuration parameters in most wok is always dependent on the experience of domain experts yet. Therefore, configuration parameters selection based on machine learning algorithms is a non-trivial research issue. In this paper, a method based on machine learning to identify Spark important parameters ISIP is proposed. By providing a relatively important subset of configuration parameters, the parameter space for performance tuning on Spark can be reduced, thereby saving the time and effort of users or researchers. ISIP uses Mean-shift algorithm to cluster the applications based on the workload characteristics of the applications from Spark MLlib. Then the relationship between the performance and the configuration parameters is modeled by Regression Algorithm. In the meanwhile, the ranked list of parameters by their importance is provided respectively for each type of applications. The subset of most important configuration parameters consists of the parameters at the front of the list. The experimental results show that the effect of adjusting the subset of relatively important configuration parameters provided by ISIP is almost the same as the complete parameters set.

1 Introduction

The performance and configuration parameters of data processing engine are closely related because configuration parameters control almost all aspects of application runtime such as memory allocation and I/O tuning [1]. Apache spark is no exception. For example, spark.serializer can set the data compression method. When this parameter is set to Java serialization, it can work on any class, but it will greatly slow down the calculation. However, when this parameter is set to Kryo serialization, the calculation speed is increased by more than 10 times, but the user is required to register the class used in the program [2].

Finding a subset of parameters that are closely related to application performance on Spark is non-trivial as Spark is a complex system with a large number of tunable options. A total of 190 configuration parameters are listed on the Apache Spark official website, including the configuration parameters' names, default values, and meanings. In addition to the 28 runtime environment parameters and 17 UI parameters, the rest are set on the application runtime shuffle, compression & serialization, and memory

Q. Zhou et al. (Eds.): ICPCSEE 2018, CCIS 901, pp. 525–538, 2018.
https://doi.org/10.1007/978-981-13-2203-7_42

management. However, it is impossible and unnecessary to optimize all the configuration parameters to meet all the requirements of the application, which is beyond human capabilities. Therefore, it is necessary to find configuration parameters that are highly related to the performance of the Spark application.

In recent years, there has been an increase in research related to achieving good performance by adjusting Spark configuration parameters [3, 4]. In these work, the reason for choosing configuration parameters is based on domain experts, and the choice of parameters based on the machine learning performance model is still blank [4]. However, the configuration parameter space is huge, and the cost of artificial learning is expensive. In addition, the correct relationship between system performance and configuration parameters depends on a large number of factors, sometimes beyond what humans can reason about [5]. This scenario is well-suited for data-driven machine learning algorithms and models are based on observations of specific workloads and actual system performance under the cluster [4]. Given this, we explored the relationship model between the performance and configuration parameters of Spark applications based on machine learning.

Spark can handle many types of applications by powering a stack of libraries including SQL, DataFrames, MLlib, and so on [2]. In this paper, we focus on analyzing applications selected in MLlib. Even so, the workload characteristics of the application are still varied. Therefore, clustering algorithm is applied to applications, which is one of the effective ways to improve the accuracy of the model. RDD can capture a wide class of computations through rich operators [6]. Different characteristics will be exhibited because different operators are used in the applications during the execution. For example, operators such as groupByKey, reduceByKey, etc. will result in a hash or range partitioned RDD which can often be large. The metrics obtained from some monitors are very large, such as the number of shuffle bytes read and written [2, 6]. We use clustering algorithm to divide applications into different types. Experimental results show that applications with similar characteristics are clustered into the same cluster.

Depending on the type of application, specific model is created by using the configuration parameters and performance values of a certain category of application. This is understandable and explainable. For example, for applications where there are many operators that can trigger the shuffle, setting the configuration parameters related to the degree of parallelism enables the application to execute efficiently [2]. Regression model is used where the coefficient reflect the weight of the importance of configuration parameters, that is, the impact of configuration parameters on performance.

The rest of the paper is organized as follows. Related work is discussed in Sect. 2. A brief introduce about Spark is outlined in Sect. 3. An applications-division algorithm and a performance model based on Machine Learning will be proposed in Sect. 4. Experimental methodology and the analysis of experimental results are given in Sect. 5. Finally, conclusions and future work are presented in Sect. 6.

2 Related Work

During tuning the configuration parameters of traditional database, there are many work involved in selecting configuration parameters on machine learning algorithms. iTuned using a technique called Adaptive Sampling that under conditions to provide appropriate data and planned experiments, to find high-performance configuration parameter [1]. OtterTune applied Factor-Analysis and K-means algorithm to select a subset of the metrics provided by database platform which identify important knobs having a significant impact on the subset [5].

In recent years, more and more attention has been paid to the work of setting configuration parameters to improve the performance of Spark [3, 4, 7]. Chiba et al. analyzed the relationship between these factors and performance from various aspects such as JVM parameters, Spark configuration, operating system parameters, and app code [7]. Wang et al. used multiple classifications and tree models to attempt to automatically adjust configuration parameters [4]. However, little research has been done on the selection of configuration parameters based on machine learning methods.

3 Background

3.1 Spark Stage

The resilience distributed data set RDD is an in-memory abstraction that can be only read. The partition is the smallest unit of RDD, that is, RDD is a partitioned collection of records. Operators that can be executed on RDD fall into two categories. One is lazy, called transformations, to define new RDDs such as map, filter, and so on. The other is actions, which are responsible for starting calculations to return values to the program or write data to an external storage such as count and save. RDD will only actually launch calculations after users run an action operator. Instead all transformations that generate RDDs will be recorded, called Lineage graphs, which describe the dependencies between RDDs. When the user executes an operator on the RDD, the scheduler will check the lineage graph of the RDD to build the DAG of the stage to be executed. The boundaries of these stages are the shuffle operations required for a wide dependency, or any possible calculations. A wide dependency means that each partition in the Parent RDD will be used by multiple partitions in the child RDD, that is, a one-to-many relationship. For example, the join operator will result in a wide dependency.

3.2 Spark Monitor

Spark provides three methods for monitoring spark applications: web UI, metrics, and external interfaces. The Spark UI shows many aspects of the application displayed on port 4040 by default. Information about various grained is described on different pages, including application level, job level, stage level, and task level. In addition to viewing metrics in the user interface, they can also be recorded in JSON format to monitor running and completed applications stored in the history server, which makes monitoring application information easier and more convenient. The endpoints are installed in /api/v1. In the API, the application is referenced by its application ID [app-id]. When running on YARN, applications in cluster mode will have attempt IDs, which can be identified by their [attempt-id]. The urls we used are listed below in Table 1.

Table 1. Urls and metrics provided by Spark monitor.

url	Metrics
/applications/[app-id]/stages	A list of all stages for a given application. ? status = [active|complete|pending|failed] list only stages in the state
/applications/[app-id]/stages/[stage-id]	A list of all attempts for the given stage
/applications/[app-id]/stages/[stage-id]/[stage-attempt-id]	Details for the given stage attempt

4 Overview

We now propose our identifying Spark important parameters algorithm (ISIP) to solve the problem we described above.

First, ISIP needs a data repertory consistent of history data, including configuration parameters, workload characteristics (I e, metrics) and performance (e g, total runtime) and so on. Next, ISIP divides the applications stored in repertory into several categories to establish reasonable and effective regression models. In this period, the Mean-shift algorithm is chose to cluster applications using the features (metrics) of the application after comparing several clustering algorithms based on scores of two evaluation indicators. At the same time, ISIP collects and stores the characteristic metrics of representative application for every category.

The ISIP algorithm respectively establishes a regression model between the configuration parameters and performance corresponding to each type. In establishing the regression model, the configuration parameters are ranked according to the importance. ISIP stores several configuration parameters at the top of the list separately corresponding to the category. At this time, ISIP constructs a new database that stores the representative application's features and the most important configuration parameters of each type.

This database can provide guidance to users. For example, the user collects the metrics of the target application and the calculation of similarity between the target and the representative application of each type is executed. The collection of metrics is very easy through running the target for a short period of time by controlling the size of input data. The database provides a list of important configuration parameters correspond to the category that is the most similar to the target application. Users can optimize the configuration parameters in this list to improve performance, instead of setting all configuration parameters blindly in the experiment. This can help users save a lot of time, because configuration parameter space is to configuration parameters.

In this section, data collection, application type division, and configuration parameter identification will be described.

4.1 Data Collection

ISIP is a data-driven machine learning algorithm that requires a repertory to store history data from previous training sessions.

Table 2. Parameters name, default and meaning

Parameter name	Default	Meaning
spark.broadcast.blockSize	4 m	Size of each piece of a block for TorrentBroadcastFactory
spark.shuffle.service index.cache.entries	1024	Max number of entries to keep in the index cache of the shuffle service
spark.files maxPartitionBytes	134217728 (128 MB)	The maximum number of bytes to pack into a single partition when reading files
spark.driver maxResultSize	1g	Limit of total size of serialized results of all partitions for each Spark action (e.g. collect)
spark.shuffle.sort bypassMergeThreshold	200	In the sort-based shuffle manager, avoid merge-sorting data if there is no map-side aggregation and there are at most this many reduce partitions.
spark.files openCostInBytes	4194304 (4 MB)	The estimated cost to open a file, measured by the number of bytes could be scanned in the same time
spark.shuffle.file.buffer	32 k	Size of the in-memory buffer for each shuffle file output stream
spark.io.encryption keySizeBits	128	IO encryption key size in bits. Supported values are 128, 192 and 256
spark.rpc.message maxSize	128	Maximum message size (in MB) to allow in "control plane" communication
spark.reducer maxReqsInFlight	Int.MaxValue	This configuration limits the number of remote requests to fetch blocks at any given point
spark.storage memoryMapThreshold	2 m	Size in bytes of a block above which Spark memory maps when reading a block from disk
spark.memory storageFraction	0.5	Amount of storage memory immune to eviction
spark.default.parallelism	8	spark.default.parallelism
spark.driver.cores	1	Number of cores to use for the driver process, only in cluster mode
spark.executor.memory	1g	Amount of memory to use per executor process, in MiB unless otherwise specified. (e.g. 2g, 8g)
spark.reducer maxSizeInFlight	48 m	Maximum size of map outputs to fetch simultaneously from each reduce task, in MiB unless otherwise specified
spark.memory storageFraction	0.5	Amount of storage memory immune to eviction
spark.driver.memory	1g	Amount of memory to use for the driver process
spark.shuffle accurateBlockThreshold	100 * 1024 * 1024	Threshold in bytes above which the size of shuffle blocks in HighlyCompressedMapStatus is accurately recorded

Three typical workloads from MLlib are chosen to train the algorithm, including K-means, logistic regression, and FP-growth. We control the input data size of the applications so that the running time of each application is about 5 min, in order to make the experimental results more practical in practice.

19 configuration parameters are selected as shown in the Table 2. We run the applications in Yarn mode, so the selected configuration parameters need to meet the following requirements: (1) they work on yarn; (2) numerical; (3) when setting default value of non-numeric parameters, they also work; (4) may be related to performance, not path parameters, security information parameters, network protocol parameters, etc.

At the beginning, the range of each configuration parameter is its minimum to maximum, in the limited of the cluster resource. The minimum value is defined as 1 unit for this configuration parameter. The maximum value is defined as the smaller value between the maximum value of the parameter allowed by the cluster resource and the maximum value of the parameter itself. Due to the limitation of cluster resources, the combination of configuration parameters that make the application fail is discarded.

Sampling technology is used at the parameter space, combined with random-sampling and grid-sampling. First of all, random-sampling is applied to the parameter space, that is, all parameters are set to pseudo-random numbers created by computer at the same time. Then, grid-sampling is performed at the "edge" of the parameter space, that is, only one parameter value is changed at a time, and the parameter value is evenly spaced within the range while other parameters are set as default values.

During each experiment, metrics were acquired in JSON format through rest API [2]. We observe the characteristics of metrics at the stage-grained, including the trend of executor CPU run-time metric, shuffle read and write bytes metrics, (hereinafter referred to as shuffle bytes metrics) and the input bytes metric at the corresponding stage. At the beginning of the application run-time when the application loads data or executes a non-shuffle operation such as map, the shuffle bytes metrics are always 0. With the start of the iterative calculation, if the application has a shuffle operator, then the shuffle bytes metrics alternately display the number of bytes; otherwise, only the executor CPU run-time metric fluctuates slightly. If the application requires to aggregate data, the application's final stage's shuffle bytes metrics are zeroed again. In order to reflect this trend, 9 metrics are chosen from these three part: before the iterative calculations, during the iterative calculations and after the iterative calculations. For each part, we records the shuffle bytes metrics, executor run-time metric and input size metric with the median of the stages during every part.

However, in fact, the shuffle bytes metrics from the first and the third part are always 0, and the input bytes metric from the second part during the iterative calculation is 0. Therefore, only the remaining 6 dimensions are stored in the database in the experiment. At the same time, the configuration parameters and total time of each experiment are also stored in the database. These data are obtained by analyzing the history logs in the history server.

4.2 Application Type Division

Clustering Evaluation. We use a variety of methods to evaluate the clusters. The comprehensive values of Calinski-harabaz index and homogeneity and completeness are used to evaluate the clustering effect.

The Calinski-harabaz CH indicator, [8] also known as pseudo F-statistics, is a measure of the between-classes and intra-class deviation matrices for all samples. This indicator is defined as follows:

$$\mathrm{CH(k)} = \frac{\mathrm{tr(B(k))/(k-1)}}{\mathrm{trW(k)/(n-k)}} \tag{1}$$

Where n denotes the number of samples, k denotes the number of clusters, B(k) and W(k) represent the between-classes dispersion matrix and the intra-class dispersion matrix, respectively. tr(B(k)) and tr(W(k)) represent the trace of intra-class dispersion matrix and the intra-class dispersion matrix. The higher the Calinski-harabaz index value, the smaller the intra-class covariance and the greater the between-classes covariance. In other words, the higher the CH indicator value is, the denser the samples within the cluster are, and the more dispersed the samples are between the clusters, which is to some extent the better clustering results [9].

The homogeneity and completeness measures are based on conditional entropy to evaluate the clustering effect. The clustering result meets homogeneity when every cluster only contains data points belonging to the members of a single class. The clustering results will satisfy completeness if all data points belonging to a given class are the elements of the same cluster. Homogeneity reflects the proportion of each cluster containing a single class, while completeness reflects the ratio of a given class to a cluster. We define v_measure as the mean of these two indicators which fully reflects the degree of confusion of the seeds in the cluster. The seed is a sample whose class is determined by prior knowledge.

Clustering Algorithm. The Mean-shift algorithm was used after multiple clustering methods were tried to divide the applications. Mean-shift algorithm is a common clustering method for nonparametric estimation. There are many advantages of the Mean-shift algorithm, including no requirement on the cluster shape and excellent robustness to initialization.

Mean-Shift clustering is a centroid-based algorithm designed to find cluster blobs in smooth density samples. It uses the center of mass in a given region of interest to update the candidate centroid. Then in the following processing stage, these candidates are filtered to eliminate the approximate duplicates and form the final set of centroids [10]. In the t + 1th iteration, the candidate centroid x_i is updated as follows:

$$\mathrm{x}_i^{t+1} = \mathrm{x}_i^t + \mathrm{m}(\mathrm{x}_i^t) \tag{2}$$

Where x_i^t is a candidate centroid for iteration t and m is the Mean-Shift vector, which can effectively update the candidate centroid to the center of mass in its neighborhood, which is calculated as follows:

$$\mathrm{m}(\mathrm{x}_i) = \frac{\sum_{\mathrm{x}_j \in \mathrm{N}(\mathrm{x}_i)} (\mathrm{x}_j - \mathrm{x}_i)}{k} \tag{3}$$

Where $N(x_i)$ means the samples within the region of interest, that is a neighborhood of x_i, where the amount of the samples is k. The distance range of the neighborhood h is determined by the bandwidth parameter which is estimated as the quantile in the Euclidean distances between pairs of all samples. The formula is as follows:

$$N(\mathrm{x_i}) = (\mathrm{x_j} | (\mathrm{x_j} - \mathrm{x_i})(\mathrm{x_j} - \mathrm{x_i})^T \leq h^2) \tag{4}$$

The input of the algorithm is the applications' metrics matrix while the algorithm returns the applications' labels and clustering centers. The algorithm is as follows:

Algorithm App Type division

Input: apps' metrics matrix X
Return: apps' label y, clusters' centers matrix C
1. **For** seed in X, do:
2. A circle with seed as center point, bandwidth as radius
3. Center_new ← mean point of the points within the circle
4. C[seed] ← Center_new
5. **Until**
6. Distance(Center_new , center) > threshold or interaction > threshold
7. **For** every pairs (c1, c2) in C:
8. If distance(c_1, c_2) ≤ threshold:
9. Remote c_i with less points in the cluster
10. **For** point in X, do:
11. Point assigned to the nearest cluster with c_{point} as center point
12. y[point] = c_{point}

We also used other clustering algorithms for comparison, including mini-batch-k-means, ward and Gaussian-mixture.

Mini-Batch-k-Means. Mini-batch is used in machine learning algorithms such as gradient descent and deep network. When the data set is large, using a full data set becomes no longer viable due to memory constraints. The use of batch processing can greatly speed up the parallelization process under conditions where the accuracy is degraded. Mini-batch-k-means is a variant of the k-means algorithm that uses part of the sample rather than all of the samples in calculating the distance between data points. Mini-batch-k-means, like the k-means algorithm, requires the number of clusters to be specified and the clusters are affected by the initial points [11, 12].

Ward Linkage Method (Hierarchical Clustering). Hierarchical clustering can avoid the problems about the number of clusters and the selection of initial points which is different from planar clustering. The output of the algorithm returns to an unstructured cluster set such as k-means. Hierarchical clustering is divided into two types, one is top-to-down, also known as agglomerative hierarchical, and the other is bottom-to-up, also known as divisive hierarchical.

The ward linkage method belongs to the agglomerative hierarchical using the smallest increase in the cluster variance in the combined population when merging the cluster pair [14, 15]. The new cluster' variance is the sum of the variance between the two clusters before merging.

Gaussian-Mixture. The Gaussian-mixture model GMM refers to a number of models based on the Gaussian probability density function (normal distribution curve). Theoretically, the GMM can fit any type of distributed data to solve the classification problem of a data set that contains several different distributions of the same set, and iteratively calculates the probability that each sample is classified into different clusters. Similar to k-means, the GMM also needs to specify the number of clusters.

4.3 Identify Spark Important Parameters

Lasso Regression Analysis. Using the coefficients of linear regression to determine the strength of the relationship is a more common method, especially when judging the relationship between one or more dependent variables and each independent variable. The coefficients in the linear model reflect the contribution of each independent variable to the prediction of the dependent variable.

The L1 norm (L1 norm) is the sum of the absolute values of each element in the vector. L1 regularization is often used to select important independent variables. By compressing smaller coefficients to 0, this method is also called least absolute shrinkage and selection operator (Lasso) regression model. The [13] Lasso regression model is similar to the least square method. It is different that it adds the L1 norm as a penalty constraint instead of only using the sum of the Residual Sum of Squares (RSS) as cost function.

If selecting a different amount of punishment, from the higher values to lower values, to establish different Lasso regression models, the variable coefficient will be one by one from zero to non-zero, where the order reflects the impact on the dependent variable which is also named the LASSO_PATH algorithm [10].

Random Forest Regression Analysis. The Random Forest regression can fit the nonlinear relation data well and make up for the problem of the over fitting of the decision regression tree. Depth of the decision tree node can be used to evaluate the relative importance of the features on the target variables when predicting, that is to say, the node location upper, the greater proportion of input data is decided (to be predicted) by the node. The proportion is the contribution of the feature which can be used to estimates the relative importance of the features [10].

5 Experiment

Our experiment is running on a cluster with 3 nodes with the spark version 2.1.0. The master node is about 15g in memory and the slave nodes is about 30g. The total disk size is 1T, and there are 8 CPU cores. We use Yarn as our cluster resource manager and all data and applications are stored in HDFS.

5.1 Cluster Results

We applied a variety of clustering methods to the history data. In order to simulate the environment in practice, we add some failed logs as outliers. The clustering results and index values of various clustering methods are shown in the Fig. 1. As you can see, the Mean-shift algorithm get a better clustering result on the dataset. The Mean-shift algorithm identifies the outliers as a single cluster, and the others are divided into 3 categories.

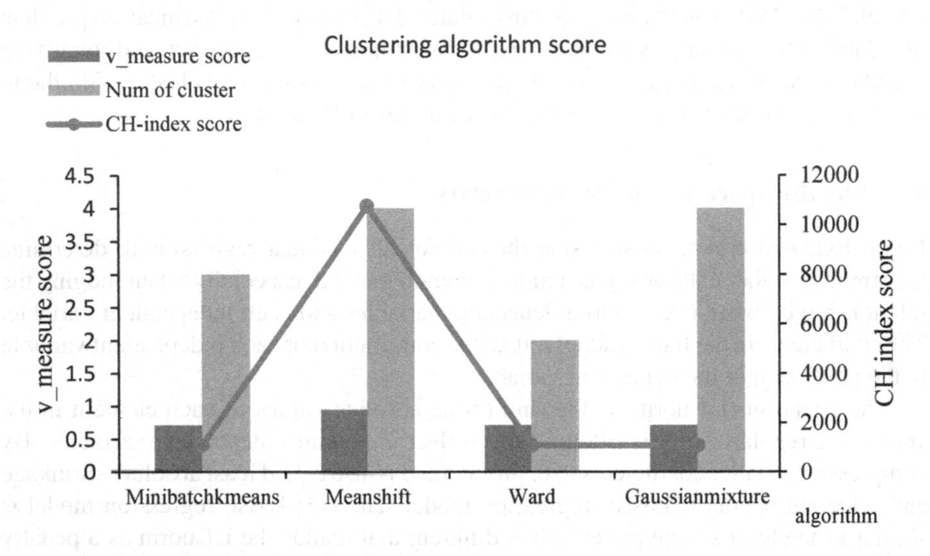

Fig. 1. Cluster algorithm score. v_measure score is the mean of the Homogeneity & Completeness. CH index score is the value of the Calinski-harabaz CH index. Performance of 4 clustering algorithms, including Mini-batch-k-means, Mean-shift, Ward and Gaussian mixture.

According to the cluster results of seed applications, we find that Mean-shift algorithm tends to cluster applications with large number of shuffle operations into one cluster, and applications with few shuffle operations into another cluster. A third cluster consists of applications with appreciate shuffle operations, not too many or too few. Mini-batch-k-means and Gaussian-mixture model need to have a priori knowledge of the number of clusters. We can get the priori-knowledge by observing the index values in experiments assigned to different number of clusters in each experiment. In the experiment, the clustering results of ward agglomerate hierarchical clustering algorithm can be divided into two clusters, most of which are applications with a lot of shuffle operations, and the other cluster contains the rest of applications whose run-time is longer compared with other algorithms.

5.2 Identify Important Parameters

We take the k-means application as an example to sort the importance of configuration parameters. By controlling the alpha parameter of the model to adjust the proportion of penalty in the cost function of Lasso, the path of the configuration parameter back to the model obtained by using lasso path is in Fig. 2.

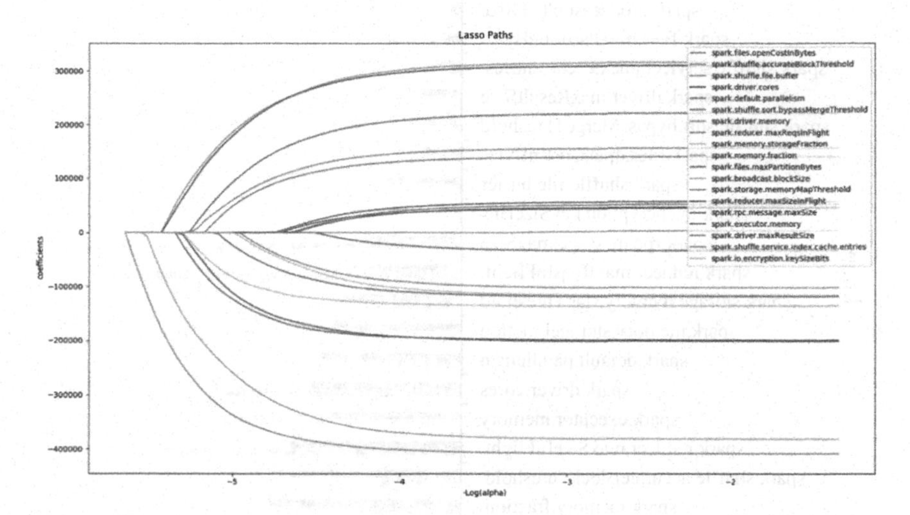

Fig. 2. Lasso path

The configuration parameters are constantly back to the linear model with the increase of alpha parameters. The order of the configuration parameter is different which represents the impact of configuration parameters on the predictive performance in the model.

The order of the configuration parameters importance from Lasso and Random Forest are compared shown in the Fig. 3. Certainly, the order from Lasso has been preprocessed. The results show the difference between the two algorithms. For example, spark.files.maxPartitionBytes, to control bytes of a partition RDD to store the data to the maximum number, changes the coefficients in the twenty-seventh iteration in lasso path, however, is located in relatively upper nodes in the Random Forest. Essentially the Lasso method is linear regression between configuration parameters and performance. However, the relationship between configuration parameters and performance may not be linear which can be solved to some content by the regression decision tree by features segmentation. For some apparently important configuration parameters, it is obvious that the relative importance is both high in the two algorithms, for example, spark.driver.memory, which controls the upper limit of memory assigned to driver.

We carried out some comparative experiments, which separately sets the values of 11 parameters which are relatively important in the two algorithms remained unchanged. The rest of the parameters were set to random values or default values. The run-time of the applications has changed only a little. In order to quantify the change rate of the application, the average relative error AARD (average absolute relative deviation) [16] is used. The results show that the effect of the Random Forest algorithm is slightly better than that of the Lasso method, which is shown as the Figs. 4, 5 and Table 3.

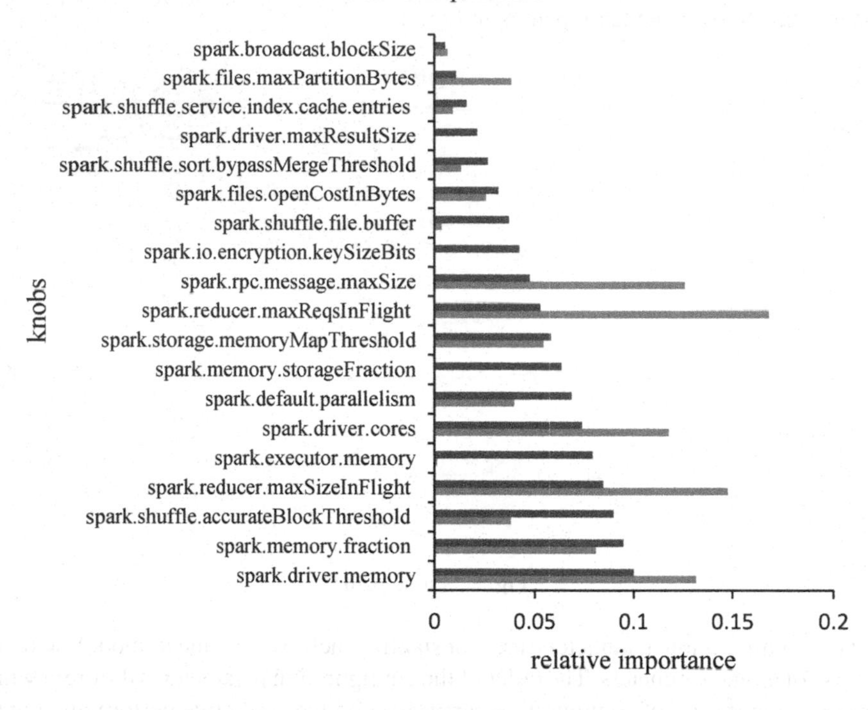

Fig. 3. Relative importance ranked by Lasso and Random Forest

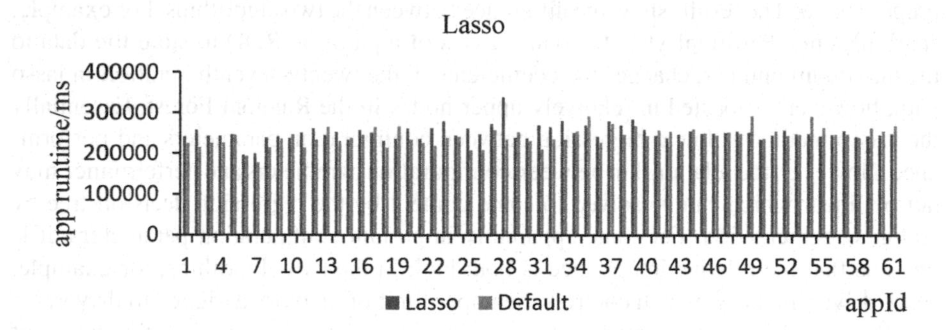

Fig. 4. Apps run-time in default configuration and setting the parameters only selected by Lasso

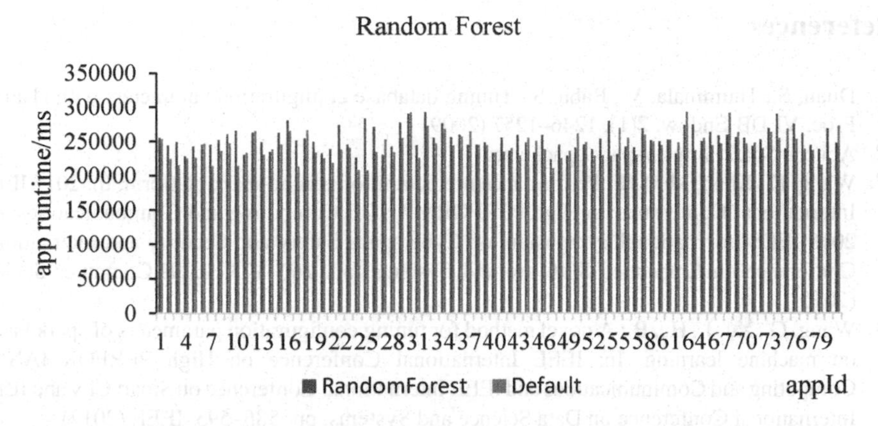

Fig. 5. Apps run-time in default configuration and setting the parameters only selected by Random Forest

Table 3. AARD between Lasso and Random algorithm

Algorithm	AARD (average absolute relative deviation)
Lasso	0.08266
Random Forest	0.07137

6 Conclusion and Future Work

In this paper, a method ISIP based on machine learning is proposed to select important configuration parameters of Spark, in which clustering and regression models are evaluated. The results of various clustering algorithms are compared. The experiments show that the effect of Mean-shift is better than others which can be applied to get a clearer division of applications. In addition, the Lasso and Random Forest regression methods are used to rank the impact on the performance of the configuration parameters. The experiment results show that the relatively important parameters selected by Random Forest play a decisive role in the run-time of the applications. The subset of configuration parameters obtained by ISIP can help users reduce the configuration parameter space by discarding the parameters out of the subset.

In the future, we will explore how other types of configuration parameters impact on the performance of Spark, such as Boolean, Category, etc. In addition, we will also consider adding more applications from the Spark's MLlib to the practice and explore new method of application type division.

Acknowledgement. This work is supported by the National Key Research and Development Program under No. 2016YFB1000703.

References

1. Duan, S., Thummala, V., Babu, S.: Tuning database configuration parameters with iTuned. Proc. VLDB Endow. **2**(1), 1246–1257 (2009)
2. Apache Spark. https://spark.apache.org
3. Wang, K., Khan, M.M.H.: Performance prediction for apache spark platform. In: 2015 IEEE International Conference on High PERFORMANCE Computing and Communications and 2015 IEEE International Symposium on Cyberspace Safety and Security and International Conference on Embedded Software and Systems, pp. 166–173. IEEE Computer Society (2015)
4. Wang, G., Xu, J., He, B.: A novel method for tuning configuration parameters of spark based on machine learning. In: IEEE International Conference on High PERFORMANCE Computing and Communications and IEEE International Conference on Smart City and IEEE International Conference on Data Science and Systems, pp. 586–593. IEEE (2017)
5. Aken, D.V., Pavlo, A., Gordon, G.J., et al.: Automatic database management system tuning through large-scale machine learning. In: ACM International Conference on Management of Data, pp. 1009–1024. ACM (2017)
6. Zaharia, M., Chowdhury, M., Das, T, et al.: Resilient distributed datasets: a fault-tolerant abstraction for in-memory cluster computing. In: Usenix Conference on Networked Systems Design and Implementation, p. 2. USENIX Association (2012)
7. Chiba, T., Onodera, T.: Workload characterization and optimization of TPC-H queries on Apache Spark. In: IEEE International Symposium on PERFORMANCE Analysis of Systems and Software, pp. 112–121. IEEE (2016)
8. Driscoll, P., Lecky, F., Crosby, M.: An introduction to statistics. **30**(10), 540 (2000)
9. Caliński, T., Harabasz, J.: A dendrite method for cluster analysis. Commun. Stat. **3**(1), 1–27 (1974)
10. Sklearn. http://scikit-learn.org
11. Feizollah, A., Anuar, N.B., Salleh, R., et al.: Comparative study of k-means and mini batch k-means clustering algorithms in android malware detection using network traffic analysis. In: International Symposium on Biometrics and Security Technologies, pp. 193–197. IEEE (2015)
12. Newling, J., Fleuret, F.: Nested mini-batch K-means (2016)
13. Tibshirani, R.: Regression shrinkage and selection via the Lasso. J. R. Stat. Soc. **58**(1), 267–288 (1996)
14. Wardjr, J.: Hierarchical grouping to optimize an objective function. Publ. Am. Stat. Assoc. **58**(301), 236–244 (1963)
15. Szekely, G.J., Rizzo, M.L.: Hierarchical clustering via joint between-within distances: extending ward's minimum variance method. J. Classif. **22**(2), 151–183 (2005)
16. Hastie, T., Tibshirani, R., Friedman, J.H., et al.: The Elements of Statistical Learning. World Publishing Corporation, New York (2015)

Design and Implementation of Dynamic Memory Allocation Algorithm in Embedded Real-Time System

Xiaohui Cheng, Yelei Guan, and Yi Zhang(✉)

School of Information Science and Engineering, Guilin University of Technology, Guilin, China
zywait@glut.edu.cn

Abstract. With the development of Internet of Things technology, embedded real-time operating system has been more and more widely used. The embedded real-time operating system has higher requirements on the real-time, fragmentation rate and reliability of dynamic memory allocation. Therefore, dynamic memory allocation has become an important research content of embedded real-time operating system. Aiming at the shortage of μC/OS-II memory management mechanism, an improved memory management algorithm is proposed. By predicting transient objects, allocating them on one side of the heap memory, and then allocating the remaining objects on the other side of the heap memory, the algorithm uses enhanced multilevel separation mechanisms and look-up tables and hierarchical bitmaps to make efficient use of memory occupy. The comparison experiment of μC/OS-II platform shows that the improved dynamic memory allocation algorithm can better improve the speed and utilization of memory allocation. The dynamic memory algorithm has better real-time performance and can effectively improve the memory management of embedded real-time operating system performance.

Keywords: Memory management · Operating system · Heap memory
Real-time

1 Introduction

Embedded systems, as an element of miniaturization and intelligence of equipment, have been widely used in various fields such as defense, industry, transportation, energy, information technology and daily life, and have played a very important role [1]. Operating system dynamic memory allocation is one of the most important components of modern software engineering. It offers maximum flexibility in software system design [2]; however, developers of real-time systems often avoid using dynamic memory allocation due to limited or unlimited response time and memory fragmentation of the embedded operating system. Modern, complex applications such as multimedia streaming and web applications make dynamic memory allocation mandatory for applications [3]. The main challenge with memory allocation algorithms is to minimize fragmentation, provide good response times, and maintain a good place between memory blocks [4].

Q. Zhou et al. (Eds.): ICPCSEE 2018, CCIS 901, pp. 539–547, 2018.
https://doi.org/10.1007/978-981-13-2203-7_43

2 Embedded Algorithm Dynamic Allocation Algorithm Analysis

In computer systems, memory space is a limited but indispensable resource, and system performance is closely related to the use and management efficiency of memory space [5]. Therefore, a reasonable memory management strategy for the overall performance of the system has a very important significance [6]. Currently more common memory management algorithms are many, such as partner system algorithm, TLSF allocation algorithm [7].

The allocation principle of the partner algorithm is to divide all free pages into 11 block lists, each containing the same size and address contiguous page box [8]. If you have a task to apply for address space, you can select the size of the block in the list. For example, if you want to apply for a block of 64 pages, find out whether there is a free block from the starting address of the list of 64 * 4K, and if not, go to 128 * 4K linked list. If found, divide the blocks of 128 pages into two 64-page blocks, one for the application space, and the other for a list of 64 page frames. If this level of the frame list is still not find free memory block, step by step to find upward when the scanning list that all frame and no free blocks, the algorithm will automatically give up the memory allocation, and eventually returns an error signal. The release of the memory block and block allocation process, to the contrary, the system will first to release the user block is inserted into the available space in the table, to meet the "partners" free blocks according to the partner system the principle of the combining integration algorithm.

TLSF (Two Level Segregated Fit) or secondary interval dynamic memory allocation algorithms, the proposed algorithm for the management of the memory pool has a fixed time, the basic principle is to list combined with a bitmap thought through appropriate adaptation strategies to get the same results as optimal adaptive strategy, seeking to exactly match the block of memory, or the most close to the requirements of the memory block. The core data structure used by the algorithm is a free list array, and an array corresponds to a free memory block of corresponding size. These empty list tables can be divided into two levels in order to make it more efficient to find the free blocks that meet the requirements, and to simplify the processing of many isolated linked lists [9]. The first level (fl) corresponding size is 2i ($i = 2, 3, \ldots$) The free block can be positioned by i. The two level(sl) is on the basis of the proceeds from the list at the next higher level is divided into several groups, all of the space groups are the same size, there are one-to-one and group bitmap, Its function is to identify whether there are empty linked lists and the list of free blocks. To avoid crossing the line, and keep the continuity of the memory area, in division 1 chain table, secondary treatment is usually divided into its serial number respectively 0 to 7 or 8 isolation linked list.

These algorithms are commonly used in embedded systems memory management algorithms, although these algorithms have many advantages, but there are also some shortcomings, should not predict the life of the memory block in advance when the system is idle [10]. In this paper, we design an algorithm for predicting the longevity of memory blocks in advance, which improves the efficiency of the system in terms of time and fragment rate.

3 Algorithm Design of Improved Memory Management

Through analysis of the above two commonly used memory management mechanisms, a memory management mechanism based on the prediction of the length of memory blocks will be proposed in this section. In the earlier microkernel architecture, the system usually re-divided the memory block in the memory partition into a corresponding interval. In this paper, we introduce an improved dynamic memory allocation algorithm. Improved memory allocation algorithms predict transient objects, allocate them on one side of heap memory, and allocate the remaining objects on the other side of heap memory to reduce memory fragmentation. Allocation algorithms are implemented using enhanced multi-level isolation mechanisms, using look-up tables and layered bitmaps, ensuring very good response times and reliable timing performance. Allocation algorithms reduce memory fragmentation by using adaptive orientation based on predicted object lifetime. Multiple parameters can be used to predict lifetime, such as block size, number of instructions executed between block allocation and deallocation, total number of bytes allocated between memory block allocation and deallocation, and allocation events between block allocations Number and its release. However, the number of instruction parameters executed is not particularly suitable for this context, since memory management events are the only interesting events in this context. The total number of bytes allocated is not valid because the size of objects varies from a few bytes to a few megabytes. Our experiments show that the combination of block size and number of allocated events is a good measure of object prediction. The proposed allocator uses the combination of object size and number of assigned events to predict the life of an object.

Table 1 shows the structure of free and used data blocks. The memory allocator inserts header information into each free and used block. The block header of a free block holds information such as BS (32 bits, the last two bits are always zero because the block size is always a multiple of 4), which specifies the size of the block, BT (1 bit), which specifies the block type, AV (1 bit) that specifies the block status, Prev_Memory_Blk and Next_Memory_Blk, which identifies the status of the Memory adjacent blocks required. Prev_FreeList and Next_FreeList need to be used to locate the last and next free blocks in the delimited free list. The block header used for the block holds all the fields of the free block header, except for the free list pointers, which are not necessary for the used blocks as they are not linked to any isolated free list. However, the header of the used block contains an extra field called BlkAllocStat, which holds the block allocation statistics of the block prediction algorithm. Head overhead is counted as internal fragmentation.

Table 1. Data structures of free and allocated blocks

BS(b)	BT(b)	AV(b)
Prev_Memory_Blk		
Next_Memoryl_Blk		
Prev_FreeList		
Next_FreeList		
Free Block		

BS(b)	BT(b)	AV(b)
Prev_ Memoru_Blk		
Next_Memory_Blk		
Used Block		

The new memory allocation algorithm uses a large number of free lists. Where each list keeps the size of free blocks within a predefined range, and the blocks belong to some particular block type. Effectively handling short-term storage blocks between short-term and long-term blocks without incurring additional search overhead, and the free list is organized into a three-level array, as shown in Fig. 1. The first level divides the free list into free list classes the free list for each class is further subdivided into different free List collections, each holding a free block of size within a predefined range (the dynamic range of a class is linearly subdivided across all free list collections). Finally, each group is divided into two free lists, one for ephemerality and the other for long-term objects.

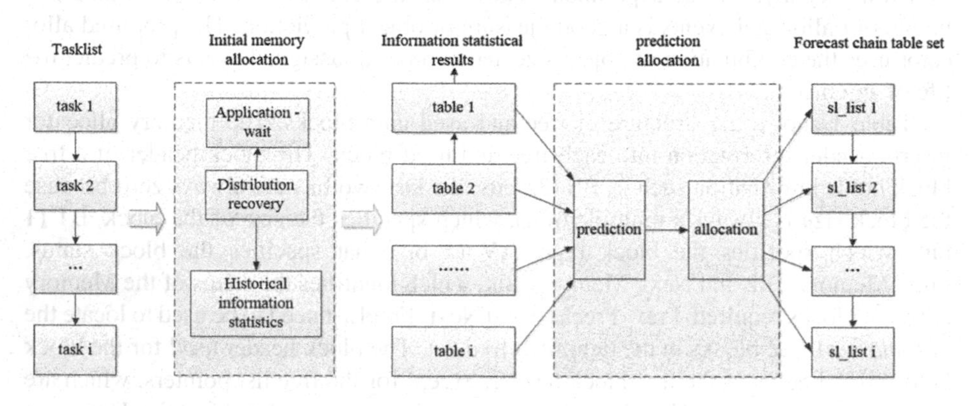

Fig. 1. Free block list chart

Figure 2 shows a two-level bit mask used in the distributor to identify available free blocks. Two 32-bit fields are used as a primary mask, one for ephemeral blocks and one for long-term blocks. Each bit of the first-level mask indicates the availability of free blocks in the corresponding free list class (32 bits correspond to 32 classes). Based on the type of memory block, one or two 32-bit fields are used to find free list classes with free blocks of the most suitable size for the block request. An 8-bit second level mask (configurable) is maintained for each bit of the first level mask. In total, 64 8-bit masks are used to identify free blocks in any free list set.

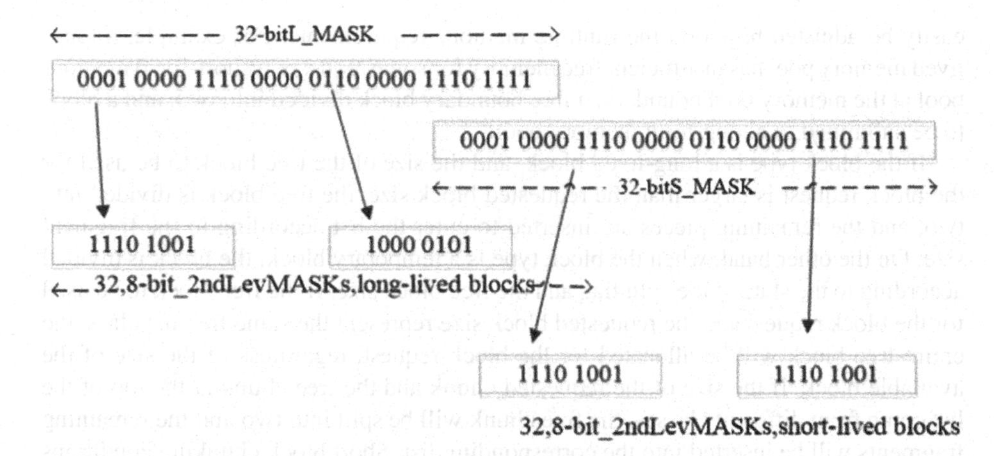

Fig. 2. Two bit mask

The distribution algorithm is to deal with long-term blocks and short-term blocks separately. As shown in Fig. 3, short blocks are allocated upward from the heap, and long-lived blocks are allocated from top to bottom. Used space heap growth from both sides. For example, let the heap size be 200 bytes and each short and long-term object pool has only one free block of size 100 bytes. In response to a memory block request (a short-term object) of size 8 bytes, the bottom 8 bytes of the free block corresponding to the short-term block are assigned to the request. Conversely, in the case of a memory request with a block size of 32 bytes (assumed to be a long-term object), the first 32 bytes long corresponding to the free block are allocated to the request. Similarly, all short blocks may be allocated from the heap from top to bottom, and long blocks may be allocated from bottom to top.

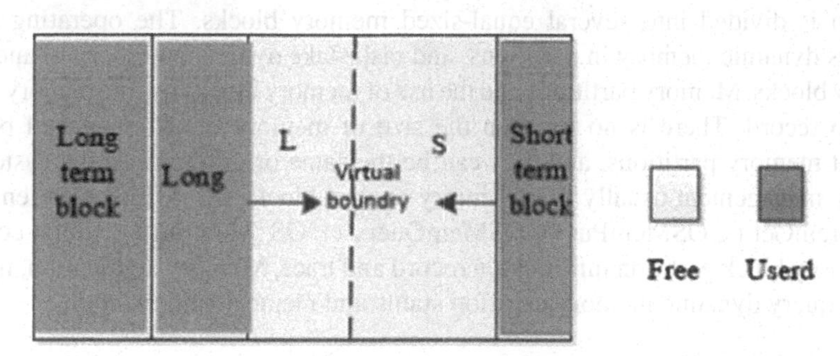

Fig. 3. Memory organization chart

Initially, the entire heap memory is free, with only one free block for each of the short-term and long-term memory pools. Heap space is initially divided into two blocks (no Memory boundaries, only virtual memory pool boundaries), predefined ratios, The one for the short term and the other for the long term. As the heap grows from both sides, the boundaries between short-term memory pools and log storage memory pools can easily be adjusted based on the runtime memory requirements. For example, a long-lived memory pool has insufficient free memory for some requests, a short-lived memory pool at the memory pool boundary, a free boundary block divided into two, and a block to be submitted to a long-term memory pool.

If the block type is a long-lived block, and the size of the free block to be used for the block request is larger than the requested block size, the free block is divided into two, and the remaining pieces are inserted to enter the list according to the fragment size. On the other hand, when the block type is a temporary block, the block is divided according to the state of the split flag and the free block size. If the free block to be used for the block request and the requested block size represent the same free list class, the entire free block will be allocated for the block request, regardless of the size of the available block. If the size of the requested chunk and the free chunk at the top of the list come from different classes, the free chunk will be split into two and the remaining fragments will be inserted into the corresponding list. Short block chunking conditions (the chunk sizes are likely to be between a few bytes and a few kilobytes) will be rescued from memory that is pinned to a very small block.

4 Experiment

4.1 μC/OS-II Memory Management Mechanism Is Analyzed

μC/OS-II is a preemptive real-time multi-task embedded operating system based on priority, which is concise and practical. It has very low system hardware requirements and can meet the needs of many projects. Its stable and reliable features have been certified by the US Airways Administration, Successfully used in medical science, aerospace engineering and other major projects. μC/OS-II two levels of memory management, that is, a continuous memory space is divided into several partitions, each partition is divided into several equal-sized memory blocks. The operating system manages dynamic memory in partitions, and tasks take dynamic memory in and out of memory blocks. Memory partitions and the use of memory blocks by the memory control block to record. There is no limit on the size of memory blocks managed between different memory partitions, and they can be the same or different. In the system, the memory management usually uses memory control blocks OS_MEM, OSMemCreate (), OSMemGet (), OSMemPut (), OSMemQuery (), OS_MemInit (), which complete the memory block partition information record and trace, Memory application, memory release, query dynamic memory partition status and memory initialization.

4.2 Experimental Results are Compared and Analyzed

Test embedded memory allocation mechanism is good or bad, mainly through the memory allocation time overhead indicators to evaluate. In order to analyze the performance of the improved algorithm before and after, all experiments were performed on a 2 GB RAM Inter (R) Core i5-4460 3.2 GHz CPU. This article selects μC/OS-II as the experimental system, and transplants this system to the VC++ environment to test before and after the improvement the system performance. By comparing before and after the algorithm to improve the dynamic memory allocation and dynamic memory release time to determine the real-time. The experiment divides the memory size requested by the user into five intervals (in bytes), which are [0, 128], [128, 256], [256, 1024], [1024, 4096], [4096, 16384] Generate random values in each range and frequently allocate and release random-size memory. Each group of test 100 times, take the average. Call μC/OS-II timing function in the system to measure memory calls and memory release time, compare the two real-time performance. Figure 4 is an experimental result diagram that distributes 4098 bytes of memory.

Fig. 4. Experimental result diagram of memory allocation

Two different mechanisms were tested in different intervals and recorded in Table 2 below:

Table 2. Comparison of allocation and release time table(us)

Operation	Algorithm	Range 1	Range 2	Range 3	Range 4	Range 5
Memory allocation	Old	5.859	6.998	9.546	9.621	10.639
	Improved	3.862	5.927	7.469	8.126	9.256
Memory release	Old	4.568	5.034	7.865	7.685	9.324
	Improved	2.651	4.236	4.865	5.186	7.365

As can be seen from Table 2, the improved memory allocation and release time overall less than before the improvement, with better real-time. The improved algorithm allocates and releases small memory more quickly than before, since small memory allocation and release use a two-level bitmap algorithm to allocate and release fixed-size blocks of memory with fewer instruction cycles.

5 Conclusion

This paper analyzes a variety of memory allocation algorithm and μC/OS-II in the memory management methods to achieve the mechanism to absorb the advantages of the original memory algorithm based on the proposed memory allocation algorithm to improve the original system Performance, so that it can be better applied to the need for dynamic distribution of the occasion. The experimental results show that the μC/OS-II system with the predictive lifetime algorithm has a higher allocation efficiency than the μC/OS-II prototype system. The experimental results show that it has achieved the expected purpose and has a good reference value for the related research work.

Acknowledgment. As the research of the thesis is sponsored by National Natural Science Foundation of China (No: 61662017, No: 61262075), Key R & D projects of Guangxi Science and Technology Program (AB17195042), Guangxi Natural Science Foundation (No: 2017GXNSFAA198223), Major scientific research project of Guangxi higher education (No: 201201ZD012), Scientific and Technological Research Program for Guangxi Educational Commission grants (#2013YB113), Guilin Science and Technology Project Fund (No: 2016010408) and Guangxi Graduate Innovation Project (No: SS201607), we would like to extend our sincere gratitude to them.

References

1. Diwase, D., Shah, S., Diwase, T., et al.: Survey report on memory allocation strategies for real time operating system in context with embedded devices. Int. J. Eng. Res. Appl. (IJERA) **2**, 1151–1156 (2012)
2. Shen, F.-Y., Zhang, Y., Lin, Y.: Design and implementation of dynamic memory management algorithm in embedded real-time system. Comput. Sci. Modern. **7**, 103–107 (2015)
3. Jabeen, Q., Khan, F., Hayat, M.N., et al.: A survey: embedded systems supporting by different operating systems. arXiv preprint arXiv:1610.07899(2016)
4. Yan, J., Xuewen, Z., Sun, P.: Low memory process management algorithm based on statistical analysis and prediction. Comput. Eng. Des. **35**(1), 107–111 (2014)
5. Mancuso, R., Dudko, R., Betti, E., et al.: Real-time cache management framework for multi-core architectures. In: 2013 IEEE 19th Real-Time and Embedded Technology and Applications Symposium (RTAS), pp. 45–54. IEEE (2013)
6. Patil, N.V., Irabashetti, P.S.: Dynamic memory allocation: role in memory management (2014)
7. Cheng, X., Gong, Y., Anming, X.: Embedded memory prediction and allocation algorithm based on markov chain. Comput. Eng. Des. **34**(8), 2727–2731 (2013)
8. Xiao, L., Kejiang, L.: Analysis and comparison of dynamic memory management mechanism of μC/OS and FREERTOS. Softw. Eng. **19**(5), 21–22 (2016)

9. Wang, C., Wong, W.F.: Observational wear leveling: an efficient algorithm for flash memory management. In: 2012 49th ACM/EDAC/IEEE Design Automation Conference (DAC), pp. 235–242. IEEE (2012)
10. Wang, X., Qiu, X., Mu., F., et al.: Research on a new dynamic memory management mechanism of embedded system. Microelectron. Comput. **34**(8), 66-6 (2017)

A Heterogeneous Cluster Multi-resource Fair Scheduling Algorithm Based on Machine Learning

Wenbin Liu, Ningjiang Chen(✉), Hua Li, Yusi Tang, and Birui Liang

School of Computer and Electronic Information, Guangxi University, Nanning 53004, China
chnj@gxu.edu.cn

Abstract. The resource scheduling of data center is a research hotspot of cloud computing. The exiting research work is concerned with the issue of fairness, resource utilization and energy efficiency, which are only applicable to the same cluster environment or specific application situations. First, the default scheduling algorithm (DRF) of Mesos is analyzed. The DRF algorithm does not consider machine performance and task types. Then, this paper presents a heterogeneous cluster multi-resource fair scheduling algorithm based on machine learning to solve the problem. The algorithm is to test the performance of the machine and use the machine learning method to classify the computing tasks and reach the goal of reasonable resource allocation. Finally, the experimental results show that the method presented in this paper not only ensures the fairness of resource allocation, but also makes the system more reasonable allocation of resources and further improves the system's resource utilization.

Keywords: Heterogeneous clustering · Resource scheduling · DRF
Machine learning

1 Introduction

With the rapid development of new technologies such as cloud computing and big data, more and more data need to be processed. The big data computing framework represented by Hadoop, Spark and Storm has been rapidly developed and applied. However, due to the continuous expansion of the company size and the increasing number of applications. Various big data computing frameworks have made the clustering environment of enterprise data centers complicating. With the continuously increase of data volume, the scale of the data center cluster is also expanding. The computing framework and applications deployed on the cluster are constantly diversified. The large-scale resource scheduling of data center clusters is still one of the hot fields.

Mesos [1] is an Apache's open source project. It is also a large-scale resource management and scheduling framework for heterogeneous clusters. It strives to optimize resource utilization through dynamic sharing of resources among multiple frameworks. Mesos has been widely used in the production environment such as Twitter, Apple et al. However, in actual production, the cluster environment of the data center is relatively complicated. The Mesos is difficult to unify the allocation of physical machine resources

Q. Zhou et al. (Eds.): ICPCSEE 2018, CCIS 901, pp. 548–559, 2018.
https://doi.org/10.1007/978-981-13-2203-7_44

in data centers. Mesos uses a unified machine configuration and resource allocation methods without distinguishing task types, which lead to the result that the resource allocation effect of Mesos in practical application is not ideal, such as the average resource utilization of Twitter is less than 20% [2]. Some experts have also conducted relevant researches. For example, fish swarm intelligent algorithm [3] is a new idea to dynamically adjust Mesos cluster resources. The method could maintain the imbalance of Mesos load and improve resource utilization. Mesos uses DRF [4] (Dominant Resource Fairness) algorithm for resource allocation. DRF is a resource fair scheduling algorithm for multi-resource application. The DRF algorithm is a generalization and improvement of the maximum-minimum fair resource algorithm in the case of multiple resources. Each computational framework's jobs can get a fair allocation of resources when it needs resources most. Although being the default scheduling algorithm of Mesos, DRF shows excellent performance in heterogeneous multi-resource allocation. The DRF still has deficiencies of no difference in physical machine performance and the task type is not classified. Therefore, this paper presents a heterogeneous cluster multi-resource fair scheduling algorithm based on machine learning to solve the problem.

Therefore, the main contributions of this paper are as follows: this paper presents a heterogeneous cluster multi-resource fairness scheduling algorithm based on machine learning to improve the original DRF algorithm without considering machine performance and task types. The experiment results prove the effectiveness of the algorithm. This method provides a new solution for data center managers to manage heterogeneous cluster resources. The rest of this paper is organized as follows: Sect. 2 analyzes the principle of Mesos default DRF fair scheduling algorithm and its shortcomings in practice. Section 3 presents a fair scheduling algorithm for heterogeneous resources that is sensitive to machine performance and task types. Section 4 is an experimental comparison to prove the effectiveness of the method. Section 5 is a summary of relevant research work. Section 6 concludes the paper.

2 Problem Analysis

The DRF algorithm does not consider machine performance and task types. No difference in physical machine performance means that there are machines purchased at different times in the actual data center cluster. So the machines have different performance and configuration. However, the machine performance of the DRF algorithm in the cluster is no difference. The DRF does not take into account the unfairness of cluster resource allocation, which is due to the no difference in machine performance. Even if the same numbers of resources are allocated, the task of getting more high quality resources is relatively fast in executing efficiency. Dominant resource share calculation of the DRF algorithm uses the method of uniform physical machine performance, the formula is as follows:

$$S_i = \left(\max_{1 \le j \le m} Ru_{i,j}/r_j \right) / w_i \tag{1}$$

Here i is the current computation task, j is the resource type, m is the total number of resource types. $Ru_{i,j}$ is the total amount of j-type resources that task i has obtained, r_j is the total amount of j type resources, w_i is the total task weight that task i occupies. Formula (1) only does a certain amount of processing on the resource type and computing task weight, but does not distinguish between machine performances. Even if different computing frameworks for the same share of resources are obtained, there is a huge difference in the efficiency of execution. In the long run, it will not only lead to unfairness of resource allocation, but also will lose the envy-freeness of the fairness algorithm [5].

The task type is not classified. It means that in the description of the original DRF algorithm. Each resource allocation of the system is the resource allocation of resource requirements for the next task of the current computing framework. However, when the actual algorithm is applied, Mesos performs coarse-grained resource allocation among various computing frameworks. Each framework also performs specific fine-grained task scheduling according to the characteristics of its own task. Therefore, Mesos is unable to know the job type of each computing framework. In order to achieve the fairness of the resource allocation in specific implementations, the DRF algorithm has made a simple strategy to allocate all the resources in the system to the framework that has the least amount of resources. This will allow the system to allocate the resources without classifying the job types on each computing framework. It will bring fragmentation of system resources and decline of resource utilization. For a distributed scheduling system such as Mesos, pre-aware system job conditions and cluster resource information can make more reasonable scheduling decisions. For example, A SD-Predictor based on cluster system configuration information was proposed before cluster scheduling to predict the termination status (success or failure) of the system tasks [6]. Therefore, this paper draws on this idea of pre-aware and presents a multi-resource fair scheduling algorithm based on machine learning to solve the problem.

3 Research Idea

At present, machine learning is the most popular method for solving such large-scale data classification problems. Common machine learning classification methods include decision tree methods, artificial neural networks, support vector machines, etc. Their adaptation scenarios are also different. This paper does not focus on the features of various machine learning methods. However, in order to adapt to the application scenarios of heterogeneous data clusters in this paper, we have chosen the following criteria for the machine learning model: (1) Supports parallel processing and distributed computing, and can adapt to large-scale data calculations; (2) Fast data processing and less system resources; (3) More accurate output. Therefore, we abandon decision-trees that is easily over-fitting, study artificial neural networks with long times and support vector machines that is difficult to train large-scale data [7]. We choose XGBoost [8] model that supports parallel processing and distributed computing. The XGBoost model can specify the default direction of the branch for missing values to improve the efficiency of the algorithm. The task type problem of each computing framework on Mesos is a typical large-scale data classification problem. Therefore, the XGBoost machine learning model is introduced in this paper to classify and predict the resources on each framework and output the next type of heavy resources for each computational framework. The system selects suitable resources according to the matching degree of computing framework and machine performance. It could further improve the utilization of system resources. Although the introduction of XGBoost model will inevitably occupy part of the system resources. It provides a new way for data center managers to manage heterogeneous clusters and multiple resources.

The improved DRF algorithm is presented in this paper, named X-DRF algorithm. It first performs performance testing and scoring for the physical machines of each cluster and sorts them according to their scores. This paper only considers memory and CPU resources for the moment. Then, the system calculates the dominant share of each computing framework, and sorts them from small to large. Next, the system predicts the task heavy type of each computing framework through the resource information collected by the XGBoost model. Finally, it uses the corresponding higher (lower) quality resources to allocate resources for the next tasks of the framework according to the task heavy type. In this way, the system would balance the dominant share of all computing frameworks as much as possible. Assume a data center using Mesos as a cluster management system. Given n is the amount of calculate node in the hypothesis, G_p is the score of the machine performance evaluation, and p is the various performance types of the system machine. The factor ξ that defines the performance of the machine is the ratio of the performance evaluation score and the average score of the machine. S_i is the dominant share of calculation task i. r_j is the total amount of j type resource on machine q. $Ru_{i,j}^{q}$ is the amount of j type resource that calculation task i gets in the machine q. $Rc_{q,j}$ is the amount of j type resource which machine q can allocate to task. w_i is the weight of calculation task i. The processing flow is as follows:

(1) Measure and score the CPU and memory performance of the physical machine of each cluster node. The test tool is Ubench [9]. The scoring mechanism is Ubench's own benchmark and score. Record the CPU and memory score G of each machine and get the average value $\bar{G}$:

$$\overline{G} = \sum_{i=1}^{n} G_{i,p}/n \tag{2}$$

(2) The ratio ξ of machine performance evaluation score with the average score of machine q is:

$$\xi_{q,p} = G_{q,p}/\overline{G} \tag{3}$$

(3) The dominant share $S_{i,q}$ of computing task i on machine q is:

$$\mathrm{S}_{i,q} = \left(\max_{1\le j\le m} Ru_{i,j}/r_j\right).\xi_{q,p}/\mathrm{w}_i \tag{4}$$

(4) The dominant share S_i of computing task i is the sum of dominant shares on each machine:

$$S_i = \sum_q^n S_{i,q} \tag{5}$$

(5) The differences dS_i in the dominant share from calculating adjacent computing tasks is:

$$dS_i = S_{i+1} - S_i \tag{6}$$

(6) The XGBoost model is trained based on the monitored historical resource information including memory and CPU usage data. Then, the trained XGBoost model is used to analyze and predict the task typeon each computing framework.
(7) The resource allocation required by the user i in the t th calculation is determined by the following issues: When predicting that the next task of user i is the CPU-intensive tasks, the system allocates CPU premium resources for its next tasks. The constraint condition is as shown in Eq. (7).

$$\begin{cases} \sum_{q=1}^n (Rc_{q,cp} \cdot \xi_{k,cp}) / \sum_{t=1}^i W_t = dS_t \\ \min k \\ k = n + 1 - q \end{cases} \tag{7}$$

If the next task of user i is the memory-intensive, it allocates memory quality resources for its next tasks. The constraints are as shown in Eq. (8).

$$\begin{cases} \sum_{q=1}^n (Rc_{q,mp} \cdot \xi_{k,mp}) / \sum_{t=1}^i W_t = dS_t \\ \min k \\ k = n + 1 - q \end{cases} \tag{8}$$

(8) Perform steps 2 to 5 in a loop until the resource allocation is complete or there is no resource request. The allocation process is finished. The X-DRF algorithm minimizes the resource gap with other computing frameworks by allocating more resources, as shown in Fig. 1.

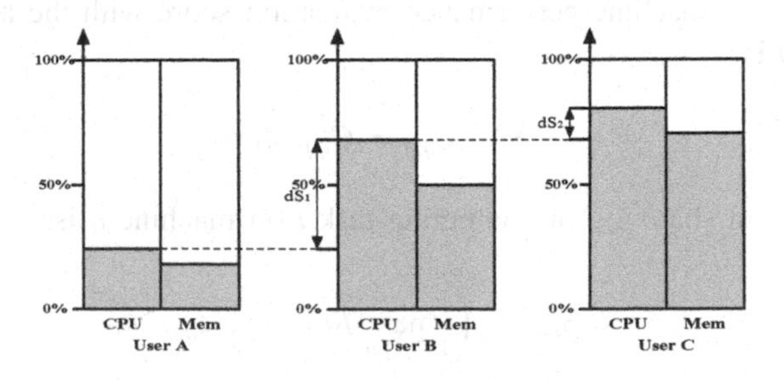

Fig. 1. Algorithm assignment diagram

As shown in Fig. 1, the gray area represents the resources of each user. When user A needs CPU as dominant share, dS1 is the difference between the dominant share of user A and user B, and dS2 is the difference between the dominant share of user B and user C. In order to balance the dominant share among different users, the first thing to do is to increase the user A's CPU resource dS1 to be equal to the user B's CPU resource. The CPU resource that user A needs to increase, is not an undifferentiated allocation of CPU performance. It is based on the historical data of the A user to train the XGBoost model. Then, the system uses the resource information consumed by the user A on the cluster as inputs to train XGBoost model. The XGBoost model predicts the heavy type of task of the user A. If it predicts that the next task type of user A is CPU resources, the system allocates high-quality CPU resources for user A. The next time the resources are allocated, if the system still has enough available CPU resources, the system will set dS2, which is the increase of the CPU resources of both user A and user B. The pseudo-code of the X-DRF algorithm is as shown in Algorithm 1:

Algorithm 1 X - DRF

X - DRF allocation()

1. $R = r_1, \ldots r_m$ — total of m resources
2. $C = c_1, \ldots c_m$ — consumed resources
3. $S_i (1 \le i \le n)$ — resource vector of user i, is initialized to 0
4. $U_i = u_{i,1}, \ldots u_{i,m} (1 \le i \le m)$ — the number of resources allocated to user i, initialized to 0
5. $MP = m_1 p, \ldots m_q p, CP = c_1 p, \ldots c_q p$ — sort machine q's performance
6. $\xi_{q,p} = n G_{q,p} / \sum_{i=1}^{n} G_{i,p}$ — calculate performance score factor
7. *select* min S_i *from* $S_i (1 \le i \le n)$ — select the least resource vector in user i
8. $D_i = d_i, \ldots d_m$ — demand of next task user i
9. *if* $C + D_i \le R$ *then*
10.   min k
11. $k = n + 1 - q$
12.   *if* $j = mp$ *then*
13. $\sum_{q=1}^{n} (Rc_{q,mp} . \xi_{k,mp}) / \sum_{t=1}^{i} W_t = dS_t$
14.   *else*
            $\sum_{q=1}^{n} (Rc_{q,cp} . \xi_{k,cp}) / \sum_{t=1}^{i} W_t = dS_t$ — determine resource types
15.   *end if*
16. $U_i = U_i + dS_i$ — update the resource vector assigned by user i
17. $C_i = C + dS_i$ — update the consumed resource vector
18. $dS_i = S_{i+1} - S_i$ — dominant share of computing tasks calculated adjacent difference dS_i
19. $S_{i,q} = \left(\max_{1 \le j \le m} u_{i,j} / r_j \right) g \xi_{q,p} / w_i$ — update leading resource vector
20. *else*
21.   *return* — allocation of resources
22. *end if*

There are main characteristics of the X-DRF algorithm. The feature is the judgment statements in lines 9–15, which classify the job types on the calculation framework, adding the training of the XGBboost model and the determination of job types. However, the training of the XGBoost model requires a certain amount of time and resources. It can be seen that the space-time complexity of the X-DRF algorithm mainly depends on the XGBoost model in the 12th line of the algorithm to train and classify the resources collected by the system. The principle of the XGBoost model shows that its time complexity is mainly caused by the maximum number of iterations n. So, the algorithm has a time complexity of about O(n).

4 Experiments and Evaluation

The heterogeneous clustering environment of this experiment is consists of five computing nodes with a total of 28 core CPUs and 146G of memory. The specific hardware parameters of the machine are shown in Table 1. The cluster experiment sets up a Hadoop framework and a Spark framework. It processes four typical tasks such as WordCount, PageRank, MergeSort and K-means. The resource utilization of the cluster and resource information of the cluster is collected for a week as historical data for classification and forecasting. The experimental configuration is shown in Table 1.

Table 1. Experiment configuration

Types	Mem	CPU	OS/Software
Inspur	96G DDR3	6 Xeon(R) E5-2620	Ubuntu14.04
Acer	8G DDR3	4 Core(TM) i7-3770	Hadoop- 2.5.0
Acer	8G DDR3	4 Core(TM) i7-3770	Spark-2.1.0
Sugon	8G DDR3	6 Xeon(R) E5-2420	Zabbix-2.2.1
Sugon	16G DDR3	8 Xeon(R) E5-620	

The experiment submits the same task to the Mesos system of the original DRF algorithm and the Mesos system of the X-DRF algorithm respectively and compares the evaluation methods of the work itself to verify the effectiveness of the X-DRF algorithm introduced into machine learning.

① **Resource utilization:** percentage of CPU usage of the system and the amount of available memory;
② **Average task execution time**: the average value of the execution time of all tasks of the system.;
③ **Average task waiting time:** The average value of all task waiting time of the system;
④ **System normalized performance:** SNP is the geometric mean of ANP. ANP is the ratio between the theoretical running time and the actual running time of the running operation, i.e. $ANP = T_{theory}/T_{\exp eriment}$.

4.1 System Resource Utilization Verification

This section is mainly to verify the system's resource utilization. We select WordCount, PageRank and MergeSort as Hadoop tasks to perform experiments. We select WordCount, PageRank, and K-means as Spark tasks to perform experiments. At the same time, we monitor and count the system CPU and memory usage. Figures 2 and 3 are the CPU utilization comparison charts and memory usage comparison charts for each group of tasks.

In Figs. 2 and 3, when the same Hadoop task was submitted separately to the X-DRF algorithm system and the original DRF algorithm system, the system's CPU utilization increased by nearly 10%, and the memory footprint is increased by about

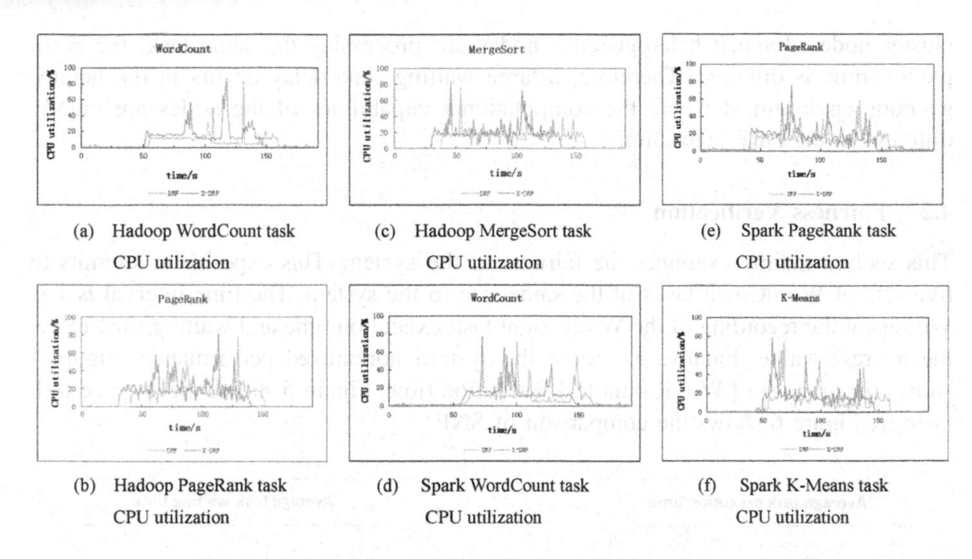

(a) Hadoop WordCount task CPU utilization (c) Hadoop MergeSort task CPU utilization (e) Spark PageRank task CPU utilization

(b) Hadoop PageRank task CPU utilization (d) Spark WordCount task CPU utilization (f) Spark K-Means task CPU utilization

Fig. 2. Comparison of CPU utilization during task execution

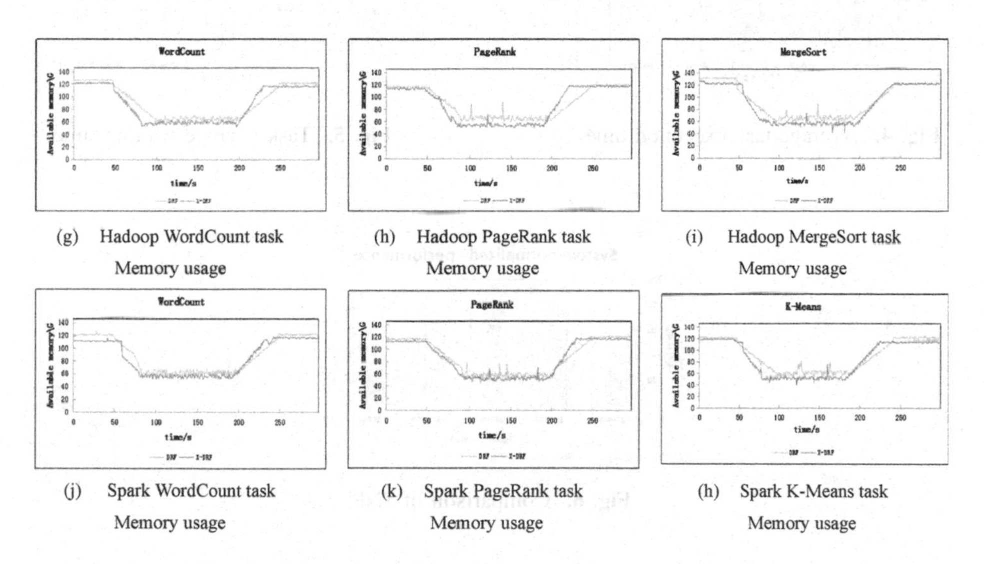

(g) Hadoop WordCount task Memory usage (h) Hadoop PageRank task Memory usage (i) Hadoop MergeSort task Memory usage

(j) Spark WordCount task Memory usage (k) Spark PageRank task Memory usage (h) Spark K-Means task Memory usage

Fig. 3. Comparison of memory usage during task execution

12G. When running the Spark task, the system CPU utilization increased by nearly 6%, and the memory footprint increased by about 8G. Because the introduction of XGBoost model will occupy a part of system resources. This part of the resource is still acceptable compared to the overall improved CPU and memory resource utilization of the Mesos cluster system of the original DRF algorithm. It can be seen that the system with the converged X-DRF algorithm is also more stable. The reason is that in the heterogeneous environment, there are differences in the computing capabilities of the cluster nodes. Even if heterogeneous nodes are processing the same task, the completion time is different. Therefore, a large waiting time delay occurs in the heterogeneous environment where the computational capabilities of the nodes are greatly different in the DRF algorithm.

4.2 Fairness Verification

This section mainly examines the fairness of the system. This experiment submits to five sets of WordCount tasks of the same size to the system. The time interval is 1 s. We repeat the recording of the WordCount task execution time and waiting time to get the average value. Finally, we count the system normalized performance. Figure 4 shows the average of WordCount task execution time. Figure 5 shows the average task latency. Figure 6 shows the comparison of SNP.

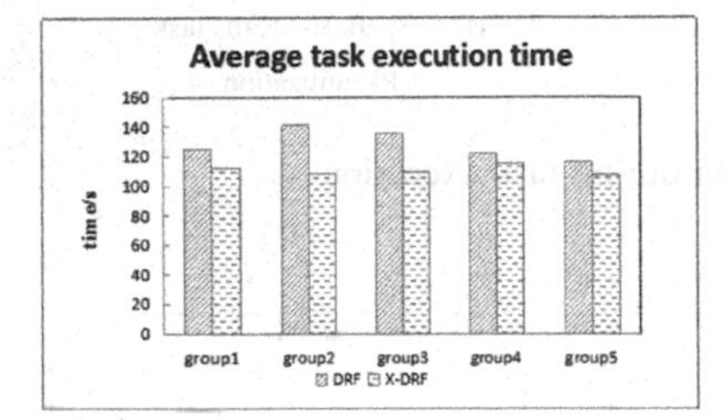

Fig. 4. Average task execution time

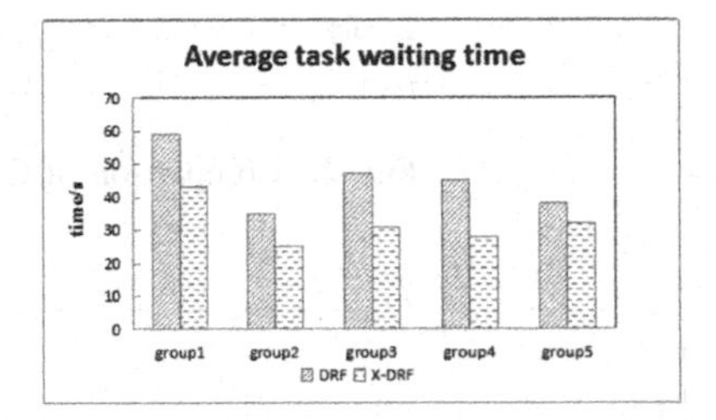

Fig. 5. Task average waiting time

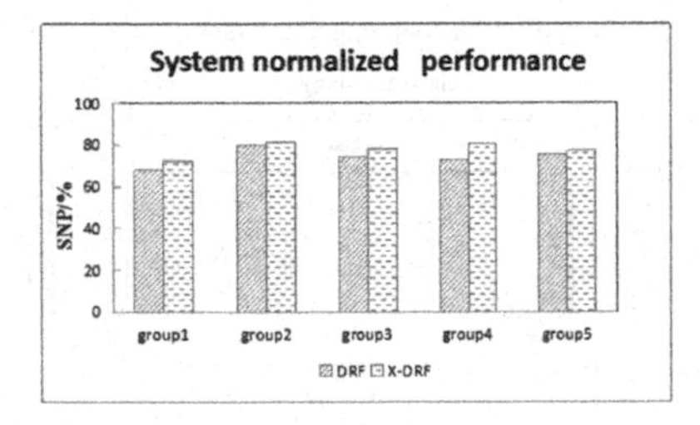

Fig. 6. Comparison of SNP

As can be seen from Fig. 4, the average execution time of each group of WordCount tasks on the X-DRF algorithm system is reduced by about 15 s than the average execution time on the system of the original algorithm. However, the average waiting time of each group of WordCount tasks on the improved X-DRF algorithm system is reduced by about 10 s than the average waiting time on the original DRF algorithm system in Fig. 5. It shows that the system tasks of the X-DRF algorithm is more efficient. Figure 6 is the normalized performance of the system. As can be seen from the figure, the system normalized performance ratio of the WordCount task of each group in the X-DRF algorithm is about 3% higher than that of the original DRF system.

This shows that the X-DRF allocation algorithm is more equitable and reasonable in resource allocation than the original DRF allocation algorithm. It's also more suitable for resource allocation of heterogeneous clusters. The X-DRF consumes part of the system resources when allocating resources. It still reduces the resource waste caused by the unreasonable resource allocation of the original DRF algorithm.

5 Related Works

The scheduling of data center resource is one of the hot topics in cloud computing and big data. This paper focuses on the two goals of data center resource scheduling.

(1) **Fairness-oriented resource scheduling.** Isard et al. proposed a Quincy [10] scheduling strategy for shared distributed clusters. It simply mapped the fair scheduling problem to the minimum cost of flow graph problem, which effectively calculated and optimized the global matching of scheduling decisions online and reached the goal of balancing task fairness and data locality. Ghodsi et al. [11] extended the Max-Min algorithm to motivate users to truthfully report their demands and share resources for sharing. It proposed a resource fair scheduling method Choosy, which implements resource sharing under placement constraints. But it lacked better Convergence time and research under heterogeneous resources. Ousterhout et al. [12] took into account fairness and data locality, who proposed a resource scheduling method named Sparrow, based on random sampling and late binding [13]. The method continuously adjusts the scheduling strategy based on historical operating data to achieve the best resource scheduling effect.

(2) **Utilization-oriented resource scheduling**. Robert et al. [14] consider the trade-off between fairness, performance and efficiency in modern cluster systems. A long-term altruistic scheduling strategy CARBYNE was proposed, which can significantly improve application performance and cluster resource utilization by regrouping and allocating the remaining resources in the cluster. Moreover, in terms of performance isolation and fairness, CARBYNE is closed to DRF. Chen et al. [15] proposed a large-scale resource scheduling modeling method based on the characteristics of large-scale resource scheduling parallel operations. The method supports multiple scheduling objectives and has flexibility, but it's the cost of parameter configuration and the efficiency of the solution of the graph requires further studying. The resource utilization and fairness of data centers are two important indicators of system resource scheduling.

6 Conclusion and Future Work

The large-scale resource scheduling of heterogeneous clusters is one of the hot issues recently studied. Firstly, this paper introduces the background of the research work of the paper and the default scheduling algorithm (DRF) of Mesos is analyzed. The DRF algorithm does not consider machine performance and task types. Then, this paper presents a heterogeneous cluster multi-resource fair scheduling algorithm based on machine learning to solve the problem. The algorithm is to test the performance of the machine and use the machine learning method to classify the computing tasks and reach the goal of reasonable resource allocation. Finally, several typical jobs such as Word-Count and PageRank are selected to experiments. The results proves the effectiveness of the method. However, in the practical application situation, there still remains room for improvement in this method. First of all, this paper only considers the CPU and memory that have a large influence on the machine performance, but it has not considered the network bandwidth and disk I/O of the machine. This is one of the works for follow-up research. Secondly, the research objectives of this paper mainly focus on fairness and cluster resource utilization. How to achieve energy efficient heterogeneous cluster large-scale resource scheduling method is the future research direction.

Acknowledgments. This work is supported by the Natural Science Foundation of China (No. 61762008), the Natural Science Foundation Project of Guangxi (No. 2017GXNSFAA198141), the Key R&D project of Guangxi (No. GuiKE AB17195014), and the R&D Project of Nanning (No. 20173161).

References

1. Hindman, B., Konwinski, A., Zaharia, M., Ghodsi, A., Joseph, A., et al.: Mesos: a platform for fine-grained resource sharing in the data center. In: Proceedings of the 8th USENIX Conference on Networked Systems Design and Implementation, pp. 429–483. USENIX Association (2013)
2. Delimitrou, C., Kozyrakis, C.: Quasar: resource-efficient and QoS-aware cluster management. ACM SIGPLAN Not. **49**(4), 127–144 (2014). ACM
3. Li, Y., Zhang, J., Zhang, W., Liu, Q.: Cluster resource adjustment based on an improved artificial fish swarm algorithm in Mesos. In: IEEE International Conference on Signal Processing, pp. 1843–1847. IEEE (2017)
4. Ghodsi, A., Zaharia, M., Hindman, B., Konwinski, A., Shenker, S., et al.: Dominant resource fairness: fair allocation of multiple resource types. In: Usenix Conference on Networked Systems Design and Implementation, pp. 323–336. USENIX Association (2011)
5. Wang, W., Liang, B., Li, B.: Multi-resource fair allocation in heterogeneous cloud computing systems. IEEE Trans. Parallel Distrib. Syst. **26**(10), 2822–2835 (2015)
6. Tang, H., Li, Y., Wang, L., et al.: Predicting misconfiguration-induced unsuccessful executions of jobs in big data system. In: Computer Software and Applications Conference, pp. 772–777. IEEE (2017)
7. Cernadas, E., Amorim, D.: Do we need hundreds of classifiers to solve real world classification problems? J. Mach. Learn. Res. **15**(1), 3133–3181 (2014)
8. Chen, T., Guestrin, C.: XGBoost: a scalable tree boosting system. In: ACM SIGKDD International Conference on Knowledge Discovery and Data Mining, pp. 785–794. ACM (2016)
9. Ubench. http://www.phystech.com/download/ubench.html
10. Isard, M., Prabhakaran, V., Currey, J., Wieder, U., Talwar, K., Goldberg, A.: Quincy: fair scheduling for distributed computing clusters. In: ACM SIGOPS, Symposium on Operating Systems Principles, pp. 261–276. ACM (2009)

11. Ghodsi, A., Zaharia, M., Shenker, S., Stoica, I.: Choosy: max-min fair sharing for datacenter jobs with constraints. In: Proceedings of the 8th ACM European Conference on Computer Systems, pp. 365–378. ACM (2013)
12. Ousterhout, K., Wendell, P., Zaharia, M., Stoica, I. Sparrow: distributed, low latency scheduling. In: Proceedings of the Twenty-Fourth ACM Symposium on Operating Systems Principles, pp. 69–84. ACM (2013)
13. Grandl, R., Chowdhury, M., Akella, A., Ananthanarayanan, G.: Altruistic scheduling in multi-resource clusters. In: 12th USENIX Symposium on Operating Systems Design and Implementation, pp. 65–80 (2016)
14. Ukidave, Y., Li, X,, Kaeli, D.: Mystic: predictive scheduling for GPU based cloud servers using machine learning. In: IEEE International Parallel and Distributed Processing Symposium, pp. 353–362. IEEE (2016)
15. Chen, X., Wu, H., Wu, Y., Lu, Z., Zhang, W.: Large-scale resource scheduling method based on minimum cost maximum flow. J. Softw. **28**(3), 598–610 (2017)

A Network Visualization System for Anomaly Detection and Attack Tracing

Xin Fan, Wenjie Luo, Xiaoju Dong(✉), and Rui Su

BASICS, Department of Computer Science and Engineering,
School of Electronic Information and Electrical Engineering,
Shanghai Jiao Tong University, Shanghai, China
xjdong@sjtu.edu.cn

Abstract. Analyzing network data is one of the important means to safeguard network security. However, how to detect anomalies and trace back the origin of attacks in the enlarging scale of network data is still a challenge now. This paper designs and implements a network visualization system, which meets three main requirements: the situation awareness of the whole network, the rapid detection of anomalies, and the track of attack source. To combine multiple visualization technologies reasonably, the system provides information from three levels. It also uses unsupervised learning methods to detect anomalies in different ways. Therefore, the system enhances the ability of identifying abnormal behaviors from network data. Its efficiency is tested by the usage of data in the ChinaVis 2016.

Keywords: Network security · Visualization · Anomaly detection

1 Introduction

With the rapid development of network technique, the Internet is widely used in all trades and fields. The increasing scale of network attacks, on the other hand, is attracting the attention of network security researchers. Although systems like IDS are effective to detect some kinds of attacks, there are still many threats that cannot be detected according to traditional methods.

Network log analysis is one of the most important means to safeguard network security. It helps find the attacking sources after the attack events happened in the method of analyzing a huge amount of network data. Nevertheless, manual work alone are not enough for getting results from network logs directly. Visualization technology, as a method of data display, taking full advantage of humans' visual system, helps users comprehend hidden meanings from a high level and analyze complicated data preferably.

However, with the challenge of big data and complicated attacking methods, traditional visualization methods for network security fall down. Whereas, in

Q. Zhou et al. (Eds.): ICPCSEE 2018, CCIS 901, pp. 560–574, 2018.
https://doi.org/10.1007/978-981-13-2203-7_45

the last decade, machine learning has been widely researched in the field of big data. By using machine learning to preprocess a large number of network data or detect trends, patterns and anomalies, the analyst is enabled to identify attacks automatically, which greatly simplifies the tracing process. When we combine the visualization technology and machine learning, we must achieve greater results in the aspect of network security.

In view of the above problems, this paper combined machine learning with visualization technology to propose a new system, which achieved abnormal identification and traceability in a large number of network data. The system tightly integrates multiple visualization methods to present data from different levels, enabling users to drill down to details needed. Besides, the system has a high flexibility for it is developed as a website and has a high scalability in supporting the user-defined configuration of data source and the mapping between visual variables and data attributes.

The contributions of our work can be summarized as follows. We combine visualization technology and machine learning to calculate abnormality degree, clustering IPs with their behavior patterns and match IPs that cause certain network traffic. We propose three goals to be met in the process of network data analysis and design an integral system that implements small tasks subdivided from the three high-level goals.

Considering the inseparable relationship between situation awareness, anomaly detection and IP trace-back, the innovative of this paper is that we combine the researches of these three points organically according to our design strategy, which can help network security analysts work more efficiently.

The rest of this paper is organized as follows: Sect. 2 discusses related works, Sect. 3 introduces the design rationales of our network visualization system and three main goals for analyze network data. Section 4 describes the framework of system and gives a detailed introduction on core algorithms of each modules, in Sect. 5 we analyze the data from challenge 1 of ChinaVis as a case study to validate the system and finally in Sect. 6, we summarize our task and propose the further work.

2 Related Work

There has been not a few research achievements in the field of network security visualization techniques. For example, Shiravi et al. have introduced category of network security data as well as made a systematic conclusion on network security visualization [14].

IP addresses and ports are indispensable parts of network security analysis. IPMatrix [7] maps the four parts of IP address into four axes of two matrix. HNMap [11] uses Treemap to express IP addresses. Bundling techniques [5] which simplify graphs, can be applied to solve the problem of confounding edges in large networks caused by force-directed algorithm. PortVis [12] uses a 256×256 matrix to express the traffic situation of 65536 ports, and a movable small window to observe detail information. PortMatrix [17] splits the ports into four groups where each 100 consecutive dynamic ports are expressed in a single grid in order to highlight specific ports and improve utility rate of space.

For complex network attacks, radar charts can do it better to identify the association between events. VisAlert [9] uses the 3W (what when where) model and radar chart to analyze network data, and many studies are then published to improve and expand it. For example, AlertWheel [4] combines three bundling techniques and changes the layout and edges between hosts and attack types. IDSRadar [18] uses a kind of concentric circle structure with five entropy functions to analyze abnormal behaviors. NetSecRadar [19] uses a hierarchical force-directed algorithm to improve the internal structure. Most of the above approaches use only radar view for analysis, though can achieve situation awareness of the whole network, they will also cause a loss of further details.

In addition, machine learning can be combined with visualization to detect anomalies. Buczak and Guven [3] introduce the use of machine learning and data mining for intrusion detection. They provide a brief introduction of various models of machine learning and the papers that have been widely cited for misuse and anomaly detection. Sommer [15] provide a set of guidelines for applying machine learning to detect network intrusion. But users cannot intuitively judge the validity of results drawn only from machine learning algorithms which may have a high false positive rate. Therefore such methods are not suitable for users except the network analysts.

In the respect of the combination of machine learning and visualization, TVi [2] proposes a PCA-based anomaly detection algorithm. Hao et al. [6] find anomalies by using dynamic time warping to get the optimized match among ensemble members and discuss the value of clustering number k in hierarchical clustering. Ma [10] mentions that associative memory neural networks can be used to handle data with noise and distortion. VizRank [8] uses k-nearest neighbor algorithm to evaluate the usefulness of a projection. EnsembleMatrix [16] utilizes visualization and human knowledge to adjust the weights of each classifier so that a better classification result can be obtained. If we combine several machine learning algorithms with visualization, we may achieve better results on network security.

3 Design Rationales

As we discussed in Introduction, for the purpose of safeguarding network security, a network analysis system should try to enhance the ability to identify abnormal behaviors from large-scale of network data.

To detect threats among a large number of network data generated by complex network, some demands should be met. First, the situation awareness of the whole network, which means acquiring and visualizing security elements that may cause a change in the network situation and further, forecasting future trend. Second, anomalies should be detected as fast as possible. For example, the flow anomaly of the entire network may be caused by DDoS, the traffic spikes of ports can be obtained from a port scan. Though it is not necessary that there are attacks hidden under anomalies, when some anomalies are observed, we should take steps at the first time to minimize the lost caused by attacks. Third, tracing the attacks sources. Tracing attacks instantly is essential for preventing further damage. Checking IPs' former behavior patterns, we can deny the access or fix them if they are attackers or infected hosts. Actually, the three demands go forward one by one. Situation awareness is for anomaly detection while the ultimate goal of anomaly detection is attack-tracing.

Further, we need to get some detail information and meet some demands if we want to achieve the above goals. For situation awareness, the necessary information contains: how the network flow changes over time, the topological structure of IPs and the traffic exchange between network segments. In order to obtain this information, we can make use of some classic views like timeline, matrix, force-directed graph and radar graph. When it comes to anomaly detection, a significant factor is the detecting speed, otherwise the system is unable to work in a real-time environment. Here we use three methods to help analysts identify anomalies: the method based on PCA, K-means cluster and genetic algorithm. Finally, for the purpose of finding out the original attacker, we may want to know what other IPs a given IP has connected with in the past period of time. The form of the IP trace view is like a tree graph in that the connected IPs of a given IP can also be tracked back and the graph can be unfolded layer-by-layer.

4 System Framework

The attributes used in the system are: timestamp, SRCIP, DSTIP, SRCPORT, DSTPORT, filename, file type, protocol and check result: the connections are grouped into safe, low-risk, middle-risk and high-risk by local detector.

Figure 1 shows the dashboard of the system. The general workflow is to select a time period of interest according to network traffic, after all views have been

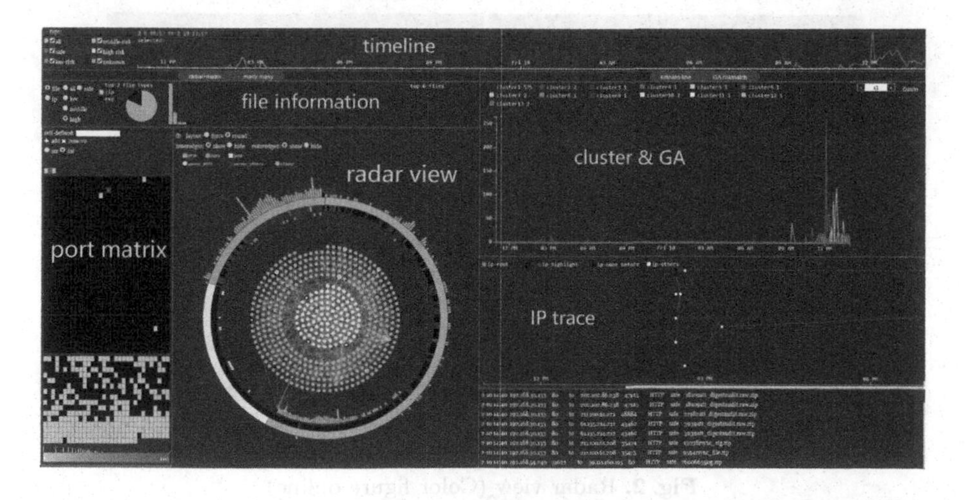

Fig. 1. System overview

updated, users can predict the condition of whole network or find abnormal IPs by filtering high-risk files, ports and cluster id, then trace these IPs to check whether they are attackers or not. Anomaly detection and IP trace-back are core parts of the system.

4.1 Situation Awareness

According to the detail tasks in Sect. 3, a timeline is placed at the top of the interface, displaying the change of network flow and alert numbers over time. When a certain IP is chosen, vertical grey bars will be drawn in the timeline at the corresponding locations where this IP occurred. If an IP fits well with the risk traffic, it may be one of the causes that led to the risk events. The timeline is also used for selecting the time range of other views. In view of the file attribute contained in the dataset, the system also provides a pie graph and histogram to display file information.

Besides, we use two methods to complete other tasks and help users to judge the situation of the whole network, users can switch between them when needed.

Radar View and Port Matrix. The radar view and port matrix are shown in the bottom-left corner of Fig. 1. The design of radar view is similar to what we have mentioned in related work. The outer ring is the distribution of protocols and within the rings are concentric or force-directed IPs. What's different is the orange middle ring that indicates the anomaly value, described in Sect. 4.2.

The radar view is applied to the system to display part of the information needed by situation awareness: The outer arcs indicate the alert data and the inner distribution of IPs can show the topological information of hosts. For example, when switch to force-directed layout (Fig. 2(a)), we can find: one of

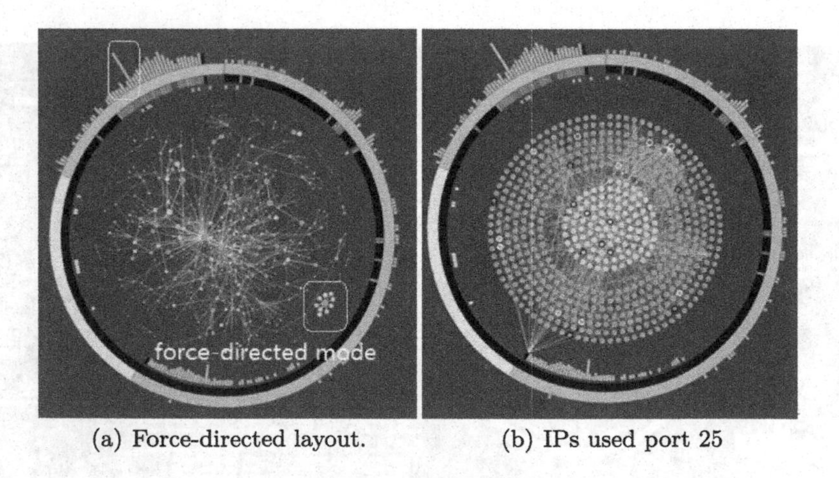

(a) Force-directed layout. (b) IPs used port 25

Fig. 2. Radar view (Color figure online)

these IPs had a high volume of traffic with several HTTP servers and there was a pair of IPs that only have communication with each other frequently.

We use the matrix form to show the network connections situation, thus users can identify attacks such as port scan. As shown in the left part of Fig. 1, the port matrix divides the ports into three parts: ports that users are interested in, well-known ports (0-1023) and other ports. In the third part, each grid indicates the aggregated traffic of 128 ports. The color of matrix indicates the volume of traffic. The histogram under the matrix is the distribution of the volume of network traffic which can be used to filter ports.

Through interactions, we can get additional information such as: The port utilization of a certain IP and which IPs have used the mail port 25. Figure 2(b) gives the IPs used port 25 which are highlighted in red.

Detail to Overview. Finally, the last information the system should show is the input traffic volume and output traffic volume between different network segments. To do this, we can sort the IPs by their decimal format and then place them on the axis from left to right and the vertical axis indicate the ports of these IPs. Such a design is similar to scatter plot but the IPs who have communications will be linked by line. However, the links can make the view cluttered and crowded. On one hand, we provide a filter function to focus on segments that users are interested in. On the other hand, the system allows users to select some IPs and manually classify them together, then points in the same class will be shown as a single circle in the view.

Based on such idea, we get the view as shown in Fig. 3. This view can be divided into two parts, in the detail part (the upper portion of Fig. 3(a)), the vertical axis shows port number and the horizontal axis shows the A part of IP addresses, sockets are linked if there are connections between them.

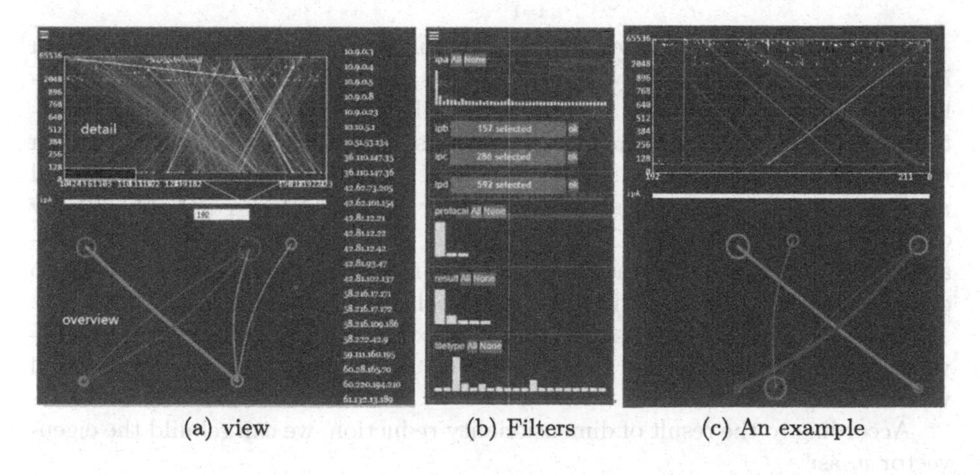

(a) view (b) Filters (c) An example

Fig. 3. Detail to overview (Color figure online)

Selecting a region in the detail view can add the inner nodes to a new class or an existing class. Circles corresponding to the classes will be drawn in the overview part (the lower portion of Fig. 3(a)). Circles are connected to each other by two lines that indicate the inbound and outbound traffic of the class. The size of circle represents the number of class members and the thickness of line represents the volume of traffic. IP addresses contained in the clicked circle will be listed on the right. Click the icon in the top left corner to expand the view for filtering, as shown in Fig. 3(b).

Such a method puts the information of IP and port together, enabling users to observe the inbound and outbound traffic between network segments and find special IPs and ports. An example is shown in Fig. 3(c), filter the segments 192 and 211, we can see that the network traffic between these two segments is mainly caused by the connections from 192.168.30.133 in segment 192 to five IPs in segment 211.

4.2 Anomaly Detection

Entropy and PCA. In order to find anomalies efficiently, we need a score to measure the abnormal degree of every period of time. We use the information entropy and PCA (Principal Component Analysis) to calculate the score.

Information entropy, referring to disorder or uncertainty of data, is conductive to anomaly detection for it changes significantly when attacks occur. PCA is one of the most commonly used dimensionality reduction methods that transforms the raw data into a set of linearly independent vectors through linear transform.

For a random variable X, its entropy H(X) can be expressed as:

$$H(x) = -\sum_{i=1}^{n} p(x_i) \cdot \log_2 p(x_i) \tag{1}$$

The anomaly value can be calculated by following steps: First, slice the data into timestamps where every timestamp has a 5 min time scan. Then calculate the entropy of SRCIP, DSTIP, SRCPORT and DSTPORT every timestamps and this will result in four time series, which can also be formed as a matrix: $M = \{v_1, v_2, ..., v_i, ..., v_K\}^T$, Where K is the total number of 5 min and $v_i = \{x_{i1}, x_{i2}, x_{i3}, x_{i4}, x_{i5}\}$ is the eigenvector. $x_{i1}, x_{i2}, x_{i3}, x_{i4}$ are the entropy of SRCIP, DSTIP, SRCPORT, DSTPORT in the ith 5 min, x_{i5} is the percentage of record number that the check result is risk. We take risk percentage into consideration as some IDS check the risk level of every connection locally.

Second, reduce the dimensionality of matrix M using PCA, then get the eigenvector of covariance matrix: $B = \{\varphi_1, \varphi_2, ..., \varphi_n\}$, where φ_i is a 5 dimensional vector and $q < 5$.

According to the result of dimensionality reduction, we can rebuild the eigenvector v_i as:

$$\tilde{v_i} = \sum_{m=0}^{q} \left(\sum_{n=0}^{5} v_i[n] \cdot \varphi_m[n] \right) \cdot \varphi_m \tag{2}$$

Finally, measure the anomaly value, namely, the deviation of origin vector, by the Euclidean distance between original eigenvector v_i and reconstructed eigenvector $\tilde{v_i}$. The anomaly value of the ith 5 min can be expressed as:

$$A_i = \sum_{m=0}^{5} (\tilde{v_i}[m] - v_i[m])^2 \tag{3}$$

The anomaly value is presented as an orange ring in the radar view, the bigger the anomaly value, the closer the color is mapped to orange.

Cluster. Regarding IPs distant from most other groups as abnormal IPs is a fast way to detect anomalies. These IPs are picked out firstly and whether they are actual attackers or not will be verified then.

We use K-means to cluster IPs by their network traffic changes so that abnormal IPs can be distinguished. K-means clustering is one of the most commonly used clustering arithmetic, its basic idea is to select k centroids in the feature space firstly, then classify the points that are nearest to these centroids and update the location of centroids according to new cluster members. The cluster results can be obtained through iterations. The K-Means++ algorithm [1] can be used to solve the problem caused by the selection of initial centroids.

In this paper, the observation of K-means is set as the network flow series of every IP, sliced by 5 min. Parameter k can be chosen automatically by Silhouette Coefficient [13]. An input box is also provided for user to adjust k manually. According to the result of test, most IPs are grouped into the same class meaning they have similar behavior while others act out especially. On closer inspection, those outlier IPs match the risk curve more or less, may be identified as anomalous IPs.

Each cluster is drawn as a curve in screen indicating the average network traffic of this cluster. When clicked, IPs contained in this cluster will be listed on the right side of the view. When double clicked, the cluster will be unfolded, the member IPs plotted respectively.

Genetic Algorithm. When analysts find something abnormal in a period of time, they may try to locate the IPs whose communications caused the anomalies. This requirement introduces the last small task of anomaly detection: discovering IPs that have similar network flow curves as alerts curves quickly.

Genetic algorithm is a method of simulating the natural evolutionary process to search the optimal solution. Assume that the number of IP is N, the definition of each individual is an array A with length of N, A[i] = 0 means that the ith IP is not selected, A[i] = 1 means selected. Here the communication behavior refers to a discrete time series composed of network traffic. The initial population size is set to 200, that is, randomly generated 200 individuals. Genetic algorithm uses a fitness function to evaluate individuals. In this paper, the individual fitness is the Euclidean distance between individual's communication mode and the target communication mode. The closer the distance, the higher the fitness.

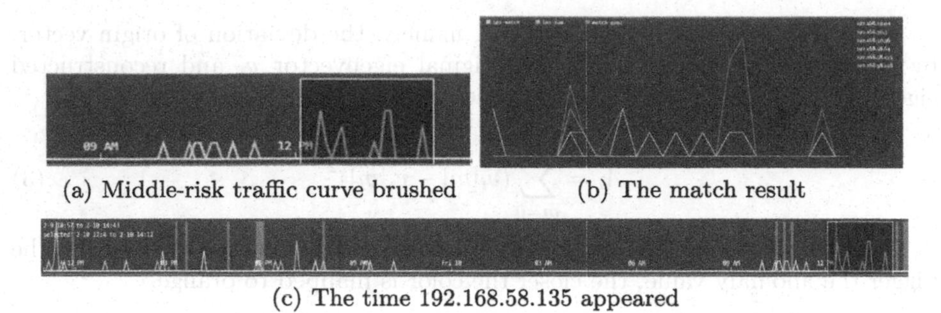

(a) Middle-risk traffic curve brushed (b) The match result

(c) The time 192.168.58.135 appeared

Fig. 4. Genetic algorithm example (Color figure online)

The selected individuals generate new individuals by crossover and mutations to form the next generation. In order to improve efficiency, we set the rule that only when fitness of new individuals is higher than that of its parents will the new one be selected into the next generation, otherwise we take the parents instead. After enough generations we can get the result IP.

We apply the Genetic algorithm to the time brushed as shown in Fig. 4(a) to find the threatening IPs. Figure 4(b) shows the result, the 6 IPs we find are listed on the right. The white line is the traffic line for each IP, the yellow one is the sum of their traffic and the orange one is the target traffic line. We can see that the similarity of result and target pattern is relatively high.

Then we choose one of the result IPs 192.168.58.135 to check its activity in other time periods, as shown in Fig. 4(c). We find a good agreement between the risk curve and the time 192.168.58.135 occurred, indicating 192.168.58.135 may be concerned with attacks.

4.3 IP Trace-Back

Situation awareness and anomaly detection are not our ultimate goal. The situation awareness is designed for the analysts to validate whether the network is healthy and find abnormal events. But we must pick out possible causes after noticing anomalies, otherwise we can do nothing. Therefore, it is of great importance to trace the origin causes of anomalies.

Chances are that after a possible anomaly IP is found, we need to trace its previous behavior to decide whether it is an attacker, a victim or a normal user. If it is indeed an attacker, we may want to who it has contacted with and confirm whether those IPs are infected hosts or attackers as well.

The system needs a view to implement the operations mentioned above so that users can use the view to observe the actions of abnormal IPs by tracing back and to find out other abnormal IPs associated with one abnormal IP. By this means, the source of attack and hidden attacks can be found.

The system provided a special view for tracing, as shown in Fig. 5. To start the procedure, a certain start time and an initial IP for tracing should be determined

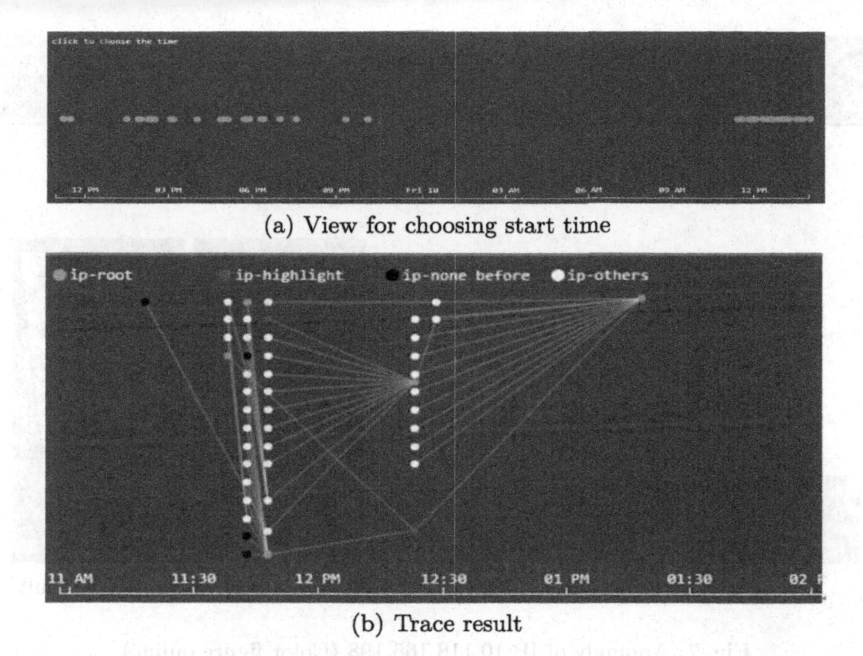

(a) View for choosing start time

(b) Trace result

Fig. 5. IP tracing example (Color figure online)

(Fig. 5(a)). After that, users can double click the IP of interest to trace-back. IP nodes are mapped in accordance with the time they occurred (Fig. 5(b)). The same IP will not appear two or more times at the same time/x-axis. An IP can be traced any times until it has no record before, which will be black colored. The orange circle represents the initial traced IP. The thickness of the lines between IPs can be used to represent the network traffic between them during that period of time.

By expanding the nodes repeatedly in the attack tracing view, the tracing process is represented in an intuitive and comprehensible way.

5 Case Study

In this section, we utilize the Challenge 1 of ChinaVis as case study to illustrate how our system works.

5.1 DDos on July 31

Since there is no attribute of check result which indicates the risk level of connections, we take the records of crack file (the file transferred is judges corrupted by monitor system) as low risk, others as safe, without regard to middle risk or high risk. Now looking at the low risk curve, there was a continuous record of cracked files during July 31 11:30 to 15:30, as shown in Fig. 6.

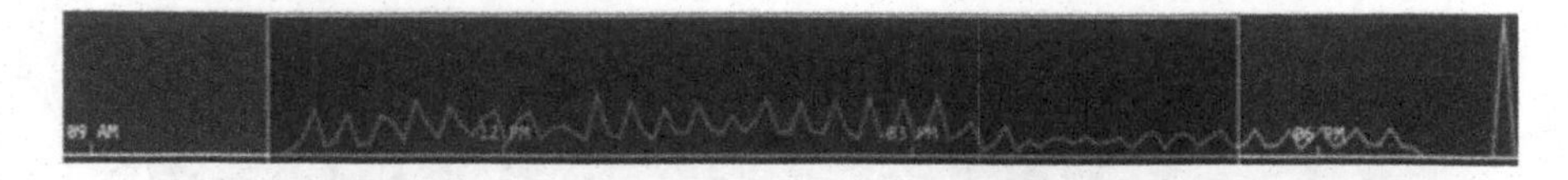

Fig. 6. The curve of file cracked

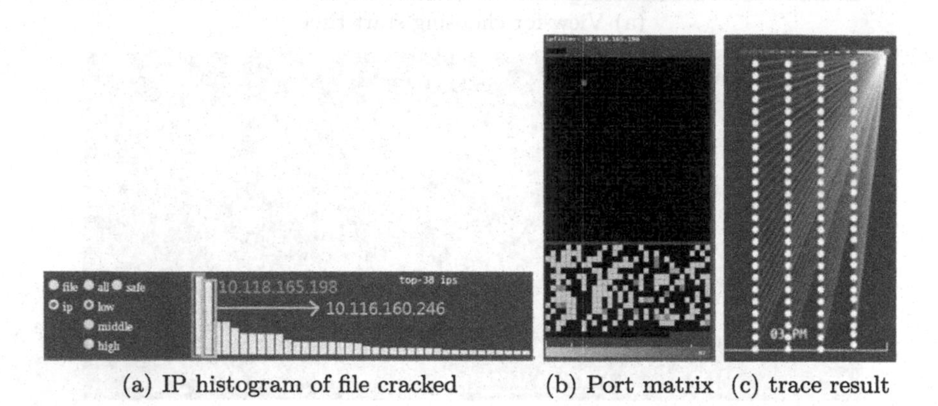

(a) IP histogram of file cracked (b) Port matrix (c) trace result

Fig. 7. Anomaly of IP 10.118.165.198 (Color figure online)

Found by Histogram and Matrix. In Fig. 7(a), the IP histogram of cracked files shows that IP 10.118.165.198 caused most of crack records, followed by 10.116.160.246. Besides, the topological structure of the network shows that these two IP shared common network contacts.

When we turn to the port matrix of 10.118.165.198 and 10.116.160.246, we can find that 10.118.165.198 was scanned, most of its dynamic ports was connected once and port 135 were connected many times, as shown in Fig. 7(b).

Then trace this anomalous IP. The process of tracing is depicted in Fig. 7(c) where the orange node is 10.118.165.198. The tracing shows a repetitive pattern: 10.118.165.198 was accessed frequently by the same 26 hosts (the sameness can be verified by highlighting), especially by 10.52.140.227. The IP addresses of these 26 hosts were basically continuous, such as 10.52.140.210-213 and 10.52.189.137-138. At this point, we can argue that this could be a DDos attack, where 10.118.165.198 was attacked by 26 hosts.

Found by Genetic Algorithm. We also use genetic algorithm mentioned in Sect. 4.2 to find the possible causes of the abnormal cracked traffic curve. The match result is shown in Fig. 8. The result traffic curve (the light yellow curve) fits the target curve (the dark yellow curve) quite well. Comparing the result with the 26 IPs we got above, we can see that these IPs are in the list of the GA match result, which proves the correctness of both methods. While these two means have similar results, genetic algorithm gives conclusion more efficient and requires less manual analysis compared with normal method.

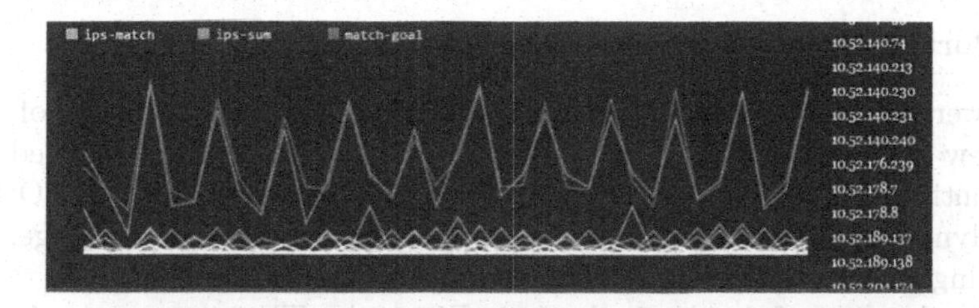

Fig. 8. Match result of GA (Color figure online)

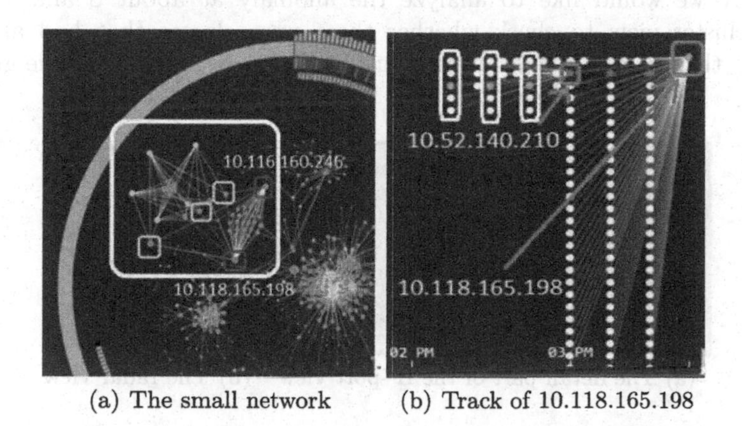

(a) The small network (b) Track of 10.118.165.198

Fig. 9. Peculiar pattern of activity (Color figure online)

Following these 26 hosts by switch the view to the force-directed layout, we notice the special relationship between these IPs. The small network as shown in the white frame in Fig. 9(a) is composed of the 26 IPs, 10.118.165.198, 10.116.160.246 (the two IPs that had the most cracked files and they were both servers) and five file-sharing servers (five light blue nodes in the upper-left corner of the white frame): 10.116.160.247, 10.118.165.199, 10.118.161.2, 10.116.160.248 and 10.118.161.1. Actually, users may find the curves of these servers also fit well with the crack curve, but they were not found by GA because in our system the algorithm only matches the source IPs with target otherwise a single connection may be calculated twice.

Back to the force-directed layout, most of the 26 IPs only have relations with 10.118.165.198 and 10.116.160.246 (in red frame) while 10.52.140.210, 10.52.140.227, 10.52.140.211 (in yellow frame) were also in contact with the five file-sharing servers.

Peculiar pattern of activity is found in Fig. 9(b) when tracing 10.52.140.210, it accessed the port 135 of servers 10.118.165.199, 10.116.160.247 and 10.118.161.1 every 15 min. Analogously, 10.52.140.211 connected to port 135 at 10.116.160.248 every 15 min. We guess that such particular resistance pattern is the result of RPC. In order to maintain the stability of remote connection to the server, clients may initiate a new connection every 15 min.

5.2 Port Scan and LAND Attack on July 28

There were also something abnormal on July 28. The detail part of the IP-port view is shown in Fig. 10(a). The ports of 10.118.165.198 (the red circles) were continuously connected just like what happened on July 31. Other IPs whose dynamic ports were continuously requested are marked in orange. What's interesting is that most of these IPs were file sharing servers.

The radar view of the day is shown in Fig. 10(b). The orange color in the two red frames indicating there were anomalies at about 8 AM and 10 AM.

Firstly we would like to analyze the anomaly at about 8 am. We turn to the cluster view to check whether there were classes that had abnormal curves at that time. The clustering result in Fig. 11(a) shows that the group 13

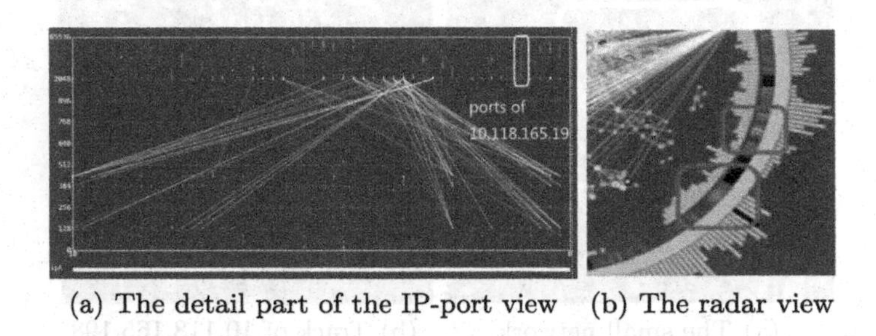

(a) The detail part of the IP-port view (b) The radar view

Fig. 10. Anomaly on July 28 (Color figure online)

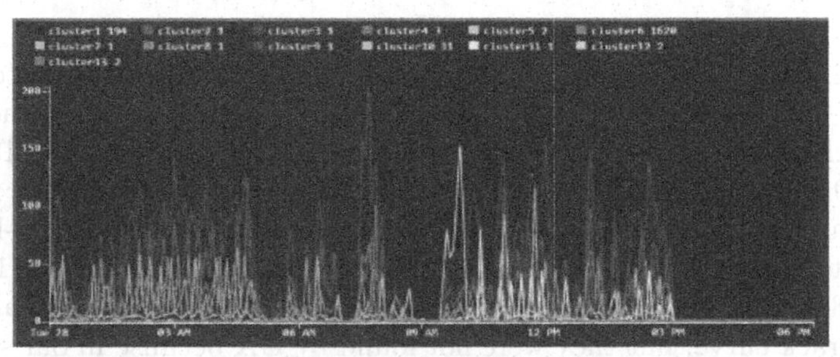

(a) Cluster view at 8 AM

STARTTIME	SRCIP	SRCPORT	DSTIP	DSTPORT	IPSMALLTYPE	FILEAFFIX	ISCRACKED
7-28 7:51	10.67.220.228	4343	10.67.216.188	3206	67	.unk	safe
7-28 7:51	10.67.220.228	4343	10.67.216.188	3205	67	.unk	safe
7-28 7:51	10.67.220.228	4343	10.67.216.188	3204	67	.unk	safe
7-28 7:51	10.67.220.228	4343	10.67.216.188	3207	67	.unk	safe
7-28 7:51	10.67.220.228	4343	10.67.216.188	3202	67	.unk	safe
7-28 7:51	10.67.220.228	4343	10.67.216.188	3201	67	.unk	safe
7-28 7:51	10.67.220.228	4343	10.67.216.188	3203	67	.unk	safe

(b) Message view at 8 AM

(c) Track of 10.67.216.226

Fig. 11. Anomalies at 8 AM and 10 AM (Color figure online) (the orange curve, containing IPs 10.67.216.188 and 10.67.220.228) had a traffic peak at about 8 AM but almost no traffic at other times which may lead to the anomaly. Focusing on the two members of group 13 by highlighting them in radar view, we notice that 10.67.216.188 only occurred at 8 AM, connected by 10.67.220.228. Furthermore, as shown in the message view in Fig. 11(b), we find that 10.67.220.228 scanned the ports of 10.67.216.188 multiple times.

In the same way, we find that IP 10.67.216.226 may cause the anomaly at 10 AM. Tracing 10.67.216.226, the pattern in Fig. 11(c) can be observed: it served as both the source IP and destination IP 3 times. It was a LAND attack which may lead to the starvation of resources. An attacker tried to bring the server down but failed, as 10.67.216.226 still provided services at 11 AM.

6 Conclusion

This paper proposes a visualization system to enhance the situation awareness of the whole network and help analysts find anomalies efficiently as well as trace them rapidly. The system is composed of several modules including situation awareness, anomaly detection, anomaly verification and IP trace-back. On one hand, we tried multiple machine learning algorithms including PCA, K-means and GA to improve the efficiency of anomaly detection. On the other hand, we emphasized the importance on attack tracing and a special view is provided in the system for tracing back IPs. The multi-views work cooperatively to present the data from different perspectives.

The system is developed as a web application so that it will not be restricted to analysts' working environment. Supporting the configuration of data source and field mapping by users, the system also has a good expansibility.

For future work, we will try other algorithms and optimization methods to enhance the performance of system in the face of the large dataset. We will also use some parameters to present the total evaluation of the system.

Acknowledgments. Supported by National Key Research and Development Program of China (Grant No. 2017YFB0701900), National Nature Science Foundation of China (Grant No. 61100053) and CCF-Venustech Hongyan Research Initiative (2016-013). Thanks Prof. Xiaoru Yuan, Peking university and unknown reviewers for instruction.

References

1. Arthur, D., Vassilvitskii, S.: k-means++: the advantages of careful seeding. In: Proceedings of the Eighteenth Annual ACM-SIAM Symposium on Discrete Algorithms. pp. 1027–1035. Society for Industrial and Applied Mathematics (2007)
2. Boschetti, A., Salgarelli, L., Muelder, C., Ma, K.L.: TVi: a visual querying system for network monitoring and anomaly detection. In: Proceedings of the 8th International Symposium on Visualization for Cyber Security, p. 1. ACM (2011)
3. Buczak, A.L., Guven, E.: A survey of data mining and machine learning methods for cyber security intrusion detection. IEEE Commun. Surv. Tutor. **18**(2), 1153–1176 (2016)

4. Dumas, M., Robert, J.M., McGuffin, M.J.: Alertwheel: radial bipartite graph visualization applied to intrusion detection system alerts. IEEE Netw. **26**(6), 12–18 (2012)
5. Fischer, F., Mansmann, F., Keim, D.A., Pietzko, S., Waldvogel, M.: Large-scale network monitoring for visual analysis of attacks. In: Goodall, J.R., Conti, G., Ma, K.-L. (eds.) VizSec 2008. LNCS, vol. 5210, pp. 111–118. Springer, Heidelberg (2008). https://doi.org/10.1007/978-3-540-85933-8_11
6. Hao, L., Healey, C.G., Hutchinson, S.E.: Ensemble visualization for cyber situation awareness of network security data. In: 2015 IEEE Symposium on Visualization for Cyber Security (VizSec) pp. 1–8. IEEE (2015)
7. Koike, H., Ohno, K., Koizumi, K.: Visualizing cyber attacks using IP matrix. In: IEEE Workshop on Visualization for Computer Security 2005 (VizSEC 05), pp. 91–98. IEEE (2005)
8. Leban, G., Zupan, B., Vidmar, G., Bratko, I.: Vizrank: data visualization guided by machine learning. Data Min. Knowl. Discov. **13**(2), 119–136 (2006)
9. Livnat, Y., Agutter, J., Moon, S., Foresti, S.: Visual correlation for situational awareness. In: IEEE Symposium on Information Visualization 2005. INFOVIS 2005, pp. 95–102. IEEE (2005)
10. Ma, K.L.: Machine learning to boost the next generation of visualization technology. IEEE Comput. Graph. Appl. **27**(5), 6–9 (2007)
11. Mansmann, F., Keim, D.A., North, S.C., Rexroad, B., Sheleheda, D.: Visual analysis of network traffic for resource planning, interactive monitoring, and interpretation of security threats. IEEE Trans. Visual. Comput. Graph. **13**(6), 1105–1112 (2007)
12. McPherson, J., Ma, K.L., Krystosk, P., Bartoletti, T., Christensen, M.: Portvis: a tool for port-based detection of security events. In: Proceedings of the 2004 ACM Workshop on Visualization and Data Mining for Computer Security, pp. 73–81. ACM (2004)
13. Rousseeuw, P.J.: Silhouettes: a graphical aid to the interpretation and validation of cluster analysis. J. Comput. Appl. Math. **20**, 53–65 (1987)
14. Shiravi, H., Shiravi, A., Ghorbani, A.A.: A survey of visualization systems for network security. IEEE Trans.Visual. Comput. Graph. **18**(8), 1313–1329 (2012)
15. Sommer, R., Paxson, V.: Outside the closed world: On using machine learning for network intrusion detection. In: 2010 IEEE Symposium on Security and Privacy (SP), pp. 305–316. IEEE (2010)
16. Talbot, J., Lee, B., Kapoor, A., Tan, D.S.: EnsembleMatrix: interactive visualization to support machine learning with multiple classifiers. In: Proceedings of the SIGCHI Conference on Human Factors in Computing Systems, pp. 1283–1292. ACM (2009)
17. Zhao, Y., Liang, X., Wang, Y., Yang, M., Zhou, F., Fan, X.: MVsec: a novel multi-view visualization system for network security. In: Proceedings of Visual Analytics Science and Technology, pp. 7–8. IEEE Computer Society Press, Los Alamitos (2013)
18. Zhao, Y., Zhou, F., Fan, X., Liang, X., Liu, Y.: IDSRadar: a real-time visualization framework for IDS alerts. Sci. China Inf. Sci. **56**(8), 1–12 (2013)
19. Zhou, F., Shi, R., Zhao, Y., Huang, Y., Liang, X.: NetSecRadar: a visualization system for network security situational awareness. In: Wang, G., Ray, I., Feng, D., Rajarajan, M. (eds.) CSS 2013. LNCS, vol. 8300, pp. 403–416. Springer, Cham (2013). https://doi.org/10.1007/978-3-319-03584-0_30

Opportunistic Concurrency Transmission MAC Protocol Based on Geographic Location Information

Jianfeng Wang[1], Dongjia Zhang[2], Haomin Zhan[3], Zhen Cao[3], and Hongbin Wang[4(✉)]

[1] China Institute of Marine Technology and Economy, Beijing 100081, China
[2] China Aerospace Science and Industry Corp, Beijing 1100048, China
[3] Beijing General Institute of Electronic Engineering, Beijing 100854, China
[4] College of Computer Science and Technology, Harbin Engineering University, Harbin 150001, China
wanghongbin@hrbeu.edu.cn

Abstract. In the wireless sensor networks, the wireless resources are very limited. In the design of the MAC protocols, how to make full use of the limited channel resources to complete more data communications must be considered as much as possible. In this paper, in view of the shortcomings of the OPC method on decisions of parallel communications, an opportunistic concurrency transmission MAC protocol based on geographic location information (OPCLI-MAC) is proposed, an improved local parallel mapping table and a parallel control algorithm are proposed. The node location information is added to the mapping table. The parallel control algorithm is carried out using the distance value to deal with the problems of the parallel transmission link's two terminals. Experiments show that the OPCLI-MAC protocol reduces the interference of the newly launched parallel transmission link to the ongoing data communications, optimizes the parallel transmission communication decision, improves the probability of the parallel transmissions, and improves the channel utilization.

Keywords: Wireless sensor networks · MAC protocol · Parallel transmission
Geographic location information · Communication decision optimization

1 Relevant Work

With the rapid development of wireless sensor networks, MAC protocol, as the channel resource controlling the network, has been a research hotpot. Due to the limit of channel resource, it is a good idea to introduce parallel transmission into the design and improvement of MAC protocol for its fuller utilization.

At present, single-channel MAC protocol focuses more on the optimization of serial transmission, however, when the improvement increases to a certain extent, no matter what actions have been taken, the improvement of MAC protocol is no longer obvious because of its limit and bottleneck. By this time, it is more reasonable to introduce parallel transmission. The so-called "parallel transmission" refers that there are two or more data link in the data communication the same time. The basic principle of parallel

Q. Zhou et al. (Eds.): ICPCSEE 2018, CCIS 901, pp. 575–588, 2018.
https://doi.org/10.1007/978-981-13-2203-7_46

transmission is that the upcoming data communication cannot interfere with the ongoing data communication [1].

The solution to parallel transmission under multiple single channels has been proposed. Reference [2] proposes a C-MAC protocol with high throughput capacity, which makes use of parallel access method based on power control and wireless channel of physical layer interference signal model to achieve the MAC layer parallel transmission control, and Signal-to-Interference-plus-Noise-Ratio (SINR) to adjust the data transmission power accordingly to ensure effective transmission of parallel transmission. In order to solve this problem, an efficient parallel transmission LACT-MAC protocol based on geographic location information is proposed in Ref. [3], which calculates the signal-to-noise ratio condition by using the coordinate data and the distance value between nodes, thus solving the problem that parallel transmission is not allowed in the traditional MAC protocol. The decision and control of the LACT-MAC for parallel transmission is completed by four parts: the acquisition of the node location information, the confirmation of the exposed terminal node, the detection and execution of the parallel transmission and the multiple parallel transmission collision avoidance mechanism. Capture effect refers that the receiving node will give priority to a greater power in the choice of access link. Capture effect is more common in the wireless sensor network, especially in the case of large deployment of the node. Reference [4] verifies that capture effect can greatly improve the feasibility of parallel transmission, on this basis, Ref. [1] proposes a new type of parallel transmission MAC protocol - NCGTPCCT (Non-Cooperative Game Theory based Power Controlled Forward Transmission MAC) protocol, thereby improving the performance of the entire network. In addition, there are also other improvement methods such as physical layer technology application, cross-layer design and control frame improvements [5–7].

Reference [8] proposes a type of opportunistic concurrency (OPC) MAC protocol. OPC utilizes the latest research progress, MIM (message in message), of the physical layer technology, which makes the OPC allow the sensor nodes to capture the available parallel transmission opportunities, replacing the mode of waiting for a completely idle channel in the traditional MAC protocol and making parallel transmission possible. However, it does not deal with the existing problems of the sender and the receiver in the establishment of a new parallel transmission link, and this article deals with the issues above.

2 Problems Analysis and Solutions

The OPCLI-MAC protocol analyzes the problems existing in the sender and receiver in the link establishment of the parallel transmission and gives the corresponding solutions.

2.1 The Potential Interference of Sender to Nodes Within Two-Hop Range and Corresponding Solutions

In the wireless sensor network, each sensor node has a fixed transmit power value, that is, the rated transmitting power, but in the actual operational environment, not only the transmitting power can be adjusted by each node in each process of data communication, the intensity of signal emitted by each node is also changed by many factors because the wireless sensor network is mainly deployed in harsh, changing and uncontrollable conditions, such as forests and marine. As showed in Fig. 1, the local parallel mapping table in the OPC method records only the information of the neighbor node within one-hop range such as the signal intensity value and the node ID. Assuming that the node that has not been in the communication coverage has entered the transmission distance due to the change of the transmission power, and there is no response record in the local parallel mapping table, the potential interference of sender to nodes within two-hop range will take place before any decision is made according to the mapping table through the parallel control algorithm. The probability of data conflict in OPC method can be effectively reduced and the success rate of parallel transmission can be improved as long as this problem is solved.

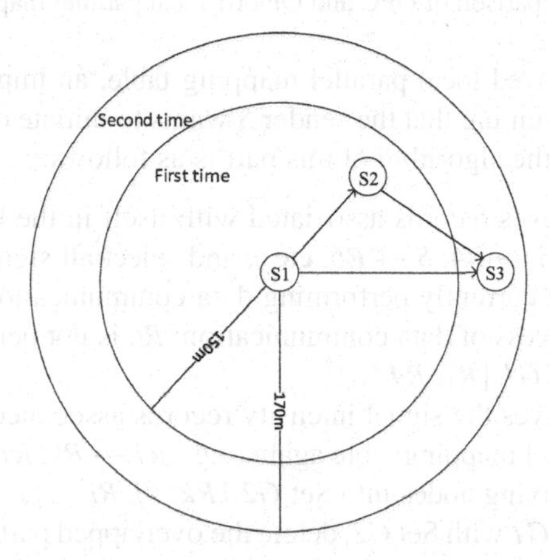

Fig. 1. The potential interference of sender to nodes within two-hop range.

This paper introduces "geographic location information" of the node into the design and parallel transmission decision of the MAC protocol. By utilizing this idea, the "geographic location information" is applied to the improvement of the local parallel mapping table, the improved local parallel mapping table is given on the base of which. Based on this, an improved parallel control algorithm is proposed to solve the potential interference of parallel transmission link sender to nodes within two-hop range.

First, add geographic location information of the receiver to the local parallel mapping table and calculate the distance between the two nodes corresponding to the existing link records in the mapping table. Figure 2 shows the comparison between the improved local parallel mapping table and the original. In addition, the wireless sensor network is a peer-to-peer network, in which each node broadcasts their own local parallel mapping table in flooding, and after a period of time, relevant information of neighbor nodes in its own one-hop range will be saved in the local parallel mapping table maintained by each node.

Concurrency Map

Signal intensity value
$S_1 \rightarrow R_1$, D7
$S_1 \rightarrow R_2$, D9
$S_2 \rightarrow R_3$, E6
...
...
$S_2 \rightarrow R_5$, DA

OPC

Concurrency Map

Signal intensity value	Node's location coordinates and distance
$S_1 \rightarrow R_1$, D7	$R_1(X_1,Y_1)$, $L(S_1 <-> R_1)$
$S_1 \rightarrow R_2$, D9	$R_2(X_2,Y_2)$, $L(S_1 <-> R_2)$
$S_2 \rightarrow R_3$, E6	$R_3(X_3,Y_3)$, $L(S_2 <-> R_3)$
...	...
...	...
$S_2 \rightarrow R_5$, DA	$R_5(X_5,Y_5)$, $L(S_2 <-> R_5)$

OPCLI

Fig. 2. Comparison of OPC and OPCLI local parallel mapping tables.

Based on the improved local parallel mapping table, an improved parallel control algorithm is given. Assuming that the sender *S* wants to initiate data communication to the target node *R*, and the algorithm of this part is as follows:

Step 1: Node *S* retrieves records associated with itself in the local parallel mapping table (e.g., $S \rightarrow R1$, $S \rightarrow R4$, $S \rightarrow R6$, etc.), and select all signal intensity records of the receivers that are currently performing data communication in the channel (e.g., *R1*, *R4* are in the process of data communication, *R6* is not performing data communication) to form Set *G1* {*R1, R4* …};
Step 2: Node *S* retrieves the signal intensity records associated with all the nodes in *G1* in the local parallel mapping table again (e.g., $R1 \rightarrow Rk$, $R1 \rightarrow Rj$, $R4 \rightarrow Ri$, etc.), and form all the receiving nodes into Set *G2* {*Rk, Rj, Ri* …};
Step 3: Compare Set *G1* with Set *G2*, delete the overlapped part, and form the rest into Set *G3* {*Rk, Rj* …};
Step 4: Calculate the distance, *Li*, between node *S* and all nodes in set *G3* and record it into local parallel mapping table;
Step 5: This time, when $S \rightarrow R$ is initiated, compare the transmission radius *Rs* of *S* with all the distances *Li*one by one, and the algorithm is Algorithm 1.

Algorithm 1: Solution of OPCLI parallel transmission sender

1: for $i = 0$ to k do //k represents the number of elements in collection *G3*
2:    if $R_s > L_i$ then //L_i represents the distance between *S* and R_i
3:    *S* sends a beacon message to R_i
4:       test the link signal intensity value of $S \rightarrow R_i$
5:       the record enters the local parallel mapping table
6:    end if
7: end for

Through the two processes of OPCLI-MAC, i.e. the improvement of the local parallel mapping table and the parallel control algorithm, the potential interference of sender to nodes within two-hop range have been greatly reduced, so has the probability of collision between nodes, thus improving the efficiency of parallel transmission.

2.2 Problem of Hidden Terminal Receiver and Solutions

Problem of hidden terminal is one of the most important problems to be solved in the MAC protocol design process of wireless sensor network. No matter how good the MAC protocol is, the hidden terminal problem will exist to a certain extent, a major approach to improving existing MAC protocol performance is to take some measures to reduce the negative effects brought by the performance of the hidden terminal. At one point, there may be multiple senders that want to initiate communication with the receiver *R*, and this problem can be discussed in two cases.

Case 1: The sender *S1* and the sender *S2* are within the communication range of each other, although this is not a hidden terminal problem, this direct conflict situation still needs to be taken into account. As shown in Fig. 3, sender *S1* and sender *S2* initiate data communication to node *S3* at the same time, resulting in multiple data collisions at the receiver *S3*.

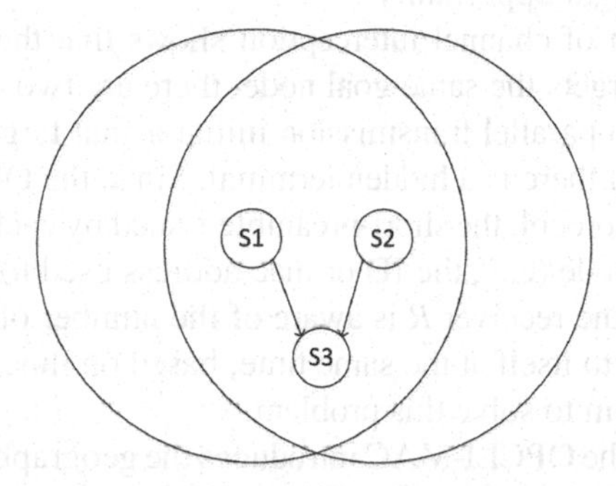

Fig. 3. Direct conflicts at the receiver caused by multiple senders.

Case 2: The sender *S1* and the sender *S2* are not within the communication range of each other, that is, the hidden terminal problem at the receiver. As shown in Fig. 4, the sender *S1* wants to initiate data communication to the receiver *S3* at a certain time, through the interception channel, sending end *S1* will determine that there is no parallel transmission initiator in the channel which targets the same node in the absence of a direct conflict situation as in Case 1, so that the *S1* performs the SINR condition judgment according to the normal parallel transmission control algorithm, and at this point, it is very likely that *S1* has passed the SINR condition judgment, and, ultimately, has the parallel transmission permission. At this point, there will be two or more data transmission conflicts at receiver *S3*.

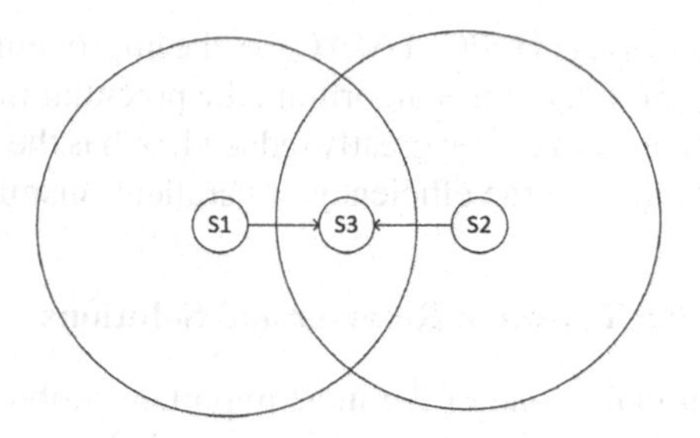

Fig. 4. Hidden terminal problem of receiver.

On the basis of the improved local mapping table, the improved parallel control algorithm of the receiving end is given to solve the hidden terminal problem of the parallel transmission link receiver.

First, the solution to Case1 is as follows:

A sender *S* needs to intercept the channel in the first place before the official transmission of data to ensure whether there is a parallel transmission initiator *S2* that targets the same goal node, if any, then *S2* is in the communication range of *S*. At this point, the sender *S* retreats according to the "binary index back-off algorithm" and selects the re-transmission in the right opportunity.

Second, if the result of channel interception shows that there is no parallel transmission initiator that targets the same goal node, there are two cases at this point. One is that there really is no parallel transmission initiator that targets the same as node *S* and the other one is that there is a hidden terminal. Since the OPCLI-MAC protocol is based on the X-MAC protocol, the short-preamble issued by each sender contains information about the goal node (e.g., the ID or mac address used to identify and notify the goal node). Therefore, the receiver *R* is aware of the number of the nodes that want to initiate communication to itself at the same time, based on this, OPCLI-MAC protocol gives a specific algorithm to solve this problem.

As shown in Fig. 2, the OPCLI-MAC introduces the geographic location information and the distance between corresponding nodes into the local parallel mapping table. Based on the specific number of the senders that targets itself, the receiver *R* can make a decision according to the "geographical location information" and the distance between the nodes in the local parallel mapping table, that is, select a sender with the shortest distance from itself because the shorter the distance, the less the attenuation of the signal sent by the sender, and the greater the success probability of data transmission between the two points. As shown in Fig. 5, after *R* determines which data transmission is to be received, *R* will reply to an early-ACK acknowledgment frame to notify the selected sender according to the X-MAC protocol, which contains ID or mac address of the selected sender, at the same time, the other potential senders can also receive the confirmation frame, so that they will know that they have not been selected, so they will retreat randomly in accordance with the "binary index back-off algorithm", and choose to retransmit in the right opportunity [9].

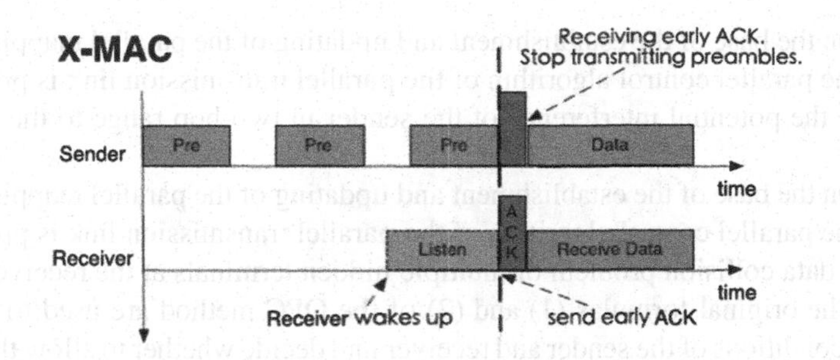

Fig. 5. Schematic diagram of X-MAC protocol.

3 Description of the Overall Communication Strategy of OPCLI-MAC Protocol

By integrating the solution to the problems above and describing the overall communication flow of the OPCLI-MAC protocol, the specific steps are as Fig. 6.

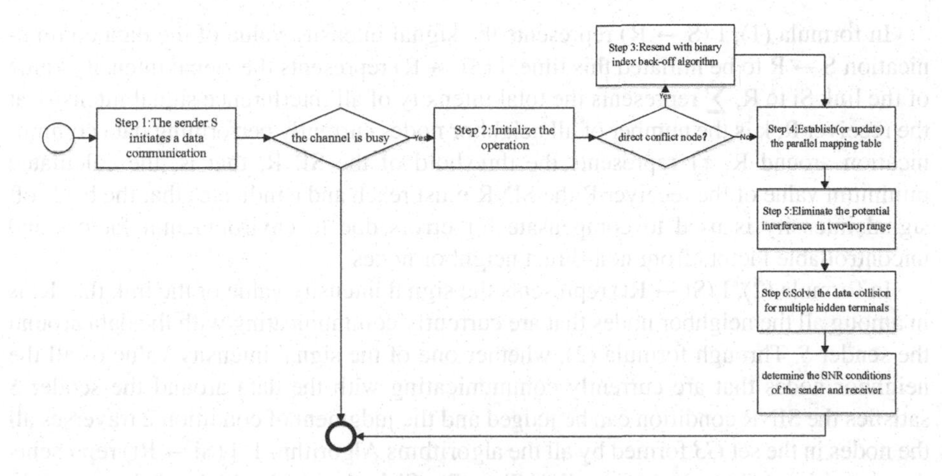

Fig. 6. The process of OPCLI-MAC.

Step 1: The sender *S* initiates a data communication and performs channel interception. If the channel is busy, the parallel transmission link is prepared to be established, and start the parallel transmission control algorithm;
Step 2: Initialize the operation: initialize the permitted maximum number of parallel transmission links Cmax and the parallel transmission decision D;
Step 3: If there is a direct conflict node, the sender uses the "binary index back-off algorithm" to retreat, and choose to retransmit in the right opportunity, otherwise, proceed to Step 4;
Step 4: Establish and update the relevant information in the local parallel mapping table, and establish the information exchange in the initial period through the network, so that the information of the mapping table maintained by each node keeps complete;

Step 5: On the base of the establishment and updating of the parallel mapping table in Step 4, the parallel control algorithm of the parallel transmission link is processed to eliminate the potential interference of the sender in two-hop range to the maximum extent;
Step 6: On the base of the establishment and updating of the parallel mapping table in Step 4, the parallel control algorithm of the parallel transmission link is processed to solve the data collision problem of multiple hidden terminals at the receiver;
Step 7: The original formulas (1) and (2) of the OPC method are used to determine the SNR conditions of the sender and receiver and decide whether to allow this parallel transmission;

$$\varepsilon + \sum_{i=1}^{k} I(Si \rightarrow R) \leq \frac{I(S \rightarrow R)}{10^{(\tau/10)}} \tag{1}$$

$$\varepsilon + I(S \rightarrow Rt) + \sum_{i \neq t}^{k} I(Si \rightarrow Rt) \leq \frac{I(St \rightarrow Rt)}{10^{(\tau/10)}} \tag{2}$$

In formula (1), I (S → R) represents the signal intensity value of the data communication S → R to be initiated this time, I (Si → R) represents the signal intensity value of the link Si to R, $\sum$ represents the total intensity of all interference signal intensity at the receiver R, k is the number of all neighbor nodes currently performing data communication around R, $\tau 1$ represents the threshold of the SINR, that is, the calculated minimum value of the receiver R the SINR must reach and ε indicates that the back-off signal intensity is used to compensate for errors due to environmental factors and uncontrollable factors from non-direct neighbor nodes.

In formula (2), I (St → Rt) represents the signal intensity value of the link that Rt is in among all the neighbor nodes that are currently communicating with the data around the sender *S*. Through formula (2), whether one of the signal intensity value of all the neighbor nodes that are currently communicating with the data around the sender *S* satisfies the SINR condition can be judged and the judgment of condition 2 traverses all the nodes in the set *G3* formed by all the algorithms Algorithm 1. I (Si → Rt) represents the signal intensity value of the link Si to Rt, Si is the sender node of the currently ongoing data communication other than the node *S*, k is the number of neighbor nodes currently communicating with each other around Rt, and ε is the back signal intensity used to compensate for the environmental factors due to the environmental factors, the left part of the formula (2) shows the combined intensity of all the interference signals formed by the interference signal of the sender S at the receiver Rt, $\tau 2$ represents the threshold of the SINR, i.e. the calculated minimum value of the receiver Rt the SINR must reach.

4 Experimental Results and Analysis

By collating and analyzing the experimental data of a series of simulation experiments, under the same conditions, the performance comparison of the basic CSMA protocol, OPC method and OPCLI-MAC protocol is presented by drawing the curve, and the performance improvement of OPCLI-MAC protocol in average terminal-to-terminal delay and average throughput capacity than OPC method and that of the parallel transmission to traditional single channel MAC protocol can both be verified, in which the basic CSMA protocol and OPC method are selected as comparisons.

4.1 Experimental Platforms

The hardware environment of the experiments: the machine is a PC, the processor is Intel (R) Core (TM) CPU 2.90 GHz and its memory is 4 GB.

The software platform of the experiments: the operating system is Linux, OPNET Modeler 14.5 network simulation software, OriginPro8 technology mapping and data analysis software.

4.2 Parameter Setting

All of the node deployment ranges of the simulation experiments are set to 1000 m * 1000 m, the rated transmission power coverage is set to 150 m, the number of re-transmission allowances is 4, the preamble length is set to 500 ms, the SINR thresholds $\tau 1$ and $\tau 2$ are set to 8 dB and 3 dB respectively, and the compensation signal intensity value ε is set to 0.5 dB and each experiment is set to 1 h, in which the node's location coordinates can be obtained in the OPNET Modeler network simulation wireless modeling module in order to simulate the actual deployment of nodes on the GPS devices.

4.3 Results and Analysis

A total of three experiments were conducted. First, under the different node sizes, the two performance index of the three protocols are compared and analyzed. As shown in Fig. 6, the average end-to-end delay of the three protocols is given under the condition of different network node deployment. There are three lines in Fig. 6, they show the changes of, from above to below, the basic CSMA protocol, OPC Method and OPCLI-MAC protocol. The three curves show an upward trend as a whole because the possibility of data transmission collision and data re-transmission between nodes in a wireless sensor network increases with the increase of node deployment, resulting in an increase in average end-to-end transmission delay.

Figure 7 shows the changes of average throughput of three types of protocols under different sizes of network node deployment. There are a total of 3 lines in Fig. 7, they are relatively, from above to below, OPCLI-MAC protocol, OPC Method and the basic CSMA protocol. The three curves show the trend of rising first and then decreasing as a whole because the initial increase in node deployment scale can increase channel utilization and throughput capacity, but when the node deployment scale is too large, the possibility of data transmission conflicts and data re-transmission between nodes in the wireless sensor network increases, resulting in a decrease in average throughput capacity. However, it can be seen that the OPCLI-MAC protocol gives the improvement measures and solutions to the problems of OPC method, which also is a parallel transmission protocol, which makes its MAC protocol performance, the average end-to-end delay, always better than the OPC method.

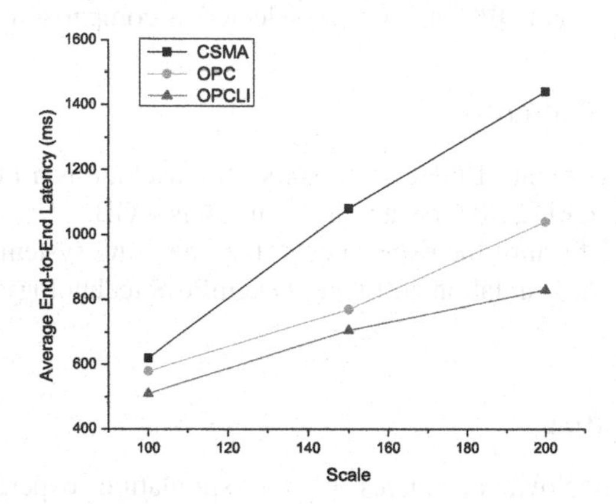

Fig. 7. Average end-to-end delay comparison under different node sizes.

There are a total of three lines in Fig. 8, respectively showing the changes of, the basic CSMA protocol, OPC method and OPCLI-MAC protocol. The three curves as a whole show an upward trend, because when the network load increases, the number of messages to be transmitted in the network within the same time period is also increasing, and the possibility of data transmission collision and data re-transmission between nodes in the network also increases, resulting in an increase in average end-to-end transmission delay. When the data transfer rate is small, the performances of the three protocols are

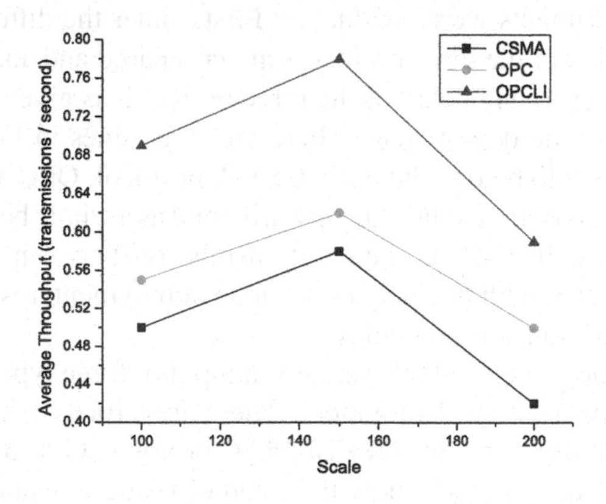

Fig. 8. Comparison of average throughput capacity under different node sizes.

very close. The average end-to-end delay of OPCLI-MAC protocol is slightly larger, but with the data transfer rate increases, the average end-to-end delay of basic CSMA protocol increases significantly.

Figure 9 shows the changes of average throughput capacity of the three types of protocols under different data transfer rates. There are a total of three lines in Fig. 9, respectively showing the changes of, OPCLI-MAC protocol, OPC method and the basic CSMA protocol. The three curves as a whole show an upward trend, but when the data transfer rate is greater than 40 packets/min they show a downward trend, because with the continuous increase of data transmission rate, the wireless sensor network channel has been more fully utilized, and the average throughput capacity also increases, but when the data transfer rate in the network is too large, the throughput capacity of the network will reach the bottleneck, and due to the increased data conflict and re-transmission, the throughput capacity decreases. Although the throughput capacity of the three protocols increases and then decreases as the data rate increases, and their performances are close when the data transfer rate is low, the parallel transmission edges of the OPCLI-MAC protocol and its improvement to OPC emerges as the data transfer rate increases.

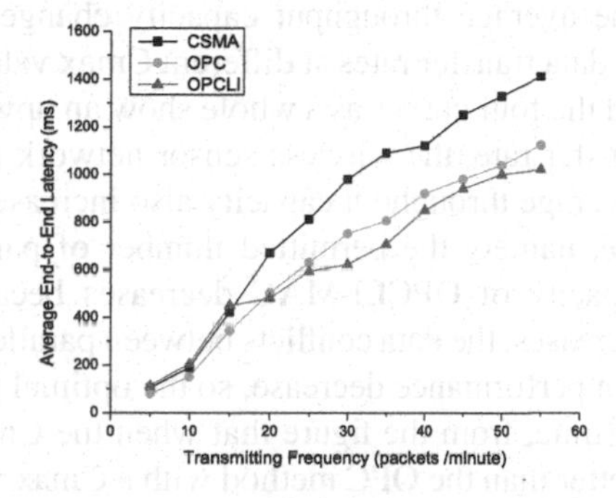

Fig. 9. Comparison of average end-to-end delay under different data transfer rates.

Finally, the two kinds of performances of the OPCLI-MAC protocol are observed by controlling the changes in Cmax value under different data transfer rates to determine the optimal permitted number of parallel transmission. Figure 10 shows the average end-to-end delay changes of the OPCLI-MAC protocol under different data transfer rates at different Cmax values. There are a total of four lines in Fig. 10, and the four curves as a whole show an upward trend because with the increase of network load, the number of messages to be transmitted in the network within the same time period is also increasing, and the possibility of data transmission collision and data re-transmission between nodes in the network also increases, resulting in an increase in average end-to-end transmission delay. Besides, it can be seen that the average end-to-end delay of the OPCLI-MAC protocol increases as the C_{max} value, i.e. the number of parallel transmission allowed, increases because with the number of parallel transmission increases, data transmission collision between parallel transmission is significantly increased, resulting in the decrease of performance, that is why the optimal Cmax value of OPCLI-MAC is 2. At the same time, from the figure that when the Cmax value is 3, OPCLI-MAC protocol is still better than the OPC method with a Cmax value of 2, and it can be seen that the improvement of OPCLI-MAC to OPC is apparent.

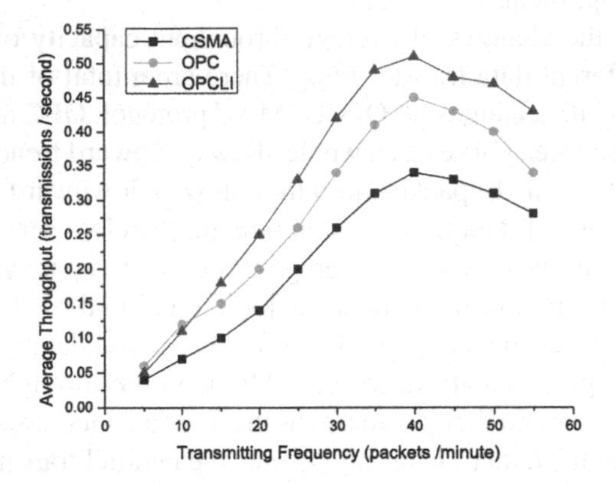

Fig. 10. Comparison of average throughput capacity under different data transfer rates.

Figure 11 shows the average throughput capacity changes of the OPCLI-MAC protocol under different data transfer rates at different Cmax values. There are a total of four lines in Fig. 10, and the four curves as a whole show an upward trend because with the increase of data transfer rate, the wireless sensor network channel has been more fully utilized, and the average throughput capacity also increases. In addition, with the increase of Cmax value, namely the permitted number of parallel transmission, the average throughput capacity of OPCLI-MAC decreases because as the number of parallel transmission increases, the data conflicts between parallel transmission increase significantly, resulting in performance decrease, so the optimal Cmax value of OPCLI-MAC is 2. At the same time, from the figure that when the Cmax value is 2, OPCLI-MAC protocol is still better than the OPC method with a Cmax value of 2, and it can be seen that the improvement of OPCLI-MAC to OPC is obvious (Fig. 12).

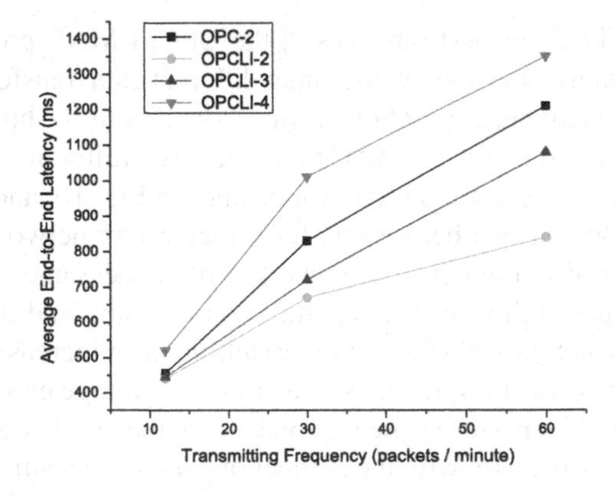

Fig. 11. Comparison of average end-to-end delay at different Cmax values.

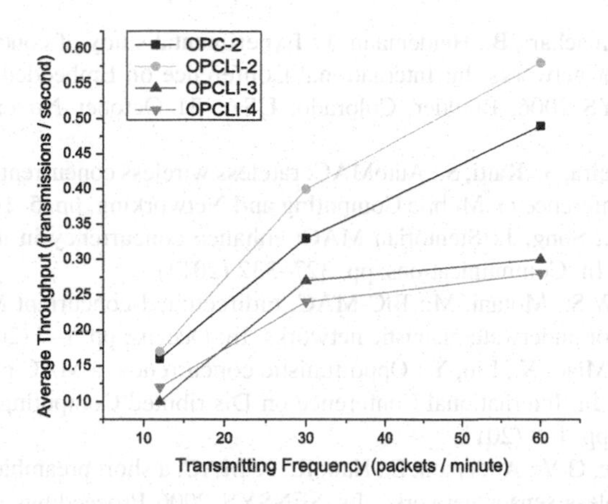

Fig. 12. Comparison of average throughput capacity at different Cmax values.

5 Conclusion

Through the analysis of the results of the whole simulation experiments, it can be seen that under the same condition, OPCLI-MAC protocol is superior to OPC method in terms of average end-to-end delay and average throughput capacity, and OPCLI-MAC has a greater advantage than traditional single-channel MAC protocol because of the enhanced opportunity of parallel transmission. The improvement of the OPCLI-MAC protocol may lead to a certain degree of extra computing costs, storage overhead and energy consumption, but the current sensor node's computing power, storage capacity and endurance are sufficient to support these requirements. In particular, the significant improvement in protocol performance brought about by OPCLI has been experimentally proven, and extra cost is negligible compared to which, or it can be said that the cost of such extra overhead is worth it. But in the course of future research, other methods can be thought of to increase the efficiency of protocol while reducing these extra costs.

Acknowledgement. This work was funded by the National Natural Science Foundation of China under Grant (No. 61772152 and No. 61502037), the Basic Research Project (No. JCKY2016206B001, JCKY2014206C002 and JCKY2017604C010), and the Technical Foundation Project (No. JSQB2017206C002).

References

1. Wei, Y., Heidemann, J.: Medium access control in wireless sensor networks. USC/ISI Technical report ISI-TR-580, vol. 51, pp. 961–994 (2004)
2. Sha, M., Xing, G., Zhou, G., Liu, S., Wang, X.: C-MAC: model-driven concurrent medium access control for wireless sensor networks. Proc. IEEE INFOCOM **11**, 1845–1853 (2009)
3. Guoyan, Y., Guoyin, Z.: Concurrent transmission MAC protocol for wireless sensor network based on nodes geographical location information. Comput. Sci. **39**, 105–108 (2012)

4. Son, D., Krishnamachari, B., Heidemann, J.: Experimental study of concurrent transmission in wireless sensor networks. In: International Conference on Embedded Networked Sensor Systems, SENSYS 2006, Boulder, Colorado, USA, 31 October–November, pp. 237–250 (2006)
5. Gudipati, A., Pereira, S., Katti, S.: AutoMAC: rateless wireless concurrent medium access. In: International Conference on Mobile Computing and Networking, pp. 5–16 (2012)
6. Shi, W., Liu, W., Song, J.: Stentorian MAC: enhance concurrency in underwater acoustic sensor networks. In: Communications, pp. 327–332 (2014)
7. Ng, H.H., Soh, W.S., Motani, M.: BiC-MAC: bidirectional-concurrent MAC protocol with packet bursting for underwater acoustic networks. In: Oceans, pp. 1–7 (2010)
8. Ma, Q., Liu, K., Miao, X., Liu, Y.: Opportunistic concurrency: a MAC protocol for wireless sensor networks. In: International Conference on Distributed Computing in Sensor Systems and Workshops, pp. 1–8 (2011)
9. Buettner, M., Yee, G.V., Anderson, E., Han, R.: X-MAC: a short preamble MAC protocol for duty-cycled wireless sensor networks. In: SENSYS 2006 Proceedings of the International Conference on Embedded Networked Sensor Systems, Boulder, Colorado, USA, pp. 307–320 (2006)

Multi-channel Parallel Negotiation MAC Protocol Based on Geographic Location Information

Jianfeng Wang[1], Hongbin Wang[2](✉), Haomin Zhan[3], Rouwen Dang[3], and Yang Bai[2]

[1] China Institute of Marine Technology & Economy, Beijing 100081, China
[2] College of Computer Science and Technology, Harbin Engineering University, Harbin 150001, China
wanghongbin@hrbeu.edu.cn
[3] Beijing General Institute of Electronic Engineering, Beijing 100854, China

Abstract. In the wireless sensor networks, the wireless resources are very limited. In the design of the MAC protocols, how to make full use of the limited channel resources to complete more data communications must be considered as much as possible. In this paper, in view of the shortcomings of the LPR-MAC protocol on conflict handing of multiple nodes in multiple channels, a multi-channel parallel negotiation MAC protocol based on geographic location information (LPRLI-MAC) is proposed, an improved neighbor table and a P-persistence algorithm are proposed. The node location information is added to the table. The P-persistence algorithm is carried out using the distance value to deal with the conflicting problem of multiple transmitters. Experiments show that the LPRLI-MAC protocol optimizes the calculation of the probability values in the P-persistence algorithm, reduces the probability of data collisions on a single channel at a certain time, optimizes the parallel negotiation strategy of multiple transmitters on multiple channels, reduces the probability of collisions between multiple parallel transmissions and improves the success rate of transmissions.

Keywords: Wireless sensor networks · MAC protocol · Parallel negotiation Geographic location information · P-persistence algorithm

1 Relevant Work

With the rapid development of wireless sensor networks, some classical multi-channel MAC protocols have encountered problems such as control signal bottleneck. How to improve the efficiency of parallel transmission on multiple channels and reduce the probability of collision among links have become a research focus.

In recent years, with the development of hardware technology and the reduction of price, many kinds of multi-channel communication chips have appeared. Researchers have also developed various types of multi-channel MAC protocols for different fields and there are more mature multi-channel MAC protocols such as MMSN, MMAC, TMMAC, TMCP, TFMAC, MC-LMAC, McMAC, TSMP and SSCH. These MAC

Q. Zhou et al. (Eds.): ICPCSEE 2018, CCIS 901, pp. 589–598, 2018.
https://doi.org/10.1007/978-981-13-2203-7_47

protocols focus on different network features according to different application scenarios because of their application-oriented features, such as channel access mode and time slicing mode. Reference [1] has described and analyzed the existing multi-channel MAC protocols in detail.

The multi-channel MAC protocol using multiple channels for data communication with the concept of "parallel transmission" itself. However, since there are multiple channels and the available space is larger, the performance of the original multi-channel MAC protocol can be greatly improved through designing and improving the channel utilization mode by introducing time slicing, distribution and different communication strategies. At this point, the introduction of the concept of parallel transmission is of practical feasibility. The so-called "parallel transmission" refers that there are two or more data links in the data communications at the same time. The basic principle of parallel transmission is that the data communication to be initiated can not interfere with the ongoing data communication [2].

There is no unified standard for multi-channel MAC protocol classification. Generally speaking, there are several ways to classify MAC protocols. They can be divided into fixed allocation channel, grading frequency hopping channel and parallel channel negotiation channel according to the channel allocation; CSMA and TDMA according to the channel access mode; static allocation, dynamic allocation, competitive time-slice and time-free slice according to the time slicing mode; time synchronization in the whole net, time synchronization in pairs and no time synchronization based on different time synchronization methods [3–6].

A Low-power Parallel Rendezvous (LPR) MAC protocol is proposed in Ref. [7]. LPR-MAC utilizes methods such as time synchronization in the whole network, multiple time slices, random allocation of time-slices and pseudo-random sequences to reduce competition in communication and energy consumption. However, no specific solutions are provided for the issue of multi-channel conflicts in Ref. [4] and this article addresses the above issue.

2 Problems Analysis and Solutions

The LPRLI-MAC protocol analyzes the conflicts among multiple senders on multiple channels and presents the corresponding solutions.

2.1 Analysis on Problems Existing in Multi-channel Parallel Negotiation

Multi-channel MAC protocol still performs in the way of a single-channel MAC protocol on a specific single channel, that is, on one channel, and only one data link is in data transmission during each time segment. Otherwise, congestion and data collisions can occur.

First of all, the problem of channel collision that multiple senders initiate communication to multiple receivers respectively is analyzed. As shown in Fig. 1, it is assumed that both node *R1* and node *R2* randomly select the time-slice 4 as their own receiving cycle. The current time-slice number is 4, and node *R1* and node *R2* respectively switch to channel 0 according to their own pseudo-random frequency hopping sequences and wait for receiving data when they are powered on. At this point, when node *S1* sends data to node *R1* and node *S2* needs to send data to node *R2*, the two nodes will generate data collision on channel 0, which is a channel collision in which multiple senders respectively initiate communication to multiple receive ends.

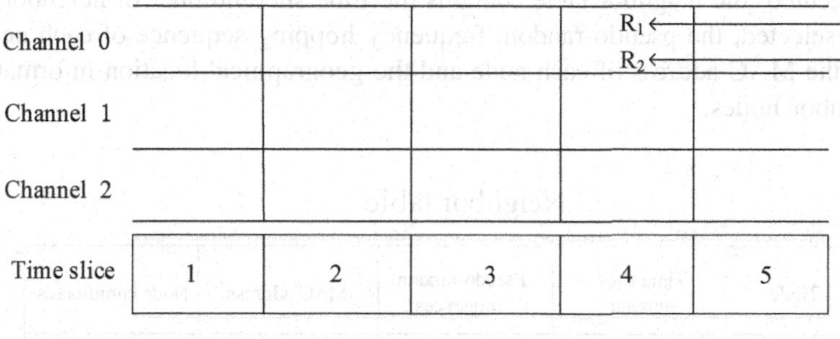

Fig. 1. Channel collision between multiple senders and multiple receivers.

Then, the problem of channel collision that multiple senders initiate communication to one receiver at the same time is analyzed. As shown in Fig. 2, assuming that node *R* randomly selects time-slice 4 as its own receiving cycle and the current time-slice number is 4, node *R* switches to channel 0 according to its own pseudo-random frequency hopping sequence and waits for receiving data when it starts up. In this case, when node *S1* sends data to node *R* and node *S2* also sends data to node *R*, the two will generate data collision on channel 0, that is, the problem of channel collision that multiple senders simultaneously initiate communication to one receiver.

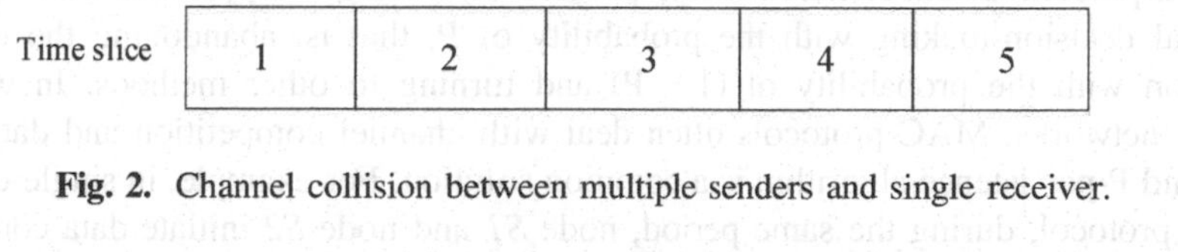

Fig. 2. Channel collision between multiple senders and single receiver.

2.2 Solution to Multi-channel Parallel Negotiation Conflict

Communication congestion and data collision are one of the three typical problems of multi-channel parallel negotiation MAC protocol. The aforementioned channel collision where multiple senders initiate communication to multiple receivers respectively and that in which multiple senders initiate communication to one receiver at the same time can be attributed to the conflict of data communication in a single channel, and it can be solved with a unified solution.

First, the improvement of the neighbor table. As shown in Fig. 3, add the geographical location information of the nodes to the neighbor table to calculate the distance between each node in the control algorithm, thus an effective back-off strategy can be executed; the neighbor table contains the time slice number of neighbor nodes randomly selected, the pseudo-random frequency hopping sequence of each neighbor node and the MAC address of each node and the geographical location information of their neighbor nodes.

Neighbor table

Node	Time slice number	Pseudo-random sequences	MAC address	Node coordinates
R_1	2	R1 sequence	R_1-mac	$R_1(X_1, Y_1)$
R_2	4	R_2 sequence	R_2.mac	$R_2(X_2, Y_2)$
R_3	2	R_3 sequence	R_3.mac	$R_3(X_3, Y_3)$
...	...	...	...	
...	...	...	...	
R_n	3	R_n sequence	R_n-mac	$R_n(X_n, Y_n)$

LPRLI-MAC

Fig. 3. Neighbor table.

Second, the improvement of the P-persistence algorithm. The idea of P-persistence algorithm is used in many cases. Simply speaking, P-persistence algorithm refers that when a problem is encountered in the process of solving the problem, adhere to the original decision-making with the probability of P, that is, abandoning the original decision with the probability of (1 − P) and turning to other methods. In wireless sensor networks, MAC protocols often deal with channel competition and data collision and P-persistence algorithm is a common solution. For example, in single-channel MAC protocol, during the same period, node *S1* and node *S2* initiate data communication at the same time and the P-persistence algorithm can be used here. Both node *S1* and node *S2* adhere to the original data transmission with the probability of P, that is, node *S1* and node *S2* both back off with the probability of (1 − P).

In LPRLI-MAC, the distance between each pair of nodes can be calculated by using the coordinates of each neighbor node in the neighbor table, and this distance information can be used to improve the P-persistence algorithm. Assuming that the data collision mentioned in (1) occurs during the execution of the LPRLI-MAC protocol, the sender *S1* obtains the distance *D1* between itself and the receiver *R1* by calculation, and the sender *S1* insists to continue sending data to the node *R1* with the probability *P1*, and the computational method of *P1* is Eq. (1).

$$Pi = \varepsilon * (\frac{Ri}{Di}) * 100\% \tag{1}$$

In this equation, *Pi* refers to the probability that one sender *Si* insists on initiating this data communication, *Di* refers to the distance between the sender *Si* and the target node, and *Ri* refers to the rated sending power radius of the sender *Si*. In particular, *Di* must be less than *Ri*, otherwise, it is impossible to carry on data communication. ε is a correction parameter. It is used to reduce the value of *Pi* according to the number of time-slices and channels when LPRLI-MAC is implemented. Otherwise, if each sender persists to send data with a larger probability, not only the data conflict problem can not be solved, but also it is meaningless to improve the P-persistence algorithm. For example, the network node deployment size is 20 nodes and each super-frame contains 5 time slices, although the node selection of time-slice is random, it is reasonable to speculate that there are 4 nodes in each time-slice, that is, 4 nodes choose the same time-slice as their own receiving cycle, then the maximum collision probability is when 4 nodes are at the point to reach the time-slice, they switch to the same channel just in accordance with their respective pseudo-random frequency hopping sequences, then there are four receiving nodes in the same channel. If, in this case, the problem in Sect. 2.1 occurs, then it is more reasonable for each sender to insist the transmission with the probability of 25%.

The size of the probability *Pi* is decided according to the geographical location information or the distance between the nodes because if the distance between the two nodes is closer, the data signal intensity sent by the transmitting end decays less when it reaches the receiving end, so that the data transmission between two nodes is more likely to be successful and the improved P-persistence algorithm is more accordant with the actual situation of each node in the implementation of MAC protocol, that is, the closer the distance, the greater the probability that the sender insists to send. This will maximize the success rate of LPRLI-MAC protocol data transmission, thereby enhancing the performance of the LPRLI-MAC protocol.

3 The Overall Communication Strategy Description of LPRLI-MAC Protocol

The solution to the above problems are integrated and the overall communication process of the LPRLI-MAC protocol is described. The specific steps are as Fig. 4.

Step one: Establish the network and initialize the relevant parameters, and each node establishes their own neighbor tables and randomly selects a time-slice as their own fixed receiving cycle and generates their own pseudo-random frequency hopping sequences according to the same pseudo-random sequence generator with its own mac address, that is, channel switching sequence. All nodes keep sending their neighbor table information to other neighboring nodes through broadcast. Through the network,

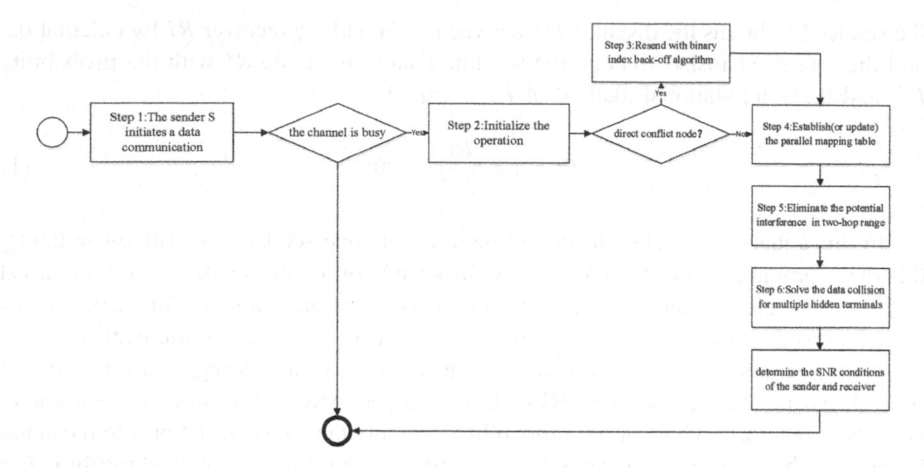

Fig. 4. The process of LPRLI-MAC.

an initial period of maintaining and updating neighbor tables is established. The neighbor table of each node contains the time-slice selection of neighboring nodes, random frequency hopping sequence and geographic location information and other related information;

Step two: Establish and update the neighbor table;

Step three: Assuming that node *S* wants to initiate data communication with node *R*, it knows that the receiving cycle time-slice number *Tr*, the pseudo-random frequency hopping sequence and the position coordinates of the target by querying the neighbor table, it calculates the channel number the next receiving cycle node *R* is in through the frequency hopping sequence of *R* and it calculates the distance between two nodes by the coordinate of its own and node *R*;

Step four: In the time-slice *Tr*, boot up the sender S and broadcast a beacon message, which contains the channel label of the target node of the data communication to be initiated this time in the time-slice *Tr*. At the same time, intercept whether there are any other sending end (for example, node *S1*) is the same as its own target node in the channel; whether there are any other target node *R2* of the sending end (for example, node *S2*), though it is different from its own target node *R*, the pseudo-random frequency hopping sequences of the node R2 and the node R exactly overlap in the time slice *Tr*, that is, when the node R2 and the node R are in the same time-slice Tr, they are in the same channel;

Step five: If neither of the two cases exist, switch the node *S* directly to the channel where the target node is located in the normal multi-channel parallel transmission.

Step six: If one of the two cases exists, calculate the probability *P* of persisting sending data according to the formula (1). If the calculated probability is greater than the given threshold, the sender insists on sending data this time and the node *S* directly switches to the channel where the target node *R* is located for normal multi-channel parallel transmission. In this case, it may collide with other transmitters on the channel. If the result is to abandon this data transmission, avoid a random time according to a random function, and then initiate the data transmission again;

Step seven: After the sender S completes this data communication, it switches back to its own pseudo-random frequency hopping sequence.

4 Experimental Results and Analysis

The experimental data obtained from a series of simulation experiments are arranged and analyzed. Under the same conditions, performance of McMAC protocol, LPR-MAC protocol and LPRLI-MAC protocol are displayed by drawing curves, and the performance improvements of LPRLI-MAC protocol compared to LPR-MAC protocol in terms of average end-to-end delay and average throughput can be verified, and the performance of parallel negotiation compared to traditional multi-channel MAC protocol performance can also be verified, in which McMAC protocol and LPR-MAC protocol are selected as the comparison object.

4.1 Parameter Setting

All of the node deployment ranges of the simulation experiments are set to 500 m * 500 m, the number of re-transmission allowances is 4, 2.4 GHz channels are used and the channel capacity is set to 2 Mbps, the number of available channels is set to 3. Each super-frame setting includes five time slices and the size of each time-slice is set to 5.5 ms. The correction parameter ε is set is 0.25, the probability threshold is set to 0.5 and the time for each experiment is set to 1 h. In addition, the node's location coordinates can be obtained in the OPNET Modeler network simulation wireless modeling module in order to simulate the actual deployment of nodes on the GPS devices.

4.2 Results and Analysis

First of all, under different network load conditions, two performance indicators of the three protocols are comparatively analyzed. The average throughput changes of the three protocols under different network load conditions are shown in Fig. 5. There are three lines in the figure, which are the changes of the three protocols of McMAC, LPR-MAC and LPRLI-MAC respectively. It can be seen that the LPRLI-MAC protocol has no obvious advantage before the network load is 2.5 Mbps. On the contrary, the McMAC protocol is slightly better because a series of algorithm control of the LPRLI-MAC protocol leads to the performance of the LPRLI-MAC protocol slightly reduced when the network load exceeds 2.5 Mbps. The network has an overall upward trend of the three curves before 4 Mbps because as the network load continues the channel in wireless sensor network is more fully utilized and the average throughput also increases. However, when the network load exceeds 4 Mbps and keeps increasing, the throughput of all three protocols drops because the load capacity of the wireless sensor network is limited, and when the network load is increasing, the overall performance of the MAC protocol will be degraded. Overall, the LPRLI-MAC protocol is significantly better than the LPR-MAC protocol over a range of network loads, which also shows that LPRLI-MAC is effective for the improvement of the original protocol.

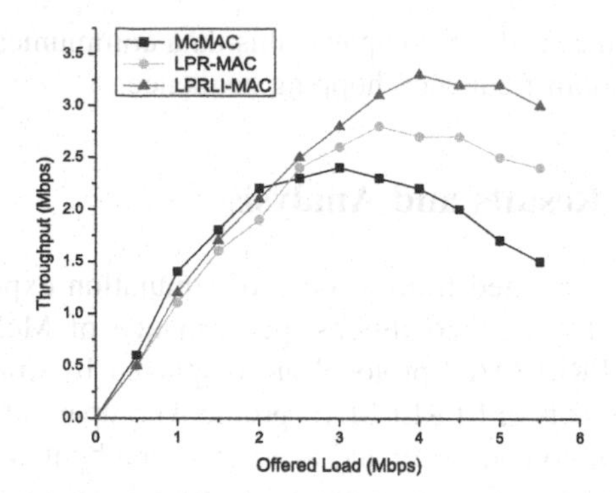

Fig. 5. Comparison of average throughput under different network loads.

Figure 6 shows the average end-to-end delay of the three protocols under different network load conditions. It shows the average end-to-end delay variation of the three protocols under different network load conditions. There are three lines in the figure, which are the changes of the three protocols of McMAC, LPR-MAC and LPRLI-MAC respectively. The three curves all show an upward trend because as the data transmission rate continues to increase, the number of data packets in the network in the same time period increases, data transmission conflicts and the possibilities of data retransmission between nodes in the wireless sensor network also increase, resulting in an increase in average end-to-end transmission delay. When the data transmission rate is small, the performance of the three protocols is very similar, and the average end-to-end delay of the LPRLI-MAC protocol is still slightly larger. However, as the data transmission rate increases, the average end-to-end delay of the McMAC protocol increases greatly, and the LPRLI-MAC protocol in this performance is better than the LPR-MAC protocol.

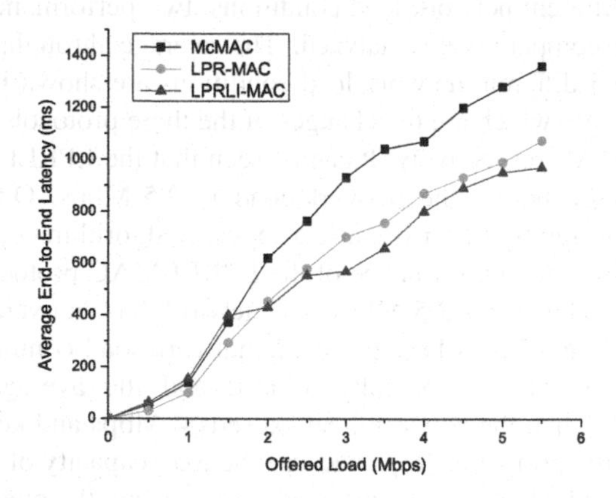

Fig. 6. Comparison of average end-to-end delay under different network loads.

Secondly, under different node sizes, two network performance indexes of the three protocols are compared and analyzed. Figure 7 shows the comparison of the average throughput of the three protocols at different node sizes. Respectively it shows the average throughput of the three protocols under different network node deployment scales. There are totally three lines in the figure, which are LPRLI-MAC protocol, LPR-MAC protocol and McMAC protocol from top to bottom. The variation trend of throughput of three kinds of protocols all firstly increase and then decrease because when the deployment size of nodes increase to a certain extent, the channel utilization rate is improved. However, as the node deployment increases, the possibility of data transmission conflicts and data re-transmission among nodes in a wireless sensor network also increases, resulting in a decrease in average throughput. It is obvious that the LPRLI-MAC protocol has given improvements and solutions to the problems of the OPC method and has always performed better than the LPR-MAC protocol in terms of the average throughput of the MAC protocol.

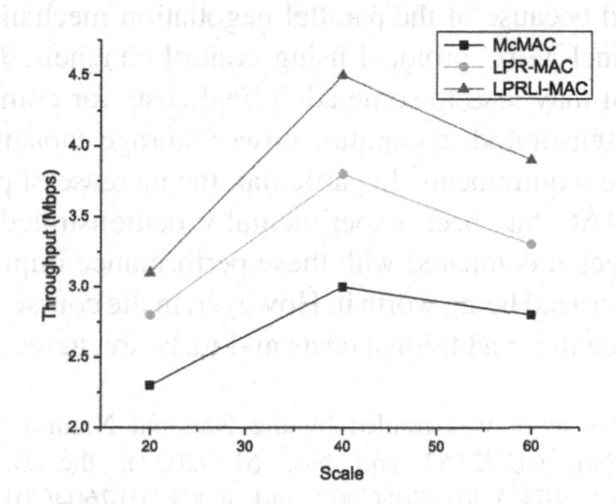

Fig. 7. Comparison of average throughput at different node sizes.

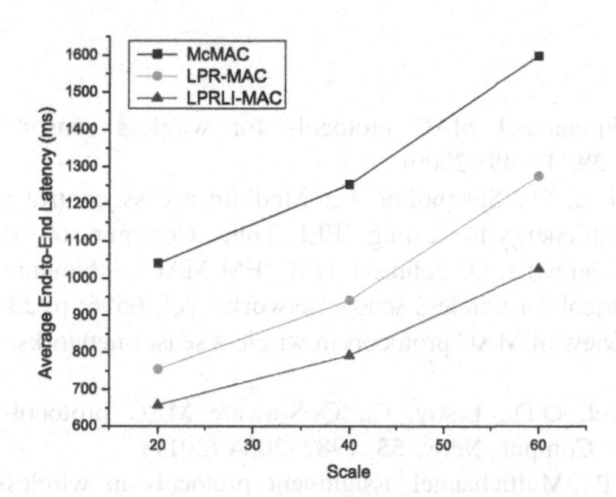

Fig. 8. Comparison of average end-to-end delay for different node sizes.

Figure 8 shows the average end-to-end delay variation of the three protocols under different network node deployment scales. There are totally three lines in the figure, which are changes of McMAC protocol, LPR-MAC protocol and LPRLI-MAC protocol. The three curves shows an upward trend as a whole, because with the increasing deployment of nodes, the possibility of data transmission conflicts and data re-transmission among nodes in a wireless sensor network also increases, as a result, the average end-to-end transmission delay increases. As the node size increases, the performance of the LPRLI-MAC protocol is always better than that of the LPR-MAC protocol.

5 Conclusion

Through the analysis of results of the overall simulation experiments, LPRLI-MAC protocol outperforms LPR-MAC protocol in terms of average end-to-end delay and average throughput under the same conditions. And the performance of LPRLI-MAC is significantly improved because of the parallel negotiation mechanism compared to the traditional multi-channel MAC protocol using control channels. Improvements of the LPRLI-MAC protocol may lead to some additional costs for computation and storage space, but the current sensor node computing power, storage capacity and battery life are sufficient to meet these requirements. In particular, the increase of protocol performance brought by LPRLI-MAC has been experimentally demonstrated, and the additional overhead can be neglected compared with these performance improvements, or we can say the cost of such overhead being worth it. However, in the course offuture research, we can find ways to reduce these additional costs and make the agreement more efficient.

Acknowledgement. This work was funded by the National Natural Science Foundation of China under Grant (No. 61772152 and No. 61502037), the Basic Research Project (No. JCKY2016206B001, JCKY2014206C002 and JCKY2017604C010), and the Technical Foundation Project (No. JSQB2017206C002).

References

1. Chen, D.Y.: Multi-channel MAC protocols for wireless sensor networks: a survey. J. Shandong Univ. **39**, 41–49 (2009)
2. Iannello, F., Simeone, O., Spagnolini, U.: Medium access control protocols for wireless sensor networks with energy harvesting. IEEE Trans. Commun. **60**, 1381–1389 (2012)
3. Tang, L., Sun, Y., Gurewitz, O., Johnson, D.B.: EM-MAC: a dynamic multichannel energy-efficient MAC protocol for wireless sensor networks, vol. 6696, p. 23 (2011)
4. Zheng, G.Q.: Overview of MAC protocols in wireless sensor networks. J. Softw. **19**, 389–403 (2008)
5. Yigitel, M.A., Incel, O.D., Ersoy, C.: QoS-aware MAC protocols for wireless sensor networks: a survey. Comput. Netw. **55**, 1982–2004 (2011)
6. Soua, R., Minet, P.: Multichannel assignment protocols in wireless sensor networks: a comprehensive survey. Pervasive Mob. Comput. **16**, 2–21 (2015)
7. Qin, S.H., Chen, D.Y., Chen, G.Y.: LPR-MAC: a low power parallel Rendezvous MAC protocol. Comput. Eng. Sci. **32**, 6–8 (2010)

PBSVis: A Visual System for Studying Behavior Patterns of Pseudo Base Stations

Haocheng Zhang, Xiang Tang, Chenglu Li, Yiming Bian, Xiaoju Dong(✉), and Xin Fan

BASICS, Department of Computer Science and Engineering, Shanghai Jiao Tong University, Shanghai, China
xjdong@sjtu.edu.cn

Abstract. Pseudo base stations do great harm to people's daily lives through cheating on mobile terminals to reside in the pseudo base station and forcing users to receive spam messages. However, conducting track detection of pseudo base stations is a challenging task because there exists not only a requirement for an aggregation of millions of spam messages, but also a requirement for effective methods to express behavior patterns and trajectory characteristics of pseudo base stations. In this paper, we put forward a visualization approach for analyzing pseudo base stations. Our approach leverages valid clustering algorithm to extract distinct pseudo base stations from millions of spam messages and provides an integrated visualization system to analysis behavior patterns and trajectory characteristics. By means of case studies of real-world data, the practicability and effectiveness of the method are demonstrated.

Keywords: Visualization · Pseudo base station · Trajectory

1 Introduction

With the rapid development of mobile Internet, almost everyone is engaging into the mobile phone network where they can communicate with each other and obtain useful messages. However, the emergence of pseudo base station has seriously influenced the experience of using mobile phones because of spam messages. Therefore, it is of great significance to conduct track detection of pseudo base stations to analysis behavior patterns and trajectory characteristics and increase sanctions against criminals.

By the identification and localization algorithm [1, 2], we can collect information such as sending phone number, locations of sending spam messages, content of spam messages and sending time. But it is difficult to determine spam messages' corresponding pseudo base stations because the same spam messages may come from different pseudo base stations and the same pseudo base station may send different messages. Therefore, there exists a request to develop an effective algorithm for extracting distinct pseudo base stations from spam messages. If we can determine spam messages' corresponding pseudo base stations, our analysis objects will become distinct pseudo base stations other than millions of spam messages, which contributes to the analyzation of behavior patterns and executing specific managements. Meanwhile, we

Q. Zhou et al. (Eds.): ICPCSEE 2018, CCIS 901, pp. 599–610, 2018.
https://doi.org/10.1007/978-981-13-2203-7_48

can be familiar with the amount and type of spam messages in different areas. Moreover, by comparing behavior patterns of different pseudo base stations, we can determine the relationship between these pseudo base stations, which is good for the joint governance.

However, given the structural complexity and rich information of spam messages, most current pseudo-base-station analytics researches try to create a new communication protocol between base station and mobile terminals which is very difficult to implement. Although there exist some approaches to detect spam messages in different fields [3–5], few of them provides an integrated visualization with the exploration and analysis ability. Therefore, it is difficult to extract behavior patterns of pseudo base stations, which is of great importance to obtain the regularity of criminals. Moreover, it is hard to identify and verify long time-span trajectory patterns of pseudo base stations by current analytics approaches. It is difficult for users to grasp the pseudo base station's daily activities intuitively and to understand the reason why they are hard to capture. Such functionalities are not supported by existing approaches.

In our work, a visual analytics method is proposed for extracting behavior patterns and trajectory characteristics of pseudo base stations. We put forward a valid clustering algorithm to extract distinct pseudo base stations from huge amounts of spam messages and then analyze the specific pseudo base station for more details. There exists a significant challenge for representing and analyzing behavior patterns and trajectory characteristics of pseudo base stations. Our method utilizes the pie matrix visualization to display the long time-span trajectory patterns and the dynamic trajectory visualization to display the detailed activities. We have exploited a adequately operational system and applied it for analyzation of the real-world spam messages. In addition, we demonstrate the effectiveness and practicability of the approach by means of case studies.

2 Related Work

2.1 Pseudo Base Station Working Principle

"Pseudo base station" is illegal radio communications equipment, which disguises as a mobile network and send RF signal to all mobile terminals within the range of the pseudo base station network for group messaging operation. Because it uses the GSM 900 standard and its transmitted power is between 40 to 43 DBM, it sends messages at a high speed and supports a variety of sending strategies. A pseudo base station is mainly composed of a mainframe, an operating terminal (laptops installing Linux system) and an antenna, which has small volume and good concealment and liquidity. "Pseudo base station" has two ways of placement. The first way is fixed, which places the mainframe at areas of good liquidity such as hotels and places the antenna nearby the window. The other way is flowing, which is able to be deployed and leave quickly and is hard to locate because the pseudo base station is placed in a van, an electric car or pull rod box.

2.2 Pseudo Base Station Positioning Algorithm

There are some localization algorithms to locate the position of the pseudo base station according to the working principle of pseudo base station [6–8]. The signal of GSM base stations is divided into 900 MHz and 1800 MHz and the maximum normal communication radius is up to 3.5 km. The localization algorithm has the ability of rapid real-time positioning by means of the relationship equation between signal attenuation and distance, which measures the mobile test signal and distance and fit calculation results with the least square method on the basis of triangulation method to rule out the influence on signal attenuation caused by buildings and terrain.

2.3 Pseudo Base Station Visualization

It is of great value to visualize the activities of pseudo base stations which makes the analyzation intuitive and efficient. However, there are few existing approaches for the analyzation of pseudo base stations because much attention is pay to creating a new communication protocol between base station and mobile terminals which does no help to the current GSM network. Huawei LTE NASTAR analysis system provides a fast passage for screening network incurable diseases in the era of big data. In the routine KPI health examination of the entire network, NASTAR is able to visualize the representations of the indicators in different geographical distributions in the condition of switch failure times. As a result, with the help of NASTAR, analysts can filter out the most possible area that has pseudo base stations and then go for field trips.

However, the visualization system should have the ability of Pseudo base station trajectory reconstruction and analysis of the behavior patterns while little work has been done for extract the trajectory and behavior patterns of pseudo base stations. The system developed by Qihoo 360 displays the detailed trajectory effectively, nevertheless, there is a lack of way to find the long time-span trajectory patterns. Li and his partners mark the positions of pseudo base station directly on the map, which makes it difficult to observe the behavior patterns in the case of large amount of data [9].

Although there exist a few researches in trajectory visualization, almost none of them could be utilized to visualize behaviors of pseudo base stations directly [10–13]. Tominski and his partners use stacked 3D trajectory bands to show the trajectories in a spatial-temporal context [14]. Such methods could hardly be adapted to visualize trajectory patterns of pseudo base stations since illegal stations might change their routes after a period of time and then stacked 3D trajectory bands are overlapped. Scheepens and his partners present a method to explore trajectory attributes using density maps [15], which could display the trajectory of one pseudo base station clearly while it would be a mess in the situation of multiple stations.

3 Data and Task Abstraction

The selected data and task abstraction is introduced in this section for a better understanding of the problem domain of visual speech analysis.

3.1 Spam Message Data

With the localization algorithm mentioned in Sect. 2.2, we can collect some valuable information of spam messages such as its source location, sending time, sending phone number and content. We can transform data as the format in Table 1.

Table 1. Data format.

Field name	Field meaning
phone	Disguised sending phone number
content	Specific text messages
md5	MD5 of text messages
recitime	Received timestamp
conntime	Connect timestamp
lng	Approximate longitude
lat	Approximate latitude

According to the survey, a pseudo base station sends out hundreds of spam messages per day on average. With the information above, we cannot extract behavior patterns and trajectory characteristics of pseudo base station because the corresponding relationship between spam messages and pseudo base stations is not available. Therefore, we put forward a valid clustering algorithm in Sect. 5.1 to extract distinct pseudo base stations from millions of spam messages, as a result of which the data such like the movement and activities of a specific pseudo base station can be obtained.

3.2 Task Abstraction

The very first task of pseudo base station analysis is finding the corresponding relationship between spam messages and pseudo base stations. Afterwards, general tasks of pseudo base station analysis involve extracting behavior patterns and trajectory characteristics. Another task of pseudo base station analysis is telling types of spam messages in different districts for the aim of the customized governance. We derive a task list for pseudo base station analysis of spam message data using a method which combines tasks of two research domains:

- **Pseudo base station extraction:** What part of spam messages come from the same pseudo base station? Which pseudo base station sends most spam messages? What is the corresponding relationship between spam messages and pseudo base stations?
- **Behavior patterns and trajectory characteristics of pseudo base station:** What are the regular activity time and area of a specific pseudo base station? What kinds of spam messages does a specific pseudo base station often send? What is the regular route of a specific pseudo base station? Where does a specific pseudo base station stop for a rest? What are the similarities and differences between pseudo base stations in behavior patterns and trajectory characteristics?

- **Regional situation:** Which districts suffer most from pseudo base stations? What are main kinds of spam messages does a specific district suffer from? How do law enforcement agencies take actions according to different districts?

Moreover, conducting a comprehensive pseudo base station analysis of spam messages requires not only exploring the above aspects in respective research fields, but also summarizing the activity rule of all pseudo base stations. For example, it is of great importance to know peaks and valleys of spam messages regardless of types of different pseudo base stations.

4 System Overview

As shown in Fig. 1, the visual analysis proposed in the paper is consisted of three modules: an extractor of pseudo base stations, a classifier of spam messages and a visualization system. The extractor of pseudo base stations uses a valid clustering algorithm to extract distinct pseudo base stations from millions of spam messages, which is described in Sect. 5.1. The classifier of spam messages utilizes the natural language processing technology to divide spam messages into different categories, which is described in Sect. 5.2. The visualization system which includes the map view, the behavior pattern view, the timeline view and the distribution view is described in detail in Sect. 5.3. With the data generated by the extractor and the classifier, the visualization system provides integrated analysis tools to make it easy for law enforcement officials to be aware of the overall situation and catch criminals as soon as possible.

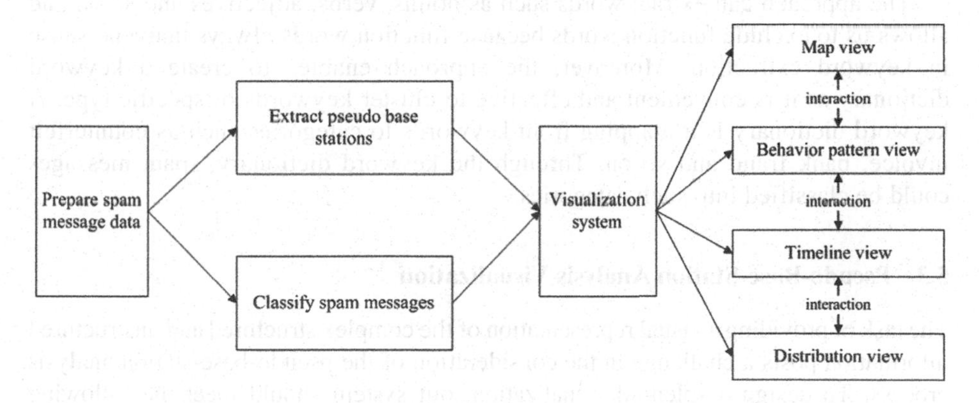

Fig. 1. System flow chart.

5 System Analysis

5.1 Pseudo Base Station Extraction

The data format used in pseudo base station extraction is described in Sect. 3.1. A pseudo base station is defined as a set of phone numbers which is used to send spam messages in turn. Therefore, different phone numbers should be extracted and classified to distinguish different pseudo base stations.

However, it is difficult to tell if two source phone numbers are from the same pseudo base station because their different sending time and area. For example, when one phone number is sending spam messages, another phone number has no activity. Then we put forward a valid clustering algorithm to address this question.

We define a variable to represent the active area of a specific phone number, which might be none if this phone number does not have any movement. If two phone numbers are consistent at all time periods when comparing their active areas, we should put them into a set which represents a pseudo base station. If at least one of the two compared phone numbers does not have any movement in the period, we continue our comparison because there is no conflict. However, if both of the two compared phone numbers have movements in the period, we should judge their consistency. Once a conflict happens in any period of time, we demonstrate that they are from different pseudo base stations. Otherwise they are considered from the same station.

5.2 Spam Message Classification

We adopt a Chinese word segmentation approach based natural language processing technology to analyze the content of the spam message, which is able to extract the key meaningful words from spam messages. Different types of spam messages have different keywords. For example, "young married woman" and "flight attendant" are marked as eroticism, while "boom" and "traders" are marked as stock.

The approach can extract words such as nouns, verbs, adjectives and so on and allows us to exclude function words because function words always make no sense in keyword extraction. Moreover, the approach enables to create a keyword dictionary as it is convenient and effective to cluster keywords of specific type. A keyword dictionary is a mapping from keywords to categories such as counterfeit invoice, bank fraud and so on. Through the keyword dictionary, spam messages could be classified into such categories.

5.3 Pseudo-Base-Station Analysis Visualization

The task of providing a visual representation of the complex structured and unstructured information posts a challenge in the consideration of the pseudo-base-station analysis process. To design a splendid visualization, out system should meet the following requirements which are based on the analytic tasks in Sect. 3.2:

- **Exhibition of distinct pseudo base stations from spam messages:** The visualization should provide an exhibition of distinct pseudo base stations which some spam messages belong to. Moreover, the visualization should exhibit characteristics of pseudo base stations and tell the difference between different pseudo base stations.
- **Exhibition of behavior pattern and dynamic trajectory:** It is important to visually display behavior patterns of pseudo base stations because users can easily find the spam message type and the discovery area of the specific pseudo base station on the

specific time. The exhibition of dynamic trajectory provides an animation display of pseudo-base-station movement which is much intuitive and enables users to make prediction.

- **Exhibition of regional distribution:** The visualization should not only provide users with the characteristics of pseudo base stations, but also the characteristics of regional distribution, which enables law enforcement agencies to take reasonable actions.

5.3.1 Visualization Design and Interface

Figure 2 indicates the pseudo-base-station analysis visualization based on the requirements above. Our visualization system uses a whole-and-part framework to settle structural complexity. Pseudo-base-station visualization is the central part that coordinates the map view, the behavior pattern view, the timeline view and the distribution view. It is encoded in coherent visual sense, but at different detail levels. Specifically, the visualization employs overall behavior pattern plus specific dynamic trajectory to describe the pseudo base station because overall characteristics are displayed by behavior pattern and specific space-time actions are displayed by dynamic trajectory.

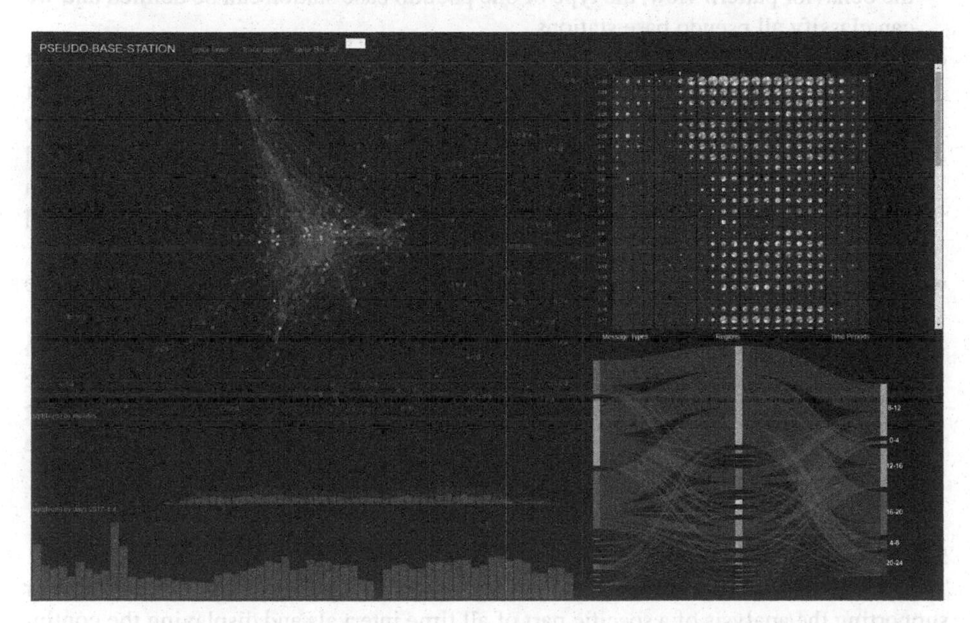

Fig. 2. The visual interface. (a) The map view exhibits the behavior of one pseudo base station. (b) The behavior pattern view shows the behavior pattern of one pseudo base station. (c) The timeline view indicates the spam message distribution of all time intervals. (d) The distribution view indicates the relationship between message type and distribution area. (Color figure online)

Specific visual encodings are presented below:

- **Map view:** The map view indicates overall actions and movement of pseudo base stations (Fig. 2(a)). The location of one circle indicates where the pseudo base station sent a spam message. The color of one circle indicates the type of the spam message.

The gray line indicates the movement of the pseudo base station between two actions. From most of the circle locations, the overall active area of this pseudo base station can be indicated. From most of the circle color, the message characteristic of this pseudo base station can be known. Moreover, to help users identify the process and detail of movement and actions, we adopt the dynamic trajectory which indicates every movement and action of the pseudo base station by continuous animation. As the result of dynamic trajectory, we can not only summarize the rules of pseudo base stations intuitively and easily, but also can we predict next steps of pseudo base stations.

- **Behavior pattern view:** The behavior pattern view shows the overall space-and-time action characteristic of one pseudo base station throughout all time (Fig. 2(b)). One row in the pattern matrix represents an active day for the station, while a row is divided into 24 blocks which represent 24 h. The pie chart in the block consists of six sectors, while the color of each sector indicates the position of the pseudo base station. Then the overall activity and rest time of one pseudo base station can be found, as well as the number and locations of overall active area. With the help of the behavior pattern view, the type of one pseudo base station can be defined and we can classify all pseudo base stations.
- **Timeline view:** The timeline view shows the number of spam messages throughout all time intervals and the number of spam messages every half an hour throughout one specific day (Fig. 2(c)). And different types of messages could be chose to show the distribution of messages of selected type and display the proportion of the total numbers of spam messages in one specific day. With the help of the behavior pattern view, the severity of pseudo-base-station attacks can be observed every day, as well as the time intervals with high frequency pseudo-base-station attacks.
- **Distribution view:** The distribution view provides the relationship of the message type, the distribution area and the time quantum in regard to spam messages (Fig. 2(d)). For example, we can identify the composition of different distribution area and time quantum in regard to one specific type of spam messages such like eroticism spam messages. Meanwhile, we can identify the composition of different message type and time quantum in regard to one specific distribution area such like Chaoyang District.

5.3.2 Interactive Exploration

The pseudo-base-station visualization includes two major interactive functions for supporting the analysis of a specific part of all time intervals and displaying the continuous animation of dynamic trajectory in regard to one day and one pseudo base station.

- **Analysis of a specific part of all time intervals:** We design a brush in timeline view to select a range of whole time intervals, as a result of which, the map view will change in response. This design is used for the deep analyzation of pseudo-base-station data because there should be a function of selecting and analyzing the specific time interval where users are interested in. Therefore, when the brush tool is used (Fig. 2(c)), users are able to extract the movement and actions of this pseudo base station during this period.

- **Displaying the continuous animation of dynamic trajectory:** We design a continuous animation display in map view to demonstrate the dynamic trajectory when one specific pseudo base station and one date are chosen. This design is used for intuitive display of details and all actions in regard to one pseudo base station. Moreover, with the help of the continuous animation of dynamic trajectory, users can summarize the rules of pseudo base stations and predict next steps of pseudo base stations.

6 Case Study

Our system was applied to the spam message data extracted in Beijing in 2017 for the evaluation of the approach raised in the paper. Here, there are two common questions to be solve. What is the general time-and-space activity rule of pseudo base stations? What are the spatial and temporal distribution while sending different types of spam messages?

6.1 The General Time-and-Space Activity Rule of Pseudo Base Stations

The biggest difficulty for law enforcement officials to seize the pseudo base stations is that they cannot determine the behavior pattern of the pseudo base station by simple spam message information. Therefore, this system designs a clustering algorithm which extract seventy-seven pseudo base stations from spam messages of more than ten thousand phone numbers.

Fig. 3. The geographical positions of pseudo base stations sending spam messages in different time periods which could be selected in timeline view.

From timeline view, we can identify the number of spam messages all the time. We can choose one date such as 2017-4-3 for deep exploration. The map view, the behavior pattern view and the distribution view will change according to the selected date and can be further filtered by brushing on the timeline (Fig. 3).

Combined with the interaction with behavior pattern view (Fig. 4), four types of pseudo base stations can be found (Table 2).

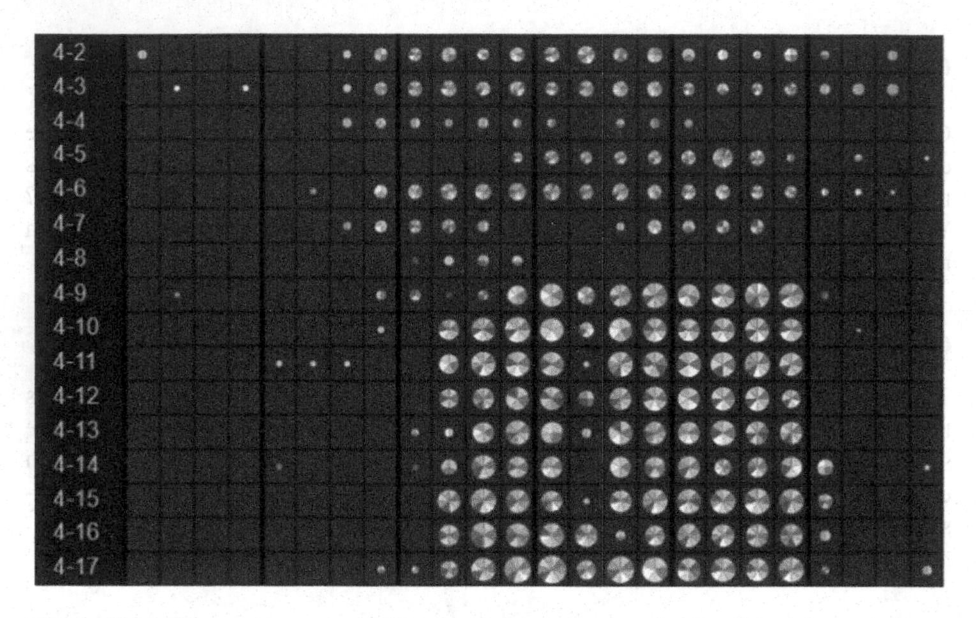

Fig. 4. The behavior pattern view. The selected pseudo base station sent spam messages mainly in Chaoyang District, Dongcheng District and Xicheng District from 9:00 to 20:00.

Table 2. Four types of pseudo base stations.

	Active regions	Time periods	Main types of spam messages
1	Dongcheng district, Chaoyang district	08:00–20:00	Banking fraud
2	Dongcheng district, Fengtai district, Dafeng district, Chaoyang district	18:00–02:00	Invoice, banking fraud, maintenance ads
3	Dongcheng district, Fengtai district, Chaoyang district	Almost all day	Invoice, banking fraud
4	Chaoyang district	20:00–04:00	Pornographic

6.2 The Spatial and Temporal Distribution of Spam Messages

Interacting with the distribution view (Fig. 5), the spatial and temporal distribution of different types of spam messages can be summarized as in Table 3. Reasons for this situation could be given. Spam messages of invoice do less harm than the others and they are not the focal point for crack down campaign, which account for higher proportion. Spam messages of banking fraud are mostly sent from 9:00 to 17:00, which is the opening time period of banks. Spam messages of porn focus on the midnight for their particularities.

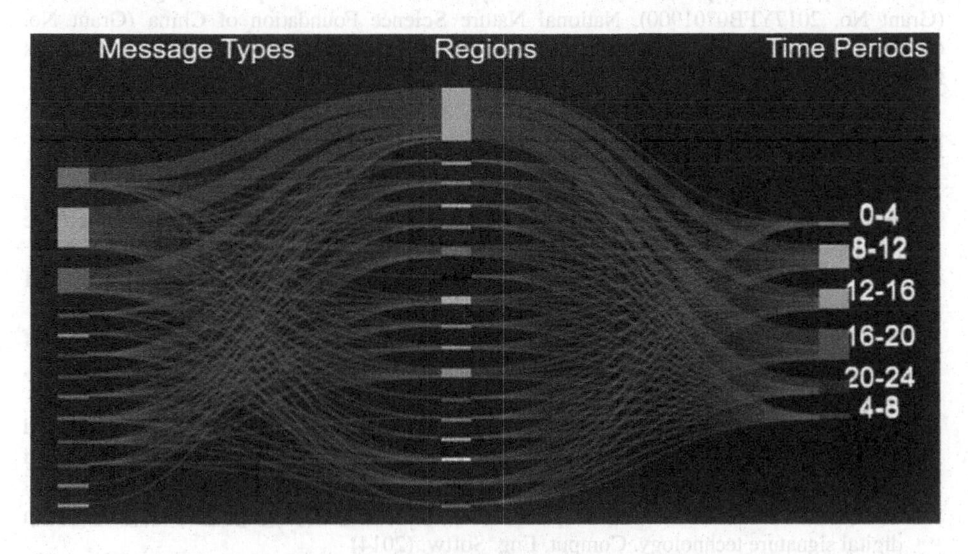

Fig. 5. The distribution view indicates the relationship between message type, distribution area and time periods.

Table 3. The administrative regions with most spam messages.

	Administrative regions	Time periods	Main types of spam messages
1	Chaoyang district	Almost all day	Pornographic, banking fraud, invoices
2	Fengtai district	8:00–11:00, 16:00–20:00	Banking fraud, invoice
3	Haidian district	8:00–12:00	Banking fraud, invoice

7 Conclusion

In this paper, a method of visual analysis of long time-span trajectory patterns of pseudo base stations is proposed for the first time. The main feature of the method is that it is capable of obtaining detailed behaviors of a specified pseudo base station multi-dimensionally with adequate interactive approaches and visually revealing the long time-span trajectory patterns of pseudo base stations. By means of case studies, the practicability and effectiveness in analyzing complicated spatial and temporal data, particularly in implementing analysis of spam messages data have been demonstrated. We are going to use the PBSVis system to analyze more complicated trajectory data from different sources and apply to larger amount of data. In future work, we will devote our effort to improve the accuracy and practicality of the approach and try to convey more comprehensive features of behaviors on a larger dataset.

Acknowledgement. Supported by National Key Research and Development Program of China (Grant No. 2017YFB0701900), National Nature Science Foundation of China (Grant No. 61100053, 61572318). Also, thanks Prof. Xiaoru Yuan, Peking university and unknown reviewers for instruction.

References

1. Zhao, M.W., Lin-Zhou, X.U., Shi, Z.F., et al.: A method for illegal pseudo base station site fast measuring and positioning. Mob. Commun. (2016)
2. Yan, H., Han, Z.H., Chuan-Zhu, L.V.: Research on quasi real time monitoring method of pseudo base station based on large data. Telecom Eng. Tech. Stand. (2016)
3. Androutsopoulos, I., Koutsias, J., Chandrinos, K.V., et al.: An experimental comparison of naive Bayesian and keyword-based anti-spam filtering with personal e-mail messages. **97**(2), 160–167 (2000)
4. Sun, X., Liu, C.: Improved algorithm identifying spam message of pseudo base station based on digital signature. In: IEEE International Conference on Electronics Information and Emergency Communication, pp. 176–179. IEEE (2017)
5. Shang, Q.W., Wei, G.Y.: Study on pseudo base-station identifying spam message based on digital signature technology. Comput. Eng. Softw. (2014)
6. Wang, D., Zhang, X.: Study the method that using mobile phone information accurate positioning pseudo base station. Microcomput. Appl. **11**, 8 (2014)
7. Xiao-Dong, F.U., Hui-Lian, L.I., Xiao, H.Q., et al.: Pseudo base station capture tracing method research and application. Mob. Commun. (2016)
8. Huang, Q., Cui, H., Yan, L.: Method and device for identifying short messages from pseudo base stations (2017)
9. Li, Y., Yang, H., Hao, A., et al.: Detecting and tracking pseudo base stations in GSM signal hijacking and frauds: a visualized approach. Inf. Secur. Comput. Fraud **5**(1), 1–8 (2017)
10. Bogorny, V., Avancini, H., Paula, B.C.D., et al.: Weka-STPM: a software architecture and prototype for semantic trajectory data mining and visualization. Trans. in GIS **15**(2), 227–248 (2011)
11. Buschmann, S., Trapp, M., Döllner, J.: Animated visualization of spatial–temporal trajectory data for air-traffic analysis. Vis. Comput. **32**(3), 371–381 (2016)
12. Liu, S., Liu, C., Luo, Q., et al.: A visual analytics system for metropolitan transportation. IEEE Trans. Intell. Transp. Syst. **14**(4), 1586–1596 (2013)
13. Liao, Z.F., Li, Y., Peng, Y., et al.: A semantic-enhanced trajectory visual analytics for digital forensic. J. Vis. **18**(2), 173–184 (2015)
14. Tominski, C., Andrienko, N., Andrienko, N., et al.: Stacking-based visualization of trajectory attribute data. IEEE Trans. Vis. Comput. Graph. **18**(12), 2565–2574 (2012)
15. IEEE: Interactive visualization of multivariate trajectory data with density maps. In: Visualization Symposium, pp. 147–154. IEEE (2011)

C2C E-commerce Credit Model Research Based on IDS System

Xiaotang Li(✉)

Computer and Information Engineering College, Harbin University of Commerce, Harbin, China
htyiguo@qq.com

Abstract. The credit problem is the bottleneck problem of e-commerce, In particular, C2C e-commerce has the user dispersion, it is not easy to manage, all credit crises are a critical issue. This paper presents an IDS system for the credit problems of C2C e-commerce. The system is the integrity platform of the C2C users in the whole network, which makes the online shopping users have a trace in the whole network transaction. In the IDS system, user ID information will be used and managed safety by means of ID abstract technology. The user's full network credit and comprehensive evaluation are obtained by combining with the website. IDS system adopts the principle of probability calculation. And combine various factors to obtain the user's objective credit value, reflect the use's credit truly. Everyone and websites, all can query user's credit value on the whole network through the port offered by any website, and through the credit curve graph the form image shows the user's credit trend vividly, it can response the user's integrity situation very intuitively. The C2C e-business credit model based on IDS is a national management credit system, it is the fundament to guarantee the healthy development of C2C e-commerce in the long term.

Keywords: Credit platform · IDS system · ID abstract technology

1 Problems Exists in C2C E-commerce Credit System

The credit problem in C2C e-commerce is the most important in several e-commerce models, as a result of the difficulty for authentication and management in retail business, and it leads the credit asymmetry problem between buyer and seller, in turn, brings a series of credit evaluation problems for the C2C transaction, for example credit speculation, cheat transaction, cash out, money laundering etc., but the origin of all the credit problems is unreasonable of user identity. At present, existing credit ratings are based on a website's credit rating for its users, this kind of rating has one-sidedness. It's more important to cannot avoid the fact that malicious users are trading on different sites, and a series of issues that deal with different identities on the same site, and these issues seriously disrupt the security of C2C electronic transaction [1].

It can be seen that establishing a unified authentication platform is very meaningful for C2C e-commerce. On this basis, the credit evaluation of both parties is carried out, can solve the effective management problem of customer in C2C transaction, makes it possible that every online transaction user can be managed in a unified way, can truly

Q. Zhou et al. (Eds.): ICPCSEE 2018, CCIS 901, pp. 611–618, 2018.
https://doi.org/10.1007/978-981-13-2203-7_49

solve the credit asymmetry problem in C2C transactions, and make the credit evaluation of the user more comprehensive and real.

2 IDS Synopsis

2.1 The Structure of the Public Security Network ID Card Management System

At present, China's id card management system belongs to a function of public security network, because of the large amount of data, it is a hierarchical management structure, each province has an independent household registration system, identity card information is in the household registration system, it was generated when the household registration business was, in the public security network, there is the function of checking the identity card information, and if you want to inquire about the id card information out of your province you need to login to the public security network, you can get it only you have the accredit though public security network.

2.2 IDS System Abstract

In the first question, it has been expounded that user identity management is very important for C2C e-business credit, previous identifiers are based solely on the one-sided management of a single site, such management cannot avoid many credit problems, the authentication system established in this paper is an extension of the existing public security system's identity card management system, named IDS system, Such proposal originates in the DNS system. At present, in China's case, there are about 130 million computers, the population of online shopping is about 430 million, it's visible that the number of online shoppers is indeed a huge number, there is no better system for managing such a vast amount of data than the DNS system. In the DNS system, the management task of computer IP is distributed to network segments at all levels, this makes a huge data management feasible. Consider the large number of online shoppers, and each user can have more than one net name for online shopping. In this system, the network name management tasks are assigned to various websites, the management of unified identity and the integrated online transaction information management of the identity are given to the IDS system management. This system has the following advantages:

(1) Id card management system of public security system is a identity authentication system with very mature technology, id card management system can be used to direct access to a large number of user information, for the IDS system provides a good data source.
(2) The current site in the real-name authentication will compare the id card system with the public security system, Just write the identity card information to the IDS database when you compare, Therefore, IDS should be established in the public security system's id card management system to facilitate the generation of data.
(3) The data of the police system's identity management system is true and reliable, which provides a safe basis for the credit of C2C e-commerce.

3 C2C E-business Credit Model Based on Authentication System

3.1 IDS Platform Characteristic

Online shopping users will need real-name certification, certification at this time is the id information, in addition to user is using the network name, thus it's convenient to gain user's personal information. IDS has the management on the real-name information and net information is as follows:

(1) Network name management: user can use different network name not only on different website but also on the same website to trade, these network names will be managed by corresponding website.
(2) Real-name information management: all the user network names of network shopping on network will be bound together in the process of real-name certification, and be managed by website, but the secret considers the secrecy of identity information, the identity information will be taken care in an encrypted form (The technology of website management identity information will be discussed in the next question).

IDS platform management: network name of network shopping is not single, but it's obviously that the user evaluation information for trading can not be gained though the network name, because the network name is non-uniqueness, so we need to bind the identity information and the net name information, the evaluation is gained though website and be bounded to identity information. The user generates an identity information in a transaction on the Internet while authenticating, in this case, the authenticated user is called an ID user, this information is imported into the IDS database, the site will then synchronize the data to the IDS platform after obtaining the credit rating of an ID user, then in the IDS system database value of comprehensive credit rating all user ID on any website. By any query interface of a trading site access IDS system database to query comprehensive credit evaluation value of the ID user online, and in order to distinguish between the buyer and the buyer, and prevent some users from buying from selling, the phenomenon such as cash out etc., IDS has accumulated the total value of the evaluation of the seller and the comprehensive value as the buyer's evaluation [2, 3].

3.2 The Technology of the IDS Platform

3.2.1 ID Hash Technology

User identity information will be authenticated by the public security network when real-name authentication on website. If you are passed your ID information will be set up in the system database. In order to record every ID user's credit evaluation on trading, the site need bound all the ID information. In order to protect the user's ID information, the information will be saved as form as cryptographic, and the encryption algorithm must be reliable entirely, based on this the principle of digital digest will be used to implement the user ID information encryption processing on website and IDS platform see in Fig. 1.

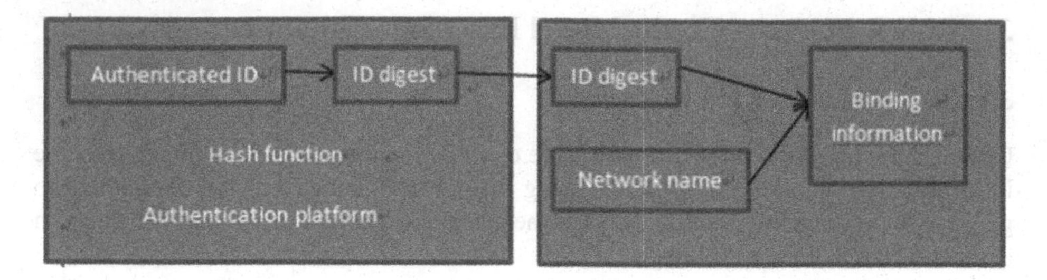

Fig. 1. Website user ID management model

The digital digest principle is: In sending the original information is encrypted to digest using HASH function, then the digital digest and prime information is sent to the receiver together, the receiver also use HASH function for the original message encryption, then the two digest will be compared if they are same. According to the original that the original should agree to generate the principle is also consistent, if they are same, it Indicates information integrity, or incomplete. Here are a lot of data encryption algorithm and technology, but digest is the best algorithm to protect the data integrity. The encryption of the HASH function is irreversible, that original information can be generated digest, but digest cannot be reverted to the original information, visible, in the absence of demand from the original, only through this digest to verify the information, digital technology is undoubtedly the best technology to ensure information integrity. In this for ID user ID information for site is invisible. On the network transmission requirements are absolutely safe. For this the model choses the digest as the information exiting form from the certification from platform profile ID to leaving the authentication platform, this ensures the website that the user ID information is invisible websites. Website only made a binding of user ID and net name information. In this site the user assessment information is gained through the net. By binding with the ID passed to the IDS platform of, so the transmission is ID digest on the Internet, so make sure the safety of the user ID information. After the data reach the IDS platform through comparing with ID digest in the IDS database. If the agreement is found the user ID, then add up the credit data. Such management model, the binding data can be generated in the user identities on the website at the same time. Shied away from he user ID information retrieval of site. The site can only gained user ID digest, this can protect user personal privacy, by using a user ID and the net name binding, this website may through the net name to evaluate user credit, and at last it will be bind with user ID digest, and transfer the value to the IDS system, accumulated the user IDS system by all sites as buyers and sellers credit rating value. We can read ID user of credit information in IDS system by any website interface, so that realize the track of users online trading credit evaluation.

The user ID digest will be transmit to IDS system database as while as a new user is authenticating on the website, From then on, the user transactions in the entire network have become trace, strengthen the comprehensive credit evaluation to the user. The management of IDS system for ID user information includes: ID summary information, ID user as the seller's credit evaluation value, ID user as the buyer's credit evaluation value. See Table 1.

Table 1. IDS database table structure

Serial number	ID digest	Seller credit	Buyer credit

3.2.2 IDS System Credit Standing Calculation Arithmetic

Id user credit information comes from every website, but it is not add up to IDS system simply arriving at IDS system it is integrant to calculate though arithmetic, and then create user credit comprehensive value on the whole network.

Issue of C2C e-commerce credit evaluation, at Current existing C2C e-commerce credit evaluation system to a certain extent, make sure the safety and fairness of online transactions, Most of C2C e-commerce credit evaluation model using simple credit accumulation form to gain a user's credit history and credit evaluation algorithm. Although reduce the credit crisis of online trading, increase the confidence of users online trading, but there are still some problems in the evaluation rules and evaluation method, etc. About website credit evaluation algorithm, the author did some research [1, 2], evaluation model is put forward in the many factors that affect e-commerce credit problems, including: two-way real-name authentication, transaction time, transaction amount, factors such as the type of the evaluation user, credit evaluation algorithm of this paper is limited to the IDS system, not easy to get trade time, trade amount, the information such as evaluate user type, so these factors call for credit evaluation to consider on the website, the IDS system of credit evaluation management important to embody the characteristics of two aspects:

(1) There are two evaluation for one user as seller and buyer, so to avoid users in different sites on the identity of the seller and the buyer to wash sale, money laundering, cash, etc., two kinds of identities for one user to credit evaluate is more comprehensive.
(2) Considered factors in the evaluation:

At present, many websites of credit evaluation mainly consider the transaction time and transaction and factors such as the evaluation user types, mainly because the credit is the result of long-term accumulation, long ago to make evaluation and recent evaluation of the effects on the credit is the same, this will make some sellers early do trade honestly, then use of the accumulation of credit fraud, some users through the accumulation of small credit and then turned to fraud, in addition many websites with no comment as good reputation customers, this is not an objective and some users in a malicious bad for evaluation, trade time, trade amount and evaluation user types very well to avoid these problems, reference sites to consider these factors combined with the particularity of IDS platform, in this paper the credit evaluation factors included: time lag and the last evaluation, credit evaluation scale factor.

The time lag between the current assessment and evaluation of the last time reflects the user transaction frequency, stable frequency reflect the stability state of the user, transaction time factors in the evaluation of similar sites, prevent credit fraud, Selection of credit scale factor added a adjustment balance factor for the credit value, when the credit value exceeds a certain value can be regulated through a small factor, and when below a certain value through big factor to adjust, in a word make credit value relatively objective and stable, don't lead to the evaluation value is not objective on account of the evaluation calculation differences in different websites, then the choice of the scaling factor is crucial, needs comprehensive credit evaluation data through a large number of websites. In addition, considering the IDS platform is different from the website, it reflect comprehensive credit of users through the network transactions, here the calculation principle of probability as the user credit calculation algorithm, as follows:

$$P(B) = \sum_{i=1}^{n} \sigma P(A_i)P(B/A_i)$$

This is a deformation of the full probability formula, the more the full probability formula is a constantσ, this constant is the scale factor mentioned in the article. The principle of full probability formula is, if the event A1, A2... Constitute a complete event group, and all have a positive probability, the arbitrary $P(A_i) > 0$, then for any event B to have the full probability formula. Here, the user credibility of any transactions web site as $P(A_i)$, obviously meet $P(A_i) > 0$, P(B) is regarded as the integrated value of the credit evaluation of the user in the IDS system, namely the known users website credit conditions to calculate comprehensive credit evaluation value, so obtained user credit evaluation is a combination of each site average, more objective. The scaling factor σ of Algorithm rise to balance the action of user credit, when the user made a malicious evaluation, the credit evaluation value should be weakened, a malicious user's information will be reflected in the site transfer credit standing, if it is the evaluation of normal users to give him a moderate proportion factor value, such adjustments that embodies the effect of user types on the calculation of credit evaluation, and the scale factor can be determined more moderate. So to define the scale factor, is to think through it to protect to the reasonable user and control malicious users appropriately. The calculation for a user is divided into two direction for sellers and buyers, more comprehensive to reflect the user's credibility.

3.2.3 Credit Chart

The IDS platform also provides a user credit chart, through two curves on a graph can be intuitive understanding to the user as buyers and sellers of credit and credit contrast, As shown in Fig. 2 [4].

Credit chart can take users the credit as seller and buyer over a period of time, facilitate comparing users credit condition as buyers and sellers different roles in the trade, a more comprehensive response user online credibility, for example, a user's sellers credit curve and buyer credit curve is very smooth and steady rise shows that the user is a good online users.

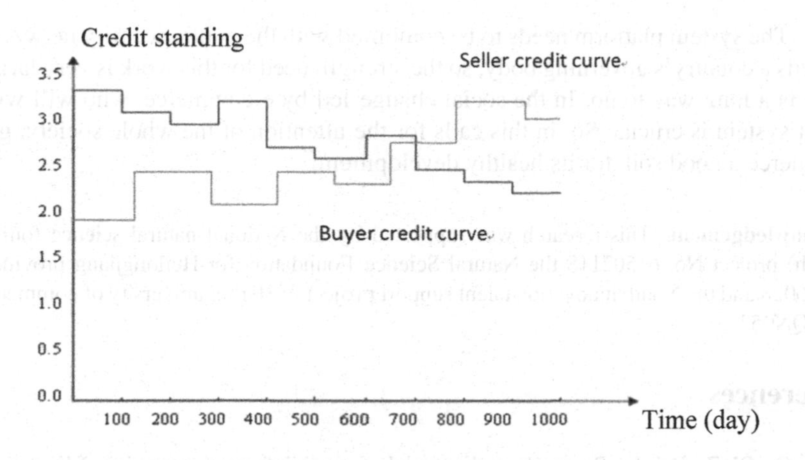

Fig. 2. Credit chart

3.3 Platform Access Way

IDS system is international credit management system independent from each site, if it allow each user login access, this will increase the weight of the IDS system, and it is not convenient for the users. Users must hope to gain the opposite site's credit evaluation on IDS system in its shopping environment, such procedures as the sequential shopping is seamless, it can greatly convenient user, also makes the shopping process more smoothly. Therefore, this paper establish a IDS system access interface of the web site provides to the user, namely in the user's shopping environment through websites provide access to remote access to IDS data in the system, direct access to the credit evaluation information of the other side, and interface is simple, easy to operate. Such access pattern is very humanization, convenient for the user's shopping.

4 Conclusion

Electronic commerce has been developing rapidly from generation to now, there is no question about the changes in the lives of e-commerce leaders, in the trend of such overwhelming no one can deny the significance of electronic commerce, However, in the rapid development, there are still many problems in e-commerce, the most important thing of these problems is the credit problems. E-commerce credit problems are the bottleneck of e-commerce, if we can't solve the credit problem very well in its development, even if no one can intercept the development of e-commerce, the future will see the mistakes of e-commerce. Therefore, in the process of e-commerce development, it is advisable to keep pace and standardize its healthy growth. Therefore, the credit research of e-commerce has become the focus of research, especially the credit of C2C e-commerce. Because the biggest disadvantage of C2C e-commerce is that the fragmented user is not easy to manage, for this purpose, the purpose of the IDS system is to establish a national unified system of integrity to consider its authority, security and unity. The system platform needs to be combined with the public security network, and it needs a country's governing body, so the strength need for this work is very large, but there is a long way to go. In the social change led by e-commerce, who will win, the credit system is crucial. So, in this calls for the attention of the whole society, give e-commerce a good soil, for its healthy development.

Acknowledgement. This research was supported by the National natural science foundation (youth) project No. 61502118 the Natural Science Foundation for Heilongjiang province No. F2016028 and the Youth innovation talent support project of Harbin university of commerce No. 2016QN052.

References

1. Bi, Q., Qi, Z., Bai, Y.: Research on the model of credit information service of E- commerce. Inf. Sci. **25**(11), 1634–1639 (2007)
2. Li, X.: Dynamic weighted trust evaluation model for C2C electronic commerce based on bidirectional authentication mechanism. Sci. Eng. Res. Support Soc. **7**, 329–338 (2014)
3. Yu, L.-A.: E-commerce credit risk early-warning with a least squares proximal support vector regression model. Syst. Eng. Theory Pract. **32**(3), 508–514 (2012)
4. Li, X.: Research for two-way real-name authentication unified platform of C2C E-commerce based on DNS. Sci. Eng. Res. Support Soc. **9**, 279–288 (2015)

An Evolutionary Energy Prediction Model for Solar Energy-Harvesting Wireless Sensor Networks

Guangya Yang[1(✉)], Xue Hu[2], and Xiuying Chen[2]

[1] School of Computer Science, Wuhan University, Wuhan 430072, Hubei, China
whuyangguangya@qq.com

[2] School of Cyber Science and Engineering, Wuhan University, Wuhan 430072, Hubei, China

Abstract. Energy harvesting plays a significance role in wireless sensor networks for it can keep the nodes surviving as long as possible, especially when the wireless sensor networks are established in somewhere that electricity is unavailable from the power station. Making use of solar energy is one solution to mitigate this problem, however, on account of the ever-changing weather conditions and the sun's cycles, the solar energy can be very unreliable and inconstant. Thus, in this paper, a new energy prediction model named RE-prediction is presented for solar energy-harvesting wireless sensor networks, which adopts current solar energy data calculated by the ASHRAE model and the mean of last days to estimate the solar energy data in future. By comparing our RE-prediction model with other existing energy prediction models, such as EWMA, WCMA, and Pro-Energy model via the experimental analysis of these four prediction models with the same datasets, the RE-prediction model is proved to be superior to the other three in accuracy, and obtains a far smaller relative average error successfully.

Keywords: Energy predictions · RE-prediction model · ASHRAE model Energy harvesting

1 Introduction

As distributed autonomous sensors, the wireless sensor networks (WSNs) [1–3] are to monitor the environmental or physical conditions, such as pressure, sound, temperature and so on, and to deliver their data through networks to one main location cooperatively. WSNs rely heavily on electrical power and they can only depend on battery power if planted in the wild. However, the battery lifetime remains a vital restriction for the development of WSN. With the development of rechargeable stored energy, such as the super capacitor or efficient battery, the problem can be addressed by applying energy-harvesting techniques to WSN.

There exist various energy sources for energy harvesting technologies, such as wind, thermoelectric, solar, vibration and so forth, all of which can be quite inconsistent and unreliable. Comparing with these various energy sources, the solar energy

Q. Zhou et al. (Eds.): ICPCSEE 2018, CCIS 901, pp. 619–627, 2018.
https://doi.org/10.1007/978-981-13-2203-7_50

harvesting is proved to be the most efficiency [4]. Moreover, the networks powered by solar energy and other sources have to handle the problem of unstable energy intake. For example, solar energy changes frequently throughout the day and possesses various patterns in different days, months, seasons, as well as weather conditions. Therefore, an energy prediction model with accuracy is required to construct the network to acquire a reliable and consistent source of energy. Many works adopt data from previous history [5] or sophisticated models of energy prediction [6] to estimate energy intake in future for the network, but there is still one relative average error in the results that cannot be ignored.

Establishing an accurate energy prediction model for WSN is of great significance to the network because it is conducive to the network to harvest as much energy as possible, to minimize energy waste, and to direct the network to turn on or off depending on the energy intake and storage with higher efficiency.

One evolutionary energy prediction model named RE-prediction is proposed in this paper for the solar energy-harvesting wireless sensor networks, which utilizes representative past-days' solar energy data and current solar energy data to offer future estimations of the solar energy data. Our RE-prediction model is compared with other prediction methods, such as EWMA [7], WCMA [8], and Pro-Energy [9], it is observed that our model's results are proved to be far better and obtain a fairly smaller error rate. Besides, the concept of ASHRAE model and this model are used to calculate the predicted value of the target slots and to calculate the mean solar energy of the target slots in the past few days. Then the final predicted value is determined by adding some weighting factors to the above two values. What's more, an update strategy is introduced for sample database and all the key parameters are analyzed to find the most efficient and appropriate values for each of them.

The organization of the rest of this paper is as below. Section 2 is the summary of relevant work of the energy prediction models; Our RE-prediction model is described in Sect. 3; Sect. 4 presents an analytic comparison of the performance of RE-prediction model and other prediction models, including analysis of the proper values for key parameters. In the final section, our conclusions are presented.

2 Related Work

Exponentially Weighted Moving-Average (EWMA) is an extensively used solar energy prediction algorithm [7–10], calculating energy value that is able to be harvested at a special time, as one weighted mean of energy that is received at the same time over prior days. The algorithm assumes that the energy available at a particular time is similar to what is observed at the same time of prior days. This method can handle both the diurnal and seasonal variations in solar energy, but when it comes to the ever-changing weather, EWMA fails to adapt to these conditions and the results can be quite inaccurate when sunny and cloudy days are mixed.

Compared to EWMA, Weather-Conditioned Moving Average (WCMA) is one different prediction algorithm for solar energy, which solves the problem that EWMA does not solve. When the weather is cloudy and sunny synchronously, the WCMA prediction algorithm considers the average energy availability in the identical time slot as the current one of the prior days. By establishing a weighting factor that is capable of demonstrating how the weather conditions of current day change concerning prior days, the average value is then scaled based on the factors. It has been proved that WCMA outperforms EWMA comparatively and acquires a far smaller error rate than EWMA especially when the weather conditions are frequently changing.

As a high-efficiency prediction model for solar energy, Pro-Energy enhances solar energy prediction's accuracy and obtains far smaller average prediction errors than WCMA. Pro-Energy is capable of selecting forecasting time frame dynamically due to application requirements and only utilizing the information that nodes themselves collect for estimating the energy intake in future. Besides, there are many other solar energy prediction models such as the method provided by Moser [11], which combines the energy that is harvested during current interval and past one, but its results can be very inaccurate. The approach proposed by Lu [12] focuses on the issue of energy harvesting prediction for the real-time embedded systems, although the results show that the regression analysis obtains the highest accuracy in the estimations in one-second time slot, this approach cannot function well for the medium-term prediction horizons. Moreover, the solution proposed by Sharma [13] is completely different methods, which employ a model to convert the forecasting data of weather into the prediction on energy harvesting. Nevertheless, these methods' performance is not convincing due to the existing flaws. Therefore, in general, all these untypical prediction models have a poor performance that cannot compete with WCMA or Pro-Energy method, even though some of these models do find a different approach, the results are unsatisfying.

3 Basic Models and Methods

In this section, a brand new energy prediction model for the solar energy-harvesting wireless sensor networks named RE-prediction is presented, which adopts representative solar energy data of current and last days for the estimation of the data of solar energy in future.

Additional solar datasets are exploited from the U.S. National Renewable Energy Laboratory [11]. The raw data is processed according to the following principles:

- Exclude abnormal data, such as data with negative exposure;
- Given the lost data, the linear interpolation method is adopted to complete the data;
- Eliminate the data whose exposure is greater than the theoretical value of radiation;
- If the relative humidity is more than 100, it is treated to 100;
- If the length of the calculated sunshine is greater than that of the day's theoretical sunshine, it is replaced by the length of the theoretical sunshine.

Each day is divided into a number of N time slots, and matrix E is used to store the energy data for the last D days. Hence E = D × N and E_{ij} is the energy that is stored in matrix for j^{th} time slot on i^{th} day, C_t is used to represent the solar energy data during the time slot t in current day:

$$E_{d\times n} = \begin{bmatrix} E_{11} & E_{12} & \cdots & E_{1n} \\ E_{21} & E_{22} & \cdots & E_{2n} \\ \vdots & \vdots & \ddots & \vdots \\ E_{d1} & E_{d2} & \cdots & E_{dn} \end{bmatrix}$$

$$C = \begin{bmatrix} c_1 & c_2 & \cdots & c_t \end{bmatrix}$$

Solar energy of *n + 1* time slot on a certain day is determined by the average energy value of the same time slot (the *n + 1* time slot) of the prior sample days and the predicted value of ASHRAE model at that time. Therefore, the solar energy data estimate model of the time slot $t+1$ is established as:

$$E(d,n+1) = GAP_k \bullet (1-\alpha) \bullet M_D(d,n+1) + \alpha \bullet E_A(d,n+1) \quad (1)$$

Where α denotes one weighting factor that is similar to EWMA algorithm, $M_D(d,n+1)$ denotes the average value of D past days at $n+1$ sample of the day:

$$M_D(d,n+1) = \frac{\sum_{i=d-1}^{d-D} E(i,n+1)}{D} \quad (2)$$

In our algorithm, predicted value $E_A(d,n+1)$ of ASHRAE model and inclusion of factor GAP_k are the main innovation.

Firstly, calculation of $E_A(d,n+1)$ in ASHRAE model are introduced. As a sunny-day solar radiation model recommended by the American Society for heating, air conditioning and refrigeration, the ASHRAE model is based on the radiation data in the United States, and the model is expressed as:

$$I_{TH} = (C + sin\alpha) Aexp(-B/sin\alpha)$$
$$I_{DN} = Aexp(-B/sin\alpha)$$
$$I_{dH} = CI_{DN}$$

Where A denotes the radiant value of the outer atmosphere measured on the normal plane. B represents the extinction coefficient of the atmosphere. C is the scattering coefficient and α denotes the solar altitude angle. Subsequently, the horizontal total radiation value (I_{TH}), direct radiation value (I_{DN}) and horizontal scattering radiant value (I_{dH}, Table 1) can be obtained.

In ASHRAE model, α is defined as the height angle of the sun which varies with the local time and the declination of the sun. The sun's declination (equal to the sun's direct point latitude) is δ, the geographic latitude within the range of observation (both

Table 1. ASHRAE model coefficient table.

Month	A	B	C
1	1 230.23	0.142	0.058
2	1 214.46	0.144	0.060
3	1 186.46	0.156	0.071
4	1 135.60	0.180	0.097
5	1 104.06	0.196	0.121
6	1 088.29	0.205	0.134
7	1 085.13	0.207	0.136
8	1 107.21	0.201	0.122
9	1 151.37	0.177	0.092
10	1 192.38	0.160	0.073
11	1 220.77	0.149	0.063
12	1 233.39	0.142	0.057

the latitude and the latitude of the sun are positive for the north and negative for the south) is φ and the local time (hour angle) is t. The formula for calculating the height angle of the sun can be written as follows:

$$\sin\alpha = \sin\varphi \bullet \sin\delta + \cos\varphi \bullet \cos\delta \bullet \cos[(t - 12) \times 15°]$$

Subsequently, the time of the $t + 1$ time slot to be predicted can be replaced by the above formula to obtain the value of the sun's height angle (α), and substitute α into the ASHRAE model formula.

Secondly, the authors try to compute the factor GAP_k, which measures the solar conditions in current day relative to the prior days. At first, a vector $V = [v_1, v_2, \cdots, v_K]$ with K elements is defined. If the value is larger, it indicates that the time of the day is a sunny day. Whereas if the value is smaller, it represents cloudy days:

$$v_k = \frac{E(d, n - K + k - 1)}{M_D(d, n - K + k - 1)} \tag{3}$$

Subsequently, to lay more emphasis on the closest values on time, these values with the distance to actual point are weighted in time by utilizing vector $P = [p_1 p_2,..p_K]$ as follows;

$$p_k = \frac{k}{K} \tag{4}$$

In the end, the weighting factor GAP_k is composed:

$$GAP_k = \frac{V \bullet P}{\sum P} \tag{5}$$

Figure 1 describes one example of the utilization of the RE-prediction model between Aug 22 and Aug 25 to make predictions on the solar energy data. As shown in the figure, the solar energy data in the box represent the selected typical days related to the current day. No matter how solar condition changes, regularly like type1 scenario or irregularly like type2 scenario, our RE-prediction model can make perfect estimations about the future condition of solar energy.

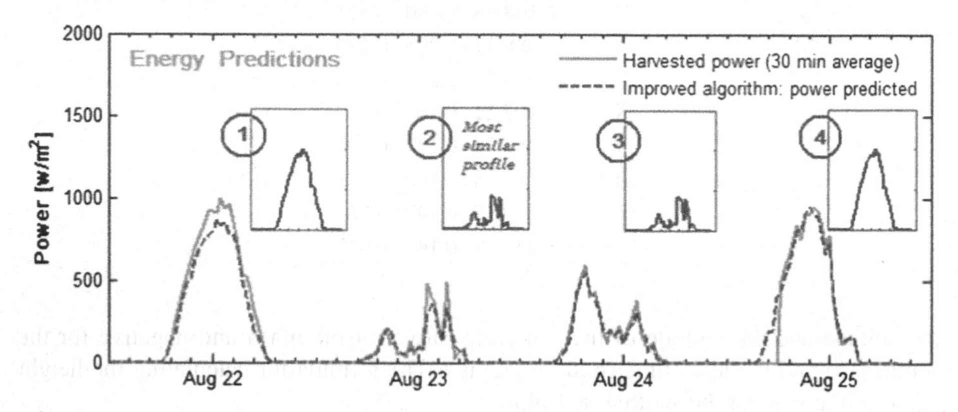

Fig. 1. RE-prediction model: typical days' profiles that are chosen from the past D days are used to make predictions on the future's solar energy data.

4 Performance Evaluation

In order to determine all the uncertain parameters in RE-prediction model, the Mean Absolute Percentage Error (MAPE) [14] function is suggested to evaluate the performance of our RE-prediction model through controlling variables. To optimize all the uncertain values of the parameters, the solar energy data of more than 50 successive days from one location on the resource network [15] are adopted, the highly effect data from July, August, September and October are selected in particular due to the sufficient sunshine as well as the rich solar energy, which is easy to analyze and helps to reduce the errors. Our data are mainly processed by Matlab2015a and the results data are obtained through it as well.

Then the predicted energy of all time slots and the actual energy are compared for determining values of the uncertain parameters through the value calculated by the equation below:

$$\mathrm{MAPE} = \frac{1}{\mathrm{T}}\sum\left|\frac{e_t - \bar{e}_t}{e_t}\right| \tag{6}$$

Where e_t denotes the actual energy that is harvested during the time slot t, $\bar{e}_t$ presents the energy that is predicted for the time slot t, T is the gross number of time slots that MAPE calculates.

In addition to minimizing the MAPE of our prediction model, attention should also be paid to the trade-offs between energy consumption and accuracy. When more data are gathered per day and more samples are used in the RE-prediction model, the predictions are expected to be more precise, while the cost of CPU, memory, and energy consumption can be a big downside for the whole prediction model. On the contrary, an extremely low sampling rate can result in an inaccurate prediction and the whole prediction model will be incapable of calibrating itself.

4.1 PE-Prediction's Parameters Optimization

Whereafter, several sets of experiments are performed to determine the most appropriate values for parameters α, D, m, k, and matrix Z. By employing control variable method, only one parameter is changed per time and the other parameters are set to minimize the overall MAPE, in addition, our experiments only take three prediction horizons: 30 min, 1 h and 2 h into account.

Figure 2 demonstrates the different influences of various α parameter's values on the prediction accuracy of RE-prediction model. Whether the prediction horizon is 30 min, 1 h or 2 h, Fig. 2 displays that α parameter exerts great influence on the overall MAPE of RE-prediction model and when $\alpha = 0.2$, the value of the total MAPE reaches the minimum.

Figure 3 shows that the D parameter does not exert much influence on the overall MAPE. When the value of D parameter changes, the overall MAPE has a small decrease of less than 5% and the MAPE nearly remain unchanged when the value of D parameter exceeds 8. Therefore, $D = 8$ is utilized to determine the D parameter. And we can conclude that when the solar energy of the future time slot is predicted, only the past 8 days' solar data should be adopted to establish the matrix E.

By comparing Figs. 2 and 3, one can find that only when the prediction horizon is 30 min, can the MAPE be far smaller than the time horizons of 1 h and 2 h. When a prediction horizon smaller than 30 min is used in our experiments, there is not an apparent change in the MAPE, while the cost of CPU, memory and energy remarkably increase. As a result, it is wise to set the prediction horizon as 30 min, so that a considerable prediction accuracy and a low cost of CPU, memory or energy consumption can be achieved.

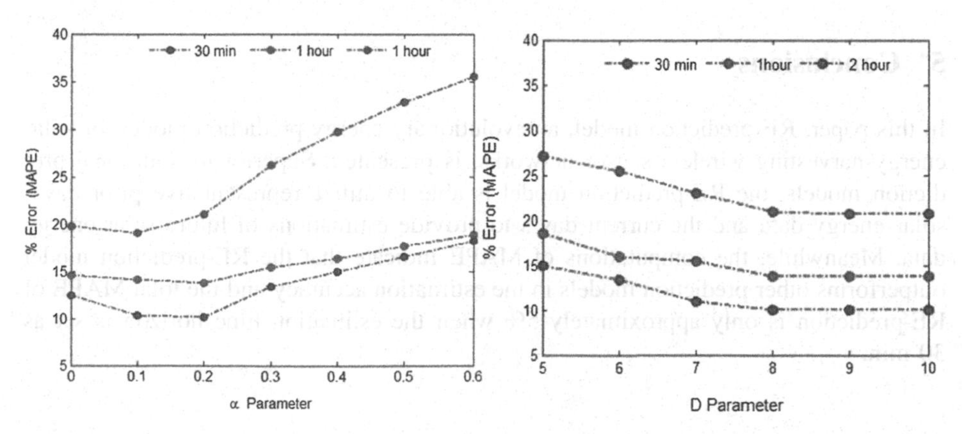

Fig. 2. Impact of varying α parameter on the overall MAPE for 30 min, 1 h and 2 h prediction horizons.

Fig. 3. Impact of varying D parameter on the overall MAPE for 30 min, 1 h and 2 h prediction horizons.

4.2 Comparison Among RE-Prediction, Pro-Energy, WCMA and EWMA

The same solar energy data are adopted to make a comparison of prediction accuracy among RE-prediction, Pro-Energy, WCMA and EWMA. Figure 4 demonstrates the overall prediction MAPE for these four prediction models in various time horizons. It is apparent that RE-prediction model outperforms the other three prediction models in all time horizons from 30 min to 3 h.

As one can observe from Fig. 4, each of the four prediction model's MAPE increases immediately and the differences of MAPE among the four prediction models become smaller when the time horizon extends. However, when the time horizon of 30 min is used, our RE-prediction model has a less than 5% MAPE, which is much smaller than the others.

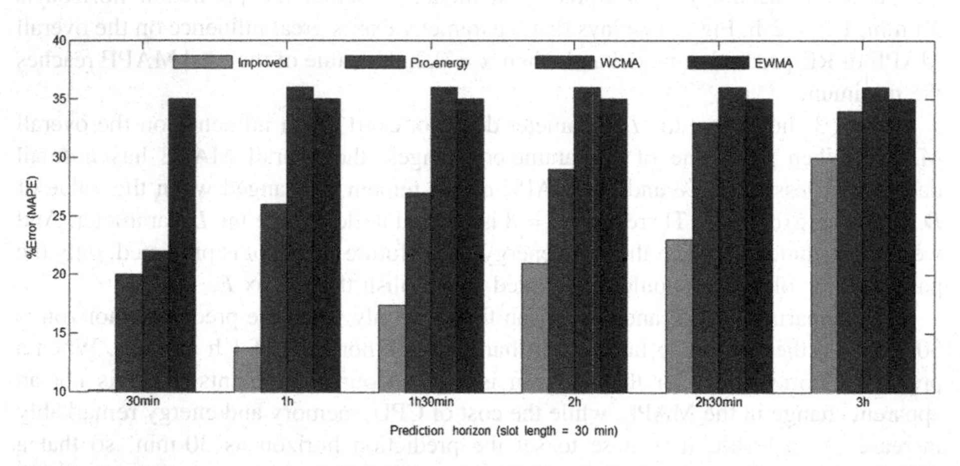

Fig. 4. Predictions accuracy under different prediction horizons: comparisons between RE-prediction model, Pro-Energy, WCMA, and EWMA.

5 Conclusions

In this paper, RE-prediction model, an evolutionary energy prediction model for solar energy-harvesting wireless sensor networks, is presented. Superior to traditional prediction models, the RE-prediction model is able to utilize representative prior days' solar energy data and the current day's to provide estimations of future solar energy data. Meanwhile, the computations of MAPE indicate that the RE-prediction model outperforms other prediction models in the estimation accuracy and the total MAPE of RE-prediction is only approximately 5% when the estimation time horizon is set as 30 min.

References

1. Park, C., Chou, P.: Ambimax: autonomous energy harvesting platform for multi-supply wireless sensor nodes. In: Proceedings of IEEE SECON 2006, Reston, Virginia, USA, 25–28 September, vol. 1, pp. 168–177 (2006)
2. Raghunathan, V., Kansal, A., Hsu, J., Friedman, J., Srivastava, M.: Design considerations for solar energy harvesting wireless embedded systems. In: Proceedings of ACM/IEEE IPSN 2005, UCLA, Los Angeles, CA, USA, 25–27 April, pp. 457–462 (2005)
3. Simjee, F., Chou, P.: Everlast: long-life, supercapacitor-operated wireless sensor node. In: Proceedings of ACM ISLPED 2006, Tegernsee, Germany, 4–6 October, pp. 197–202 (2006)
4. Raghunathan, V., Kansal, A., Hsu, J., Friedman, J., Srivastava, M.: Design considerations for solar energy harvesting wireless embedded systems. In: Proceedings in Sensor Networks (IPSN 2005), pp. 457–462 April 2005
5. Moser, C., Brunelli, D., Thiele, L., Benini, L.: Lazy scheduling for energy harvesting sensor nodes. In: Kleinjohann, B., Kleinjohann, L., Machado, R.J., Pereira, C.E., Thiagarajan, P.S. (eds.) DIPES 2006. IIFIP, vol. 225, pp. 125–134. Springer, Boston, MA (2006). https://doi.org/10.1007/978-0-387-39362-9 14
6. Moser, C., Chen, J.-J., Thiele, L.: Power management in energy harvesting embedded systems with discrete service levels. In: Proceedings Of ACM/IEEE ISLPED 2009, San Francisco, CA, USA, 19–21 August, pp. 413–418 (2009)
7. Cox, D.R.: Prediction by exponentially weighted moving averages and related methods. R. Stat. Soc. **23**(2), 414–422 (1961)
8. Piorno, J., Bergonzini, C., Atienza, D., Rosing, T.: Prediction and management in energy harvested wireless sensor nodes. In: Proceedings Of Wireless VITAE 2009, Aalborg, Denmark, 17–20 May, pp. 6–10 (2009)
9. Cammarano, A., Petrioli, C., Spenza, D.: Pro-energy: a novel energy prediction model for solar and wind energy-harvesting wireless sensor networks. In: Proceedings of IEEE 9th International Conference on MASS, pp. 75–83 (2012)
10. Kansal, A., Hsu, J., Zahedi, S., Srivastava, M.B.: Power management in energy harvesting sensor networks. ACM Trans. Embed. Comput. Syst. **6**(4), 1–38 (2007). article 32
11. Moser, C., Thiele, L., Brunelli, D., Benini, L.: Adaptive power management in energy harvesting systems. In: Proceedings of IEEE DATE 2007, Nice, France, 16–20 April, pp. 773–778 (2007)
12. Lu, J., Liu, S., Wu, Q., Qiu, Q.: Accurate modeling and prediction of energy availability in energy harvesting real time embedded systems. In: Proceedings of IEEE IGCC 2010, Chicago, IL, USA, 15–18 August, pp. 469–476 (2010)
13. Sharma, N., Gummeson, J., Irwin, D., Shenoy, P.: Cloudy computing: leveraging weather forecasts in energy harvesting sensor systems. In: Proceedings of IEEE SECON 2010, Boston, Massachusetts, USA, 21–25 June, pp. 1–9 (2010)
14. Ali, M., Al-Hashimi, B., Recas, J., Atienza, D.: Evaluation and design exploration of solar harvested-energy prediction algorithm. In: Proceedings of IEEE DATE 2010, Dresden, Germany, 8–12 March, pp. 142–147 (2010)
15. Yang, C., Chin, K.W.: Novel algorithms for complete targets coverage in energy harvesting wireless sensor networks. IEEE Commun. Lett. **18**(1), 118–121 (2014)

A Cooperative Indoor Localization Method Based on Spatial Analysis

Qian Zhao[1], Yang Liu[2], Huiqiang Wang[2], Hongwu Lv[2], Guangsheng Feng[2(✉)], and Mao Tang[3]

[1] Harbin University of Commerce, Harbin 150028, China
zhaoqian@hrbcu.edu.cn

[2] Harbin Engineering University, Harbin 150001, China
fengguangsheng@hrbeu.edu.cn

[3] Science and Technology Resource Sharing Service Center of Heilongjiang, Harbin 150001, China

Abstract. We study the cooperative localization in indoor multipath environments considering the larger localization error and insufficient LOS (LOS) localization signals. Using the surrounding nodes as reference points, some existing localization schemes can cooperatively localize the candidate terminals when there are insufficient LOS localization signals. But the localization error arising from the reference points is unavoidable. We first propose a space-partitioning method based on ray tracing technology, in which the indoor area is divided into a direct localization area with LOS conditions (DLA) and a cooperative localization area with NLOS (NLOS) conditions (CLA). In a DLA, there are sufficient LOS localization signals that can be used for localization, while in a CLA, the number of LOS localization signals is insufficient. Then, we develop a cooperative localization scheme based on the space partitioning, in which the reference points can assist the candidate terminals to accomplish their localization. Finally, extensive experiments verify the effectiveness of the proposed cooperative localization scheme.

Keywords: Cooperative localization · Space analysis · Ray tracing

1 Introduction

According to the statistics [1], most of people spend 70% of their time indoors, e.g., shopping and fitness, which far exceeds that of the outdoor activities. In a complicated indoor environment, losing way or incapable of finding some specific spot often occur, especially for elderly or children, where the GPS is usually unusable for its larger signal attenuation while passing through the buildings [2]. Therefore, the demand for high-precision indoor localization is more imminent than any period in history [3–5].

To solve the problem of indoor localization, the wireless localization techniques, e.g., WiFi [6], Bluetooth [7], RFID [8] and UWB [9], have been widely

Q. Zhou et al. (Eds.): ICPCSEE 2018, CCIS 901, pp. 628–637, 2018.
https://doi.org/10.1007/978-981-13-2203-7_51

studied, which can be grouped into two classes, namely, non-cooperative localization and cooperative localization [1,10–14]. In a non-cooperative localization system, the terminal position can be calculated according to the signals received from different transmitters. Due to the higher cost and lower coverage, however, such techniques are unsuitable for large-scale localization in a complicated indoor environment [15]. On the other hand, the cooperative localization scheme is more promising in the mixed environment, especially with the development of the fifth generation mobile communications (5G) [12]. In such scenarios, the indoor localization could be achieved by the cooperative heterogeneous networks, i.e., the terminals that have been localized can transmit localization signals to other candidate terminals to be localized via the device-to-device (D2D) communications [14]. However, the localization precision is severely affected by the NLOS and multipath effects arise from the complicated indoor environment [16–19].

To address above concerns, this paper proposes a collaborative localization scheme based on spatial analysis, where the existing cellular base station (CBS) or localization base station (LBS) are used to locate the candidate user equipments (UEs), e.g., smartphones. In the cooperative localization scheme, the distance between two neighboured UEs could be obtained accurately, which can somehow offset the effects of the NLOS and multipath. Grounding on the more accurate distance between neighbours, the localization precision could be improved. With the cooperative localization scheme, if at least one UE to be localized under LOS conditions, the shadow effect could be improved.

Table 1. Parameter definitions

Parameters	Definitions
N_D	The number of spatial partitions
P_{LOS}	The position sampling rate
P_{min}	The minimum coverage in one DLA
RSS_{min}	The threshold of the localization signal
D_i	Grid i's number
$P(x_i, y_i, z_i)$	The coordinates of each LOS sampling point
X_F	The ranging error of the reference points
C_R	The reference point set
C_L	The distance set
D_A	The common coverage area by all reference points according to C_R and C_L
C_D	The spatial partition covered by the area D_A
P_D	The partition coverage rate of each partition
$N(\mu_F^2, \sigma_F^2)$	The Gaussian distribution of variable X_F, where μ_F denotes the mean and σ_F denotes the standard deviation
$Noise$	The variance of X_F

2 A Spatial Analysis-Based Collaborative Localization Approach

Given that a terminal in an LOS area is localized with a cooperative localization approach, the localization precision could not be improved when the cooperative terminals, namely the reference points, is located in the NLOS environments. The reason is that the ranging error of the reference points in the NLOS environments cannot be guaranteed. The mainly used notations are summarized in Table 1.

To guarantee the localization precision under complicated environments, we develop a spatial analysis-based collaborative localization approach (SACP), which includes a ray tracing-based decision algorithm (RTSRD) and a two-stage cooperative localization algorithm (TSCL). The RTSRD algorithm is used to determine one area is a direct localization area with LOS conditions (DLA) or a cooperative localization area with NLOS conditions (CLA). In the DLA, there are sufficient LOS signals such that the terminal can be localized by the LOS signals, and otherwise to localize a terminal requires the cooperation from the reference points. The basic idea of RTSRD is summarized in Algorithm 1.

Algorithm 1. RTSRD: Ray Tracing-based Spatial Region Decision

---

Input: the number of spatial partitions N_D, the position sampling rate P_{LOS}, the minimum coverage P_{min} in one DLA, and the threshold of the localization signal RSS_{min}

Output: DLA set and CLA set

1 Partition the localization area into N_D rectangular three-dimensional grids, and each grid is numbered as D_i;
2 Calculate the NLOS sampling points according to P_{LOS} in each grid;
3 Randomly generate the coordinates of each LOS sampling point, denoted by $P(x_i, y_i, z_i)$;
4 **for** *the strengths of all signals received by the sampling point is greater than* RSS_{min} **do**
5   Calculate the signal propagation based on the ray tracing;
6   **if** *the signal reaches to the sampling point directly* **then**
7   This signal w.r.t. the sampling point is recorded;
8   **end**
9 **end**
10 /*All the sampling points in the sub-area have completed the LOS signal track calculation*/;
11 **if** *the number of recorded sampling points is no less than 4 and* $P_{D_i} \geq P_{min}$ **then**
12   The candidate area is a DLA;
13 **end**
14 **if** *the number of recorded sampling points is less than 4 or* $P_{D_i} < P_{min}$ **then**
15   The candidate area is a CLA;
16 **end**
17 **return** DLA set and CLA set;

The TSCL algorithm can be divided into two stages, which are given as follows:

- In the first stage, the target terminal which is to be localized performs initial localization according to the localization signals received from all reference points, and then determines the sub-area according to the localization information.
- In the second stage, according to the partition in the first stage, the target terminal determines whether to add the surrounding mobile reference nodes into its localization reference nodes.

The two stages of TSCL are described as Algorithms 2 and 3. The initial localization in the first stage is employed to determine the spatial partition of the target terminal. Theoretically, the localization error is existed in both stages, and the value in first stage is generally larger than the second one. Given a inaccurate determination of the partition, the collaborative localization in the second stage could be affected. To avoid a lower localization precision, the target terminal can select appropriate reference points for cooperative localization, where the reference points have lower ranging error.

As shown in Fig. 1, the target terminal receives the localization signals from reference points R1, R2 and R3, and the distances between the target terminal and the reference points are L1, L2 and L3, respectively. The areas covered by the three reference points are D1, D2, D3 and D4. According to geometric analysis, it can be found that the common coverage area ratio D3 > D1 > D2 > D4. Therefore, the probability that the target terminal is located in D3 is larger than that in other areas.

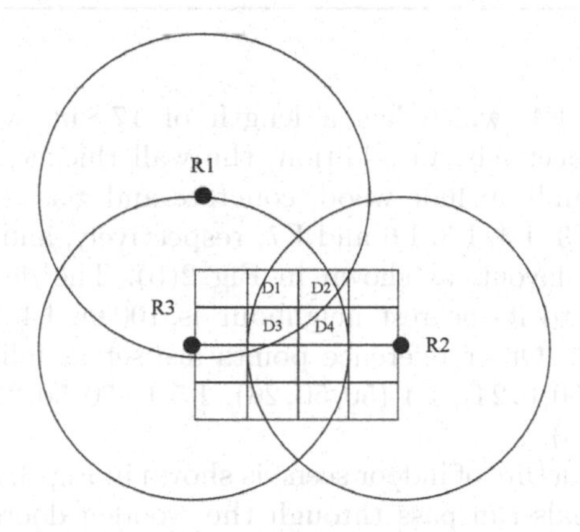

Fig. 1. The coverage of the reference points.

3 Performance Evaluation

The experimental scenario is a two-floor building, and each floor consists of six rooms, as shown in Fig. 2(a). The building is localized in the centre of reference

Algorithm 2. The first stage of TSCL

Input: The target terminal and all candidate reference points
Output: The target partition

1 The target terminal obtains a reference point set C_R, each of which can receives the localization signal;
2 By estimating the distance between the target terminal and each reference point in C_R, obtain the distance set C_L;
3 Determine the common coverage area D_A by all reference points according to C_R and C_L;
4 Obtaining spatial partition according to RTSRD, denoted by set C_D that is covered by the area D_A;
5 Calculate the coverage area of each partition in C_D and calculate the partition coverage rate P_D;
6 **if** the partition with the highest coverage rate in the collection C_D is unique **then**
7   Select the partition with the highest coverage as the partition where the target terminal is located;
8 **end**
9 **if** the partition with the highest coverage rate in the collection C_D is not unique **then**
10   Calculate the centroid coordinates, including the common coverage area D_A and the multiple partitions with the largest coverage;
11   According to the Euclidean distance formula, the partition where the centroid with the shortest distance from the common coverage area D_A is considered as the target partition of the target terminal;
12 **end**
13 **return** The target partition;

points F1, F3 and F4, which has a length of 17.8 m, width of 6.4 m, and height of 8.7 m, respectively. In addition, the wall thickness of 0.2 m, and the materials indoor mainly include wood, concrete and glass. The fixed reference points are F1, F2, F3, F4, F5, F6 and F7, respectively, and their locations follow the honeycomb layout, as shown in Fig. 2(b). The default distance from one reference point to its nearest neighbour is 100 m. F1 is located with the coordinates $(0, 0, 20)$. Other reference points are set as follows (in meter): F2 $(50, -50, 22)$, F3 $(100, 0, 24)$, F4 $(50, 50, 26)$, F5 $(-50, 50, 21)$, F6$(-100, 0, 23)$ and F7 $(-50, -50, 25)$.

The two-floor structure of indoor scene is shown in Fig. 3, it can be found that the localization signals can pass through the wooden doors between adjacent rooms. The coordinates of each point are N1 (42.1, 32.0675, 6.2), N2 (47.3, 32.0675, 6.2), N3 (52.5, 32.0675, 6.2), N4 (57.9, 32.0675, 6.2), N5 (57.9, 27.1675, 6.2), N6 (52.5, 27.1675, 6.2), N7 (47.3, 27.1675, 6.2), N8 (42.1, 27.1675, 6.2), N9 (59.9, 32.0675, 6.2), N10 (59.9, 25.6675, 6.2) and N11 (42.1, 25.6675, 6.2).

In the experiments, we use Taylor algorithm in both the proposed scheme SACP and the classical scheme CP, and we compare their localization precision under different cases, including different noises, different number of reference

Algorithm 3. The second stage of TSCL

Input: the number of iterations k, the target terminal set
Output: the positions of the target terminals

```
1  while the current times of iterations is no larger than k do
2      Partition the indoor space according to Algorithm RTSRD, and determine
       each area is a DLA or CLA;
3      Perform the first stage of TSCL to determine the partition where the target
       terminal is localized;
4      The target terminal receives the localization signals from the moving
       reference points and the fixed reference points, and determine the target
       area according to the TDOA-based localization scheme;
5      if the target terminal is in the DLA then
6          The second-stage localization is re-performed based on the reference
           information of the fixed node;
7      end
8      if the target terminal is in the CLA then
9          The second-stage localization is not performed;
10     end
11     /*The target terminal finishes localization*/ The target terminal sends its
       own position and orientation to its neighbour nodes and the mobile
       reference points;
12     for the neighbour nodes that receives updated position information from the
       surrounding mobile reference nodes do
13         if the neighbour node in a CLA then
14             The localization of the neighbour node should be re-localized
               according to the supplemental reference information;
15         end
16     end
17     Increase the iteration times by 1;
18 end
19 if the number of localization iterations in space is greater than the specified k or
   the global position information no longer changes then
20     The localization is terminated;
21 end
```

points and the candidate terminals that required to be localized. The ranging error X_F of the reference points satisfies the Gaussian distribution with $X_F \sim N(\mu_F^2, \sigma_F^2)$ and we define $Noise = \sigma_F^2$ for convenience. Based on the spatial analysis, T1 is located in a DLA, T2 is located in a CLA, and partitions of T1 and T2 are adjacent with each other. T1 and T2 are localized in the square centre $(50, \frac{50\sqrt{3}}{3}, 4.2)$ and $(50, 5\frac{50\sqrt{3}}{3} - 2, 4.2)$, respectively, with the side length of 2m. Ranging error follows Gaussian distribution with the mean $\mu_F = 0$ and the standard deviation $\sigma_F = \sqrt{5}$. The ranging error of the target terminal follows the Gaussian distribution with mean $\mu_M = 0$ and standard deviation $\sigma_M = 1$. The simulation are repeated 10000 times.

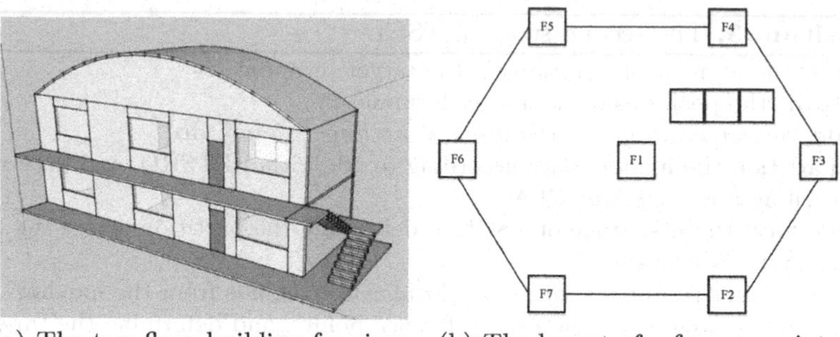

(a) The two-floor building for simulation. (b) The layout of reference points

Fig. 2. The sketch map of the experimental scenario.

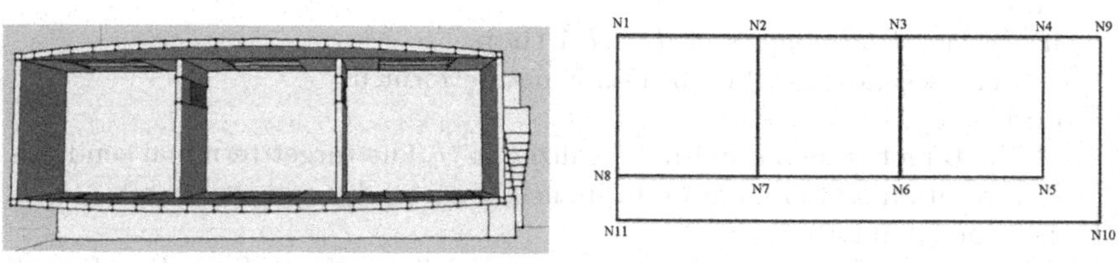

(a) The sectional view of the building. (b) The plane view of the building.

Fig. 3. The different views of the two-floor structure.

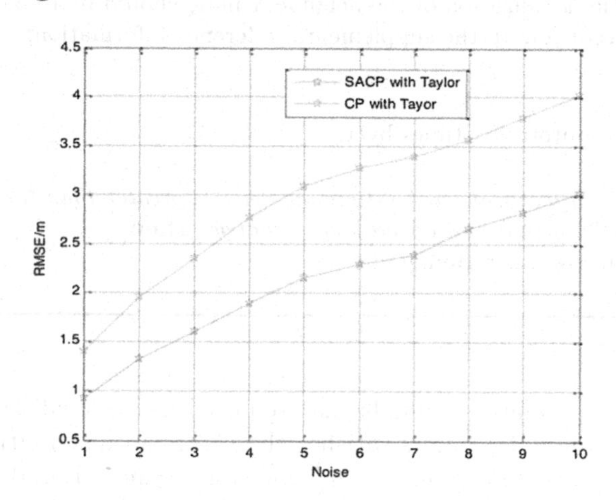

Fig. 4. Collaborative localization results.

Figure 4 shows the RMSE versus the measurement error, denoted by $Noise$, which varies from 1 to 10 m. It can be found that the $Noise$ has a large affect on the RMSE for both of the two schemes. It also can be found that the proposed SACP scheme is less affected by the $Noise$ in comparison to the traditional CP scheme. When $Noise = 1$ m, the precision of SACP is increased by 34.6%.

Figures 5 and 6 show the ranging errors of T1 and T2, where the reference point follows Gaussian distribution with mean is 0 and standard deviation is $\sqrt{5}$. The ranging error of the target terminal follows Gaussian distribution with mean is 0 and standard deviation is 1. According to the statistics in the experiments, the localization precision of the proposed SACP is increased by 23.0% in comparison to the traditional CP scheme. The average RMSE of SACP is 2.0695 m, that of CP is 2.6917 m. Compared with the CP scheme, the proposed SACP increased by 20.4% in $[0, 1]$ and 2.1% in $[1, 2]$, respectively. The localization error of the first 80% of the SACP is 0.8910 m, while the localization error of the first 80% of the CP is 2.3626 m.

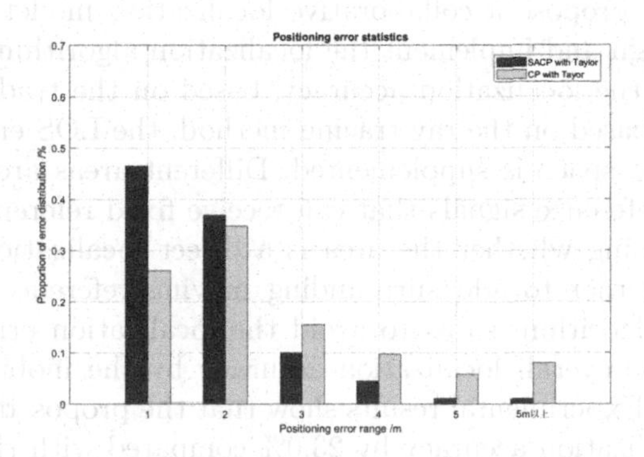

Fig. 5. The comparison of the collaborative localization error statistics.

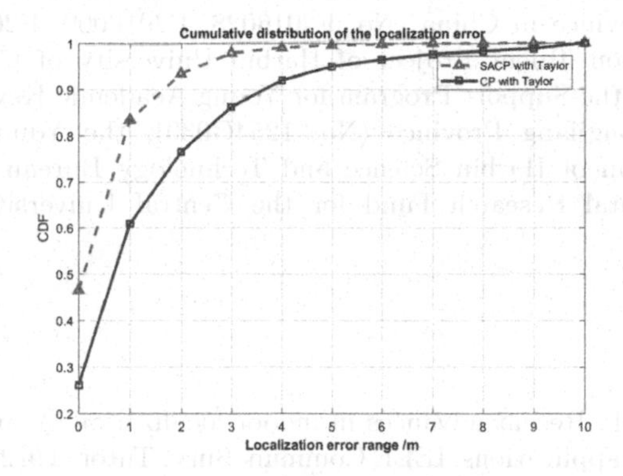

Fig. 6. The CDF comparison of the collaborative localization.

From above experiments, the performance of localization scheme is highly affected by the reference points. In general, given a constant precision of the reference points, the more number of the reference points, the higher precision of the localization scheme. Similarly, given a constant number of the reference points, the higher precision of the reference points, the higher precision of the localization scheme. According to above experiments, the localization errors is partly caused by the reference points, which is positively correlated to the ranging errors of the reference points. In other words, when the ranging errors of reference points are large, the localization precision is decreased due to the accumulation of errors. In the proposed SACP, the target terminals in DLA will little affected by the reference points in CLA with large localization errors, and meanwhile the target terminals in CLA can accomplish the localization.

4 Conclusion

In this paper, we propose a collaborative localization model based on spatial analysis, and design and implement the localization algorithm of the model. In order to improve the localization accuracy, based on the traditional collaborative localization, based on the ray tracing method, the LOS environment of the indoor localization space is supplemented. Different areas are divided and the number of LOS reference signals that can receive fixed reference points in each area. By determining whether the area is a direct localization area or not, it is determined whether to add surrounding moving reference signal data into the localization algorithm, so as to avoid the localization error being lowered and improving the overall localization accuracy by the mobile reference node with larger error. Experimental results show that the proposed SACP algorithm improves the localization accuracy by 23.0% compared with the traditional CP algorithm in a typical indoor scenario.

Acknowledgements. This work is supported by the Natural Science Foundation of Heilongjiang Province in China (No. F2016028, F2016009, F2015029, F2015045), the Youth Innovation Talent Project of Harbin University of Commerce in China (No. 2016Q N052), the Support Program for Young Academic Key Teacher of Higher Education of Heilongjiang Province (No. 1254G030), the Young Reserve Talents Research Foundation of Harbin Science and Technology Bureau (2015RQQXJ082), and the Fundamental Research Fund for the Central Universities in China (No. HEUCFM180604).

References

1. Yassin, A., et al.: Recent advances in indoor localization: a survey on theoretical approaches and applications. IEEE Commun. Surv. Tutor. **19**(2), 1327–1346 (2016)
2. Kang, W., Han, Y.: Smartpdr: smartphone-based pedestrian dead reckoning for indoor localization. IEEE Sens.S J. **15**(5), 2906–2916 (2015)
3. Chintalapudi, K., Padmanabha Iyer, A., Padmanabhan, V.N.: Indoor localization without the pain. In: Proceedings of the Sixteenth Annual International Conference on Mobile Computing and Networking, pp. 173–184. ACM (2010)
4. Rai, A., Chintalapudi, K.K., Padmanabhan, V.N., Sen, R.: Zee: zero-effort crowdsourcing for indoor localization. In: Proceedings of the 18th Annual International Conference on Mobile Computing and Networking, pp. 293–304. ACM (2012)

5. Yang, Z., Wu, C., Liu, Y.: Locating in fingerprint space: wireless indoor localization with little human intervention. In: Proceedings of the 18th Annual International Conference on Mobile Computing and Networking, pp. 269–280. ACM (2012)
6. Yang, C., Shao, H.-R.: Wifi-based indoor positioning. IEEE Commun. Mag. **53**(3), 150–157 (2015)
7. Faragher, R., Harle, R.: An analysis of the accuracy of Bluetooth low energy for indoor positioning applications. In: Proceedings of the 27th International Technical Meeting of the Satellite Division of the Institute of Navigation (ION GNSS+14), pp. 201–210 (2014)
8. Jin, G., Lu, X., Park, M.-S.: An indoor localization mechanism using active RFID tag. In: 2006 IEEE International Conference on Sensor Networks, Ubiquitous, and Trustworthy Computing, vol. 1, p. 4. IEEE (2006)
9. Tiemann, J., Schweikowski, F., Wietfeld, C.: Design of an UWB indoor-positioning system for UAV navigation in GNSS-denied environments. In: 2015 International Conference on Indoor Positioning and Indoor Navigation (IPIN), pp. 1–7. IEEE (2015)
10. Li, S., Hedley, M., Collings, I.B.: New efficient indoor cooperative localization algorithm with empirical ranging error model. IEEE J. Sel. Areas Commun. **33**(7), 1407–1417 (2015)
11. Mendrzik, R., Bauch, G.: Constrained stochastic inference for cooperative indoor localization. In: 2017 IEEE Global Communications Conference, GLOBECOM 2017, pp. 1–6. IEEE (2017)
12. Witrisal, K., et al.: High-accuracy localization for assisted living: 5G systems will turn multipath channels from foe to friend. IEEE Signal Process. Mag. **33**(2), 59–70 (2016)
13. Chen, Z., Zou, H., Jiang, H., Zhu, Q., Soh, Y.C., Xie, L.: Fusion of WiFi, smartphone sensors and landmarks using the Kalman filter for indoor localization. Sensors **15**(1), 715–732 (2015)
14. Qiu, J.-W., Lin, C.-P., Tseng, Y.-C.: BLE-based collaborative indoor localization with adaptive multi-lateration and mobile encountering. In: 2016 IEEE Conference on Wireless Communications and Networking Conference (WCNC), pp. 1–7. IEEE (2016)
15. Otsason, V., Varshavsky, A., LaMarca, A., de Lara, E.: Accurate GSM indoor localization. In: Beigl, M., Intille, S., Rekimoto, J., Tokuda, H. (eds.) UbiComp 2005. LNCS, vol. 3660, pp. 141–158. Springer, Heidelberg (2005). https://doi.org/10.1007/11551201_9
16. Li, F., Zhao, C., Ding, G., Gong, J., Liu, C., Zhao, F.: A reliable and accurate indoor localization method using phone inertial sensors. In: Proceedings of the 2012 ACM Conference on Ubiquitous Computing, pp. 421–430. ACM (2012)
17. Chenshu, W., Yang, Z., Liu, Y., Xi, W.: Will: wireless indoor localization without site survey. IEEE Trans. Parallel Distrib. Syst. **24**(4), 839–848 (2013)
18. Martin, E., Vinyals, O., Friedland, G., Bajcsy, R.: Precise indoor localization using smart phones. In: Proceedings of the 18th ACM International Conference on Multimedia, pp. 787–790. ACM (2010)
19. Jung, S.-Y., Hann, S., Park, C.-S.: TDOA-based optical wireless indoor localization using led ceiling lamps. IEEE Trans. Consum. Electron. **57**(4) (2011)

Phishing Detection Research Based on LSTM Recurrent Neural Network

Wenwu Chen[1(✉)], Wei Zhang[2], and Yang Su[1,2]

[1] Key Laboratory for Network and Information Security of Chinese Armed Police Force, Engineering University of Chinese Armed Police Force, Xi'an, Shaanxi, China
Chenwen5abc@163.com
[2] Department of Electronic Technology, Engineering University of the Chinese Armed Police Force, Xi'an, Shaanxi, China

Abstract. In order to effectively detect phishing attacks, this paper designed a new detection system for phishing websites using LSTM Recurrent neural networks. LSTM has the advantage of capturing data timing and long-term dependencies. LSTM has strong learning ability, has strong potential in the face of complex high-dimensional massive data. Experimental results show that this model approach the accuracy of 99.1%, is higher than that of other neural network algorithms.

Keywords: Phishing detection · LSTM · RNN · Deep learning Cyberspace security

1 Introduction

Phishing attacks are growing threats to cyber security in worldwide. According to the Phishing Activity Trends Report (the first half year of 2017 and the third quarter of 2017) [1] released by the Anti-Phishing Working Group (APWG), from the first quarter of 2017 to the third quarter of 2017 with an increase of 65%, targeting a month more than 420 brands. This is the most frequent attack found since phishing was started in 2004 to track and report (Fig. 1).

In order to obtain the user name, password, ID number, bank card number and other private information, the attackers attract unknown victims to click the fake websites and deceptive E-mails [2]. These criminals are usually profitable using phishing, so their goal usually is online banking, online payment platform, and mobile commerce applications. Researchers firstly developed the blacklist technology to combat phishing attacks [3]. Although URL blacklists have been somewhat effective, the attacker can bypass the blacklist system by slightly modifying the characters in the URL string, and the time of blacklist suspicious sites is relatively delayed and cannot effectively identify new phishing websites.

To make up for the shortcomings of blacklist technology, researchers have tried heuristic detection methods, such as CANTINA [4] and CANTINA+ [5], and the visual similarity test [6]. Recently, the use of machine learning algorithms to identify phishing links becomes the mainstream of current research [7–9]. Long Short-Term Memory

Q. Zhou et al. (Eds.): ICPCSEE 2018, CCIS 901, pp. 638–645, 2018.
https://doi.org/10.1007/978-981-13-2203-7_52

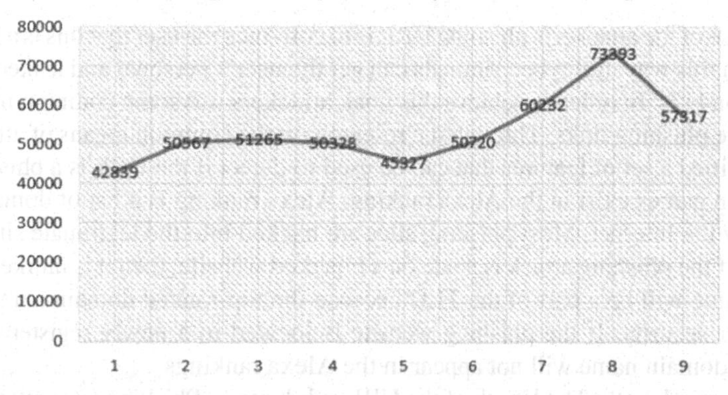

Fig. 1. Phishing activity trends (2017.1–2017.9)

(LSTM) is an architecture proposed by Hochreiter and Schmidhuber [10]. LSTM is a recurrent neural network (RNN), but it differs from the RNN mainly in that it incorporates an LSTM cell that determines the usefulness of information in the algorithm. LSTM has already had many applications in the field of science and technology. The LSTM-based system can learn the tasks of translating languages, controlling robots, image analysis, document summaries, speech recognition image recognition, handwriting recognition, controlling chatbots, predicting diseases, click-through rates and stocks, and synthesizing music.

2 URL Feature Extraction and Analysis

2.1 Uniform Resource Locator Standard Format

The Uniform Resource Locator is a standard resource address on the Internet, and the entrance to a website. Uniform Resource Locator confusing is very common to phishing, to lure users to click on the URL to visit their phishing website is an important part of phishing. To increase the likelihood of users visiting phishing sites, phishing attackers often use deceptive URLs that are visually similar to the fake ones. The format of a standard URL is as follows:

Protocol://hostname[:port]/path/[;parameters][?query]#fragment

The common way to confuse URLs is to construct a phishing URL by partially modifying and replacing the host name part and the path part based on the target URL in order to confuse the user.

For example, the attackers using "www.amaz0n.com" as a fake Amazon website (the real URL is "www.amazon.com"), or using the "www.interface-transport.com/www.paypal.com/" as a fake PayPal website (the real URL is "www.paypal.com") and so on.

2.2 Extract the Features of the Uniform Resource Locator

The purpose of the attacker's phishing URL is to convince the user that this is a legitimate website. In this way, the cybercriminals can get the user's personal and leaked financial information [12]. In order to achieve this goal, attackers use some common methods to camouflage phishing links. Through the research on the common means of attacker, we have identified a set of features that can be used to detect if the URL is a phishing link:

Domain names exist in the Alexa ranking: Alexa ranking is a list of domain names ordered by the Internet. Most phishing sites are hacked into the legitimate sites or new domains. If the phishing attack is made on a hijacked website, then it is unlikely that the domain name will be a part of the TLD because the top-ranked domain names tend to have better security. If the phishing website is located in a newly registered domain name, the domain name will not appear in the Alexa rankings.

Subdomain length: The length of the URL subdomain. Phishing sites attempt to use their domain as their subdomain to mimic the URL of a legitimate website. Legitimate websites tend to have a short subdomain name.

URL length: Phishing URLs tend to be longer than legitimate URLs. Long URLs increase the likelihood of confusing users by hiding the suspicious part of the URL, which may redirect user-submitted information or redirect uploaded web pages to suspicious domain names.

Prefixes and suffixes in URLs: Phishers trick users by remodeling URLs that look like legitimate URLs.

Length ratio: Calculate the ratio between the length of the URL and the length of the path; phishing sites often have a higher proportion of legitimate URLs.

The "@" and "-" counts: The numbers of "@" and "-" in the URL. In the URL, the symbol "@" causes the browser to ignore inputs of previous and later redirects the users to the typed links.

Punctuation counts: The number of "! # $% &" in the URL. Phishing URLs usually have more punctuation.

Other TLDs: The number of TLDs displayed in the URL path. Phishing web links emulate legitimate URLs by using domain names and TLDs in the path.

IP address: The host name - part of the URL uses an IP address instead of a domain name.

Port Number: If a port number exists in the URL, verify that the port is included in a list of known HTTP ports, such as 21, 70, 80, 443, 1080 and 8080. If the port number is not in the list, mark it as a possible phishing URL.

URL Entropy: Calculate URL Entropy. The higher the entropy of the URL, the more complicated it is. Because phishing URLs tend to have random text, so we can try to find them by their entropy.

3 Algorithm Model

3.1 Long Short-Term Memory Cell

Long-term short-term memory (LSTM) is a neural network architecture proposed by Hochreiter and Schmidhuber [10] in 1997. It differs from the RNN mainly in that it incorporates an LSTM unit that determines the usefulness of information in the algorithm. Figure 2 shows a single LSTM cell.

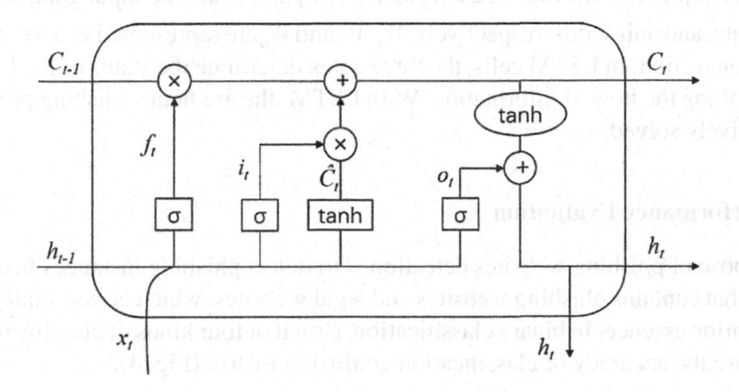

Fig. 2. LSTM cell

At time t, the components of the LSTM cell are updated as follows:

(1) Forgotten information from the cell state, determined by the Sigmoid layer of the Forgotten Gate, with the input xt of the current layer and the output ht − 1 of the previous layer as input, and the cell state output at the time t − 1 is formula (1)

$$f_t = \sigma(W_f \cdot [h_{t-1}, x_t] + b_f) \quad (1)$$

(2) Store information in the cell state, consisting mainly of two parts:
(a) Results of the Sigmoid layer is i_t entering the gate as information to be updated;
(b) Vector C_t newly created by the tanh layer, to be added in the cell state. The old cell state C_{t-1} is multiplied by f_t to forget the information, and the new candidate information $i_t * \hat{C}_t$ is summed to generate an update of the cell state.

$$i_t = \sigma(W_i \cdot [h_{t-1}, x_t] + b_i) \quad (2)$$

$$\hat{C}_t = \tan h(W_C \cdot [h_{t-1}, x_t] + b_C) \quad (3)$$

$$C_t = f_t * C_{t-1} + i_t * \hat{C}_t \quad (4)$$

(c) Output information, determined by the output gate. First use the Sigmoid layer to determine the part of the information to output the cell state, and then use tanh to process the cell state. The product of the two parts of the information yields the output value.

$$o_t = \sigma(W_o[h_{t-1}, x_t] + b_o) \quad (5)$$

$$h_t = o_t * \tan h(C_t) \quad (6)$$

Among them σ is the sigmoid function, h_{t-1} represents the hidden state of the $t-1$ moment, b represents the bias of each gate, i_t, f_t, o_t and C_t are the input gate, forget gate, output gate, and unit status, respectively. W_f, W_i and W_o are represented as a weight matrix for the connection. In LSTM cells, the three gates determine the status of the LSTM cell by controlling the flow of information. With LSTM, the gradient vanishing problem can be effectively solved.

3.2 Performance Evaluation

The purpose of phishing websites detection is to detect phishing instances from the test data set that contains phishing websites and legal websites, which is essentially a binary classification essence. In binary classification, a total of four kinds of classification, used to measure the accuracy of classification confusion matrix (Fig. 3).

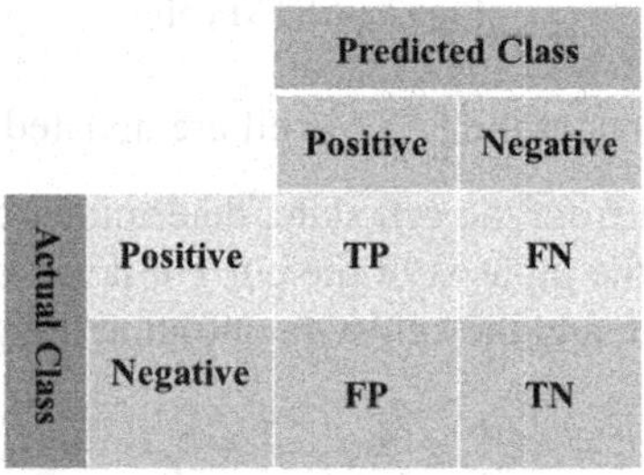

		Predicted Class	
		Positive	Negative
Actual Class	Positive	TP	FN
	Negative	FP	TN

Fig. 3. Confusion matrix

Each URL falls into one of the four possible categories: true positive (TP, correctly classified phishing URL), true negative (TN, correctly classified as non- Phishing URL), false positives (FP, non-phishing URLs are incorrectly classified as phishing) and false negatives (FN, phishing URLs are incorrectly classified as non-phishing). Standard measures, such as accuracy, precision, recall, false negative rate, were determined using the following equation:

$$Accuracy = \frac{TN + TP}{TN + FP + TP + FN} \quad (7)$$

$$Precision = \frac{TP}{FP + TP} \quad (8)$$

$$Recall = \frac{TP}{FN + TP} \tag{9}$$

$$FNR = \frac{FN}{TP + FN} \tag{10}$$

4 Experimental Methods

The experiment uses the Python programming language. The LSTM model is implemented by a deep learning class such as Keras. It contains 5 LSTM layers with 128 nodes each. The model uses a stochastic gradient descent (SGD) optimization method with an initial learning rate of one thousandth and a batch size of 128. The objective function of the least-squares fit is a quadratic polynomial function. The dataset used consisted of 2000 legitimate websites collected from Yahoo Directory (http://dir.yahoo.com/) and 2,000 phishing websites collected from Phishtank (http://www.phishtank.com/). Collected data sets carry label values, "legal" and "phishing". In this data set randomly selected 70% for training, 30% for the test. The training dataset is used to train the neural network and adjust the weight of the neurons in the network, while the test dataset remains unchanged and used to evaluate the performance of the neural network. After training, run the test data set on the optimized neural network.

Predict phishing websites using LSTM Recurrent neural network. The above ten features are taken as input, that is, the number of input layer nodes in the LSTM network is 10 and the number of output layer nodes is one. Training network to choose a strong adaptability of the three-layer LSTM network, incentive function is sigmoid function:

$$f(x) = \frac{1}{1 + e^{-x}} \tag{11}$$

LSTM neural network for classifying phishing URLs based on LSTM units. When entering a URL into an RNN, one-hot encoding is performed on each URL first. Since the characters composing the URL are all contained in ASCII characters (128 characters in total), each URL becomes one-hot encoded. An input vector with a dimension of (len_of_URL)*128 and then brings the input vector into the RNN. So each input character is translated by an 128-dimension embedding. The translated URL is fed into a LSTM layer as a 100-step sequence. Finally, the classification is performed using an output sigmoid neuron. The learning rate of LSTM neural network is 0.1.

In order to better illustrate the accuracy of the algorithm in this paper, an ordinary CNN is used to test the experimental data set. By experimenting with the selected data set, the results show that LSTM network are better than normal CNN, and their prediction accuracy is higher than that of CNN (Table 1).

Table 1. Evaluations of LSTM RNN and CNN

Method	Accuracy	Precision	Recall	FNR
CNN	0.9742	0.9648	0.9723	0.0591
LSTM	0.9914	0.9874	0.9891	0.0212

5 Conclusion and Discussion

In the phishing site testing process, many factors affect the test results, with a certain degree of non-linearity, this paper implements a LSTM-based phishing detection method, which solves the problem that it is difficult for other machine learning methods to extract valid features from the data. It is proved that the prediction method is effective in practice and can solve the problems that traditional methods are difficult to solve. At the same time, this paper adopts LSTM deep learning method and optimizes the training method of the model in combination with the characteristics of RNN. The training time of deep learning model is generally possible from hours to days, and the optimization convergence time is strict on the timeliness of power dispatching and other issues. It is of great significance.

References

1. Phishing Activity Trends Report: Phishing Activity Trends Report 1st Half. Methodology (2017)
2. PHISHTANK: Free community site for anti-phishing service. http://www.phishtank.com/
3. Sinha, S., Bailey, M., Jahanian, F.: Shades of grey: on the effectiveness of reputation-based "blacklists". In: International Conference on Malicious and Unwanted Software, pp. 57–64. IEEE (2008)
4. Zhang, Y, Hong, J.I., Cranor, L.F.: Cantina: a content-based approach to detecting phishing web sites. In: International Conference on World Wide Web, WWW 2007, Banff, Alberta, Canada, May, pp. 639–648. DBLP (2007)
5. Xiang, G., Hong, J., Rose, C.P., et al.: CANTINA+: a feature-rich machine learning framework for detecting phishing web sites. ACM Trans. Inf. Syst. Secur. **14**(2), 21 (2011)
6. Wenyin, L., Huang, G., Xiaoyue, L., et al.: Detection of phishing webpages based on visual similarity. In: Special Interest Tracks and Posters of the 14th International Conference on World Wide Web, pp. 1060–1061 (2005)
7. Ma, J., Saul, L.K., Savage. S., et al.: Beyond blacklists: learning to detect malicious web sites from suspicious URLs. In: ACM SIGKDD International Conference on Knowledge Discovery and Data Mining, Paris, France, 28 June–July, pp. 1245–1254. DBLP (2009)
8. Choi, H., Zhu, B.B., Lee, H.: Detecting malicious web links and identifying their attack types. In: Usenix Conference on Web Application Development, p. 11 (2011)
9. Ma, J., Saul, L.K., Savage, S., et al.: Identifying suspicious URLs: an application of large-scale online learning. In: International Conference on Machine Learning, pp. 681–688. ACM (2009)
10. Hochreiter, S., Schmidhuber, J.: Long short-term memory. Neural Comput. **9**(8), 1735–1780 (1997)
11. Sadeghi, B.H.M.: A BP-neural network predictor model for plastic injection molding process. J. Mater. Process. Technol. **103**(3), 411–416 (2000)
12. Ma, J., Saul, L.K., Savage, S., et al.: Learning to detect malicious URLs. ACM Trans. Intell. Syst. Technol. **2**(3), 1–24 (2011)
13. Sahoo, D., Liu, C., Hoi, S.C.H.: Malicious URL detection using machine learning: a survey (2017)
14. Kim, D., Achan, C., Baek, J., et al.: Implementation of framework to identify potential phishing websites, p. 268. IEEE (2013)

15. Garera, S., Provos, N., Chew, M., et al.: A framework for detection and measurement of phishing attacks. In: ACM Workshop on Recurring Malcode, pp. 1–8. ACM (2007)
16. Olivo, C.K., Santin, A.O., Oliveira, L.S.: Obtaining the threat model for e-mail phishing. Appl. Soft Comput. J. **13**(12), 4841–4848 (2013)
17. Herzberg, A., Jbara, A.: Security and identification indicators for browsers against spoofing and phishing attacks. ACM Trans. Internet Technol. **8**(4), 1–36 (2008)
18. Pan, Y., Ding, X.: Anomaly based web phishing page detection. In: 2006 Computer Security Applications Conference, ACSAC 2006, pp. 381–392. IEEE (2006)
19. Fu, A.Y., Liu, W., Deng, X.: Detecting phishing web pages with visual similarity assessment based on earth mover's distance (EMD). IEEE Trans. Dependable Secur. Comput. **3**(4), 301–311 (2006)

Performance Evaluation of Queuing Management Algorithms in Hybrid Wireless Ad-Hoc Network

Ertshag Hamza[1], Honge Ren[1,2](✉), Elmustafa Sayed[3], and Xiaolong Zhu[1,2]

[1] College of Information and Computer Engineering, Northeast Forestry University, Harbin, China
ertshag@yhoo.com, nefu_rhe@163.com, zhuxiaolonglong@sina.cn

[2] Heilongjiang Forestry Intelligent Equipment Engineering Research Center, Harbin, China

[3] Faculty of Engineering, Red Sea University, Port Sudan, Sudan
elmustafasayed@gmail.com

Abstract. Congestion is one of the problems that will reduce the performance of the networks. In wireless ad-hoc networks connected to the wired backbone through gateways. The performance will be affected due to the type of the routing protocol, and the queuing interfaces were used in the link between the wireless ad-hoc network and the wired backbone network through gateways, which will lead to delay in packet delivered ratio and packet loss in addition to reduce the network throughput. The technique of queue management is used to reduce the congestion rate to have successfully transferred data between any sources and destinations. In this paper, the effect of four queuing management algorithms, fairness queue (FQ), stochastic fairness queuing (SFQ), random early detection {RED) and random exponential marking (REM) will be evaluated on FTP application, based on IEEE802.11 MAC protocol. The algorithms will be simulated in the Hybrid wireless ad-hoc network using NS-2, and the results will be discussed in three scenarios; hybrid network, queuing area (gateways) and different bandwidth with dedicated simulation time, and packet size by measuring the packet delay, packet delivery ratio, and packet losses.

Keywords: Hybrid wireless ad-hoc network · Queuing management · RED REM · FQ · SFQ · Drop-tail algorithm · DSDV · FTP

1 Introduction

Wireless ad hoc network (WANET) means connection between the points of the wireless network (access points) without a central medium or router [1]. Wireless ad hoc network can be connecting two computers or more with each other without a central point of contact and without any central server of the network and each point the role of passing packets to other nodes of the network [2]. It also connects wireless access points without a central moderator or router [3]. Peer-to-Peer is also a direct link between wireless clients. This technology uses emergency networks and military applications that want fast wireless network solutions without creating a central data server or router. Such applications are requiring to connect the wireless ad hoc network to the central sever by wired backbone network through gateways for purposes of decision making and

Q. Zhou et al. (Eds.): ICPCSEE 2018, CCIS 901, pp. 646–655, 2018.
https://doi.org/10.1007/978-981-13-2203-7_53

operation managements [4]. The connectivity through gateways should be interfaced due to routing protocol in wireless ad hoc network and gateway queuing management in wired backbone network to the server for ensuring easy flowing data between the networks, that because long waiting is not desirable in wireless networks [5]. Queuing management algorithms such as RED, REM, SFQ, and FQ can reduce waiting when the gateways queue buffer becomes full, and therefore improve the throughput and reduce delays and loss that occurs for the packets in haul hybrid network [6].

2 Related Works

The present work evaluated the performance of different types of queuing management algorithms in hybrid wireless ad hoc network to wired backbone network through gateways using DSDV ad hoc routing protocol and based on IEEE802.11 MAC protocol and FTP Application. The transmission of packet over a medium at any instance of time requires a packet processing routine. Thus, to maintain a proper processing of the packets over a node an interface must be deployed. This interface object must be able to accept the request from node objects to transmit a packet, even when the medium is busy transmitting a previous packet. Performance of Queues over Multi-Hop Networks were analyzed by Agrawal et al. [7], authors evaluated the performance of Drop tail, DRR, RED, SFQ, and FQ by varying the number of hops, the paper concluded that the performance of the queues is mostly affected due to increment in number of hops. Average Delay, Throughput and Packet loss parameters were considered and the study showed that the parameters were directly or indirectly influenced by the variation in number of hops. Also the study showed that the average delay and simulation time were increased due to the increment in hops. There are two popular types of queue management algorithm; Passive queue management algorithms, which in it the packets are dropped only when the buffer is full [8]. These algorithms are easy to implement in real networks, such algorithms are Drop Tail, FQ and SFQ [9]. Active queue management algorithms, which make to avoid congestion and lock-out. Active Queue Management algorithms reduce the average queue length of buffer. Therefore, end to end delay is also reduced as well. Such active algorithms are RED and REM [10].

3 Simulation Model and Parameters

The Hybrid wireless ad hoc Network environment is consisting of two wired nodes as a switch and server, two gateways and ten fixed wireless ad hoc nodes randomly distributed as shown in Fig. 1 below, using the IEEE802.11 wireless protocol to communicate the ad hoc nodes with each other and to the wired backbone through gateways. The wireless ad hoc network uses the DSDV routing protocol and FTP traffic application in connection to the wired backbone network. The queuing management algorithms which will be evaluated in the hybrid network are RED, REM, SFQ, and FQ. The simulated performance results obtained using different performance metrics such as packet delivery ratio (PDR), packet delay, and data loss. The performance metric equations show as follow:

$$Throughput = \frac{\sum Number\ of\ all\ packets\ delivered}{Receiving\ time\ interval\ length} \quad (1)$$

$$Packetdelay = \frac{TX\ time - RX\ received}{Number\ of\ all\ packets\ delivered} \quad (2)$$

$$PDR = \frac{\sum Number\ of\ all\ packets\ received}{Number\ of\ all\ packets\ sent} * 100 \quad (3)$$

$$Packetlosses = \sum Number\ of\ dropped\ data\ packets\ at\ all\ nodes \quad (4)$$

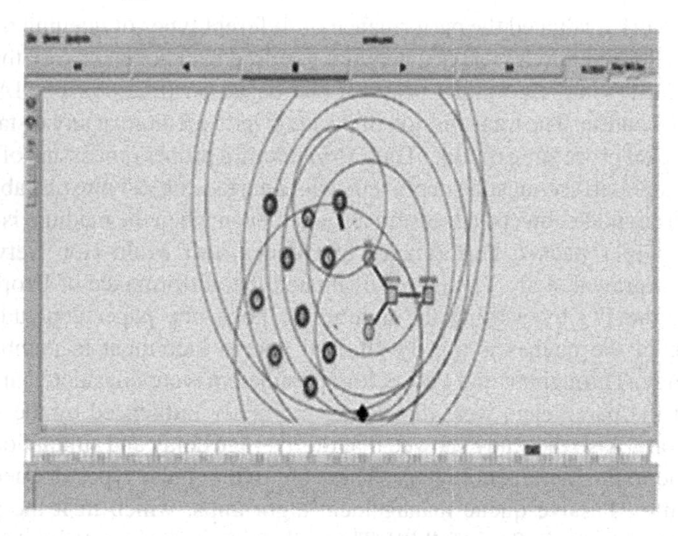

Fig. 1. Hybrid wireless ad-hoc network environment.

Table 1. Simulation environment.

Parameters	Environment
Simulation area	200 × 200
MAC protocol	IEEE802.11
Routing protocol	DSDV
Packets size (byte)	1400 bytes
Queuing bandwidth (Mb)	5, 10, 15, 20
Simulation time (sec)	20, 40, 60, 80, 100
Application Traffic	FTP
Queuing management algorithm	RED, REM, SFQ and FQ
No. of fixed wireless ad hoc nodes	10
No. of wired nodes	2
No. of gateways	2

The following Table 1 shows the values of the various parameters used during simulation to evaluate the performance of queuing management algorithm on the network in different Scenarios.

4 Experimental Results and Discussion

After performing the simulation of hybrid wireless ad hoc network to evaluate the performance of queuing management algorithms, the results of the simulation are divided into three scenarios as follows:

(1) SCENARIO 1: QUEUING ALGORITHMS PERFORMANCE IN HYBRID NETWORK

In the scenario 1, the queuing algorithms were applied in hybrid wireless ad hoc network, and the results obtained for the hybrid network with different simulation time 20, 40, 60, 80, 100 s in a packet size of 1400 byte. The following three figures show the performance metrics considered for evaluating the performance of queuing management algorithms in the network.

In scenario 1, we obtained the results from the simulated network which reviewing the performance of different queuing algorithms and their impact in Wireless ad hoc to wired network. For hybrid network packet delay, Fig. 2 shows that RED and REM algorithms have a stable packet delay compared to the SFQ and FQ but in high simulation time all algorithms remain to have a low packets delay. FQ has a high delay due to the problem of allocate resources.

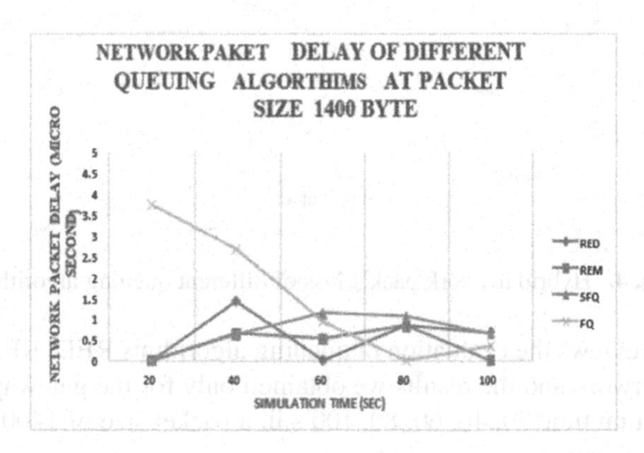

Fig. 2. Hybrid network packet delay of different queuing algorithms

In case of packet delivered ratio (PDR) as in Fig. 3 noticed that all PDR of algorithms increased due to simulation time increases. FQ gives the lower PDR due to their poor network throughput. For network packet losses, as shown in Fig. 4 the RED and SFQ algorithms give a semi matched performance in packet loss, where they give high packet loss compared to REM and FQ, but RED gives higher loss due to its randomly packets dropping mechanism. REM gives lower loss compared to other algorithms.

(2) SCENARIO 2: QUEUING ALGORITHMS PERFORMANCE IN GATEWAYS

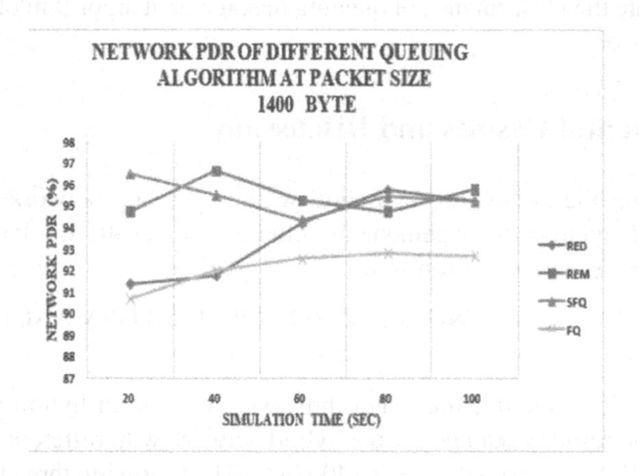

Fig. 3. Hybrid network PDR of different queuing algorithms

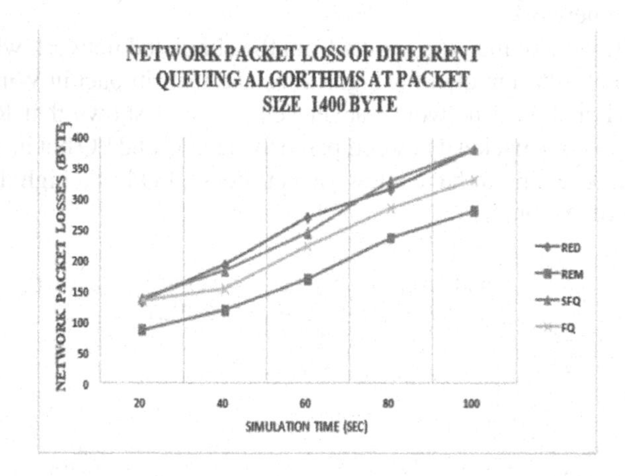

Fig. 4. Hybrid network packet loss of different queuing algorithms

Scenario 2, reviews the evaluation of queuing algorithms RED, REM, SFQ and FQ in the hybrid network and the results we obtained only for the gateways queuing with different simulation time 20, 40, 60, 80, 100 s in a packet size of 1400 bytes as shown in the following figures:

In scenario 2, the gateways queuing packet delay according to the simulation results is presented as shown in Fig. 5, the observed performance shows that the delay of all algorithms is low in 1 micro sec except FQ because it depends to allocate resources. In PDR results as in Fig. 6, all the algorithms performance is stable and consistent with the change in packets size. For gateways queuing packet loss as shown in Fig. 7 all algorithms increase performance while increasing the simulation time. SFQ gives the higher packet loss because of the increase in traffic due to increase throughput of the queuing area. REM gives lower packets loss compared with others algorithms.

(3) Scenario 3: Queuing Algorithms Performance based on Queuing Bandwidth

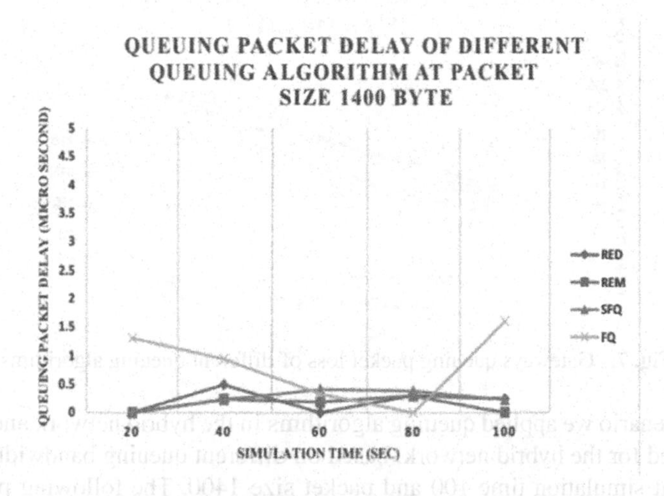

Fig. 5. Gateways queuing packet delay of different queuing algorithms

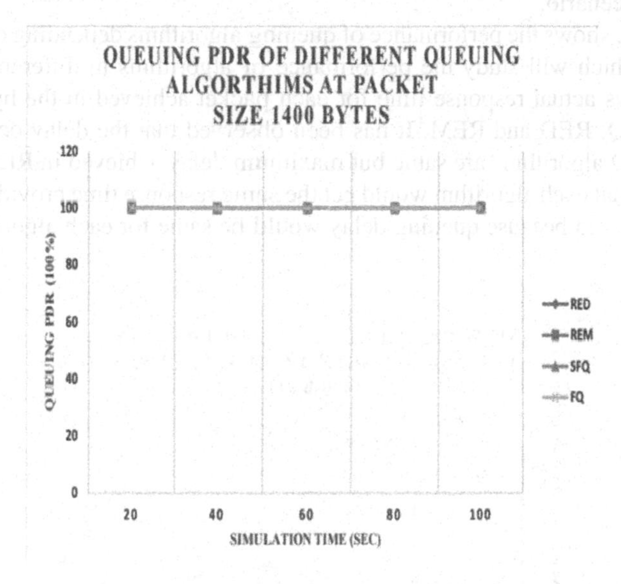

Fig. 6. Gateways queuing PDR of different queuing algorithms

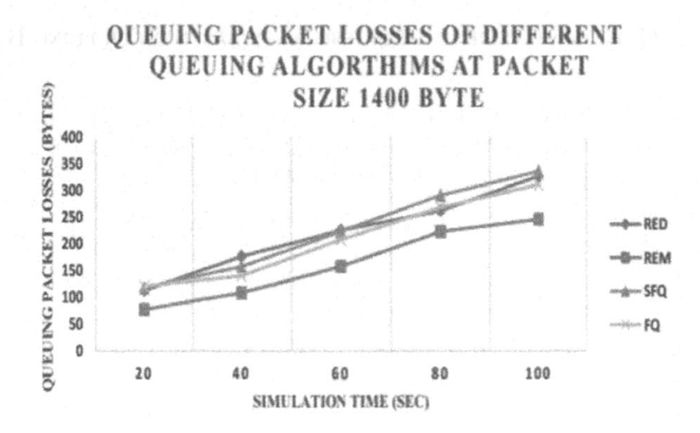

Fig. 7. Gateways queuing packet loss of different queuing algorithms

In this scenario we applied queuing algorithms in the hybrid network and the results were obtained for the hybrid network, based on different queuing bandwidth 5, 10, 15, 20 Mbytes at simulation time 100 and packet size 1400. The following performance metrics are considered for evaluating the performance of queuing management algorithm in this scenario.

Scenario 3, shows the performance of queuing algorithms depending on the queuing bandwidth, which will study the performance of algorithms in different bandwidths. Figure 8 shows actual response time for each packet achieved in the hybrid network using SFQ, FQ, RED and REM. It has been observed that the delay occurred in FQ, SFQ, and RED algorithms are same but maximum delay achieved in REM. Therefore, we conclude that each algorithm would get the same response time provided congestion has been observed because queuing delay would be same for each algorithm in higher bandwidths.

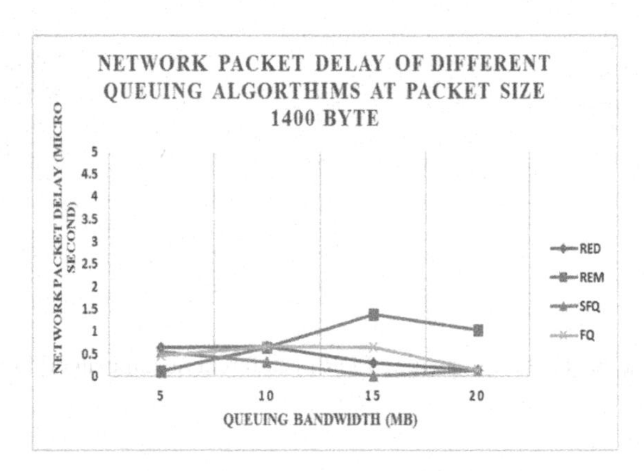

Fig. 8. Hybrid network packet delay of queuing algorithms based on queuing bandwidths

The packet delivered ratio (PDR) of all queuing algorithms as showed in Fig. 9, which it shows that the PDR of the REM decreases with increased queuing bandwidth.

RED, SFQ, and FQ give same performance at queuing bandwidth 20 Mbytes. Figure 10 showed that in case of REM, loss rate decreased when the bandwidth of bottleneck increased. There is a drastic change in loss rate at higher queuing bandwidths in 20 Mbytes in case of SFQ, and FQ because of unfairness achieved at this bandwidth. Otherwise, it has been concluded that FQ, SFQ, and RED could achieve higher loss rate at higher bandwidth and reflected more in case of FQ. At the overall performance we noticed that RED and SFQ performances are semi-matched and same in queuing bandwidth 20 Mbytes. FQ performance in queuing bandwidth 5 Mbytes is low, and with increased queuing bandwidth the performance of FQ algorithm be nearing the performance of SFQ and RED. REM gives the lower packet loss compared with another algorithm.

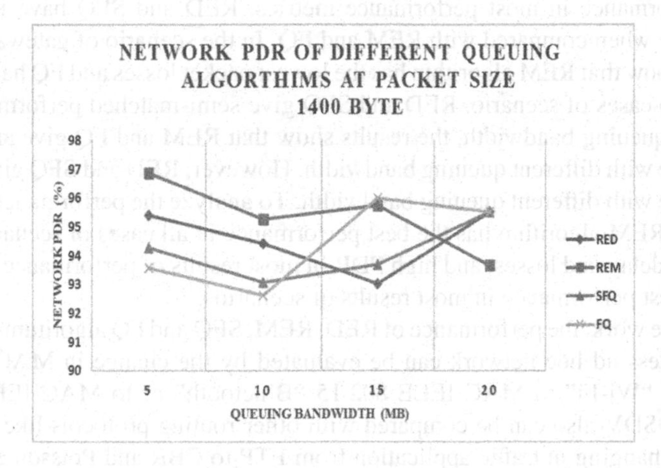

Fig. 9. Hybrid network PDR of queuing algorithms based on queuing bandwidths

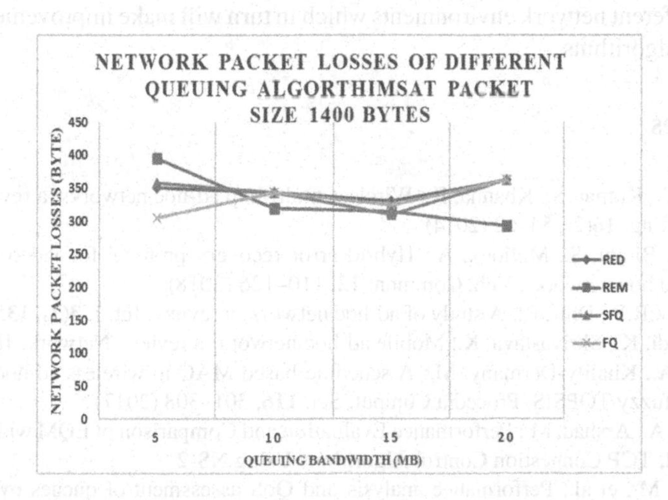

Fig. 10. Hybrid network packet loss of queuing algorithms based on queuing bandwidths

5 Conclusion

This paper presents the comparative analysis of the FQ, RED, SFQ and REM algorithms performance in the hybrid wireless ad hoc network using IEEE802.11 based on FTP application, basis of various performance parameters; packet loss, PDR and delay measured for all hybrid network data and also only to the queuing in gateways area with different simulation time 20, 40, 60, 80, 100 s and packetsize1400 bytes in addition to varying bandwidths for a configured network.

As observed in scenarios A, B, and C of the simulated hybrid wireless network, the performance results of queuing management algorithms show that the REM has the best performance when it compared with RED, SFQ, and FQ algorithms, and FQ has the lowest performance in most performance metrics. RED and SFQ have intermedium performance when compared with REM and FQ. In the scenario of gateways queuing, the results show that REM algorithm has the lowest packet losses and FQ has the highest delay in two cases of scenario. RED and SFQ give semi-matched performance. In the scenario of queuing bandwidth, the results show that REM and FQ give an oscillating performance with different queuing bandwidth. However, RED and SFQ give a Regular performance with different queuing bandwidth. To analyze the performance of queuing algorithms, REM algorithm has the best performance in all cases of scenarios because it gives low delay and losses, and high PDR in most results of performance metrics. FQ has the lowest performance in most results of scenarios.

For future work, the performance of RED, REM, SFQ and FQ algorithms in the same hybrid wireless ad hoc network can be evaluated by the change in MAC type, from IEEE802.11 “Wi-Fi” to MAC IEEE 802.15 “Bluetooth” or to MAC IEEE 802.15.4 “ZigBee”. DSDV also can be compared with other routing protocols like AODV and DSR, with changing in traffic application from FTP to CBR and Poisson applications. This extensive analysis can bring up more features and performance of queuing algorithms in different network environments which in turn will make improvements in these considered algorithms.

References

1. Prabha, C., Kumar, S., Khanna, R.: Wireless multi-hop ad-hoc networks: a review. IOSR J. Comput. Eng. **16**(2), 54–62 (2014)
2. Zaidi, S., Bitam, S., Mellouk, A.: Hybrid error recovery protocol for video streaming in vehicle ad hoc networks. Veh. Commun. **12**, 110–126 (2018)
3. Student, V.R.P., Dhir, R.: A study of ad-hoc network: a review. Int. J. **3**(3), 135–138 (2013)
4. Chaturvedi, K., Shrivastava, K.: Mobile ad-hoc network: a review. Network, **1**(1) (2013)
5. Rahdar, A., Khalily-Dermany, M.: A schedule based MAC in wireless ad-hoc network by utilizing fuzzy TOPSIS. Procedia Comput. Sci. **116**, 301–308 (2017)
6. Nauman, A., Arshad, M.: Performance Evaluation and Comparison of EQM with RED, SFQ, and REM, TCP Congestion Control Algorithms Using NS-2
7. Agrawal, M., et al.: Performance analysis and QoS assessment of queues over multi-hop networks. In: International Symposium on Computing, Communication, and Control Proceedings of CSIT (2009)
8. Kochher, S., et al.: A Review on Active and Passive Queuing Techniques

9. Reddy, T.B., Ahammed, A.: Performance comparison of active queue management techniques. J. Comput. Sci. **4**(12), 1020 (2008)
10. Kumar, A., et al.: Comparison and analysis of drop tail and RED queuing methodology in PIM-DM multicasting network. RED **3**(10), 11 (2012)

Selection of Wavelet Basis for Compression of Spatial Remote Sensing Image

Meishan Li[1], Jiamei Xue[1](✉), and Hong Zhang[2]

[1] Information and Electronic Technology Institute, Jiamusi University, Jiamusi 154007, Heilongjiang, China
xuejiameixzy@sina.com

[2] Graduate School Jiamusi University, Jiamusi 154007, China

Abstract. Characteristics of spatial remote sensing images are studied. Based on their characteristics, remote sensing images are compressed through wavelet transformation. Wavelet properties are first analyzed. The relationship of these properties with image compression is then obtained. Simulations on wavelet properties are performed via MATLAB. Influence of individual properties on compression of spatial remote sensing images is obtained. Wavelet decomposition and reconstruction of several images are simulated in the experiment. With peak signal to noise ratio (PSNR) as the performance metric, the d9/7 wavelet is finally determined as the optimal choice for compression of spatial remote sensing image.

Keywords: Spatial remote sensing images · Wavelet basis
Image compression

1 Introduction

Remote sensing is an integrated subject of science. It bears a close relationship with electronics, optics, geology, computer science and space science, making it an important part of modern science. Theoretically, remote sensing refers to the technology of recognizing, classifying and analyzing the object after collecting its data without directly contacting it [1]. With rapid progress in remote sensing, the number of remote sensing images is growing dramatically, thereby highlighting the need for efficient storage and transmission through image compression. The traditional compression methods, however, cannot operate as effectively as expected. The complexity of the vector quantification-based compression method increases exponentially with the dimensionality. The compression rate of difference pulse code modulation is poor. The Joint Photographic Experts Group (JPEG) method produces obvious block artifacts. Wavelet transformation is a new mathematical tool for time-frequency analysis. Its advent enables many images to be compressed, transmitted and analyzed more efficiently. Due to its suitability for image compression, wavelet transformation provides desirable PSNR and compression ratio. More importantly, it is able to reconstruct images effectively and reduce the time consumption of image decomposition,

Q. Zhou et al. (Eds.): ICPCSEE 2018, CCIS 901, pp. 656–665, 2018.
https://doi.org/10.1007/978-981-13-2203-7_54

reconstruction and coding. For the wavelet-based compression algorithm, the first step is to find a wavelet basis suited for image compression. This is the main focus of this paper.

2 Typical Classes of Wavelet Basis [2]

2.1 Haar Wavelet

The Haar function is the most common orthogonal wavelet function with compactness for wavelet analysis. It is also the simplest wavelet function, and a rectangulare wave with support region in $t \in [0, 1]$. The Haar function can be defined as:

$$\psi(t) = \begin{cases} 1 & 0 \leq t \leq 1/2 \\ -1 & 1/2 \leq t \leq 1 \\ 0 & other \end{cases} \tag{1}$$

Due to its discontinuity in the time domain, the Haar wavelet is not a good choice as a basic wavelet.

2.2 Daubechies (dbN) Wavelet

The Daubechies function [3, 4] was constructed by Ingrid Daubechies, a world-famous wavelet analyzer. It is often written as dbN for short, where N denotes the order of wavelet. It makes it possible to perform discrete wavelet analysis and can be classified into two categories, orthogonal and biorthogonal. The Daubechies biorthogonal wavelet system strongly resembles the orthogonal wavelet system. The only difference is that the former consists of two wavelet functions, i.e. ψ and its dual wavelet $\tilde{\psi}$.

The word biorthogonal means that the low-pass decomposition filter h is orthogonal to the high-pass reconstruction filter $\tilde{g}$, and that the low-pass reconstruction filter $\tilde{h}$ is orthogonal to the high-pass decomposition filter g, where h and g correspond to the wavelet ψ, while $\tilde{h}$ and $\tilde{g}$ correspond to the wavelet $\tilde{\psi}$. Usually, a wavelet function ψ is used for decomposition, and its dual wavelet $\tilde{\psi}$ is used for reconstruction. As we know, decomposing and reconstructing an image using the same filter makes it impossible to guarantee the accuracy of symmetry and reconstruction at the same time. On the contrary, adopting two functions provides an effective approach to this problem. The major difference with the orthogonal wavelet system is that the biorthogonal wavelet system is partly orthogonal and completely symmetric.

2.3 Symlet (symN) Wavelet

The Symlet function system is an approximately symmetric wavelet function proposed by Daubechies, and can be written as symN, where $N = 2, 3, \ldots, 8$, representing an improvement in the db function. According to Daubechies, symN enhances its symmetry while maintaining remarkable simplicity.

2.4 Coiflet (coifN) Wavelet

Daubechies constructed the Coiflet wavelet which is compliant with the requirements of R.Coifman. It includes the series of coif N, where $N = 1, 2, 3, 4, 5$. The 2Nth moment of the Coiflet wavelet function $\psi(t)$ and the (2N − 1)th moment of the scale function are zero. The support of $\psi(t)$ and $\varphi(t)$ has a length of 6N − 1. Moreover, $\psi(t)$ and $\varphi(t)$ of Coiflet are superior to dbN in terms of symmetry.

3 Evaluation Metrics of Wavelet Basis

PSNR [5] is an important measure in evaluating the quality of image reconstruction and the performance of image compression algorithms. Experiments were performed to compress remote sensing images using different wavelet bases. The experimental results were analyzed to find an optimal wavelet for image compression. With other conditions being the same, a higher value of PSNR leads to a higher quality of image reconstruction and it indicates that the selected wavelet is more appropriate for remote sensing images. And the wavelet is more suited for compression of remote sensing images. PSNR in db can be computed as:

$$PSNR = 10\lg\frac{255^2}{MSE} \tag{2}$$

And we have:

$$MSE = \frac{1}{M \times N}\sum_{i=0}^{M-1}\sum_{j=0}^{N-1}(x(i,j) - \hat{x}(x,j))^2 \tag{3}$$

where M and N denote the length and width of an image in pixels, respectively, $x(i, j)$ denotes the value of the original image at (i, j), $\hat{x}(x, j)$ denotes the value of the reconstructed image at (i, j).

4 Relationship of Wavelet Basis Properties with Image Compression

4.1 Wavelet Basis Properties

Major properties of wavelet basis include orthogonality and biorthogonality [6, 7], compactness, symmetry, regularity [8] and high-order vanishing moments. The biorthogonal wavelet is superior to the orthogonal wavelet at the cost of orthogonality. For the property of compactness, a small width of carried set corresponds to a low level of computational complexity and facilitates rapid implementation. The symmetric wavelet is instrumental in effectively avoiding phase distortion and alleviating distortion of some edges in the reconstructed image. The highly regular wavelet enables the signal or image to be reconstructed more effectively, reducing the impact of quantification or round-off error on visual effect. A higher order of vanishing moment, corresponds to a higher compression ratio.

4.2 Simulation Experiment

The dbN wavelet contains orthogonal and biorthogonal categories, being characterized by compactness, regularity and vanishing moment. These properties are closely related to image compression. Thus it is selected in this paper for experiment. Note that other wavelets (e.g., Morlet, Mexican, Hat and Meyer) are devoid of compactness. The Coiflet and Symlets wavelets are variants of dbN.

Basic properties of dbN are shown in Table 1, where the support width is 2N − 1, vanishing moment is N, and the regularity is about 0.2N for large N. The wavelet function and the scale function belong to C 0.2N, where N denotes the serial number of wavelet base.

Table 1. Basic properties of dbN

Wavelet basis	Orthogonal	Support width	Vanishing moments	Regularity
db2	yes	3	2	1.0
db3	yes	5	3	2.75
db4	yes	7	4	5.1
db5	yes	9	5	7.98
db6	yes	11	6	11.33
db7	yes	13	7	15.11
db8	yes	15	8	19.32
db9	yes	17	9	23.95
db10	yes	19	10	29.02

Three-level wavelet decomposition is done on a large number of spatial remote sensing images in the experiment. Consider the "Taiyuan Airport" and "Resolution Crowd" image with a size 256 * 256, as in Fig. 1 Extensive simulations were performed via MATLAB. Tables 2 and 3 present the data on PSNR, decomposition and reconstruction time. The optimal wavelet for compression of a remote sensing image is obtained from comparing, plotting, summarizing and analyzing the experimental data on orthogonality, compactness, vanishing moment and regularity of the dbN wavelet.

After analyzing the three tables, Figs. 2, 3 and 4 show the relationship of image compression with Compactness, regularity and vanishing moment.

It can be observed from Fig. 2 that the compression and decompression time increases with the support width. PSNR increases accordingly. Hence, the more compact the wavelet basis, the better the reconstructed image. From analysis of the large amount of experimental data, it has been shown that the image compression and decompression time as well as PSNR increase with compact width. Therefore, the more compact it is, the higher the quality of reconstructed image. Figure 3 shows that the

a)Taiyuan Airport

b) Resolution Crowd

Fig. 1. Original image

Table 2. The test results of remote sensing image (Taiyuan airport)

Wavelet basis	PSNR (db)	Decomposition time (s)	Reconstruction time (s)
db2	37.39	0.16	0.12
db3	38.21	0.18	0.13
db4	38.74	0.20	0.14
db5	38.95	0.23	0.14
db6	38.99	0.26	0.16
db7	39.00	0.30	0.17
db8	39.05	0.37	0.18
db9	39.09	0.42	0.18
db10	39.11	0.48	0.20

Table 3. The test results of remote sensing image (resolution crowd)

Wavelet basis	PSNR (db)	Decomposition time (s)	Reconstruction time (s)
db2	35.40	0.12	0.15
db3	35.87	0.12	0.15
db4	35.90	0.14	0.17
db5	35.74	0.14	0.17
db6	35.74	0.15	0.17
db7	35.67	0.15	0.19
db8	35.65	0.17	0.21
db9	36.09	0.17	0.22
db10	36.21	0.18	0.24

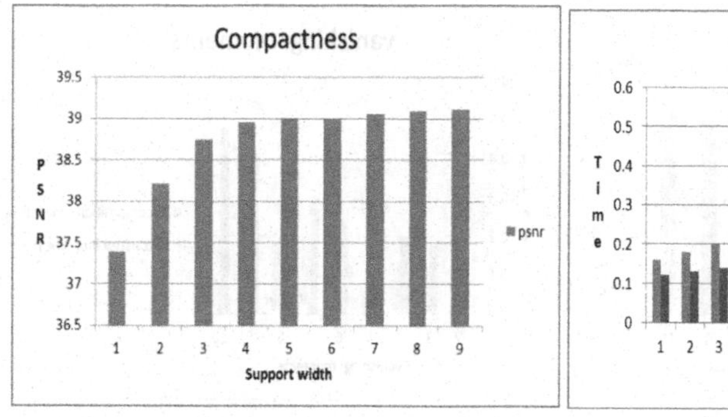

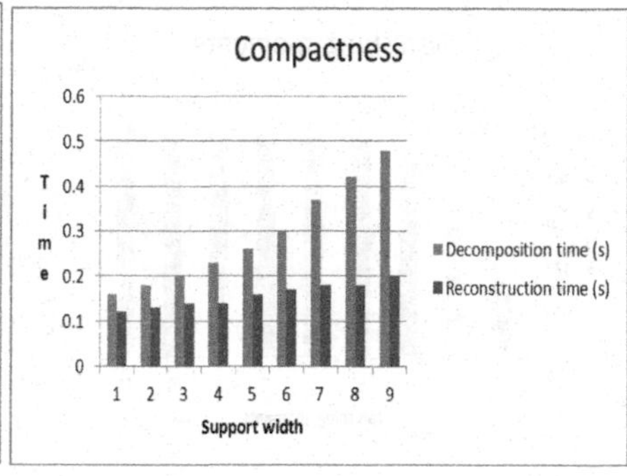

a) PSNR b) decomposition time and reconstruction time

Fig. 2. Compactness vs. image compression

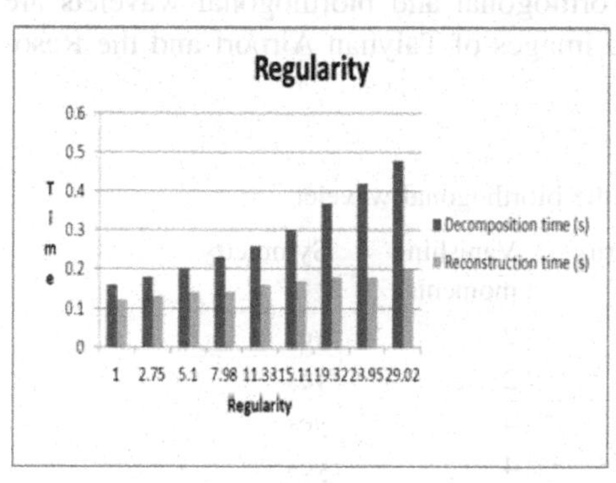

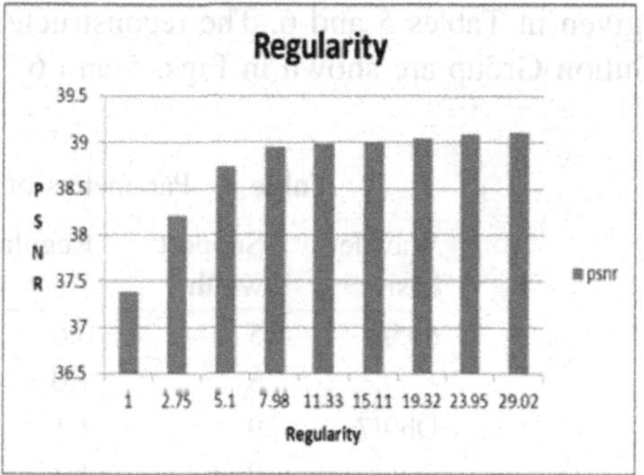

a) PSNR b) decomposition time and reconstruction time

Fig. 3. Regularity vs. image compression

PSNR increases with the order of regularity. But the PSNR varies slightly after the order exceeds 7. Therefore, an appropriate order is recommended. Analysis of Fig. 4 shows that the PSNR increases with the vanishing moment. But the histogram tends to be stable after the vanishing moment reaches the level of 5. That is, setting the vanishing moment to an extremely high value is discouraged. Note that the compression and decompression time increases with the vanishing moment, underscoring the need to choose an optimal vanishing moment by finding a balance between operation time and PSNR. Based on results of extensive experiments on remote sensing images for this paper, setting the vanishing moment to approximately 5 is recommended for compression of remote sensing images. Recommended for compression of remote sensing images.

Symmetry is another property of wavelet which is essential to image compression. The db2–db10 wavelet bases are all orthogonal, but only biorthogonal wavelets are symmetric. The db5/3 and db9/7 biorthogonal wavelets are commonly used at present,

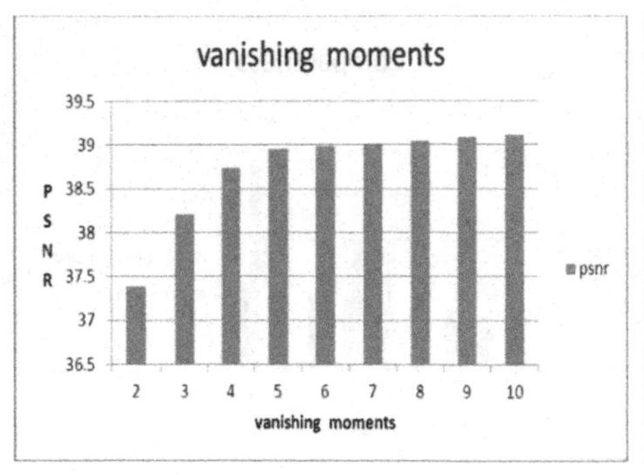

a) PSNR

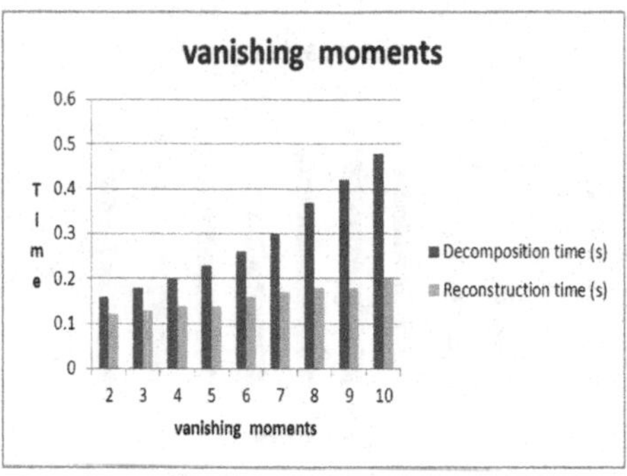

b) decomposition time and reconstruction time

Fig. 4. Vanishing moment vs. image compression

as shown in Table 4. Results of tests on orthogonal and biorthogonal wavelets are given in Tables 5 and 6. The reconstructed images of Taiyuan Airport and the Resolution Group are shown in Figs. 5 and 6.

Table 4. Parameters of the biorthogonal wavelet

Wavelet basis	Support width	Regularity	Vanishing moments	Symmetry
db5/3	5	0.0	2	yes
	5	1.0	2	yes
Db9/7	9	1.1	4	yes
	9	1.7	4	yes

Table 5. Results of test on wavelet symmetry (Taiyuan airport)

Wavelet Basis	PSNR (db)	Decomposition time (s)	Reconstruction time (s)
Db4	38.74	0.20	0.14
Db5/3	38.98	0.35	0.17
Db9/7	39.86	0.42	0.20

Table 6. Results of test on wavelet symmetry (resolution crowd)

Wavelet basis	PSNR (db)	Decomposition time (s)	Reconstruction time (s)
Db4	35.9	0.14	0.17
Db5/3	35.98	0.14	0.17
Db9/7	36.86	0.17	0.21

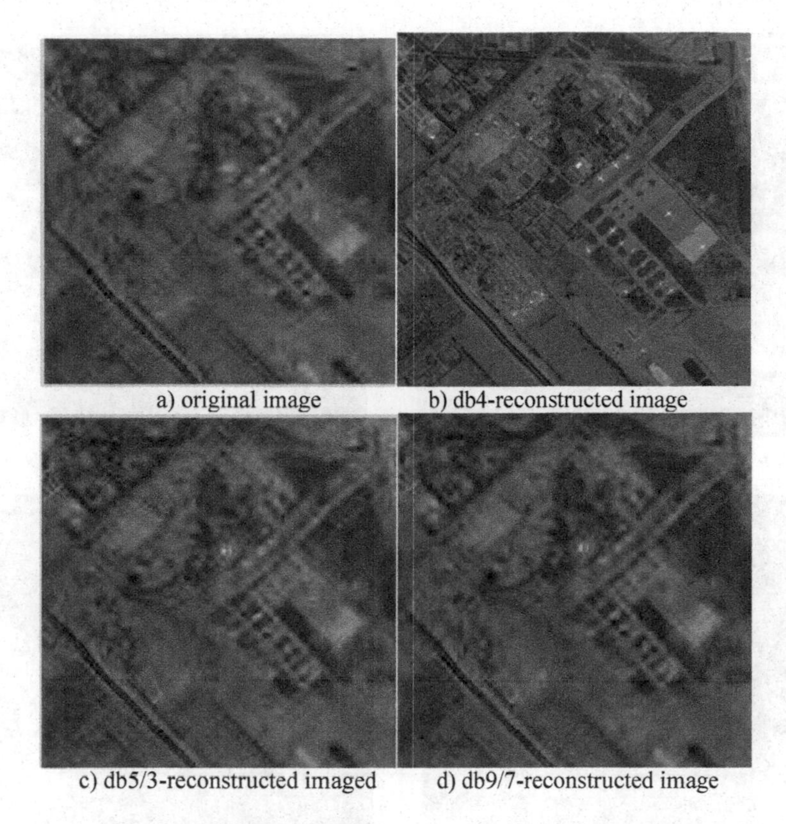

a) original image b) db4-reconstructed image

c) db5/3-reconstructed imaged d) db9/7-reconstructed image

Fig. 5. Reconstructed image of Taiyuan airport

4.3 Analysis of Experimental Results

The following conclusions can be drawn from analysis of experimental data in Tables 1 2, 3, 4, 5 and 6.

(1) A high support width corresponds to a high PSNR and a protracted length of compression and decompression. A support width of 10 is recommended. Wavelets with this property include db5, db6 and db9/7;
(2) The increase in PSNR slows down after the order of regularity exceeds 7. Hence, wavelet with a regularity order of 6 or less is recommended. Wavelets with this property include db2, db3, db4, db5/3 and db9/7;
(3) The increase in PSNR slows down after the vanishing moment exceeds 5, but the compression and decompression time still rises. Vanishing moment of around 5 is thus recommended. Wavelets with this property include db4, db5 and db9/7;
(4) Analysis of Tables 5 and 6 indicates that despite increase in the compression and decompression time, the symmetric wavelet is superior to non-symmetric wavelet in terms of PSNR.

To sum up, the d9/7 wavelet provides desirable compactness, regularity, vanishing moment, symmetry and high PSNR. It is learned from Figs. 5 and 6 that the images

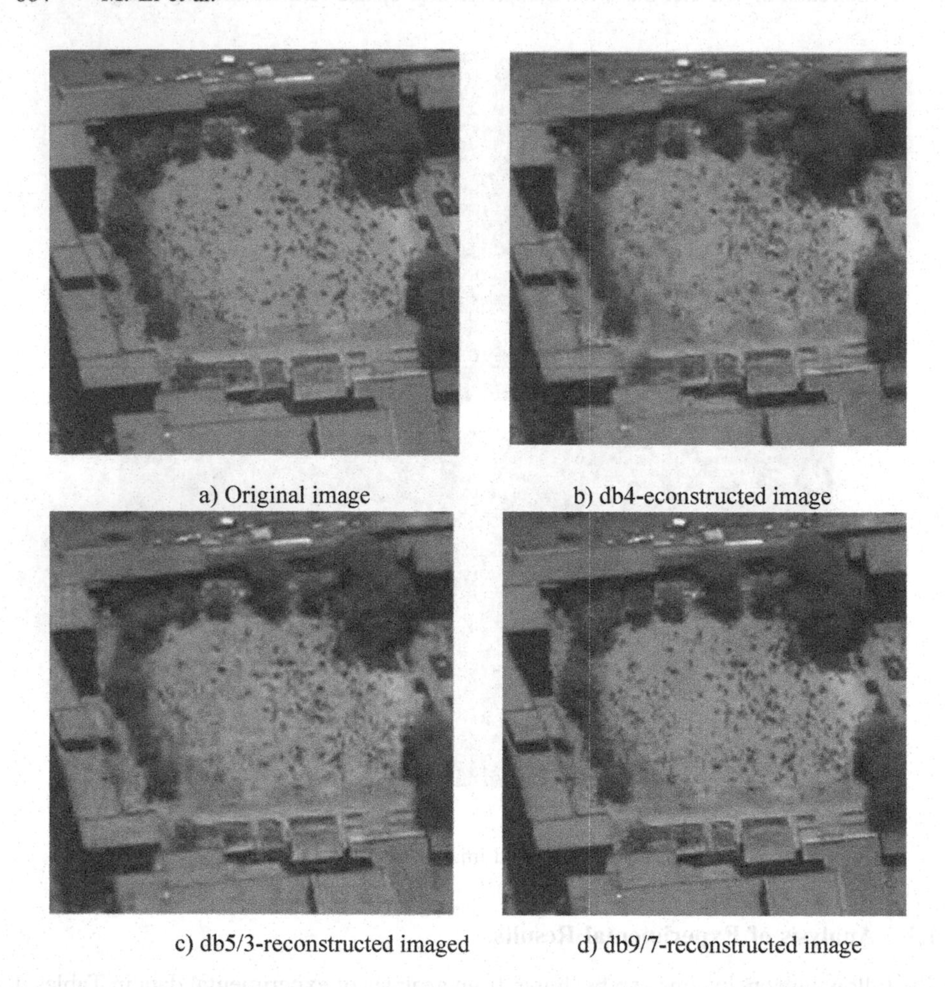

a) Original image b) db4-econstructed image

c) db5/3-reconstructed imaged d) db9/7-reconstructed image

Fig. 6. Reconstructed image of resolution crowd

reconstructed via the d9/7 wavelet are clearer. Hence, the d9/7 wavelet is selected as the wavelet basis for real-time compression of remote sensing images.

5 Conclusions

The characteristics of remote sensing images are described. Relationship of wavelet basis properties with spatial remote sensing images is analyzed. Simulations are performed on each wavelet basis propertiey through MATLAB. Experimental results are analytically summarized. Observation of the reconstructed image in the experiment shows that PSNR and image sharpness of the d9/7 wavelet are superior to other choices. Hence, the d9/7 wavelet is selected for real-time compression of remote sensing images.

Acknowledgements. The work was financially supported by the surface scientific and research projects of Jiamusi University (Grant No. 13Z1201576) and the basic research projects of Jiamusi University (Grant No. JMSUJCMS2016-009).

References

1. Chen, S.: Remote sensing image compression and DSP implementation based on wavelet transform. Ph.D. dissertation from Chinese Academy of Science, pp. 1–3 (2006)
2. Changhua hu, G., Tao liu, Z.: System analysis and design based on MATLAB 6.x., 2nd edn. Xi'an Electronic Science and Technology University Press (2004)
3. Daubechies, I., Sweldens, W.: Factoring wavelet transforms into lifting steps. J. Fourier Anal. Appl. **4**, 247–269 (1998)
4. Taswell, C.: A spectral-factorization combinatorial-search algorithm unifying the systematized collection of Daubechies wavelets. In: ICSSCC, Durban, South Africa, pp. 1–3 (1998)
5. Li, K.: Research on remote sensing image real time compression based on wavelet transform. Ph.D. dissertation from Chinese Academy of Science, pp. 69–74 (2005)
6. Kim, H.O., Kim, R.Y.: Characterizations of biorthogonal wavelets which are associated with biorthogonal multiresolution analyses. Appl. Comput. Harmon. Anal. **11**, 263–272 (2001)
7. Long, R., Chen, D.: Biorthogonal wavelet bases on Rd. Appl. Comput. Harmon. Anal. **2**, 230–242 (1995)
8. Cooklev, T., Nishihara, A., Sablatash, M.: Regular orthonormal and biorthogonal wavelet filters. Signal Process. **57**, 121–137 (1997)

Recognition of Tunnel Cracks Based on Deep Convolutional Neural Network Classifier

Min Yang[1(✉)], Qing Song[1,2], Xueshi Xin[1,2], and Lu Yang[1,2]

[1] Laboratory of Pattern Recognition and Intelligent Vision, Beijing University of Posts and Telecommunications, Beijing, China
hangman@bupt.edu.cn

[2] Beijing Songze Technology Co., Beijing, China
http://songzeai.com

Abstract. It is of vital importance to obtain timely identification and treatment of tunnel cracks in ensuring railway traffic safety. As the amount of tunnel images collected by high-speed cameras equipped on the train is extremely large, manual method is unable to meet the actual needs in recent years while traditional image processing methods become powerless to achieve a satisfying speed and accuracy. In this paper, we propose a novel algorithm based on deep learning to solve this problem. We first use Simple Linear Iterative Clustering super-pixel algorithm to segment the original tunnel data and then artificially annotate the images. And eventually establish a dataset called CLS-CRACK for training and testing the deep convolutional neural network classification model. We design our classification network on the basis of ResNet18 and finish the experiments with one NVIDIA GeForce GTX Titan X on a framework named caffe. Finally, results demonstrate that the new algorithm performs well in the validation set with the result of a 94.0% accuracy and an 83.0% recall and time cost for each original tunnel image is within 0.025 s.

Keywords: Tunnel cracks · Super-pixel segmentation
Deep learning · CNN classification model

1 Introduction

The appearance of cracks is inevitable during construction and use of the tunnels, while the existence of these cracks will affect the stability of the tunnel and have a negative impact on the train running. As a result, tunnel cracks must be effectively identified in time. Unfortunately, currently existed detection methods are unsatisfactory in efficiency and lack of intelligence. Most of them are manual methods, so the test results are unavoidably influenced by human factors and it is difficult to obtain a uniform standard. On the other hand, due to the large amount of data, the process of manual method is very slow and it also costs a

Q. Zhou et al. (Eds.): ICPCSEE 2018, CCIS 901, pp. 666–678, 2018.
https://doi.org/10.1007/978-981-13-2203-7_55

lot of money. However, there are also some different detection methods, including ultrasonic detection method, optical fiber sensor detection method, acoustic emission detection method and image processing detection method. Among them, the image processing detection method has the advantages of non-contact, high efficiency, convenience and intuitiveness, which make the kind of method become the main direction [1] of research in this field as the computer science and digital image processing technology develops continuously.

The tunnel image data has complex characteristics such as water stains, pollution and other structural seams. Combined with the influence of the environment including uneven lighting, noise, and irregular distribution of cracks, traditional methods based on image processing are faced with a bottleneck. In recent years, the technology of artificial intelligence is developing rapidly. Particularly, the deep learning method has fully penetrated into the field of computer vision and has made great achievements.

In this paper, we aim to apply the deep learning technology in the traditional tunnel cracks recognition field creatively to propose a new automatic algorithm to realize the fast and accurate recognition for tunnel cracks. And some experiments will be done to verify the algorithm.

2 Related Work

As mentioned above, the image processing detection method has been a main direction of research in this field and has made a lot of achievements at home and abroad. Another breakthrough is the development of deep learning which greatly contributes to computer vision.

2.1 Algorithm Based on Image Processing

Yusuke Fujita et al. proposed a two-step algorithm [2], which can effectively remove the noise in the image caused by uneven illumination, shadows, stains and so on. Later, a fractal monitoring algorithm [2] based on morphological processing and logistic regression using the statistical classification was proposed. This algorithm had more than 80% accuracy for crack extraction, but it still exists some defects, for example, it will miss some small cracks and the calculation is extremely large and as a result it is inefficient. In China, methods based on image processing have also developed at a fast speed. Li Gang et al. proposed an image segmentation algorithm based on Sobel operator and maximum entropy method [2] and Chu Yanli proposed a crack feature extraction algorithm based on gray-scale image and its texture characteristics [2]. Wang Xiaoming, Feng Xin et al. proposed a multi-image and multi-resolution pavement crack detection method [2] with image fusion technique, which preserves the collection characteristics of the image well and greatly improves the reliability and precision of crack detection. With the rapid development of machine learning, the algorithm of crack recognition combined with machine learning method and image processing is constantly emerging.

2.2 Deep Learning for Image Classification

The concept of deep learning was proposed creatively by Hinton et al. [3] in 2006, and later Lecun et al. [4] proposed a more intuitive multi-layer network learning algorithm named Convolution Neural Networks. CNN mainly uses image space information to reduce the number of training parameters and thus improve the performance of model training.

The purpose of image classification is to identify whether the image contains a specific class of object and the main content of the process is characterizing the image. Expectedly, CNN has great superiority in feature representation. As network's depth increases, the characteristics that model extracts are becoming more and more abstract and better to represent semantic topic of the image so that there exists less uncertainty in identification. Above all, the CNNs based on deep learning are stronger than other methods. 2012 was the year of deep learning, success of AlexNet [5] proves that the CNN can improve the effect of image classification. AlexNet uses an 8-layer network structure and won the champion in 2012 ImageNet [6] image classification competition. The paper also contributes in putting forward the skills including Relu function, local response normalization, overlapping pooling, data augmentation and dropout to prevent over-fitting. All of these provided us with a reference in training deep convolutional neural network model. In ILSVRC2013, the top 20 teams all used deep learning method under the great influence of AlexNet. In 2014, GoogLeNet [7] developed a new way to improve the recognition effect from the perspective of re-designing network structure. The main contribution of this paper is to design the inception module structure to capture the feature of different scales, and to reduce the dimension through the convolution of 1*1. Since then, many updated versions have been proposed based on GoogLeNet, which put forward more improved inception modules and provide valuable experience in network design for researchers. Another work in 2014 is VGG [8], the article further demonstrates the importance of network depth in enhancing the effectiveness of the model. The most important work of 2015 is the deep residual network [9] (ResNet), which proposed a method of fitting the residual network for better training the deeper network. This method is the big winner of the ILSVRC2015, which won the champion of the classification, segmentation, detection and many other tasks in the ImageNet and COCO datasets. The development of subsequent classification networks all followed the design ideas of ResNet such as DenseNet [10], which won the best paper award in CVPR2017. The model of our paper is also designed based on the basic network structure of ResNet.

3 Methodology

The scheme of our algorithm for tunnel cracks recognition includes image preprocessing, super pixel segmentation, dataset establishment, classification model design, training and model testing analysis. The main framework of the algorithm is shown in Fig. 1.

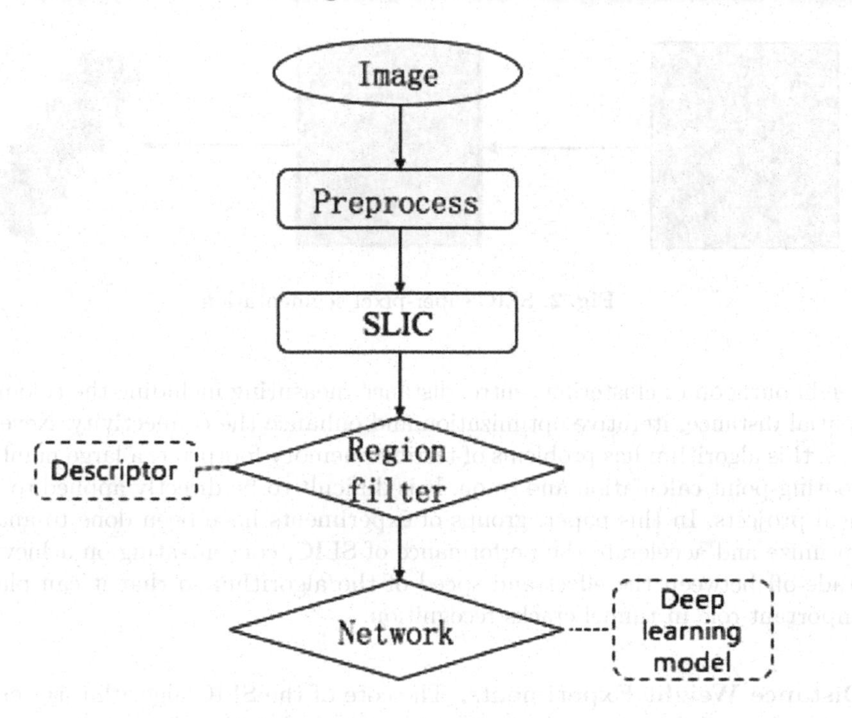

Fig. 1. Framework of system.

3.1 SLIC Segmentation

The size of original tunnel data obtained through camera is too large to handle. To adjust the deep CNN model, it is necessary to segment the image in some way. However, due to the non-uniformity of the distribution of cracks, the common segmentation method is very likely to make the cracks appear on the edge of the divided patches, which will adversely affect the subsequent classification model training. Therefore, our paper uses the SLIC segmentation method [11] to divide the original image data according to the generated super pixels. In this paper, several groups of experiments are performed to optimize and accelerate the SLIC method to adjust to the tunnel data and improve the segmentation effect. The overview of this process is shown in Fig. 2.

SLIC (Simple Linear Iterative Clustering) first transforms the RGB image to the 5-dimensional feature vectors in CIELAB colour space and XY coordinates and then constructs the standard of the distance measuring for these feature vectors. After the process, local clustering o f the image pixels is performed according to the measured distance. SLIC segmentation method can generate compact and uniform super pixels, which has got a comprehensive evaluation in terms of computing speed, object counter preserving and generated shape of super pixels among researchers.

Steps of SLIC algorithm include initializing the clustering centre, reselecting the clustering centre in n*n neighbourhood, assigning class labels to each pixel in

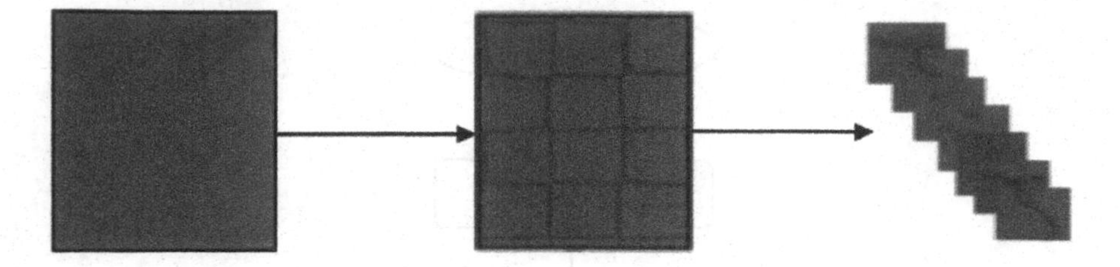

Fig. 2. SLIC super-pixel segmentation.

neighbourhood of clustering centre, distance measuring including the colour and spatial distance, iterative optimization and enhance the connectivity. Nevertheless, this algorithm has problems of terrible memory footprint, a large number of floating-point calculation and so on. It is difficult to be directly applied to practical projects. In this paper, groups of experiments have been done to analyse, optimize and accelerate the performance of SLIC, concentrating on achieving a trade-off between the effect and speed of the algorithm so that it can play an important role in tunnel cracks recognition.

Distance Weight Experiments. The core of the SLIC algorithm is iterative computing, while the core of iteration is calculating the distance. Because the source of clustering is a complex of LAB and XY, it will bring a very chaotic result if we directly use the European Distance measurement. Hence, the original paper proposed a new formula which coordinates the distribution of the two distance ratio with m and s parameters. The weight decay of m which stands for LAB colour space is 0.3, 0.6, 0.9 from left to right in Fig. 3. We can conclude that the edge between segmented super pixels is not so smooth when the weight decay is 0.3. However, if the weight decay is set very large, the result will lose the measure of XY coordinates. Taking the analysis into consideration, we finally set value to 0.6. The results are shown in Fig. 3.

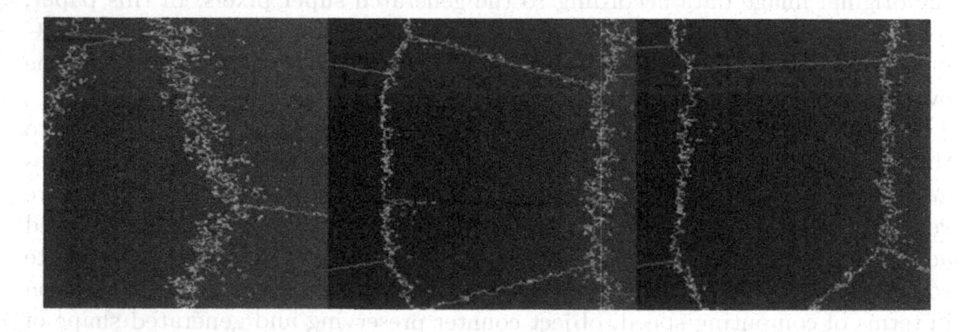

Fig. 3. Experiments with the change of parameter distance weight.

Acceleration Experiments. We both use the Python and C++ programming language to realize the SLIC segmentation algorithm. Considering the efficiency, we choose the C++ as the main language and carry out a series of optimization to accelerate the process. We finally adopt cuda [12] programming to run this algorithm on GPU. Experimental methods and results are shown in Table 1.

Table 1. SLIC acceleration experiments.

Method	Total time/s	Segment time/s
CPU common method	14.820	1.536
CPU acceleration	2.600	0.130
GPU running	0.050	

As shown in Table 1, despite considerably optimization in programming skills, the segmentation process on CPU is still very slow. Taking into consideration that subsequent CNN classification model training process is deployed on GPU (NVIDIA GeForce GTX Titan X) and GPU has obvious advantage in graphics operation than CPU, therefore, we apply cuda programming to transfer the SLIC segmentation algorithm into GPU in this paper. The final speed of the process is 0.05 s per image. Compared with CPU, the speed of segmentation process increases significantly in the case of similar effect.

All irregular super pixels segmented from the original images are saved in the same format to establish the dataset for tunnel cracks recognition task. Since SLIC algorithm helps to find the cracks potentially, we can control the divide process to avoid the cracks. Saved super pixel patches are shown in Fig. 4. These irregular super pixels maintain the shape of cracks well and effectively prevent the cracks appearing on the edge. This is in line with our expectation.

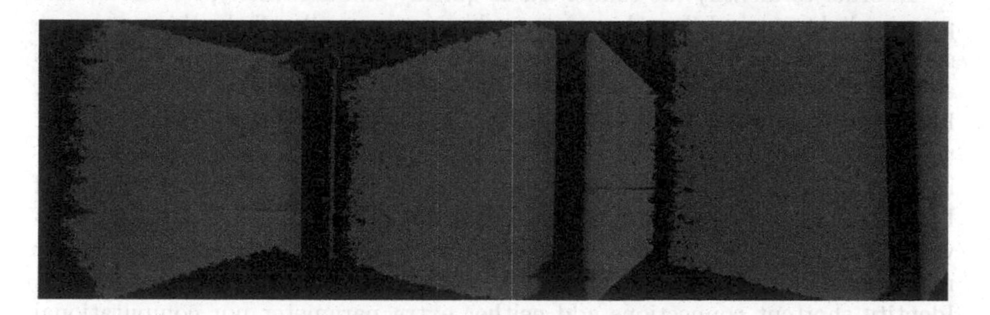

Fig. 4. Saved image from super pixels.

3.2 Dataset

The establishment of an appropriate dataset is critical for training deep CNN model in deep learning field. In order to solve the problem of tunnel cracks recognition, we construct a specific dataset named CLS-CRACK. The dataset selects three complete tunnel image data obtained originally from the camera as the source covering different road sections, lighting conditions and tunnel types and eventually has a total of 2000 image data. Among them, we choose 1600 for the training set and the other 400 for the validation set. Each image is segmented to about 100 super pixel patches by SLIC algorithm and these patches compose the CLS-CRACK dataset. After a series of data cleaning process, all images in the dataset are annotated by professional people and the label for classification is to determine whether the image contains cracks. At present, the training set has 6550 images and relevant classification labels and the validation set has 2000 images and the corresponding labels. The positive and negative sample ratio is basically 2:1. In the future, CLS-CRACK dataset can be extended by supplying detection and segmentation wise label to adapt to more diversified needs in this field.

3.3 Model Design

A complete classification network includes input, image preprocessing, inference process and output. In the training phase, the input includes the image and the corresponding label while input contains only the image in testing phase. To feed into the deep convolutional neural network, data preprocessing is required in training process including mean value subtraction, variance division, operation of image shuffling and so on. Our model is trained by sending images into model through mini-batch and then test images are sent into the trained model to get the inference results. Finally, the model outputs the probability whether the image contains cracks.

In order to identify the tunnel cracks quickly and accurately, we test several popular light network to achieve a speed/accuracy trade-off. In this paper, we choose the ResNet18 network to construct our classification model. ResNet is a kind of deep residual learning network instead of plain ones such as VGG. Formally, using H(X) to represent the optimal solution mapping, and then let the stacked nonlinear layer fit out another map: $F(X) = H(X) - X$. So, the original optimal solution H(X) can be rewritten as $F(X) + X$. The formulation of $F(X) + X$ can be realized by making a shortcut connection in the forward network, which can skip one or more layers. The shortcuts simply perform identity mapping, and their outputs are added to the outputs of the stacked layers. Identity shortcut connections add neither extra parameter nor computational complexity. The entire network can still be trained end-to-end by SGD with back-propagation, and can be easily implemented using common libraries without modifying the solvers.

Experiments are finished with one NVIDIA GeForce GTX Titan X on caffe [13]. We first test the speed of ResNet18 by tools in caffe and the time cost of a forward and back process is respectively 4.48 ms and 5.07 ms with an image of approximately 300 * 300 resolution. At the meantime, it is able to achieve a 0.29 top-1 error and a 0.1 top-5 error in the classification task of ImageNet, better than the state-of-the-art traditional method. Obviously, ResNet18 contains 18 convolution layers, among which the first layer is a 7 * 7 convolution layer followed by eight similar block module, each contains two 3 * 3 convolution layers and the last layer is a 1000 fully-connected layer. The network structure is shown in Fig. 5.

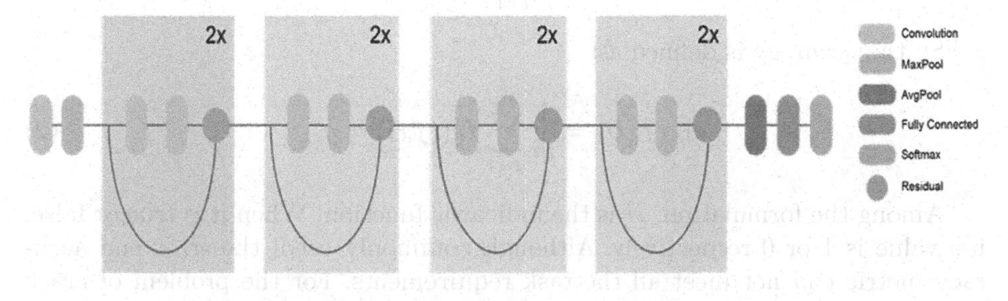

Fig. 5. Network structure of ResNet18.

4 Experiments and Analysis

Parameters of our network are found with straightforward SGD using caffe and are fine-tuned from the ImageNet classification pretrained model rather than random initialization. Since our work is a dichotomous task, we replace the last 1000-softmax fully connected layer with 2-softmax fully connected layer. Results show that fine-tuning the network is able to accelerate the convergence of training process and get a more satisfying result compared with training the network from scratch.

4.1 Implementation Details

Training is performed over 100 epochs, each consisting of full training image samples. The gradients for each iteration are estimated using mini-batches of size 15, and the learning rate steps down from 1e−3 to 1e−5 for 3 stages. All images are resized to 600 * 600 to feed the network and normalization process is necessary. In testing phase, the data preprocess is almost the same with that in training phase. Differently, we adopt a 'multi-crop' strategy to each test image so that each image will produce several crops of certain size. And apparently, our trained model will output a result to these crops after the inference. We simply sum the probability of crops produced by every image as the final result. As a result, this strategy outperforms the basic method.

4.2 Evaluation Methodology

In the classification task, the most common used performance metrics are error and accuracy. The error is the ratio of the number of misclassified samples to the total number of samples while the accuracy is the ratio of the number of correctly classified samples to the total number of samples [14]. Let's consider a set D, the classification error is given by following:

$$E(f;D) = \frac{1}{m}\sum_{i=1}^{m} \Pi((fx_i) \neq y_i) \tag{1}$$

So the accuracy is defined as:

$$acc(f;D) = \frac{1}{m}\sum_{i=1}^{m} \Pi((fx_i) = y_i) \tag{2}$$

Among the formulation, Π is the indicator function. When it is true or false, it's value is 1 or 0 respectively. Although commonly used, the error and accuracy metric can not meet all the task requirements. For the problem of crack recognition, the error and accuracy metrics measure the percentage of recognized images containing the cracks. However, if one is concerned about how many of the picked pictures contain cracks or how many of all pictures including cracks have been picked out, the error rate and accuracy are no longer enough. At this point, the precision and recall metrics are more appropriate for such requirements. For the dichotomous tasks, the situation combined by real categories and model predictions can be divided into four cases: true positive (TP), false positive (FP), true negative (TN) and false negative (FN). The precision (P) and recall (R) are respectively defined as:

$$P = \frac{TP}{TP + FP} \tag{3}$$

$$R = \frac{TP}{TP + FN} \tag{4}$$

Precision and recall are a pair of contradictory measure. In many cases, the samples can be ranked according to the prediction results of the model. And as a result, at the top of the list are most likely to be positive examples. Taking the sample as a positive prediction one by one according to the ranked sequence, we can calculate the current recall and precision and get the P-R curve by employing the precision as vertical axis and recall as horizontal axis. In this paper, we not only analyze these metrics but also evaluate the model's performance on the P-R curve.

4.3 Results and Analysis

Model uses softmax classifier whose final output of the network is probability value that if cracks exists in current image. In our experiment, we think it is true

Table 2. Experimental parameters and results.

Number	Crop method	Crop num	Accuracy %	Precision %	Recall %
1	600/600	1	92.7	40.0	81.0
2	600/632	12	94.0	46.0	83.0
3	600/632	12	82.8	22.5	95.3

if the probability value is greater than 0.5. As mentioned above, we perform a group of experiments using different crop strategy. The experimental parameters and results are shown in Table 2.

Taking into consideration that the single-crop (600/600) experiment as the baseline, we can conclude from Table 2 that the 12-crop (600/632) experiment has a significant improvement in accuracy, precision and recall compared with the first experiment with the result of 94.0, 46.0, 83.0 respectively. The reason is that a sample is cropped to 12 patches which contain different information of original data and our model will make a inference to all 12 patches, then a final result is combined with these probabilities with an averaging operation in the second experiment. The difference between the third experiment and the second one is the strategy to combine the final results of the 12 patches from the same one image. In the third experiment, we find the max probability among 12 values to define the classification result. This is reasonable because in this recognition task, people often hope to find all cracks despite some of them are not. While the tunnel data does not have so many true samples and the shape of cracks are elongated which are difficult to find, so it can improve the recall to consider the existence of cracks once identified in some state. Just as we can see from the Table 2 that the final recall is 95.3, a significant improvement than that in the second experiment.

We also compare our model with the state-of-the-art algorithm SVM [15] classifier using features extracted by this CNN model on our CLS-CRACK dataset and test it on the same validation set. The accuracy and speed performance is shown in Table 3.

Table 3. Experimental comparison results.

Method	Average accuracy/%	Time cost/s
Our algorithm	94.0	0.025
SVM	83.2	0.150

We adopt SVM classifier in 'scikit-learn' which is a popular machine learning package and set the best parameter to perform experiments on CLS-CRACK. The results are shown in Table 3 and we can conclude that our deep CNN classifier outperforms previous traditional methods.

4.4 Evaluation on P-R Curve

P-R curve can intuitively depicts an overall precision and recall of one classification model. The performance of classifier can be evaluated by analyzing the distribution of P-R curve and comparing the area under it. P-R curves of three groups of experiments are shown in Fig. 6.

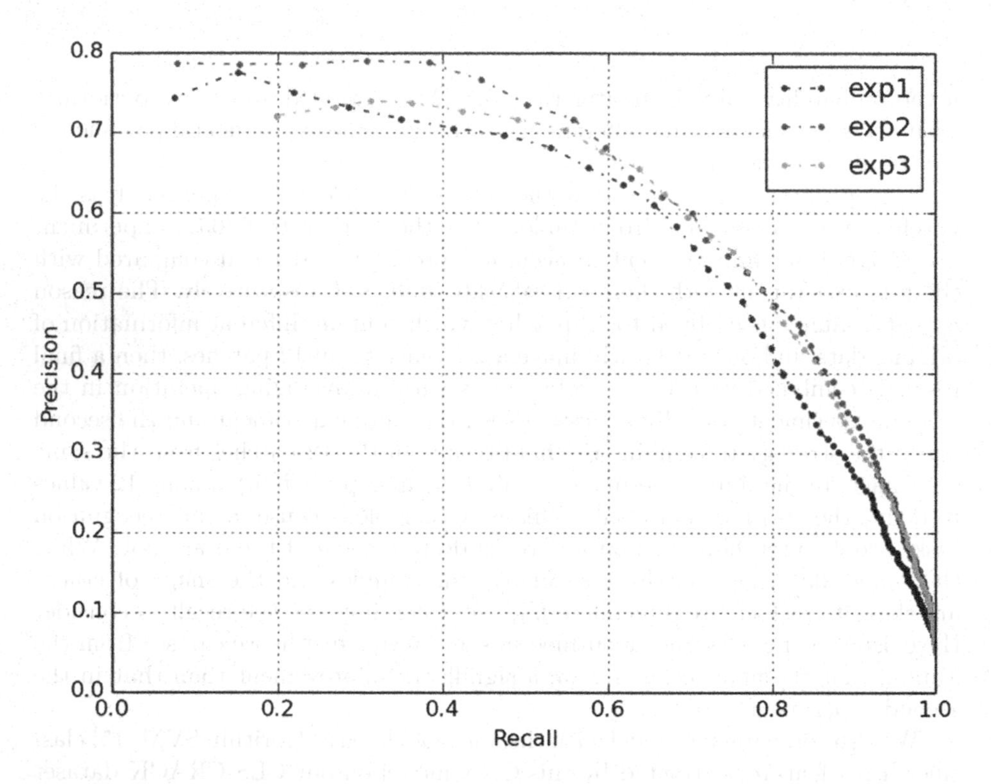

Fig. 6. P-R curves of three groups of experiments.

It can be seen from Fig. 6 that P-R curves of the second and third experiment completely covers that of basic experiment. That is, at any point on the curve, precision and recall of the last two experiments are higher than the first one so we can conclude the multi-crop strategy adopted in testing phase has a good effect to improve the performance of classifier. The P-R curves of the second and the third have an obvious overlap so it is difficult to simply evaluate the performance. Despite the recall of experiment 3 is higher than that of experiment 2 under the same score thresh as can be seen from Table 2, area under curve 2 is larger than that under curve 3 so that the possibility to get a high precision and recall at meantime on curve 2 is higher than on curve 3. Hence, performance of the CNN model can be appropriately evaluated by analyzing it's P-R curve. In addition, we can find the best thresh to distinguish the true or false examples according to the curve.

5 Conclusion and Further Work

We propose an algorithm based on deep learning to automatically solve the problem of tunnel cracks recognition in this paper. First, we specifically optimize the SLIC segmentation method to divide the original tunnel image to a proper size to simplify the preprocess and adjust to the CNN model. Then, we present a standard dataset named CLS-CRACK to train and test deep classification model. This project builds a classification network with ResNet18 and uses caffe to perform training and optimization. Experimental results demonstrate that the network can work on the dataset well and finally get a result of a 94.0% accuracy, an 83.0% recall and a 46.0% precision. Our model can recognize the tunnel cracks accurately and quickly and for new unknown tunnel images, it can be used offline to make satisfying predictions.

At present, the dataset has a small number of pictures, which limits the learning ability of the model so the scale of the dataset should be further improved. In addition, the bounding box annotation for detection and polygon annotation for segmentation can be complemented to meet a variety of mission needs. For the tunnel cracks recognition, we can further analyze the location and shape of the cracks after the cracks are correctly identified. This requires another important technique in deep learning: semantic segmentation. Semantic segmentation technology is a pixel-wise classification that enables the complete representation of cracks by their outline. Since the shape of crack is different from other common objects, the segmentation network for this task needs to be redesigned, and all of these are the directions for future work.

References

1. Wang, Y., Bai, B., Xu, X.: Research on metro tunnel crack identification algorithm based on image processing. J. Instrum. (2009)
2. Bai, B.: Research on image crack identification algorithm of subway tunnel surface. Master thesis (2015)
3. Hinton, G.E., Salakhutdinov, R.R.: Reducing the dimensionality of data with neural networks. Science **313**(5786), 504–507 (2006)
4. Sermanet, P., Chintala, S., LeCun, Y.: Convolutional neural networks applied to house numbers digit classification. In: 21st International Conference on Pattern Recognition (ICPR), pp. 3288–3291. IEEE (2012)
5. Krizhevsky, A., Sutskever, I., Hinton, G.: ImageNet classification with deep convolutional neural networks. In: Neural Information Processing Systems (2012)
6. Deng, J., Berg, A., Satheesh, S., Su, H., Khosla, A., Fei-Fei, L.: ILSVRC-2012 (2012). http://www.image-net.org/
7. Szegedy, C., et al.: Going deeper with convolutions. In: Computer Vision and Pattern Recognition (2015)
8. Simonyan, K., Zisserman, A.: Very deep convolutional networks for large-scale image recognition. In: International Conference on Learning Representations (2015)
9. He, K., Zhang, X., Ren, S., Sun, J.: Deep residual learning for image recognition. In: Computer Vision and Pattern Recognition (2016)

10. Huang, G., Liu, Z., Weinberger, K.Q.: Densely connected convolutional networks. arXiv preprint arXiv:1608.06993 (2016)
11. Achanta, R., Shaji, A., Smith, K., Lucchi, A., Fua, P., Susstrunk, S.: SLIC superpixels compared to state-of-the-art superpixel methods. IEEE Trans. PAMI **34**(11), 2274–2282 (2012)
12. Sermanet, P., Chintala, S., LeCun, Y.: Convolutional neural networks applied to house numbers digit classification. In: 2012 21st International Conference on Pattern Recognition (ICPR), pp. 3288–3291. IEEE (2012)
13. Jia, Y., et al.: Caffe: convolutional architecture for fast feature embedding. arXiv preprint arXiv:1408.5093 (2014)
14. Mensink, T., Verbeek, J., Perronnin, F., Csurka, G.: Metric learning for large scale image classification: generalizing to new classes at near-zero cost. In: Fitzgibbon, A., Lazebnik, S., Perona, P., Sato, Y., Schmid, C. (eds.) ECCV 2012. LNCS, pp. 488–501. Springer, Heidelberg (2012). https://doi.org/10.1007/978-3-642-33709-3_35
15. Bengio, Y., Goodfellow, I.J., Courville, A.: Deep Learning. MIT Press, Cambridge (2015)

Quality of Geographical Information Services Evaluation Based on Order-Relation

Yi Cheng, Wen Ge(✉), and Li Xu

Zhengzhou Institute of Surveying and Mapping, Zhengzhou 450052, China
lvwen1016@163.com

Abstract. With the widespread of web based geospatial services, more services are creating and publishing. There will be a large number of geographic information services with the same function which have different quality. Quality of Web Services (QoS) has become an important. Order-relation is introduced to study the quality of geographical information service evaluation and geographical information service matchmaking based on QoS in this paper. The calculating method of QoS indicators weights is discussed by three steps, which are confirming order-relation, comparing the relative importance between indicators and calculating weight factor. The quality of geographical information service indictors' matrix is built and normalized. The quality of geographical information service evaluation model is built by using simple additive weighted method. Geographical information services matching methods of QoS preference model based on order-relation and weight are discussed. Finally, experimental prototype is developed to practice and analyze the method of QoS evaluation and service matchmaking based on QoS. Analysis result shows that the proposed algorithm can improve service matching accuracy.

Keywords: Geographical information service · Quality of Service
Order-relation · QoS evaluation · Service matching · Preference model

1 Introduction

With the widespread of web based geospatial services, the availability and reliability of geospatial services have become the focus of attention. Quality of Web Services (QoS) has become an important factor to distinguish reliable services from faulty ones [1]. QoS makes it possible to discover, filter, acquire and interact with different web services in a reliable, efficient and easy-to-use manner. It is for this reason that the research of quality evaluation methods of geospatial services has been paid more and more attention nowadays.

In 2016, OGC Quality of Service and Experience Domain Working Group (QoSE DWG), which is concerning evaluating and improving the Quality of Service and the Quality of Experience of the Spatial Data Services, data sets and applications implementing and leveraging OGC standards, published its chapter, aiming to provides an open forum for the discussion and presentation of interoperability requirements, use cases, pilots, and implementations of OGC standards in this domain [2]. It gives a

Q. Zhou et al. (Eds.): ICPCSEE 2018, CCIS 901, pp. 679–688, 2018.
https://doi.org/10.1007/978-981-13-2203-7_56

strong signal that the study of geospatial services quality has become an important issue that needs to be raised to an OGC standards level.

2 Related Works

2.1 Conceptual Framework of Geospatial Information Service Quality

In fact, QoS is a fundamental concern in the field of SOA [3, 4]. WS-Quality Model (WS-QM), a conceptual model for web Services quality, which defines the concept, role and reciprocal action between quality associates, quality activities and quality factors in the lifecycle of Web Services, does not concern about the details of the web services [5]. In the other words, WS-QM can be applied in geospatial information service field with few modifications [6]. But in the level of implement, WS-Quality Factors (WS-QF), as an implementation specification of WS-QM for common web services which defines a set of attributes used to represent and evaluate the quality of a web service, is related with services fields closely [5]. For different Web service in special field, the quality factors will be different more or less due to the characteristics of the field. It is generally accepted that the common WS-QF cannot fulfill the quality of geospatial services [6–9, 12–14, 20]. Liu et al. indicated that there are two factors need to be modified: WS-QF does not make any descriptions of the inherent characteristics of geospatial data, and many problems can be simplified from web services to geospatial information service, especially for measurement methods [6]. Based on the analyze of differences between quality factors of geospatial services and common services, Wu et al. proposed a new concept of QoGIS (Quality of geospatial information service) [7–11]. Most of the conceptual frameworks including QoGIS are refer to WS-QM, so that they can be considered as a specific implementation of WS-QM in geospatial fields. This can not only reflect the characteristics of geographic information, but also keep the consistency with the common QoS models.

2.2 Evaluation Method of Geospatial Information Service Quality

In terms of evaluation methods of geospatial QoS, import and expand the existing common QoS evaluation model in the context of geospatial information application is a general solution of idea [13–16]. Cheng et al. established an extensible synthesis evaluation model to realize comprehensive evaluation of geographic information QoS through fuzzy comprehensive evaluation method [17]. Focused on the quality of location based services, Machaj et al. put forward parameters of quality and a criterion for QoS evaluation [18]. Liu et al. proposed a heterogeneous quality values to evaluate quality of geo-services [19]. A software tool GeoQoS supporting QoS-aware geoprocessing on the Web was developed by Yue et al. [20, 21].

2.3 Applications of Geospatial Information Service Quality

As mentioned above, Quality of geospatial information service plays important roles in applications. The quality of services is needed to be considered in most large complex systems like SDIs [23–26]. The quality of geographic information service has been proved to be able to significantly improve the efficiency of service discovery and selection [27–31]. In order to combine different services, quality factors were imported to service composition to optimize the service chain [32–34]. With the rapid growth of geospatial data services, geographic information services quality can also be developed as a special service for the data processing, integrating and analyzing [35–38].

In general, the quality of geographic information services has received enough attention and made considerable process in recent years. However, although mature tools for measuring and analyzing the QoS of geospatial services do exist, but there is little to support for a standardized way of communicating the expected QoS level of the services to the end users and various analyzing tools [2]. Likewise, we can say that quality features of geographic information services do exist, but there is little to evaluate by a standard method for varied user demands of it. Measuring and evaluating the quality of geographic information services needs to take full account to the needs of different users. There is thereby an urgent need but it is still a significant challenge to find a QoS evaluating method therefore.

In this paper, we present a kind of order relation analysis method for indicators weight calculation and quality evaluation of quality of geographic information services. Further study of the quality matching of geographic information services is conducted on the basis of it.

3 Methods

3.1 Weighting of Geographic Information Services Quality Indicators

Quality indicators of Geographic Information Services can be considered as a combination of geographic information quality indicators and web service indicators. According to ISO/TC 211 (Technical Committee) standard 19138, geographic information quality indicators are: completeness, consistency, positional accuracy, temporal accuracy, and thematic accuracy. Meanwhile, according to WS-QM, the major indicators of web services are: availability, accessibility, integrity, performance, reliability, regulatory and security. Therefore, as the combination of these indicators, quality of geographic information services is described by more than 8 independent indicators normally. It is obviously that the weight of these indicators is varied for different users. In response to this question, we introduced an order relationship [39] to weighting the indicators. Compared with Delphi method, Analytic Hierarchy Process, Subjective experience and Expert scoring method, order relationship method has the advantage of strong logically, small computation and sequencing. The workflow of the method is as follows.

Order Relationship Determination

1. If the importance degree of evaluation indicator q_i is bigger (or not less) than that of q_j, recorded as $q_i \succ q_j$.
2. If evaluation indicators set $q_1, q_2, \cdots, q_m$ has the sequence of $q_1^* \succ q_2^* \succ \cdots \succ q_m^*$, then considered the sequence relationship has established by "$\succ$" among these evaluation indicators. q_i^* is the i-th indicator sorted by "$\succ$".

The steps to establish the sequence relationship for $\{q_1, q_2, \cdots, q_m\}$ are as follows:

- Step 1: select the most importance indicator and recorded as q_1^* from $\{q_1, q_2, \cdots, q_m\}$;
- Step 2: select the most importance indicator and recorded as q_2^* from the rest m − 1indicators;
- ……
- Step k: select the most importance indicator and recorded as q_k^* from the rest m − (k − 1)indicators;
- ……
- Step m: recorded the last indicator as q_m^*.

Comparison of Relative Importance Between Evaluation Indicators. The relative importance given by evaluators between evaluation indicator q_{k-1} and q_k are recorded as $w_{k-1}/w_k = r_k$, $k = m, m-1, \cdots, 3, 2$, the value of r_k can refer to Table 1.

Table 1. Table reference value of r_k.

r_k	Description
1.0	q_{k-1} and q_k has the same importance
1.3	q_{k-1} is slightly important than q_k
1.5	q_{k-1} is obviously important than q_k
1.7	q_{k-1} is strong important than q_k
1.9	q_{k-1} is extremely important than q_k

There are the following theorems on the numerical constraints of r_k: if evaluation indicators set $q_1, q_2, \cdots, q_m$ has the sequence of $q_1^* \succ q_2^* \succ \cdots \succ q_m^*$, then $r_{k-1} > 1/r_k$, $k = m, m-1, \cdots, 3, 2$.

Weighting Factor Calculation. Weighting factor w_k of geographic information services qualities can be calculated according to the following formula:

$$\begin{cases} w_m = (1 + \sum_{k=2}^{m} \prod_{i=k}^{m} r_i)^{-1} \\ w_{k-1} = r_k w_k \end{cases} \quad (1)$$

3.2 Quality Evaluation of Geographical Information Services

A multi-attribute comprehensive evaluation method is needed to evaluate the quality of services based on the QoS indicator set. In this paper, we select simple linear weighting model for it is effective and easy to implement. The main idea is "synthesize" each attribute into a composite value in a linear weighted manner. Detailed steps are as follows:

Quality Index Matrix Building and Normalizing. Set the Geographic information services needed to be evaluated as $S = \{s_1, \cdots, s_n\}$, evaluation matrix and normalizing proceeding according to the following formula:

$$Q = \begin{pmatrix} q_1 \\ \vdots \\ q_n \end{pmatrix} = \begin{pmatrix} q_{11} & \cdots & q_{1m} \\ \vdots & \ddots & \vdots \\ q_{n1} & \cdots & q_{nm} \end{pmatrix} \Rightarrow \begin{pmatrix} q'_{11} & \cdots & q'_{1m} \\ \vdots & \ddots & \vdots \\ q'_{n1} & \cdots & q'_{nm} \end{pmatrix} \quad (2)$$

q_i is evaluate indicators of i-th service, q_{ij} is the j-th(j = 1, 2, …, m) indicator value of it, q'_{ij} is the normalized value of q_{ij} and m is the number of indicators. Due to the inconsistency between the dimension and the value range of the service quality indicators, different indicators cannot be compared and calculated in the same way. Therefore, it is necessary to normalize the service quality indicator value.

From the induction, we can see that there are two kinds of quality indicators: positive indicators and negative indicators. For positive indicators, higher value means higher quality, i.e. reliability, availability, reputation, integrity, logical consistency and topic accuracy etc. For negative indicators, higher value means lower quality, i.e. execution time, cost, location accuracy and timeliness etc. The normalized formulas for these two quality indicators are as follows:

- For positive indicators:

$$q'_{ij} = \begin{cases} \frac{q_{ij} - q_j^{\min}}{q_j^{\max} - q_j^{\min}} & (q_j^{\max} - q_j^{\min} \neq 0) \\ 1 & (q_j^{\max} - q_j^{\min} = 0) \end{cases} \quad (3)$$

- For negative indicators:

$$q'_{ij} = \begin{cases} \frac{q_j^{\max} - q_{ij}}{q_j^{\max} - q_j^{\min}} & (q_j^{\max} - q_j^{\min} \neq 0) \\ 1 & (q_j^{\max} - q_j^{\min} = 0) \end{cases} \quad (4)$$

$q_j^{\max}$ and $q_j^{\min}$ are the maximum and minimum values of a QoS indicators in the quality evaluation matrix Q.

Evaluation Model Building and Composite Value Calculation. Composite value is the ultimate criteria for measuring service satisfaction for evaluator. In order to obtain the composite value, we need to construct a linear weighted evaluation function as:

$$Score(s_i) = q'_i \cdot w = \sum_{j=1}^{n} (q'_{ij} \cdot w_j) \quad (5)$$

That is, the value obtained by normalizing each quality index is multiplied by the weight coefficient and then summed. $w = (w_1, \cdots, w_m)^T$ is the weight of indicators given by the evaluator and $Score(s_i)$ is the composite value of the QoS of geographic information service to be evaluated.

3.3 Geographical Information Services Matching Based on QoS

Services matching based on QoS is to find the best quality service according to user demands from all services to be selected. Different users will have their own preference on QoS indicators so that the best service is not the same one for different applications. The matching processes are as follows (Fig. 1):

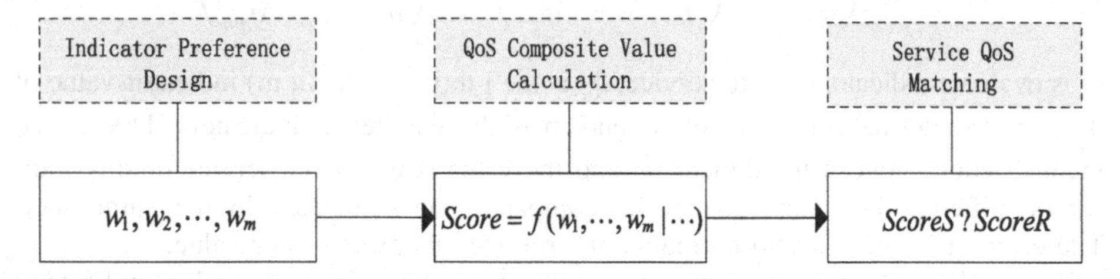

Fig. 1. Matching process based on QoS preference.

- Step 1: establish the weight of geographic information services quality indicators, thus establishing the user's preference for quality indicators.
- Step 2: calculate QoS composite values of geographic information services with Formula 5.
- Step 3: ranking services by their QoS *Score*, if $ScoreS \geq ScoreR$ means service S is better than R to meet the requirements.

4 Experiments

We collected 8 web feature gazetteer services from different web providers. All of these services have the same function but because of different sources, their services QoS attributes are different also. As showed on **错误!未找到引用源。**, QoS attributes are represented by 10 indicators, which is $\{q_1, q_2, \cdots, q_{10}\}$ = {executing time, reliability, usability, credibility, fee, integrity, logical consistency, position accuracy, timeliness, thematic accuracy} (Table 2).

Table 2. QoS indicators of 8 web feature gazetteer services.

Indicators	q_1	q_2	q_3	q_4	q_5	q_6	q_7	q_8	q_9	q_{10}
S1	80	0.9	0.9	7.8	30	0.8	0.9	1.5	1.2	8.0
S2	120	0.9	1.0	8.5	50	0.8	0.9	2.0	2.5	6.0
S3	160	0.5	0.4	5.0	80	0.4	0.6	4.5	1.5	7.5
S4	110	0.6	0.3	3.5	60	0.9	0.5	6.0	3.3	9.5
S5	25	0.8	0.9	8.5	20	0.5	0.8	1.0	1.1	6.0
S6	30	0.7	0.8	8.0	40	0.6	0.7	2.5	0.5	7.0
S7	45	0.9	1.0	9.2	20	0.5	0.8	6.0	1.5	5.5
S8	60	0.8	0.8	7.8	40	0.6	0.7	4.0	2.0	6.5

Using the simple linear weighting method described in this paper to evaluate the quality of geographic information service, construct the quality index matrix and normalize it as follows:

$$Q = \begin{pmatrix} q_1 \\ \vdots \\ q_{10} \end{pmatrix} = \begin{pmatrix} 80 & 0.9 & 0.9 & 7.8 & 30 & 0.8 & 0.9 & 1.5 & 1.2 & 8.0 \\ 120 & 0.9 & 1.0 & 8.5 & 50 & 0.8 & 0.9 & 2.0 & 2.5 & 6.0 \\ 160 & 0.5 & 0.4 & 5.0 & 80 & 0.4 & 0.6 & 4.5 & 1.5 & 7.5 \\ 110 & 0.6 & 0.3 & 3.5 & 60 & 0.9 & 0.5 & 6.0 & 3.3 & 9.5 \\ 25 & 0.8 & 0.9 & 8.5 & 20 & 0.5 & 0.8 & 1.0 & 1.1 & 6.0 \\ 30 & 0.7 & 0.8 & 8.0 & 40 & 0.6 & 0.7 & 2.5 & 0.5 & 7.0 \\ 45 & 0.9 & 1.0 & 9.2 & 20 & 0.5 & 0.8 & 6.0 & 1.5 & 5.5 \\ 60 & 0.8 & 0.8 & 7.8 & 40 & 0.6 & 0.7 & 4.0 & 2.0 & 6.5 \end{pmatrix}$$

$$\Rightarrow \begin{pmatrix} 0.59 & 1.0 & 0.86 & 0.93 & 0.83 & 0.8 & 1.0 & 0.9 & 0.75 & 0.63 \\ 0.3 & 1.0 & 1.0 & 0.88 & 0.5 & 0.8 & 1.0 & 0.8 & 0.29 & 0.13 \\ 0.0 & 0.0 & 0.14 & 0.26 & 0.0 & 0.0 & 0.25 & 0.3 & 0.64 & 0.5 \\ 0.37 & 0.25 & 0.0 & 0.0 & 0.33 & 1.0 & 0.0 & 0.0 & 0.0 & 1.0 \\ 1.0 & 0.75 & 0.86 & 0.88 & 1.0 & 0.2 & 0.75 & 1.0 & 0.79 & 0.13 \\ 0.81 & 0.5 & 0.71 & 0.79 & 0.67 & 0.4 & 0.5 & 0.7 & 1.0 & 0.5 \\ 0.85 & 1.0 & 1.0 & 1.0 & 1.0 & 0.2 & 0.75 & 0.0 & 0.64 & 0.0 \\ 0.74 & 0.75 & 0.71 & 0.75 & 0.67 & 0.4 & 0.5 & 0.4 & 0.46 & 0.25 \end{pmatrix}$$

Based on the surveying of 10 indicators, the order relations of them are:

$q_3 \succ q_1 \succ q_8 \succ q_2 \succ q_4 \succ q_5 \succ q_9 \succ q_7 \succ q_{10} \succ q_6 \Rightarrow q_1^* \succ q_2^* \succ q_3^* \succ q_4^* \succ q_5^* \succ q_6^* \succ q_7^* \succ q_8^* \succ q_9^* \succ q_{10}^*$

Importance comparisons of these indicators are:

$r_2 = \frac{w_1^*}{w_2^*} = 1.3$, $r_3 = \frac{w_2^*}{w_3^*} = 1.9$, $r_4 = \frac{w_3^*}{w_4^*} = 1.7$, $r_5 = \frac{w_4^*}{w_5^*} = 1.3$, $r_6 = \frac{w_5^*}{w_6^*} = 1.3$, $r_7 = \frac{w_6^*}{w_7^*} = 1.7$, $r_8 = \frac{w_7^*}{w_8^*} = 1.7$, $r_9 = \frac{w_8^*}{w_9^*} = 1.5$, $r_{10} = \frac{w_9^*}{w_{10}^*} = 1.3$.

According to Formula 1 we can calculate the indicator weights:

$w_1 = w_2^* = 0.263$, $w_2 = w_4^* = 0.081$, $w_3 = w_1^* = 0.342$, $w_4 = w_5^* = 0.063$, $w_5 = w_6^* = 0.048$, $w_6 = w_{10}^* = 0.009$, $w_7 = w_8^* = 0.017$, $w_8 = w_3^* = 0.138$, $w_9 = w_7^* = 0.028$, $w_{10} = w_9^* = 0.011$.

Firstly, we calculate the geographic information service QoS synthesis value which does not take the user's quality indicators preference into account, as shown in Fig. 2.

Then, we add the weight of user QoS preference, that is, the weight of geographic information service quality indicators, and calculate the comprehensive value of geographic information service QoS as shown in Fig. 3.

As shown in Figs. 2 and 3, it is clear to find S5 replace S1 to become the best result. That is to say considering the user's preference, S5 is more user-friendly than S2.

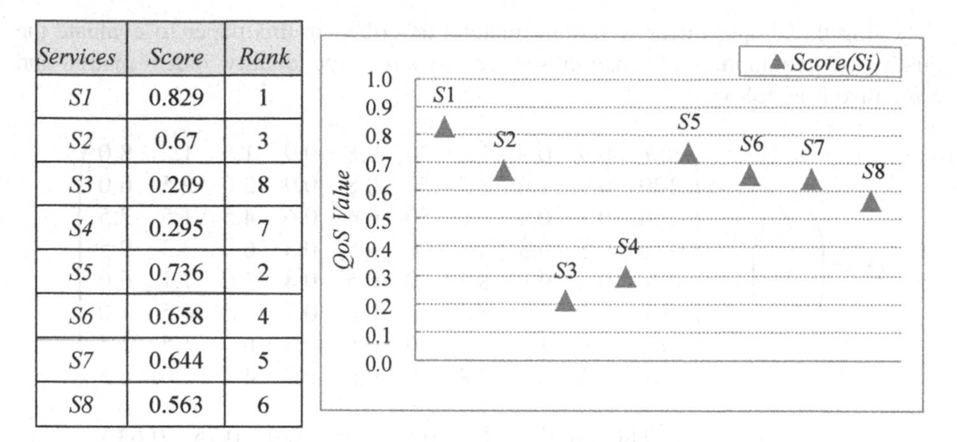

Services	Score	Rank
S1	0.829	1
S2	0.67	3
S3	0.209	8
S4	0.295	7
S5	0.736	2
S6	0.658	4
S7	0.644	5
S8	0.563	6

Fig. 2. QoS value and ranking without user's quality indicators preference

Services	Score	Rank
S1	0.805	2
S2	0.725	4
S3	0.133	8
S4	0.153	7
S5	0.897	1
S6	0.720	5
S7	0.790	3
S8	0.640	6

Fig. 3. QoS value and ranking with user's quality indicators preference

5 Conclusion

More and more geospatial web services are creating and publishing every day. There will inevitably be a large number of geographic information services with the same function but different QoS levels. Based on this background, we studied the evaluation method of geographic information services qualities based on order relation, and discusses the following three questions: (a) introduce the relationship between the index of geographic information service quality and put forward the method of determining the weight of geographic information service quality index based on order relation. (b) constructed and normalized a geographic information service quality indicator matrix and the evaluation model by simple linear weighting method. And (c) proposed a method of geographic information service matching based on QoS relation model.

References

1. Mani, A., Nagarajan, A.: Understanding quality of service for Web services (2002). https://www.ibm.com/developerworks/library/ws-quality/
2. OGC Project Document: OGC Quality of Service and Experience Domain Working Group (QoSE DWG) Charter (2016)
3. Xiao, X., Lionel, M.N.: Internet QoS: a big picture. IEEE Netw. **13**(2), 8–18 (1999)
4. Menssсé, D.A.: Qos issues in web services. IEEE Internet Comput. **6**(6), 72–75 (2002)
5. Kim, E., Min, D., Lee, Y., Kang, G.: Web services quality model v1.0. http://www.oasis-open.org/committees/tc-home.php?wg-abbrev=wsqm. Accessed 15 Nov 2008
6. Liu, Z., Wu, H., Chen, Y.: An introduction of WS-QF and the specialized factor set based on WS-QF for QoGIS. In: The 17th International Conference on Geoinformatics (Geoinformatics 2009), Fairfax, USA (2009)
7. Wu, H., Zhang, H., Liu, X., et al.: Adaptive architecture of geospatial information service over the internet with QoGIS embedded. In: Proceedings of the International Society of Photogrammetry and Remote Sensing (ISPRS) Workshop on Service and Application of Spatial Data Infrastructure, Hangzhou, China, vol. XXXVI (4/W6), pp. 53–57 (2005)
8. Zhang, H., Yueming, H., Huayi, W.: QoGIS supported OWS framework extension. Sci. Surv. Mapp. **4**, 51 (2011)
9. Wu, H., Li, D.: QoS: quality of geospatial information service. In: Proceedings of 6th International Conference on ASIA GIS, Malaysia (2006)
10. Huayi, W., Zhang, H.: QoGIS: concept and research framework. Geomat. Inf. Sci. Wuhan Univ. **32**(5), 385–388 (2007)
11. Wu, H., Zhang, H., et al.: Theory and Technology of Geographic Information Service Quality. Wuhan University Press, Wuhan (2011)
12. Gui, Z., Wu, H., Liu, W., Chen, Y.: The research on QoS assessment and optimization for geospatial service chain. In: Proceedings 17th International Conference on Geoinformatics, Fairfax, USA, pp. 1–5 (2009)
13. Onchaga, R.: Extending the Quality Concept in Geo-Information Process. Map Asia, Kuala Lumpur, Malaysia (2003)
14. Onchaga, R.: On quality of service and geo-service compositions. In: Proceedings of AGILE 2005 8th Conference on Geographic Information Science, Estoril, Portugal, pp. 519–528 (2005)
15. Subbiah, G., Alam, A., Khan, L., Thuraisingham, B.: Geospatial data qualities as Web Services performance metrics. In: Proceeding of the 15th International Symposium on Advances in Geographic Information Systems, ACM GIS 2007, Washington, USA (2007)
16. Moses, R.: Enabling quality of geospatial web services. In: Geospatial Web Services: Advances in Information Interoperability, pp. 33–63. IGI Global, Hershey (2011)
17. Cheng, D., Han, G., Chen, Y.: A research of extensible synthesis evaluation model for geographic information service of quality. In: 18th International Conference on Geoinformatics. IEEE, Beijing (2010)
18. Machaj, J., Brida, P., Majer, N.: Novel criterion to evaluate QoS of localization based services. In: Pan, J.-S., Chen, S.-M., Nguyen, N.T. (eds.) ACIIDS 2012. LNCS (LNAI), vol. 7197, pp. 381–390. Springer, Heidelberg (2012). https://doi.org/10.1007/978-3-642-28490-8_40
19. Liu, L., Fang, J., Liang, D.: A model for heterogeneous quality evaluation of Geographic web service. In: 21st International Conference on Geoinformatics. IEEE, Kaifeng (2013)

20. Yue, P., Tan, Z., Zhang, M.: GeoQoS: delivering quality of services on the Geoprocessing Web. In: Proceedings of OSGeo's European Conference on Free and Open Source Software for Geospatial (FOSS4G-Europe 2014), Europe, pp. 15–17 (2014)
21. Yue, P., et al.: GeoPW: laying blocks for the geospatial processing web. Trans. GIS **14**(6), 755–772 (2010)
22. Simonis, I., Sliwinski, A.: Quality of service in a global SDI. In: Proceedings of Form Pharaohs to Geoinformatics FIG Working Week 2005 and GSDI-8, Cairo, Egypt (2005)
23. Gao, S., Mioc, D., Yi, X.: The measurement of geospatial Web service quality in SDIs. In: 17th International Conference on Geoinformatics, pp. 1–6. IEEE, Fairfax (2009)
24. Luo, Y., Liu, X., Wang, W., et al.: QoS analysis on web service based spatial integration. J. Commun. Comput. **2**(7), 70–81 (2005)
25. He, L., Yue, P., Di, L., et al.: Adding geospatial data provenance into SDI—a service-oriented approach. IEEE J. Sel. Top. Appl. Earth Obs. Remote Sens. **8**(2), 926–936 (2015)
26. Liu, C., Liu, D.: QoS-oriented web service framework by mixed programming techniques. J. Comput. (Finland) **8**(7), 1763–1770 (2013)
27. Yang, S., Shi, M.: A model for web service discovery with QoS constraints. Chin. J. Comput. **28**(4), 589–594 (2005)
28. Guo, D., Ren, Y., Chen, H., et al.: A QoS-guaranteed and distributed model for web service discovery. J. Softw. **17**(11), 2324–2334 (2006)
29. Zheng, X., Wang, J.: Research on web service discovery model based QoS. Inf. Sci. **2**, 18 (2007)
30. Guo, G., Yu, F.: A method for semantic web service selection based on QoS ontology. J. Comput. **6**(2), 377–386 (2011)
31. Xin, M., Jiang, T., Zhang, R.: A QoS constraints location-based services selection model and algorithm under mobile internet environment. Int. J. Grid Distrib. Comput. **7**(2), 127–138 (2014)
32. Zhang, X., Li, Y., Wang, H.: A kind of QoS-sensitive web service composition method based on genetic algorithm. J. Shandong Univ. (Nat. Sci.) **9**, 11 (2007)
33. Song, X., Liu, J.: Optimization of GIS web service chaining based on QoS. J. Univ. Electron. Sci. Technol. China **2**, 34 (2010)
34. Du, W., Fan, H.: An automatic service composition algorithm for constructing the global optimal service tree based on QoS. In: 2010 IEEE International Geoscience and Remote Sensing Symposium (IGARSS), pp. 3976–3979. IEEE, Hawaii (2010)
35. Subbiah, G., Alam, A., Khan, L., et al.: Geospatial data qualities as web services performance metrics. In: Proceedings of the 15th Annual ACM International Symposium on Advances in Geographic Information Systems, p. 66. ACM, Washington (2007)
36. Xin, M., Qian, Q., Chen, Y., et al.: An approach to QoS-driven engine for spatial knowledge processing. In: Eighth ACIS International Conference on Software Engineering, Artificial Intelligence, Networking, and Parallel/Distributed Computing, 2007, SNPD 2007, pp. 33–38. IEEE, Qingdao (2007)
37. Hamdi, S., Bouazizi, E., Faiz, S.: A new QoS management approach in real-time GIS with heterogeneous real-time geospatial data using a feedback control scheduling. In: Proceedings of the 19th International Database Engineering & Applications Symposium, IDEAS 2015, pp. 174–179. ACM, Yokohama (2015)
38. Hamdi, S., Bouazizi, E., Faiz, S.: QoS management in real-time spatial big data using feedback control scheduling. ISPRS Ann. Photogramm. Remote Sens. Spat. Inf. Sci. **II-3/W5**, 243–248 (2015)
39. Gierz, G., Hofmann, K.H., Keimel, K., Mislove, M., Scott, D.S.: Continuous Lattices and Domains. Encyclopedia of Mathematics and its Applications, vol. 93. Cambridge University Press, Cambridge (2003)

High Precision Self-learning Hashing for Image Retrieval

Jia-run Fu[1], Ling-yu Yan[1(✉)], Lu Yuan[1], Yan Zhou[2], Hong-xin Zhang[1], and Chun-zhi Wang[1]

[1] School of Computer Science, Hubei University of Technology, Wuhan 430068, Hubei, China
lingboli1997@126.com

[2] Hubei Entry-Exit Inspection and Quarantine Bureau, Wuhan 430050, Hubei, China

Abstract. At present, hashing algorithm has been combined with deep learning to accelerate image retrieval. Against this background, there are many ways to construct hashing, but most of the methods do not show excellent performance in reducing semantic loss. At the same time, the vast majority of cases that adopt hashing algorithm and obtain successful cases involve the identification model requiring labels. So we propose a high precision with the combination of self-learning hash algorithm (HPSLH) to conduct experiments, the algorithm can not only through the analysis of the data itself, and construct a set of false label, then using the data from the identification model of deep learning can also avoid enormous semantic loss in the process of our hash. Through experiments on traditional datasets, this method can achieve the desired goal.

Keywords: Self-learning · Deep learning · Hashing · Image retrieval

1 Introduction

With the advent of the era of big data, data volume compared with the previous grew exponentially, extracted from a mass of complex image data we want to be associated with similar image data, is the core of this background attention. In this context, the content based image hashing efficient image retrieval has attracted the attention of researchers. Hashing maps high-dimensional features to the compact hash code, and then computes the hamming distance between two binary hash codes via a PC. For example, use the XOR operation to complete the operation. In the field of fast similarity search, hashing algorithm shows great advantages.

Based on this, more and more people began to use hashing to solve problems, and put their eyes on the direction of further improving operation speed and improving accuracy. For now, most of the hash algorithms are based on manual label, but manual label is quite limited. In order to solve the above problems, deep learning is introduced into the study of hashing algorithm. The first hash algorithm based on deep learning is the Semantic Hashing method proposed by Hinton research group [1]. After that, the combination of deep learning and hash is widely studied. The emergence of CNNH [2] opened a new chapter in the study. In the experiment, CNNH has achieved remarkable

Q. Zhou et al. (Eds.): ICPCSEE 2018, CCIS 901, pp. 689–697, 2018.
https://doi.org/10.1007/978-981-13-2203-7_57

performance improvement compared with traditional methods based on manual design feature. However, this method is still not an end-to-end learning method, and the learned image representation can't react to the update of binary encoding, so it is not able to fully develop the ability of deep learning. After that, there are many new methods, such as NINH [3] based on triples. DSRH [4] which directly optimizes the final evaluation index. And insert a new full connection layer to provide scope constraints for DLBHC [5]. To sum up this type of method, it is generally used as a deep model to learn the image representation + sigmoid/tanh function to limit the output range + different loss functions, and this type of study also has some representative articles [6–9]. The focus of this approach is on the quantified part, but as before, it has not escaped the dependency on labels. Also, several researchers combined CNN with hash algorithm [15–19], and found that the experimental results were very satisfactory. But these studies, like the past, rely on labels to make them poorly adapted to the sample, and the loss of semantic meaning has never been resolved.

For the above problems, we propose high precision self-learning hashing (HPSLH) algorithm. HPSLH can adapt to the situation without a label, and automatically generated hash label for end-to-end learning. At first, the specific process to use machine learning to construct the objective function, to minimize, save the hash code of neighborhood structure is obtained. After that, we can further minimize the target function value and get the exact hash code. Then, at the second part, using a relatively simple deep learning network structure, with generated hash labels. The generated hash label is used for end-to-end learning. In the following article, we will introduce and demonstrate the method in the second section. In the third section, we used traditional data sets to test our algorithm. Finally, we reached the conclusion in Sect. 4.

2 Related Work

Hashing is a technique that maps feature data to compact hash code, similar objects are assigned to a tight hamming code. Since hashing based on deep learning [1], data conversion to hash code has become the focus of attention. Let's start with the data transformation section. Shallow hashing provides the basic concepts of data transformation to hash code. In order to maintain the similarity in the hashing process, the shallow learning algorithm always combines the various shallow features extracted by different mapping methods. Due to the inherent defects of shallow feature extraction, these methods are surpassed by deep hashing. In order to adapt to the situation without labels, the structure of self-study in shallow learning is noteworthy. For example STH [13], STH [13] focuses on the local similarity structure, namely, for each data point, it is a nearest-neighbors k. This method maintains the semantic relation between data and the ability to control the length of hash code by mapping the characteristic data relative to Euclidean distance. Therefore, it provides a strong correlation hash code and shows good performance. In this most typical example, the difference between our algorithm and STH is as follows: (1) HPSLH is based on automatic feature extraction rather than manual production; (2) HPSLH adopts the depth model; (3) we adopt iterative optimization to achieve better semantic preservation.

In order to make the hash code has more efficiency and accuracy, the researchers proposed some data-aware hash method of machine learning techniques to hash area, to improve the effectiveness of the hash code – [23–25]. Due to the characteristics of the end-to-end between tasks and interactive learning, deep learning algorithm makes the extracted image features better suited to the task itself. A typical example is CNNH algorithm [15]. Although CNNH [15] also needs to rely on the classification tag, it maps the classification tag to hash label, and finally realizes the end-to-end deep hash learning. CNNH solves the problem of insufficient labels to generate hash code through deep learning. However, the label is used to generate hash labels that are independent from the label, ignoring the relevance of the image semantics. In general, this method simply departs the extracted features, resulting in the loss of semantics. In addition, these simple hash methods are not much improved in terms of compactness.

Through the above examples, we realize that better feature extraction methods generate feature vectors, thus producing more accurate hash labels to improve the accuracy of hashing functions. So, we introduced our algorithm.

3 High Precision Self-learning Hashing

The custom hash framework proposed in this paper is composed of hash label creation and hash function learning phase. First, we use CNN (such as AlexNet or GoogLeNet) to obtain image features. In the process, you can choose whether to use the classification label. On this basis, KNN algorithm is used to create the feature graph model. Then the calculation method proposed in this paper is used to map it. Finally, we apply the binarization method to map the result to hash code as the newly created hash label. This method takes advantage of deep learning in feature extraction. And local iteration is used to reduce the loss of semantics in the mapping phase. In order to speed up the construction of hash code and minimize the loss of accuracy, we chose a complex network during the label creation phase and chose a simple network during the function learning phase. Next, we'll look at HPSLH in detail.

A. Label Creating Stage

Given training feature $X = [x_1, x_2, \ldots, x_n] \in R^{p \times n}$, we want to get hash codes denoted by $H = [h_1, h_2, \cdots, h_n]^T \in \{1, -1\}^{n \times c}$ (c is the length of the hash codes). We construct an n-by-n similarity matrix S by using the structural characteristics of local iteration:

$$S_{ij} = \begin{cases} 1, & \text{if } x_i \in N_{k(x_j)} \text{ } or \text{ } x_j \in N_{k(x_i)} \\ 0, & otherwise \end{cases} \quad (1)$$

where $N_k(x)$ represents the set of k-nearest-neighbors of feature vector x.

The hamming distance between two binary code h_i and h_j (corresponding to the feature x_i and x_j is given by the difference between them). We seek to minimize the weighted average hamming distance, representing semantic loss and making the results more accurate.

$$\sum_{i=1}^{n}\sum_{j=1}^{n} S_{ij}\|h_i - h_j\|^2 \tag{2}$$

To achieve the implementation of a large dataset, we need to find a way to convert the new feature data into a binary hash code. Here we use the linear transformation as a hash function to implement and optimize the simplicity.

Given a feature data x_i, the $l-th(1 \leq l \leq c)$ hash function h_l which is defined as:

$$h_l(x_i) = sign(w_l^T x_i + b_l) \tag{3}$$

where $w_l \in R^{p\times 1}$ is the transformation matrix and $b_l \in R$ is the bias term.

Because the constraint $H = [h1, h2, \ldots, hn] \in \{1, -1\}^{c\times n}$ makes the objective function to be an NP-hard problem, we relax that constraints to make the problem computationally solvable. After that, we construct a joint framework which aims to minimize semantic loss and empirical error simultaneously. The final objective function turns out to be:

$$\begin{aligned} &\arg\min_{H,W,b} \sum_{i,j=1}^{n} S_{ij}\|h_i - h_j\|^2 \\ &+\phi\left(\|X^T W + 1b - H\| + \gamma\|W\|_F^2\right) \\ &s.t\, HH^T = I \end{aligned} \tag{4}$$

W = [w1, w2, …, wc] ∈ R b = [b1, b2, …, bc] ∈ R (c is the length of the hash codes), and 1 is a vector of all ones, F is a regularization function φ and γ are parameters.

To get the an optimal solution of the objective function, we need to first minimize the objective function with respect to W and b. Set the derivative of (4) with respect to b to zero, we have

$$\begin{aligned} &1^T(X^T W + 1b - H) = 0 \\ &\Rightarrow b = \frac{1}{n}(1^T H - 1^T X^T W) \end{aligned} \tag{5}$$

Set the derivative of W to 0. We have:

$$X(X^T W + 1b - H) + \gamma W = 0 \tag{6}$$

Using the obtained result of (5), we transform (6) to be as follows:

$$\begin{aligned} &XX^T W + X1\left(\frac{1}{n}(1^T H - 1^T X^T W)\right) - XH + \gamma W = 0 \\ &\Rightarrow W = (XAX^T + \gamma I)^{-1} XAH \end{aligned} \tag{7}$$

Where $C = A - AX^TBXA, B = (XAX^T + \gamma I)^{-1}$

So we get:

$$\begin{aligned} &\|X^TW + 1b - H\|_F^2 + \gamma\|W\|_F^2 \\ &= trH^TCH \end{aligned} \tag{8}$$

Apart from that, the first part of the objective function can be transformed to be follows:

$$\sum_{i,j=1}^{n} S_{ij}\|h_{i\_}h_j\|^2 = tr(H^T(N - S)H) \tag{9}$$

where $N_{ii} = \sum_{j=1}^{n} S_{ij}\Big)$ and other elements are zero. Combine the (8) and (9), the objective function turns to be:

$$\begin{aligned} &\arg \min_{H,W,b} tr(H^T(N - S + \phi C)H) \\ &s.t. HH^T = I \end{aligned} \tag{10}$$

In the end, we take the minimum n of eigenvalues of $(N - S + \varphi C)$ (the length of the hash code) to be the hash label we want. After that, H is binarized to be $\{-1, 1\}$ for further optimization. The hash label as final result is H.

B. Hash Function Learning Stag

In this stage, we implement an end-to-end hashing deep learning to learn hash function.

Firstly, using hash labels acquired in hash label generating stage, we employ CNNs again to receive fine grained features. After that we used MLP to approximate the hash label. The MLP includes an input layer, an output layer, and several hidden layers. We can use Backprop (backward propagation of errors) [21] algorithm to realize its modeling, the algorithm has the characteristics of simple structure, easy to implement.

We use an artificial neural network (ANN [20]) which consists of input layer, hidden layer and output layer. It can adjust the input according to different weights of different nodes in the hidden layer. After CNNs, the reason why we chose the single hidden-layer MLP to learn the hash tag is that CNNs is also a conversion model of MLP. Therefore, we can construct the end to-end deep learning framework of ANN under multi-output conditions.

4 Experiment

In this part, we designed experiments to test traditional datasets. Specifically, we try to use this algorithm to perform performance tests on traditional datasets. Then, we compare our methods with other hash algorithms, including some hand-labelling monitoring hash algorithms and some unlabeled unsupervised hash algorithms. Finally, different data sets are analyzed and summarized. The results show that the algorithm is not only advantageous, but also has some problems.

In the data set selection, we chose the mnist data set. MNIST is a classic primer on deep learning. MNIST is composed of 60,000 training images and 10,000 test images, each of which is 28 * 28 and is made of black and white. These images are collected by different people from 0 to 9. These pictures are not in the traditional sense of the PNG or JPG format images, because the PNG or JPG image format, will have a lot of interference information, these images will be processed into a simple two dimensional array. Taking mnist data set as the test model of the algorithm, it is beneficial for us to get the intuitive result and modification.

At the beginning, let's look at the results of other algorithms on the mnist.

We choose contrast algorithm is as follows: using hyperplane to separate the data points to two different sets of SPH [11], the first data set of original space dimension of itq [12], fast similarity search STH [13], the optimization of the final result of CNNH [14], as well as the depth of self-study DSTH [22]. For these algorithms, we can see that the exact rate is basically the same. The difference is the process of reaching the final result.

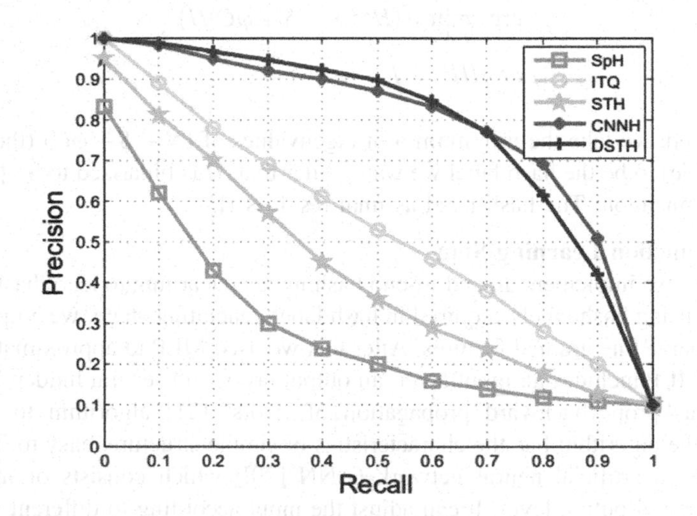

Fig. 1. Results of the comparative algorithm

Next, let's look at the parameter Settings of the HPSLH algorithm (Table 1).

Table 1. Parameter settings

Step	Type	Kernel size	Pad	Stride	Output	Pool style
1	conv	5 × 5	2	1	32	
2	pool	3 × 3	0	2	1	Max
3	conv	5 × 5	2	1	32	
4	pool	3 × 3	0	2	1	Ave
5	conv	5 × 5	2	1	64	
6	pool	3 × 3	0	2	1	Ave

We convert mnist into the form of mnis_unit8, with the following parameters:

test_x	10000x784	7840000	uint8
test_y	10000x10	100000	uint8
train_x	60000x784	47040000	uint8
train_y	60000x10	600000	uint8

Fig. 2. mnis_unit8

Finally, let's take a look at the HPSLH test. 1. Finish the training time and test time after the label creation.

Time Train,	Time Test
0.0377	0.0035

The MNIST database of handwritten digits, has a training set of 60,000 examples, and a test set of 10,000 examples.

Different from the other algorithms shown earlier. In the final accuracy experiment, we randomly selected the scale of the test and the sample of the test, which is more conducive to our analysis of the performance of this algorithm (Fig. 3).

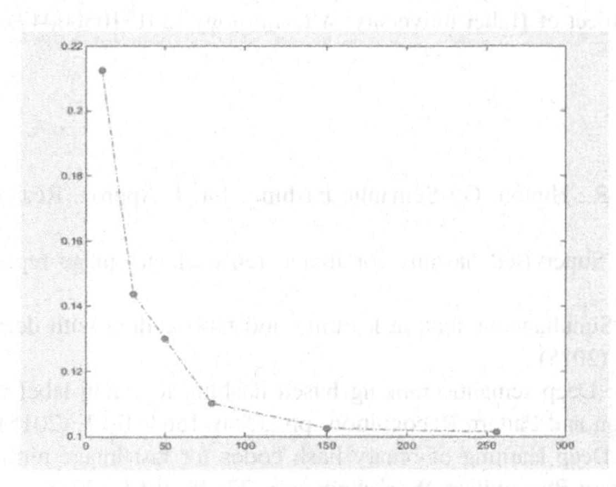

（Y: precision X: recall）

Fig. 3. Result of HPSLH

From Fig. 2, we can see that the final result of HPSLH is not very different from the final result of Fig. 1. However, in the process of change, we can see that the experimental process of HPSLH is more stable and rapid compared with the comparison algorithm. Compared with the results of the algorithm rely on the labels, HPSLH can obtain the same exact result without relying on the original labels.

Since the algorithm is not fully mature, our experiment is only conducted here so that we can improve and adjust it.

5 Conclusion

This paper presents a self-learning hashing algorithm based on deep learning. It is a unified model for obtaining hash codes for training data. In the hash label creation and hash function learning phase, it can not only generate the hash label with guaranteed accuracy under unlabeled conditions, but also save time. While no other data sets have been tested, the performance of a small range of test results in a traditional dataset is ideal. So we believe the algorithm is very promising. Comparative experiments show that our scheme has advantages in the stability of the proposed scheme. The final accuracy is not much different from other algorithms, but the process is relatively stable. In the same way, the algorithm still needs to improve the test and algorithm of the larger data set. However, we conclude from the experimental results that our algorithm is superior to other algorithms in the current test range. Especially in terms of accuracy and stability.

Acknowledgement. Project supported by the National Natural Science Foundation (61502155, 61772180); Education cooperation and cooperative education project (201701003076); Research start-up fund of Hubei university of technology (BSQD029); University student innovation and entrepreneurship project of Hubei university of technology (201710500047).

References

1. Salakhutdinov, R., Hinton, G.: Semantic hashing. Int. J. Approx. Reason. **50**(7), 969–978 (2009)
2. Xia, R., et al.: Supervised hashing for image retrieval via image representation learning (2014)
3. Lai, H., et al.: Simultaneous feature learning and hash coding with deep neural networks, pp. 3270–3278 (2015)
4. Zhao, F., et al.: Deep semantic ranking based hashing for multi-label image retrieval. In: Computer Vision and Pattern Recognition, pp. 1556–1564. IEEE (2015)
5. Lin, K., et al.: Deep learning of binary hash codes for fast image retrieval. In: Computer Vision and Pattern Recognition Workshops, pp. 27–35. IEEE (2015)
6. Zhang, R., et al.: Bit-scalable deep hashing with regularized similarity learning for image retrieval and person re-identification. IEEE Trans. Image Process. **24**(12), 4766–4779 (2015)
7. Liu, H., et al.: Deep supervised hashing for fast image retrieval. In: IEEE Conference on Computer Vision and Pattern Recognition IEEE Computer Society, pp. 2064–2072 (2016)

8. Zhu, H., et al.: Deep hashing network for efficient similarity retrieval. In: Thirtieth AAAI Conference on Artificial Intelligence, pp. 2415–2421. AAAI Press (2016)
9. Li, W.J., Wang, S., Kang, W.C.: Feature learning based deep supervised hashing with pairwise labels, pp. 1711–1717 (2015)
10. Gardner, M.W., Dorling, S.R.: Artificial neural networks (the multilayer perceptron)—a review of applications in the atmospheric sciences. Atmos. Environ. **32**(14–15), 2627–2636 (1998)
11. Heo, J.P., et al.: Spherical hashing: binary code embedding with hyperspheres. IEEE Trans. Pattern Anal. Mach. Intell. **37**(11), 2304–2316 (2015)
12. Gong, Y., et al.: Iterative quantization: a procrustean approach to learning binary codes for large-scale image retrieval. IEEE Trans. Pattern Anal. Mach. Intell. **35**(12), 2916 (2013)
13. Zhang, D., et al.: Self-taught hashing for fast similarity search. In: International ACM SIGIR Conference on Research and Development in Information Retrieval, pp. 18–25. ACM (2010)
14. Liu, Y., et al.: FP-CNNH: a fast image hashing algorithm based on deep convolutional neural network. Computer Science (2016)
15. Xia, R., Pan, Y., Lai, H., Liu, C., Yan, S.: Supervised hashing for image retrieval via image representation learning. In: AAAI, pp. 2156–2162 (2014)
16. Lin, K., Yang, H.-F., Hsiao, J.-H., Chen, C.-S.: Deeplearning of binary hash codes for fast image retrieval. In: CVPR Workshops, pp. 27–35 (2015)
17. Lai, H., Pan, Y., Liu, Y., Shuicheng, Y.: Simultaneous feature learning and hash coding with deep neuralnetworks. In: CVPR, pp. 3270–3278 (2015)
18. Dong, Z., Jia, S., Wu, T., Pei, M.: Face video retrieval via deep learning of binary hash representations. In: 12th IEEE Transactions on Multimedia, pp. 3471–3477. AAAI (2016). **XX**(X), December 2014
19. Guo, J., Zhang, S., Li, J.: Hash learning with convolutional neural networks for semantic based image retrieval. In: Bailey, J., Khan, L., Washio, T., Dobbie, G., Huang, J.Z., Wang, R. (eds.) PAKDD 2016. LNCS (LNAI), vol. 9651, pp. 227–238. Springer, Cham (2016). https://doi.org/10.1007/978-3-319-31753-3_19
20. Gardner, M.W., Dorling, S.: Artificial neural networks (the multilayer perceptron)ła review of applications in the atmospheric sciences. Atmos. Environ. **32**(14), 2627–2636 (1998)
21. Sinyavskiy, O., Polonichko, V.: Apparatus and methods for backward propagation of errors in a spiking neuron network. US9489623 (2016)
22. Zhou, K., et al.: Deep self-taught hashing for image retrieval, pp. 1215–1218 (2015)
23. He, K., Wen, F., Sun, J.: K-means hashing: an affinity-preserving quantization method for learning binary compact codes. In: IEEE Conference on Computer Vision and Pattern Recognition, pp. 2938–2945. IEEE Computer Society (2013)
24. Heo, J., Lee, Y., He, J., Chang, S.F., Yoon, S.: Spherical hashing. In: Proceedings of the IEEE Computer Society Conference on Computer Vision and Pattern Recognition, pp. 2957–2964 (2012)
25. Lai, H., et al.: Sparse learning-to-rank via an efficient primal- dual algorithm. IEEE Trans. Comput. **62**(6), 1221–1233 (2013)

Face Detection and Recognition Based on Deep Learning in the Monitoring Environment

Chaoping Zhu[1,2(✉)] and Yi Yang[1,2]

[1] School of Computer Science and Information Engineering, Chongqing Technology and Business University, Chongqing 400067, China
jsjzcp@163.com

[2] Detection and Control of Integrated Systems Engineering Laboratory, Chongqing 400076, China

Abstract. In the construction of smart cities and public safety, the number of cameras has increased dramatically, the artificial video management cannot meet the demand of urban construction. Therefore, intelligent video monitoring technology has become a research hotspot. In the actual video monitoring,the monitoring environment is very complex, Face recognition in video is often difficulty. This paper research the face recognition problem of monitoring video in the city, propose a mothed of detection and recognition method based on deep learning. The detection part adopts the fast speeding YOLO2 algorithm, the recognition part adopts high accuracy ResNet algorithm. Using WIDERFACE face detection database as training data set, Use the CASIA_Webface database for validation experiments. The experimental data show that the mothod YOLO2 algorithm and ResNet algorithm can complete detection and recognition of human face in monitoring video. the result is better. To further verify the recognition effect, the data set collected by the actual camera was tested. the test results show that the real time and accuracy of the system can meet the needs of practical engineering application.

Keywords: Deep learning · Rapid detection · Face detection · YOLO2 algorithm ResNet algorithm

1 Introduciton

Face recognition is a non-contact biometric identification technology, and has been widely used in commercial, security and many other aspects. Compared with the fingerprint and other contact identification technology, the face recognition information acquisition channels are more extensive, more hidden.

Face recognition is divided into two steps from the natural background: face detection and face recognition. Many traditional detection methods such as AdaBoost, SVM, DPM etc., the multi-scale deformation component model DPM [1] effect is more prominent among them, it has been VOC(visual object class) [2] detection champion from 2007 to 2009 continuously. Deep learning took place the traditional detection methods by showing obvious advantages in performance later, but the effect promoted slowly at

Q. Zhou et al. (Eds.): ICPCSEE 2018, CCIS 901, pp. 698–705, 2018.
https://doi.org/10.1007/978-981-13-2203-7_58

the initial stage. The detection effect of deep learning is significantly improved until Ross Girshick proposed R-CNN [3] in 2014. The mAP is up to 48% on the VOC 2007 test set. Ross Girshick proposed Fast R-CNN [4] algorithm and Faster R-CNN [5] algorithm in year 2015 and 2016 in succession The mAP of VOC 2007 was increasing to 70% and 73.2% respectively after that. The detection effect of Faster R-CNN was good so far. The effect of detection of faster RCNN has been very good, However, due to its substep detection strategy of "extraction the candidate box first, then classification based on the candidate box", the FPS only reaches 7 frames per second, which is far from satisfying real-time demand.

YOLO algorithm made a series of improvements based on the slow speed of Faster R-CNN. It converted the detection problem into the strategy of classification problem different from that of R-CNN series, YOLO algorithm speeded the calculation to a large extent by converting target detection problem into regression problem, FPS can reach 45 frames per second. However, the accuracy decreased while the speed increased as the selection scale and proportion of the candidate box were relatively simple, the mAP of VOC 2007 is only 63.4%. YOLO2 algorithm continued to improve based on YOLO algorithm, which is different from the manual selection of multi-scale candidate boxes in the Faster R-CNN algorithm. YOLO2 algorithm improved the accuracy greatly while increasing the speed at the same time, by adopting k-means clustering as the rules of the candidate box to select, and its mAP reached 78.6% in VOC 2007.

In this paper, method of face detection and recognition based on YOLO2 [6] algorithm and ResNet algothrim, aimed to solve the problem of face recognition in video monitoring. YOLO2 algorithm detected each frame of face video rapidly, then got the facial features by inputing the result into ResNet [7] algorithm, compared the target face feature and local face library, calculate the cosine similarity of face feature, take the highest similarity as the recognition result. The whole recognition speed improved because of the detection speeding up.

2 Algorithm Implementation

2.1 YOLO2 Algorithm

Yolo2 and YOLO [8] algorithm is a series of excellent detection algorithm after the Faster R-CNN. Faster R-CNN algorithm adopted "extraction from candidate frame plus classification" (RPN plus fast R-CNN) steps to achieve multiple objectives detection, The essence of this is to convert the detection problem into a classification problem, the precision is high, but the speed is poor. Yolo Used a method of classifying and detecting joint training for targets, directly regress the position and type of the detected target frame in the output layer, converting the detection problem into regression problem, greatly reduced the speed of detection.

The pre-training classification network of YOLO is the first 20 convolution network of GoogleNet [9] algorithm adding the average pooling layer and the full connection layer. The input image size is 224 × 224. The detection training network added four convolutional layers and two fully connected layers based on it, and changed the input image size to 448 × 448. The image of 448 × 448 is divided into a 7 × 7 grid, with each grid corresponding to two candidate borders, and borders are used to predict the position of objects in the grid center. The grid is used to predict object categories. After entering the network, the border of the object is selected, and the overlapped border is eliminated by NMS, and the location and category of the target in the input image are obtained (Fig. 1).

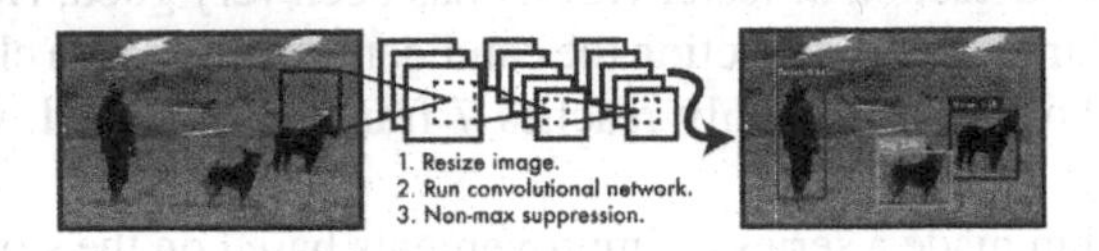

Fig. 1. Schematic diagram of YOLO treatment [6]

Compared with the Faster R-CNN algorithm, YOLO algorithm has a clear advantage in speed, but the accuracy rate is not high with a single proportional candidate box. For the, YOLO2 algorithm made a series of improvements to settle low accuracy of YOLO algorithm. Aiming at the differences between the training data and the distribution of test data, YOLO2 added a "batch normalization" step at the back of each convolution layer to improve the training speed of the network. In addition, YOLO2 improved the resolution of the pre-training network to 448 × 448. and gained some improvements on mAP. In the field of candidate box selection, YOLO2 algorithm used k-means clustering to obtain the size, proportion and quantity of candidate box, and the clustering results and computational complexity are balanced on this basis, then got the reasonable parameters of candidate box (Fig. 2).

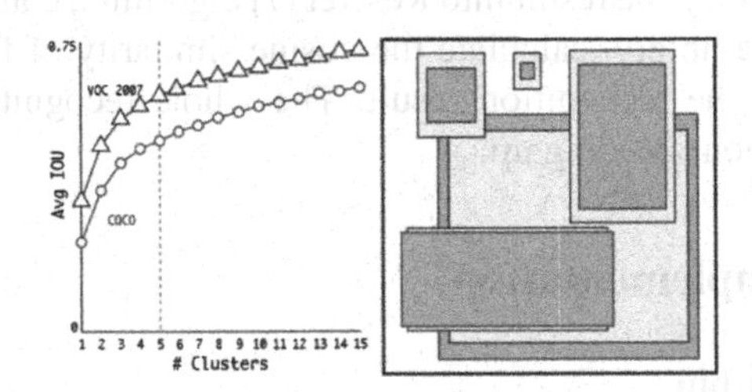

Fig. 2. Results of the clustering results of VOC and COCO [6]

Each candidate box outputs the corresponding prediction position, confidence, and the probability of the object in the border. In order to adapt YOLO2 algorithm to multi-scale features, YOLO2 algorithm has added a "passthrough layer", which was used to connect the characteristics of shallow and depth character, multi-scale training was adopted in training, and the scale of the new images was randomly selected after 10 rounds of training.

In terms of speed, YOLO2 also made some improvements. Many detection networks used VGGNet-16 algorithm as feature to extract network, although its accuracy is very high, but it is too complex. YOLO2 algorithm used the GoogleNet network as the feature to extract network, the number of operations that was transmitted one time is only about a quarter of that of VGG-16. Although using GoogleNet network, the accuracy is slightly lower than that of vggnet-16, but the speed is significantly improved.

In addition to the above improvements, many details have been optimized for YOLO2 algorithm. The test results showed that when the input image size was 228×228, the frame rate reached 90 FPS, and mAP nearly reached the same level as that of Faster R-CNN. When the input image resolution is higher as 554×554, the mAP of YOLO2 algorithm reached 78.6% at VOC 2007. and the frame rate can reach 40 FPS, which fully meets the real-time requirement.

2.2 ResNet Deep Residual Networks Algorithm

The depth residual network (ResNet) is a deep learning network proposed by Kaiming He, which obtained the first of ImageNet target detection, target positioning, COCO target detection and COCO image segmentation in 2015.

Many image processing related tasks can achieve better results thanks to the deeper network structure. The deeper the layer of the network means, the more features can be extracted, which can reflect the semantics of the image. But simple stacking networks can cause serious gradient disappearance problems. This problem may be initialized by a standard (normalized initialization) and regularization (intermediate normalization) to some extent, dozens of layer of the network can ensure the normal convergence, but on a deeper network, precision has reached the saturation, the effect has become worse.

In response to this problem, ResNet introduced the residual learning to solve the problem that the depth network is difficult to optimize. Formally, H of x is the optimal mapping, Let the stacked nonlinear layer fit another mapping $F(x) = H(x) - x$, the optimal mapping can be expressed as H of x is equal to F of x plus x at this moment. suppose the residual mapping is more optimized than the original mapping, then it is more easier to push the residuals to 0 in extreme cases, which is much easier than mapping the mapping to another mapping.

$F(x) + x$ can be represented by adding a "shortcut connection" to the feedforward network, "Quick connection" skips one or more layers and perform simple identity mapping without adding additional parameters or adding computational complexity. And the entire network can still be used for end-to-end training with SGD and reverse propagation (Fig. 3).

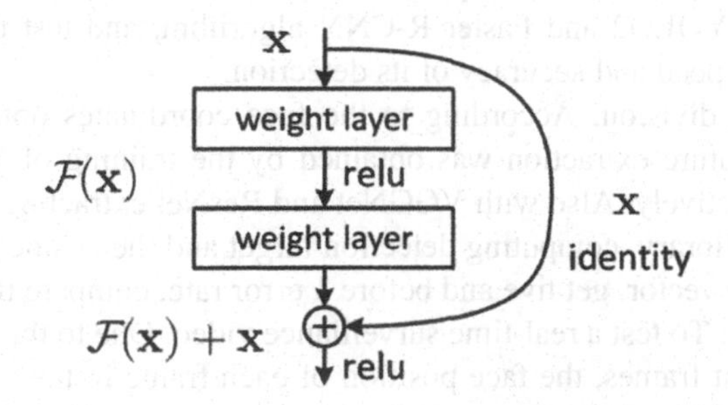

Fig. 3. Schematic diagram of residual network [7]

The specific approach is as follows: each group of network of the residual network mapping is treated as a construction block, and each constructed block is defined as:

$$y = F(x, \{W_i\}) + x \quad (1)$$

The input and output vectors of the construction blocks are x and y respectively. $F(x, \{W_i\}0)$ is the residual mapping to be trained. And x and y dimension need to be consistent in (one). If the dimensions are inconsistent, a linear projection is added on "shortcut connection" to match the dimension, which is expressed as follows:

$$y = F(x, \{W_i\}) + W_s x \quad (2)$$

In the ImageNet classification data set, the deep residual network has achieved excellent results. Residual network of one hundred and fifty-two layers is the deepest network on ImageNet at present. and the complexity is lower than the VGG network, the composite model has a error rate only 3.57% on the ImageNet top five test set. In the end, ResNet obtained the first place in ImageNet detection, ImageNet positioning, COCO detection and COCO segmentation in the competition of ILSVRC and COCO 2015.

The introduction of residual learning enables the performance of the basic network to be further optimized, whether it is detection or recognition, and it can achieve better results.

2.3 Face Detection and Recognition Based on Deep Learning

In this paper, a method of face detection and identification is proposed for the city monitoring video. Method includes detection and identification two parts. The detection part adopts YOLO2 algorithm, which has a good effect on speed and accuracy. The recognition part adopts ResNet algorithm, and the residual learning introduced makes the model more accurate than the original network.

In order to verify the effectiveness of the method, a set of tests were designed for testing, identification and comprehensive performance testing.

(1) Testing part: Compared with the detection performance of YOLO2 and Faster R-CNN algorithm, a series of comparative experiments were designed. Use the same data to train YOLO2 and Faster R-CNN algorithm, and test the same video to compare the speed and accuracy of its detection.
(2) Identification division: According to the face coordinates obtained by YOLO2 algorithm, feature extraction was obtained by the training of VGGNet [10] and ResNet respectively. Also with VGGNet and ResNet extracting the feature vector of local face library, computing detection target and the cosine similarity of local image feature vector, get five and before 1 error rate, compare their performance.
(3) Synthetic test: To test a real-time surveillance video. Due to the smaller change of video adjacent frames, the face position of each frame in the video was detected every 1 s (25 FPS). According to the location to extract the facial feature detection, and compared the facial features of local database and and videos appeared, then

to calculate the cosine similarity. If the similarity threshold is reached, recognition shall be deemed a success. Record the total number of actual human face in video, as well as the total number of correct detection of human face, the total number of correct recognition, the average time to complete, the final assessment of the overall effect is done.

3 Experimental Results and Analysis

The experimental platform adopts intel's e5-2620v4 model processor, GPU adopts NVIDIA tesla K80, and the operating system is CENTOS 6.5, face detection adopts FDDB data set, face recognition adopts CASIA_Webface data set.

(1) The first group of human face detection experiments using FDDB dataset as a test, each frame detected one time, compared to the Faster R-CNN and YOLO2 detection effect.

From Table 1, it can be seen that YOLO2 algorithm detection recall rate is slightly lower than Faster R-CNN algorithm, both can reach about 90%, because of the phenomenon of motion blur and occlusion for monitoring video existed, accuracy is within the acceptable range.

Table 1. Comparison of the results of faster R-CNN and YOLO2 face detection

Detect algorithm	Face amount	Total detected	AVG. processing time/ frame	Recall rate
Faster R-CNN	248	226	181.264 ms	91.13%
YOLO2	248	224	22.421 ms	90.32%

In terms of speed, the speed of YOLO2 algorithm reaches about eight times that of Faster R-CNN. The result of video test is missing and without error detection.

(2) In the second group of face recognition, the error rate of top1 and top5 similarity for VGGNet and ResNet obtained by extracting the characteristic vectors of the last layer of the network, which is used to compute the cosine similarity, and the results are as shown in Table 2.

Table 2. Comparison of VGGNet and ResNet facial recognition results

Detect algorithm	Top1 error	Top5 error
VGGNet	2.68%	0.45%
ResNet	1.79%	0%

From Table 2., the error rate is lower than that of VGGNet, ResNet top1 and top5. The effect of using ResNet Recognition is better.

(3) The third group of Comprehensive test section, using a section of human face monitoring video to test. The test results are as follows (Table 3):

Table 3. Synthetic results of face detection and recognition

Face amounts	Correctly detected	Correct ID	Recognition accuracy	AVG. recognition time
248	224	220	90%.71	47 ms

The overall performance of the test, the speed is faster which can meet the real-time application requirements. The overall recognition rate is reduced on the based on individual recognition, and the most influential is the detection part.

(4) To verify the results of the test, The CHUK' s Wider-Face detection benchmark data set was adopted. The WIDER FACE is organization by 61 event-based categories, Select 40% of each event category from them as a training set. 10% for cross validation and 50% for test set. The data collected by the DS-2cd3t45d-i3 network camera is used as the validation data set. By extracting the key frames from video, the key frames are used as the test data. Verify model identification results. The experimental results are shown in Fig. 4. The identification result is not given in the paper due to the problem of portrait right.

Fig. 4. Human face detection results

4 Conclusion

In order to alleviate the pressure of city monitoring and supervision, in this paper a method of face detection based on YOLO2 and ResNet algorithm is proposed, which is used to identify the face in urban monitoring. Three groups of tests show that the overall recognition effect of YOLO2 and ResNet algorithm is very good, which can meet the actual demand in terms of real time and accuracy. Of course, there are still some deficiencies in this scheme, when the face is tilted at a higher Angle or the face area is smaller, it is easy to leak detection phenomenon. And because of the motion blur, the recognition effect is worse even if the face is detected in this situation. In subsequent studies, more advanced detection, more advanced recognition algorithms and higher performance GPU can be used, the overall performance of the algorithm will be enhanced, which will meet the application of more complex scenarios.

References

1. Felzenszwalb, P.F., Girshick, R.B., McAllester, D.: Object detection with discriminatively trained part-based models. IEEE Trans. Pattern Anal. Mach. Intell. **9**(32), 1627–1645 (2010)
2. Sermanet, P., Eigen, D., Zhang, X., Mathieu, M., Fergus, R., LeCun, Y.: OverFeat: integrated recognition, localization and detection using convolutional network. In: ICLR (2014)
3. Girshick, R., Donahue, J., Darrell, T., Malik, J.: Rich feature hierarchies for accurate object detection and semantic segmentation. In: ImageNet Large-Scale Visual Recognition Challenge Workshop, ICCV (2013)
4. Girshick, R.: Fast R-CNN. Comput. Sci. (2015)
5. Ren, S., He, K., Girshick, R., Sun, J.: Faster R-CNN: towards real-time object detection with region proposal networks. In: Advances in Neural Information Processing Systems (NIPS), vol. 28 (2015)
6. Redmon, J., Farhadi, A.: YOLO9000: Better, Faster, Stronger, pp. 332–328 (2016)
7. He, K., Zhang, X., Ren, S., Sun, J.: Deep residual learning for image recognition. In: IEEE Conference on 2016 Computer Vision and Pattern Recognition (CVPR), Las Vegas, NV, pp. 770–778 (2016)
8. Redmon, J., Divvala, S., Girshick, R., et al.: You only look once: unified, real-time object detection. In: Computer Vision and Pattern Recognition, pp. 779–788. IEEE (2016)
9. Szegedy, C., Liu, W., Jia, Y., et al.: Going deeper with convolutions, pp. 1–9 (2014)
10. Simonyan, K., Zisserman, A.: Very deep convolutional networks for large-scale image recognition, CoRR abs/1409.1556 (2014)

Localization and Recognition of Single Particle Image in Microscopy Micrographs Based on Region Based Convolutional Neural Networks

Fang Zheng(✉), FuChuan Ni, and Liang Zhao

College of Informatics, Huazhong Agricultural University, Wuhan 430079, China
zhengfang@mail.hzau.edu.cn

Abstract. Single-particle Cryo-electron microscopy (Cryo-EM) is an important tool to study the structure and function of biological macromolecules. In order to acquire higher resolution macromolecular structure, large number of molecular particles should be extracted from micrograph. Particle selection is a very important step for single particle reconstruction. Manual selection of particle is a bottleneck of 3D reconstruction, even with the assistance of computer. The low signal-to-noise ratio and low contrast of the micrograph pose great difficulty for automatic identification of particles. This has attracted the attention of researchers and many methods for particle identification were proposed, such as the template matching, edge detection, image segmentation method, neural network etc. In this paper we propose an automatically picking particles method in Cryo-EM micrographs. In this paper, a method of automatically particle extraction based on Faster-RCNN is proposed. The experimental results show that the accuracy of particle image recognition is 82.7% and F_1-score is 84.2%. The detection time of single picture was 0.3 s.

Keywords: Faster-RCNN · Deep learning · Cryo-electron microscopy

1 Introduction

Structural biology studies the molecular mechanism by analyzing the structure of biological molecules. The biological sample (such as the virosome, the macromolecule and etc.) itself is a three-dimensional (3D) object, whereas conventional electron microscopy can only observe the two-dimensional ultrastructure of it [1]. With the development of computer technology, reconstruction algorithm can be used to restore 3D structure of the sample with its 2D electron microscope images. Single-particle Cryo-electron microscopy (Cryo-EM) is an important tool for study the structure and function of biological macromolecules [2]. However, the large number of images produced by these devices require efficient and timely processing. Therefore, improving the computing efficiency of electron microscopy image processing becomes an urgent problem in computational biology.

Q. Zhou et al. (Eds.): ICPCSEE 2018, CCIS 901, pp. 706–720, 2018.
https://doi.org/10.1007/978-981-13-2203-7_59

In recent years, with the development of cryo-electron microscopy technology [3, 4], molecular structure which biologist obtained is near atomic resolution [5, 6]. However it needs ten thousands of protein particles, these particles are usually manually or partial manually selected from thousands of microscopic photo. In order to improve efficiency, automatic or semi-automatic particle recognition algorithms have been developed. The identification algorithms have been divided into two parts. One is template matching [7–12], this method identify particles in a micrograph according to the similarity of the reference template. In [13], they use feature vectors instead of pixels to represent templates. The inherent disadvantage of this method is the assumption that the 3D reference structure of the particle has already known, but in many cases, this is impossible. The other method is based on machine learning [14–17], user needs to select some real particles to train the classifier [18–23]. Once the classifier is trained, it can automatically identify the particles.

Deep learning is one of the most potential branches of artificial intelligence (AI) [35]. The deep learning algorithm can automatically extract the raw feature from a large data set such as images, genomes and etc., they are applied in speech recognition, automatic drive and so on. Its unique advantage is that the multidimensional data can be used directly as the input of the model to avoid the complexity of feature extraction in the model training. In biology, deep-learning algorithms dive into data to detect feature in ways that human can't to catch. Researchers are using the algorithms to classify cellular images, make genomic connections, advance drug discovery and even find links across different data types, from genomics, imaging to electronic records [36, 37]. In 2016, firstly a deep learning algorithm is proposed for particle recognition and the time to recognize a single picture is 1.5 min [12], but the recognition efficiency and accuracy can still be improved.

For the excellent computation performance of CNN (Convolutional Neural Network) in computer vision field, it has become the preferred algorithm for object recognition. In 2014, Ross et al. [24] proposed a simple, scalable algorithm which significantly improved the result of target detection, called RCNN (Region based Convolutional Neural Network). RCNN is used to identify target by extracting region that may contain target in the image by using CNN. However, due to computational complexity, it is difficult to deal with a large number of electron microscope images. The next year, inspired by SPPnets [25], Ross et al. proposed a faster RCNN algorithm —Fast-RCNN. Xiao et al. [26] used Fast-RCNN to pick the particle in three cryo-microscopy datasets. Comparing with CNN for recognizing particles [12], its computation efficiency improved, but the accuracy was still low.

Faster RCNN uses a regional proposed network instead of a time-consuming selection search method to calculate the target area, it uses CNN network to get raw features and a regional proposed network (RPN) to search the area of the targets. So that feature extraction and region selection are both integrated into the deep network framework [27–29]. Therefore, it greatly improved the detection efficiency of object recognition. Although faster-RCNN has made breakthrough achievements in many fields for the past few years, how to use faster-RCNN to realize the identification of particles in Cryo-EM is still a blank. This paper attempts to apply the algorithm in this field.

Therefore, this paper proposed a regional generation network based on Convolutional Neural Networks algorithm for single particle recognition of Cryo-EM. The organization of this paper is as following. In the Sect. 2 will introduce the Faster-RCNN algorithm, training scheme, datasets and data preprocessing that we used. The evaluation procedure, experiment results are presented in Sect. 3. Finally, conclusion and discussion are given in Sect. 4.

2 Materials and Methods

Comparing with Spatial pyramid pooling Net (SPPNet) and Fast-RCNN, Faster-RCNN not only breaks the calculation time bottleneck of region proposed, but also ensures the accuracy of target recognition. This paper we use Faster-RCNN to extract particle features and identify target particles. The algorithm structure is shown in Fig. 1 [30]. The Faster-RCNN contains two CNN networks: Regional Proposal Network (RPN) and the Fast-RCNN detection network. As shown in Fig. 2, the algorithm contains four stages.

(1) Inputting the image to the shared convolution network;
(2) Extracting the feature of the image and generating the feature map;
(3) RPN generates region proposed boxes of the feature map, and then sends these candidate boxes to the ROI pool of Fast-CNN model. Finally, the feature is sent to the full connection layer after ROI processing;
(4) Full connection layer output candidate box and classification results.

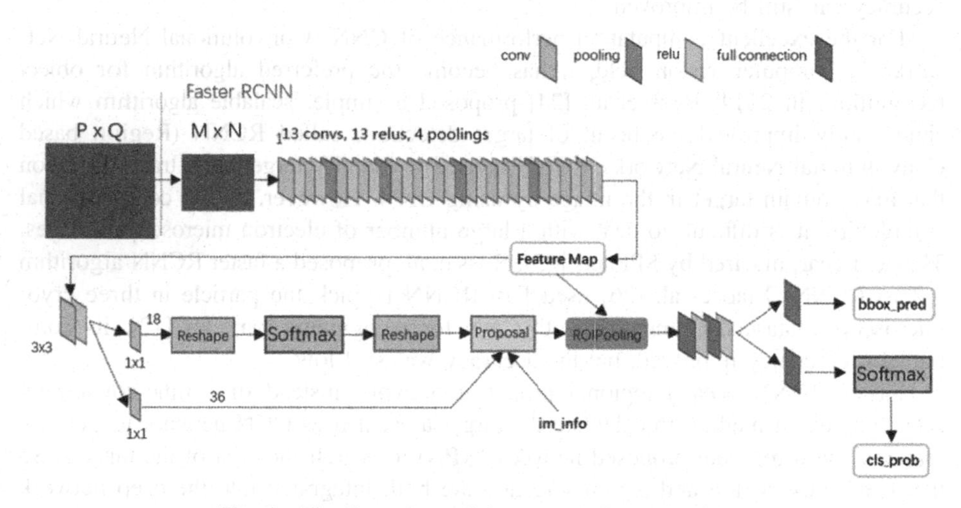

Fig. 1. The flow chart of the Faster-RCNN algorithm

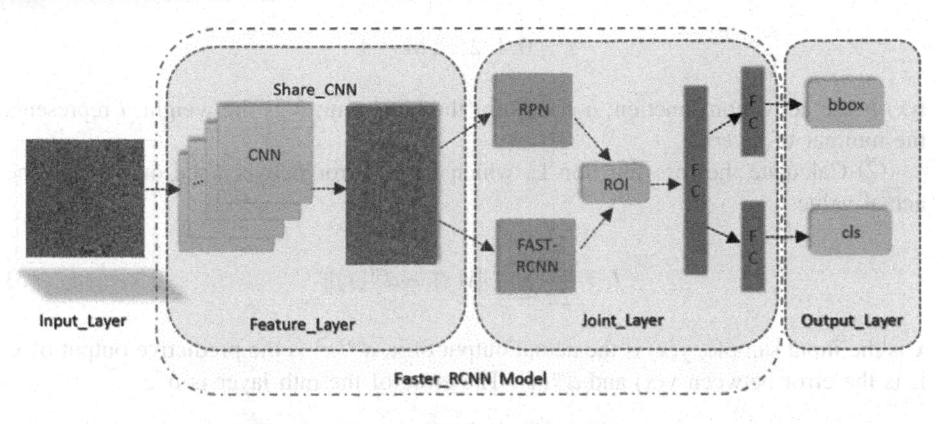

Fig. 2. The simplified Faster_RCNN model

2.1 Particle Recognition and Location Based on Faster-RCNN

The particle detection algorithm is composed of two parts: the network training and the particle detection.

1. Network Training for Particle Recognition

The network training is to learn the feature of particle image. In the process of training, the network update the weight parameter, and the cost function of the network is gradually reduced to 0.

The Faster-RCNN model is composed of RPN and Fast-RCNN, and RPN get the region proposed candidate box as input to the subsequent Fast-RCNN network. Therefore, the model is based on RPN, the training process is divided into two steps. RPN is optimized firstly, then combined with Fast-RCNN joint training to get the complete training model.

(1) Optimization of RPN

RPN is a full convolution network, which is trained using end-to-end reverse propagation algorithm and stochastic gradient descent method.

Back propagation algorithm (BP algorithm) contain three stages. ① performing forward propagation algorithm; ② the error between the actual value and the estimated value obtained from the forward propagation algorithm is used to guide reverse propagation; ③ adjusting each parameter according to the error in the reverse propagation and iterating to convergence.

① Input training set to the network and perform the forward propagation algorithm. The output of the m layer z^m in the network is the input of $m+1$ layer, and d^{m+1} represents the output of $m+1$ layer.

$$z^m = \sum\nolimits_k a^m \cdot w^m + b^m \tag{1}$$

$$d^{m+1} = f^{m+1}\left(\sum\nolimits_k a^m \cdot w^{m+1} + b^{m+1}\right) \tag{2}$$

$$z = 0, 1, 2..., \mathrm{M} - 1$$

f(x) is the activation function; b stands for the bias term; w is the weight; l represents the number of layers.

② Calculate the loss function L, which is the error between the output and the actual value.

$$L = \frac{1}{2n} \sum_x \|y(x) - d^m(x)\|^2 \tag{3}$$

x is the input sample, y(x) is the actual output of x, $d^m(x)$ is the predictive output of x. L is the error between y(x) and $d^m(x)$. The error of the mth layer is δ^m.

$$\delta^m = \frac{\partial L}{\partial z^m} \tag{4}$$

The error gradient of the mth layer weight and bias can be calculated according to δ^m.

$$\frac{\partial L}{\partial w^m} = a^{m-1} \delta^m \tag{5}$$

$$\frac{\partial L}{\partial b^m} = \delta^m \tag{6}$$

③ Finally, the updating weight $w^m(k+1)$ and bias $b^m(k+1)$ are obtained by using stochastic gradient descent algorithm.

$$w^m(k+1) = w^m(k) - \eta \delta^m \left(a^{m-1}\right)^T \tag{7}$$

$$b^m(k+1) = b^m(k) - \eta \delta^m \tag{8}$$

In the formula (7) and formula (8), $\eta(0 < 1)$ is the learning rate, the cost function L is reduced to zero by iteration.

(2) Training model

The training of CNN and RPN network model requires initialization of network parameters. The ZFnet [31] and VGG16 [32] are the most commonly used networks, we adopted VGG16 network for training to improve the detection accuracy.

① Pre-training of RPN

Firstly, the training result of ImageNet was used as the initialization weight to train RPN, and RPN-1 was obtained, and the candidate region proposed box was generated.

② Training of Fast-RCNN

The convolution layer of Fast-RCNN (object recognition) network was initialized with the training result of ImageNet, then training the Fast-RCNN network into Fast-RCNN-1with the candidate box region proposed generated by step ①.

③ The RPN was initialized with Fast-RCNN-1, the shared convolution layer was fixed. Training the full connection layer of RPN, then RPN-2 was generated.

④ The Shared convolution layer was the same as step ③, the region proposal box generated by step ③ is used to train the full connection layer of fast RCNN to get the Fast-RCNN-2.

⑤ Finally, RPN-2 and Fast-RCNN-2 formed a unified network with the Shared convolution layer.

In this way, the two networks share the convolutional layer and form a joint network that reduce training parameters and improve training efficiency, the network training algorithm is as following.

Algorithm flow of training phase

Procedure

Input : Electron microscope image data and the XML file of identifying particle object;

Output: Weight parameters of trained Fast-RCNN and RPN network model;

step1: Setting model parameters, learning parameters and etc.;

step2: Specifying the network model type and selecting the GPU number;

step3: Adopting an end to end approximation joint training method, and initializing the weights of the pre-training model W_0;

while (training loss< threshold) **do**

step4: RPN and Fast RCNN networks are iterative training , $i=0$;

repeat

(1) Train RPN with W_i parameters. Extraction of candidate regions on training sets using RPN;

(2) Starting with the W_i, the candidate region is trained with Fast RCNN, and the parameter is denoted as W_{i+1};

End

step5: Using Softmax Loss (detection classification probability) and Smooth L1 Loss (check border regression) to optimize the training.

End

2. Detection of particle image

The electron microscope particle image has the characteristics of high resolution and small target. The angle of single particle is diversity and randomness [33]. The process of detection and identification is shown in Fig. 3.

(1) Convolution features were generated by CNN convolution operation on the input image;
(2) The region proposed candidate box were produced by the RPN (a sliding window operation) on the convolution features of step (1);
(3) The feature mapping layer mapped the feature of region proposed candidate box to a lower dimension vector;
(4) The output layer is divided into a box-regression layer and a box-classification layer, which shows the boundary box in which the particles are located, and the detection ended.

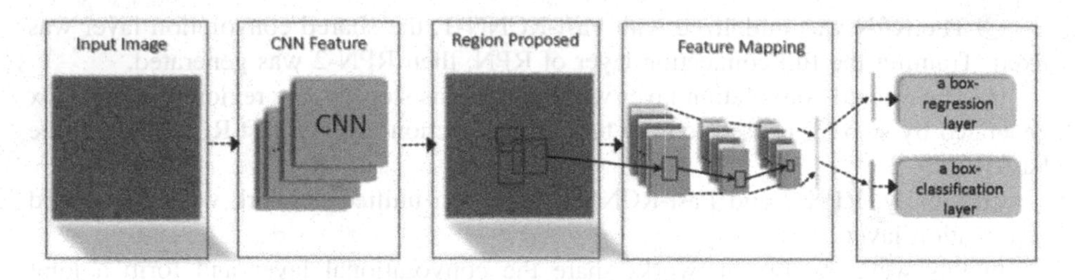

Fig. 3. Detection of particle image

2.2 Datasets and Data Preprocessing

1. Datasets

We used the KLH II (Keyhole Limpet Hemocyanin) [34] dataset, it is a single particle image dataset of cryo-electron microscope imaging, and also a standard benchmark [33] for many particle recognition. In the dataset, the image size is 2048 * 2048. KLH II includes 655 defocus pairs of micrographs, coordinate centers of 11309 side-view particles from the FFF image (Farther from Focus, image is acquired at farther from focus conditions (−3 μm)) and NFF image(Near from Focus, image is acquired at near to focus conditions (−0.6 to −1.5 μm)) in MRC and JPEG format. Images were acquired with a FEI Tecnai F20 equipped with a 2Kx2K CCD Tietz camera, as defocal pairs at a nominal magnification of 62,000 x and a voltage of 120 keV. In this study, the whole data set contains 655 micrographs, including 11309 particle objects. In order to improve the accuracy of recognition, the original image is rotated by 900, 1800 and 2700, so that the number of samples is expanded to 2620 images, and the number of particles is 45236. We trained the algorithm with 70% of the particles in the dataset, tested the algorithm with 15% of the particles, the remaining 15% is validation set.

In the test and validation sets, the same number of background images were taken as negative samples. The data set structure is shown in the following Table 1.

Table 1. The details of datasets

Dataset	Whole dataset	Training set	Test set	Validation set
Number of micrograph	2620	1834	393	393
Number of particle	45236	31666	6785	6785

2. Preprocessing of Electron Micrograph

For the low contrast and low signal to noise ratio of the original electron microscope image, we preprocess the image to improve the efficiency and accuracy of particle recognition. A histogram equalization algorithm was used to improve the image contrast, a Gauss low-pass filtering algorithm was to improve the SNR (SIGNAL-NOISE RATIO) of image.

(1) Histogram equalization. As shown in the following Fig. 4, a micrograph which processed using the histogram equalization algorithm has a wider dynamic range and higher contrast.

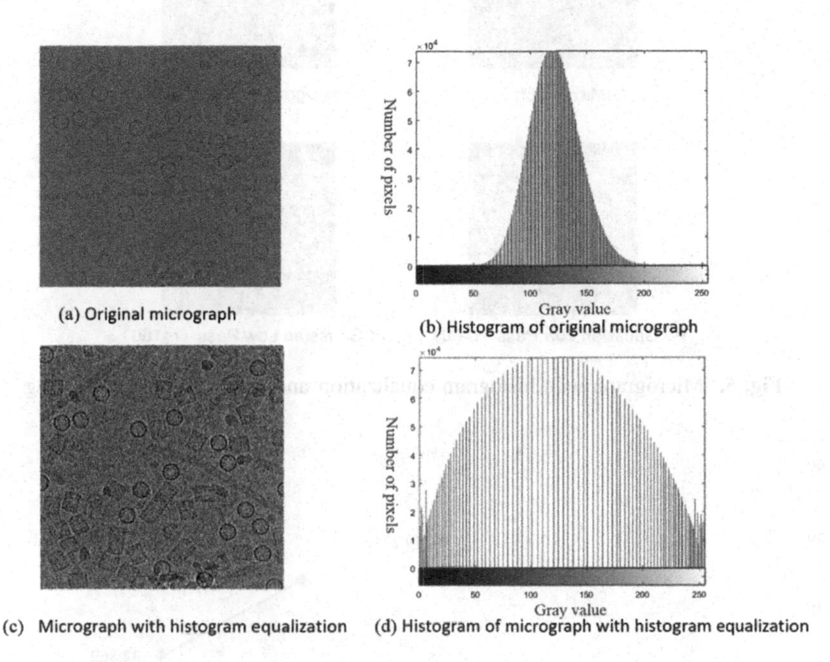

Fig. 4. Micrograph and Micrograph processed with histogram equalization

(2) Gauss low-pass filtering processing. We compare the original micrograph, micrograph processed with histogram equalization and micrograph processed with Gauss low-pass filtering under different radius. The results are shown as below Fig. 5. After equilibration, the contrasts between the particles and the background is larger, but the SNR reduce nearly 50%. In order to promote the accuracy of training and recognition, we use the Gauss low-pass filtering algorithm on micrograph which have been processed with histogram equalization to improve the SNR. The results show that the micrograph is clearer, and at the same time SNR of the image increased nearly doubled. In Fig. 6, we compared SNR of the micrograph, histogram equalization micrograph, Gauss low-pass filtering micrograph e and smoothing micrograph. The following experiments show that training on the filtered image greatly improved the accuracy of the test data.

3. Identification of Particle Object in Training Set

In the data set of KLHII, the central coordinate *p (x, y)* of the particle object in world coordinate system are given. But in Faster-RCNN, it is necessary to adopt the rectangle box (upper left and lower right coordinates of target object) for target training and recognition. So a transformation coordination from center coordination to rectangle box coordination are needed. Since the XML calibration is based on the screen

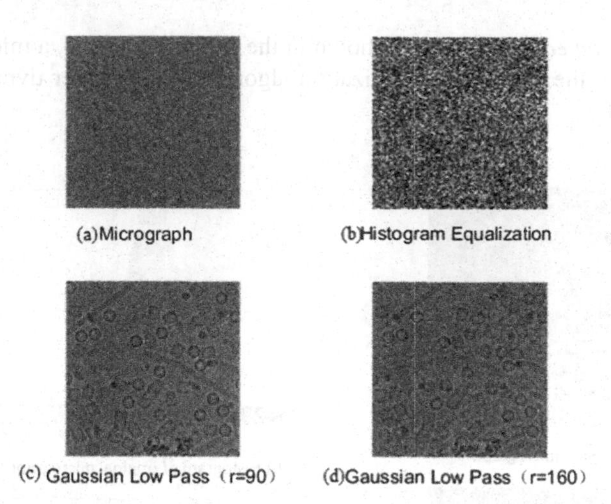

Fig. 5. Micrograph with histogram equalization and Gauss Low-Pass filtering

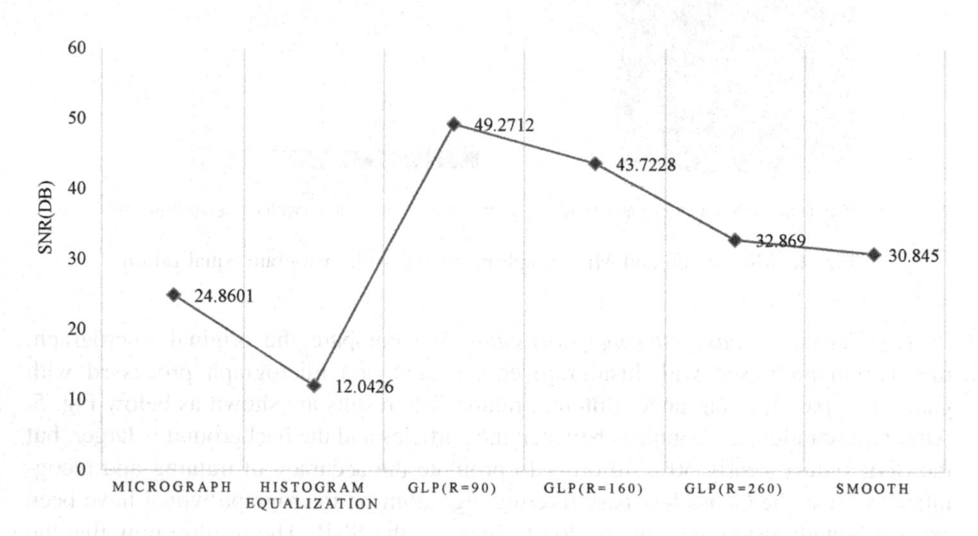

Fig. 6. SNR of the Micrograph, Histogram Equalization Micrograph, Gauss low-pass filtering Micrograph and Smoothing Micrograph

coordinate system, so the world coordinate *(x, y)* is converted into screen coordinate. Then the x and y are added or subtract 100 to obtain the rectangle box coordinates *(x − 100, y + 100, x + 100, y − 100)*, that the particle target in original image can be calibrated with rectangle box coordinates. In order to increase the training set, we also rotate the image at 90, 180 and 270 angle degree, coordinate of the image need to be recalculated for the rotation. After the rotation, it is also necessary to identify the particles, and the concrete steps are as follows.

(1) In order to calibrate the rotated image, the screen coordinates are converted into image coordinates. As shown in the Fig. 7, the black line represents the screen coordinate system, and the green line represents the image coordinate system. *P (a, b)* is the coordinates of an image pixel in the screen coordinate system. The *P (x, y)* is the coordinates of that point *P (a, b)* under the image coordinate system after converting. The transformation formula is as Formula (9).

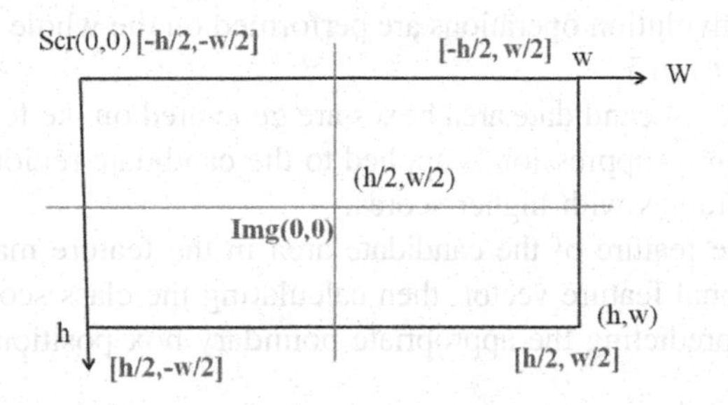

Fig. 7. Screen coordinate and image coordinate (Color figure online)

$$\begin{cases} x = a - \frac{h}{2} \\ y = b - \frac{w}{2} \end{cases} \tag{9}$$

(2) Rotation transformation of image coordinates, P (x′, y′) is coordinate after rotation of P (x, y), the formula (10) and formula (11) are as follows.

$$\begin{cases} h' = hcos\theta + wsin\theta \\ w' = hsin\theta + wcos\theta \end{cases} \tag{10}$$

$$\begin{pmatrix} x' \\ y' \end{pmatrix} = \begin{pmatrix} +cos\theta, +sin\theta, h' - hcos\theta - wsin\theta \\ -sin\theta, +cos\theta, w' + hsin\theta - wcos\theta \end{pmatrix} \times \begin{pmatrix} a \\ b \\ 1/2 \end{pmatrix} \tag{11}$$

(3) Inverse transformation of the rotated image, the formula (12) is as follows.

$$\begin{pmatrix} a \\ b \end{pmatrix} = \begin{pmatrix} +cos\theta, -sin\theta, -h'cos\theta + w'sin\theta + h \\ +sin\theta, +cos\theta, -h'sin\theta - w'cos\theta + w \end{pmatrix} \times \begin{pmatrix} x' \\ y' \\ 1/2 \end{pmatrix} \tag{12}$$

(4) XML calibration of the target which is carried out on the rotated image

The detection target is divided into two categories: the particle image and the non-particle image. In the test set and the verification set, the same number of background images are randomly selected as non-particle images.

2.3 Detection and Recognition Process

The above training shows that the two networks can share the same 5 layers of the convolutional neural network, which makes the whole detection process only need to complete a series of convolution can complete the detection and recognition process. It completely solves the time bottleneck problem of region proposed network. The process of detection and identification is as followings.

(1) A series of convolution operations are performed on the whole image to obtain the feature graph *Conv5*;
(2) A large number of candidate area boxes are generated on the feature map by RPN;
(3) A non-maximum suppression is applied to the candidate region box, and keeping the first 300 frames with higher scores;
(4) Taking out the feature of the candidate area in the feature map and forming the high-dimensional feature vector, then calculating the class score by the detection network and predicting the appropriate boundary box position of target.

3 Experiments

In this section, we conducted a series of experiments to measure the effectiveness of the particle recognition algorithm. The details of these experiments are described below.

3.1 Experiment Setting

The experimental platform of this project is a servers with K40 and K80 Graphics processing units of NVIDIA.

3.2 Experiment Results

After iterative training, the loss function and detection results of the training set and validation set are shown in Fig. 8.

In order to evaluate the effectiveness of the algorithm and the model, the test results are analyzed respectively from the PR (Precision/Recall) index, training efficiency and etc. Using recall rate (γ) and accuracy (ρ) as a measure evaluate the detection algorithm, γ.and ρ is defined as formulas (13) and (14).

$$\rho = \frac{P_t}{P_t + P_f} \tag{13}$$

$$\gamma = \frac{P_t}{P_t + N_f} \tag{14}$$

Among them, P_t is the number of particle which was correctly detected, P_f is the number of particles detected by mistake, and N_f is the number of be wrongly detected as background image. However, it is still not enough to evaluate the algorithm only

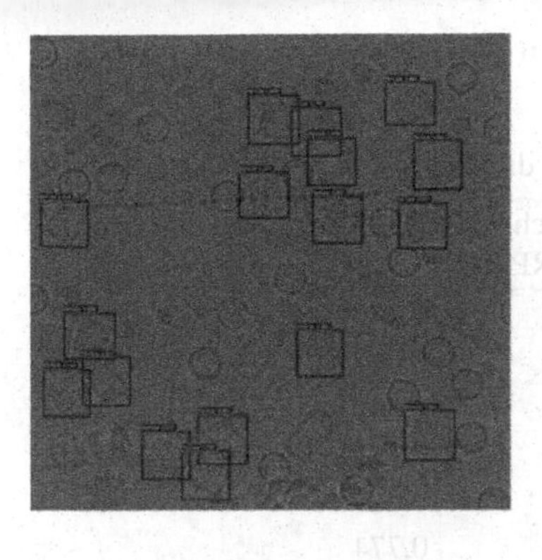

(a) Detection results

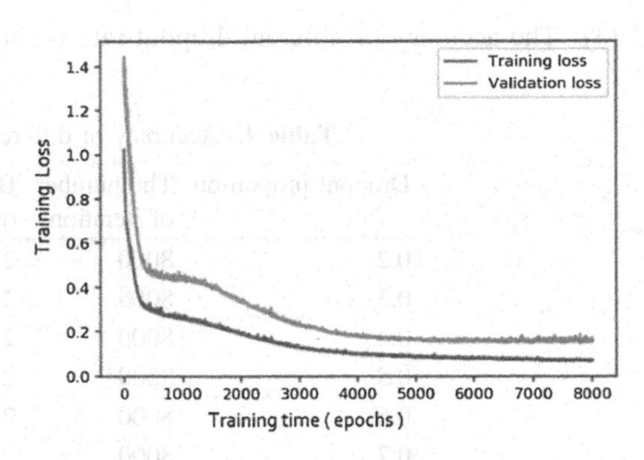

(b) Training loss with the epochs

Fig. 8. The loss function and detection results

with two indexes of recall and accuracy. Therefore, we also use F_1-score (F_1 score also F-score or F-measure) is a measure of a test's accuracy. The F_1-score is the harmonic average of the precision and recall, where a F_1-score reaches its best value at 1 (perfect precision and recall) and worst at 0 to evaluate the algorithm which is defined as formula (15).

$$F_1 = 2 \times \frac{\rho * \gamma}{\rho + \gamma} \tag{15}$$

(1) The Precision and Recall rates in different preprocessing datasets is shown in Table 2.

Table 2. Precision and Recall

Dataset	Recall	Precision	F1-score
Histogram equalization	64.2%	55.2%	69.1%
Gaussian low pass (r = 90)	83.7%	82.7%	84.2%
Gaussian low pass (r = 160)	78.4%	73.2%	77.5%
Gaussian low pass (r = 260)	70.2%	64.3%	64.1%

(2) The training time and accuracy under different models are shown in Table 3.

Table 3. Execution efficiency and execution accuracy

	ZFnet	VGG16
Training time(h)	6	12.5
Detecting time(s)	0.3	0.3
Accuracy	65.4%	82.7%

(3) The accuracy of different dropout rate is shown in Table 4.

Table 4. Accuracy of different dropout rate

Dropout proportion	The number of iterations	Batch size of RPN	Precision
0.2	8000	256	0.686
0.3	8000	256	0.663
0.4	8000	256	0.742
0.5	8000	256	0.785
0.6	8000	256	0.827
0.7	8000	256	0.774
0.8	8000	256	0.698

4 Discussion and Conclusion

This study selected identification of protein particles in electron microscope images, the results can pave the way for subsequent protein 3D reconstruction. Picking electron microscope particles is an important process of electron microscope data processing. For the large number of targets (level one hundred thousand, or even millions), the automatic selection process is imperative. Because of the low signal-to-noise ratio of the electron microscope image, the recognition accuracy is low. At the same time, large micrograph with many particles lead to low recognition efficiency. Therefore, improving the accuracy and efficiency of identification is the two main objective of the study.

This research adopts Faster-RCNN for the automatic identification of particles, the recognition accuracy rate is 82.7%. In order to improve the accuracy of recognition, we used the method of image rotation to increase training set samples, and used image preprocessing to improve the quality of data sets. These measures greatly improved the accuracy of recognition. Because the detection target is a small object in the whole image, it also improves the difficulty of detection, which is also the main reason for the relatively low detection precision. In the latter experiment, we hope to use the small target object to pre train the network parameters to improve the detection effect. The follow-up is going to continue to test and compare the data sets in different identification algorithms, and to test the accuracy and efficiency of the algorithms in different data sets.

Acknowledgments. The authors thank R. F. Zhai for helpful discussions, This study is supported by Natural Science Foundation of Hubei Province of China (Program No. 2015CFB524 and Program No. 2016CKB705) and the Fundamental Research Funds for the Central Universities (Program No. 2015BQ023 and Program No. 2014QC008).

References

1. Guo, F., Jiang, W.: Single particle cryo-electron microscopy and 3-D reconstruction of viruses. In: Kuo, J. (ed.) Electron Microscopy. MMB, vol. 1117, pp. 401–443. Humana Press, Totowa, NJ (2014). https://doi.org/10.1007/978-1-62703-776-1_19
2. Russo, C.J., Passmore, L.A.: Ultrastable gold substrates: properties of a support for high-resolution electron cryomicroscopy of biological specimens. J. Struct. Biol. **193**(1), 33 (2016)
3. Cheng, Y., et al.: A primer to single-particle cryo-electron microscopy. Cell **161**(3), 438–449 (2015)
4. Bai, X.-C., et al.: How cryo-EM is revolutionizing structural biology. Trends Biochem. Sci. **40**(1), 49 (2015)
5. Yan, C., et al.: Structure of a yeast spliceosome at 3.6-angstrom resolution. Science **349** (6253), 1182 (2015)
6. Liao, M., et al.: Structure of the TRPV1 ion channel determined by electron cryo-microscopy. Nature **504**(7478), 107–112 (2013)
7. Xu, X.-P., Page, C., Volkmann, N.: Efficient extraction of macromolecular complexes from electron tomograms based on reduced representation templates. In: Azzopardi, G., Petkov, N. (eds.) CAIP 2015. LNCS, vol. 9256, pp. 423–431. Springer, Cham (2015). https://doi.org/10.1007/978-3-319-23192-1_35
8. Mao, Y., Castillomenendez, L.R., Sodroski, J.: Dual-target function validation of single-particle selection from low-contrast cryo-electron micrographs. Quantitative Biology (2013)
9. Wu, X., Wu, X.: A review of automatic particle recognition in Cryo-EM images. J. Biomed. Eng. **27**(5), 1178 (2010)
10. Abrishami, V., et al.: A pattern matching approach to the automatic selection of particles from low-contrast electron micrographs. Bioinformatics **29**(19), 2460 (2013)
11. Chen, Y., Ren, F., Wan, X., Wang, X., Zhang, F.: An improved correlation method based on rotation invariant feature for automatic particle selection. In: Basu, M., Pan, Y., Wang, J. (eds.) ISBRA 2014. LNCS, vol. 8492, pp. 114–125. Springer, Cham (2014). https://doi.org/10.1007/978-3-319-08171-7_11
12. Wang, F., et al.: DeepPicker: a deep learning approach for fully automated particle picking in cryo-EM. J. Struct. Biol. **195**(3), 325 (2016)
13. Kumar, V., et al.: Robust filtering and particle picking in micrograph images towards 3D reconstruction of purified proteins with cryo-electron microscopy. J. Struct. Biol. **145**(1–2), 41–51 (2004)
14. Proença, M.C., Nunes, J.F.M., De Matos, A.P.A.: Development of an algorithm for automatic image detection of biological particles in transmission electron microscopy images. Microsc. Microanal. **18**(S5), 3–4 (2012)
15. des Georges, A., et al.: High-resolution cryo-EM structure of the *Trypanosoma brucei* Ribosome: a case study. In: Herman, Gabor T., Frank, J. (eds.) Computational Methods for Three-Dimensional Microscopy Reconstruction. ANHA, pp. 97–132. Springer, New York (2014). https://doi.org/10.1007/978-1-4614-9521-5_5
16. Roseman, A.M.: Particle finding in electron micrographs using a fast local correlation algorithm. Ultramicroscopy **94**(3–4), 225 (2003)
17. Volkmann, N.: An approach to automated particle picking from electron micrographs based on reduced representation templates. J. Struct. Biol. **145**(1–2), 152 (2004)
18. Arbeláez, P., et al.: Experimental evaluation of support vector machine-based and correlation-based approaches to automatic particle selection. J. Struct. Biol. **175**(3), 319–328 (2011)

19. Yin, Z., Doerschuk, P.C., Gelfand, S.B.: Cryo electron microscopy of mixed ensembles: simultaneous pattern recognition and 3-D reconstruction, p. 421 (2003)
20. Min, W.K., et al.: Cryo-electron microscopy single particle reconstruction of virus particles using compressed sensing theory. In: Electronic Imaging (2007)
21. Langlois, R., Pallesen, J., Frank, J.: Reference-free particle selection enhanced with semi-supervised machine learning for cryo-electron microscopy. J. Struct. Biol. **175**(3), 353 (2011)
22. Guo, F., Jiang, W.: Single Particle Cryo-Electron Microscopy and 3-D Reconstruction of Viruses, p. 401. Humana Press (2014)
23. Sorzano, C.O., et al.: Semiautomatic, high-throughput, high-resolution protocol for three-dimensional reconstruction of single particles in electron microscopy. Methods Mol. Biol. **950**, 171–193 (2013)
24. Girshick, R., et al.: Rich Feature Hierarchies for Accurate Object Detection and Semantic Segmentation, pp. 580–587 (2013)
25. He, K., et al.: Spatial pyramid pooling in deep convolutional networks for visual recognition. IEEE Trans. Pattern Anal. Mach. Intell. **37**(9), 1904–1916 (2014)
26. Xiao, Y., Yang, G.: A fast method for particle picking in cryo-electron micrographs based on fast R-CNN. In: Applied Mathematics and Computer Science: International Conference on Applied Mathematics and Computer Science (2017)
27. https://github.com/rbgirshick/py-faster-rcnn.git
28. Ren, S., et al.: Faster R-CNN: towards real-time object detection with region proposal networks. IEEE Trans. Pattern Anal. Mach. Intell. **39**(6), 1137–1149 (2017)
29. Jia, Y., et al.: Caffe: convolutional architecture for fast feature embedding. In: ACM International Conference on Multimedia (2014)
30. https://github.com/rbgirshick/py-faster-rcnn/blob/master/models/pascal_voc/VGG16/faster_rcnn_alt_opt/faster_rcnn_test.pt
31. Zeiler, M.D., Fergus, R.: Visualizing and understanding convolutional networks. In: Fleet, D., Pajdla, T., Schiele, B., Tuytelaars, T. (eds.) ECCV 2014. LNCS, vol. 8689, pp. 818–833. Springer, Cham (2014). https://doi.org/10.1007/978-3-319-10590-1_53
32. Simonyan, K., Zisserman, A.: Very deep convolutional networks for large-scale image recognition. Computer Science (2014)
33. Sorzano, C.O., et al.: Automatic particle selection from electron micrographs using machine learning techniques. J. Struct. Biol. **167**(3), 252–260 (2009)
34. Stoschek, A., Hegerl, R.: Automated detection of macromolecules from electron micrographs using advanced filter techniques. J. Microsc. **185**(1), 76–84 (2010)
35. Angermueller, C., Pärnamaa, T., Parts, L., et al.: Deep learning for computational biolog. Mol. Syst. Biol. **12**(7), 878 (2016)
36. Webb, S.: Deep learning for biology. Nature **554**, 555–557 (2018)
37. Jones, W., Alasoo, K., Fishman, D., et al.: Computational biology: deep learning. Emerg. Top. Life Sci. **1**(3), 257–274 (2017)

A Novel Airplane Detection Algorithm Based on Deep CNN

Ying Wang[1], Aili Wang[1(✉)] and Changyu Hu[2]

[1] Higher Education Key Lab for Measure and Control Technology and Instrumentations of Heilongjiang, Harbin University of Science and Technology, Harbin 150080, China
aili925@hrbust.edu.cn

[2] Department of Communication Engineering, Nanjing University of Aeronautics and Astronautics, Nanjing 210016, China

Abstract. Driven by the massive natural scene image and computer's high speed computation ability, deep CNN becomes the mainstream method for completing the computer vision task by the power of feature representation capability. Because object detections based on deep CNN are unable to achieve optimization due to lack of training set of remote sensing images, a novel method based on transfer learning and deep CNN combined with SVM is proposed in this paper. Firstly, we obtain the deep CNN model AlexNet trained on a large-scale data set. Then, we truncate the first five convolution layers and get the initial parameters through transfer learning. Then, Deep CNN is used as feature extractor to extract the depth feature for training SVM, and the final remote sensing image airplane detection model is obtained. Experimental results show that the average precision of the proposed algorithm outperforms other traditional algorithms.

Keywords: Remote sensing · Airplane detection
Deep Convolutional Neural Network (CNN) · Transfer learning

1 Introduction

Aircraft detection in optical remote sensing images has always been an important task in computer vision. It is found that in recent years, there are many types of aircraft detection algorithms for remote sensing images, and the performance is slowly improved. Most detection algorithms use feature extraction + classifier (i.e. SVM, Adaboost, Hofferlin). Due to training set of high-resolution remote sensing images is few, many algorithms are not fully trained and cannot obtain the optimal model parameters which limit the performance of the algorithm to further improve.

For a long period of time before the deep learning was re-emphasized, there were many people proposed the target detection algorithms combined with the feature extraction and the classifier. For example, Dalal et al. [1] proposed using the Histogram of Oriented Gradient (HOG) feature to detect the target. A lot of target detection algorithms are improved on the basis of HOG features and obtain a good performance. Felzenszwalb et al. [2] proposed the most classic model of deformable components, which firstly used the root filter on the target for global rough detection, and then used the local filter for

Q. Zhou et al. (Eds.): ICPCSEE 2018, CCIS 901, pp. 721–728, 2018.
https://doi.org/10.1007/978-981-13-2203-7_60

local fine detection of the target. Because the local filter can be deformed according to the structure of the spring, the effective combination of the contour of the target can obtain good detection results. Liang et al. [3] used two features plus the multikernel support vector machine (MKLSVM), in which the multi-core SVM was parallel-trained in multithreading. The multikernal SVM was not suitable for large scale data sets, because it takes a lot of time for training.

Of course, in addition to feature extracted from remote sensing images can be replaced, the classifiers can also be replaced. Such as Grabner et al. [4] fused HOG, LBP and wavelet characteristics to train Adaboost classifier. Hof Lei [5] used forest classifier to achieve the aircraft target detection in remote sensing images. Inglada [6] proposed using the geometric features of the image to describe different types of targets. Hu et al. [7] proposed using saliency map to get the target. Sun [8] proposed using sparse encoding to detect the image plane under complex environment, and the model of geometric component information used to describe the target change and rotation has good robustness. Zhang [9] proposed to use the rotation invariant component model for detecting objects with complex shapes in remote sensing images.

Compared with the characteristics of the artificial design, deep CNN can learn rich features from the training data set in the depth learning framework. Krizhevsky [10] designed a 8-layer depth CNN network model, using convolution layer to continuously extract the underlying characteristics of image and get 4096 dimensional feature vector combination through the connection layer of feature vector. Finally, the output layer discriminated the feature vector and output the label probability. Nowadays, a variety of high resolution remote sensing image target detection algorithms based on CNN have been proposed. For example, Chen [11] proposed a hybrid DNN (HDNN) approach to achieve target detection in remote sensing images. Cheng [12] used CNN for target detection of remote sensing images. Chen [13] proposed Deep Belief Net (DBN) network to realize remote sensing image plane detection, which takes into account the target's color, scale and complex background.

The above methods are based on the deep CNN and require a large number of image data sets. The amount of image data in the natural scene is very large, so the deep CNN can be trained for a long time to obtain the optimal model parameters. Because the remote sensing image training set is small, deep CNN can not be fully trained to obtain the optimal model parameters. Therefore, if a small amount of remote sensing image is used to train the deep CNN, the effect is not ideal. Taking into account the deep CNN, the shallow learning obtains the general characteristics of the image, the deep learning is for the unique characteristics. Therefore, this paper proposes a method based on transfer learning and deep CNN combined with SVM to detect the aircraft targets of remote sensing images.

2 Structure of AlexNet Deep CNN

Deep CNN is a multi-layer deep learning framework, which is connected by several different layers: convolution layer, pool layer, full connectivity layer and non linear operators. For convolution layer, each layer is connected with the previous layer by convolution kernel, and each convolution kernel carries out the convolution operation with feature map in local receptive fields. At the same time the convolution kernel is expressed as weights of neurons. Shared weights in Deep CNN greatly reduce the number of parameters needed to be trained. It has been shown that convolution operation is used for feature extraction, the feature map of each layer are obtained by convolution with the feature map of the upper layer Eq. (1) gives the definition of the convolution layer:

$$x_j^l = f(\sum_{i=1}^{M} x_i^{l-1} \times k_{ij}^l + b_j^l) \quad (1)$$

Where x_i^{l-1} is the i-th feature map of the (l−1)-th convolution layer and x_j^l the j-th feature map of the l-th convolution layer respectively. M is the number of feature map of the current layer. k_{ij}^l and b_j^l are weight parameters and deviation of the convolution layer respectively. $f(\cdot)$ is nonlinear function.

The maximum pool operation is implemented after convolution which uses nuclear function to non repeated traversal on local receptive fields of each feature map. The feature map is divided into a plurality of non overlapping areas, then calculate the maximum value of regional characteristics in each region for subsequent training, this is max-pool. As can be seen in Fig. 1, the feature map of aircraft target in the remote sensing images after convolution operation and max-pool respectively.

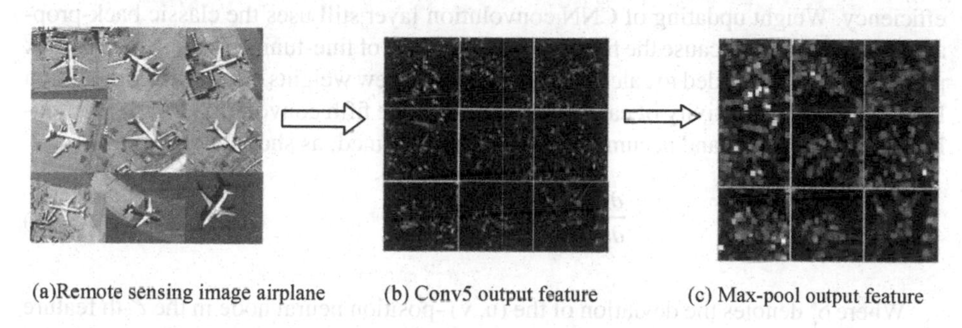

(a)Remote sensing image airplane (b) Conv5 output feature (c) Max-pool output feature

Fig. 1. The basic framework of max-pooling

3 Transfer Learning

The remote sensing image training data set in the real environment is very small (only a few thousand), and unable to meet the requirements of the ImageNet training set. Therefore, this paper introduces the concept of transfer learning. Through the analysis of hidden layer in Deep CNN, it can be found that the bottom of the CNN can achieve the extraction of the general features of the image, while the high-level generation is the unique characteristics of the image. That is to say, CNN can be used in other visual task, the bottom of which is regarded as a feature extractor, also high-level can be optimized using fine-tuning based on target data set, then obtains parameters for generating high-level features.

Based on the above analysis, this paper first gets a pretrained Deep CNN model on ImageNet with 5 convolution layers and 3 full connection layers. Then delete the last three connection layers, namely fc6, fc7, fc8, to obtain truncated Deep CNN. At last get the initial parameters through the transfer learning to generate the depth feature extractor before fine tuning.

4 Fine-tuning of the Convolutional Layers

The remote sensing image sets are used for the retaining the fifth convolution layer of truncated deep CNN in CAFFE library. The training of the fifth layer consists of forward and backward propagation. For forward propagation, this paper uses squared-error loss function as cost function. According to N training samples and 2 classes detection problems, squared-error loss function form is shown as Eq. (2):

$$E^N = \frac{1}{2}\sum_{n-1}^{N}\sum_{k=1}^{2}\left(t_k^n - y_k^n\right)^2 \tag{2}$$

Where t_k^n is the k-th dimension of the n-th training sample class. y_k^n is similar as t_k^n, and it represents the response value of the k-th output to the n-th input sample of the output layer. For the activation function of the convolution layer in each layer, RELU function can effectively shorten the training cycle to improve the learning rate and efficiency. Weight updating of CNN convolution layer still uses the classic back-propagation algorithm. Because the fifth convolution layer of fine-tuning is connected to max pooling layer, it is needed to calculate every neurons new weights of the fifth convolution layer. When the sensitivity of each neural node for the fifth convolution layer are calculated, then execution and accumulation can be performed, as shown in the Eq. (3):

$$\frac{\partial E}{\partial b_j} = \sum_{u,v}(\delta_j^\ell)_{u,v} \tag{3}$$

Where δ_j^ℓ denotes the deviation of the (u, v) -position neural node in the ℓ-th feature map in the j-th convolution layer. The sum of these deviations is equal the partial derivative of the output error to the current layer deviation. And finally the weight and the gradient of the convolution kernel can be solved by using Eqs. (4) and (5):

$$\frac{\partial E}{\partial W^\ell} = x^{\ell-1}(\delta^\ell)^T \tag{4}$$

$$\Delta W^\ell = -\eta\frac{\partial E}{\partial W^\ell} \tag{5}$$

In Eq. (4), we obtain the input of the ℓ-th convolution layer, and then use δ for scaling, where x is the output of the (ℓ−1)-th convolution layer. According to the Eq. (5), the partial derivative of Eq. (4) is multiplied by a negative learning rate to obtain the updated weight of the current layer neuron.

5 Experimental Results and Analysis

This section first introduces the data set use for the fine tuning and SVM training, and then explains how to evaluate the fifth layer convolution layer characteristics and fine-tuning. The experiments are divided into two parts, the first part is to determine which convolution layer needs to be fine tuned; the second part is to retrain the selected convolution layer. In the first part of the experiment, the data set of Google Earth v7.1.5.1557 is used which obtained in Hamburg Airport in Germany, Amsterdam Holland, International Airport, London Heathrow International Airport, Paris Charles De Gaulle Airport. The aircraft data set contain 1000 positive samples, 300 negative samples without the aircraft, among which 800 positive samples and 240 negative samples are used as the training set, and the remaining 200 positive samples and 60 negative samples are used as the test set.

In the second part experiment, we use 600 positive samples and 300 negative samples. Because training set and test set are both needed for the retaining of the convolution layer, the positive and negative samples were divided into two groups. 500 positive samples are used as training set and the remaining 100 samples are used as validation set. 250 negative samples are used as training set, the remaining 50 are used as validation set. The size of the data set will be adjusted to 227 * 227 pixels.

The truncated CNN obtains the initial parameters by transfer learning in ImageNet. Combine the feature map of the upper convolution layer to get the feature map of the current layer. The feature map forms feature vector used to train the SVM detection model. The first layer feature map is put into the SVM to get the final detection model. Obtain the detection result on the test set, i.e. the average precision, then each layer performs the first layer operation.

Precision - recall curve (P-R) and average precision (AR) are used to evaluate the detection results of different convolution layer. It can be seen from Fig. 3, the performance of SVM detection model are different based on output depth characteristics of different convolution layers. With the increase of the number of convolution layers, the average accuracy of the trained SVM model is increasing. It can be inferred that the performance of the SVM detection model should be better trained by depth feature of the fifth than that trained by depth feature of the fourth layer, however, the reality is just on the opposite. And among five convolution layers training, the average precision of deep feature decreases which model is trained by the fifth convolution layer. Take into account the deep structure of CNN, and obtain the initial parameters in ImageNet. The fifth layer is deep convolution layer, whose weight parameters and deviation of neurons have get the optimal value in the training of natural scene images. Figure 2 shows that each layer of the truncated 5-layer CNN combining with SVM constitutes a detection system, and uses the training set to train the SVM, and then tests the remote sensing image to detect aircraft. The average accuracy from Conv5+SVM to Conv1+SVM are 0.712, 0.762, 0.643, 0.564, 0.484 respectively.

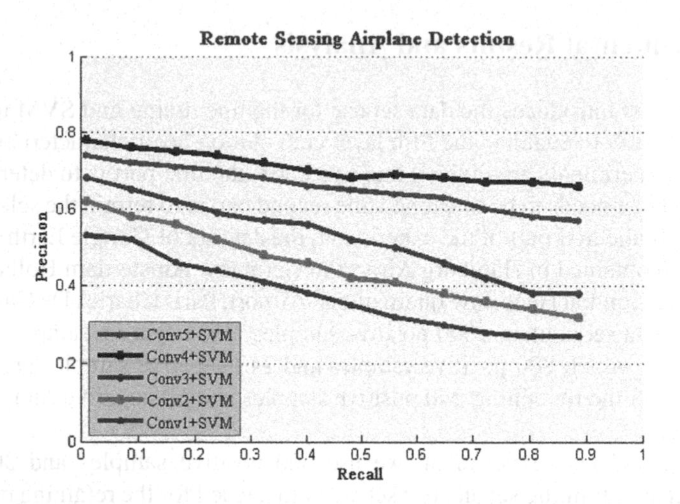

Fig. 2. P-R comparison of different convolution layer

It can be seen that the average accuracy of the fifth layer has decreased, so it needs to be fine tuned. The fine-tuning model is the fifth convolution layer of deep CNN which is pretrained on 1 million 500 thousand natural scene images (including 1000 classes). The same is as the pretraining process, and CAFFE library is also used for fine-tuning. The dimension of the last full connection layer (fc8) is the same as the number of classes in Image Net. In order to make CNN suitable for remote sensing image plane detection, the output of the fc8 needs to be changed to 2 dimensions. The pretrained deep CNN cannot match the original fifth convolution layer, whose parameters can not be initialized, so the process of fine-tuning is actually re-training the fifth convolution layer, so that the parameters can not be initialized. After fine-tuning, the parameters of the fifth layer are the best parameters fitting the remote sensing images, and the Deep CNN constitutes a feature extractor suitable for extracting the deep features of remote sensing images.

In order to evaluate the performance of CNN+FT5+SVM proposed in this paper, several classical target detection algorithms are used to compare. Figure 3 shows a variety of different algorithms in the detection of P-R curves which contain 200 positive samples and 60 negative samples of the remote sensing images. These algorithms are presented including CNN+FT5+SVM, DPM (the fifth version VOC5.1), HOG-LBP +SVM and HOG+SVM. The average accuracy of the four methods are 0.864, 0.762, 0.553, 0.426, which can be seen that the detection performance of our method is better than other methods. Figure 4 gives the airplane detection results on high resolution remote sensing image used by CNN+FT5+SVM proposed in this paper. It can be concluded that our proposed method has excellent detection performance.

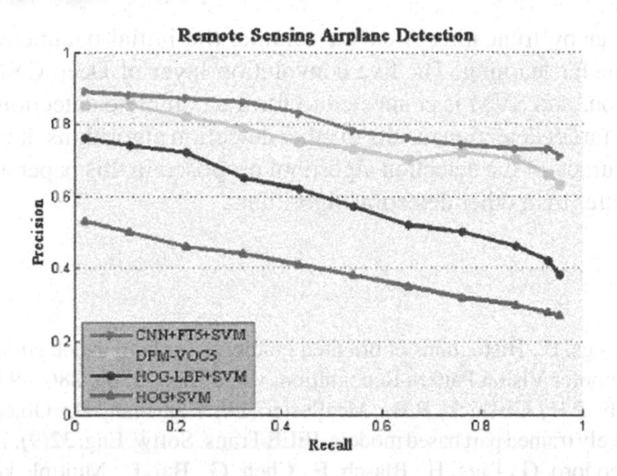

Fig. 3. Detection performance of different methods

Fig. 4. The airplane detection results used CNN+FT5+SVM

6 Conclusions

In this paper, a plane detection algorithm based on transfer learning and deep CNN combined with SVM is proposed for high-resolution remote sensing images. Firstly, deep CNN model AlexNet is trained through large data sets ImageNet, then get five convolution layer by truncating AlexNet volume, the initial parameters of which are obtained by transfer learning. The five convolution layer of Deep CNN is used as the feature extraction, and SVM is connected to form a complete detection system set. By comparison the target detection results to other detection algorithms, it can be found that the average accuracy of the detection algorithm proposed in this paper achieves to 86%, significantly better than other detection algorithms.

References

1. Dalal, N., Triggs, B.: Histograms of oriented gradients for human detection. In: Proceedings of IEEE Computer Vision Pattern Recognition, vol. 1, no. 12, pp. 886–893 (2005)
2. Felzenszwalb, P.F., Girshick, R.B., Mcallester, D., Ramanan, D.: Object detection with discriminatively trained part based models. IEEE Trans. Softw. Eng. **32**(9), 1627–1645 (2014)
3. Liang, P., Teodoro, G., Ling, H., Blasch, E., Chen, G., Bai, L.: Multiple kernel learning for vehicle detection in wide area motion imagery. In: Proceedings of 15th International Conference on Information Fusion, pp. 1629–1636 (2012)
4. Grabner, H., Nguyen, T., Gruber, B., Bischof, H.: On-line boosting based car detection from aerial images. Remote Sens. **63**(3), 382–396 (2008)
5. Lei, Z., Fang, T., Huo, H., Li, D.: Rotation-invariant object detection of remotely sensed images based on texton forest and hough voting. IEEE Trans. Geosci. Remote Sens. **50**(4), 1206–1217 (2012)
6. Inglada, J.: Automatic recognition of man-made objects in high resolution optical remote sensing images by SVM classification of geometric image features. ISPRS J. Photogram. Remote Sens. **62**(3), 236–248 (2007)
7. Hu, X., Shen, J., Shan, J.: Local edge distributions for detection of salient structure textures and objects. IEEE Geosci. Remote Sens. Lett. **10**(3), 466–470 (2013)
8. Sun, H., Sun, X., Wang, H.Q., Li, Y.: Automatic target detection in high-resolution remote sensing images using spatial sparse coding bag-of-words model. IEEE Geosci. Remote Sens. Lett. **9**(1), 109–113 (2012)
9. Zhang, W., Sun, X., Fu, K., Wang, C., Wang, H.: Object detection in high-resolution remote sensing images using rotation invariant parts based model. IEEE Geosci. Remote Sens. Lett. **11**(1), 74–78 (2014)
10. Krizhevsky, A., Sutskever, I., Hinton, G.E.: ImageNet classification with deep convolutional neural networks. In: Proceedings of Advances in Neural Information Processing, pp. 1097–1105 (2012)
11. Chen, X.Y., Xiang, S., Liu, C.L., Pan, C.H.: Vehicle detection in satellite images by hybrid deep convolutional neural networks. IEEE Geosci. Remote Sens. **11**(10), 1797–1800 (2014)
12. Cheng, G., Zhou, P., Han, J.: Learning rotation-invariant convolutional neural networks for object detection in VHR optical remote sensing images. IEEE Geosci. Remote Sens. Soc. **54**(12), 7405–7415 (2016)
13. Chen, X., Xiang, S., Liu, C.L., Pan, C.H.: Aircraft detection by deep belief nets. In: Proceedings of 2nd IAPR Asian Conference on Pattern Recognition, pp. 54–58 (2013)

Object Tracking Based on Hierarchical Convolutional Features

Aili Wang[1(✉)], Haiyang Liu[1], Yushi Chen[2], and Yuji Iwahori[3]

[1] Higher Education Key Lab for Measure and Control Technology and Instrumentations of Heilongjiang, Harbin University of Science and Technology, Harbin 150080, China
aili925@hrbust.edu.cn
[2] Institute of Image and Information Technology, Harbin Institute of Technology, Harbin 150001, China
[3] Department of Computer Science, Chubu University, Aichi, Japan

Abstract. A novel object tracking algorithm based on hierarchical convolutional features was proposed in this paper. Firstly, the tracking algorithm uses the hierarchical networks of VGG-Net-19 to extract the hierarchical convolutional features of image, having a greater improvement than using only one layer to do that. Secondly, the algorithm obtains features by using correlation filtering method with weighted fusion, so as to determine the real position of the target according to the characteristics of different layers. The experimental results show that, compared with the current four popular object tracking algorithms, the proposed algorithm achieves better accuracy and success rate, and the results are consistent in OPE (one-pass evaluation), SRE (spatial robustness evaluation) and TRE (temporal robustness evaluation).

Keywords: Object tracking · Hierarchical convolutional features · Feature fusion
Correlation filters

1 Introduction

Object tracking is one of the most popular topic in computer vision field, which has been wide applied in the military field for missile guidance and the civil field for telemedicine or visual reality [1]. In recent years, with the high-speed development of computer performance, as well as the appearance of Graphics Processing Unit (GPU), the accuracy and efficiency of tracking algorithms have been improved in a significant degree [2]. But with the effect of occlusion, rotation, illumination variation, etc., the robustness and accuracy of tracking algorithms are still facing serious challenges [3].

Recently, the most popular tracking algorithms based on deep learning [4] usually draw positive sample and negative sample in the target position, by using incremental learning to exploit feature classifier from the Convolutional Neural Network (CNN) [5, 6]. The traditional method is to choose the last layer of CNN for feature representation, which has an excellent ability in semantic inference. Whereas it is insensitive to variable changes and accurate positioning. So, it cannot accomplish the mission of accurate object representation with only the last layer of CNN. The kernel concept of Deep Learning is training and

Q. Zhou et al. (Eds.): ICPCSEE 2018, CCIS 901, pp. 729–737, 2018.
https://doi.org/10.1007/978-981-13-2203-7_61

learning, which need a large quantity of positive samples and negative samples. In order to obtain a well-robust classifier, all of the samples are integrant.

The rest of the paper is organized as follows. Section 2 reviews related work of target tracking, and Sect. 3 introduces the hierarchical convolutional features and the network structure. Section 4 is the proposed tracking algorithms and Sect. 5 gives the qualitative estimation and quantitative estimation of our algorithms. We conclude the future works in the last Sect. 6.

2 Related Work

All of the present object tracking algorithms are mainly divided into three fields. First of all, the tracking algorithm based on binary classifiers which is mainly taking use of tracking-by-detecting. It regards object tracking as a repetition detection within a local detector, and collects a series of positive samples and negative samples from all frames of video sequences. Incremental learning is used to separate target form background, such as semi-supervised learning [7], which is the combination of supervised learning and unsupervised learning, and uses both a large quantity of unlabeled data and some labeled data to recognize the target. It can not only reduce the manpower but also increase the accurate to a large degree.

The second one is the tracking algorithm based on correlation filters, which mainly cares about its high-speed computing efficiency of Fourier Transform. Bolme et al. [8] proposed a tracking algorithm based on minimum output of luminance channel and mean square error filter. While tracking, it will need to keep adjusting the filter according to the current frame. Some expand methods improve the accuracy of tracking, such as Multi-dimensional Features [9, 10], Context Learning [11] and Scale Estimation [12] etc.

The third one is the tracking algorithm based on Convolutional Neural Network, which is the most popular in recent years. The traditional tracking algorithms use the manual extracted features to represent the target, but the result cannot be satisfied by us. However CNN can represent the target's feature perfectly be by using the positive samples and negative samples. The earlier tracking algorithms based on CNN extracted features using the last fully connected layer, but it cannot extract all of the feature to represent the target. Few years later, Nam et al. proposed HCF (Hierarchical Convolutional Features) algorithm [13], which took use of three convolutional layers together to extract features and gave us a better result than the former.

In this paper, we improve the VGG-Net [14] to extract the convolutional features with the combination of three convolutional layers and one fully connected layer, and accomplish the tracking target with correlation filters. The proposed tracker can solve the tracking failures caused by occlusion and overmuch high frequency information.

3 Hierarchical Convolutional Features

With the increasing of the layer of CNN, it tends to code semantic towards the target, and its output is well robust to the appearance changes. In contrast, the lower layer tends to represent space details, which is more helpful for accurate positioning. Recently, with the improvement of computer performance, many models of deep convolutional neural network with excellent performance are proposed such as AlexNet and VGG-Net etc. They can receive multidimensional image as the input, which can avoid the procedure of complex feature extractions and data reconstructions by contrasting to the traditional object tracking algorithms. So, Convolutional Neural Network is wide applied to the field of image processing and computer vision.

CNN is consists of two levels, which is a multilayer perceptron and we can obtain a sires of different feature representation from every layer while inputting an image to the net. In this paper, we choose the VGG-Net-19 as the basic of deep convolutional network, and the number “19” means the weight of the network learning. The structure of our network is shown in the Fig. 1. In Fig. 1, there are five groups of convolutional layers (totally 16 layers) and three fully connected layers (recorded as FC1, FC2 and FC3), which consists of two fully connected feature layers (FC-4096) and one fully connected classification layer (FC-10). Among the whole network, the convolutional group Conv1-2 consist of two convolutional layers and Conv3-5 consist of four convolutional layers, which all use the convolution kernel with the size of 3 * 3. By training on the labeled large scale training dataset (ImageNet), we can obtain hierarchical feature representation from every layer of VGG-Net-19.

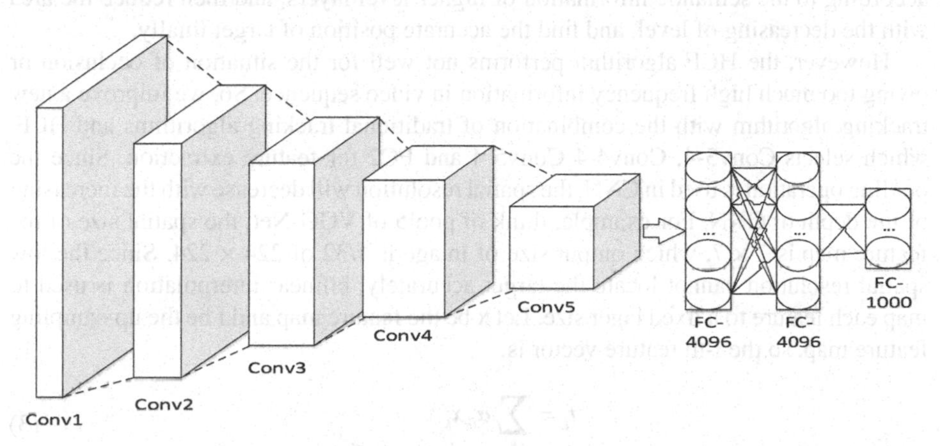

Fig. 1. The structure of VGG-Net-19 convolutional neural network which including 5 convolutional layer groups and three fully-connected layers.

To the VGG-Net-19, let x be the i-th layer feature vector whose size is $M \times N \times O$, and M, N, O separately represent width, height and channel numbers. So, the multi-channel convolutional feature map of the i-th layer is $f_i \in \Re^{M \times N \times O}$, and every characteristic channel o ($o \in \{1, 2, \ldots, O\}$) will be used as the positioning filter:

$$H_i^O = \frac{G_I \odot \overline{F}_i^O}{\sum_{k=1}^{O} F_i^k \odot \overline{F}_i^k + \lambda_p} \quad (1)$$

Where λ_p is the regularization parameter of the positioning filter, F_i is the DFT transformation of f_i, $\overline{F_i}$ and F_i are conjugate complex numbers, and the operator $\odot$ is the expression of Hadamard product. Let g_i represent the expectation of the cyclic displacement sample of $f_i(m, n)$, which is a two-dimensional Gauss kernel function, and its expression is Eq. (2):

$$g_i(m,n) = e^{-\frac{\left(m-\frac{M}{2}\right)^2+\left(n-\frac{N}{2}\right)^2}{2\sigma_p^2}} \tag{2}$$

Where $(m, n) \epsilon \{0, 1, \ldots, M-1\} \times \{0, 1, \ldots, N-1\}$, and σ_p is the width of the Gaussian kernel.

Considering about the features abstracted by each layer, clear contour can be seen in Conv3, and cannot be recognized with the increasing of layer levels, such as in Conv5. However, we can distinguish the semantic information from the highlighting information section, and the highlight area is the position that the target located. So, the process of determining the location of target is as follows, roughly estimate the position of target according to the semantic information of higher level layers, and then reduce the area with the decreasing of level, and find the accurate position of target finally.

However, the HCF algorithm performs not well for the situation of occlusion or owing too much high frequency information in video sequence. So, we improve a new tracking algorithm with the combination of traditional tracking algorithms and HCF, which selects Conv3-4, Conv4-4 Conv5-4 and FC2 for feature extraction. Since the pooling operation is used in CNN, the spatial resolution will decrease with the increasing of the depth of CNN. For example, think of pool5 of VGG-Net, the spatial size of the feature map is 7×7, which output size of image is 1/32 of 224×224. Since the low spatial resolution cannot locate the target accurately, bilinear interpolation is used to map each feature to a fixed lager size. Let x be the feature map and f be the up-sampling feature map, so the i-th feature vector is:

$$f_i = \sum_k \alpha_{ik} x_k \tag{3}$$

From the Eq. (3), the interpolating weight α_{ik} relies on the feature vector of (i, k) neighborhood.

4 Object Tracking Algorithm Based on Correlation Filters

A typical correlation filter accomplish the object estimation by learning from discriminate classifiers and searching the maximum of correlative feature maps. In this paper, we mainly choose the output of the last layer of the four groups. According to the parameter settings in the Sect. 3, simplify x^i to x and ignore the dependency of M, N and O about the layer index i. We set all the cyclic shifts of x along to M and N dimensions as training samples.

By calculating the minimum of Eqs. 2 and 3, a correlation filter u can be obtained, whose size is the same to x:

$$u^* = \underset{u}{\operatorname{argmin}} \sum_{m,n} \|u \cdot x_{m,n} - y(m,n)\|^2 + \lambda \|u\|_2^2 \tag{4}$$

Where λ is a regularization parameter ($\lambda \geq 0$), and the inner product is a linear kernel function in Hilbert space, such as $u \cdot x_{m,n} = \sum_{o=1}^{O} u^T_{m,n,o} x_{m,n,o}$. The samples with threshold will not be needed, since y(m, n) is not binary. According to the Eq. (2), Fast Fourier Transform (FFT) can be used to get the minimum value through the feature of every channel. The Capital is used to represent the signal which is transformed by FFT. The frequency spectrum of the filter in the o-th channel ($o \in \{1, 2, \ldots, O\}$) is:

$$U^O = \frac{Y \odot \overline{X}^O}{\sum_{l=1}^{O} X^i \odot \overline{X}^i + \lambda} \tag{5}$$

Where Y is the frequency spectrum of $\{Y(m, n)|(m, n)\{0, 1, \ldots, M-1\} \times \{0, 1, \ldots, N-1\}\}$. The feature vector of the j-th layer is set as z, and the size of the image is $M \times N \times O$. The correlation feature map of the j-th layer can be calculated by:

$$f_i = \mathcal{F}^{-1}\left(\sum_{o=1}^{O} U^o \odot \overline{Z}^o\right)$$

And the j-th convolutional layer can be estimated by finding the maximum of the feature map f_j.

Finally, by giving the size $M \times N$ (e.g. 1.8 times the size of the target) of the searching window of each image frame, set a fixed spatial size $\frac{M}{4} \times \frac{N}{4}$ to adjust the feature size of each convolutional layer. Set the regularization parameter $\lambda = 10^{-4}$ in Eq. (1), and generate the Gauss function label with 0.1 kernel width. In order to remove the discontinuity of the boundary, use cosine window to weigh each feature extraction channel. We set Conv4-4, Conv3-4, Conv5-4 and FC2 with the value of 1, 0.5, 0.02 and 0.01, which is equivalent to simply summing and deducing the target location from the weighted response maps of multiple levels.

5 Experimental Results and Analysis

In this paper, we evaluate the proposed method on OTB100 (a large benchmark dataset, including 100 annotated video sequences), and compare tracking performance with several advanced tracking methods. We use the distance accuracy rate, the overlap rate and the center position error to quantitatively evaluate the tracker.

5.1 Quantitative Evaluation of Experimental Results

The experimental results of this paper and the other four most advanced trackers are shown in Fig. 2. These trackers can be roughly divided into three categories: i. The algorithm based on deep learning, such as HCF and the proposed algorithm (Mine); ii. The algorithm based on correlation filter, such as KCF; iii. A representative tracker using a single or multiple online classifier, such as MEEM and Struck.

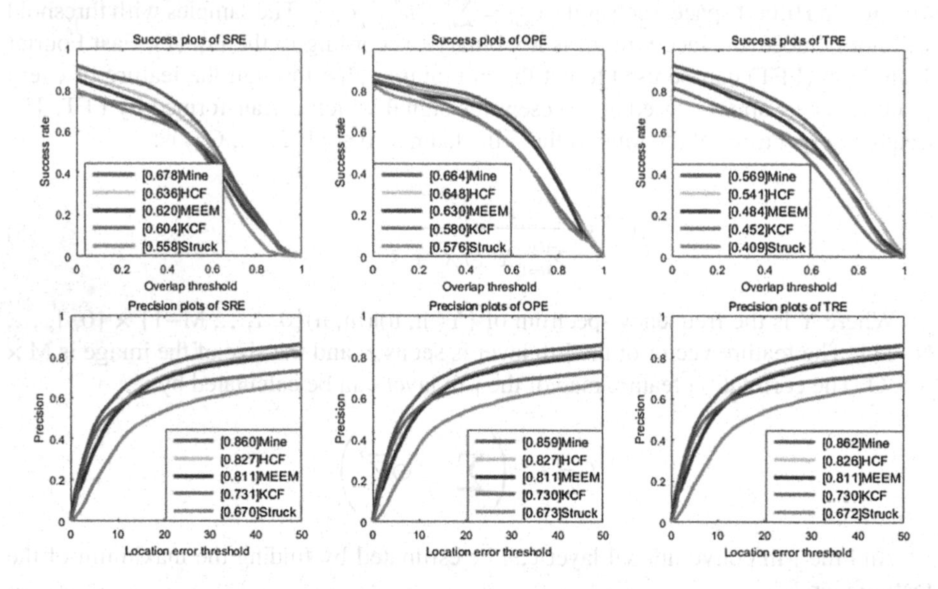

Fig. 2. Success plots and precision plots of test sequences by comparing with four popular tracking algorithms, HCF, MEEM, KCF and Struck.

Figure 2 shows the evaluation of three indexes respectively about spatial robustness evaluation (SRE), one-pass evaluation (OPE) and temporal robustness evaluation (TRE) in the two aspects of the accuracy and overlap of the distance. The algorithm we proposed performs better than any other methods evaluated on the three indexes.

Table 1. Quantitative comparison of several trackers in the three aspects.

	Distance precision (DP, %)		Overlap success (OS, %)		Center location error (CLE, pixel)	
Benchmark	OTB 50	OTB 100	OTB 50	OTB 100	OTB 50	OTB 100
Mine	91.3	85.2	78.4	67.1	14.4	20.0
HCF	89.1	83.7	74.0	65.5	15.7	22.8
MEEM	83.0	78.1	69.6	62.2	20.9	27.7
KCF	74.1	69.2	62.2	54.8	35.5	45.0
Struck	65.6	63.5	55.9	51.6	50.6	47.1

By comparing with the most advanced tracking methods based on two tracking benchmark of OTB50 and OTB100, a quantitative comparison is made in three aspects:

distance precision (DP), overlap success (OS) and center location error (CLE). The specific results are shown in Table 1. From the Table 1, the method we proposed has good consistency within the three aspects.

5.2 Qualitative Evaluation of Experimental Results

In the two tracking benchmark of OTB, all video sequences have different video attributes, including Illumination variation (IV), Scale variation (SV), Occlusion (OCC), Deformation (DEF), Motion blur (MB), Fast motion (FM), In-plane rotation (IPR), Out-of-plane rotation (OPR), Out of view (OV), background clutters (BC) and Low resolution (LR). The Table 2 lists part of the video sequences and their attributes used in the experiments.

Table 2. The parameters of the selected video sequences.

Sequences	Total frames	Attributes
Basketball	725	IV, OCC, DEF, OPR, BC
MotorRolling	164	IV, SV, MB, FM, IPR, BC, LR
Shaking	400	IV, SV, OCC, DEF, OPR, BC
Coke	291	IV, OCC, FM, IPR, OPR, BC
Tiger2	365	IV, OCC, DEF, MB, FM, IPR, OPR, OV

Figure 3 gives the tracking results of different video sequences. We respectively select the 1st, 30th, 50th and 100th frame to evaluate the proposed algorithm. Especially for the sequence of Shaking, it can be founded, the objected is lost in the 30th frame and re-found in 50th frame and so on. The experimental results sufficiently proved that the proposed algorithm can achieve target tracking in complex scene.

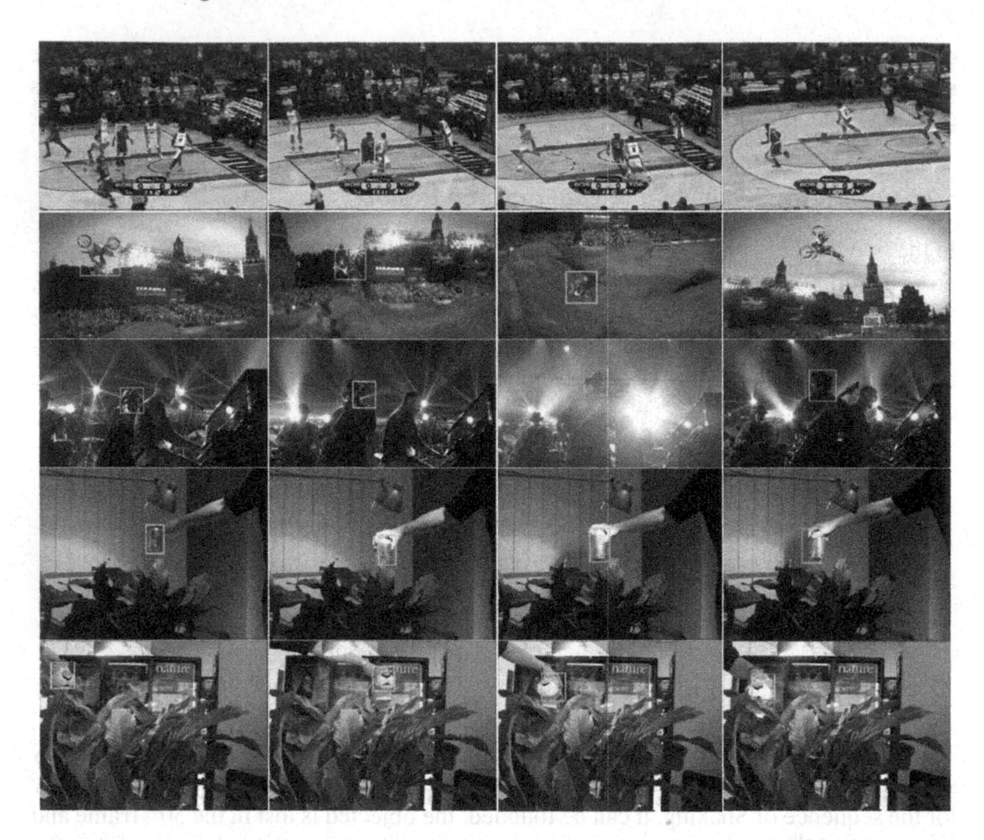

Fig. 3. Tracking results of different video sequences, concluding basketball, motorrolling, shaking, coke and tiger2.

6 Conclusions

The proposed algorithm in this paper mainly optimizes the part of feature extraction, which uses Conv4-4, Conv3-4, Conv5-4 and FC2 together for object representation with the hierarchical convolutional features. The obtained feature has more extensive practicality, and can solve the tracking failures caused by occlusion to a large degree. By comparing with the excellent tracking algorithms proposed in recent years, our method has better tracking performances in DP, OS and CLE three aspects.

References

1. Gao, L., Pan, H., Xie, X., Zhang, Z., Li, Q., et al.: Graph modeling and mining methods for brain images. Multimedia Tools Appl. **75**(15), 9333–9369 (2016)
2. Wang, Y., Wang, H., Li, J., Gao, H.: Efficient graph similarity join for information integration on graphs. Front. Comput. Sci. **10**(2), 317–329 (2016)
3. Smeulders, A.W.M., Chu, D.M., Cucchiara, R., Calderara, S., Dehghan, A., Shah, M.: Visual tracking: an experimental survey. TPAMI **36**(7), 1442–1468 (2014)

4. Henriques, J.F., Caseiro, R., Martins, P., et al.: High-speed tracking with kernelized correlation filters. Pattern Anal. Mach. Intell. IEEE Trans. **37**(3), 583–596 (2015)
5. Danelljan, M., Hager, G., Khan, F.S., et al.: Learning spatially regularized correlation filters for visual tracking. In: IEEE International Conference on Computer Vision, pp. 4310–4318 (2015)
6. Danelljan, M., Hager, G., Khan, F.S., et al.: Learning spatially regularized correlation filters for visual tracking. In: IEEE International Conference on Computer Vision, pp. 4310–4318 (2015)
7. Buccoli, M., Bestagini, P., Zanoni, M., et al.: Unsupervised feature learning for bootleg detection using deep learning architectures. In: IEEE International Workshop on Information Forensics and Security, pp. 131–136 (2015)
8. Yin, Z., Liu, J.: Introduction of SVM algorithms and recent applications about fault diagnosis and other aspects. In: Industrial Informatics (INDIN), pp. 550–555 (2015)
9. Henriques, J.F., Caseiro, R., Martins, P., Batista, J.: Highspeed tracking with kernelized correlation filters. TPAMI **37**(3), 583–596 (2015)
10. Park, E., Ju, H., Jeong, Y.M., et al.: Tracking-learning-detection adopted unsupervised learning algorithm. In: Seventh International Conference on Knowledge and Systems Engineering, pp. 234–237 (2015)
11. Sun, H., Yu, T.: Crane tracking and monitoring system based on TLD algorithm. In: IEEE International Instrumentation and Measurement Technology Conference Proceedings, pp. 1–5 (2016)
12. Min, W.P., Jung, S.K.: TLD based vehicle tracking system for AR-HUD using HOG and online SVM in EHMI. In: IEEE International Conference on Consumer Electronics, pp. 289–290 (2015)
13. Varfolomieiev, A., Lysenko, O.: An improved algorithm of median flow for visual object tracking and its implementation on ARM platform. J. Real-Time Image Proc. **11**(3), 1–8 (2016)

A Volleyball Movement Trajectory Tracking Method Adapting to Occlusion Scenes

Ting Yu, Zeyu Hu, Xinyu Liu, Pengyuan Jiang, Jun Xie, and Tianlei Zang(✉)

Southwest Jiaotong University, Chengdu 610031, Sichuan, China
zangtianlei@126.com

Abstract. This paper proposes a volleyball trajectory tracking method adapting to occlusion scenes. Firstly, the target volleyball is obtained in the video manually and the Kalman filter algorithm combined with Continuously Adaptive Mean Shift (CAMSHIFT) algorithm is used to track and determine the size and position of the volleyball in each frame of video, and then it is determined whether there exists an occlusion. If there is no occlusion, positions of the volleyball in each frame of video are connected to obtain the trajectory of the volleyball. If there is occlusion, the Kalman filter algorithm is used to predict the positions of the volleyball in the occlusion section, and the size remains unchanged. Finally, the positions of the volleyball in each frame of video is connected to a line to obtain the trajectory of the volleyball motion. The proposed approach solves the problem of more complicated video background in volleyball movement. When the volleyball is blocked, it can accurately predict the volleyball movement trajectory so as to accurately track the volleyball movement trajectory under dynamic and occlusion scenes.

Keywords: Kalman filter algorithm
CAMSHIFT algorithm · Trajectory tracking · Occlusion

1 Introduction

At present, the Hawkeye system is used in sports games such as tennis, badminton and volleyball. However, the Hawkeye system is still in the exploratory stage. In the volleyball game with athletes as background, the Hawkeye system has difficulty tracking volleyball trajectories. In addition, if the volleyball is blocked during the movement, it is difficult for the Hawkeye system to predict the trajectory of the volleyball, which means the tracking of volleyball trajectories under dynamic and occlusion scenes by the Hawkeye system is not perfect [1, 2]. The method studied in this paper belongs to the problem of target tracking, and the data processing is complex. It is a hot issue in data science research.

The existing algorithms for the trajectory tracking of volleyball such as Gaussian mixture algorithm [3] and the Codebook algorithm [4] separate foreground (moving objects) and from background (static objects or motions) in each frame of image of the video, which is based on the color components. The positions of foreground in each frame of the image finally match the moving objects' trajectory. According to the above

Q. Zhou et al. (Eds.): ICPCSEE 2018, CCIS 901, pp. 738–749, 2018.
https://doi.org/10.1007/978-981-13-2203-7_62

description, these two methods do not adapt to the trajectory tracking of objects having a complex background which includes intense motion. For volleyball videos, there are athletes with intense motion, due to which the two methods are not applicable.

The algorithm used in this paper is Kalman filter algorithm combined with CAMSHIFT algorithm [5, 6]. This algorithm uses prior knowledge to track the volleyball trajectory. Firstly, the position of the volleyball in the first frame is circled manually and the parameters of the volleyball are obtained. Next, it is the volleyball in each frame of the image to be tracked. When the volleyball encounters occlusion, the Kalman filter predicts the volleyball trajectory to fit the volleyball trajectory. Compared with the first two algorithms, this algorithm has higher accuracy, faster operation, and stronger adaptability.

2 A Volleyball Trajectory Tracking Method Adapting to Occlusion Scenes

In this paper, a volleyball trajectory tracking method adapting to the occlusion scenes is designed, which includes the following steps.

Firstly, the parameters of the tracked volleyball are obtained manually in the video, including the size and the position. Secondly, the CAMSHIFT algorithm is used in combination with the Kalman filter algorithm to track and determine the size and position of the volleyball in each frame of the video image [7]. Next, whether the volleyball is blocked is judged in the video. If there is no occlusion, the positions of the volleyball in each frame of the video are connected to get the volleyball motion trajectory. If there is occlusion, the Kalman filter algorithm is used to predict the positions of the volleyball in the occlusion section. Then the positions of the volleyball in each frame of the image determined by the algorithm are connected to a line to get the trajectory of the volleyball [8, 9].

2.1 Obtain Parameters Manually

Semi-automatic method is adopted to track the volleyball, namely gain its size M (a, b), location $\boldsymbol{N}$ (cc, cr) and target matrix $\boldsymbol{Z}$ which consists of circled target pixel block in the first frame of video image by manually circling the volleyball, to obtain the parameters of the volleyball finally (Figs. 1 and 2).

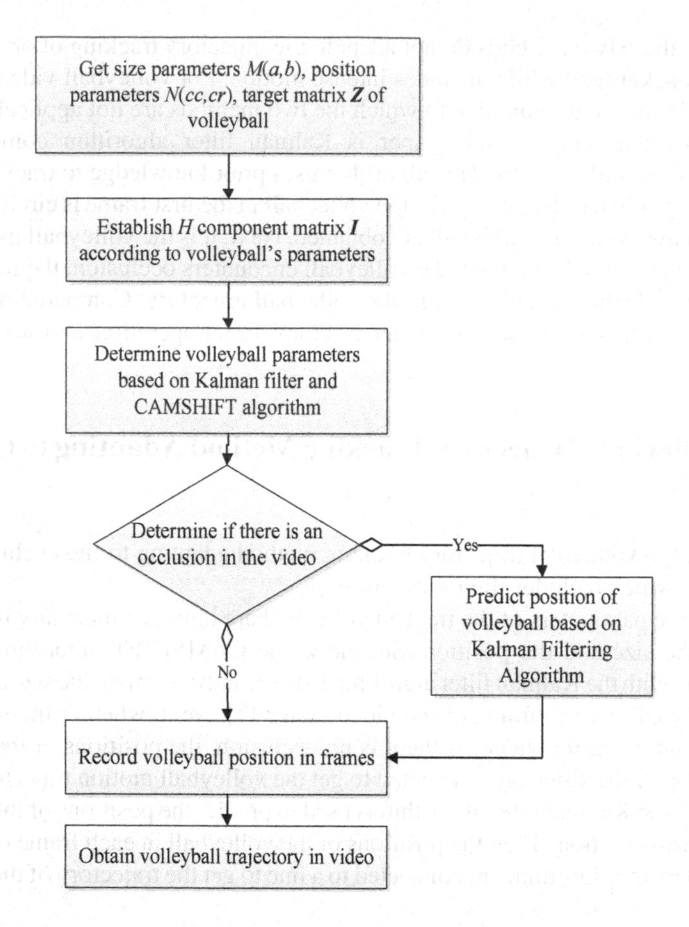

Fig. 1. Algorithm flowchart

Fig. 2. Manually circle the volleyball

2.2 Track the Volleyball Using CAMSHIFT

The CAMSHIFT algorithm is used to track and determine the size and position of the volleyball in each frame of the video. The obtained position parameters are used as location measurement in the Kalman filter algorithm [10]. Then, according to the Kalman filter algorithm, a more accurate position parameter is obtained.

A: Obtained Color Histogram Matrix

The target matrix **Z** of volleyball is transformed from RGB color model to the hexagonal cone model (HSV), and color histogram matrix is established using *H* component matrix ***I***. Color histogram matrix is also known as target histogram matrix.

The pixel value is equally divided into m intervals. The pixel value range of the r interval is $\left(\frac{225(r-1)}{m}, \frac{255r}{m}\right)$. If a certain pixel' value belongs to interval r, r is the index value corresponding to the pixel.

B: Get Color Index Function Value

Get the target volleyball model according to the *H* component matrix ***I***, namely the probability density function is:

$$q_u(x) = C\sum_{i=1}^{n}\left\{k\left(\left\|\frac{x-x_i}{h}\right\|^2\right)\delta\left[b(x_i)-u\right]\right\} \tag{1}$$

where, x is the center of matrix $\boldsymbol{I}$, x_i is the pixel location of matrix $\boldsymbol{I}$, whose value range is $1, \ldots, n$ (pixel location is defined from line to column, from left to right, and from top to bottom); functions $b{:}T^2 \to \{1, \ldots, m\}$ is defined as color index function at pixels x_i; $b(x_i)$ is the color index value corresponding to the pixel x_i; u is the index subscript, and its value range is $1, \ldots, m$; $k\left(\|x\|^2\right)$ is the contour function of the kernel function; h is the bandwidth, namely the sum of squares between the half of target length and the width [11–13].

Because of $\sum_{u=1}^{m} q_u = 1$, C can be obtained as follow:

$$C = \frac{1}{\sum_{i=1}^{n} k\left(\left\|\frac{x-x_i}{h}\right\|^2\right)} \tag{2}$$

Color index function of target volleyball is:

$$\delta\left[b(x_i)-u\right] = \begin{cases} 1 & b(x_i) = u \\ 0 & b(x_i) \neq u \end{cases} \tag{3}$$

C: Estimate the Candidate Probability Density Function

In each frame of volleyball video image, the identified target center in previous frame is used as the present center, and the same method in steps of section B is adopted to establish the candidate model of the same size as the target, namely the candidate probability density function:

$$p_u(f) = C\sum_{i=1}^{n}\left\{k\left(\left\|\frac{f-x_i}{h}\right\|^2\right)\delta[b(x_i)]-u\right\} \tag{4}$$

where, f is the volleyball position in each frame of video.

D: Identify the Volleyball Center with a Similar Function

The Bhattachariya coefficient is used as a similar function, which is defined as:

$$\rho(f) = \sum_{u=1}^{m}\sqrt{p_u q_u} \tag{5}$$

In the current frame, tracking the new volleyball position begins from the estimated volleyball location in the previous frame, and search the maximum in the surroundings. Get the Taylor expansion for function $\rho(f)$ on the target location f_0 in the previous frame:

$$\begin{aligned}\rho(f) &\approx \frac{1}{2}\sum_{u=1}^{m}\sqrt{p_u(f_0)q_u} + \frac{1}{2}\sum_{u=1}^{m}p_u(f)\sqrt{\frac{q_u}{p_u(f_0)}}\\ &= \frac{1}{2}\sum_{u=1}^{m}\sqrt{p_u(f_0)q_u} + \frac{C}{2}\sum_{i=1}^{n}w_i k\left(\left\|\frac{f-x_i}{h}\right\|^2\right)\end{aligned} \tag{6}$$

where, $w_i = \frac{1}{2}\sum_{u=1}^{m}\delta[b(x_i)-u]\sqrt{\frac{q_u}{p_u(f_0)}}$.

In the above equation, only the second term is changed with the f, and its maximization process is completed by the Mean shift algorithm iterative equation from the candidate regional center to the real regional center.

The iterative equation is:

$$f_{k+1} = f_k + \frac{\sum_{i=1}^{n}w_i(f_k-x_i)g\left(\left\|\frac{f_x-x_i}{h}\right\|^2\right)}{\sum_{i=1}^{n}w_i g\left(\left\|\frac{f_k-x_i}{h}\right\|^2\right)} \tag{7}$$

The new volleyball location is calculated as follow:

$$f_1 = \frac{\sum_{i=1}^{n}x_i w_i g\left(\left\|\frac{f_0-x_i}{h}\right\|^2\right)}{\sum_{i=1}^{n}w_i g\left(\left\|\frac{f_0-x_i}{h}\right\|^2\right)} \tag{8}$$

After the new volleyball location is obtained, the similar function is expanded at the new volleyball location, repeating the iteration process until the number of iterations is reached and the final volleyball position is obtained.

E: Adjust the Target Size
Each location update is calculated by using $h \pm h \times 10\%$ (the target length and width scaling) for three times. The minimum bandwidth is used to determine the size of the new target window.

F: Initialize the Kalman Filter
The state transition matrix $\boldsymbol{E}$, the control matrix $\boldsymbol{B}$, the observation matrix $\boldsymbol{G}$, the measurement noise covariance matrix $\boldsymbol{R}$, the process noise excitation covariance matrix $\boldsymbol{Q}$, and the prediction estimation covariance matrix $\boldsymbol{PP}$ are determined in this section. Then the state matrix $\boldsymbol{l}$ is initialized (including the target coordinates and speed in the X and Y directions) based on the volleyball parameters obtained according to Sect. 2.1.

G: Obtain the Final
Firstly, according to the Kalman filter, the prior system state (the position and speed of the target volleyball in the X and Y directions) estimate is updated with the estimate of the trustworthy system state obtained in the previous frame, and so is the prior covariance. Secondly, calculate the Kalman gain based on the updated covariance, and then the Kalman gain and the prior system state estimate are used to calculate the trustworthy system state estimate of the current frame [14, 15]. Finally, repeat the above steps to obtain the trustworthy system state estimate of each frame. Specifically:

(1) *Update the prior system state estimate*

Update the prior system state estimate according to $\bar{\boldsymbol{l}}_i = \boldsymbol{E} * \boldsymbol{l}_{i-1}$, where, $\boldsymbol{l}_{i-1}$ is the posterior state matrix of the previous frame and $\bar{\boldsymbol{l}}_i$ is the current frame prior state estimation matrix.

(2) *Update the prior covariance*

Update the prior covariance according to $\overline{\boldsymbol{PP}_i} = \boldsymbol{E} * \boldsymbol{PP}_{i-1} * \boldsymbol{E}^{-1} + \boldsymbol{Q}$, where, $\boldsymbol{PP}_{i-1}$ is the previous frame's posterior estimate covariance matrix, $\overline{\boldsymbol{PP}_i}$ is the current frame's a priori estimate covariance matrix.

(3) *Calculate the Kalman gain*

$$K = \overline{\boldsymbol{PP}_i} * \boldsymbol{G}^{-1} * inv\left(\boldsymbol{G} * \overline{\boldsymbol{PP}_i}\boldsymbol{G}^{-1} + \boldsymbol{R}\right) \tag{9}$$

(4) *The system state estimate of the current frame*

Calculate the system state estimate of the current frame $\boldsymbol{l}$.

$$\boldsymbol{l}_i = \bar{\boldsymbol{l}}_i + K * \left([cc(i), cr(i)]^{-1} - \boldsymbol{G} * \bar{\boldsymbol{l}}_i\right) \tag{10}$$

(5) *The covariance of the current frame*

Calculate the covariance PP_i of the current frame: $PP_i = (eye(4) - K * G) * \overline{PP_i}$

2.3 Determine the Existence of Occlusions

It is determined whether there exists an occlusion. If there is no occlusion in the video, positions of the volleyball in each frame of video are connected to obtain the trajectory of the volleyball. If there is occlusion, jump directly to Sect. 2.4 [5, 16].

2.4 Predict Volleyball Trajectory Using Kalman Filter

According to the prior state prediction equation in the Kalman filter [17], the state parameters of the occlusion section are predicted as follow:

$$l_i = E \times l_{i-1} \tag{11}$$

Connect the positions of volleyball in each frame into a line to obtain the trajectory of the volleyball.

3 Example Test

In this paper, two scenarios are used to verify the proposed method, in which volleyball is blocked by objects or not.

3.1 Volleyball not Obstructed by Objects

In volleyball competitions, since volleyballs have a wide range of sports space on the field, most volleyball sports are in an unobstructed condition. Therefore, the trajectory of volleyball not obstructed by objects is tracked firstly.

The actual trajectory and tracking trajectory of volleyball are shown in Fig. 3, where the black line is the actual trajectory, and the red line is the tracking trajectory. Some of the specific coordinates are shown in Table 1. Analysis of coordinates is shown in Table 2.

From Fig. 3, it is obvious that the volleyball tracking trajectory obtained by the proposed method can reflect its actual trajectory. In order to quantitatively analyze the tracking effect, the Manhattan distance, Euclidean distance and Pearson correlation coefficient between the two curves are calculated. It can be seen that the tracking of the proposed method is effective.

Fig. 3. The trajectory of volleyball not obstructed by objects (Color figure online)

Table 1. Specific coordinates

Point	1	2	3	4
Tracked pixel	(746.58, 769.41)	(845.56, 794.27)	(939.06, 816.76)	(1028.78, 841.08)
Actual pixel	(767.13, 786.35)	(843.56, 810.25)	(924.88, 820.12)	(1005.12, 855.20)
Point	5	6	7	8
Tracked pixel	(1120.14, 865.63)	(1208.13, 873.58)	(1286.68, 905.98)	(1370.34, 933.77)
Actual pixel	(1105.55, 872.16)	(1184.23, 876.23)	(1252.65, 908.56)	(1324.70, 835.24)

Table 2. Analysis of coordinates

Analysis	Average Manhattan distance	Average Euclidean distance	Pearson correlation coefficient
Index value	42.40 pixel	33.45 pixel	0.9214

3.2 Volleyball Blocked by Objects

The difference between volleyball and other ball games lies in the fact that there are many players in the field and they are densely distributed. Therefore, it is inevitable that volleyball is blocked by athletes in the process of camera acquisition which is also the difficulty in tracking volleyball.

In order to better verify the proposed method's ability to deal with occlusion, two cameras are used to track the same trajectory of volleyball to verify whether the proposed method can solve the problem of occlusion occurring.

Figure 4 shows the actual trajectory and tracking trajectory of volleyball in the situation which volleyball is blocked by an orange area. In Fig. 4 the blue line is the actual trajectory, and the red line is the tracking trajectory. Some of the specific coordinates are shown in Table 3. Analysis of coordinates is shown in Table 4.

Fig. 4. The trajectory of volleyball obstructed by objects (Color figure online)

Table 3. Specific coordinates

Point	1	2	3	4
Tracked pixel	(478.06, 489.57)	(569.27, 492.87)	(651.66, 495.23)	(729.06, 495.25)
Actual pixel	(499.64, 506.89)	(567.42, 508.88)	(637.19, 510.20)	(705.44, 509.36)
Point	5	6	7	8
Tracked pixel	(789.31, 499.83)	(863.49, 501.61)	(937.67, 503.39)	(1011.86, 505.17)
Actual pixel	(774.93, 508.36)	(841.31, 507.24)	(907.27, 508.33)	(975.88, 509.58)

Table 4. Analysis of coordinates

Analysis	Average Manhattan distance	Average Euclidean distance	Pearson correlation coefficient
Index value	24.84 pixel	31.30 pixel	0.9651

Three indexes are calculated to quantitatively analyze the tracking effect. According to the results, the proposed method has high accuracy.

4 Algorithm Comparison

In this section, the algorithm proposed in this paper is compared with other algorithms related to trajectory tracking. Each algorithm is used to track the same volleyball video, and the tracking results are compared in the same coordinate system. As shown in Fig. 5, the blue track is the actual track of volleyball and is the reference track.

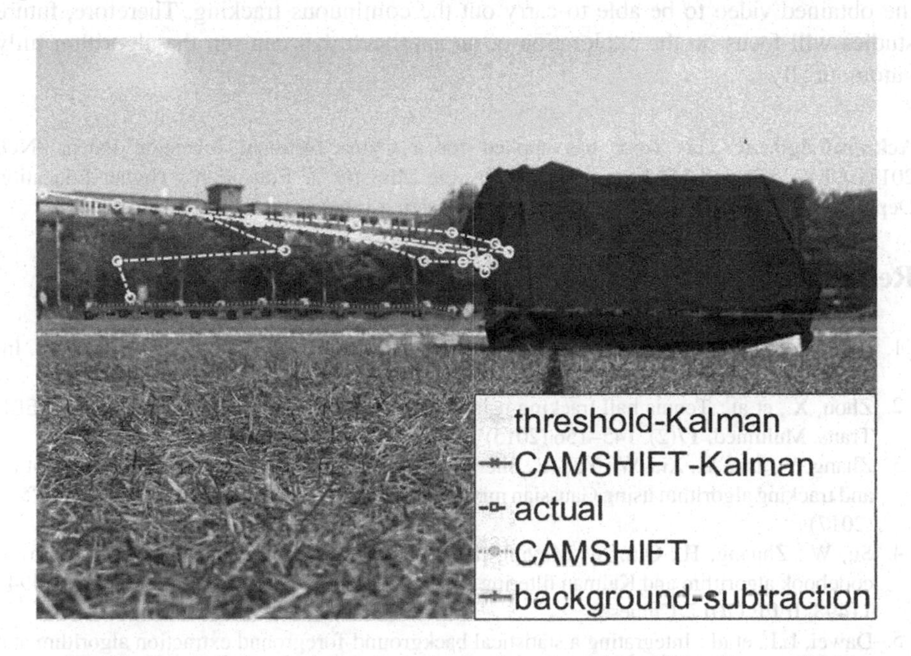

Fig. 5. Comparison of track trajectories by four algorithms (Color figure online)

From Fig. 5 it can be seen intuitively that the algorithm proposed in this paper is superior to other algorithms for the tracking of volleyball trajectories in occlusion scenarios. Compared with other algorithms, the proposed method can track the volleyball trajectory in the occlusion scene. At the same time, the tracking time of the algorithm proposed in this paper is very fast.

In summary, this paper proposes a volleyball motion tracking method for processing occlusion scenes, which has higher accuracy than other trajectory tracking algorithms.

5 Conclusion

Due to the characteristics of multiple occlusion of volleyball, the existing trajectory tracking algorithms do not handle the occlusion problem well. This paper proposes a ball trajectory tracking method adapting to occlusion scenes. According to the example test, it can be seen that the tracking trajectory obtained by the proposed method is strongly correlated with the actual trajectory of volleyball in the two cases. So, the method proposed in this paper has an effective tracking effect for situations where volleyball is blocked by objects. The Kalman filter algorithm combined with CAMSHIFT algorithm used in this paper has the characteristics of high tracking accuracy, strong anti-interference ability, and high speed. At the same time, it can handle the occlusion problem more accurately. In the future, it can be applied in the case of occlusion or no occlusion to track objects which have color features. The disadvantage of the method proposed in this paper is that it is necessary to manually circle volleyball from the obtained video to be able to carry out the continuous tracking. Therefore, future studies will focus on the exploration of an approach that can run the algorithm fully automatically.

Acknowledgment. This work has applied for a China National Invention Patent (NO: 201710981725.7) and has been supported by the Ministry of Education's Higher Education Department's Industry-Science Collaborative Education Innovation Fund (201601030018).

References

1. Chakraborty, B., Meher, S.: Real-time position estimation and tracking of a basketball. In: 2012 IEEE International Conference, pp. 1–6. IEEE (2012)
2. Zhou, X., et al.: Tennis ball tracking using a two-layered data association approach. IEEE Trans. Multimed. **17**(2), 145–156 (2015)
3. Zhang, Y., Zhu, Y., Xia, W., Yan, F., Shen, L.: Semidefinite programming-based localisation and tracking algorithm using Gaussian mixture modelling. IET Commun. **11**(16), 2514–2523 (2017)
4. Su, W., Zhuang, H., Qiu, X.: Moving targets detection and tracking based on improved codebook algorithm and Kalman filtering. In: 36th Chinese Control Conference, pp. 11494–11498. IEEE (2017, Chinese)
5. Dawei, L.I., et al.: Integrating a statistical background-foreground extraction algorithm and SVM classifier for pedestrian detection and tracking. Integr. Comput.-Aided Eng. **20**(3), 201–216 (2013)
6. Gao, S.: Research on target tracking algorithm based on color features. Master, Sun Yat-sen University (2009)
7. Kim, J.-Y., Kim, T.-Y.: Soccer ball tracking using dynamic Kalman filter with velocity control. In: 6th International Conference on Computer Graphics, Imaging and Visualization, pp. 367–374. IEEE (2009)
8. Ahmad, A., Lawless, G., Lima, P.: An online scalable approach to unified multirobot cooperative localization and object tracking. IEEE Trans. Robot. **33**(5), 1184–1199 (2017)
9. Dikairono, R., et al.: Visual ball tracking and prediction with unique segmented area on soccer robot. In: 2017 International Seminar on Intelligent Technology and Its Applications, pp. 362–367. IEEE (2017)
10. Chen, W., Zhang, Y.: Tracking ball and players with applications to highlight ranking of broadcasting table tennis video. In: 2006 IMACS Multiconference on Computational Engineering in Systems Applications, pp. 1896–1903. IEEE (2006)
11. Tsoi, J.K.P., Patel, N.D., Swain, A.K.: Real-time object tracking based on colour feature and perspective projection. In: 9th International Conference on Sensing Technology, pp. 665–670. IEEE (2015)

12. Fitriana, A.N., Mutijarsa, K., Adiprawita, W.: Color-based segmentation and feature detection for ball and goal post on mobile soccer robot game field. In: International Conference on Information Technology Systems and Innovation, pp. 1–4. IEEE (2017)
13. Triamlumlerd, S., et al.: A table tennis performance analyzer via a single-view low-quality camera. In: 2017 International Electrical Engineering Congress, pp. 1–4. IEEE (2017)
14. Chakraborty, B., Meher, S.: A trajectory-based ball detection and tracking system with applications to shot-type identification in volleyball videos. In: 2012 International Conference on Signal Processing and Communications, pp. 1–5. IEEE (2012)
15. Kao, S.-T., Wang, Y., Ho, M.-T.: Ball catching with omni-directional wheeled mobile robot and active stereo vision. In: 26th International Symposium on Industrial Electronics, pp. 1073–1080. IEEE (2017)
16. Lyu, C., et al.: High-speed object tracking with its application in golf playing. Int. J. Soc. Robot. **9**(3), 449–461 (2017)
17. Feng,Y., Guo, G., Zhu, C.: Object tracking by Kalman filtering and recursive least squares based on 2D image motion. In: 2008 International Symposium on Computational Intelligence and Design, pp. 106–109. IEEE (2008)

Author Index

Printed in the United States
By Bookmasters